Lecture Notes in Electrical Engineering

Volume 944

The book series *Lecture Notes in Electrical Engineering* (LNEE) publishes the latest developments in Electrical Engineering—quickly, informally and in high quality. While original research reported in proceedings and monographs has traditionally formed the core of LNEE, we also encourage authors to submit books devoted to supporting student education and professional training in the various fields and applications areas of electrical engineering. The series cover classical and emerging topics concerning:

- Communication Engineering, Information Theory and Networks
- Electronics Engineering and Microelectronics
- Signal, Image and Speech Processing
- Wireless and Mobile Communication
- Circuits and Systems
- Energy Systems, Power Electronics and Electrical Machines
- Electro-optical Engineering
- Instrumentation Engineering
- Avionics Engineering
- Control Systems
- Internet-of-Things and Cybersecurity
- Biomedical Devices, MEMS and NEMS

For general information about this book series, comments or suggestions, please contact leontina.dicecco@springer.com.

To submit a proposal or request further information, please contact the Publishing Editor in your country:

China

Jasmine Dou, Editor (jasmine.dou@springer.com)

India, Japan, Rest of Asia

Swati Meherishi, Editorial Director (Swati.Meherishi@springer.com)

Southeast Asia, Australia, New Zealand

Ramesh Nath Premnath, Editor (ramesh.premnath@springernature.com)

USA, Canada:

Michael Luby, Senior Editor (michael.luby@springer.com)

All other Countries:

Leontina Di Cecco, Senior Editor (leontina.dicecco@springer.com)

**** This series is indexed by EI Compendex and Scopus databases. ****

Wuhong Wang · Jianping Wu · Xiaobei Jiang ·
Ruimin Li · Haodong Zhang
Editors

Green Transportation and Low Carbon Mobility Safety

Proceedings of the 12th International Conference on Green Intelligent Transportation Systems and Safety

Springer

Editors
Wuhong Wang
Department of Transportation Engineering
Beijing Institute of Technology
Beijing, China

Jianping Wu
School of Civil Engineering
Tsinghua University
Beijing, China

Xiaobei Jiang
Beijing Institute of Technology
Beijing, China

Ruimin Li
School of Civil Engineering
Tsinghua University
Beijing, China

Haodong Zhang
Beijing Institute of Technology
Beijing, China

ISSN 1876-1100 ISSN 1876-1119 (electronic)
Lecture Notes in Electrical Engineering
ISBN 978-981-19-5614-0 ISBN 978-981-19-5615-7 (eBook)
https://doi.org/10.1007/978-981-19-5615-7

This Springer imprint is published by the registered company Springer Nature Singapore Pte Ltd.
The registered company address is: 152 Beach Road, #21-01/04 Gateway East, Singapore 189721, Singapore

Contents

A Scheduling Plan Model for Metro Crew Incorporating Fatigue and Biological Rhythms

Yueyuan Chen, Weining Fang, Si Li, and Jianxin Wang

Abstract Crew scheduling is one of the critical planning decisions in railway transportation. The existing scheduling and rostering methods usually take the lowest cost as the objective, ignoring the metrzzo crew members' fatigue and biological rhythms. This paper proposed an optimization approach considering fatigue's impact on solving real-world metro crew scheduling and rostering problems. The shift work characteristics of the metro crew were analyzed firstly. The usability of the Ikeda formula for fatigue evaluation was verified and applied to the metro crew. Then the metro crew scheduling and rostering model were described, and the process of incorporating fatigue factors into the model was demonstrated. Moreover, using the genetic algorithm to solve the problems. Finally, this model was applied to the Beijing Metro Yanfang Line. The results illustrated that the method could significantly reduce the metro crew members' fatigue value with optimized operating costs.

Keywords Crew scheduling · Biological rhythms · Fatigue evaluation · Set covering model · Metro crew

1 Introduction

The scheduling plan of the metro crew involves two significant problems: crew scheduling and crew rostering [6, 7, 20]. Crew scheduling refers to dividing the planned train timetable into several tasks according to the daily train operation plan and pairing the crew members' tasks. It is a short-term schedule, typically one day.

Y. Chen · W. Fang (✉)
State Key Laboratory of Rail Traffic Control and Safety, Beijing Jiaotong University, Beijing 100044, China
e-mail: wnfang@bjtu.edu.cn

S. Li
Haidian District, Beijing 100044, China
e-mail: 16116372@bjtu.edu.cn

J. Wang
China Institute of Marine Technology and Economy, Beijing 100081, China

W. Wang et al. (eds.), *Green Transportation and Low Carbon Mobility Safety*, Lecture Notes in Electrical Engineering 944,
https://doi.org/10.1007/978-981-19-5615-7_1

Crew rostering refers to a set of daily plan arrangements for drivers based on crew scheduling [8]. Many scholars have done studies on crew scheduling plans. The goals are generally the lowest total operating cost [6, 7, 20], the most economical time cost [13, 17, 46], and the minimum number of duties [21, 49]. Metro crew scheduling plan is not just a matter of cost and time, but also directly related to the members' health and operation safety. Several studies have proved that the driver's physiological state is significantly related to driving safety [39, 54]. Shift work induces fatigue, and irrational scheduling can exacerbate the situation [43]. It is significant to make a humanized scheduling plan to enhance the metro operation's efficiency and service level.

An investigation by the Rail Safety and Standards Board found that drivers suffering from fatigue accounted for 21% [4]. McGuffin found that the effect of continuous driving on fatigue was relatively stable in the first 4 h of continuous driving but increased rapidly after that. The fatigue risk doubles when continuous driving lasted for more than 6 h [36]. Spencer proposed that only sustained sleep can sustained sleep fatigue and suggested that train drivers should ensure at least 16 h sleeping between consecutive night shifts [48]. Simultaneously, the risk of driving fatigue caused by shift work is more prominent under high automation [39]. With the universal application of automated driving technology, the metro crew members are often unconsciously extended working hours by scheduling personnel in the practice schedule due to simplified operating mode [57]. The long-time monitoring in automation will not reduce the fatigue effect and even cause problems such as distraction and lower situational awareness, which will affect driving safety [42].

There are few scheduling studies combined with fatigue caused by the drivers' biological rhythm, which is a critical factor that cannot be ignored. Crum and Morrow [10] investigated the influence of carrier scheduling on driver fatigue. However, it focuses on the fatigue model, not the scheduling and rostering model incorporating fatigue and biological rhythms. Bowden researched the optimal schedule problem and used the alertness performance model, not the fatigue evaluation model [3].

The metro crew scheduling plan is a synthesis NP-hard problem involving the train, human, and timetable [24]. The crew scheduling problems are mainly described in the form of a set covering problem [9, 17, 25] and set partition problem [12, 37]. A set partitioning problem determines how the items in one set can be partitioned into smaller subsets, all items in set S must be contained in one and only one partition [29]. The set covering problem is selecting a minimum set to contain all the elements in any put sets [53]. The most commonly used model to solve the scheduling problem is set covering models, which good at dealing with complex non-linear problems and are also widely used in aviation and bus fields [19]. In terms of algorithm, it is mainly for an exact solution at the beginning. With the continuous expansion of the problem scale, the combination explosion will occur when the data volume is enormous [20]. Some heuristic algorithms were used to solve these problems, such as Shen used a hybrid genetic algorithm to solve the public transport crew scheduling problem[44], Khmeleva presented a fuzzy-logic controlled genetic algorithm for the rail-freight crew scheduling problem [28].

This paper focuses on developing a new model and analyzing scheduling plans for the metro crew using the set covering model and genetic algorithm. Based on that, a scheduling plan model for the metro crew incorporating fatigue caused by biological rhythm is proposed. The scheduling plan has two steps: crew scheduling problem and crew rostering problem. First, it is verified that the Ikeda formula for evaluating the drivers' fatigue. Then the set covering model is used for crew scheduling with fatigue as one of the constraints. On this basis, the crew rostering model is established with fatigue minimization as the objective function. Finally, taking the Beijing Subway Yanfang Line as an example, the scheduling plan model is applied. The results are compared with the actual results to verify the proposed model's effectiveness in reducing the risk of metro drivers' fatigue.

2 Characteristic Analysis of Metro Crew Scheduling

2.1 General Metro Crew Scheduling Problem

The metro crew scheduling plan refers to the formulation of a crew duty scheduling plan based on the specific time and space matching relationship determined in the train diagram. For the sake of clarification, a few terminologies of the crew scheduling plan should be introduced. "The trips of a train operated in this cycle runs in a terminal-to-terminal pattern, running from the origin terminal to the destination terminal and stopping at each of the intermediate stations" [21]. And "duty is the work to be performed by a crew during one day from sign-on until sign-off at the depot and consists of pieces of trips" [33], shown as Fig. 1. The most common form is the "three-duty system", that is three duties within 24 h a day, each duty lasts eight hours, and every duty changes at regular intervals. Figure 2 shows an example of a crew rostering for duties.

The scheduling plan problem has two phases combining with the train diagram [5]. Firstly, the daily duty is scheduled according to the transfer station and train diagram. Then, the rostering is made combining the duties work in one cycle to complete a rostering plan.

2.2 The Crew Scheduling Problem

The crew scheduling problem of a single duty can be described as follows. The operational tasks are divided into trips and combined to form a duty. All duties of one group drivers can cover the train numbers in the train operation diagram. And the working hours of one duty need to follow the labor laws and regulations [24]. The problem can is a Set Covering Problem (SCP) [24], which solution is a set that can cover all train numbers. Usually, the objective function is to minimize the cost, that is, based on satisfying labor laws and regulations, to arrange the duties for as few drivers as possible to meet train operation planning. The known variable is the set of split driving segments $P = \{p_1, p_2, \ldots, p_m\}$ and all the available train number $S = \{S_1, S_2, \ldots, S_n\}$. Decision variables are integer variables of 0–1. The mathematical model formulas show in (1) to (3).

$$\min G(X) = \sum_{j=1}^{n} (C + c_j)x_j \tag{1}$$

$$s.t. \sum_{j=1}^{n} a_{ij}x_j \geq 1, i \in \{1, 2, \ldots, m\} \tag{2}$$

$$x_j = 0 \text{ or } 1, j \in \{1, 2, \ldots, n\} \tag{3}$$

where c_j is the operating cost of the train S_j of a trip, a_{ij} is a 0–1 variable. If S_j covers p_i, the value is 1, otherwise, 0; x_j is a 0–1 variable of the corresponding decision. If train number j is used, it is counted as 1; otherwise, it is counted as 0. C is a large constant to ensure the priority minimization of the total train numbers. The constraint (2) provides that each driving segment is covered by only one train number. According to the selection of specific constraint and objective function, model parameters can be modified.

2.3 *The Crew Rostering Problem*

The crew cycle rostering problem can be described as follows. After the crew scheduling, all duties are assigned to drivers. Drivers perform periodic work in a predetermined sequence. And, the rostering plan is required to satisfy the basic labor laws and regulations as well [24]. In general, rostering is designed to minimize the number of crew members or to balance the workload [14]. Establish mathematical models according to the different needs of operating companies flexiblly. The known variable is the set of train numbers $T = \{t_1, t_2, \ldots, t_n\}$ and duty date $D = \{d_1, d_2, \ldots, d_m\}$. The decision variable is an integer variable of 0–1. The mathematical model is as follows:

$$\min F(X) = \sum_{o=1}^{l} \sum_{p=1}^{m} \sum_{j=1}^{n} f(y_{opq}) \tag{4}$$

$$s.t. \sum h_i(y_{opq}) = 1 \tag{5}$$

$$y_{opq} = 0 \text{ or } 1, o \in \{1, 2, \ldots, l\}, p \in \{1, 2, \ldots, m\}, q \in \{1, 2, \ldots, n\} \tag{6}$$

where f represents the objective function of the corresponding model; y_{opq} is a 0–1 variable; if a crew o perform a task q on the day p, the value is 1; otherwise, the value is 0. The notation h_i is the corresponding function of constraint.

According to the operating company regulations, the constraint conditions of the model are different. Necessary constraints include as follows: each duty must be assigned to a driver; each driver is allowed to perform only one driving task at one time; a driver should have enough rest time between two continuous duties (see Sect. 2.2.3); the rest days and distribution in a cycle of one driver should satisfy the labor laws and regulations; the working hours of a driver during the cycle shall be within a specific range to avoid driving fatigue caused by excessive working hours or personnel waste caused by excessively low working hours (see Sect. 2.2.3). Moreover, other constraints, such as fairness of crew work and rostering mode.

3 Impact Factors of Metro Crew Scheduling

3.1 Complex Time Constraints

Different countries and cities have various operating institutions in crew mode, dining custom, duty type, and shifting mode [55]. Therefore, the scheduling plan is elaborate and should consider specific time constraints [15]:

(1) Train operation timetable: the metro crew scheduling should cover all the train operation timetable to ensure train driving continuity.
(2) Preparation time for departure: reserve train preparation and inspection time for drivers, which is usually 30 min earlier before the train entered the mainline;
(3) Break and dining time: the time for break and dining should satisfy the rest requirement of drivers. The lower time bound could be generally obtained from local working time standards.
(4) Commuting time: most metro lines bind the night shift and day shift (the driver who was on duty the night before should continue the morning shift after a rest in the metro vehicle depot). It is necessary to arrange the driver's arrival time in the morning reasonably, and some drivers need to take the outgoing train to the transfer point to take the first train of the day.

3.2 Objective

The objectives are diverse for metro crew scheduling. In terms of cost, the number of crew members and the total crew cost needs to a minimum. In terms of workload, there is the objective of fairness or the prescribed shift system and vacation mode [26]. It is impossible to consider all the beneficial objectives, and there are conflicts in cost, load, vacation, and other factors [24]. So, it is necessary to proceed from the practical needs of the operating line and optimize the target as much as possible under the condition of guaranteeing the most favorable factors.

3.3 Biological Rhythms and Fatigue

Metro drivers are always working underground and driving in a monotonous and closed cab, who need to keep standing for a long time and maintaining a high concentration of attention. Driving fatigue caused by continuous work is a problem that must be paid attention to solve. Meanwhile, the shift work is inevitably for the metro operation. The dysfunctional circadian rhythm will exacerbate fatigue [24]. Therefore, the fatigue in this paper refers to sleep deprivation-related fatigue caused by irregular biological rhythms and prolonged work.

Furthermore, the unreasonable working, dining, and rest time have adverse effects on drivers' physiology rhythm, which destroys existing human biological rhythms and causes health problems, such as insomnia [38], heart disease [1]. Therefore, measuring the fatigue value of drivers from the perspective of biological rhythm is critical to ensure the rationality of scheduling. The scheduling problem should consider the allocation of working time and sleeping time and the adaptability of the human body. When evaluating the influence of biorhythm on the metro crew scheduling plan, it is usually necessary to consider the working time domain, working hours, work break, rest time, rostering mode.

(1) Working time domain

Studies have shown that for the work from 13:00 to 24:00, the average sleeping time of metro drivers is 7.4 h, the average sleeping time is 5.3 h with work from 00:00 to 02:00, and the average sleeping time is 4.4 h with work from 06:00 to 08:00 [36]. The reason is that the works at night and in the early hours of the morning conflict with the fixed sleeping time, resulting in reduced sleeping time and sleeping quality. The long-term night work destroyed the human body clock, and the sleeping quality and time are affected, which caused cumulative fatigue [14].

(2) Working hours

Working hours refers to the duration time of driving. Drive requires a high level of concentration, which can easily lead to high cognitive load. The long working hours make the fatigue of drivers aggravate, which makes them sleepy quickly [16]. Although the effect of continuous driving on fatigue was relatively stable in the first 4 h of continuous driving, increased rapidly after that. When continuous driving lasts for more than 6 h, the fatigue risk doubled [36]; therefore, the working hours of drivers and the short breaks on duty should be reasonably arranged, decreasing fatigue caused by long working hours.

(3) Work break

Work break means a short rest time on duty, which can help drivers restore energy and reduce fatigue. A study of pilots' breaks showed that in the 15 min after the break, the slow eye movement of pilots decreased significantly, and the energy recovered from a break will vanish after working for 40 min [40]. Compared with drivers who have only one short rest, the driver who takes a regular break is less likely to sleepy. Therefore, the proper break is arranged to avoid working long hours can effectively reduce drivers' fatigue.

(4) Rest time

Rest time refers to the rest between two continuous duties. The human body can only offset fatigue through constant sleeping, and enough time should be set aside for the driver to sleep before the next duty. Drivers who work at night often have sleep problems because the effects of light and noise during the day are more severe than at night. To make sure the drivers work in the night getting adequate rest in the day after a night work, the rest time of two continuous duties should increase, which lasts at least 16 h [48].

(5) Rostering mode

Rostering mode involves rostering orientation and rostering speed. The rostering orientation can be divided into clockwise and counter-clockwise, both of which have no difference in usually sleeping [45]. However, considering the sleep deprivation caused by night work, counter-clockwise rostering is prone to cause the amplitude reduction and phase delay of biological rhythm, which leads to the disorder of biological rhythm. Therefore, give priority to the clockwise rostering [41]. The rostering speed usually includes fast (1–4 days), medium (1–2 weeks), and slow (3–4 weeks). The fast rostering can reduce the cumulative effect of sleep loss, making the drivers keep alert during the rostering period. The slow rostering has a minimal impact on sleeping duration. It may be better to keep the same rostering (such as permanent night mode) at all times, but it is often challenging to carry out such a rostering because of social and family factors [2].

4 Method

4.1 Fatigue Evaluation Model

Most of the fatigue evaluation methods [23] can not be used in crew scheduling and rostering model because of the complex time constraints. Furthermore, different fields have their operation characteristics and relevant laws and regulations. At present, the two most common methods used in the fatigue evaluation of the metro crew are the Ikeda formula [51] and Fatigue Index (FI) [35]. Both of them considered fatigue evaluation's input factors, including the working time domain, the working hours, and work contents. But compared with the Ikeda formula calculate in minutes, FI is more affected by the time domain and calculate in hours [26]. Ikeda formula has high precision and can distinguish different duties in the same working time domain. The changes in fatigue of the same task in different time domains can be accurately measured. Therefore, the Ikeda formula was used to evaluate the fatigue value of the metro crew in this paper.

Japanese scholar Oshima Masahiro has established a "workload rating list" [34] for job fatigue based on the human daily rhythm, evaluating the fatigue of the continuous work and the work that starts in different time domains. Based on the "workload rating list", Ikeda Toshihisa introduced a weighted correction factor according to the characteristics of work and rest time in practice, established a dynamic formula for assessing the fatigue of the metro crew, namely the Ikeda formula [51].

$$W = \alpha \times L - \beta \times R \tag{7}$$

$$L = \sum_{i=1}^{n_i} \sum_{j=1}^{n_j} \left(C_{ij} \times \left(\frac{L_{abi}}{b \times 60} \times T_{ij} \right) \right) \tag{8}$$

$$R = \sum_{k=1}^{m_k} \sum_{l=1}^{m_l} \left(C_{kl} \times \left(\frac{R_{cdk}}{d \times 60} \times T_{kl} \right) \right) \tag{9}$$

Table 1 shows the definitions of index, parameters, and variables. The fatigue value of continuous work from 9:00 a.m. to 5:00 p.m. is regarded as 100. Similar to fatigue evaluation, the sleep effect from 10:00 p.m. to 6:00 a.m. in the next morning is regarded as 100. Based on this, the fatigue value or sleep effect at any other time is evaluated quantitatively compared with that. The daily fatigue loss on duty is subtracted from the restored amount of the rest on that day. Then the fatigue value at any time point in the roster and the total fatigue consumption value can be obtained according to the time series, which can evaluate the metro crew rostering fatigue.

Many researchers have verified the Ikeda formula's feasibility for evaluating the fatigue of metro drivers [50, 56]. This study used subjective questionnaires to verify its usability further. Japanese scholars have developed the "Subjective Fatigue Symptoms Questionnaire" in the 1970s and proposed a revised version in 2002 [27]. The questionnaire is widely used in Japan and many other countries, which measures 25 the subjective fatigue items from five aspects: sleepiness, restlessness, discomfort, fatigue, and ambiguity. 1–5 points for each item, the lowest score is 25, and the highest score is 125. The higher the score, the higher the fatigue level. We conducted a questionnaire survey of 60 metro train drivers aged 22–35 (Mean = 26.92, SD = 3.08), all of whom were driving for more than two years. The fatigue of all single duties and roster work were measured. The correlation test between subjective fatigue scores and Ikeda formula fatigue results of single duties and roster work are shown in

Table 1 Index, parameters used in the fatigue evaluation model

Notations	Description
W	The value of workload evaluation (When W is less than 0, it counts 0)
α	Weighting coefficient of fatigue effect
β	Weighting coefficient of rest effect
L	Cumulative value of fatigue effect
R	Cumulative value of rest effect
n_i	The i-th work; n_j is the j-th period in the work
m_k	The k-th rest; m_l is the l-th period in the rest
$\frac{L_{abi}}{b\times 60}$	Average fatigue score in the i-th work (starting at point a clock, lasting b hours)
$\frac{R_{cdk}}{d\times 60}$	Average fatigue score in the k-th work (starting at point c clock, lasting d hours)
C_{ij}	Fatigue weighting coefficient of the j-th period in the i-th work
T_{ij}	Time of duration of the j-th period in the i-th work (min)
C_{kl}	Weighting coefficient of the k-th period in the l-th work
T_{kl}	Time of duration of the k-th period in the l-th work (min)

Table 2 Correlation analysis of Ikeda formula and subjective fatigue

Type	Single duty			Rostering
	Day	Middle	Night	
N	27	13	13	60
Pearson Correlation	0.750	0.604	0.687	0.600
p	0.000	0.029	0.01	0.000

Table 3 Index, parameters and variables used in metro crew scheduling model

Notations	Description
j	j-th trip in a duty
$T = \{T_1, T_2, \ldots, T_n\}$	Train number
$D = \{D_1, D_2, \ldots, D_m\}$	Driver number
$TS = \{TS_1, TS_2, \ldots, TS_n\}$	Departure time for each train
$TF = \{TF_1, TF_2, \ldots, TF_n\}$	Arrival time for each train
$TFi = \{TFi_1, TFi_2, \ldots, TFi_n\}$	The fatigue value of each trip
$Size$	Maximum number of trips in a duty
$BTMin$	Minimum time length of the break
$BTMax$	Maximum time length of the break
$Fimax$	Acceptable maximum fatigue value of drivers

Table 4 Index, parameters and variables used in metro crew rostering model

Notations	Description
j	j-th duty in a roster
$T = \{T_1, T_2, \ldots, T_n\}$	Train number
$D = \{D_1, D_2, \ldots, D_m\}$	Driver number
$TDr = \{TDr_1, TDr_2, \ldots, TDr_m\}$	The daily rest time of drivers
$TDs = \{TDs_1, TDs_2, \ldots, TDs_m\}$	Daily sleeping time of drivers
$W = \{W_1, W_2, \ldots, W_l\}$	The number of duty
$WFi = \{WFi_1, WFi_2, \ldots, WFi_l\}$	The fatigue value of each duty
$Wty = \{Wty_1, Wty_2, Wty_3\}$	The type of each duty
$Ws = \{Ws_1, Ws_2, \ldots, Ws_l\}$	The attendance time of each duty
$We = \{We_1, We_2, \ldots, We_l\}$	The get off work time of each duty
T_C	Normal commuting time
T_Dc	Commuting time in the depot
$Date$	The date of a roster
Day_size	The days of a roster
Day_type	The sequence requirements of duty in a roster. If there is no shift sequence requirement, the set is empty
Day_Fi	Accumulated fatigue value of a driver D_i on j-th duty of a roster

Table 5 The parameter inputs of the model

Notations	Description	Parameters
$Size$	Maximum number of trips in a duty	Day duty: 10
		Middle duty: 7
		Night duty: 7
D	Driver number of a group	14
$BTMin$	Minimum break time	5
$BTMax$	Maximum break time	30
$Fimax$	Acceptable maximum fatigue value	100
Com_time	Commute time between home and depot	60(min)
$Dcom\_time$	Commute time in the depot	20(min)
Day_size	Calculated days	8
Day_type	Duty sequence	[2,3,1,0,2,3,1,0]
P_size	Initial population	100
I	Iterations	500
M	Mutation rate	0.5

Table 6 The results of metro crew scheduling plan of a group

Date of roster Driver number	1	2	3	4	5	6	7	8
1	N-9	M-9	R	D-3	N-9	M-3	R	D-7
2	N-10	M-7	R	D-4	N-7	M-6	R	D-3
3	N-4	M-3	R	D-7	N-3	M-2	R	D-9
4	S	M-6	R	D-6	N-10	M-5	R	D-4
5	N-6	S	R	S	N-8	S	R	S
6	N-7	M-5	R	S	N-4	M-11	R	S
7	N-5	M-4	R	D-9	N-5	M-4	R	D-10
8	N-12	M-1	R	D-10	N-11	M-12	R	D-5
9	N-3	S	R	D-2	N-2	M-9	R	D-8
10	N-11	M-2	R	D-5	S	M-1	R	D-2
11	N-1	M-12	R	S	N-1	M-8	R	S
12	N-8	M-11	R	D-8	N-12	M-10	R	D-1
13	S	M-10	R	D-11	S	S	R	D-6
14	N-2	M-8	R	D-1	N-6	M-7	R	D-11

Note 1. The duty type- trip number, for example, "N-9" means the 9th trip of the night duty
2. The day duty-D; The middle duty-M; The night duty-N; Rest-R; Standby-S

Table 2. The fatigue values calculated by the Ikeda formula for all types are significant correlative with subjective fatigue (95% confidence interval). The results proved the validity of the Ikeda formula for evaluating the fatigue of the scheduling plan.

4.2 Metro Crew Scheduling Plan Model

The metro crew scheduling model includes two phases that crew scheduling model and crew cycle rostering model. The fatigue evaluation model results was applied to both models as constraints and objective functions, respectively.

4.2.1 Metro Crew Scheduling Model

A proper metro crew scheduling is favorable for reducing the number of drivers and operation costs. The scheduling plan of the metro crew should support reducing the total number of duties as far as possible based on meeting various constraints, to reduce staff costs and operation costs. Therefore, the objective function of the metro crew scheduling model was to minimize the number of duties. The developed optimization model was described as follows:

$$\min \sum_{i=0}^{m} \sum_{k=0}^{n} X_{i1k} \tag{10}$$

where the variable X_{ijk} was a 0–1 variable:

$$X_{ijk} = \begin{cases} 1, & if\ the\ driver\ D_i\ drive\ the\ train\ numbered\ T_k\ for\ the\ jth\ trip \\ 0, & otherwise \end{cases} \tag{11}$$

The index, parameters, and variables used in the crew scheduling model were defined shown in Table 3.

Each train was drived by one and the only one driver in one trip, and the metro drivers should cover all the trains in the train diagram:

$$\sum_{j=0}^{size} \sum_{i=0}^{m} X_{ijk} = 1, k \in T \tag{12}$$

Each driver can drive only one car at a time:

$$\sum_{i=0}^{m} \sum_{j=0}^{size} \sum_{k=0}^{n} X_{ijk} = 1, i \in D, j = 1, 2, 3, \ldots, Size \tag{13}$$

Every driver must have the n-th trip before the (n + 1)-th trip:

$$\sum_{k=0}^{n} X_{ijk} - \sum_{k=0}^{n} X_{ij+1k} = 1, i \in D, j = 1, 2, 3, \ldots, (Size - 1) \tag{14}$$

For each driver, ensure that the time interval between the (n + 1)-th trip and the n-th trip was within the specified time range, which is longer than the minimum time length of the break and less than the maximum time length of the break:

$$\sum_{k=0}^{n} X_{ij+1k} \cdot TS_k - \sum_{k=0}^{n} X_{ijk} \cdot TF_k \geq BTMin, i \in D, j = 0, 1, 2, \ldots, (Size - 1) \tag{15}$$

$$\sum_{k=0}^{n} X_{ij+1k} \cdot TS_k - \sum_{k=0}^{n} X_{ijk} \cdot TF_k \leq BTMax, i \in D, j = 1, 2, 3 \ldots, (Size - 1) \tag{16}$$

In addition to the above constraints, the model incorporated the biological rhythm evaluation model into the constraints. The acceptable maximum fatigue value of drivers was constrained in one duty, that is, the maximum fatigue of each driver after completing a duty does not exceed the prescribed upper limit:

$$\sum_{i=0}^{m}\sum_{j=0}^{size}\sum_{k=0}^{n} X_{i1k} \cdot TFi_k \le Fimax, i \in D \tag{17}$$

4.2.2 Metro Crew Rostering Model

The metro crew rostering model proposed was a periodic scheduling plan model that takes the fatigue caused by biological rhythm disorder as the objective function and considers the other necessary constraints. The purpose was to find out the best match between each duty and the drivers.

Because of the periodicity of human biological rhythm, it should reduce the risk of fatigue in the whole work cycle rather than one working day. Use the fatigue evaluation model described in Sect. 3.1 to calculate the driver's fatigue value. The fatigue peak of each driver in one rostering was selected to sum and minimize it to establish the objective function. The developed optimization model was described as follows:

$$\min\left(\sum \max Day\_Fi\right) \tag{18}$$

where the variable Y_{ijk} was a 0–1 variable:

$$Y_{ijk} = \begin{cases} 1 thedriver D_i drivethetrainnumbered T_k forthejthtask \\ 0 otherwise \end{cases} \tag{19}$$

The index, parameters, and variables used in the metro crew rostering model were defined as shown in Table 4.

Make sure that the drivers' tasks can cover a train diagram, and each duty must be in charge of a driver every day:

$$\sum_{i=0}^{m} Y_{ijk} \ge 1, j \in Date, k \in W \tag{20}$$

Ensure that the number of drivers' tasks is constrained. A driver can and only can take one duty in one day, and regard the break day as a duty into the model:

$$\sum_{k=0}^{l} Y_{ijk} = 1, i \in D, j \in Date \tag{21}$$

The duty sequence in the crew rostering needed to constrain. Some rail transit operators have defined duty sequence rules. For example, a driver cannot take the same type of duty for two consecutive days. If there were no rules for that, the Day_type was an empty set.

$$\sum_{k=0}^{l} Wty * Y_{ijk} = Day\_type \tag{22}$$

4.3 Model Solution

The solving process of the model included two steps, shown in Fig. 3. The first was to solve the crew scheduling problem. Combine different trips to form a duty based on the train diagram, relevant laws, and regulations constraints. Duty was one shift, which included all tasks of a driver in a day. To ensure the rationality of the workload and working hours of one duty. The fatigue evaluation model calculated each driver's fatigue values on every duty, and the results as constraints. The crew scheduling model was used to generate multiple duty types, ensuring that the duty set covers all the trips in the train diagram. In the second part, based on crew scheduling fatigue value, the fatigue evaluation results of crew rostering were obtained. Regarding each driver's fatigue value as the objective function, calculate crew rostering results based on the laws and regulations constraints for reducing the risk of rhythm disturbance and fatigued driving for every driver on duty.

It is an NP-hard problem to optimize the scheduling of the metro crew [24]. Because of its large computation scale and complex constraints, this problem is challenging to solve, which meta-heuristic algorithm is used to solve the problem in most studies effectively. Meta-heuristic algorithm refers to searching the optimal solution in a random population according to specific search strategies [31], including simulated annealing algorithm, ant colony algorithm, genetic algorithm, and so on. The genetic algorithm can be applied to solve the traditional programming model. The algorithm can quickly jump to another region when searching locally and approach the optimal solution quickly in the search space. By comparing several meta-heuristic methods, the optimization result of the genetic algorithm is better when solving the problem of crew scheduling [47]. The solution method of genetic algorithm is to establish coding and optimize the population according to the fitness function value, which is more suitable for solving periodic shifts. Therefore, this paper adopted the genetic algorithm to solve the model of the crew scheduling problem.

5 Verification

This section empirically validated the metro crew scheduling approach by solving real-world instances of the Beijing Subway Yanfang line. The calculation results were compared with the existing two schemes, and the optimization results are analyzed.

5.1 *Instances*

The Beijing Subway Yanfang Line has a total length of 14.4 km, a fully automatic subway line has nine stations and one depot. The depot and the first station near the train depot are the transfer stations, which have a driver transfer room and duty room. On weekdays, seven trains are running simultaneously, and each train has a driver on duty. A total of 302 trains are operated one day. The first train leaves at 5:05 in Yancun East, and the last train leaves at 22:55 in Yanshan.

Based on the existing train diagram, the staff makes a scheduling plan for the metro crew according to their working experience. Sixty-eight drivers are divided into four groups, each with 17 drivers. In one group, 15 drivers on duty for the driving task, and 2 drivers are on standby. There are three types of metro crew duty: the day duty (08:00–17:00), the middle duty (17:00–24:00), the night duty (04:00–08:00). These three duties cover all the running tasks in the train diagram.

The work and rest time of drivers, train diagrams were acquired by interview. The maximum number of trips on the line during the day duty, the middle duty, and the night duty are 10, 7, and 7, respectively. Set the maximum allowable fatigue value to 100 according to the Ikeda formula (Sect. 3.1). The rostering scheduling adopts the cycle mode of "day duty—middle duty—night duty–rest" to calculate 8 days (2 rostering). The parameter inputs for the crew scheduling plan model see Table 5.

5.2 *Results*

Using the crew scheduling plan model proposed, and the genetic algorithm solved the crew scheduling plan problem of 8 days. The results of the crew scheduling model showed that the minimum number of trains to complete the train operation plan included: 11 day duties, 12 middle duties, and 12 night duties. Correspondingly, 35 drivers on duty for the ordinary driving task and two drivers on standby for each group are needed. Therefore, a total of 41 crew members were required for regular operation. Compared with the original metro crew of 51 drivers (17 drivers in each group), 10 drivers were reduced for the Beijing Subway Yanfang Line, which reduced workforce costs. Based on the results of single duty scheduling, considering the metro crew's daily schedule, the crew rostering model was used to determine which duty each driver performs each day. The results of one group were shown in Table 6.

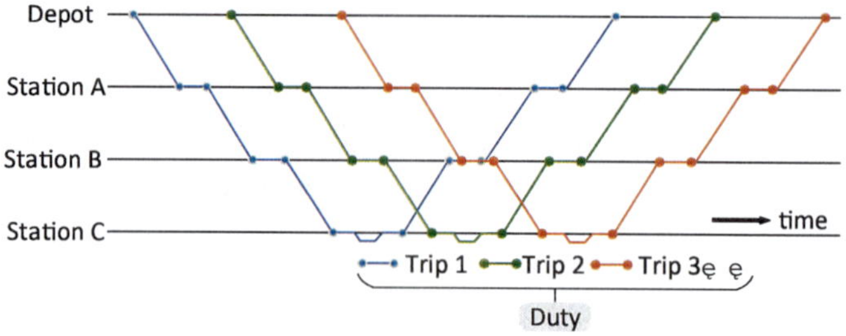

Fig. 1 Driving trips and duty

Fig. 2 Crew rostering for duties

Driver 1	Duty1	Duty2	Duty3	Holiday
Driver 2	Holiday	Duty1	Duty2	Duty3
Driver 3	Duty3	Holiday	Duty1	Duty2
Driver 4	Duty2	Duty3	Holiday	Duty1
Driver 5	Duty1	Duty2	Duty3	Holiday

After obtaining the crew rostering results, the fatigue evaluation model using the Ikeda formula was applied to verify the results of the crew scheduling plan model. The fatigue values of the 14 drivers in the group after work and after the rest were calculated. The drivers' fatigue results of using the model of metro crew scheduling plan were compared with the original scheduling plan, as shown in Figs. 4 and 5. The fatigue peak value is not more than 100, and the average fatigue value of crew members decreased by 39%.

Using the paired sample T-test to analyze the fatigue values of two crew scheduling plans. The results showed significantly different ($M = -6.62$, $SD = 8.71$, $t = -3.043$, $p < 0.05$). Compared with the original scheduling plan, the metro crew scheduling model proposed can significantly improve the work fatigue of the drivers. It proved the effectiveness of the optimization model.

6 Discussion

The key to the metro crew scheduling plan is to meet the necessary metro operation constraints and find a solution that meets the expected requirements. Our model incorporated fatigue and biological rhythms within the metro crew scheduling domain to enforce a minimum fatigue level while including the necessary variables and constraints. The fatigue value of drivers was 0 after the end of one roster cycle from Fig. 4 (the timing of before R and after R), which means that the rostering fatigue is low. Before each duty, the fatigue value was close to 0, which indicated that the crew

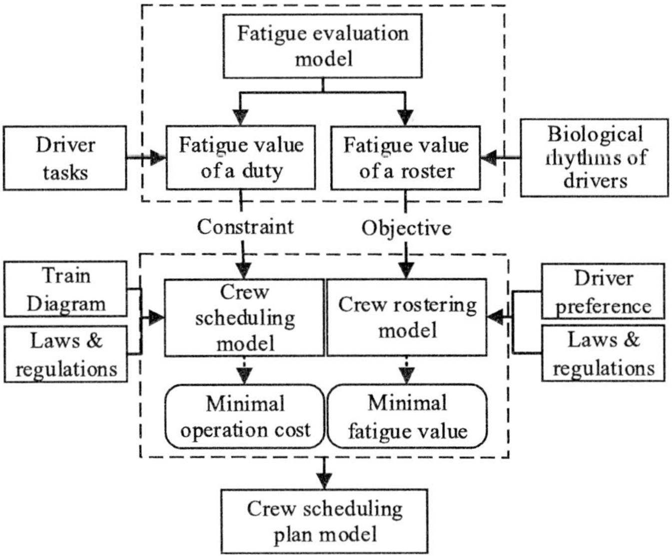

Fig. 3 A framework for the metro crew scheduling problem

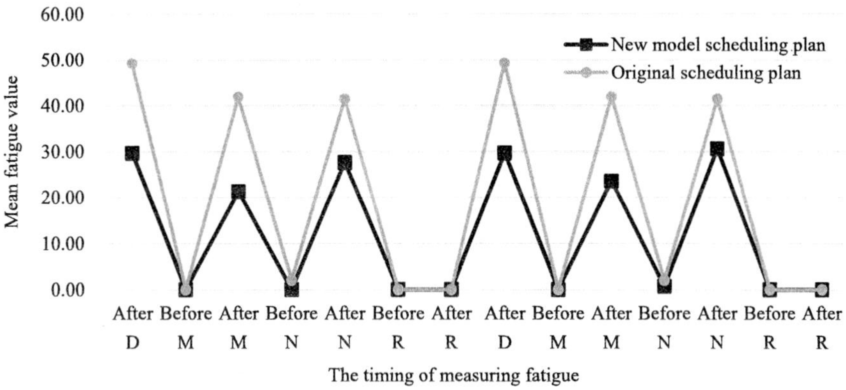

Fig. 4 The mean fatigue value of drivers in 8 days. *Note* The day duty-D; The middle duty-M; The night duty-N

scheduling results have a significant effect on relieving the risk of rhythm disorder caused by sleep interruption when the crew is on the night duty.

The maximum fatigue values of the new scheduling plan were higher than the original scheduling plan after day duty (Fig. 5). The reason for this phenomenon could be that the number of drivers in each group was reduced, then the work arrangement was more compact. The driver's workload increased slightly during the day duty, increasing the peak value of fatigue. However, each driver's peak fatigue value was less than 70, which avoids the fatigue risk of a single duty and proves the rationality of

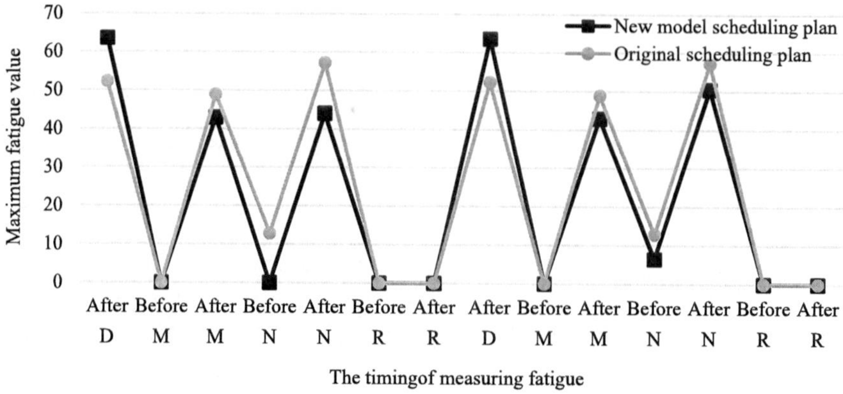

Fig. 5 The maximum fatigue value of drivers in 8 days. *Note* The day duty-D; The middle duty-M; The night duty-N

the metro crew scheduling model. And because the new model took into account the driver's biological rhythms, the trips have been more reasonably arranged, reducing the risk of cumulative fatigue caused by the disorder of biological rhythm [17].

The results showed that this model facilitates the production of schedules that meet existing constraints while also reduced the fatigue value of the metro crew. Our model generated the metro crew scheduling plan can better maintain a functional working status of drivers. It could enhance their work efficiency and the ability to deal with emergencies [11] and improve the service quality for passengers and operation safety.

There are some limitations to our model. Interviews identified the biological rhythms of drivers, and a subjective questionnaire survey verified the method of fatigue evaluation. In future work, physiological monitoring methods such as collecting heart rate and heart rate variability to evaluate fatigue. Furthermore, there are biological differences between individual drivers. For some drivers, short sleep at night can not relieve his fatigue because of sleep interruption [52]. It may increase subjective fatigue. The Ikeda formula uses the same biological rhythm by default and ignores sleep interruption. In the next research, we will improve the Ikeda formula considering the biological rhythm between individuals and sleep interruption.

7 Conclusions

This paper proposed a modeling approach to scheduling plan optimization of metro crew incorporating fatigue risk factor effects and drivers' biological rhythm. First, the model starts from the fatigue evaluation model to evaluate the fatigue value for the metro crew of crew scheduling and crew rostering. Second, it considered the prescribed driver fatigue limit as the constraint to minimizing the labor costs of each duty in the crew scheduling model. Finally, it optimized the total fatigue value in the whole metro crew rostering into account of the driver's biologic rhythm to establish the metro crew rostering model.

Taking the Yanfang Line of Beijing Metro as a case study, the model was established, and the genetic algorithm solved the crew scheduling plan of the line. The case application results using real-world data showed that the model reduced the 51 people to 41 people for the metro crew; 19.6% of human resource costs were saved. Simultaneously, the fatigue factor caused by the driver's biological rhythm was considered, which dramatically reduces the fatigue value during the period of being on duty, and reduced the average fatigue value by 39% for rostering. Therefore, the model achieved lowering fatigue risk and rhythm disorder of the metro crew while optimizing operating costs. The case proved that the model could optimize staffing and reduce workload for urban rail transit enterprises.

Acknowledgements This work was supported by the Natural Science Foundation of Beijing Municipality Grant L191018. The authors are very grateful to all the participants who contributed to the study. The authors have no conflict of interest to declare.

Data Availability Statement Some or all data, models, or code that support the findings of this study are available from the corresponding author upon reasonable request.

References

1. Arendt J (2010) Shift work: coping with the biological clock. Occup Med 60(1):10–20. https://doi.org/10.1093/occmed/kqp162
2. Rogers AS, Spencer MB, Stone BM (1999) Validation and development of a method for assessing the risks arising from mental fatigue. Health Safety Executive, Sudbury
3. Bowden ZE, Ragsdale CT (2018) The truck driver scheduling problem with fatigue monitoring. Decis Support Syst 110:20–31. https://doi.org/10.1016/j.dss.2018.03.002
4. Bowler N, Gibson H (2015) Fatigue and its contributions to railway incidents. Rail Safety and Standards Board, London
5. Caprara A, Fischetti M, Toth P, Vigo D, Guida PL (1997) Algorithms for railway crew management. Math Prog 79(1–3):125–141. https://doi.org/10.1007/BF02614314
6. Caprara A, Monaci M, Toth P (2001) A global method for crew planning in railway applications. Comput Aided Scheduling Public Transport, Springer 17–36. https://doi.org/10.1007/978-3-642-56423-9_2
7. Ceder AA, Hassold S (2015) Applied analysis for improving rail-network operations. J Rail Trans Planning Manag 5(2):50–63. https://doi.org/10.1016/j.jrtpm.2015.06.001
8. Chu SC (2007) Generating, scheduling and rostering of shift crew-duties: applications at the Hong Kong International Airport. Eur J Oper Res 177(3):1764–1778. https://doi.org/10.1016/j.ejor.2005.10.008
9. Clement R, Wren A (1995) Greedy genetic algorithms, optimizing mutations and bus driver scheduling. Comput Aided Transit Scheduling, Springer 213–235. https://doi.org/10.1007/978-3-642-57762-8_14
10. Crum MR, Morrow PC (2002) The influence of carrier scheduling practices on truck driver fatigue. Transp J 42(1):20–41. https://doi.org/10.2307/20713513

11. Dorrian J, Roach GD, Fletcher A, Dawson D (2007) Simulated train driving: fatigue, self-awareness and cognitive disengagement. Appl Ergon 38(2):155–166. https://doi.org/10.1016/j.apergo.2006.03.006
12. Dos Santos AG, Mateus GR (2009) General hybrid column generation algorithm for crew scheduling problems using genetic algorithm. In: Proceedings of IEEE congress on evolutionary computation. IEEE 1799–1806. https://doi.org/10.1109/cec.2009.4983159
13. Elizondo R, Parada V, Pradenas L, Artigues C (2010) An evolutionary and constructive approach to a crew scheduling problem in underground passenger transport. J Heuristics 16(4):575–591. https://doi.org/10.1007/s10732-009-9102-x
14. Ernst A, Jiang H, Krishnamoorthy M, Nott H, Sier D (2001) Rail crew scheduling and rostering optimization algorithms. Computer-Aided Scheduling of Public Transport, Springer 53–71. https://doi.org/10.1007/978-3-642-56423-9_4
15. Ernst AT, Jiang H, Krishnamoorthy M, Sier D (2004) Staff scheduling and rostering: a review of applications, methods and models. Eur J Oper Res 153(1):3–27. https://doi.org/10.1016/s0377-2217(03)00095-x
16. Filtness AJ, Naweed A (2017) Causes, consequences and countermeasures to driver fatigue in the rail industry: the train driver perspective. Appl Ergon 60:12–21. https://doi.org/10.1016/j.apergo.2016.10.009
17. Freling R, Huisman D, Wagelmans AP (2001) Applying an integrated approach to vehicle and crew scheduling in practice. Comput Aided Scheduling Public Transport, Springer 73–90. https://doi.org/10.1007/978-3-642-56423-9_5
18. Freling R, Lentink RM, Odijk MA (2001) Scheduling train crews: a case study for the Dutch railways. Comput Aided Scheduling Public Transport, Springer 153–165. https://doi.org/10.1007/978-3-642-56423-9_9
19. Fuentes M, Cadarso L, Marín Á (2019) A hybrid model for crew scheduling in rail rapid transit networks. Transport Res Part B Methodol 125:248–265. https://doi.org/10.1016/j.trb.2019.05.007
20. García J, Altimiras F, Peña A, Astorga G, Peredo O (2018) A binary cuckoo search big data algorithm applied to large-scale crew scheduling problems. Complexity. https://doi.org/10.1155/2018/8395193
21. Han AF, Li EC (2014) A constraint programming-based approach to the crew scheduling problem of the Taipei mass rapid transit system. Ann Oper Res 223(1):173–193. https://doi.org/10.1007/s10479-014-1619-1
22. Van Dongen HP, Dinges DF (2005) Circadian rhythms in fatigue, alertness, and performance. In: Meir TR, Kryger R, William R, Dement W (eds) pp 435–443. Principles and Practice of Sleep Medicine (Fourth Edition). https://doi.org/10.1016/B0-72-160797-7/50042-2
23. Hartley L, Horberry T, Mabbott N, Krueger GP (2000) Review of fatigue detection and prediction technologies. National Road Transport Commission Virginia
24. Heil J, Hoffmann K, Buscher U (2019) Railway crew scheduling: models, methods and applications. Eur J Oper Res. https://doi.org/10.1016/j.ejor.2019.06.016
25. Hoffmann K, Buscher U (2019) Valid inequalities for the arc flow formulation of the railway crew scheduling problem with attendance rates. Comput Ind Eng 127:1143–1152. https://doi.org/10.1016/j.cie.2018.05.031
26. Jütte S, Müller D, Thonemann UW (2017) Optimizing railway crew schedules with fairness preferences. J Sched 20(1):43–55. https://doi.org/10.1007/s10951-016-0499-4
27. Kazuhiro S (2002) Revised work of the Japanese society for occupational health industry fatigue study group "Subjective symptoms" 2002. (Japanese). Digest of science of labour 57(5):295–298
28. Khmeleva E, Hopgood AA, Tipi L, Shahidan M (2018) Fuzzy-logic controlled genetic algorithm for the rail-freight crew-scheduling problem. KI-Künstliche Intelligenz 32(1):61–75. https://doi.org/10.1007/s13218-017-0516-6
29. Korf RE (1998) A complete anytime algorithm for number partitioning. Artif Intell 106(2):181–203. https://doi.org/10.1016/s0004-3702(98)00086-1

30. Lee C-K (2004) The integrated scheduling and rostering problem of train driver using Genetic algorithm. In: Proceedings of 9th international conference on computer-aided scheduling of public transport (CASPT), San Diego–California, Citeseer
31. Lee KS, Geem ZW (2005) A new meta-heuristic algorithm for continuous engineering optimization: harmony search theory and practice. Comput Methods Appl Mech Eng 194(36–38):3902–3933. https://doi.org/10.1016/j.cma.2004.09.007
32. Li Si FW, Wu, Jie, Bin W (2019) A comparison between two methods of evaluating shift subway drivers fatigue (Chinese). China Safety Sci J 029(002):7–12
33. Ma J, Ceder A, Yang Y, Liu T, Guan W (2016) A case study of Beijing bus crew scheduling: a variable neighborhood-based approach. J Adv Transp 50(4):434–445. https://doi.org/10.1002/atr.1333
34. Masahiro O (1979) Fatigue research (Second Edition), Tong Wen College, Tokyo
35. McGuffog A, Spencer MB, Stone BM, Turner C (2005) Guidelines for the management and reduction of fatigue in train drivers. Report T059. London: RSSB.
36. McGuffog A, Spencer MB, Stone BM, Turner C (2004) Working patterns of train drivers: implications for fatigue and safety. QinetiQ Centre for Human Sciences Report. KI/CHS/CR043098
37. Mingozzi A, Boschetti MA, Ricciardelli S, Bianco L (1999) A set partitioning approach to the crew scheduling problem. Oper Res 47(6):873–888. https://doi.org/10.1287/opre.47.6.873
38. Monk TH, Kennedy KS, Rose LR, Linenger JM (2001) Decreased human circadian pacemaker influence after 100 days in space: a case study. Psychosom Med 63(6):881–885. https://doi.org/10.1097/00006842-200111000-00005
39. Naweed A (2014) Investigations into the skills of modern and traditional train driving. Appl Ergon 45(3):462–470. https://doi.org/10.1016/j.apergo.2013.06.006
40. Neri DF, Oyung RL, Colletti LM, Mallis MM, Tam PY, Dinges DF (2002) Controlled breaks as a fatigue countermeasure on the flight deck. Aviat Space Environ Med 73(7):654–664. https://doi.org/10.1016/S1352-2310(02)00273-X
41. Nesthus T, Cruz C, Boquet A, Detwiler C, Holcomb K, Della Rocco P (2001) Circadian temperature rhythms in clockwise and counter-clockwise rapidly rotating shift schedules. J Human Ergol 30(1–2), 245–249. https://doi.org/10.11183/jhe1972.30.245
42. Parasuraman R, Sheridan TB, Wickens CD (2008) Situation awareness, mental workload, and trust in automation: viable, empirically supported cognitive engineering constructs. J Cogn Eng Decision Making 2(2):140–160. https://doi.org/10.1518/155534308x284417
43. Shen J, Botly LC, Chung SA, Gibbs AL, Sabanadzovic S, Shapiro CM (2006) Fatigue and shift work. J Sleep Res 15(1):1–5. https://doi.org/10.1111/j.1365-2869.2006.00493.x
44. Shen Y, Peng K, Chen K, Li J (2013) Evolutionary crew scheduling with adaptive chromosomes. Trans Res Part B Methodol 56:174–185. https://doi.org/10.1016/j.trb.2013.08.003
45. Shiffer D, Minonzio M, Dipaola F, Bertola M, Zamuner AR, Dalla Vecchia LA, Solbiati M, Costantino G, Furlan R, Barbic F (2018) Effects of clockwise and counterclockwise job shift work rotation on sleep and work-life balance on hospital nurses. Int J Environ Res Public Health 15(9):2038. https://doi.org/10.3390/ijerph15092038
46. Sodhi MS, Norris S (2004) A flexible, fast, and optimal modeling approach applied to crew rostering at London Underground. Ann Oper Res 127(1–4):259–281. https://doi.org/10.1023/b:anor.0000019092.76669.a1
47. Souai N, Teghem J (2009) Genetic algorithm based approach for the integrated airline crew-pairing and rostering problem. Eur J Oper Res 199(3):674–683. https://doi.org/10.1016/j.ejor.2007.10.065
48. Spencer M, Robertson K, Folkard S (2006) The development of a fatigue/risk index for shiftworkers. Health and Safety Executive, Sudbury
49. Suyabatmaz AÇ, Şahin G (2015) Railway crew capacity planning problem with connectivity of schedules. Transport Res Part E Logistics Transport Rev 84:88–100. https://doi.org/10.1016/j.tre.2015.10.003
50. Toshiaki T, Shinzo Y (1991) The validity of the workload evaluation by driver's subjective symptoms. Jpn J Ergonom 27(Suppl):248–249. https://doi.org/10.5100/jje.27.Supplement_248

51. Toshihisa I, Hisae O (1986) Quantification of workload in power car operation (Japanese). Jpn J Ergonom 22(Supplement):182–183
52. Tsukasaki K, Kido T, Makimoto K, Naganuma R, Sunaga K (2006) The impact of sleep interruptions on vital measurements and chronic fatigue of female caregivers providing home care in Japan. Nurs Health Sci 8(1):2–9. https://doi.org/10.1111/j.1442-2018.2006.00261.x
53. Vazirani VV (2013) Approximation algorithms, Springer Science & Business Media
54. Wang Q, Yang J, Ren M, Zheng Y (2006) Driver fatigue detection: a survey. In: Proceedings of 6th world congress on intelligent control and automation. IEEE 8587–8591. https://doi.org/10.1109/wcica.2006.1713656
55. Wilson J, Farrington-Darby T, Cox G, Bye R, Hockey GRJ (2007) The railway as a socio-technical system: human factors at the heart of successful rail engineering. In: Proceedings of the institution of mechanical engineers, Part F: J Rail Rapid Transit 221(1):101–115. https://doi.org/10.1243/09544097jrrt78
56. Zhang Y, Fang W, Er M-C, Wang J, Guo B-Y (2010) Analysis on fatigue risk of subway divers on shift (Chinese). Railway Trans Econ 32(4):90–94. https://doi.org/10.3969/j.issn.1003-1421.2010.04.023
57. Zhao Y, Yue WJ (2017) Cognitive radio networks with multiple secondary users under two kinds of priority schemes: performance comparison and optimization. J Industr Manag Optimizat 13(3):1449–1466. https://doi.org/10.3934/jimo.2017001

Evaluation of Park and Ride Effect Based on Fuzzy Analytic Hierarchy Process

Yan Xing, Xin Rao, Weidong Liu, and Wenhao Song

Abstract In order to reduce the complexity of the park-and-ride effect evaluation, enhance practicability, combining with the fuzzy hierarchy analysis model to evaluate park-and-ride effect analysis based on multiple park-and-ride point of field survey and access to relevant data, select more concerned about when the passengers to transfer to the three level indicators and nine secondary indicators, quantify the indexes, and the weight of each index is obtained by analytic hierarchy process, using the fuzzy analysis method for comprehensive evaluation In the end, this paper selects the parking and transfer point of Shengli South Street in Shenyang as an example to evaluate its implementation effect. The evaluation result is better, which is consistent with the results of the on-site passenger transfer satisfaction survey. Meanwhile, Suggestions are proposed for further optimization and improvement of the transfer point.

Keywords Traffic engineering · Effect evaluation · Analytic hierarchy process · Park-and-ride

1 The Introduction

Park-and-ride plays an important role in alleviating traffic congestion and improving travel efficiency. In the study of the evaluation of driving effect, Jinushan et al. [1] constructed the index system of park-and-ride, and evaluated the efficiency of park-and-ride by using the data envelope method. Gang et al. [2] evaluated the efficiency of stop and ride by establishing the evaluation structure equation. However, in the process of evaluation, more indicators are needed and difficult to obtain, and the methods used are also more complex. Hou et al. [3] analyzed and studied the influence of community opening on road capacity through fuzzy analytic hierarchy process (AHP). Peng et al. [4] evaluated passenger satisfaction in bus transfer system through fuzzy analytic hierarchy process (AHP). Guo [5] evaluated the transfer connection of

Y. Xing (✉) · X. Rao · W. Liu · W. Song
Shenyang Jianzhu University, Shenyang, China
e-mail: 770859389@qq.com

W. Wang et al. (eds.), *Green Transportation and Low Carbon Mobility Safety*, Lecture Notes in Electrical Engineering 944,
https://doi.org/10.1007/978-981-19-5615-7_2

urban rail transit hub through analytic hierarchy process. Peihong and Mingxin [6] investigated and analyzed transfer factors by means of in-depth interview. Yang et al. [7] comprehensively evaluated the rationality of spatial structure layout of urban road network through fuzzy evaluation method. Ting [8] evaluated the economy of public transportation in Tianjin through analytic hierarchy process. The fuzzy analytic hierarchy process (FAHP) is relatively simple and practical. In the actual evaluation, there are few evaluation indexes required, and the evaluation indexes are easy to obtain, and the obtained results are relatively reliable. Therefore, this paper uses the FAHP to evaluate the effect of park-and-ride. Focusing on the implementation effect, this paper constructs a multi-level evaluation index system for the implementation effect of park-and-ride, and establishes a fuzzy evaluation model for the implementation effect of park-and-ride based on the fuzzy analytic hierarchy process. Based on the evaluation and analysis of the example of Shengli South Street in Shenyang, this paper provides a theoretical basis for the effective implementation of park-and-ride and provides a basis for the government to formulate effective policies.

2 Establish the Evaluation Index System for the Effect of Park-and-Ride

There are many indicators to be considered when the fuzzy analytic hierarchy process is used to evaluate the implementation effect of park-and-ride. However, all indexes cannot be taken into account in the actual evaluation. Therefore, the selection of evaluation indexes is the key to determine the feasibility of this method. In the selection of evaluation indicators, mainly from the aspect of passengers, through research and access to relevant information to determine the issues that passengers are more concerned about during transfer, these issues are taken as evaluation indicators. However, when these problems are used as evaluation indicators, they should also have the following characteristics: (1) It is easy to obtain, and the data needed in the evaluation process is easy to obtain and check. (2) Representativeness. The indicators adopted can represent the overall effect of the implementation of park-and-ride. (3) Comparability. The indicators adopted have uniform measurement standards, which can be used for effective comparison. Based on the above index characteristics, the evaluation index system of transfer effect is obtained, as shown in Fig. 1.

3 An Evaluation Model for the Effect of Park-and-Ride Based on Fuzzy Analytic Hierarchy Process is Established

Fuzzy Analytic Hierarchy Process (FAHP) is a systematic analysis method combining qualitative and quantitative methods. Firstly, the evaluation indexes are selected for quantification of the problems to be analyzed, and then the evaluation

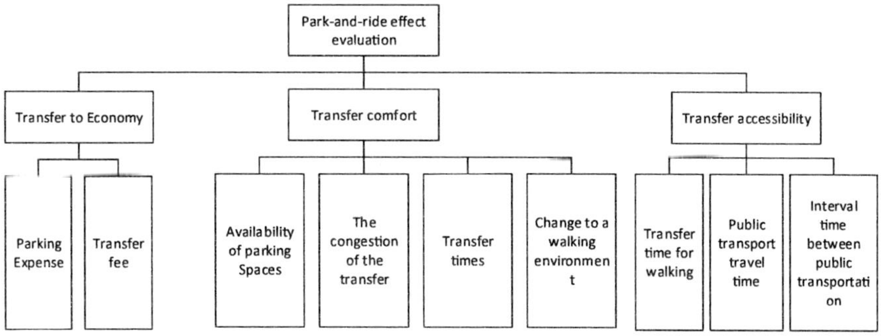

Fig. 1 Transfer effect evaluation index system

results are obtained by comprehensive evaluation. When fuzzy analytic hierarchy process is used to evaluate the effect of park-and-ride, the selection of evaluation indexes is mainly based on the consideration of passengers. The weight of each evaluation index was obtained by analytic hierarchy process, and then the grade of stop-and-ride effect was determined by fuzzy evaluation method.

3.1 *Analytic Hierarchy Process (AHP) Determines the Weight of Indicators*

Analytic Hierarchy Process (AHP) is a decision-making method that decomposes the elements related to decision-making into levels such as goals, criteria and programs, and carries out qualitative and quantitative analysis on this basis. In this paper, AHP is used to solve the weight of each index.

3.2 *Construction of Judgment Matrix*

In the evaluation index system of transfer effect, the indexes of each layer are compared pairwise with each other. According to the 9 importance levels proposed by American operations researcher Saaty and their assigned values are shown in Table 1. The matrix formed by pairwise comparison results is called judgment matrix, and the value of the elements of judgment matrix reflects people's understanding of the relative importance of various factors.

Table 1 Evaluation scale table

Scale	Meaning
1	The two indicators are equally important
3	The former indicator is slightly more important than the latter
5	The former indicator is more important than the latter
7	The former indicator is strongly more important than the latter
9	The former is extremely important than the latter
2, 4, 6, 8	The middle value of two adjacent judgments

3.3 Hierarchical Single Sort

After constructing the judgment matrix, the maximum eigenroot λ_{max} of the judgment matrix is solved first. Then solve the eigenvector corresponding to λ_{max}which is normalized and denoted as W. The element of W is the ranking weight of the relative importance of factors at the same level relative to factors at the upper level, which is called hierarchical single sorting. In this paper, the root method is used to calculate the initial weight of each index. Firstly, the product M_i of each row of the judgment matrix is calculated. Then calculate the NTH root of M_i to get $\overline{w}_i = \sqrt[n]{M_i}$, and then normalize vector $\overline{W} = (\overline{w}_1\, \overline{w}_2 \ldots \overline{w_n})^{\mathrm{T}}$ to get weight vector $W = (w_1\, w_2 \ldots w_n)^{\mathrm{T}}$. Calculate the maximum eigenroot of λ_{max} to get $\lambda_{max} = \sum_{i=1}^{n} \frac{(AW)_i}{nw_i}$.

3.4 Consistency Check

Consistency test is to determine the allowable range of inconsistency with λ_{max}, and the consistency index is calculated with CI. Consistency indicators are defined as:

$$CI = \frac{\lambda_{max} - n}{n - 1} \tag{1}$$

where: λ_{max}—the maximum eigenvalue of the judgment matrix; n—The order of the judgment matrix.

CI= 0. Complete consistency. The closer CI is to 0, the greater the consistency is.CI The bigger the discrepancy, the less it is. It is always considered consistent when the judgment matrix is first or second order. However, when the judgment matrix is of other orders, the judgment of CI is very qualitative and cannot accurately judge whether the consistency reaches the allowable range. Therefore, for the purpose of quantitative judgment, the random consistency index ratio CR is introduced. The random consistency index is defined as:

Table 2 RI value

Order number	3	4	5	6	7	8	9	10
RI	0.58	0.90	1.12	1.24	1.32	1.41	1.45	1.49

$$CR=\frac{CI}{RI} \tag{2}$$

where:CI—consistency index;RI—Random consistency index, which is related to the order of the judgment matrix, is shown in Table 2.

When $CR < 0.1$, the judgment matrix is considered to pass the consistency test; otherwise, the judgment matrix needs to be readjusted until it passes the consistency test.

3.5 *Fuzzy Evaluation of Park-and-Ride Effect*

Fuzzy evaluation method is a comprehensive evaluation method based on fuzzy mathematics. Firstly, the evaluation language set of the effect of stop and ride is determined, and then the evaluation index system of the effect of stop and ride is established, as shown in Fig. 1. In this paper, the two-level fuzzy comprehensive evaluation is adopted to evaluate, and finally an evaluation vector is obtained, and then the evaluation level of the stop-and-ride effect is determined according to the maximum membership degree principle.

4 Case Application

4.1 *Determine the Weight of Each Index*

Through the field research and access to relevant materials, determine the indicators for the evaluation of stop and ride. Then a questionnaire was conducted at the park-and-ride station on Shengli South Street in Shenyang, where passengers were asked to rate each indicator. A total of 200 questionnaires were sent out, and 153 valid answers were screened in the end. The relative importance of each index was determined by the results of the questionnaire survey. The indicators in the evaluation system are compared pairwise, their relative importance is quantified, and the judgment matrix is constructed. Then solve the weight of each index, and carry out consistency test. See Tables 3, 4, 5 and 6.

Table 3 Comprehensive evaluation judgment matrix table

A	B	C	D	The weight	Consistency check
B	1	1/5	1/6	0.078	$\lambda_{max} = 3.093$ 0.047
C	5	1	1/3	0.287	CR = CI/RI = 0.081 < 0.1
D	6	3	1	0.635	Satisfy consistency test

Table 4 Transfer economic judgment matrix table

B	B1	B2	The weight	Consistency check
B1	1	3	0.750	$\lambda_{max} = 2$ $CI = 0$
B2	1/3	1	0.250	$CR = CI/RI = 0 < 0.1$

Table 5 Transfer comfort matrix table

C	C1	C2	C3	C4	The weight	Consistency check
C1	1	3	6	4	0.535	$\lambda_{max} = 4.146$
C2	1/3	1	5	3	0.275	$CI = 0.049$ $RI = 0.9$
C3	1/6	1/5	1	1/3	0.060	$CR = CI/RI = 0.054 < 0.1$
C4	1/4	1/3	3	1	0.130	Satisfy consistency test

Table 6 Transfer accessibility judgment matrix table

D	D1	D2	D3	The weight W_E	Consistency check
D1	1	2	3	0.540	$\lambda_{max} = 3.009$ $CI = 0.005$
D2	1/2	1	2	0.297	$CR = CI/RI = 0.009 < 0.1$
D3	1/3	1/2	1	0.163	Satisfy consistency test

4.2 *Effect Rating Evaluation*

After the weight of each index is obtained by analytic hierarchy process, the grade is evaluated by fuzzy evaluation method. Firstly, the evaluation language set V = {good (V1), good (V2), average (V3), poor (V3), poor (V4)} was established for the effect evaluation of park-and-ride.

(1) Build evaluation matrix

The fuzzy evaluation matrix of each evaluation index is obtained by evaluating the different indexes of a level by experts and combining with the results of field research. The following is the evaluation matrix of each index in the evaluation system of the implementation effect of park-and-ride on Shengli South Street in Shenyang:

The evaluation matrix of parking and riding economy:

$$R_B = \begin{bmatrix} 0.2 & 0.2 & 0.5 & 0.1 & 0 \\ 0.1 & 0.2 & 0.2 & 0.4 & 0.1 \end{bmatrix}$$

Evaluation matrix of park-and-ride comfort:

$$R_C = \begin{bmatrix} 0.1 & 0.3 & 0.4 & 0.1 & 0.1 \\ 0.2 & 0.7 & 0.1 & 0 & 0 \\ 0.1 & 0.4 & 0.2 & 0.1 & 0.2 \\ 0.3 & 0.4 & 0.3 & 0 & 0 \end{bmatrix}$$

Evaluation matrix of park-and-ride accessibility:

$$R_D = \begin{bmatrix} 0.2 & 0.6 & 0.2 & 0 & 0 \\ 0 & 0 & 0.2 & 0.4 & 0.4 \\ 0.1 & 0.1 & 0.2 & 0.5 & 0.1 \end{bmatrix}$$

(2) First-level comprehensive evaluation

The weight of each index in the middle layer was multiplied with its corresponding evaluation matrix, and the evaluation results of the middle layer transfer economy, transfer comfort and transfer accessibility were obtained, as shown below.

$$R_B^* = \begin{bmatrix} 0.750 \\ 0.250 \end{bmatrix}^{\mathrm{T}} \begin{bmatrix} 0.2 & 0.2 & 0.5 & 0.1 & 0 \\ 0.1 & 0.2 & 0.2 & 0.4 & 0.1 \end{bmatrix} \begin{bmatrix} 0.125 & 0.150 & 0.400 & 0.125 & 0.200 \end{bmatrix}$$

$$R_C^* = \begin{bmatrix} 0.535 \\ 0.275 \\ 0.060 \\ 0.130 \end{bmatrix}^{\mathrm{T}} \begin{bmatrix} 0.1 & 0.3 & 0.4 & 0.1 & 0.1 \\ 0.2 & 0.7 & 0.1 & 0 & 0 \\ 0.1 & 0.4 & 0.2 & 0.1 & 0.2 \\ 0.3 & 0.4 & 0.3 & 0 & 0 \end{bmatrix} = \begin{bmatrix} 0.154 & 0.428 & 0.292 & 0.060 & 0.066 \end{bmatrix}$$

$$R_D^* = \begin{bmatrix} 0.540 \\ 0.297 \\ 0.163 \end{bmatrix}^{\mathrm{T}} \begin{bmatrix} 0.2 & 0.6 & 0.2 & 0 & 0 \\ 0 & 0 & 0.2 & 0.4 & 0.4 \\ 0.1 & 0.1 & 0.2 & 0.5 & 0.1 \end{bmatrix} = \begin{bmatrix} 0.124 & 0.340 & 0.200 & 0.200 & 0.136 \end{bmatrix}$$

(3) Second-level comprehensive evaluation.

By combining the evaluation results of transfer economy, transfer comfort and transfer accessibility obtained from the first-level comprehensive evaluation, the evaluation matrix of stop and ride effect is obtained as follows:

$$R_A = \begin{bmatrix} 0.125\ 0.150\ 0.400\ 0.125\ 0.200 \\ 0.154\ 0.428\ 0.292\ 0.060\ 0.066 \\ 0.124\ 0.340\ 0.200\ 0.200\ 0.136 \end{bmatrix}$$

Calculate the second-level comprehensive evaluation results:

$$R_A^* = \begin{bmatrix} 0.078 \\ 0.287 \\ 0.635 \end{bmatrix}^{\mathrm{T}} \begin{bmatrix} 0.125\ 0.150\ 0.400\ 0.125\ 0.200 \\ 0.154\ 0.428\ 0.292\ 0.060\ 0.066 \\ 0.124\ 0.340\ 0.200\ 0.200\ 0.136 \end{bmatrix}$$
$$= \begin{bmatrix} 0.133\ 0.350\ 0.242\ 0.154\ 0.121 \end{bmatrix}$$

4.3 Evaluation Results

According to the fuzzy analytic hierarchy process, the relative membership degrees of Shengli South Street in Shenyang with respect to the evaluation language set are respectively 0.133, 0.350, 0.242, 0.154 and 0.121. It can be seen that the maximum membership degree is 0.350, and its corresponding grade is good. Therefore, it can be considered that the implementation effect of Shengli South Street in Shenyang is good. From the actual situation, Shengli South Street transfer point in Shenyang is indeed very convenient, no matter to transfer by bus or subway is very close, but the transfer of parking costs are more expensive, the departure time and ride time of public transport is longer, these are the current problems. The government can introduce corresponding policies to reduce the parking cost of interchange and encourage people to use public transportation for park-and-ride. Shorten the time between departures, improve the efficiency of departures, reduce people's waiting time. At the same time, speed up the construction of the subway, form a complete subway network system, shorten the ride time, improve the efficiency of travel, but also can reduce the transfer cost.

5 Conclusion

Strengthening the effect evaluation research of park-and-ride is conducive to improving the efficiency of park-and-ride and easing the problem of urban traffic congestion. On the basis of constructing the multi-level evaluation index system of the effect of stop and ride, the initial weight of each evaluation index is determined by the analytic hierarchy process and fuzzy evaluation method, and then the comprehensive evaluation is carried out. Through example analysis and application, this method can quickly and accurately evaluate the implementation effect of park-and-ride strategy, improve the rationality and efficiency of the evaluation results, and provide a theoretical basis for the government departments to introduce the corresponding policies and improvement measures. However, the evaluation indexes selected in the evaluation are only the representative ones that passengers are more concerned about, and the consideration of the indexes is not complete enough. At the same time, in the process of evaluation, we refer to more artificial subjective feelings and judgments, and there are more qualitative factors. Therefore, it has a certain influence on the evaluation results, but the evaluation results are more consistent with the actual survey results, and the evaluation method has practical application significance.

References

1. Jinshuan PE, Lei ZH, Lei XU (2019) J Chongqing Jiaotong Univ (Natural Science Edition) 38(01):97–101+112
2. Gang C, Heping Z, Kefei Y (2006) A structural equation model for public evaluation of transfer efficiency of park-and-ride facilities in urban periphery. Syst Eng 03:29–34
3. Hou W, Qiao F, Shen Y, Zhang X, Fan X (2019) Research on the influence of community opening on road capacity based on AHP-FCE model. J Sci Technol Econom 27(20):207–208
4. Wu, Peng Han, Jinmei YI, (2014) Fuzzy evaluation of passenger satisfaction in bus transfer system based on AHP-DEMATEL method. J Chongqing Jiaotong Univ (Social Science Edition) 14(01):19–22
5. Guo P (2019) Fuzzy evaluation of transfer connections in urban rail transportation hubs based on Analytic Hierarchy Process (AHP). Automobile Pract Technol (19):206–207+220
6. Peihong C, Mingxin S (2019) Analysis on the influencing factors of park-and-ride behavior in Beijing. Beijing Jiaotong University (Social Science Edition) 18(01):38–47
7. Li, Yang, Liheng Z, Xu, Hou (2019) Fuzzy comprehensive evaluation of spatial structure layout rationality of urban road network. Highway Eng 44(03):102–106
8. Ting W (2015) Research on economic evaluation of public transportation in Tianjin based on AHP. China Business Rev 22:177–180

Exploiting Multi-Dec Net for Detecting Traffic Congestion in the Surveillance System

Jie Li, Jiaojiao Sun, Jun Wang, Yedi Zhuo, and A. N. D. Yinli Jin

Abstract Congestion recognition is very important for traffic flow control. The current traffic monitoring system provides a large number of videos and pictures, which are not fully utilized. With the widespread application of deep learning in the transportation field, we proposed a congestion classification method based on Convolutional Neural Network. First, we obtain representative data from different camera angles, weather, and light conditions from the traffic monitoring system in Shaanxi Province, China. Drawing on the idea of dense blocks, we established a Multi-Dec Net to classify the congestion state of the images captured from the monitoring system. The Network consists of the first-stage identification Multi-Dec Net 1 and the second-stage identification Multi-Dec Net 2. We compare Multi-Dec Net with Alexnet, VggNet, and Resnet, the test results show that Multi-Dec Net has achieved an accuracy rate of 95.96% and an error rate of 4.78%. The model will also be deployed in the highway monitoring center in Shaanxi Province in the future, which will help early warning of traffic conditions, and rapid emergency response.

Keywords Dense block · Multi-Dec Net · Congestion classification

1 Introduction

Highway transportation is an indispensable part of the national economic construction. With the increase in the number of motor vehicles, the highway characterized by high speed and large capacity has achieved rapid development [1]. As of the end of the "Thirteen-tenths Five-Year Plan" period, the total mileage of expressways in our country has exceeded 169,000 km. Besides, traffic congestion in the highway system starts to emerge frequently, which is likely to cause traffic accidents, waste

J. Li · J. Sun (✉)
School of Electronics and Control Engineering, Chang'an University, 710064 Xian, China
e-mail: 1021690370@qq.com

J. Wang · Y. Zhuo · A. N. D. Yinli Jin
School of Transportation Engineering, Chang'an University, Xi'an, China

W. Wang et al. (eds.), *Green Transportation and Low Carbon Mobility Safety*,
Lecture Notes in Electrical Engineering 944,
https://doi.org/10.1007/978-981-19-5615-7_3

transportation resources, pollute the environment, reduce residents' traffic efficiency, and seriously affect the physical and mental health of travelers [2, 3]. Therefore, it is urgent to solve the problem of traffic congestion.

The current highway construction has entered a period of management, control, and optimization from large-scale construction [4, 5]. Through real-time and accurate access to traffic flow, vehicle type, speed, lane occupancy, traffic density and other road traffic information. The artificial intelligence method is used to distinguish the road traffic conditions, and the early warning information is issued timely by means of road information board and app. That can help road managers to guide the diversion more effectively and ensure the smooth traffic.

The way to acquire highway traffic information is mainly fixed detector collection and vehicle-mounted global positioning system (GPS) collection. The fixed detectors are mainly electromagnetic, ultrasonic, radar, infrared and other detectors. These methods need to perceive the surrounding environment of the vehicle, the signal is very susceptible to interference, and maintenance is complicated [6]. The vehicle-mounted GPS collection method requires the installation of GPS equipment on the running vehicle, which has high initial investment cost and needs to collect a large amount of data (such as Google Maps). The applicability is poor and it is difficult to perform the real-time calculation of massive data [7].

With the development of computer vision and the realization of video surveillance for all sections of expressways, it is possible to detect congestion based on video images [8]. Real-time analysis of traffic video captured by surveillance cameras to obtain the required traffic data. [9] used the background difference method and the frame difference method to analyze the texture change of the image, and then obtain the road space occupancy rate and time occupancy rate to judge the road congestion. Zhu selected the data processing technology based on the YOLO-v3 algorithm to realize the detection and tracking of moving targets on the highway, and it can realize the alarm function when traffic congestion and traffic accidents occur [10]. But this method faces the problem that it is difficult to track the target under high occlusion and low frame rate.

Nowadays, a large number of surveillance cameras have been installed on highways [11]. These large-scale cameras have been replaced with cameras with pan-tilts. Traditional fixed-focus cameras cannot rotate freely and it is difficult to capture distant congested scenes. The method of using video to identify congestion has good performance in real-time and accuracy. However, due to the limitation of transmission and processing capabilities, in large-scale road networks, applications based on high frame rate video often consume a lot of resources. A large-scale highway network congestion detection system requires a lot of resource investment [12].

Image-based classification methods can be roughly divided into two categories: traditional machine learning-based methods and deep learning-based methods. In traditional machine learning methods, k-nearest neighbor (KNN) and support vector machine (SVM) are usually used to classify images. Reference [13] taking the traffic parameter data collected on the South Second Ring Expressway in Xi'an as an example, it compared and verified the effectiveness of the traffic congestion identification method under different SVM separators, and the recognition accuracy reached

91%. Traditional image processing methods are difficult to be used for traffic image classification in consideration of various scenes and interference [14]. Deep learning methods overcome the shortcomings of tradition machine learning algorithms that rely on manual feature extraction. As a special type of deep neural network, CNN has been successfully applied to image classification tasks [13]. Pamula distinguished four traffic condition levels corresponding to the LOS category with an average recognition accuracy rate of 89% [14]. Reference [15] using ResNet for congestion data classification, the test accuracy was 95%. These methods can achieve high accuracy. Using the first-stage network model to identify congestion can achieve high accuracy, but there is a high rate of misjudgment. Aiming at the problem of a high false alarm rate in congestion identification, the manual review method will produce a lot of manpower redundancy.

To sum up, to achieve the development goal of saving management, we propose a multi-stage network model, which can effectively reduce the misjudgment rate under the condition of high accuracy. In this paper, a method based on Convolutional Neural Network was proposed for traffic classification. With the idea of Dense Block, an optimization and anti-reversal mechanism were formed through a multi-stage recognition model, and we obtained better detection accuracy by implementing fine-tuning the network.contributions of this work are listed as follows:

- We established a monitoring perspective sample database and obtained six sections of Shaanxi Expressways (Raocheng, Xibao, Xiyu, Xiyao, Baomao, and Xihan) under different weather and lighting conditions, and the pictures were taken from 7 am to 7 pm. There are more than 30,000 pictures from August to December 2020.
- We employed a "multi-stage recognition" model to build the road congestion detection model. The model can effectively reduce the misjudgment rate of congestion recognition.
- We deployed the model in Raocheng Expressway of Xi'an to identify road information and convey this information to operators and travelers.

2 Related Work

Currently, highway monitoring equipment has been popularized, and a lot of information can be obtained by this equipment. With the continuous development of deep learning, it has been successfully applied in the field of transportation. Reference [16] a bi-directional partition method using link density data was proposed to identify the best-restricted zone (RZ) in a congested single-center city to assist in the implementation of regional congestion management strategy. Reference [17] proposed a short-term traffic flow prediction model based on deep learning. The model used CNN at the bottom of the network to extract traffic flow features, and input the extraction results into the Support Vector Regression (SVR) regression model for traffic prediction. Reference [18] proposed an improved single-shot detector (SSD) embedded license plate segmentation and recognition system based on feature extractors with deep separable convolution and linear bottleneck. Accurate and real-time identification of traffic conditions can meet people's needs for efficient travel. Good traffic conditions can reduce the risk of traffic accidents.

Among the many algorithms of deep learning, convolutional neural networks have attracted much attention. In 1998, Yann Lecun formally proposed the LeNet based on predecessors. LeNet used a complete network structure and laid a model for the theoretical basis of modern CNN [19]. In 2012, the AlexNet proposed by Alex established the position of Deep Learning in the field of image recognition, and CNN ushered in a period of vigorous development [20]. The VggNet proposed in 2014 achieved a classification error of 7.3% [21]. The ResNet network structure proposed in 2015 can be regarded as a milestone event in the history of CNN network development [22]. Subsequently, based on ResNet, DenseNet was proposed in the best paper of CVPR 2017, which is a brand new network structure, surpassing ResNet comprehensively in the CIFAR index [23]. DenseNet has absorbed the most essential parts of ResNet and has done more innovative work on it, which has further improved the network performance.and further improved the network performance. Thus, the CNN network has begun to be widely applied in various fields.

Convolution neural network is particularly prominent in image classification. Reference [24] used a semi-supervised convolutional neural networks to classify vehicle types. Reference [25] proposed a road congestion detection method based on multi-dimensional visual features and convolutional neural networks (CNN). These methods are based on the first-stage model of congestion classification, convolution neural network is not only reflected in the application of the first-stage model but also gradually promoted the use of a multi-stage model. Reference [26] studied a multi-stage classification construction algorithm, which separated the easily distinguishable emotions first, and then classified the easily confused emotions, and judged the emotion type of the speech to be recognized step by step. Reference [27] proposed an automatic segmentation and recognition method of dental jaw model based on multi-stage hierarchical three-dimensional convolutional neural network, which was used to realize the segmentation of teeth and gums. Reference [28] a cascade classifier was constructed by using the fusion of output category information of all levels of classifiers, which has achieved good results in the experiment of recognizing 3755 categories of national standard first-stage Chinese characters.

Currently, there are many congestion detection methods based on deep learning. Reference [1] used traffic data from the highway control center and weather data from local monitoring stations to predict traffic flow in severe weather. Reference [29] proposed a combined prediction method for short-term traffic flow based on the autoregressive integral moving average (ARIMA) model and long short-term memory (LSTM) neural network, it can make short-term predictions of future traffic flow based on historical traffic data. Reference [30] used a novel long short-term memory (LSTM) network for more accurate traffic flow prediction. A two-way long and short-term memory model that can learn the upper and lower correlations of traffic flow is constructed, and each part of the model is used to predict the corresponding lane and aggregated traffic flow on the same day [31]. These data generally need to be collected by sensors. Based on the highway image data set, in our previous work [12, 15, 32], the first-stage network model was used to identify congestion. While achieving high accuracy, there was a problem with a high misjudgment rate. For the problem of high misjudgment rate, the paper explored a multi-stage recognition traffic congestion classification model based on deep learning algorithms to identify traffic congestion. We tried to find the best solution to the problem to avoid the impact of abnormalities and obtain a lower misjudgment rate. Finally, the optimal result was obtained by adjusting the network parameters.

3 Methodology

First of all, our method is proposed based on the idea of dense blocks [33]. Each convolutional layer in the dense block will get the output of all previous convolutional layers in this dense block to realize the reuse of features. The difference between networks based on dense blocks and other networks is that it improves the transmission efficiency of information and gradients in the network. It uses a more radical dense connection (connecting all layers) to reuse features, which greatly improves its performance. We establish a multi-stage model named Multi-Dec Net based on dense blocks to achieve the purpose of traffic congestion image classification.

3.1 Build Dense Blocks

As shown in Fig. 1, this kind of dense connection structure is inspired by the cross-layer connection design of the residual network. Each layer will be connected with all the previous layers in the channel dimension (the feature map size of each layer is the same), and serve as the input of the next layer, which can realize feature sharing and improve the efficiency of information flow transmission. Its characteristic transmission mode is as follows:

$$X_l = H_l([X_0, X_1, \cdots, X_{l-1}]) \tag{1}$$

$$X_l = X_0 * H_0 \tag{2}$$

$$X_2 = (X_0 + X_1) * H_1 \tag{3}$$

Refer to (1), (2), and (3), $[X_0, X_1, \cdots X_{l-1}]$ represents the output from 0 to L-1 layer, H_1 represents the nonlinear transformation function, which is a combination (three consecutive operations are BN, ReLU and Conv, namely batch normalization, ReLU function and 3*3 convolution). Input X_0, change H_0 to get X_1, input X_0 and X_1 in turn and change H_1 to get X_2.

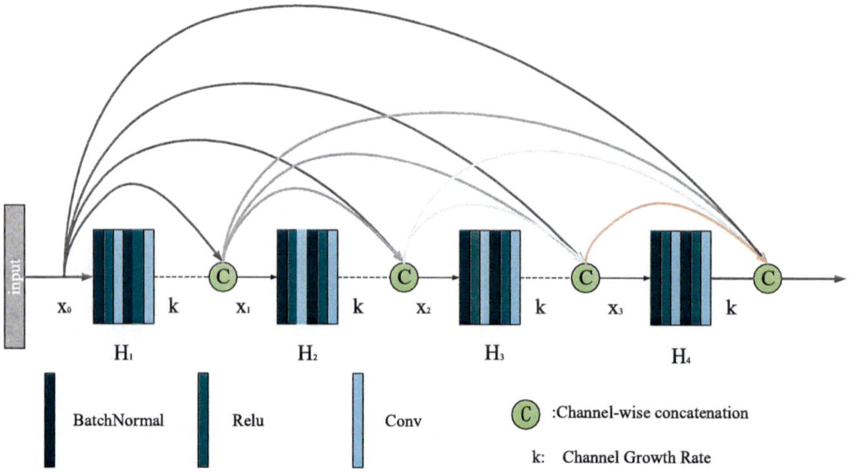

Fig. 1 Structure diagram of dense block

3.2 *Construction of Multi-Dec Net 1/2*

Based on the DenseNet, we have constructed two networks of different depths. There are two different layers 121 and 169 named Multi-Dec Net 1/2, which are used to establish the multi-stage model we proposed. The specific parameters of the two networks are shown in Table 1.

3.2.1 Construction of Multi-stage Mode

The integration of multiple classifiers has a certain complementarity, and the integration of the recognition results of multiple classifiers can make up for the deficiencies of each classifier. A system composed of multiple classifiers based on extracting different features can improve overall performance. Under the influence of the DenseNet network, a "multi-level recognition" model is established to form an optimization and anti-reversal mechanism. The structure of Multi-Dec Net 1/2 is shown in Fig. 2.

After fine-tuning the original DenseNet network, we proposed a multi-stage classification network of Multi-Dec Net. Based on the two different depths of Multi-Dec Net 1/2, the features between images with higher similarity in different categories are further extracted. The multi-stage model proposed in this paper is mainly composed of two parts: (i) the first-stage recognition network Multi-Dec Net 1, (ii) the second-stage recognition network Multi-Dec Net 2, both of which are based on the DenseNet network. At the end of the first-stage recognition network, a specific threshold probability output function is added to further filter the data. The data that the first-stage model is judged as congested and the probability is less than 100% are used as the

Table 1 The parameter of Multi-Dec Net 1/2

Layers	Multi-Dec Net 1	Multi-Dec Net 2
Convolution	7*7 Conv	7*7 Conv
Pooling	3*3 max pool, stride 2	3*3 max pool, stride 2
Dense Block(1)	DB*6	DB*6
Trans Layer(1)	1*1 Conv	1*1 Conv
	2*2 average pool, stride 2	2*2 average pool, stride 2
Dense Block(2)	DB*12	DB*12
Trans Layer(2)	1*1 Conv	1*1 Conv
	2*2 average pool, stride 2	2*2 average pool, stride 2
Dense Block(3)	DB*24	DB*32
Trans Layer(3)	1*1 Conv	1*1 Conv
	2*2 average pool, stride 2	2*2 average pool, stride 2
Dense Block(4)	DB*16	DB*32
Classification Layer	7*7 global average pool 3-D fully-connected, softmax	

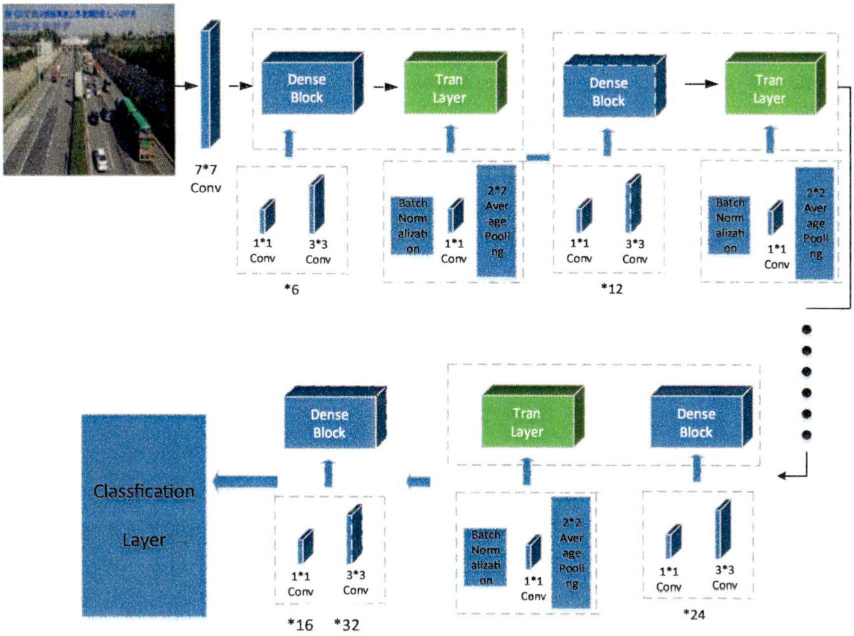

Fig. 2 The structure diagram of Multi-Dec Net 1/2

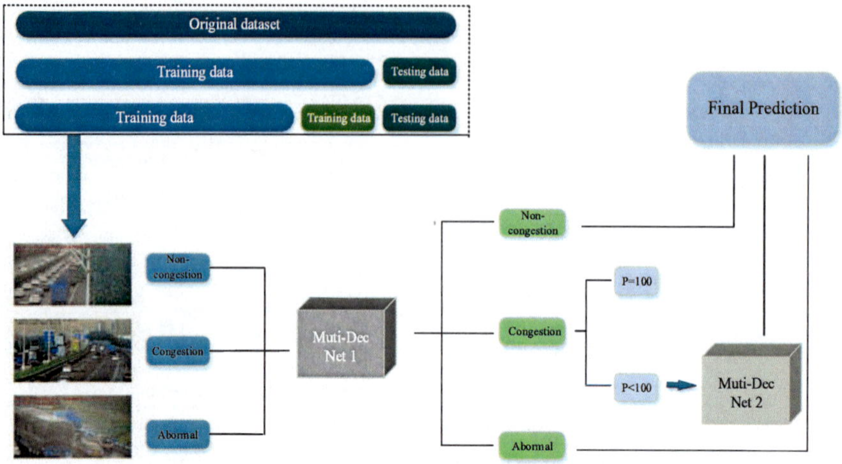

Fig. 3 The identification diagram of the multi-stage model

input of the Multi-Dec Net 2, and the final classification result is obtained through the test of the two-stage recognition model, as shown in Fig. 3.

Using the samples with known characteristics as training sets to build a mathematical model, and then using the established model to predict the unknown samples, the model can be inferred from the tagged training data set [34]. The result of our judgment on the classification problem comes from probability. The final classifier is connected with the output of the last full connection layer of the network and calculates the probability of each picture as different categories.

In this article, first, we set the probability of the congestion category to a specific threshold. Then, the data is judged using the first-level model, and when the probability of congestion is less than the set threshold, it will be filtered out as the input of the second-level model, and count the final classification results.

4 Experiments Methods

The experiment is based on Tensorflow's deep learning framework, using a computer with a hardware configuration of the Intel i7 CPU processor, 16 GB of memory, and 1080 GPU. The experiment uses a local notebook programming environment to test the model and remotely calls the server for training. The server uses the Linux operating system, Intel(R) Xeon(R) CPU E5-2678 v3 @2.5 GHz, GTX 1080Ti * 4, Python3.6, and CUDA programming environment are used to train and test our traffic congestion dataset.

4.1 Traffic Congestion Dataset

Based on the road surveillance system, we obtain traffic videos, set different time intervals, and automatically intercept the video as a picture. In this paper, to avoid the repetitiveness of the scene, we set 30 s as a fixed time interval and intercept the video obtained from the road monitoring network as a picture. Congestion conditions classifier could predict the congestion conditions using the captured images. Then dense congestion information is processed through related applications and informed to operators and travelers. Figure 4 illustrates the framework of this vision-based congestion detection system.

Our data is provided by the Shaanxi Province, China Transportation Administration. The raw image data is obtained through the camera in the outdoor transmission line environment. We established the congestion data set based on the existing monitoring system, the system includes no more than 14,700 different camera monitoring perspectives from the six sections of the Shaanxi highway (Raocheng, Xibao, Xiyu, Xiyao, Baomao, and Xihan) to obtain the images from 7 am to 7 pm between August and December 2020. The data was divided into non-congested, congested, and abnormal. As shown in Fig. 5, our data covers traffic conditions under different weather (cloudy, sunny, rain, snow, and fog) and light conditions at the same time, which is helpful to improve the robustness of the model. To increase the diversity of samples, we added data of special holidays (Labor Day, Tomb-sweeping Day, Spring Festival, National Day, New Year's Day) to build a larger traffic image database. To verify the validity and correctness of the Multi-Dec Net proposed in this paper, the sample set is randomly divided into training data, testing data, and verifying data according to the ratio of 8:1:1.

We selected more than 30,000 pictures of different weather conditions, different scenes, different seasons, and lighting conditions as the training set for this experiment. In this experiment, we marked the data as 0: non-congestion, 1: congested, 2: abnormal. When the vehicles are driving on the highway, we define that vehicles stop driving, small distance between workshops, and congestion on one or both sides of the road as congestion. In the process of driving, there are few vehicles, and the

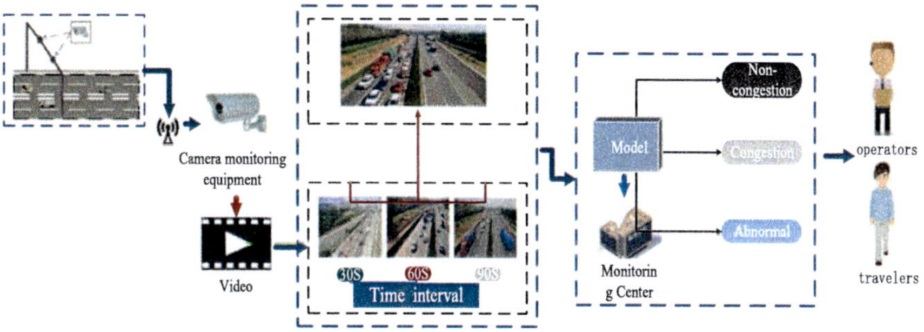

Fig. 4 Framework of vision-based congestion detection

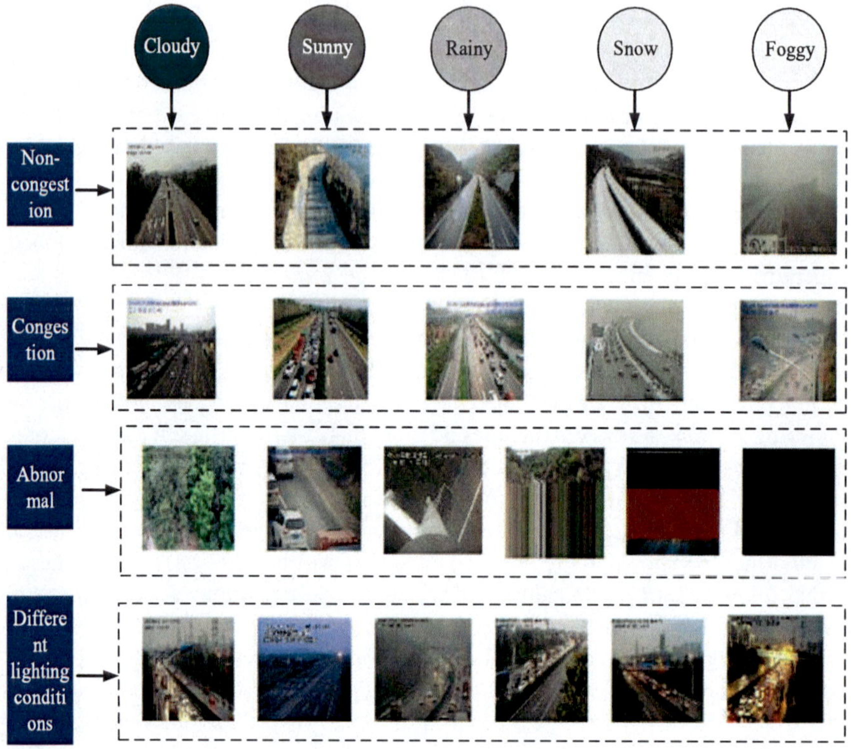

Fig. 5 Various light and weather conditions of the original image

speed of vehicles is slow, which is non-congestion. To avoid the influence of cameras and weather factors, we defined abnormal camera angle (no road condition, a small proportion of road surface, etc.), camera malfunction (blue screen, black screen, snowflake screen, et al.), and special scenes in the service area and toll station as abnormal (Fig. 6).

4.2 Transfer Learning and Fine-Tuning

Considering that our original data is smaller than the ILSVRC2012 data set, we pass the structure and parameters of the network pre-trained on the ILSVRC2012 data set to the Multi-Dec Net 1 and Multi-Dec Net 2. In this paper, Multi-Dec Net 1/2 is used as a classification network to study the properties of the features learned by each layer of the network [35]. In the training process, we used 0.1% of the data as the testing data, and the results are shown in Fig. 7.

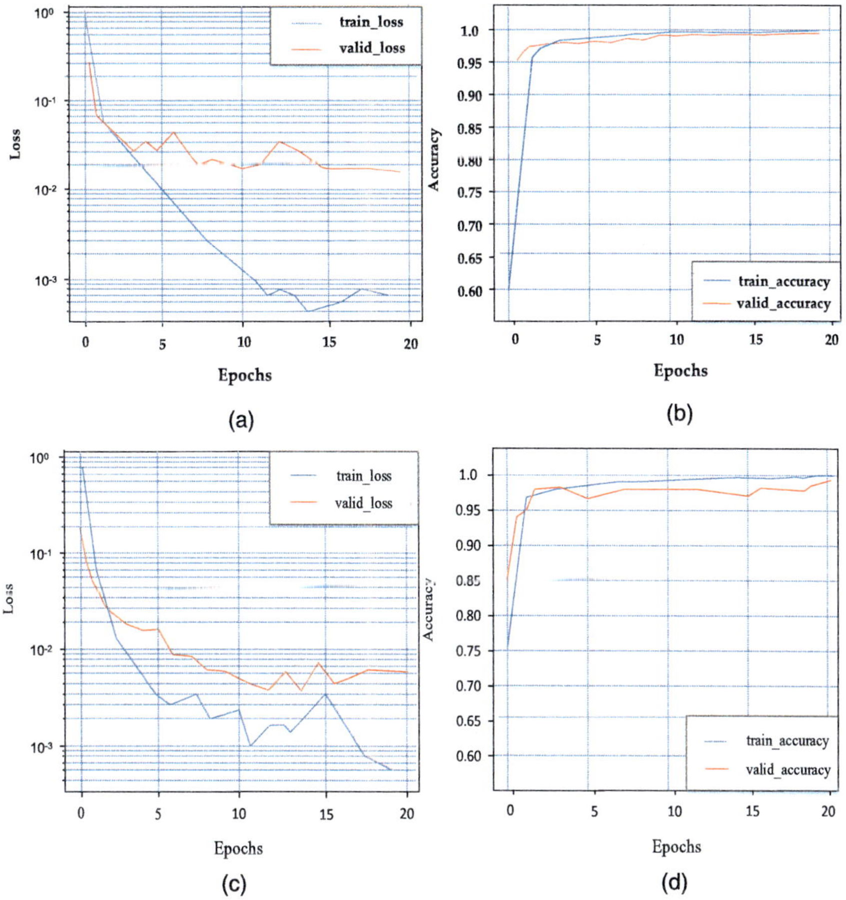

Fig. 6 The loss and accuracy curves of Multi-Dec Net 1/2: **a** The Multi-Dec Net 1 of train_loss and valid_loss; **b** The Multi-Dec Net 1 of train_accuracy and valid_accuracy; **c** The Multi-Dec Net 2 of train_loss and valid_loss; **d** The Multi-Dec Net 2 of train_accuracy and valid_accuracy

The loss represents the value of the loss function. As the training process progresses, the loss will become smaller and smaller. Accuracy represents the accuracy of the model prediction. The loss curve and accuracy curve are used to reflect the training process and the performance of the model. As shown in Fig. 6, due to the difference in the number of data sets used by Multi-Dec Net 1 and Multi-Dec Net 2, and during the pre-training process, the shuffle function randomly selects 98% of the data for random training, the same number of iterations used will lead to the starting point of the result Different from the end point loss and the accuracy in the graph. The verification loss decreases as the training accuracy decreases, and the verification accuracy increases as the training accuracy increases.

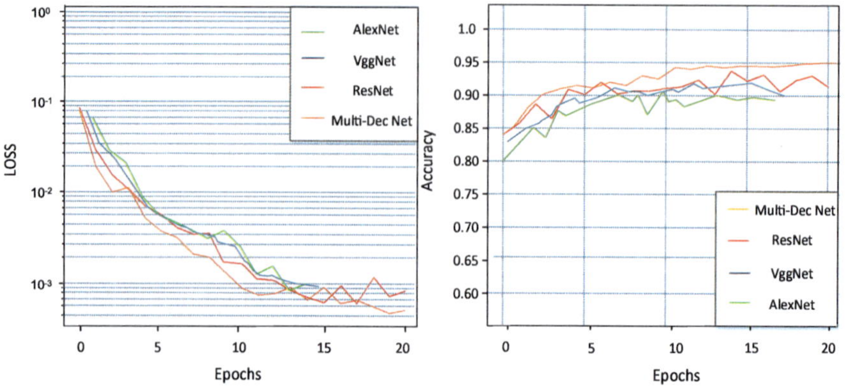

Fig. 7 Comparison of loss and accuracy of different networks

4.3 Evaluation Criteria

In classification results, Accuracy, Recall, Precision, and P-R curve are generally used to evaluate the performance of the model. In this paper, to verify the effect of the multi-stage recognition model (Multi-Dec Net) in practical applications, the accuracy and misjudgment rate of the model are used to evaluate the performance of the model classification. According to the classification result, a confusion matrix M is obtained. Row i represents the actual label, and column j represents the predicted label. Among them, $m, n = \{0, 1, 2\}$ represent non-congested, congested, and abnormal, respectively. $M_{i,j}$ means the actual label is i and the predicted label is the number of j, then the accuracy and misjudgment rate of the congestion class is calculated by the formula (4) and (5).

$$precision_1 = \frac{M_{11}}{\sum_{m=0}^{2} M_{m,1}} \tag{4}$$

$$error_1 = \frac{M_{11}}{\sum_{m=0}^{2} M_{1,m}} \tag{5}$$

4.4 Result

Aiming at the problems encountered by the team in the previous work and considering the high rate of model misjudgment, in this article we propose to use congestion data as the main research object of the experiments. The output of the model is the probability that each picture is judged as each category. We choose the congestion class as the research object, and set different thresholds for the congestion probability output by the model. The probability is greater than a certain probability as the congestion category, otherwise, it is another category.

Table 2 shows the experimental results of the congestion image classification based on different thresholds. In the experiment, we set the threshold $P = \{100, 98, 96, 94, 92, 90\}$ to get the congestion misjudgment rate under different threshold conditions. The results show that when P = 100, the error rate is the lowest.

Using the data of the actual deployment model of the team, a new test set of 17,999 was created, and the first-stage model was used for testing. The results are shown in Table 3. The data that the Multi-Dec Net 1 model is judged to be congested and the probability is less than 100 is used as the input of the Multi-Dec Net 2 model, and the misjudgment rate and accuracy rate of the final congestion class are counted. Comparing the misjudgment rate and accuracy rate of the first-stage model, the multi-stage model without the threshold, and the multi-stage model with the threshold, and the results are shown in Table 4.

The accuracy of the first-stage model test is higher, but the misjudgment rate is much higher than that of the multi-stage model. The method of threshold processing and multi-stage model testing can solve the problem of the high misjudgment rate of models to a great extent, and at the same time maintain the accuracy rate within a certain range. To evaluate the performance of the method in this paper, the Multi-Dec Net network proposed in this paper is compared with other networks. The results are shown in Table 5.

Table 5 lists three parameters for the training network model: Batch_Size, Epochs, and Learning rate, Batch_Size and Epochs may represent the training cost to obtain a reasonable model to a certain extent. Multi-Dec Net used a dropout operation with a probability of 0.2 to randomly reduce branches and avoid overfitting. For the

Table 2 Experimental results of congestion image classification based on different thresholds

Threshold-P(%)	Number of congestion	Number of error	Error rate	Number of others
100	3326	69	2.06%	6474
98	8148	212	2.60%	1652
96	8555	241	2.82%	1245
94	8743	267	3.05%	1059
92	8891	309	3.47%	911
90	8982	358	3.98%	811

Table 3 The test results of Multi-Dec Net 1

Model	Confusion Matrix	Misjudgment rate	Accuracy rate
Multi-Dec Net 1	[2652 932 431] [32610,36979] [23 48 3139]	8.64%	96.24%

Table 4 The test results of Multi-Dec Net 2

Model	P	Confusion matrix	Misjudgment rate	Accuracy rate
Multi-Dec	None	[2652 932 431] [32610,36979] [23 48 3139]	4.94%	93.8%
Net	100	[3077 496 442] [354 10,339 81] [32 24 3154]	4.79%	95.9%

Table 5 Compare test results of different models

Model	Batch_Size	Epochs	Learning rate	Validation accuracy (%)	Testing accuracy (%)	Error accuracy (%)
AlexNet	16	20	0.001	96.13	88	14.56
VggNet	16	20	0.001	98.33	90.21	12.45
ResNet	16	20	0.001	99.21	93.18	9.46
Multi-Dec-Net (Ours)	16	20	0.001	99.45	95.96	4.78

mini-batch gradient descent, we used 16 batch size and 20 epoch. Hence, it took (number of train sample) / (batch size) iterations to complete an epoch, and for each iteration, a batch of 16 images was presented to the CNN model and then the weights were updated by backpropagation. The Multi-Dec Net has a broader architecture than AlexNet, VggNet, and ResNet. Moreover, the depth of the Multi-Dec Net architecture is deeper than that of AlexNet, which may indicate that the network architecture has more functions of extracting more features. As can be seen from Table 5, the error rates of AlexNet VggNet, and ResNet are 14.56, 12.45, and 9.48%. In Multi-Dec Net, the verification accuracy and test accuracy are both higher than AlexNet, VggNet, and ResNet, and the error rate is only 4.78%. Therefore, it can be concluded that the Multi-Dec Net proposed in this paper can effectively reduce the misjudgment rate of the model, improve the accuracy of model identification and better realize the effect of congestion detection by constructing a two-stage network with the help of the idea of dense blocks.

We use Alexnet, VggNet, and ResNet as comparison networks, and use different networks to train the model. The training loss and verification accuracy of different models are shown in Fig. 6. From Fig. 6a, we can see that from the first iteration, the effect of the Multi-Dec Net is better than other networks. During the training process, the Multi-Dec network has the least loss, more stable than the other three networks. From Fig. 6b, we can see that for the comparison of different network accuracy, Multi-Dec Net also achieved the best results. The accuracy rate can reach more than 95%. To obtain better practical results and ensure the reliability of the model. Therefore, we choose a multi-level recognition network for training.

Fig. 8 Application legend of Xi'an Raocheng expressway

4.5 Application

To test the effect of the Multi-Dec Net proposed in this paper, we can deploy the existing model to the expressway in Shaanxi Province and put it into an application. Since Raocheng Expressway is prone to traffic congestion, we can choose this section as a case study, as shown in Fig. 8. Choosing 31 sets of monitoring systems currently in use can achieve real-time coverage of the entire road section and real-time transmission of traffic status videos. In future applications, we will devote ourselves to researching methods with low model misjudgment rates and high accuracy rates.

5 Conclusion

To promote the application of deep learning in the field of traffic and realize the accurate identification of congestion scenes, we propose a multi-stage classification detection structure algorithm Multi-Dec Net for image classification tasks. The algorithm is based on the dense block and uses a two-stage neural network structure. First of all, pre-training is carried out on the ILSVRC-2012 dataset, the weight of pre-training is used to train our Multi-Dec Net, and the network parameters are constantly adjusted to achieve the best performance of the network. To fine-tune the network, we obtain videos of different road sections from 14,700 cameras in the highway system of Shaanxi Province. When constructing the data set, we fully consider the diversity of data and add different weather conditions, different lighting conditions, different scenes, and different quality pictures, which can be applied to a variety of scenes. In the large-scale monitoring system, the Multi-Dec Net proposed in this paper has obtained satisfactory judgment ability, which shows the strong robustness of the algorithm. A large number of experiments show that we have found the best classification scheme for the current classification task which is different from other networks. It has a very important practical significance in traffic monitoring. It is worth extending its application to other traffic monitoring systems to enhance ITS. In the future, we will conduct research on how to reduce the false alarm rate while greatly improving the accuracy rate to build a more powerful congestion detector.

References

1. Qi L, Zhou M, Luan W (2018) A dynamic road incident information delivery strategy to reduce urban traffic congestion. IEEE/CAA J Automatica Sinica
2. Ke X (2019) Multi-dimensional traffic congestion detection based on fusion of visual features and convolutional neural network. IEEE Trans Intell Transp Syst 20(6):2157–2170
3. Wang Q, Wan J, Yuan Y (2018) Locality constraint distance metric learning for traffic congestion detection. Pattern Recogn 75:272–281
4. Liu D (2019) Research on solving the problem of provincial boundary congestion based on canceling provincial boundary toll station – Taking Liyuan toll station in Jiangxi Province as an example. World Transport Convention (WTC)., vol 9
5. Li X (2020) Construction scheme of Expressway video cloud networking. Technology Innovation and Application. 30:133–134
6. Lam CT, Gao H, Ng B (2017) A real-time traffic congestion detection system using on-line images. In: IEEE international conference on communication technology IEEE, pp 1548–1552
7. Yong C, Zuo X, Zhang L (2011) Traffic congestion detection based On GPS floating-car data. Procedia Eng 15:5541–5546. https://doi.org/10.1016/j.proeng.2011.08.1028
8. Wang P, Di J (2018) Deep learning-based object classification through multimode fiber via a CNN-architecture SpeckleNet. Appl Opt 57(28):8258–8263
9. An J, Lv X, Ji L (2019) Congestion prediction method based on spatiotemporal correlation under incomplete data. Comput Eng Appl 55(4):96–100
10. Zhu J (2020) Research on highway traffic incident detection system based on YOLO-v3. China University of Mining and Technology. M.S. thesis, Dept. Electron. Eng
11. Impedovo D, Balducci F, Dentamaro V (2019) Vehicular traffic congestion classification by visual features and deep learning approaches: a comparison. Sensors (Basel, Switzerland), 19(23)
12. Sun Z, Wang P, Wang J, Peng X, Jin Y (2020) Exploiting deeply supervised inception networks for automatically detecting traffic congestion on freeway in china using ultra-low frame rate videos. IEEE Access, 8:21226–21235
13. Dong Z, Wu Y, Pei M, Jia Y (2015) Vehicle Type classification using a semisupervised convolutional neural network. IEEE Trans Intell Transp Syst 16(4):2247–2256
14. Pamula T (2018) Road traffic conditions classification based on multilevel filtering of image content using convolutional neural networks. IEEE Intel. Transp. Syst. 10(3):11–21

15. Wang P, Li L, Jin Y, Wang G (2018) Detection of unwanted traffic congestion based on existing surveillance system using in freeway via a CNN-architecture trafficnet. In: IEEE conference on industrial electronics and applications (ICIEA), Wuhan, pp 1134–1139. https://doi.org/10.1109/ICIEA.2018.8397881
16. Luo W, Dong B, Wang Z (2017) Short-term traffic flow prediction based on CNN-SVR hybrid deep learning model. J Transp Syst Eng Informat Technol 17(5). https://doi.org/10.16097/j.cnki.1009-6744.2017.05.010.
17. Castro-Zunti RD, Yépez J, Ko SB (2020) License plate segmentation and recognition system using deep learning and OpenVINO. IET Intell Transp Syst 14:119–126
18. Gu Z, Saberi M (2019) A bi-partitioning approach to congestion pattern recognition in a congested monocentric city. Transport Res Part C Emerg Technologies. 109:305–320
19. LeCun Y, Bottou L, Bengio Y, Haffner P (1998) Gradient-based learning applied to document recognition. In: Proceedings of the IEEE, vol 86, no. 11, pp 2278–2324
20. Krizhevsky A, Sutskever I, Hinton GE (2012) ImageNet classification with deep convolutional neural networks. In: Proceedings of advances in neural information processing systems, pp 1097–1105
21. Simonyan K, Zisserman A (2015) Very deep convolutional networks for large-scale image recognition. in: Proceedings of international conference learning representation (ICLR), San Diego, CA, USA, pp 1–14
22. He K, Zhang X, Ren S (2016) Deep residual learning for image recognition. In: Proceedings of the IEEE conference on computer vision and pattern recognition (CVPR), Las V egas, NV, USA, pp 770 778
23. Huang G, Liu Z, Van Der Maaten L (2016) Densely connected convolutional networks. In: Proceedings of the IEEE conference on computer vision and pattern recognition (CVPR), pp 4700–4708
24. Dong Z, Wu Y, Pei M, Jia Y (2015) Vehicle type classification using a semisupervised convolutional neural network. IEEE Trans Intel Trans Syst 16(4): 2247–2256
25. Ke X, Shi L, Guo W (2018) Multi-dimensional traffic congestion detection based on fusion of visual features and convolutional neural network. IEEE Trans Intell Transp Syst 20(6):2157–2170
26. Ren H (2016) Speech emotion recognition based on multilevel classification. M. S. thesis, Department of Electron Engineering, Harbin Institute of Technology, Heilongjiang, China
27. Tian S, Dai N, Yuan F (2020) Segmentation and recognition of dental model based on multi-level hierarchical 3D convolution neural network. J Comput Aided Design Comput Graph 32(8):1218–1227
28. Chen J (2020) Research on the design of multilevel classifier for large scale data. M.S. thesis, Department of Computer Science., Fuzhou University, Fujian, China, 2020.
29. Ke X, Shi L, Guo W (2019) Multi-Dimensional traffic congestion detection based on fusion of visual features and convolutional neural network. IEEE Trans Intell Transp Syst 20(6):2157–2170
30. Wang Q, Wan J, Yuan Y (2017) Locality constraint distance metric learning for traffic congestion detection. Pattern Recogn:S0031320317301401
31. Liu D (2019) Research on solving the problem of provincial boundary congestion based on canceling provincial boundary toll station – taking toll station in Jiangxi Province as an example. World Transport Convention (WTC)
32. Wang P, Hao W, Sun Z, Wang S (2018) Regional detection of traffic congestion using in a large-scale surveillance system via deep residual trafficnet. IEEE Access 6:68910–68919
33. Huang G, Liu Z, Van Der Maaten L, Weinberger KQ (2016) Densely connected convolutional networks. IEEE Comput Soc
34. Du B, Liu C, Zhou W, Hou Z, Xiong H (2019) Detecting pickpocket suspects from large-scale public transit records. IEEE Trans Knowl Data Eng
35. Girshick R, Donahue J, Darrell T, Malik J (2014) Rich feature hierarchies for accurate object detection and semantic segmentation. in: Proceedings of the IEEE conference on computer vision and pattern recognition, Columbus, OH, USA, pp. 580–587

36. Li Y (2018) Traffic congestion identification method of Urban Expressway Based on support vector machine. Traffic Eng 18(01):43–47
37. Fu J, Zheng H, Mei T (2017) Look closer to see better: recurrent attention convolutional neural network for fine-grained image recognition. In: IEEE conference on computer vision & pattern recognition IEEE, USA, pp 4438–4446

Signal Control Method for Modern Roundabouts with Waiting Areas to Meet High Traffic Demands

Xiancai Jiang, Qingpeng Shang, and Yao Jin

Abstract This paper aims to address a problem at signalized roundabouts, where the queuing overflow of left-turn vehicles can block vehicles that are travelling through the roundabout when the signal control method of Two-Stopline-for-Left-Turn control (TSLT) is adopted. A method for setting up the waiting area and coordinating the traffic signal between the approach lanes and the loop lanes is proposed. With the help of the loop lane space, the through and left-turn waiting areas are set up, and the traffic flow in each direction at the roundabout is separated by a coordinated traffic signal, in order to improve the utilization rate of the space resources of the roundabout. Under saturation constraints, delay calculation models are established based on the traffic states of the approach lanes and the loop lanes, and a signal control parameter optimization model is established with the goal of minimizing the delay. The simulation results show that the proposed method can satisfy the required conditions for roundabouts with high traffic volume.

Keywords Traffic engineering · Roundabout · Waiting area · Signal control · Capacity · Delay

X. Jiang (✉)
School of Traffic and Transportation, Northeast Forestry University, Harbin 150040, China
e-mail: jxc023@126.com

Q. Shang
School of Transportation Science and Engineering, Harbin Institute of Technology, Harbin 150090, China
e-mail: sqp378@163.com

Y. Jin
Heilongjiang Transportation Information and Planning Research Center, Harbin 150001, China
e-mail: 303088610@qq.com

W. Wang et al. (eds.), *Green Transportation and Low Carbon Mobility Safety*, Lecture Notes in Electrical Engineering 944,
https://doi.org/10.1007/978-981-19-5615-7_4

1 Introduction

Many countries have adopted the modern roundabout as a common intersection form since the modern roundabout was developed in the U.K. in the 1960s [16]. Roundabouts have unique characteristics that warrant consideration by developers and managers of the road system [29]. However, the bottleneck of a roundabout lies in its weaving sections, where the vehicles enter or leave the roundabout, and one typical disadvantage of roundabouts is its failure to handle high traffic demands [1]. Thus, there need an improved method to meet the operational requirement for high traffic volume roundabouts. Replacing a roundabout by an ordinary signalized intersection may be a feasible alternative at this situation but usually at the risk of high construction costs and loss of safety and efficiency during nonpeak hours [21]. One of the low-cost solutions is adding traffic signals on the roundabout. Now, many problems begin to appear at roundabouts, especially large roundabouts with high traffic demands, such as environmental pollution, traffic safety, traffic operation evaluation, and traffic signal optimization control, etc.

To address the problem of environmental pollution at roundabouts, Shaaban et al. [31] used VISSIM and MOVES to assess the environmental effect of converting four three-lane roundabouts to signalized intersections along a heavily congested urban corridor in Qatar [26] analyzed the emissions changes of CO_2, NO_X and CO on a road intersection where a roundabout had replaced a traffic signal. All the research results showed that environmental pollution at roundabouts could be improved to some extent by changing the form and signal control mode of the roundabout.

At the same time, traffic safety at roundabouts has attracted more and more attention, and mainly focuses on two aspects: influencing factors and evaluation models. The evaluation models include a crash prediction model [3, 4], a lateral clearance model [12], and a gap acceptance model [2]. Other studies have shown that safety at roundabouts is related to vehicle speed, vehicle composition, and the inscribed circle diameter of the roundabout [23, 33]. Traffic safety at roundabouts is mainly attributed to the conflicts of weaving sections, which can be eliminated by traffic signals. Therefore, in the roundabout with high traffic demands, signal control is an effective measure to solve the problem of traffic safety [36].

Operation evaluation of roundabouts is another research hotspot. The existing research focuses on benefit evaluation [17, 20, 25] and travel time evaluation [10, 27] under conditions with no signal control, capacity calculation in various traffic environments [5, 18], and benefit evaluation and block mechanism under traffic signal control [6, 7, 9, 28, 30, 35] The main research methods used were computer simulation and multiple non-linear regressions. The results showed that traffic signal control was one of the most effective measure to improve the capacity of a roundabout when the traffic volume increases.

The first traffic signal-controlled roundabout in the West Midlands, UK came into operation in Wolverhampton, which operated well during peak hours [8]. The traffic signals are usual installed on the approaches of a roundabout. But when the traffic flow of each approach is released in turn, the utilization rate of the loop lanes is low, when opposite traffic flows release at the same time, there are conflicts between the traffic flows on the approaches and the left-turn traffic flows on the loop lanes. To solve it, [36] proposed the signal control method of Two-Stopline-for-Left-Turn control (TSLT) to eliminate the conflict points. A case study of a roundabout in Xiamen, China found that the TSLT was effective to increase the roundabout capacity by 72.1% and decrease the average delay of each vehicle by 20 s. However, TSLT is limited by the capacity of left-turn storage space. When the left-turn traffic demand is higher than the capacity, the queuing overflow of left-turn vehicles will block the passage of vehicles travelling through the roundabout. In addition, the traffic state at loop lanes and approaches (unsaturated or oversaturaed) was not considered, which could result in inaccuracy in calculating the delays.

As new application requirements have emerged, multiple induction control methods [34] and adaptive signal control methods [11, 32] have been developed for roundabouts. Xu et al. [33] proposed an improved signal control method for roundabouts considering the multi-level pedestrian signaling. At the same time, some research methods from other fields were also adopted for signal control of roundabouts, for example a group model [19], a cross-entropy method [24], fuzzy control [15], and particle swarm optimization (Gökçe et al. 2015), etc. Although these methods have improved the signal control flexibility and efficiency of roundabouts, they are also difficult to satisfy the required conditions for roundabouts with high traffic volume.

In this paper, a method of combining waiting area and coordinated signal control is proposed to satisfy the required conditions for roundabouts with high traffic demands. Compared to existing work, the novel contributions of our work are as follows:

1. The through and left-turn waiting areas are configured with the help of the loop lane space at the roundabout to improve the utilization rate of the loop lanes.
2. The through vehicles and left-turn vehicles are separated by a coordinated traffic signal to avoid conflicts and improve the roundabout capacity.
3. The delay calculation models are established based on the traffic states of approach lanes and loop lanes to calculate the delays in accuracy.

The remainder of this paper is organized as follows. Section 2 introduces a mathematical program to operate the roundabout with a waiting area. The numerical experiments are conducted in Sect. 3, and the results are analysed in Sect. 4. Finally, some conclusions and future research are discussed in Sect. 5.

2 Notations

The following symbols are used in this paper, the definitions can be seen in Table 1.

Table 1 The definition of variable

Variable	Meanings
C	Cycle (s)
c_{max}	Maximum cycle (s)
c_{min}	Minimum cycle (s)
$D_{w,i,j}$	Delay in a cycle (s)(i are 1, 2, 3 and 4 for east, west, south and north respectively, j is 1 and 2 for through traffic flow and left-turn traffic flow respectively, with the same definition used below. s)
d	Average delay of the roundabout (s/pcu)
$g_{a,i,j}$	Green time at approach lanes (s)
$g_{c,i,j}$	Green time at loop lanes (s)
$g_{c,k}$	Green time of the kth phase at loop lanes (s)
$g_{c,k,max}$	Maximum green time of the kth phase at loop lanes (s)
$g_{c,k,min}$	Minimum green time of the kth phase at loop lanes (s)
I_{ew}	Inter-green time of the east–west through phase (s)
$I_{i,j}$	Inter-green time (s)
I_{sn}	Inter-green time of the south-north through phase (s)
$L_{1,2}$	Distance between the stop line of approach lanes and the one of loop lanes (m)
$O_{ew,1}$	Lag time of the green time opening of the east–west through phase (s)
$O_{ew,2}$	Lag time of the green time opening of the east–west left-turn phase (s)
$O_{i,j}$	Lag time of the green light opening (s)
$O_{sn,1}$	Lag time of the green time opening of the south-north through phase (s)
$O_{sn,2}$	Lag time of the green time opening of the south-north left-turn phase (s)
$q_{i,j}$	Vehicle arrival rate in the j-direction of the i-approach
$r_{a,i,j}$	Red time at approach lanes (s)
$S_{i,j}$	Saturation flow rate (pcu/s)
$t_{a,i,j}$	Time difference of green turned-on time between the approach traffic signal and the loop traffic signal (s)
$t_{d,i,j}$	Time difference of red turned-on time between the approach traffic signal and the loop traffic signal (s)
$t_{d,1}$	Pre-closing time of green time at the approach lanes of the east–west through phase (s)
$t_{d,2}$	Pre-closing time of green time at the approach lanes of the east–west left-turn phase (s)
$t_{d,3}$	Pre-closing time of green time at the approach lanes of the south-north through phase (s)
$t_{d,4}$	Pre-closing time of green time at the approach lanes of the south-north left-turn phase (s)
$t_{w,i,j}$	Average time for the first car entering the waiting area from the approach stop line to the loop stop line (s)
v	Average travel speed (m/s)

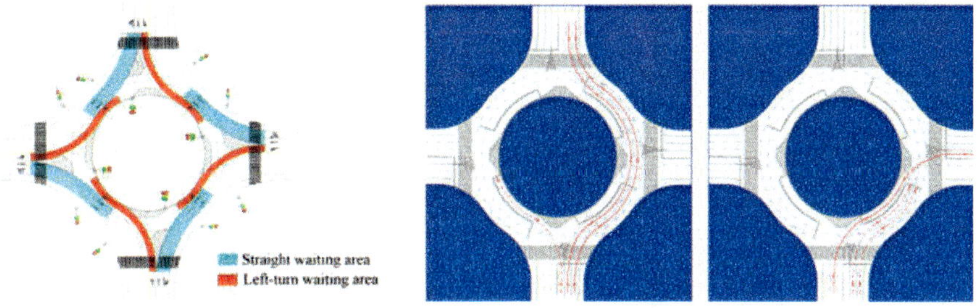

(a) Arrangement of waiting area (b) Traffic signs to guide vehicles into the waiting area

Fig. 1 Arrangement and guidance sign of waiting area at a roundabout

3 Methodology

3.1 Traffic Organization at a Roundabout with a Waiting Area

3.1.1 Configuration of the Waiting Area at a Roundabout

The waiting area at a roundabout is configured using stop lines that are set up on both approaches and loop lanes of the roundabouts, which are called the approach stop line and the loop lane stop line. The area between the two stop lines is the waiting area for left-turn vehicles and vehicles travelling through, shown in Fig. 1a. Compared with the normal layout of roundabout, the proposed layout in this paper will increase the conflict points in roundabout and result in a reduction in security levels. But we can use traffic signals to eliminate these conflict points. The conflicting traffic flow is separated by coordinating the traffic signal of the approach lanes and the loop lanes, which determines whether the vehicles in the approach lanes can enter the waiting area of the loop lanes and whether the vehicles in the loop lanes can leave the roundabout.

At the same time, the drivers are familiar with the running rules of counter-clockwise roundabout. In order to make the drivers follow the new rules proposed in this paper, two kinds of traffic signs should be set up at the roundabout facing the entrance road to remind when the left-turn and through vehicle can enter the waiting area. The layout of traffic signs is shown in Fig. 1b.

3.1.2 Release Arrangement at the Roundabout with Waiting Area

When a waiting area is set up at a roundabout, the traffic flow is organized in a similar way to that of an intersection. The traffic organization of through traffic flow and left-turn traffic flow in all directions is shown in Fig. 2.

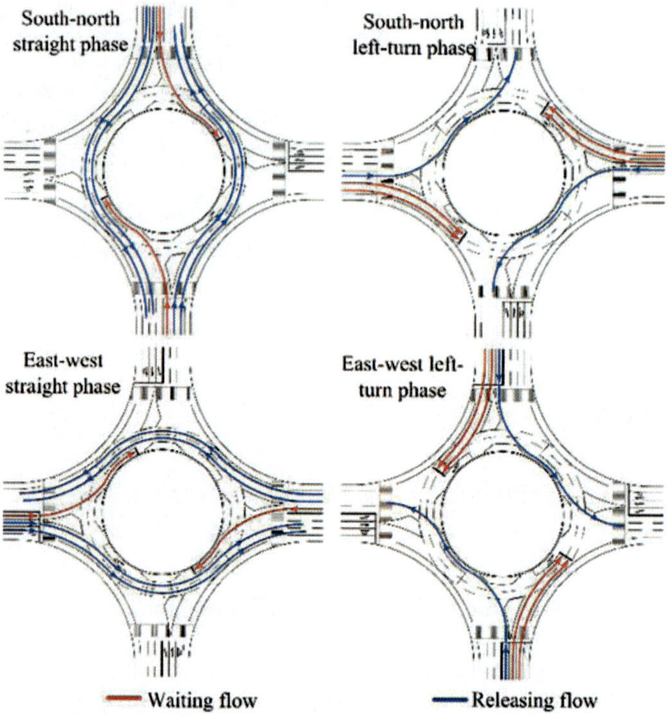

Fig. 2 Traffic organization at a roundabout with waiting area

After turning left to enter the loop lanes, it is necessary for left-turn vehicles to drive clockwise along the island, and then leave the roundabout; vehicles travelling through the roundabout (including right-turn vehicles) turn right into the loop lanes, then run counter-clockwise along the outer loop lanes, and then leave the roundabout.

Therefore, different loop lanes have different functions, so it is necessary to mark the direction of traffic flow on the loop lanes. The inner loop lane of a roundabout is set as a left-turn lane, while the outer loop lanes are set as through lanes. The traffic flow of the two types of lanes runs in opposite directions. The left-turn vehicles enter the left-turn waiting area when vehicles travelling through in a North–South direction are released; vehicles travelling through in an East–West direction enter the through waiting area when the North–South left-turn vehicles are released; the left-turn vehicles enter the left-turn waiting area when the vehicles travelling through in an East–West direction are released; the vehicles travelling through in a North–South direction enter the through waiting area when the East–West left-turn vehicles are released. The above process repeats in turn and makes the best use of space resources of a roundabout under the constraint of separating conflicting traffic flows.

3.1.3 Design of Signal Phases

As seen in Fig. 2, the traffic organization of a roundabout in this paper is similar to an ordinary intersection. Therefore, a four-phase symmetrical release can be used to separate the conflicting traffic flow. Symmetrical release can be divided into two types - through front or left front. When the left-turn vehicles are released first, the second phase is the through phase. Since vehicles travelling through the roundabout occupy the outer loop lanes, it is difficult for both vehicles travelling through and left-turn vehicles that are crossing directions to enter the waiting area in the third phase. In contrast, when the vehicles travelling through are released first, the vehicles in the previous phase will not prevent vehicles in the next phase from entering the waiting area. Therefore, in order to make full use of the loop lane space resource, a through front phase design is adopted. The pedestrian phase is the same as the through phase of loop lanes. The design of signal phases at roundabout is shown in Fig. 3. In Fig. 3, C1, C2, C3, C4 are the east–west through phase, the east–west left-turn phase, the north–south through phase, and the north–south left-turn phase of the loop lanes, respectively; E1, E2, E3, E4 are the east–west through phase, the east–west left-turn phase, the north–south through phase, and the north–south left-turn phase of the approach lanes, respectively; P1 and P2 are the east–west through phase and the north–south through phase for pedestrians, respectively.

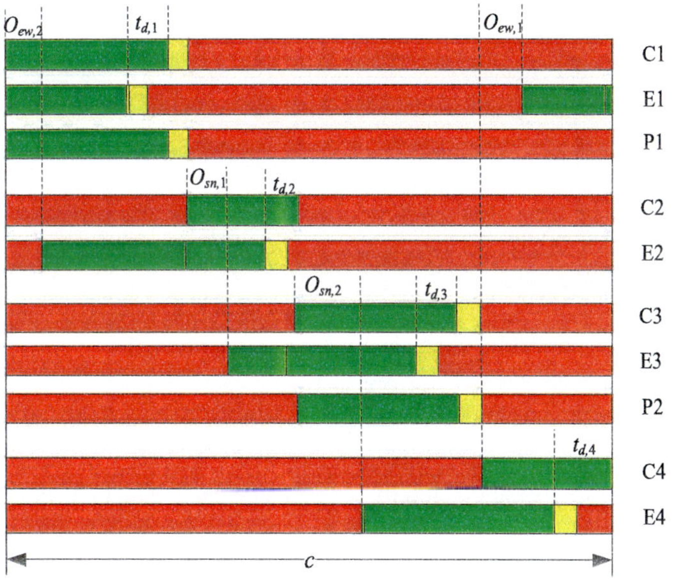

Fig. 3 Signal phase design

3.2 Optimization Models of Signal Timing Parameters for the Roundabout with Waiting Area

3.2.1 Optimization Objective Function

The objective is to minimize the average vehicle delay. The optimization model for the signal control parameters in this paper is as follows:

$$\min d = \frac{\sum_{i=1}^{n} \sum_{j=1}^{m} D_{w,i,j}}{\sum_{i=1}^{n} \sum_{j=1}^{m} q_{i,j} c} \tag{1}$$

3.2.2 Constraint Conditions

(1) General conditions

Because pedestrians and through traffic flow in the waiting area are released at the same time, the green time of the through phase should be sufficient to meet the time required by pedestrians to cross the street. The basic constraints of the roundabout signal control parameters are as follows:

$$g_{c,k,\min} \le g_{c,k} \le g_{c,k\,\max} \tag{2}$$

$$c{=}I_{ew} + I_{sn} + \sum_{k=1}^{4} g_{c,k} \tag{3}$$

(2) Saturation constraint

In practice, when the saturation at an intersection is greater than 0.9, there will be prone to congestion. Therefore, 0.9 is chosen as the practical saturation value limit, i.e. the phase saturation should not be higher than 0.9:

$$\frac{q_{i,j} c}{S_{i,j} g_{c,i,j}} \le 0.9 \tag{4}$$

(3) Coordination constraint of approach traffic signal and loop lane traffic signal.

If the green light of the loop lanes is turned on too early, the vehicles entering the waiting area will not have not reached the loop lane stop line, therefore green time will be wasted. Similarly, the green light should not be turns on later than the closing time of the approach green time. Therefore, the time difference between the green light turning on at the loop lanes and the approaches should meet the requirement of Formula (5) and should usually be the minimum value. In order to ensure that the vehicles in the waiting area are cleared before the green light changes in the loop lanes, the time difference between the green light turning off at the loop lanes and the approaches must meet the requirement of Formula (6) and should usually be the minimum value. The interval between the opening time of the approach green time compared to the loop land green time can be calculated according to Formula (7).

$$g_{a,i,j} > t_{a,i,j} \geq t_{w,i,j} \tag{5}$$

$$t_{d,i,j} \geq \frac{L_{1,2}}{v} \tag{6}$$

$$t_{a,i,j} = c + t_{d,i,j} - g_{c,i,j} \tag{7}$$

Signal control can be used to separate traffic flow to avoid many traffic conflicts, but potential traffic conflicts still exist between the tail vehicles and the first vehicles in adjacent phases when the phase switches. There are also potential traffic conflicts between the tail vehicles of traffic flow leaving, the starting vehicles in the waiting area and the vehicles entering the waiting area.

For example, if the left-turn traffic flow of the east approach is considered (similar applies to other approaches), traffic conflicts exist between the tail vehicles of the left-turn traffic flow, the first vehicles of the through traffic flow in a clockwise direction in the through waiting area and the left-turn vehicles entering the waiting area in the crossing direction, as shown in Fig. 4a. Point A is the conflict point between the tail vehicles of the left-turn traffic flow and the left-turn vehicles entering the waiting area, while point B and C are the conflict points between the tail vehicles of the left-turn traffic flow and the vehicles at the front of the through traffic flow in the waiting area.

Due to the requirement of Formula (6), the tail vehicles of left-turn traffic flow have passed through conflict points B and C before the green time of the loop lane finishes, and will not conflict with the front vehicles of the through traffic flow in the waiting area. In this case, inter-green time is not required at the end of the left-turn phase.

At the end of the green time, the tail vehicles of the left-turn traffic flow have not reached conflict point A, and the distance from conflict point A to the second stop line is L_1. In order to avoid a conflict at conflict point A, it is necessary to either delay the time of the left-turn traffic flow entering the waiting area at the south approach or increase the inter-green time of the left-turn phase of the loop lane. Since an increase in inter-green time will increase intersection delay, a conflict should be avoided by delaying the left-turn traffic flow from the south approach into the waiting area. The delay time $O_{3,2}$ is shown in formula (8).

$$O_{3,2} = \frac{L_1}{v} \tag{8}$$

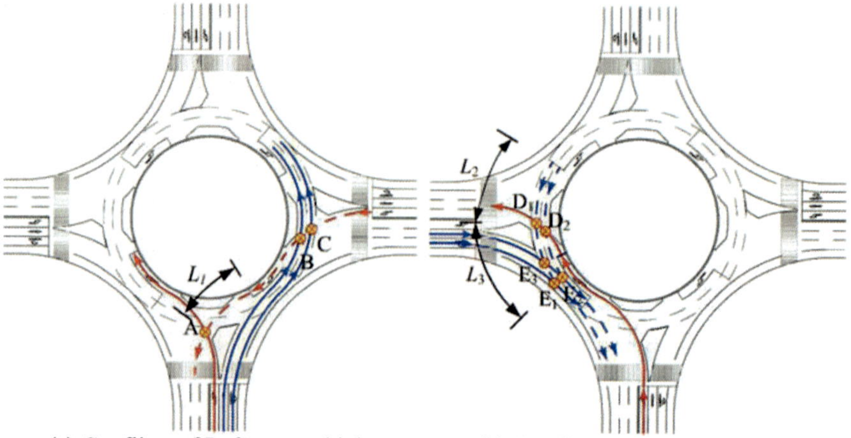

(a) Conflicts of Left-turn vehicles (b) Conflicts of through vehicles

Fig. 4 Potential traffic conflicts

According to the above analysis, the delay time $O_{sn,2}$ of the north–south left-turn green time is shown in Formula (9). The calculation method for the delay time of the east–west left-turn green time is similar.

$$O_{sn,2} = \max\{ O_{3,2}, O_{4,2}\} \tag{9}$$

Taking the through traffic flow of the north approach as an example (similar will apply to other approaches), traffic conflicts exist between the tail vehicles of the through traffic flow, the first vehicles of the left-turn traffic flow in the waiting area and the through vehicles entering the waiting area in the crossing direction, as shown in Fig. 4b. In Fig. 4b, points D_1 and D_2 are the conflict points between the tail vehicles of the through traffic flow and the first vehicles of left-turn traffic flow in the waiting area, while E_1, E_2 and E_3 are the conflict points between the tail vehicles of the through vehicle flow and the through vehicles entering the waiting area in the crossing direction. The inter-green time should be increased to avoid traffic conflicts after the through phase of the loop lanes is completed. The inter-green time required after the through phase of the loop lanes follows Formula (10):

$$I_{4,1} = \max\{ \frac{L_2}{v}, 3\} \tag{10}$$

In addition, conflict points E_1, E_2 and E_3 still exist between the through vehicles entering the waiting area of the south approach and the through vehicles going forward. The conflicts can be avoided by delaying the time of the through traffic flow into the waiting area of the west approach. The delay time $O_{2,1}$ is shown in formula (11).

$$O_{2,1}=\frac{L_3}{v} \tag{11}$$

According to the above analysis, it can be seen that the inter-green time of the north–south through phase is expressed buy Formula (12). The calculation method of the east–west through phase is similar. The delay time of the east–west through phase can be expressed by Formula (13), and the calculation method of the delay time of the north–south through phase is similar.

$$I_{sn} = \max\{ I_{4,1}, I_{3,1}\} \tag{12}$$

$$O_{ew,1} = \max\{ O_{1,1}, O_{2,1}\} \tag{13}$$

3.3 Delay Analysis

3.3.1 Basic Assumptions

1. The influence of the lane location and waiting area location are not considered; all vehicles choose approach lanes and waiting areas equally so that both waiting areas and approach lanes can be fully utilized.
2. The vehicles arrive evenly and follow the traffic signal control scheme and the traffic sign marking completely.
3. The right-turn vehicles follow the through signal phase. To simplify the description, the control model only considers vehicles travelling straight through ("through vehicles") and left-turn vehicles.
4. The saturated flow rate of vehicles in the approach lanes is the same as that in the loop lanes.

3.3.2 State of Traffic Flow

When there is no waiting area at the intersection, the intersection delay analysis can be conducted based on unsaturated and oversaturated traffic conditions. There will be two stop lines at the intersections when the waiting area is set up. Delay analysis can be conducted based on an ordinary intersection delay analysis method.

As there are two stop lines in each phase of the roundabout, there are two queues that form and dissipate. The vehicles have two types of starting waves and two types of stopping waves: the starting wave of the approach stop line and the starting wave of the loop lane stop line, and the stopping wave of the approach stop line and the stopping wave of the loop lane stop line.

When the traffic flow in the *j*-direction of the *i*-approach satisfies the constraints of Formula (14) below, it can be considered that the queuing vehicles will dissipate completely in that cycle, the starting wave when the green light of the loop lanes is turned on will be transmitted to the end of the queue, and the roundabout is in an unsaturated state. However, there are two potential situations in this case, that is, whether the queue can dissipate completely before the green light at the approaches is closed, and if so, the traffic condition of Formula (15) is called "state I", otherwise, the traffic condition of Formula (16) is called "state II".

$$q_{i,j}(r_{a,i,j} + g_{a,i,j} - t_{w,i,j}) < S_{i,j}g_{c,i,j} \tag{14}$$

$$q_{i,j}(r_{a,i,j} + g_{a,i,j} - t_{w,i,j}) < S_{i,j}(g_{c,i,j} - t_{d,i,j}) \tag{15}$$

$$q_{i,j}(r_{a,i,j} + g_{a,i,j} - t_{w,i,j}) \geq S_{i,j}(g_{c,i,j} - t_{d,i,j}) \tag{16}$$

When the traffic flow in the j-direction of the i-approach satisfies the constraints of Formula (17), it can be considered that the queuing vehicles cannot dissipate completely in one cycle, the starting wave when the green light of loop lanes is turned on cannot be transmitted to the end of the queue in one cycle, the vehicles will be waiting for two or more cycles, and the roundabout is in a saturated state. In this case, there are three possible states based on the arrival and departure of vehicles at the two stop lines after the green light turns on at the approach lanes: (1) the starting wave can be transmitted to the end of the queue and the saturated flow rate of the loop lanes returns to a normal arrival flow rate before the green light of the approaches changes; (2) the starting wave can be transmitted to the end of the queue, but because of a long queue length, the arrival flow rate of the loop lanes reaches saturated flow before the green light of the approaches changes; (3) due to the long queue length, the starting wave cannot be transmitted to the end of the queue before the green light of the approaches changes.

The three states described above are called "state III", "state IV" and "state V". In addition to satisfying the constraints of formula (17), these states should also satisfy the constraints of formula (18), formula (19) and formula (20), respectively.

$$q_{i,j}(r_{a,i,j} + g_{a,i,j} - t_{w,i,j}) \geq S_{i,j}g_{c,i,j} \tag{17}$$

$$\frac{q_{i,j}r_{a,i,j}}{S_{i,j}} + t_{w,i,j} < g_{a,i,j} \tag{18}$$

$$g_{a,i,j} \leq \frac{q_{i,j}r_{a,i,j}}{S_{i,j}} + t_{w,i,j} < g_{a,i,j} + t_{w,i,j} \tag{19}$$

$$g_{a,i,j} + t_{w,i,j} \leq \frac{q_{i,j}r_{a,i,j}}{S_{i,j}} + t_{w,i,j} \tag{20}$$

3.3.3 Delay Models

A delay analysis has been conducted based on the traffic state classifications described above.

The arrival and departure of traffic flow at the roundabout is shown in Fig. 5, when the traffic flow of the j-direction of the i-approach is in state I.

In Fig. 5, the slope of OA and BC is the vehicle arrival rate, the slope of FA, GB and HC is the saturated flow rate, the area of the triangle OAF is the delay of the approach stop line, and the area of quadrilateral region $BCHG$ is the delay of the loop lane stop line. According to Fig. 5, the delay in traffic flow of the j-direction of the i-approach in one cycle is shown in formula (21).

$$D_{w,i,j} = \frac{S_{i,j} q_{i,j} (r_{a,i,j} + t_{a,i,j} - t_{w,i,j})^2}{2(S_{i,j} - q_{i,j})} \tag{21}$$

The arrival and departure of traffic flow at the roundabout is shown in Fig. 6, when the traffic flow of the j-direction of the i-approach is in state II.

In Fig. 6, the slope of OA and CD is the vehicle arrival rate, the slope of FA, GC and HE is the saturated flow rate, the area of the triangle OAF is the delay of the approach stop line, and the area of the polygon $GCDEH$ is the delay of the loop lane stop line. According to Fig. 6, the delay in traffic flow of the j-direction of the i-approach in one cycle is shown in formula (22).

$$\begin{aligned} D_{w,i,j} &= \frac{1}{2} q_{i,j} \left(r_{a,i,j} + g_{a,i,j} - t_{w,i,j}\right) \\ &\cdot \left(\frac{(S_{i,j} - q_{i,j})(r_{a,i,j} + g_{a,i,j} - t_{w,i,j})}{S_{i,j} q_{i,j}} - g_{c,i,j} + t_{d,i,j}\right) \end{aligned} \tag{22}$$

Fig. 5 State I

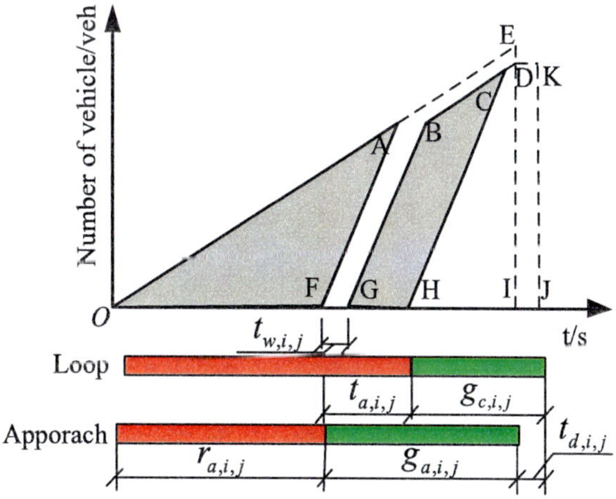

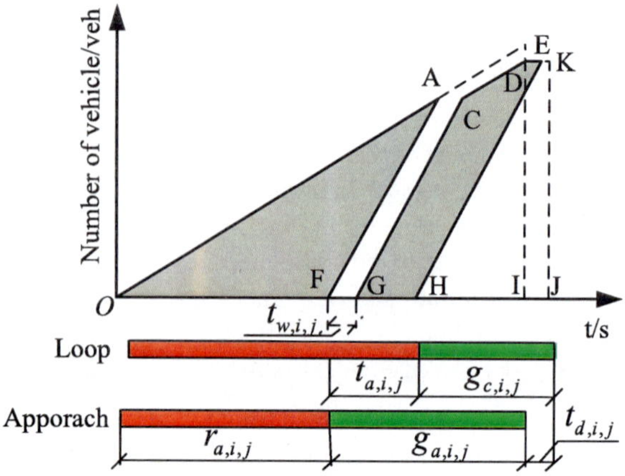

Fig. 6 State II

The arrival and departure of traffic flow at the roundabout is shown in Fig. 7, when the traffic flow of the *j*-direction of the *i*-approach is in state III.

In Fig. 7, the slope of *OA* and *BC* is the vehicle arrival rate, the slope of *FA*, *GB* and *HK* is the saturated flow rate, the area of the triangle *OAF* is the delay of the approach stop line, and the area of polygon *GBCDKH* is the delay of the loop lane stop line. According to Fig. 7, the delay in traffic flow of the *j*-direction of the *i*-approach in one cycle is shown in formula (23).

$$D_{w,i,j} = \frac{q_{i,j}(r_{a,i,j} + g_{a,i,j} - t_{w,i,j})^2}{2} - \frac{S_{i,j}(g_{c,i,j} - t_{d,i,j})^2}{2} + \frac{S_{i,j}t_{d,i,j}^2}{2} \tag{23}$$

The arrival and departure of traffic flow at the roundabout is shown in Fig. 8, when the traffic flow of the *j*-direction of the *i*-approach is in state IV.

In Fig. 8, the slope of *OA* is the vehicle arrival rate, the slope of *FA*, *GB* and *HD* is the saturated flow rate, the area of the triangle *OAF* is the delay of the approach

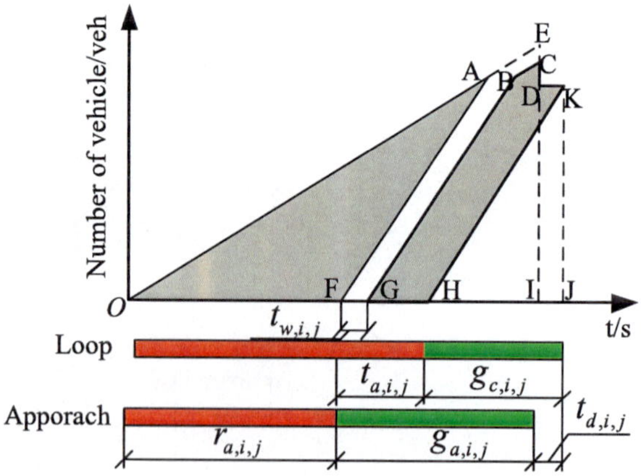

Fig. 7 State III

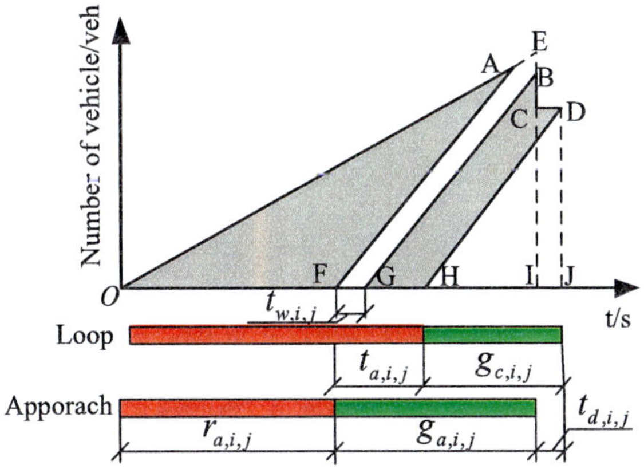

Fig. 8 State IV

stop line, and the area of the polygon *GBCDH* is the delay of the loop lane stop line. According to Fig. 8, the delay in traffic flow of the *j*-direction of the *i*-approach in one cycle is shown in formula (24).

$$D_{w,i,j} = \frac{S_{i,j} q_{i,j} r_{a,i,j}^2}{2(S_{i,j} - q_{i,j})} + \frac{S_{i,j}(g_{a,i,j} - t_{w,i,j} + 2t_{d,i,j})(g_{a,i,j} - t_{w,i,j})}{2} - \frac{S_{i,j} g_{c,i,j}^2}{2} - t_{d,i,j} S_{i,j}(g_{a,i,j} - g_{c,i,j} - t_{w,i,j}) \quad (24)$$

The arrival and departure of the traffic flow at the roundabout is shown in Fig. 9, when the traffic flow of the *j*-direction of the *i*-approach is in state V.

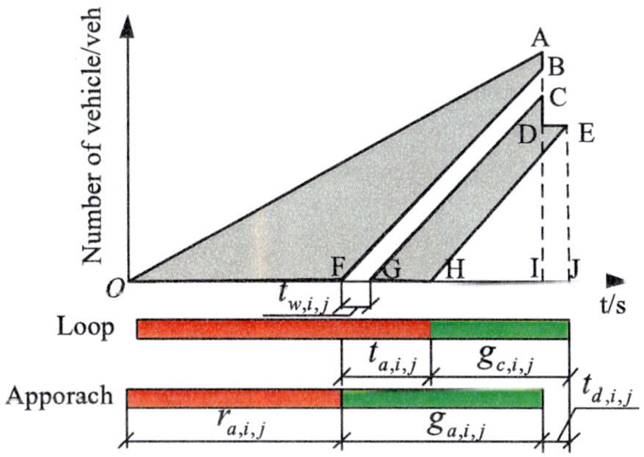

Fig. 9 State V

In Fig. 9, the slope of *OA* is the vehicle arrival rate, the slope of *FB*, *GB* and *HE* is the saturated flow rate, the area of the quadrilateral region *OABF* is the delay of the approach stop line, and the area of the polygon *GCDEH* is the delay of the loop lane stop line. According to Fig. 9, the delay in traffic flow of the *j*-direction of the *i*-approach in one cycle is shown in formula (25).

$$\begin{aligned} D_{w,i,j} &= \frac{q_{i,j}(r_{a,i,j} + g_{a,i,j})^2 - S_{i,j} g_{a,i,j}^2}{2} \\ &+ \frac{(g_{c,i,j} + t_{a,i,j} - t_{w,i,j} + t_{d,i,j}) S_{i,j} (g_{a,i,j} - t_{w,i,j})}{2} \\ &- \frac{S_{i,j} g_{c,i,j}^2}{2} - t_{d,i,j} S_{i,j} (g_{a,i,j} - g_{c,i,j} - t_{w,i,j}) \end{aligned} \quad (25)$$

3.4 Solution Algorithms

As $g_{c,k}$ is usually an integer in the time interval $[g_{c,k,\min}, g_{c,k,\max}]$, a finite searches can be used to solve the optimization problem of Eq. (1) quickly. The solution algorithm is given in Fig. 10.

4 Numerical Experiments

4.1 Basic Conditions of the Simulation Intersection

The roundabout of Xinhua Street and Xihe Road in Hailin City, Heilongjiang Province, China is taken as an example, shown in Fig. 11. The diameter of the central island is 24 m, the width of approach lane in the direction of Xinhua Street is 3.75 m, the one in the direction of Xihe road is 3.5 m, the width of inner loop lane is 5 m, and the width of the outer two loop lanes is 4.5 m. This particular roundabout is chosen as a case study owing to the capacity of storage space for left-turn vehicles in the loop lanes being less than the actual demand for left-turn vehicles, and there being a large volume of left-turn traffic. The TSLT is not applicable, so a signal control scheme to release each approach in turn is currently adopted for the roundabout, but the utilization rate of the loop lanes is low.

The right-turn, through and left-turn traffic volumes of the roundabout during peak and off-peak hour are shown in Table 2. And the current signal control scheme of the roundabout is shown in Table 3.

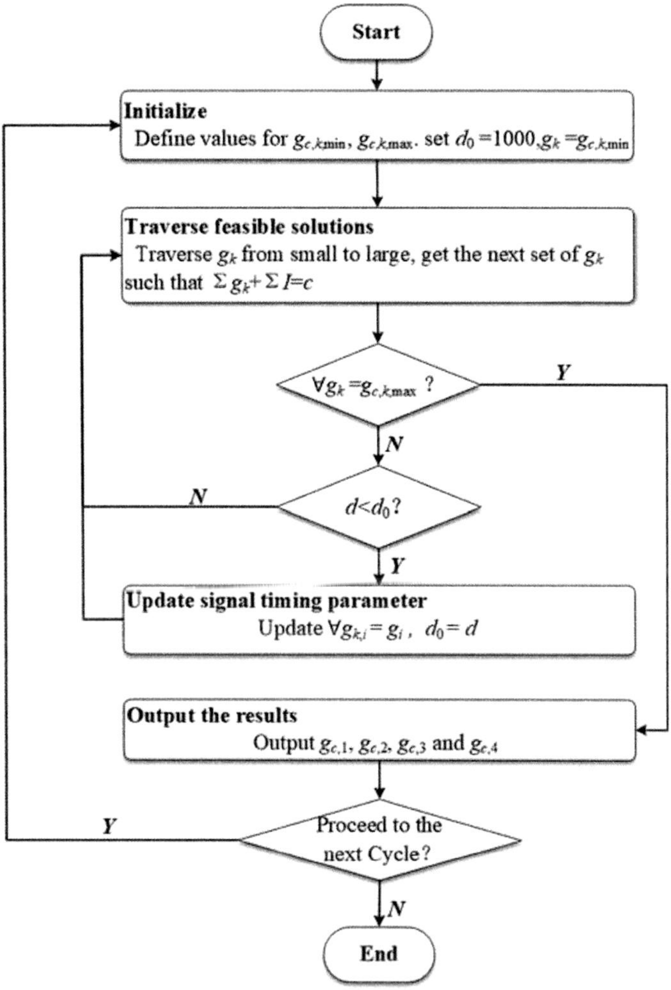

Fig. 10 Solution procedure

4.2 Simulation Process

4.2.1 Calibration of VISSIM Simulation Parameters

In order to keep the simulation results consistent with the actual operation results, the model parameters in VISSIM need to be corrected. First, the current layout model of the roundabout is built in VISSIM simulation software. Then, the default values of VISSIM parameters are temporarily kept unchanged, the peak hour traffic volume in each direction, the mean speed and current signal control scheme are used as the inputs to simulate for one hour. It is found that the mean error of delay and queue

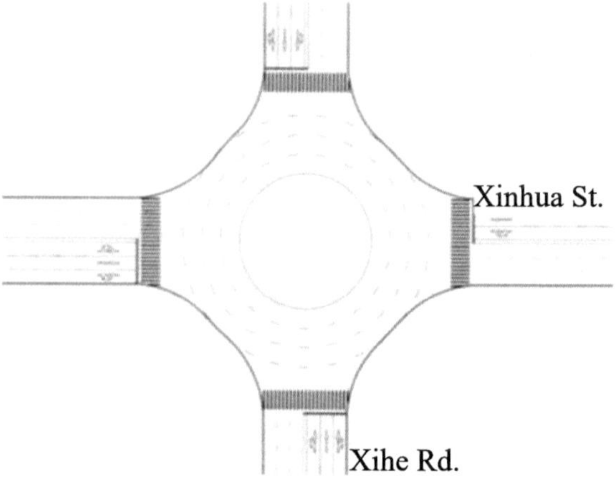

Fig. 11 Current layout of the roundabout

Table 2 Traffic volume (pcu/h)

Approach time	East	West	South	North
Peak	106/593/264	92/741/242	168/647/410	181/634/327
Off-peak	81/385/171	73/518/145	128/462/246	153/412/196

Table 3 Current signal control scheme of the roundabout (peak hour)

Items	The first phase	The second phase	The third phase	The fourth phase
Phase scheme				
Pedestrian phase	West	North	East	South
Green time (s)	33	30	37	36
Yellow time (s)	3	3	3	3
Red time (s)	112	115	108	109

Traffic survey shows that the mean speed is 23 km/h in the roundabout

length between the simulation results and the actual investigation results is approximately 13%. Next, the lane saturation flow rate, car-following model parameters, minimum spacing, and values of acceleration and deceleration are adjusted by small steps until the simulation results are essentially consistent with the actual investigation results with a mean error within 1%. At this point, the calibration of the VISSIM simulation parameters is considered complete.

The calibration results are as follows: the saturation flow rates of the left-turn and through lane, through lane and through and right-turn lane are 1280 pcu/h, 1350 pcu/h and 1320 pcu/h, respectively, the minimum spacing is 1.2 m, the maximum acceleration (deceleration) is 5 m/s^2, the expected acceleration (deceleration) is 3.0 m/s^2, the average standstill distance is 1.6 m, the minimum headway is 0.8 m, the additive part of the safety distance is 2.65, and the multiplicative part of the desired safety distance is 4.5. The investigation and simulation results of current signal control scheme at the roundabout are shown in Tables 4. The figures between the brackets in Table 4 represent the actual investigation results. Based on this, a comparative analysis is performed.

4.2.2 Establishment of the Simulation Environment

The improvement layout and optimized signal control scheme of the roundabout are shown in Fig. 11. The minimum green time of the through phase and the left-turn phase are 27 s and 12 s, respectively, and the maximum green time of the through phase and the left-turn phase is 70 s.

The improvement layout model of the roundabout is also built in VISSIM. And the traffic volume in each direction, the average speed and the optimized signal control scheme are used as the inputs to the model. Under the condition that the calibrated simulation parameters and the input traffic volume in each direction are kept unchanged, a traffic simulation of the roundabout is conducted repeatedly in accordance with the layouts and signal control plans before and after improvement. Each simulation lasts for 60 min in peak hour and off-peak hour, respectively, and the results are shown in Tables 5 and 6.

5 Results and Discussion

5.1 Results Analysis

With the improvement, the delay for vehicles travelling through the roundabout was found to be lower than the one of current signal control scheme. The current signal control scheme to release each approach in turn only achieves optimal performance when the through and left-turn traffic volumes per lane are similar, since there is sufficient green time for both the through lanes and the left-turn lanes. Since traffic volumes per lane are seldom balanced, the proposed method can better allocate green time to the left-turn and through traffic flow in order to improve the utilization rate of the green time and reduce delays compared with the current signal scheme.

The proposed method to improve the roundabout also leads to lower delays than the current signal control scheme for the right-turn vehicles, even lower than the through vehicles. Similar to the delay reduction for the through vehicles, the reduction in delay is partially due to the change of release mode, and partially due to the through waiting area. Before the waiting area was set, right-turn vehicles in the queue could leave the roundabout only when the traffic lights were green; but with the proposed method, right-turn vehicles at the front of the queue can leave the roundabout directly when right-turn vehicles are entering the waiting area. Therefore, the through waiting

Table 4 Simulation and investigation results of the current signal control scheme at the roundabout (peak hour)

Items	East approach		West approach		South approach		North approach	
	Delay(s)	Queen length(m)	Delay(s)	Queen length(m)	Delay(s)	Queen length(m)	Delay(s)	Queen length(m)
Left-turn	57.1(58.3)	41(43)	72.5(73.4)	62(63)	56.7(57.5)	34(35)	54.4(55.7)	52(51)
Through	55.3(55.2)	36(37)	57.8(58.2)	37(37)	51.0(52.1)	21(22)	62.5(63.0)	54(53)

Table 5 Simulation results before and after improvement in off-peak hour

Approach	Direction	Delay(s)	Throughput(veh)	Average queue length (m)	Maximum queue length (m)	Average number of stops
East	Through	38.0/29.4	345/332	16/7	72/51	0.82/1.31
	Left-turn	32.8/38.9	185/174	11/5	52/38	0.78/1.46
	Right-turn	38.8/26.1	97/79	16/7	72/51	0.87/0.98
West	Through	38.2/27.7	472/498	20/9	78/56	0.80/1.33
	Left-turn	39.1/39.8	158/142	11/4	55/43	0.78/1.52
	Right-turn	33.2/24.7	77/72	20/9	78/56	0.78/1.04
North	Through	40.0/23.7	369/443	21/7	92/57	0.84/1.15
	Left-turn	40.2/28.8	206/209	15/2	72/42	0.83/1.20
	Right-turn	41.9/20.5	169/165	21/7	92/57	0.87/1.12
South	Through	36.8/26.2	419/429	19/11	84/62	0.83/1.08
	Left-turn	33.9/32.9	243/253	15/7	98/46	0.77/1.30
	Right-turn	34.4/23.1	125/126	19/11	84/62	0.80/1.01

Table 6 Simulation results before and after improvement in peak hour

Approach	Direction	Delay(s)	Throughput(veh)	Average queue length (m)	Maximum queue length (m)	Average number of stops
East	Through	55.3/45.7	549/53	37/21	134/83	0.87/1.40
	Left-turn	56.7/63.2	262/257	33/17	129/69	0.85/1.63
	Right-turn	57.8/39.99	97/98	37/21	134/83	0.95/1.16
West	Through	74.6/51.7	667/704	59/35	164/111	1.03/1.57
	Left-turn	56.1/64.5	247/231	38/16	159/97	0.86/1.61
	Right-turn	74.7/47.4	87/82	59/35	164/111	1.01/1.27
North	Through	54.6/52.9	565/675	43/30	141/90	0.95/1.58
	Left-turn	52/45.8	331/318	40/16	128/83	0.85/1.34
	Right-turn	52.9/50.1	166/155	43/30	141/90	0.87/1.12
South	Through	53.2/46.8	571/585	51/37	187/148	0.83/1.43
	Left-turn	61.7/63.8	423/421	55/46	191/161	0.94/1.64
	Right-turn	53.7/45.5	148/152	51/37	187/148	0.90/1.23

area not only increases the capacity and reduces the delay of through traffic flow, but also helps to reduce the delay of right-turn vehicles.

After setting up the waiting area, the delay for left-turn vehicles in the other approach was found to be higher than the current signal control scheme for all approaches except for the north approach. The reason for this is that the through

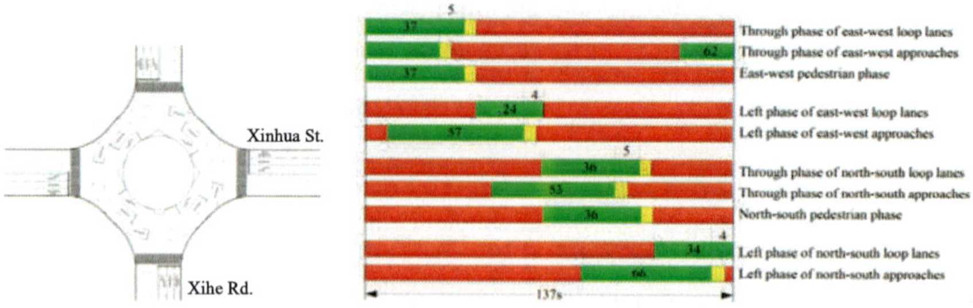

(a) improved layout (b) optimized signal plan of the roundabout

Fig. 12 Improvement of the roundabout

traffic volumes per lane of the north approach were higher than the left-turn traffic volumes per lane, and thus more green time is allocated to the through phase, which leads to higher left-turn delays at the east, south and west approaches. The lower delay at the north approach is due to the higher left-turn traffic volume at the south approach. In order to achieve the lowest overall delay, the north–south left-turn phase is allocated more green time. When the left-turn traffic volume of the north approach is low, the delay is decreased.

After setting up the waiting area, the average queue length and the maximum queue length of roundabout both decrease significantly, which not only improves the capacity, but also closes the stop line of the approach to the exit position to some extent, which can effectively reduce the queuing length of the approach. When the spacing at intersections is short, this can help to reduce the probability of "overflow".

Since there are two stop lines in the roundabout with the waiting area, vehicles that are stopped at the approach stop line need to stop again after entering the waiting area, which increases the total number of stops.

The throughput of the roundabout before and after the improvement is similar with no significant change, which cannot explain the improvement in capacity after setting up the waiting area. Therefore, on the premise that the proportion of through, left-turn and right-turn of each approach remains unchanged, the original traffic volume in peak hour was increased by 2.5% in turn, and the trend of the total throughput of the roundabout is shown in Fig. 13.

As can be seen from Fig. 13, when the increment of traffic volume reaches 15%, the capacity of the roundabout before improvement has reached its limit, and the throughput is maintained at approximately 4600 vehicles. However, when the increment reaches 27.5%, the improved roundabout reach its capacity limit, and the throughput is maintained at approximately 5100 vehicles. Based on the trend of the throughput in Fig. 13, it can be concluded that the proposed method can effectively improve the capacity of the roundabout.

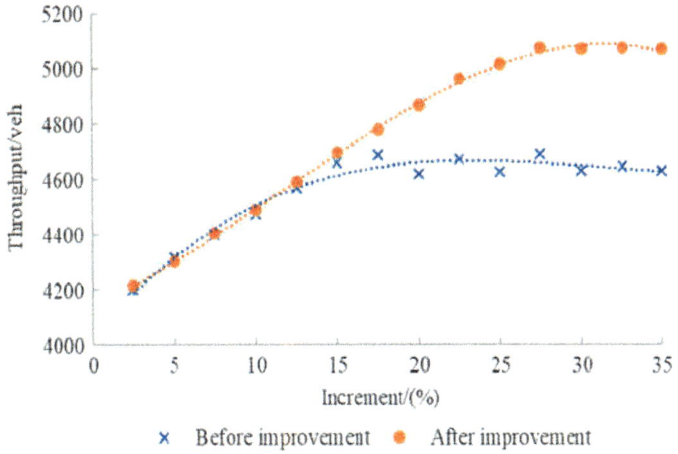

Fig. 13 Trend of throughput

5.2 *Sensitivity Analysis*

5.2.1 Traffic Volume on the Delay of Proposed Method

Under the condition that the calibrated simulation parameters are kept unchanged, the left-turn traffic volume is varied from 50 veh/h to 850 veh/h, in step-size of 50 veh/h, and the through traffic volume is varied from 250 veh/h to 1450 veh/h, in step-size of 100 veh/h. The left-turn and through traffic volumes of each approach keep growing at the same time. Simulation result is shown in Fig. 14.

As seen in Fig. 14, as the traffic volume increases the delay also increases, with slower increases in delay at lower traffic volumes, and rapid increases when the left-turn traffic volume is close to 550 veh/h or through traffic volume is close to 1350

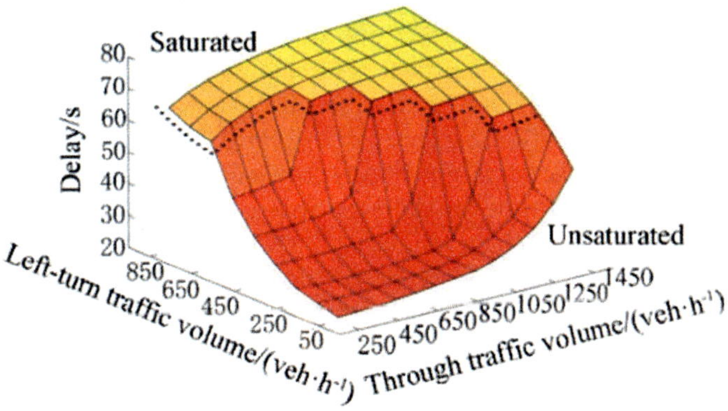

Fig. 14 Traffic volume on the delay of proposed method at roundabout

Fig. 15 Critical value of left-turn vehicle ratio

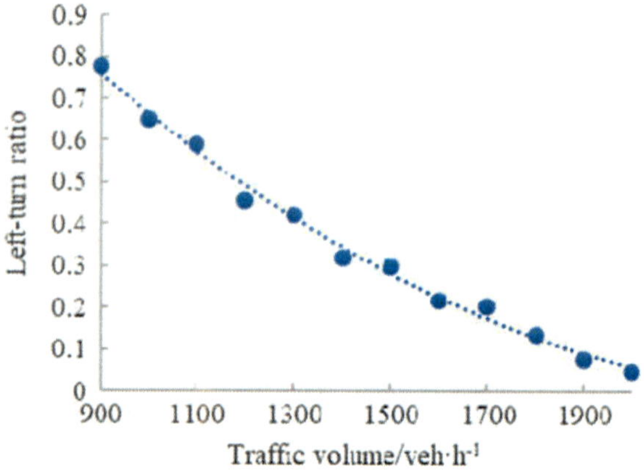

veh/h. The reasons for the rapid increase in delays are the same as those at grade signal intersections, the excessive traffic demand leads to the increase of delay, queue and stops in the approach lane. Due to the existence of the waiting area, the method proposed in this paper for the roundabout will lead to more increase in the number of stops.

In Fig. 14, when the improved roundabout is close to saturation, there is a relationship between the left-turn traffic volume and through traffic volume. The critical left-turn ratio at the improved roundabout decreases as through traffic volume increases. When the left-turn ratio is lower than the critical value, the improved roundabout is in an unsaturated state and can operate well. That is to say, the improved roundabout has an application range for left-turn and through traffic volume, beyond the application range, the improved roundabout cannot improve the operation condition.

5.2.2 Number of Loop Lanes on the Delay of Proposed Method

Under the condition that the calibrated simulation parameters are kept unchanged, the left-turn traffic volume is varied from 50 veh/h to 1450 veh/h, in step-size of 100 veh/h, and the through traffic volume is varied from 450 veh/h to 1650 veh/h, in step-size of 100 veh/h. The left-turn and through traffic volumes of each approach keep growing at the same time. Analysis is performed to assess the impact for the two cases of increasing the number of through loop lanes and the number of left-turn loop lanes. The delay under the two different cases is shown in Fig. 16.

As seen from Fig. 16, when the left-turn traffic volume is low, the delay when the number of left-turn loop lanes is increased is higher than that when the number of through loop lanes is increased. The main reason is that the through traffic volume is larger than the left-turn traffic volume. Increasing the number of through loop lanes can reduce the time for through vehicles to pass through the roundabout during the green time, so as to improve the roundabout delay as a whole. Although the increase in the number of left-turn loop lanes can improve the delay of left-turn vehicles,

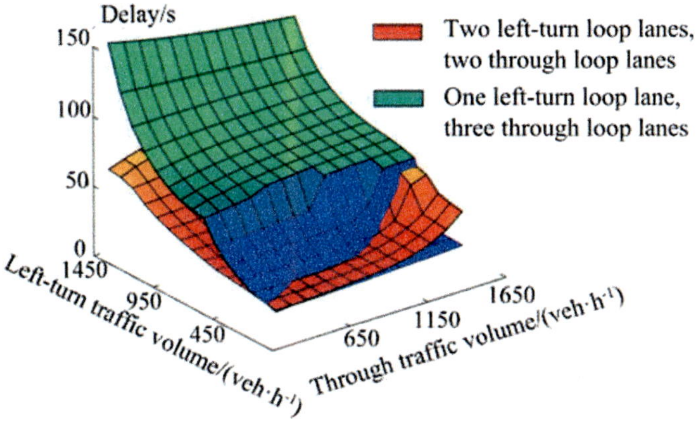

Fig.16 Number of loop lanes on the delay of proposed method

the improvement in the delay of the whole roundabout is very limited. When the left-turn traffic volume is more than 450 veh/h, the effect of increasing the number of left-turn loop lanes is better than that of increasing the number of through loop lanes. The main reason is that the increase of the number of left-turn loop lanes makes the decrease of the ratio of traffic volume to saturation flow rate of left-turn phase much larger than that of through phase caused by the increase of through loop lanes, which can solve the same traffic demand with shorter cycle.

In Fig. 16, the intersecting line of the delay surface is the critical value to determine which type of loop lanes should be increased. When the traffic volume of the improved roundabout is to the left side of the critical value, it is recommended to increase the number of left-turn loop lanes; when the traffic volume of the improved roundabout is on the right side of the critical value, it is recommended to increase the through loop lanes. In addition, when the left-turn traffic volume is less than 450 veh/h and the through traffic volume is less than 1150 veh/h, the difference in delay between the two cases is not significant, although the delay due to increasing the number of through loop lanes is lower than that of increasing the number of left-turn loop lanes.

5.2.3 The Central Island Radius on the Delay of Proposed Method

Under the condition that the calibrated simulation parameters are kept unchanged, the total traffic volume of each approach is varied from 200 veh/h to 1400 veh/h, in step-size of 100 veh/h, and the rate of left-rurn traffic volume to through traffic volume is kept 1:3. The total traffic volume of each approach keeps growing at the same time. The radius of the central island of the roundabout is changed from 20 to 50 m, in step-size of 10 m. The delay of the improved roundabout for different central island radiuses is shown in Fig. 17.

As seen from Fig. 17, when the total traffic volume of each approach is less than 800 veh/h, the influence of the central island radius on the delay of the improved

Fig.17 The central island radius on the delay of proposed method

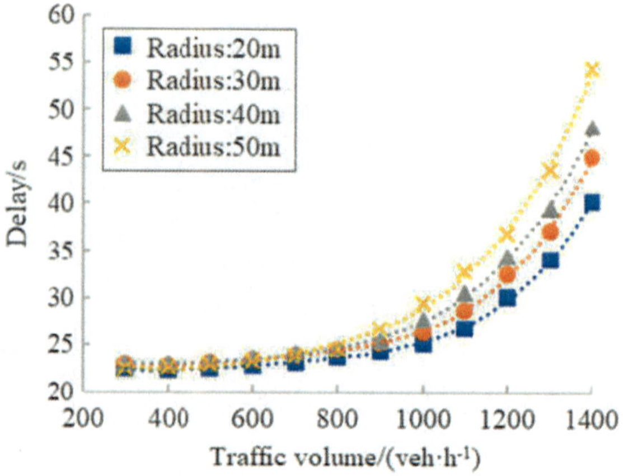

roundabout is not significant; while when the total traffic volume of each approach is higher than 800 veh/h, the influence of the central island radius on the delay of the improved roundabout is more significant with the increase of traffic volume in each approach. The larger the central island radius is, the higher the delay of the improved roundabout is. It can be seen from Fig. 4 that the conflict distance increases as the central island radius in increased, which leads to more loss time and higher delays.

6 Conclusions

1. In this paper, a method of combining waiting area and coordinated signal control is proposed to satisfy the required conditions for roundabouts with high traffic demands. The through and left-turn waiting areas are configured with the help of the loop lane space at the roundabout. When the through vehicles pass, the left-turn vehicles enter the left-turn waiting area. And the through vehicles and left-turn vehicles are separated by a coordinated traffic signal to avoid conflicts and improve the roundabout capacity.
2. The delay calculation models are established based on the traffic states of approach lanes and loop lanes to calculate the delays in accuracy, and an optimization model of signal control parameters for the improved roundabout is established with the objective of minimizing the average delay.
3. Case study shows that although the average number of stops is higher than usual due to the waiting areas, the proposed method has lower delays, queue length and higher capacity. The influence of traffic volume, number of loop lanes and central island radius on the delay of proposed method is analyzed. Results show that as the traffic volume increases the delay also goes up, with slow-motion increases in delay at lower traffic volumes, and rapid increases when the left-turn traffic volume is close to 550 veh/h or through traffic volume is close to 1350

veh/h. When the left-turn traffic volume is more than 450 veh/h, the effect of increasing the number of left-turn loop lanes is better than that of increasing the number of through loop lanes; otherwise not. When the total traffic volume of each approach is less than 800 veh/h, the influence of the central island radius on the delay is not significant; while when the total traffic volume of each approach is higher than 800 veh/h, the influence of the central island radius on the delay is more significant with the increase of traffic volume in each approach.
4. The traffic flow at the entrance of the roundabout is bifurcated in the proposed method of this paper. Compared with the usual traffic organization of the roundabout, new traffic conflicts will be added. Although these traffic conflicts can be eliminated by traffic signals, it is different from the normal running rules, and the adaptability of drivers and traffic safety deserve further study.

Acknowledgements This work is supported by the National Key R&D Program of China (2020YFB1600400).

References

1. Akçelik R (2003) A roundabout case study comparing capacity estimates from alternative analytical models. In: The 2nd Urban Street Symposium, Anaheim, California, USA
2. Akhtar MW, Ahmad A, Rastogi R (2018) Gap acceptance behavior and vehicle speed on roundabouts: case study in India. Eur Transp Trasporti Eur 68
3. Ambros J, Novak J, Borsos A, Hoz E, Kiec M, Machcinik S, Ondrejka R (2016) Central European comparative study of traffic safety on roundabouts. Transp Res Procedia 14:4200–4208. https://doi.org/10.1016/j.trpro.2016.05.391
4. Anjana S, Anjaneyulu MVLR (2015) Development of safety performance measures for urban roundabouts in India. J Transp Eng 141(1). https://doi.org/10.1061/(asce)te.1943-5436.0000729
5. Arroju R, Gaddam HK, Vanumu LD, Rao KR (2015) Comparative evaluation of roundabout capacities under heterogeneous traffic conditions. J Mod Transp 23(4):310–324. https://doi.org/10.1007/s40534-015-0089-8
6. Arshi AN, Alhajyaseen WKM, Nakamura H, Zhang X (2018) A comparative study on the operational performance of four-leg intersections by control type. Transp Res Part A Policy Pract 118:52–67. https://doi.org/10.1016/j.tra.2018.08.039
7. Bai Y, Xue K, Yang X (2009) Block mechanism of left-turned flow at signal-controlled roundabout. In 2009 WRI global congress on intelligent systems, vol 3, pp 443–449. Xiamen, China: IEEE. https://doi.org/10.1109/GCIS.2009.178
8. Belcher M (1984) Roundabout control is signal success. Surveyor 163(4780)
9. Bie Y, Mao C, Yang M (2016) Development of vehicle delay and queue length models for adaptive traffic control at signalized roundabout. Procedia Eng 138:141–150. https://doi.org/10.1016/j.proeng.2016.01.244
10. Bugg Z, Schroeder BJ, Jenior P, Brewer M, Rodegerdts L (2015) Methodology to compute travel time of a roundabout corridor. Transp Res Record 2483:20–29. https://doi.org/10.3141/2483-03
11. Chaudhary N A, Songchitruksa P (2009) Traffic adaptive signal control at roundabouts. In: Proceedings of the 4th international gulf conference on roads, pp 141–150

12. Easa SM (2019) Modeling of unsymmetrical single-lane roundabouts based on stopping sight distance. KSCE J Civ Eng 23(2):800–809. https://doi.org/10.1007/s12205-018-0344-8
13. Giuffre O, Grana A, Tumminello ML, Sferlazza A (2018) Capacity-based calculation of passenger car equivalents using traffic simulation at double-lane roundabouts. Simul Model Pract Theory 81:11–30. https://doi.org/10.1016/j.simpat.2017.11.005
14. Gokce MA, Oner E, Isik G (2015) Traffic signal optimization with particle swarm optimization for signalized roundabouts. Simul Trans Soc Model Simul Int 91(5):456–466. https://doi.org/10.1177/0037549715581473
15. Gong YJ, Zhang J (2012) Real-time traffic signal control for roundabouts by using a PSO-based fuzzy controller. In: 2012 IEEE congress on evolutionary computation, pp 1–8. Brisbane, QLD, Australia: IEEE
16. Hallworth MS (1992) Signalling roundabouts. Traffic Eng. Control. 33(6):46–51
17. Hatami H, Aghayan I (2017) Traffic efficiency evaluation of elliptical roundabout compared with modern and turbo roundabouts considering traffic signal control. Prome 29(1):1–11
18. Hummer JE, Milazzo JS II, Schroeder B, Salamati K (2014) Potential for Metering to help roundabouts ,manage peak period demands in the United States. Transp Res Record 2402:56–66. https://doi.org/10.3141/2402-07
19. Kaczmarek MK (1994) Group control of traffic at roundabouts. In: Third international conference on road traffic control, pp 119–123. London, UK: IET
20. Khan MA, Ectors W, Bellemans T, Janssens D, Wets G (2018) Unmanned aerial vehicle-based traffic analysis: a case study for shockwave identification and flow parameters estimation at pignalized intersections. Remote Sens 10(3). https://doi.org/10.3390/rs10030458
21. Ma W-J, Xie H-Z, Bai Y, Zhao J, Yang X-G (2013) Signal timing optimization model based on dual-ring phase scheme for roundabout. J Cent South Univ 20(2):563–571. https://doi.org/10.1007/s11771-013-1519-6
22. Ma W, Liu Y, Head L, Yang X (2013) Integrated optimization of lane markings and timings for signalized roundabouts. Transp Res Pt C-Emerg Technol 36:307–323. https://doi.org/10.1016/j.trc.2013.08.013
23. Mahdalova I, Krivda V, Skvain V (2013) Influence of roundabout inscribed circle diameter to the traffic safety. Appl Mech Mater 409–410:1122–1125. https://doi.org/10.1109/10.4028/www.scientific.net/AMM.409-410.1122
24. Maher M (2008) The optimization of signal settings on a signalized roundabout using the cross-entropy method. Comput Aided Civil Infrastruct Eng 23(2):76–85. https://doi.org/10.1111/j.1467-8667.2007.00524.x.
25. Martin-Gasulla M, Garcia A, Moreno AT, Llorca C (2016) Capacity and operational improvements of metering roundabouts in Spain. Transp Res Procedia 15:295–307. https://doi.org/10.1109/10.1016/j.trpro.2016.06.025
26. Meneguzzer C, Gastaldi M, Rossi R, Gecchele G, Prati MV (2017) Comparison of exhaust emissions at intersections under traffic signal versus roundabout control using an instrumented vehicle. Transp Res Procedia 25:1597–1609. https://doi.org/10.1016/j.trpro.2017.05.204
27. Obsu LL, Delle Monache ML, Goatin P, Kassa SM (2015) Traffic flow optimization on roundabouts. Math Meth Appl Sci 38(14):3075–3096. https://doi.org/10.1002/mma.3283
28. Patnaik AK, Agarwal LA, Panda M, Bhuyan PK (2020) Entry capacity modelling of signalized roundabouts under heterogeneous traffic conditions. Transp Lett 12(2):100–112. https://doi.org/10.1080/19427867.2018.1533160
29. Robison BW (2000) Roundabouts: an informational guide. ITE J Inst Transp Eng 70(4):16
30. Salter RJ, Okezue OG (1988) Simulation of traffic flow at signal-controlled roundabouts. Traffic Eng Control 29(3):142–147. http://worldcat.org/issn/00410683
31. Shaaban K, Abou-Senna H, Elnashar D, Radwan E (2019) Assessing the impact of converting roundabouts to traffic signals on vehicle emissions along an urban arterial corridor in Qatar. J Air Waste Manage Assoc 69(2):178–191. https://doi.org/10.1080/10962247.2018.1526137
32. Simmonite H (2008) Fixed Time v Single Stream MOVA Control on a signalled roundabout. Traffic Eng Control 49(10):391–394+396

33. Wilson C, Dilgir R, Zein, SR (2006) Safety risk management in large diameter modern roundabout applications. In: Conference & exhibition of the transportation association of Canada.
34. Xu H, Zhang K, Zhang D (2016) Multi-level traffic control at large four-leg roundabouts. J Adv Transp 50(6):988–1007. https://doi.org/10.1002/atr.1385
35. Yang H (2013) A traffic capacity research for five-way roundabouts with signal controls. In: 2013 The International conference on remote sensing, environment and transportation engineering (RSETE 2013). Vol 31, pp 484–487. Atlantis Press
36. Yang XG, Li XG, Xue K (2004) A new traffic-signal control for modern roundabouts: method and application. IEEE Trans Intell Transp Syst 5(4):282–287. https://doi.org/10.1109/tits.2004.838181

Short-Term Traffic Flow Prediction on a Freeway with Multiple Spatial Toll Data Via Temporal Convolutional Network

Xu Wei, Qiang Jing, YinLi Jin, and HaoChen Wang

Abstract Accurate short-term traffic flow prediction plays an important role in traffic guidance and traffic safety. In order to solve the problem that traffic flow is difficult to predict because of the sharp nonlinearity and randomness, a deep learning framework, i.e., the temporal convolutional network (TCN) was explored to capture the nonlinear spatiotemporal characteristics of traffic flow. By adjusting the hyper-parameters of TCN, a traffic flow predictor is proposed. Besides, this paper adopts a method to calculate the designated cross-section traffic volume of freeways from toll data, which makes up for the inability to obtain traffic volume at some cross-section locations due to damage or lack of traffic detection equipment. Through data merging, data cleaning, data reconstruction, data filtering, a traffic flow data set which was selected as the data input with a time interval of 15 min was constructed based on the toll data of Shaanxi Province freeway network from December 2018 to April 2019 as the data source. The TCN model was compared with the SVM, SAE, LSTM and GRU models in terms of mean absolute error (MAE) and root mean square error (RMSE). The prediction results of multiple cross-sections showed the TCN model performs best with superior prediction accuracy, which indicate the TCN model has good robustness and generalization ability. The traffic volume calculation algorithm may provide a practical method for deriving the traffic volume without installing any additional regularly maintained detectors and equipment on the freeway. And the

This paper was supported in part by the National Key Research and Development Program of China under Grant 2019YFB1600703.

X. Wei (✉) · Y. Jin · H. Wang
Institute for Transportation Systems Engineering Research (ITSER), Chang'an University, Xi'an 710064, China
e-mail: 2985317302@qq.com

X. Wei · H. Wang
School of Electronics and Control Engineering, Chang'an University, Xi'an 710064, China

Q. Jing
Hong Kong - Zhuhai - Macau Bridge Authority, Zhuhai Guangdong 519015, China

Y. Jin
School of Transportation Engineering, Chang'an University, Xi'an 710064, China

W. Wang et al. (eds.), *Green Transportation and Low Carbon Mobility Safety*, Lecture Notes in Electrical Engineering 944,
https://doi.org/10.1007/978-981-19-5615-7_5

prediction results of TCN model can provide strong support for traffic control and traffic induction.

Keywords Temporal convolutional network · Freeway · Traffic flow prediction · Short-term · Toll data

1 Introduction

Freeways have become an important channel for cargo transportation and people's daily travel. Unlike driving conditions on urban roads [1], there are more large and medium-sized vehicles on the freeways. They generally carry more loads, with higher driving speed, and drivers often need to drive continuously for a long time [2]. When these vehicles are densely distributed on the freeways, traffic congestion and traffic accidents are more likely to occur, which often cause large property losses and irreparable consequences [3]. Therefore, balancing road loads to ensure smooth roads is essential to ensure the safe operation of freeways. And the short-term traffic flow forecast is introduced to better grasp the traffic conditions of freeways [4]. On the one hand, by combining accurate short-term prediction results with some navigation products, internet companies such as Baidu can provide drivers with more scientific, intelligent and practical targeted route guidance. On the other hand, obtaining the traffic conditions of important traffic facilities such as bridges and tunnels in the future by traffic flow prediction will also help the traffic management department to take charge of the traffic flow in a timely manner, avoiding the continuous dense distribution of heavy-duty vehicles from damaging the structure of these facilities. Moreover, freeways are also an important part of the intelligent transport system (ITS), high-accuracy and real-time traffic flow prediction is one of the most important works of ITS [5]. Therefore, accurate prediction of short-term traffic flow on freeways is vital for people's travel, traffic control and management, and promoting the development of ITS.

Generally, traffic flow prediction can be divided into short-term and long-term prediction. Long-term traffic flow prediction mainly studies the general change trend of monthly or yearly traffic flow [6]. The short-term traffic flow prediction mainly studies the real-time changes in traffic flow, and the prediction time is generally a few minutes [7]. This paper focuses on how to achieve accurate short-term traffic flow prediction to meet the needs of freeway traffic control and the development of ITS. Therefore, in order to achieve the best prediction results, the existing mainstream prediction models are studied to help us make the best choice. The methods of traffic flow prediction are roughly based on the following three categories: statistical theory, traditional machine learning, and deep learning.

The statistical theory prediction methods based on the time series [8] have been deeply researched, which have developed autoregressive integrated moving average (ARIMA) [9], Kalman filtering [10] and other classic theories. These models use statistical theory to process historical traffic data. However, due to the continuous expansion of data samples and the non-linearity and complexity of traffic data, traditional prediction models have been difficult to meet the accuracy requirements of prediction.

Later, methods based on traditional machine learning were proposed for traffic flow prediction. Such as Bayesian networks, Support Vector Regression (SVR) and K-Nearest Neighbor (k-NN), etc. [11]. The SVR model maps historical traffic flow data from a low-dimensional space to a high-dimensional space. The k-NN model uses the prediction interval constructed by the K value to be used in the prediction of special road conditions, and has obtained obvious improvement effects. But it is still necessary to manually extract traffic-related functions the nonlinear time and the nonlinear temporal-spatial characteristics of traffic flow in complex freeway networks cannot be captured [12]. Therefore, in complex forecasting tasks, these methods generally cannot obtain satisfactory forecasting results.

In recent years, many models based on deep learning have been proposed for traffic flow prediction [13]. Such as stacked auto-encoder (SAE) [14], the recurrent neural network (RNN) [15] and its variants. The most classic of which are long and short-term memory (LSTM) [16] and gated recurrent neural network (GRU) [17]. In addition, Mou, Zhao and Xie, et al. proposed a temporal information enhancing LSTM (T-LSTM) to predict traffic flow of a single road segment and improved the prediction performance compared with other state-of-the-art methods [18]. Guo et al. combined one-dimension convolution with GRU and conducted a large-scale search of the hyper-parameter space to handle changes in hyper-parameters. However, for the SAE model, each layer of the stacked autoencoder needs to learn the weight matrix greedily, which brings excessive training time in traffic flow prediction [19]. For the LSTM and GRU models, the time complexity of the model training is relatively high, and there will be problems in training difficulties and poor fitting effects due to the disappearance of gradients.

Recently, some research results have shown that TCNs outperform RNNs in a series of sequence modeling tasks such as polyphonic music modeling, word- and character-level language modeling [20]. And Chen et al. [21], have presented a probabilistic forecasting framework based on TCN for multiple related time series forecasting. The results show that the framework compares favorably to the state-of-the-art in both point and probabilistic forecasting. Besides, a novel deep learning traffic forecasting framework based on graph attention network (GAT) and TCN has been presented in [22]. Results have shown that the proposed model outperforms other models. Therefore, considering that the traffic data is highly non-linear, dynamic and random, and traffic flow prediction is essentially a complex sequence modeling task [6], the TCN model is introduced to capture the complex nonlinear spatial -temporal characteristics of traffic flow in this paper. There are mainly three advantages of TCN in the traffic flow prediction task:

(1) As a kind of convolutional network, TCN has the amazing advantages of convolutional networks in effectively extracting spatial features of data compared with other deep learning models such as RNNs and its variants. In addition,

by combining causal convolution, TCN can model every input data to prevent "information leakage or loss".

(2) TCN can process the input sequence, i.e., the traffic flow data arranged in time sequence in parallel, which means TCN can reduce the time in training and evaluation compared with RNNs. And higher prediction accuracy often requires a long-time span of historical traffic data as input. Therefore, TCN can have a very fast processing speed while guaranteeing the prediction accuracy.
(3) By combining dilated convolutions and residual connections, TCN has a larger receptive field when using the same number and size of convolution kernels, which enables TCN to extract more detailed features compared to CNN and won't become difficult to train because the network is too deep.

Different from the commonly used methods for obtaining traffic flow data in the past, we adopted a method that by converting the toll data to traffic flow data of the designated road segments [23]. With this method, the original toll data of Shaanxi freeway network for a total of 5 months were collected and the historical traffic flow data of multiple cross-sections were calculated and selected as the data source to test the TCN model. The prediction results show that TCN model performs best with superior prediction accuracy which can provide strong support for traffic control and traffic induction. The contribution of this paper is as follows:

(1) A systematic method that can derive the traffic volume of designated cross-section location on a freeway from the toll data is presented, which makes up for the inability to obtain traffic volume at some cross-section locations due to damage or lack of traffic detection equipment.
(2) A traffic flow predictor with a deep learning architecture based on TCN is proposed, where the stacked casual dilated convolutions are used to enable enough receptive fields and model the entire input sequence such that the spatial–temporal characteristics of traffic flow can be captured. And the TCN is utilized for freeway traffic flow prediction for the first time as far as we know.
(3) The effectiveness of the TCN model compared with other prediction models was verified through extensive experiments. The experimental results show that our framework achieves the state-of-art performance with superior accuracy in short-term traffic flow prediction and ca provide strong support for traffic control and traffic induction.

2 Methodology

The methodology includes three parts: the acquisition of traffic flow data, the basic structure of the TCN model, and the evaluation index of the experimental results. Each part will be introduced separately.

2.1 Acquisition of Traffic Flow Data

The overall data preprocessing process is shown in Fig. 1.

Generally, the calculation of traffic volume is done through manual counters from onsite observations or surveillance video systems [24] or electronic information equipment such as vehicle detectors (VDs), radars and so on [25, 26]. But these methods inevitably require a lot of manpower or equipment installation and maintenance costs. More importantly, for road segments where no VDs are installed and VDs are broken, the traffic conditions will be unknown. Taking VDs as an example, because they are fixed in position, they can only capture traffic conditions within the detection range, as explained in Fig. 2 [23]. In order to avoid the above-mentioned problems, this paper adopts a method of using toll data to calculate the traffic volume at designated cross-section locations on a freeway based on [24], which corresponds to the data reconstruction in Fig. 1.

In our research, the toll data of 425 toll stations on the Shaanxi freeway network from December 2018 to April 2019 were collected as the raw data, examples of partial toll data at entrance and exit of toll stations are shown in Tables 1 and 2. For Tables 1 and 2, "ENDEALID" and "EXDEALID" represent the label of the toll station; the "ENTIME" and "EXTIME" indicate the vehicle entry time and exit time, respectively; the "VEHICLETYPE" represents the vehicle type; "VEHICLEID" represents the license plate number. Through the data preprocessing, the traffic flow data set including a total of 828 different cross-section locations on a freeway is obtained.

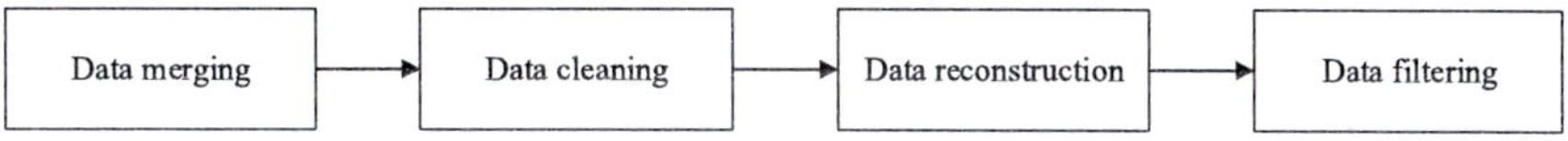

Fig. 1 Data preprocessing process

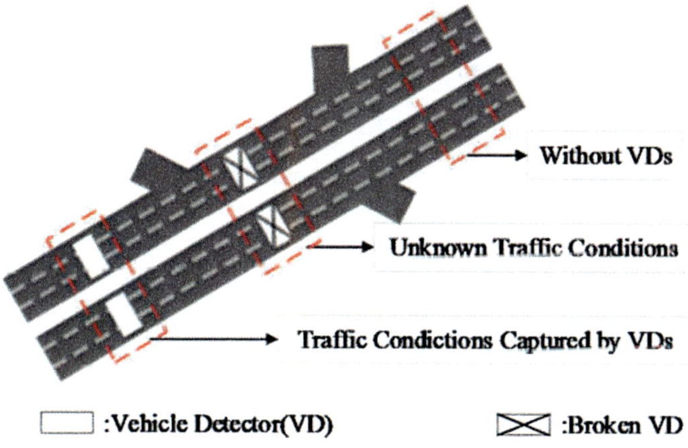

Fig. 2 Detection of traffic conditions in a road segment

Table 1 Partial data at the entrance of a toll station

ENDEALID	Entime	Vehicle type	Vehicle ID	…
G065W610030060	2019/3/22 00:08:08	1	(X) AYS253	…
G065W610030060	2019/3/22 00:06:15	16	(S) B52807	…
G065W610030060	2019/3/22 00:06:35	16	(S) A3Z7D6	…
…	…	…	…	…

Table 2 Partial data at the exit of a toll station

EXDEALID	Extime	Vehicle type	Vehicle ID	…
G065W610030040	2019/3/31 22:35:33	1	(X) BX0139	…
G065W610030040	2019/3/31 22:33:44	1	(S) A9NH79	…
G065W610030040	2019/3/31 22:35:53	1	(S) A9W10H	…
…	…	…	…	…

The detailed process of obtaining traffic flow data from the original toll collection data will be introduced in Chaps. 2–4.

2.1.1 Data Merging

From the toll collection data, the entry time, exit time, the entry toll gate and the exit toll gate, vehicle type, license plate number of each individual vehicle can be obtained. Then merge these fields to form a new table to get a complete record of each vehicle entering and exiting the freeway, and expand the toll data records for consecutive months under each field as needed as shown in Table 3.

Table 3 Partial data of vehicles entering and exiting the freeway

ENDEALID	ENTIME	EXDEALID	EXTIME	VEHICLETYPE	VEHICLEID
10,030,060	2019/03/22 06:31:06	10,030,040	2019/03/22 06:43:58	1	(S) W7K66
10,030,050	2019/03/22 06:31:14	10,030,060	2019/03/22 06:39:07	1	(S) A66C7G
10,030,040	2019/03/22 06:31:18	10,020,040	2019/03/22 07:11:33	11	(S) A7K5S6
…	…	…	…	…	.

2.1.2 Data Cleaning

Data cleaning is to filter and delete the unreasonable data in Table 2, for which we need to perform the following steps:

Step 1: Delete a row of data where the entry time of the vehicle is greater than the exit time, i.e., when the time for "ENTRYTIME" is greater than the time for "EXITTIME", this row of data is deleted from Table 2.

Step 2: Delete a row of data where the entry toll gate is the same as the exit toll gate, i.e., when the toll gate entrance number for "ENDEALID" is the same as the toll gate exit number for "EXDEALID", this row of data is deleted from Table 2.

2.1.3 Data Reconstruction

The data reconstruction is to calculate the traffic volume at the designated cross-section of on a freeway from Table 2. This process is divided into the following steps:

Step 1: Calculate the average speed of each vehicle from entering to leaving the freeway, as Eq. 1. Assume that the driving distance [27] of each vehicle from the entry toll station to the exit toll station is L, as shown in Fig. 3a.

Step 2: Calculate the average travelling speed and average travelling time of each type of vehicle on every closed road segment as Eqs. 2 and 3. According to the basic information of toll data, the vehicle types are divided into cars and trucks. Take the distance between two adjacent toll stations as a closed road segment. Assume that the number of closed road segments that each vehicle passes from the entry toll gate to the exit toll gate is n, as shown in Fig. 3a.

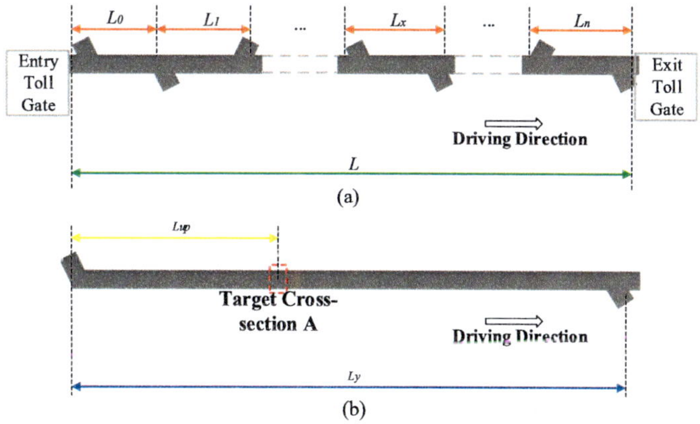

Fig. 3 A schematic of the flow estimation process. **a** Freeway structure. **b** Structure of the target cross-section

Table 4 Partial results of traffic flow data aggregated at 15-min interval

Time intervals	Traffic flow (vehicle/15 min) of different cross-sections					
	…	730,101	730,102	730,103	730,104	…
01:15:00	…	50	45	33	16	…
01:30:00	…	56	79	34	19	…
01:45:00	…	60	114	37	27	…
…	…	…	…	…	…	…

Step 3: Estimate the travelling time of each vehicle on each closed road segment as in Eq. 4.

Step 4: Estimate the time for each vehicle to reach the designated cross-section of the freeway. The length of each closed road segment is already known on a freeway. For a certain target cross section A, as shown in Fig. 3b, the distance between A and the upstream adjacent toll station can be calculated. The time at which each vehicle reaches the target cross-section of the freeway can also be calculated using Eqs. 5 and 6.

Step 5: Calculate the traffic flow of the target cross-section in the demand time period. Assume the demand time period is $T_n \in (t_0, t_0 + \Delta t)$, for the target cross-section A, if the time a single vehicle arrives at A belongs to T_n, then the traffic volume of cross-section A in time period T_n increases by 1. After numbering each target cross-section, we aggregated the traffic volume of each target cross-section with a 15-min time interval and construct a traffic flow data table, part of the data of this table is shown in Table 4.

$$\overline{V}_{(s)}=\mathrm{L}_{(s)}/\left(\mathrm{T}'' - \mathrm{T}'\right) \tag{1}$$

where $L_{(s)}$ is the travelling distance from the entry toll station to the exit toll station of each vehicle s. T' is the entry time of vehicle s. T'' is the exit time of vehicle s. $\overline{V}_{(s)}$ is the average travelling speed of vehicle s.

$$\overline{V}_{x,w} = \sum_{s=1}^{u_w} \overline{V}_{(s)}/\mathrm{u_w} \tag{2}$$

$$\overline{t}_{x,w} = L_{(x)}/\overline{V}_{x,w} \tag{3}$$

where $\overline{V}_{(s)}$ is the average traveling speed of vehicle s, w is the type of vehicle s, x represents the x-th closed road segment. u_w is the number of the toll collection data records where vehicle type is w and the travelling path includes the x-th closed road segment. $\overline{V}_{x,w}$, $\overline{t}_{x,w}$ represents the average travelling speed and average travelling time of the w-type vehicle on the x-th road segment. $L_{(x)}$ is the length of the x-th road segment.

$$t_{x,w} = T*\left(\bar{t}_{x,w}/\sum_{x=1}^{n}\bar{t}_{x,w}\right) \tag{4}$$

where $t_{x,w}$ is the estimated traveling time in the x-th road segment of each vehicle, w represents the type of vehicle, and T is the traveling time from the entry toll gate to the exit toll gate of each vehicle. n is the number of closed road segments from the entry toll gate to the exit toll gate of each vehicle.

$$\Delta t_A = \sum_{i=1}^{y-1} t_{x,w} + t_y * (L_{up}/L_y) \tag{5}$$

$$T_A = T' + \Delta t_A \tag{6}$$

where Δt_A is the traveling time from the entry toll gate to the target cross-section A. y is the number of closed road segments from the entry toll gate to the target cross-section. t_y is the estimated traveling time on the y-th closed road segment. L_{up} is the distance between the upstream toll gate at the y-th closed road segment and the target cross-section. L_y is the length of the y-th closed road segment. T_A is the estimated time of a single vehicle reaches the target cross-section, and $T\prime$ is the entry time of each vehicle.

2.1.4 Data Filtering

In this paper, the moving average filtering is used to process the traffic flow data of every cross-section and the filtering step is 2. The comparison of the flow changes before and after filtering is shown in Fig. 4. As shown in Fig. 4, through data filtering, a large part of the noise value in the traffic flow data can be well filtered out, and the values close to the real traffic state can be obtained.

2.2 *TCN Model for Traffic Flow Prediction*

TCN is a generic convolutional network architecture designed for sequence modeling tasks, which is equally suitable for the sequence modeling task of traffic flow prediction. Therefore, before introducing the TCN model, we firstly describe the inputs and outputs of a sequence modeling task.

Suppose there is a chronologically distributed input sequence $\{s_0, s_1,..., s_N\}$, and according to the sequence of output results, the results of each prediction can be combined together and recorded as $\{g_0, g_1, ..., g_N\}$. This process can be described as the mapping relationship FF $S^N \rightarrow G^N S^N \rightarrow G^N S^N \rightarrow G^N$ and the operation can be expressed as:

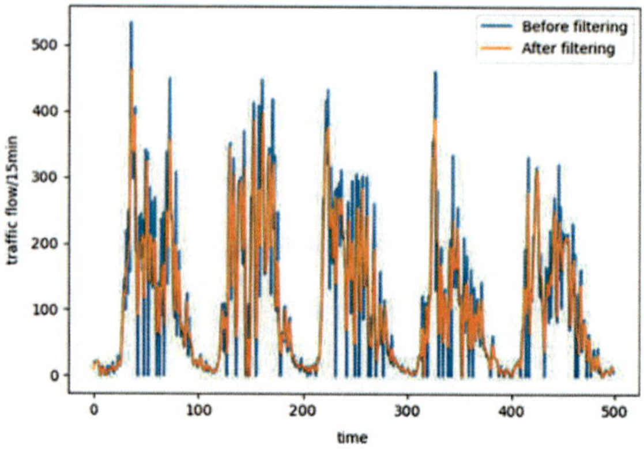

Fig. 4 The change trend of the traffic flow of cross-section *A* before filtering

$$\hat{g}_0, \hat{g}_1, ..., \hat{g}_N = F(s_0, s_1, ..., s_N) \tag{7}$$

However, in order to solve the traffic flow prediction task, when predicting an output value $g_n g_n$ between g_0 and $g_N g_N$, we can only use the values $s_n s_n$ and before $s_n s_n$ s_n as the inputs for this prediction, which means F obeys the constraint that $g_n g_n g_n$ depends only on $(s_0, s_1, ..., s_{N-1})$ the function F can be described as:

$$\hat{g}_n = F(s_0, s_1, ..., s_{n-1}) \tag{8}$$

$\hat{g}_n = F(s_0, s_1, \ldots, s_{n-1})\hat{g}_n = F(s_0, s_1, \ldots, s_{n-1})$ For the traffic flow prediction task, the learning target of is to minimize the error between $\hat{g}_n\hat{g}_n\hat{g}_n$ and $g_N g_N$ through entering sequence data distributed in chronological order.

In order to achieve the above target, TCN uses causal convolutions which means an output at time t is convolved only with elements before time t in the previous layers. Besides, 1D fully-convolutional network (FCN) architecture [28] is used in TCN to keep each hidden layer the same length as the input layer, and zero padding is added to keep subsequent layers the same length as previous ones. In this way, TCN can capture the information of input sequences without leaving nothing out in theory. An illustration of TCN model architecture whose filter size k is 3 from [20] is given in Fig. 5.

It seems that TCN has already meet the needs of the traffic flow prediction task in combination with the above two structures without considering additional factors. However, when the dimensions of the input data in temporal and spatial are very large, we need to increase the size and number of filters or the level of the network to meet the needs of model training, which often results in an extremely deep network. Therefore, dilated convolutions are combined in TCN to solve this problem.

Formally, for a filter and a $1-D$ sequence input SS $S \in R^n$, the dilated convolution operation G on element x of the sequence is defined as:

$$G_{(x)} = (S *_d f)(x) = \sum_{i=0}^{k-1} f(i) \cdot S_{x-d \cdot i} \tag{9}$$

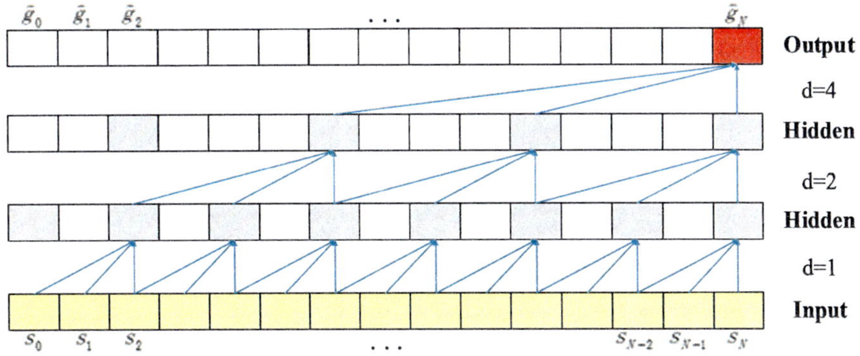

Fig. 5 Architecture of TCN

$G_{(x)} = (S *_d f)(x) = \sum_{i=0}^{k-1} f(i) \cdot S_{x-d \cdot i}$ where d is the dilation factor, kk k is the filter size, and $x - d \cdot ix - d \cdot i$ indicates the direction of the past.

By increasing the dilation factor dd or filter sizes k, TCN can obtain a sufficiently large receptive field to enable the final output to represent enough wide range of inputs. Simultaneously, the stabilization of deeper and larger TCNs becomes important when adjusting the dilation factor dd and filter sizes k. Therefore, residual block [29] is introduced to TCNs. Within a residual block, the outputs of residual connections are added to the input x of the block by a series of transformations F, which can be described as

$$o = \text{Activation}\left(x + F_{(x)}\right) \tag{10}$$

As shown in Fig. 6, TCN has two layers of dilated causal convolution in a TCN residual block. The rectified linear unit (ReLU) [30] is used to solve the non-linearity in a residual block. Weight normalization [31] is used to the convolutional filters and a spatial dropout [32] is added after each dilated convolution for normalization. Moreover, an $1 * 1$ convolution is added in case the residual input and output have different dimensions. An example of residual connection in a TCN is shown in Fig. 7, the black lines are filters in the residual function and the yellow lines are identity mappings.

Through the organic combination of causal convolution, dilated convolution, and residual block, compared with RNNs, TCN has a faster processing speed under the same length of traffic flow sequence input because it can process data in parallel. This means that when modeling the traffic flow of the same cross-section, the processing time required for TCN is shorter. And the shorter the time required for model training and evaluation, the better it can meet the needs of practical engineering applications.

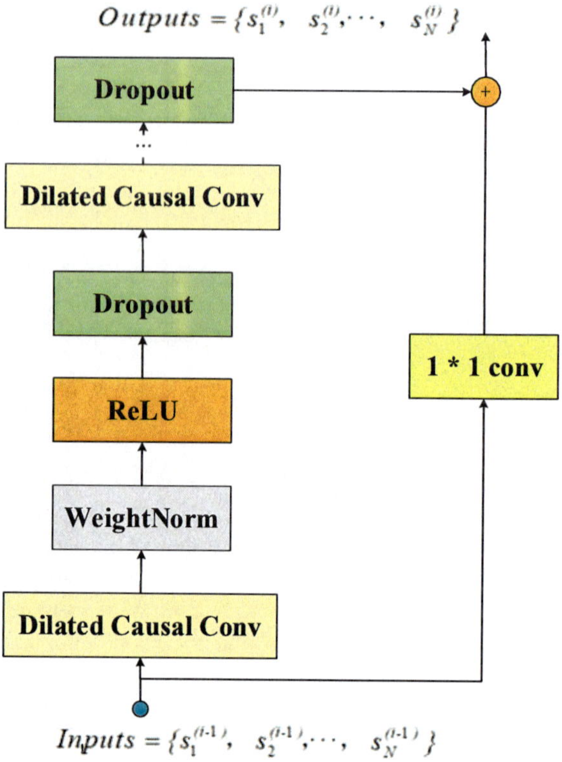

Fig. 6 Residual block of TCN

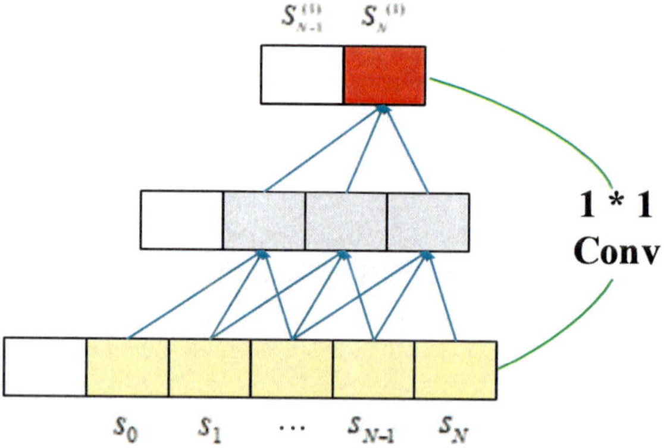

Fig. 7 An example of residual connection of TCN

2.3 Evaluation Index

This paper selects the mean absolute error (MAE) and root mean squared error (RMSE) to evaluate the performance of TCN model in the traffic flow prediction task, which are defined as:

$$MAE = \frac{1}{n}\sum_{i=1}^{n}\left|\hat{y}_i - y_i\right| \tag{11}$$

$$RMSE = \sqrt{\frac{1}{n}\sum_{i=1}^{n}\left(\hat{y}_i - y_i\right)^2} \tag{12}$$

where y_i is the observed traffic flow data, and $\hat{y}_i$ is the predicted traffic flow data.

3 Experiments

The data description, model parameters design, and comparison and analysis of experimental results are mainly introduced in this section.

3.1 *Data Description*

By aggregating historical flow data of multiple target cross-sections for a certain period of time, we can obtain the traffic flow data with multiple temporal characteristics. Then, we got a total of 828 different cross-sections of traffic data from 425 toll stations which also have multiple spatial characteristics. Finally, they were divided into two parts: traffic flow data from January 1, 2019 to April 23, 2019 were split into the training set, and the other were used as the test set.

3.2 *Identification of Design Factors*

For deep learning models, the prediction results may vary greatly with different parameters. Therefore, under the same data input, in order to obtain better prediction results, the design factors of the TCN model are designed as follows:

(1) The number of filters usually used in TCN is 6, 12, 24. In order to get the optimal results, we set the number of filters to 24.
(2) Due to the existence of dilated convolution, TCN no longer only needs to increase the size of the convolution kernel to obtain sufficient receptive fields. Considering the computational burden and the size of the data sets, the kernel size is set as 8 in this paper.
(3) By adjusting the appropriate list of the dilation, it can be ensured that TCN has enough receptive field. In order to get the optimal performance, we set three

levels, i.e., list1, list2 and list3 which are $\{2^0, 2^1, ..., 2^9\}$, $\{2^0, 2^1, ..., 2^5\}$ and $\{2, 4, 6, .., 12\}$ for comparison.

(4) For residual blocks, without a very deep network, a small residual layer is sufficient and following the minimal size recommended by [33], we set the number of stacks as 2.

We selected a cross-section with a peak flow of about 400 and used MAE and RMSE as evaluation benchmarks to test the performance of TCN at different levels of dilated convolution factors. The average values of MAE and RMSE of these tests results are shown in Table 4, from which we can conclude that the prediction results have the minimum MAE and RMSE values when the list of the dilation is list1 (i.e., $\{2^0, 2^1 \ldots 2^9\}$) using traffic flow data aggregated at 15-min, 30-min and 60-min intervals, respectively.

3.3 *Experimental Results and Analysis*

After compared the prediction results of the TCN model with the obtained traffic flow of the target cross-section, the results indicate that TCN performs well regardless of the prediction time is 15-min, 30-min or 60-min, as shown in Fig. 8.

However, judging from the fitting effect of the prediction results and the actual values on the curve alone, it is hard to determine the advantages of the TCN model in the short-term traffic flow prediction task. Therefore, using MAE and RMSE as evaluation indicators and using LSTM and other models as reference models, we conducted a series of comparative experiments. A list of average MAE and RMSE values for each model to evaluate the performance based on the prediction results are shown in Table 6.

By comparing the MAE and RMSE values predicted by each model, TCN performs best regardless of whether the single-step prediction time interval is 15-min, 30-min, or 60-min. For the 15-min ahead traffic flow prediction task, the MAE value of SAE is 16.127 which is lower than other models except for TCN, but is still higher than 14 in error compared to TCN. Although the MAE value of the LSTM model is in the penultimate position at 17.045, the MAE value of the SVM model is

Table 6 MAE and RMSE performance comparison of 15-min, 30-min and 60-min ahead traffic flow prediction for the TCN, GRU, LSTM, SAE and SVM models

Models	15-min		30-min		60-min	
	MAE	RMSE	MAE	RMSE	MAE	RMSE
TCN	**1.550**	**1.662**	**4.249**	**5.231**	**15.524**	**16.610**
GRU	16.284	21.673	60.599	77.000	95.107	123.922
LSTM	17.045	24.342	49.901	71.305	90.817	124.077
SAE	16.127	23.280	64.307	92.378	104.663	144.967
SVM	26.151	30.091	71.018	83.643	110.569	132.288

still about 9 higher than the LSTM model. This shows that as a traditional time-series prediction model, SVM performs worse than other deep learning models including TCN, GRU, SAE, and LSTM on the 15-min ahead traffic flow prediction task. When the time intervals are 30-min and 60-min, the MAE values of the TCN model are 4.249 and 15.524, which are about 45 and 75 lowers in values than other models, respectively. With the increase of prediction time intervals, although the MAE value of the TCN model has increased, its growth rate is far lower than other models. When the prediction time interval is 60-min, the TCN has the maximum MAE value, i.e.,15.524. Compared with the 15-min ahead prediction task, the increase in the MAE value still does not exceed 14, which means that even though the prediction time has doubled, the prediction accuracy of the TCN model has hardly decreased. This illustrates that TCN can perform well in the prediction of short-term traffic flow on freeways.

The RSME comparison results are also shown in Table 6. The RMSE of TCN for the 15-min, 30-min and 60-min cases are 1.662, 5.231, and 16.610, respectively. Similarly, the RMSE values of the TCN model at different prediction time intervals are still much lower than those of its counterpart. Moreover, when the prediction time interval is 30-min and 60 min, respectively, the RMSE value of the SVM model is lower than the SAE model, which indicates that the stability of the prediction results of the SVM model in these two-time intervals is better. Moreover, from the MAE and RMSE performance of Table 6, each model achieves the best performance when 15-min is the single-step prediction time interval. Therefore, we visualize the prediction results of the models compared with TCN in 15-min ahead prediction task, as shown in Fig. 9. It can be seen from the prediction results of Fig. 8a and Fig. 9 that TCN exhibits a larger advantage than other models when the traffic flow values are around “peak values” and “valley values”.

In addition, we infer that the reason that the traffic flow fluctuates once an hour or so is because the highway network has many entrances and exits and they are unevenly distributed in space. This causes the traffic flow to fluctuate during the cycle while following the cyclical changes, which also exacerbates the non-linearity of the highway traffic flow. However, in terms of the prediction results of short-term traffic flow, no matter what causes the non-linearity of the freeway traffic flow, TCN can well simulate the change of traffic flow, which is unmatched by traditional time series models. Because it is difficult for the time series prediction models to analyze the spatial connection of traffic flow. This also shows that TCN can well excavate the spatial connection of traffic flow data, which helps TCN achieve higher prediction accuracy when applied to the traffic flow prediction task.

4 Conclusions

In this paper, by adjusting the hyper-parameters of the original TCN model, an optimized freeway short-term traffic flow prediction model is proposed. At the same time, we introduced a method of extracting traffic flow data from toll data, and carried out

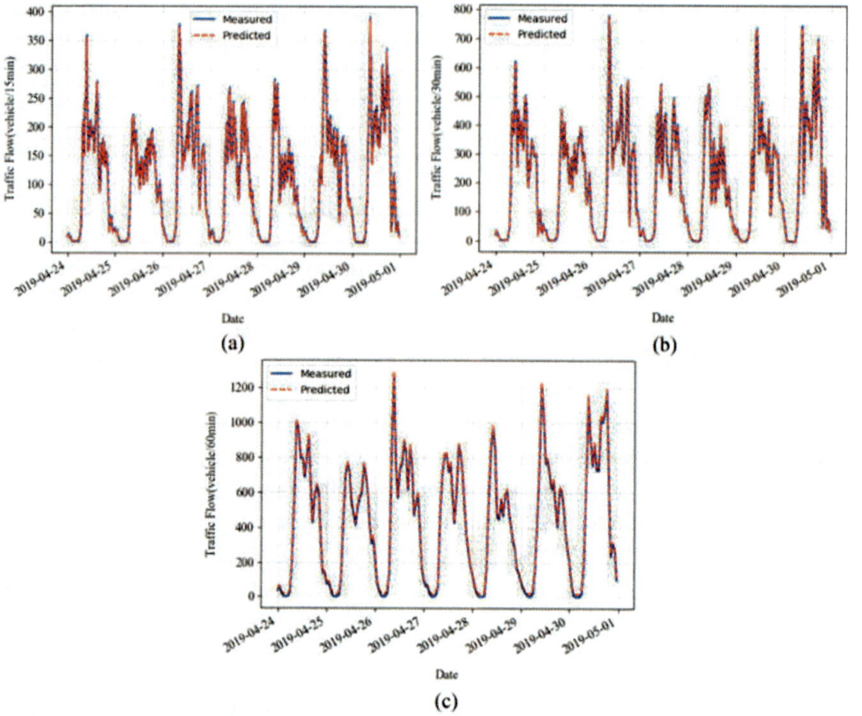

Fig. 8 Performance evaluation of 15-min, 30 min and 60 min ahead traffic flow prediction results for TCN model. **a** 15-min. **b** 30-min. **c** 60-min

practical experiments and verified the reliability of the method, which has a good driving effect on expanding the way to obtain traffic flow data and reducing the installation and maintenance costs of electronic equipment such as vehicle inspection devices. Through the comparison of the MAE and RMSE performance of different models, TCN exhibits surprising advantages, especially on the 15-min ahead traffic flow prediction task. In addition, the prediction results of multiple cross-sections at several time intervals ahead prediction task indicate the TCN model has good robustness and generalization ability. Furthermore, we found that all the prediction models used in our research had the best prediction results on the 15-min ahead prediction task. We explored the cause of this phenomenon, and initially concluded that there are many entrances and exits on the freeway, causing frequent traffic fluctuations within a changing cycle, resulting in uneven distribution of traffic flow even in a short time. So within a certain time range, the shorter the prediction time interval, the higher the prediction accuracy.

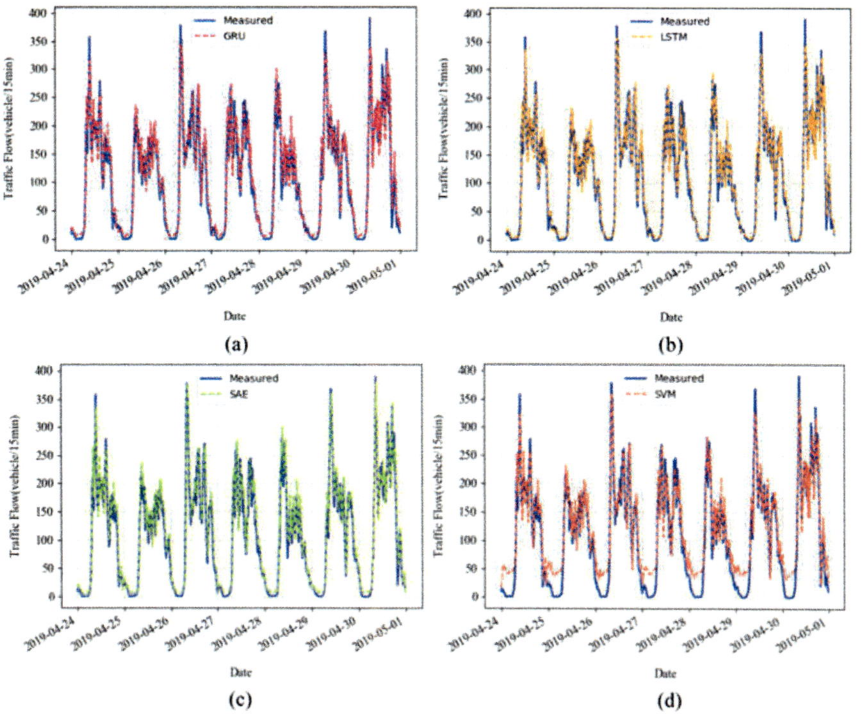

Fig. 9 Performance evaluation of 15-min ahead traffic flow prediction results for GRU, LSTM, SAE and SVM models. **a** GRU model. **b** LSTM model. **c** SAE model. **d** SVM model

Acknowledgements We are grateful for the original toll data provided by Shaanxi traffic management bureau, P. R. China. In addition, the authors would like to thank Mr. Wanrong XU, Mr. Hao ZHANG for their help in data collation and presenting work in various aspects.

Conflicts of Interest The authors declare that there are no conflicts of interest regarding the publication of this paper.

Funding Statement This paper was supported in part by the National Key Research and Development Program of China under Grant 2019YFB1600703.

Data Availability The traffic flow data used to support the findings of this study are available from the corresponding author upon request.

References

1. Wang J, Chen R, He Z (2019) Traffic speed prediction for urban transportation network: a path based deep learning approach. Transp Res Part C 100:372–385
2. Gu Y, Lu W, Qin L et al (2019) Short-term prediction of lane-level traffic speeds: a fusion deep learning model. Transp Res Part C 106:1–16
3. Zhou Y, Chung E, Bhaskar A et al (2019) A state-constrained optimal control based trajectory planning strategy for cooperative freeway mainline facilitating and on-ramp merging maneuvers under congested traffic. Transp Res Part C 109:321–342
4. Zhao Z, Chen W, Wu X et al (2017) LSTM network: a deep learning approach for short-term traffic forecast. IET Intel Transport Syst 11(2):68–75
5. Zhu L, Yu FR, Wang Y et al (2019) Big data analytics in intelligent transportation systems: a survey. IEEE Trans Intell Transp Syst 20(1):383–398
6. Zhao W, Gao Y, Ji T, Wan X, Ye F, Bai G (2019) Deep temporal convolutional networks for short-term traffic flow forecasting. IEEE Access 7:114496–114507
7. Polson NG, Sokolov VO (2017) Deep learning for short-term traffic flow prediction. Transp Res Part C Emerg Technol 79:1–17
8. Lippi M, Bertini M, Frasconi P (2013) Short-term traffic flow forecasting: an experimental comparison of time-series analysis and supervised learning. IEEE Trans Intell Transp Syst 14(2):871–882
9. Zhang H, He J, Bao J, Hong Q, Shi X (2020) A hybrid spatiotemporal deep learning model for short-term metro passenger flow prediction. J Adv Transp 2020 Article ID 4656435
10. Gong Y, Zhang Y (2013) Research of short-term traffic volume prediction based on Kalman filtering. In: 6th international conference on intelligent networks and intelligent systems (ICINIS), pp 99–102
11. Weerasekera R, Sridharan M, Ranjitkar P (2020) Implications of spatiotemporal data aggregation on short-term traffic prediction using machine learning algorithms. J Adv Transp 2020:1–21
12. Luo X, Li D, Yang Y et al (2019) Spatiotemporal traffic flow prediction with KNN and LSTM. J Adv Transp 2019:537–546
13. Yu L, Zhao J, Gao, Lin W 2019) Short-term traffic flow prediction based on deep learning network. In: 2019 international conference on robots & intelligent system (ICRIS), pp 466–469
14. Lv Y, Duan Y, Kang W, Li Z, Wang F (2015) Traffic flow prediction with Big Data: a deep learning approach. IEEE Trans Intell Transp Syst 16(2):865–873
15. Zhang Y, Mingming L, Haifeng L (2020) Urban traffic flow forecast based on fastGCRNN. J Adv Transp 2020, Article ID 8859538
16. Abbas Z, Al-Shishtawy A, Girdzijauskas S, Vlassov V (2018) Short-term traffic prediction using long short-term memory neural networks. In: 2018 IEEE international congress on Big Data (BigData Congress), pp 57–65
17. Fu X, Luo W, Xu C, Zhao X (2020) Short-term traffic speed prediction method for urban road sections based on wavelet transform and gated recurrent unit. Math Prob Eng Article ID 3697625
18. Mou L, Zhao P, Xie H, Chen Y (2019) T-LSTM: a long short-term memory neural network enhanced by temporal information for traffic flow prediction. IEEE Access 7:98053–98060
19. Zhang Z, Yang W, et al. (2020) Traffic accident prediction based on LSTM-GBRT model. J Control Sci Eng Article ID 4206919
20. Bai S, Kolter JZ Koltun V (2020) An empirical evaluation of generic convolutional and recurrent networks for sequence modeling. https://arxiv.org/abs/1803.01271
21. Chen Y, Kang Y, Chen Y et al (2020) Probabilistic forecasting with temporal convolutional neural network. Neurocomputing 399:491–501
22. Guo G, Yuan W (2020) Short-term traffific speed forecasting based on graph attention temporal convolutional networks. Neurocomputing 410:387–393
23. Wang P et al (2019) Forecasting traffic volume at a designated cross-section location on a freeway from large-regional toll collection data. IEEE Access 7:9057–9070

24. Wang P, Li L, Jin Y, Wang G (2018) Detection of unwanted traffic congestion based on existing surveillance system using in freeway via a CNN-architecture trafficnet. In: 2018 13th IEEE conference on industrial electronics and applications (ICIEA), pp 1134–1139
25. Lana I, Ser J D, Velez M, Vlahogianni IE (2018) Road traffic forecasting: recent advances and new challenges. IEEE Intell Transp Syst Magazine, 10(2): 93–109
26. Jia Y, Wu J, Ben-Akiva M, Seshadri R, Du Y (2017) Rainfall-integrated traffic speed prediction using deep learning method. IET Intell Transp Syst 11(9):531–536
27. Sun J, Dong H, Kong Y, Fang Y (2019) Solution to shortest path problem using a connective probe machine. Math Problems Eng 2019, Article ID 8709042
28. Du Z, Wu S, Huang D, Li W, Wang Y (2019) Spatio-temporal encoder-decoder fully convolutional network for video-based dimensional emotion recognition. IEEE Trans Affective Comput: 1–1
29. Li D, Lin C, Gao W, et al, (2020) "Capsules TCN network for urban computing and intelligence in urban traffic prediction. Wirel Commun Mob Comput 2020, Article ID 6896579
30. Wang G, Giannakis GB, Chen J (2019) Learning ReLU networks on linearly separable data: algorithm, optimality, and generalization. IEEE Trans Signal Process 67(9):2357–2370
31. Alur P, Asuri H, Cirelli J et al (2018) Postnatal growth restriction is reduced if birth weight is used for nutritional calculations in ELBW infants. Int J Pediatr 2018, Article ID 2045370
32. Maroli J, Özgüner Ü, Redmill K (2019) Nonlinear system identification using temporal convolutional networks: a silverbox study. IFAC-PapersOnLine 52(29):186–191
33. Chen Z, Sun S, Wang Y et al (2020) Temporal convolution-network-based models for modeling maize evapotranspiration under mulched drip irrigation. Comput Electron Agricul 169

Fast Path Planning for Fixed-Wing Unmanned Aerial Vehicle with Multiple Constraints

Gang Zhong, Yi Mao, Liandong Zhang, Shangwen Yang, and Hao Liu

Abstract Aiming at the problem of fixed-wing unmanned aerial vehicle (UAV) path planning, considering the actual flight conditions and flight performance of UAV, a multi-constraint UAV path planning model is constructed with the minimum flight range and correction times as the objective function. The improved A* algorithm is used to solve the problem: in order to adapt to the model, the objective function of the model is used as the evaluation function; in order to speed up the search efficiency, the branch and bound method is used for iterative search. The simulation results show that: the model can achieve bi-objective optimization, and it is reasonable. Compared with the traditional A* algorithm, the improved algorithm can better balance the optimization flight range and correction times, and save the algorithm planning time, and effectively complete the fast path planning of fixed-wing UAV with multiple constraints.

Keywords Fixed-wing UAV · Path planning · Positioning error · Improved A* algorithm · Branch and bound method

1 Introduction

Unmanned aerial vehicle is a kind of unmanned aircraft operated by radio remote control equipment and self-contained program control device. Because of its low cost, good flexibility, strong adaptability and other advantages, it has been widely used in military and civil fields. UAV path planning is an important topic in intelligent UAV control research, which can be defined as: planning an optimal flight path for UAV

G. Zhong (✉) · L. Zhang
College of Civil Aviation, Nanjing University of Aeronautics and Astronautics, Nanjing, China
e-mail: zg1991@nuaa.edu.cn

Y. Mao · S. Yang
State Key Laboratory of Air Traffic Management System and Technology, Nanjing, China

H. Liu
College of Science, Nanjing University of Aeronautics and Astronautics, Nanjing, China

W. Wang et al. (eds.), *Green Transportation and Low Carbon Mobility Safety*,
Lecture Notes in Electrical Engineering 944,
https://doi.org/10.1007/978-981-19-5615-7_6

from the mission starting point to the mission target point on the premise of meeting various constraints and mission objectives. Research on UAV path planning method can provide technical supports for the future promotion of UAV with autonomous flight capability, which has important practical significance.

At present, many scholars at home and abroad study the UAV path planning. The literature [1] aims at the problem of UAV formation path planning, and a formation path planning algorithm based on leader follower structure and APF algorithm is designed. The effectiveness and rationality of the proposed algorithm for UAV formation path planning are verified by simulation experiments. In literature [2], a new method of on-line path planning based on rolling time domain control is proposed. The simulation results show that the proposed method can effectively improve the planning speed, stability and real-time obstacle avoidance ability in the fixed-wing UAV path planning, and overcome the disadvantage that the traditional artificial potential field method is easy to fall into local optimal. In literature [3], a 4D trajectory generation technology based on tau theory is designed, and then the particle swarm optimization algorithm is used to optimize the 4D trajectory. The simulation results further verify the effectiveness of the method. The literature [4] aims at the shortcomings of the current path planning method, and designs the automatic path planning method in different task scenarios, and studies the optimization method of local path during climb process based on the initial path generation. Meanwhile, a new optimization model of flight path planning based on UAV with fixed-wing in threat is proposed in paper [5]. When the flight path is smoothed, a constraint improvement strategy based on differential geometry principle is proposed. In literature [6], in the background of transmission line patrol inspection, a Goal-bias Bidirectional Connection RRT*(GBCRRT*) algorithm is proposed based on the traditional RRT algorithm. The effectiveness of the algorithm is verified by simulation experiments. In literature [7], an iterative strategy is proposed to improve the performance of task assignment and path planning for distributed UAVs. The algorithm controls the task estimation error in a certain range by calculating the finite complexity, which has the advantages in the stability and complexity of the algorithm. The literature [8] proposes a strategy of UAV risk-aware path planning in urban environment. The path planning is divided into two stages: first, a temporary path is calculated according to the static risk factors; Then, the dynamic risk map is used to adjust the initial planning path and adapt it to the dynamic conditions. The simulation results verify the applicability of the method. In literature [9], the feasibility and optimal trajectory of fixed wing UAV in complex three-dimensional environment are calculated by using genetic algorithm and particle swarm optimization algorithm. A path planning algorithm is also proposed in literature [10]. The algorithm combines the rapid exploring random tree search method with formal verification algorithm, which enables UAV to perform specific tasks in the condition of meeting the constraints of real-time traffic and geographical fence. The literature [11] presents an environment model based on artificial potential field method and grid method for the application scene of forest fire extinguishing with UAV. A real-time task assignment algorithm based on improved particle swarm optimization algorithm and a real-time path planning algorithm based on improved A* algorithm is proposed. The above researches have carried out the studies on UAV path planning from different angles, but the factors are considered to be single and the flight state of the UAV is ideal, and the situation of UAV positioning error leading to yaw is not considered.

In this paper, the fixed-wing logistics UAV is taken as the research object. Considering the UAV positioning error, load limit and other factors, the problem of UAV path planning in complex environment is studied. On the premise of meeting the requirements of UAV's safe obstacle avoidance, a fast flight path planning model is established to minimize the path length and error correction times. A path planning algorithm combining improved A* and branch and bound method is proposed to plan a reasonable, economic and safe flight path for fixed-wing UAV.

2 Fast Path Planning Model for UAV

2.1 Problem Description and Related Hypotheses

At present, due to the limitations of UAV manufacturing level and system structure, its positioning system can't accurately locate itself, and errors will appear in the flight process. Once the positioning errors accumulate to a certain extent, the mission may fail. Therefore, in order to achieve accurate path planning and ensure the quality of the mission, it is necessary to correct the errors in the flight process of UAV.

Assuming that there is a demand for logistics distribution in a certain area and the location coordinates of the starting point and the target point of logistics distribution are known. Meanwhile, the fixed-wing UAV with load limit and range constraint is used for distribution. There is no positioning error during the take-off process of the fixed-wing UAV, and the error gradually occurs after the taking-off. Considering the system positioning error of the UAV, it needs to go through the horizontal correction point and the vertical correction point respectively to correct the horizontal error and vertical error. After the fixed-wing UAV takes off, its flight path is fixed and no longer accepts new mission assignment. In order to ensure the safe and efficient transportation of goods from the logistics distribution center to the logistics demand point, a safe and reasonable path planning must be carried out before the flight. Figure 1 is a schematic diagram of UAV path planning for positioning error correction.

2.2 Modeling of UAV Path Planning

First, a directed network graph is defined: $G = (N, R)$. Where N represents the set of nodes in the directed network graph; R is the set of edges. Let (i, j) denotes the link from node i to node j, and $(i, j) \in R$, d_{ij} represents the Euclidean distance from node i to node j. (D_i, V_i, H_i) indicates the state of the UAV when it is in node I. D_i indicates the cumulative flight path length of UAV at node I. The x_{ij} command

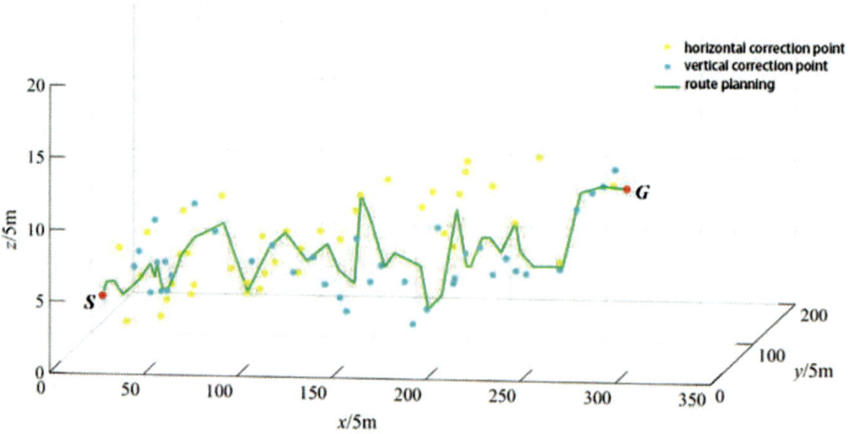

Fig. 1 Schematic diagram of fixed-wing UAV path planning considering positioning error correction

indicates whether the UAV passes through the road section (i, j), and $x_{ij}= 1$ means the UAV passed by, while $x_{ij}= 0$ means that the UAV does not pass by. According to the error constraints that UAV needs to meet when it reaches the error correction point, the connectivity between nodes is established, and the directed network graph is completed G.

Now the problem is transformed into solving the UAV path planning problem under multiple constraints. The optimization objective is to plan the shortest path from the starting point to the end point which satisfies the error constraints, and to minimize the number of nodes (correction points).

2.2.1 Objective Function

The objective function of the path planning model of fixed-wing UAV constructed in this paper is as follows:

(1) Flight range

The significant advantage of UAV is economy, so the path length should be as small as possible in path planning. The length of UAV's path from start point S to target point G is defined as 1, and its expression is as follows:

$$L = \sum_{i \in J} \sum_{j \in J} d_{ij} x_{ij} \tag{1}$$

where, x_{ij} indicates whether the UAV passes through the road section (i, j), and $x_{ij}= 1$ means the UAV passed by, while $x_{ij}= 0$ means that the UAV does not pass by. d_{ij} denotes the Euclidean distance from the correction point i to the correction point j; J represents the set of all calibration points that the UAV passes through in the planned track.

(2) Correction times

Due to the positioning error of UAV, it needs to be corrected, so the correction times should be reduced as much as possible in the process of path planning. The correction times of UAV from start point S to target point G are defined as C. according to the relationship between the segment number of UAV flight path and the number of correction points, the expression is as follows:

$$C = \sum_{i,j \in J} x_{ij} - 1 \tag{2}$$

2.2.2 Constraints

There are many factors that affect the flight of fixed-wing UAV. In this paper, considering the UAV performance constraints, planning requirements, error impact and other factors, a logistics UAV path planning model under multiple constraints is established. The constraints of the planned path from task starting point S to task target point G are as follows:

(1) Load limit

The UAV has the upper limit of load constraint:

$$q \le q_{\max} \tag{3}$$

where, q is the mass of goods and $q_{\max}$ the maximum load.

(2) Horizontal correction point constraint

When the UAV enters the vertical error correction point, it cannot exceed the upper limit of the relevant error:

$$V_j = \sum_{i,j \in J} [V_i(1 - \lambda_i) + d_{ij}\delta]x_{ji} \begin{cases} \le \alpha_1, \ \lambda_i = 1 \\ \le \beta_1, \ \lambda_i = 0 \\ < \theta, \ \lambda_i = \lambda_G \end{cases} \tag{4}$$

where, V_i represents the cumulative vertical error of UAV at the correction point i; δ is the increment of vertical error when UAV flies per 1 m; d_{ij} denotes the Euclidean distance from the correction point i to the correction point j; θ represents the maximum allowable error; λ_i indicates the correction type of the correction point i; $\lambda_i = 0$ indicates the horizontal correction point; $\lambda_i = 1$ indicates the vertical correction point; λ_G indicates the end point G; α_1 indicates the maximum vertical error allowed by UAV when correcting at the vertical correction point; β_1 indicates the maximum vertical error allowed by UAV when correcting at the horizontal correction point.

(3) Vertical correction point constraint

When the UAV enters the horizontal error correction point, it cannot exceed the upper limit of the relevant error:

$$H_j = \sum_{i,j \in J} (H_i \lambda_i + d_{ij}\delta) x_{ji} \begin{cases} \leq \alpha_2, \ \lambda_i = 1 \\ \leq \beta_2, \ \lambda_i = 0 \\ < \theta, \ \lambda_i = \lambda_G \end{cases} \tag{5}$$

where, H_i is the cumulative horizontal error of UAV at the correction point i; α_2 indicates the maximum horizontal error allowed by UAV when correcting at the vertical correction point; β_2 indicates the maximum horizontal error allowed by UAV when correcting at the horizontal correction point.

(4) Link connectivity constraints

In the shortest path network diagram with constraints, UAV shall meet the equilibrium flow constraints, that is, to ensure that the selected road section constitutes a complete connecting path from the start to the end point, and the constraints are as follows:

$$\sum_{j \in adj(i)} xij - \sum_{j:i \in adj(j)} xji = \begin{cases} 1 \, i = S \\ 0 \, other \\ -1 \, i = G \end{cases} \tag{6}$$

(5) Maximum range constraint

UAV has the constraint of farthest flight range:

$$L \leq L_{\max} \tag{7}$$

where, L is the actual flight range of UAV and $L_{\max}$ is the farthest flight range of UAV.

To sum up, the mathematical model of the problem is established as follows:

$$\min F = \eta_1 L + \eta_2 C \tag{8}$$

$$
s.t.\begin{cases}
q \le q_{\max} \\
L \le L_{\max} \\
\sum\limits_{j \in adj(i)} x_{ij} - \sum\limits_{j:i \in adj(j)} x_{ji} = \begin{cases} 1\, i = S \\ 0 \text{ other} \\ -1\, i = G \end{cases} \\
V_j = \sum\limits_{i,j \in J} [V_i(1-\lambda_i) + d_{ij}\delta]x_{ji} \begin{cases} \le \alpha_1,\ \lambda_i = 1 \\ \le \beta_1,\ \lambda_i = 0 \\ < \theta,\ \lambda_i = \lambda_G \end{cases} \\
H_j = \sum\limits_{i,j \in J} (H_i\lambda_i + d_{ij}\delta)x_{ji} \begin{cases} \le \alpha_2,\ \lambda_i = 1 \\ \le \beta_2,\ \lambda_i = 0 \\ < \theta,\ \lambda_i = \lambda_G \end{cases} \\
x_{ij} = 0, 1\, i, j \in J
\end{cases} \tag{9}
$$

3 Fast Path Planning Algorithm for UAV

3.1 *Principle of Traditional A* Algorithm*

A* algorithm is a typical heuristic search algorithm with simple structure and easy implementation. The design of its heuristic function can avoid blind search and has high search efficiency. It can find a path to the target point faster. It is one of the best algorithms for global path planning [11]. The core of A * algorithm is the design of the cost function $f(x)$. In traditional A * algorithm, the cost of path planning is usually considered as flight range:

$$f(x) = g(x) + h(x) \tag{10}$$

where $f(x)$ is the actual flight range cost from the starting point S to the extended path point C_x, which is called the actual cost function; $h(x)$ is the estimated flight range cost from the extended path point C_x to the mission target point G, which is called heuristic function.

3.2 *Improved A* Algorithm Design*

When using traditional A * algorithm for fixed-wing UAV path planning, there are the following problems: ① the evaluation function $f(n)$ only calculates the range, and does not consider the flight error, so it needs to be corrected; ② In urban environment, the buildings are dense, the problem solving scale is large, and the search time is long. Therefore, the direct use of the traditional A* algorithm is not completely suitable for practical application.

In order to solve the model built in this paper, the following design schemes are adopted: ① redesign the evaluation function $f(x)$ according to the objective function of the model built; ② Combined with the branch and bound method, the iterative search can reduce the solution time.

Because the problem belongs to the classical bi-objective optimization problem under multi-constraint conditions, this paper chooses to transform the bi-objective optimization problem into a single objective optimization problem to simplify the problem. A * algorithm combined with branch and bound iterative search is used to solve the problem:

$$f(x) = \eta_1 L + \eta_2 C = \eta_1 \sum_{i \in J} \sum_{j \in J} d_{ij} x_{ij} + \eta_2 \left(\sum_{i,j \in J} x_{ij} - 1 \right) \tag{11}$$

where, η_1 and η_2 are the factors used to adjust the weight of the objectives in the algorithm.

3.3 Algorithm Flow

The flow chart of the improved A * algorithm is shown in Fig. 2. The specific steps of the algorithm are as follows:

Step 1: Initialize and generate OPEN and CLOSED lists.

Step 2: Load the initial value and load the information of the starting point into the OPEN list generated in the first step. The generated OPEN list only contains the starting correction point.

Step 3: Search the OPEN list. If there is no value in the OPEN list, the path planning fails and exits; If the correction point of path planning has been loaded in the OPEN list, the next step will be performed.

Step 4: Select the current correction point; use the cost function to calculate the value of each correction point in the OPEN list; find the correction point with the smallest value in the list; record it as the current correction point; remove the correction point from the OPEN list; load it into the closed list, and carry out the next operation.

Step 5: The constraints of error correction limit the search range of fixed-wing UAV. According to the error state of fixed-wing UAV at the current correction point, all the reachable points of the current correction point are searched.

Step 6: Find the set of reachable points. If the reachable point is in the OPEN list, judge whether the actual cost of reaching the reachable point through the current correction point is less than the actual cost of reaching the point through the original parent correction point. If it is less, the current correction point is set as the parent

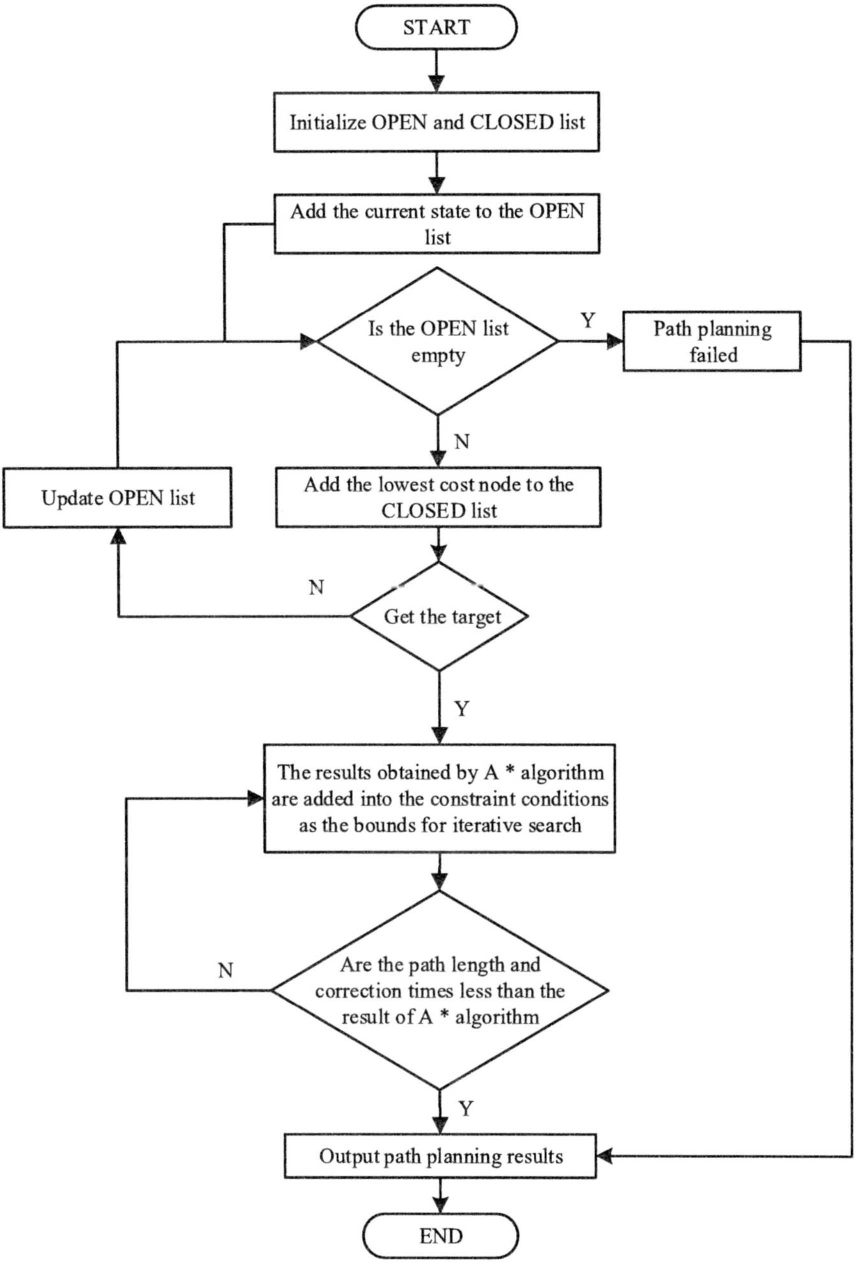

Fig. 2 The flow of improved A* algorithm

correction point of the point, and the actual cost and cost function value of the point are updated. If the reachable point is not in the OPEN list and is not in the CLOSED list, the reachable point is added to the OPEN list, and the current correction point is set as its parent correction point, and the actual cost and cost function value of the point are calculated.

Step 7: The initial path is output. If the OPEN list contains the end point, the path planning ends. The path is output in turn along the parent correction point of the end point.

Step 8: In the process of branch and bound, the path length and correction times calculated by A * algorithm are taken as the bounds and added into the constraints to carry out iterative search.

Step 9: Comparing and pruning. If the result of each branch exceeds the specified path and correction times, the branch will be pruned, and the branch will not be considered later. If the track length and correction times in the branch result are less than the result of A * algorithm, the specified boundary is updated again, and Step 8 is repeated until no better solution is found than the specified boundary, which is the optimal solution considering the path length and correction times.

4 Simulation Verification of UAV Path Planning

4.1 Parameter Setting of Simulation Experiment

The simulation conditions are as follows: the computer CPU is i7-6700; the main frequency is 3.4 GHz; the memory is 8 G, and the experimental tool is Matlab R2017a. The purpose of the simulation experiment in this paper is to verify the rationality of the proposed model and the effectiveness of the algorithm from the theoretical level. Therefore, combined with the existing research methods, this paper selects logistics transportation as the UAV operation scenario, and sets the required parameter values according to the test data of the 16th "Huawei Cup" Chinese graduate mathematical modeling competition in 2019. The specific settings are shown in Table 1.

4.2 Analysis of Simulation Results

According to the simulation experiment parameters in Table 1, the traditional A * algorithm and the improved A * algorithm are respectively used to plan the path from the starting point S to the target point G for the fixed-wing UAV, and the planned path is obtained as shown in Fig. 3 (the yellow dot in the figure represents the horizontal error correction point; the blue dot represents the vertical error correction

Table 1 Simulation parameters

Parameter	Numerical value	Parameter	Numerical value
Coordinate of starting point S/ km	(0,50,5)	Coordinate of target point G/ km	(100,60,5)
Flight range weight η_1	0.5	Correction times weight η_2	0.5
Maximum vertical error of vertical correction point α_1/ m	25	Maximum vertical error of horizontal correction point β_1/ m	20
Maximum horizontal error of vertical correction point α_2/ m	15	Maximum horizontal error of horizontal correction point β_2/ m	25
Maximum allowable error of normal flight θ/ m	30	Cumulative error per 1 m δ/ m	0.001
maximum load $q_{\max}$/ kg	10	Cargo mass q/ kg	5
Maximum flight range $L_{\max}$/ km	150		

point, and the purple and blue curves represent the tracks planned by the traditional algorithm and the improved algorithm respectively). The simulation experimental data of traditional A * algorithm and improved A * algorithm are recorded in Table 2.

Combined with Fig. 3 and Table 2, it can be found that both algorithms can complete the fixed-wing UAV path planning. However, through the comparison of the data results in Table 2, it can be found that compared with the traditional A * algorithm which only focuses on the shortest flight distance, while the path length of the improved algorithm is slightly long, the correction points and the algorithm running time are reduced by 20% and 25.86% respectively, which indicates that in actual flight, the number of error correction of UAV can be reduced, and the track planning time can be saved. The results show that the model and improved algorithm can effectively achieve the bi-objective optimization, and the algorithm runs efficiently, which reflects the rationality and effectiveness of the model and algorithm.

In order to further analyze and compare the performance of the proposed algorithm, the Dijkstra algorithm, a commonly used path planning algorithm, is used to carry out simulation experiments under the same parameter settings. The experimental data of the planned path are recorded in Table 3. It can be found that the results of each index of the improved A * algorithm are significantly better than that of Dijkstra algorithm, especially the speed of the algorithm is reduced by 52.24%. This shows that Dijkstra algorithm has low efficiency, high memory cost and large time consumption in complex search environment. It further shows that the improved A * algorithm is close to the actual needs and has superior performance, which is more suitable for solving the fast path planning problem of fixed-wing UAV in complex airspace environment.

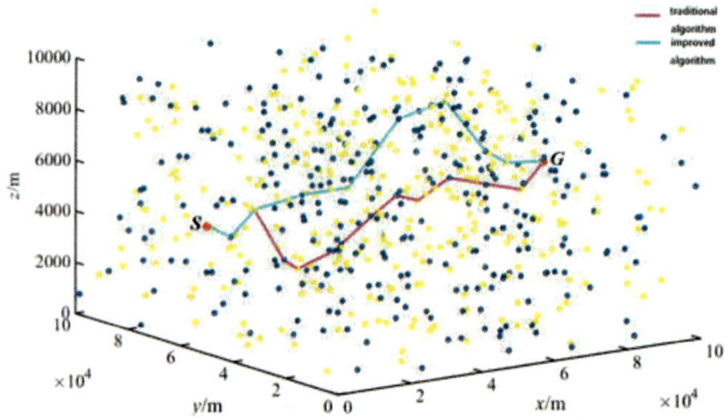

(a) Three dimensional drawing

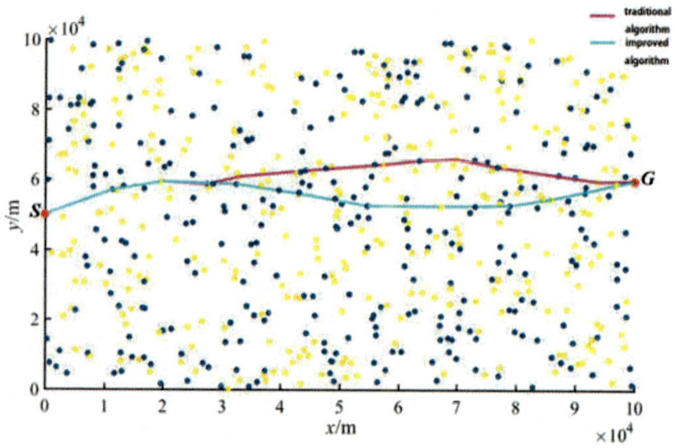

(b) Top view

Fig. 3 Comparison of two algorithms for UAV path planning

Table 2 Comparison of path planning results between traditional algorithm and improved algorithm

Comparison items	Path length/km	Correction points/piece	Planning time/ms
Traditional A * algorithm	104.6	10	298.2
Improved A * algorithm	104.9	8	221.1
Rate of change	0.28%	−20.00%	−25.86%

Table 3 Comparison of path planning results between Dijkstra algorithm and improved A*algorithm

Comparison items	Path length / km	Correction points / piece	Planning time / ms
Dijkstra algorithm	105.8	9	462.9
Improved A* algorithm	104.9	8	221.1
Rate of change	−0.85%	−11.11%	−52.24%

5 Conclusion

With the rapid development of 5G communication, automatic control and other technologies, the application of UAV is more and more widely, and the path planning technology is one of the key technologies to realize the wide application of UAV. This paper fully considers the error factors in the actual flight, and designs the correction point elements, and takes the minimum flight range and correction times as the optimization objectives of the model, which is in line with the actual flight of UAV. Based on the traditional A* algorithm, the evaluation function is redesigned and the branch and bound method is used to improve it. Simulation results show that, compared with the traditional A* algorithm, the improved algorithm can better take into account the two objectives of flight range and flight error correction times, and the running time of the algorithm is shorter, which indicates that the algorithm can effectively plan the flight path of fixed-wing UAV. This method can be extended to other fields, and provide reference for the follow-up research of real-time dynamic logistics UAV path planning and multi-UAVs path planning in three-dimensional space.

References

1. Bai WB (2019) Route planning algorithm for fixed-wing UAV. Harbin Engineering University, Harbin
2. Liu J, Qin XL, Xu Y et al (2019) On-line path planning method of fixed-wing unmanned aerial vehicle. Comput Appl 39(12):3522–3527
3. Liao WJ, Han SC, Li W et al (2020) 4D conflict-free trajectory planning for fixed-wing UAV. Trans Nanjing Univ Aeronautics Astronautics 37(02):209–222
4. Zhang YX (2016) Research on mission flight path planning and optimization method for fixed-wing UAV. Zhejiang University, Hangzhou
5. Tian J (2017) Trajectory planning optimization model based on fixed wing UAV. J Northwest Univ National (Natural Science Edition) 38(01):7–10
6. Liu MW (2018) Research and application of UAV path planning method for electric patrol line. Northeast Petroleum University, Daqing
7. Yao WR, Qi NM, Wan N et al (2019) An iterative strategy for task assignment and path planning of distributed multiple unmanned aerial vehicles. Aerosp Sci Technol 86:455–464
8. Stefano P, Giorgio G, Alessandro R (2019) A risk-aware path planning strategy for UAVs in urban environments. J Intell Rob Syst 95(2):629–643
9. Roberge V, Tarbouchi M, Labonte G (2013) Comparison of parallel genetic algorithm and particle swarm optimization for real-time UAV path planning. IEEE Trans Industr Inf 9(1):132–141
10. Balachandran S, Narkawicz A, Muñoz C, Consiglio M et al (2017) A path planning algorithm to enable well-clear low altitude UAS operation beyond visual line of sight. In: Twelfth USA/Europe Air Traffic Management Research and Development Seminar (ATM2017). America: EUROCONTROL Press
11. Chen K (2019) Research and verification of multi-UAV cooperative control technology in dynamic environment. Beijing University of Posts and Telecommunications, Beijing

Urban Rail Transit and Economic Agglomeration: A Case Study in China

Zhibin Tao, Xuesong Feng, Kemeng Li, and Ruolin Shi

Abstract This research examines the impact of urban rail transit (URT) on economic agglomeration. Initially, forty cities that have opened URT by the end of 2019 in mainland of China are selected as the study subjects. Subsequently, this study gathers panel data of the cities from 2010 to 2019 including the extent of URT, economic agglomeration and control variables. Finally, a non-spatial panel data model and four spatial econometric models are established. The estimate results indicate that URT exerts a positive impact on economic agglomeration. Furthermore, the ability of URT to promote economic agglomeration is greater in the tertiary industry than in the secondary industry.

1 Introduction

The development of urban rail transit (URT) is closely related to city evolution. URT brings grand influence on both the interior and integrity of the city [6]. URT is regarded as an efficient approach to expand the scale of the city [5], increase employment opportunities [8] and improve environmental quality [3, 10, 11]. Understanding the impact of URT is essential for policymakers to formulate relevant policies.

URT can affect urban economy from many aspects. Reference [12] points out that URT exercises a positive influence on urban economy through increasing Gross Domestic Product (GDP). Additionally, the employment rate can be risen by URT [2, 8]. URT can also reshape the employment structure of the city [4]. By applying a propensity score matching model combining difference-in-differences (DID) items, Zhang [13] finds a positive relationship between URT and GDP growth. This research mainly concentrates on the impact of URT on economic agglomeration, which lacks extensive research at present.

According to First Law of Geography put forward by [11], everything is relevant to everything else and near things are more related than distant things. As shown in

Z. Tao · X. Feng (✉) · K. Li · R. Shi
School of Traffic and Transportation, Beijing Jiaotong University, Beijing 100044, China
e-mail: xsfeng@bjtu.edu.cn

W. Wang et al. (eds.), *Green Transportation and Low Carbon Mobility Safety*,
Lecture Notes in Electrical Engineering 944,
https://doi.org/10.1007/978-981-19-5615-7_7

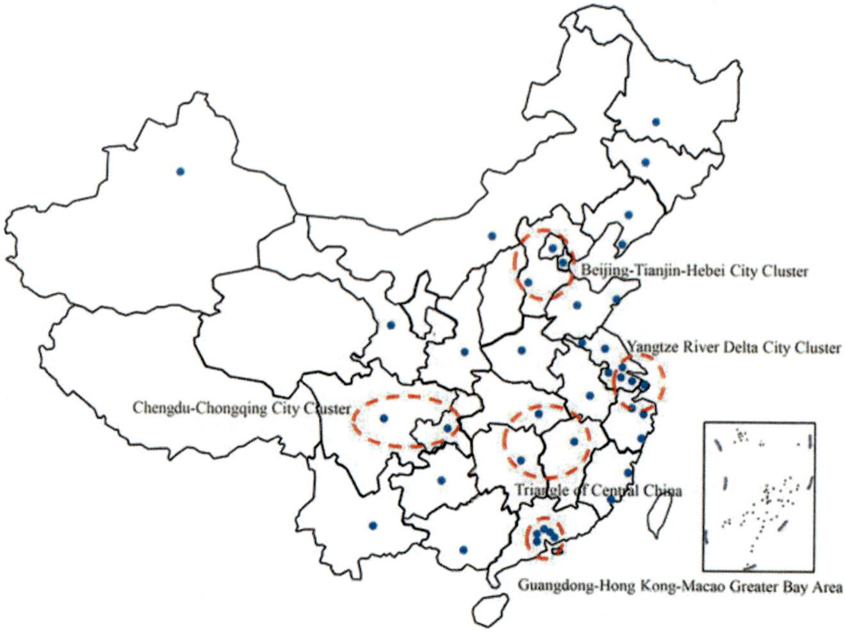

Fig. 1 The distribution of studied cities and city clusters

Fig. 1, several city clusters have been formed due to the advantages of geographical location and policy support in China. The spatial correlation among cities should not be ignored. Therefore, spatial econometric models are established in this study in order to take the spatial association of cities into consideration.

The article is organized as follows. The panel data are introduced in Sect. 2. Section 3 presents the non-spatial panel data model and spatial econometric models. The estimate results are explained in Sect. 4. Section 5 concludes the research and proposes policy recommendations.

2 Data Collection

In this study, forty cities in mainland of China that have opened URT by the end of 2019 are selected as research objects. The panel data are gathered manually from 2010 to 2019. Therefore, $40 \times 10 = 400$ observations are collected.

For the sake of measuring the extent of URT, operation kilometers are chosen to investigate the relationship between URT and economic agglomeration. Generally speaking, the complexity of URT networks is positively correlated to the values of operation kilometers. The values of operation kilometers derive from China Urban Rail Transit Annual Report. Table 1 shows the descriptive statistics for the extent of URT at the end of 2019. The mean value (Mean) of operation kilometers is

Table 1 Descriptive statistics for URT and economic agglomeration variables in 2019

Variables	Mean	Std. Dev	Max	Min	Unit
Operating kilometers (u)	168.410	192.556	809.900	8.800	Kilometers
GDP's agglomeration (GDPag)	0.108	0.080	0.351	0.108	Billion yuan per square kilometer
Secondary industrial agglomeration (Sec_inag)	0.036	0.015	0.081	0.017	Billion yuan per square kilometer
Tertiary industrial agglomeration (Thi_inag)	0.058	0.038	0.162	0.017	Billion yuan per square kilometer

168.410 km and the standard deviation (Std. Dev.) is 192.556. The maximum value (Max) is 809.000 km while the minimum value (Min) is 8.800 km.

Economic agglomeration is able to facilitate enterprises to assemble in cities and then expand the scale of cities, which is an important reason and motive force for the formation and development of cities. The agglomeration degree of a city's GDP can reflect the level of the city's economic agglomeration, which is shown in Eq. (1). Where $GDPag_{it}$ denotes the agglomeration of GDP of city i in year t, while $Area_{it}$ presents the area of city i in year t. Table 1 also presents the descriptive statistics for economic agglomeration variables in 2019.

$$GDPag_{it}=\frac{GDP_{it}}{Area_{it}} \tag{1}$$

Furthermore, some other factors including education expenditure, the number of higher education institution, proportion of employees in primary industry, proportion of employees in secondary industry and expenditure on science and technology are taken into consideration as the control variables that might affect economic agglomeration. Because the data are not complete, some missing data are filled by expectation maximization algorithm.

3 Modeling Study

With the reference of Zhang [13], a non-spatial panel data model is established to preliminarily investigate the impact of URT on the development of economy agglomeration, which is shown in Eq. (2).

$$y_{it} = \alpha u_{it} + X_{it}^{T}\beta + \mu_i + \lambda_t + \varepsilon_{it} \tag{2}$$

where y_{it} represents the level of economic agglomeration of city i in year t. u_{it} denotes the extent of URT, which is measured by operating kilometers. post_{it} is a dummy variable which equals one if city i operates URT in year t. Otherwise, post_{it} equals zero. X_{it} presents the vector consisting of control variables. μ_i denotes city fixed effect. λ_t represents year fixed effects, and ε_{it} is random interference. The coefficient of interest is α_1, which shows the average effect of URT on economy.

Nevertheless, Eq. (2) ignores the spatial relationship among explained variables. Therefore, spatial econometric models are established in order to avoid the estimation bias. First of all, it is vital to set up a spatial weight matrix W that is able to reflect the spatial connections of a certain variable. Each element w_{ij} in the matrix W presents the connectivity between city i and j. At present, the commonly used distance is the reciprocal of Euclidean distance [1]. However, a more appropriate approach to describe connectivity between cities is to measure the convenience of transportation between cities. In this study, the reciprocal of the length of the shortest highway connecting cities i and j is selected and initial matrix is formed. Each row of the initial matrix is normalized to obtain the spatial weight matrix W in the research.

Subsequently, two basic spatial econometric models, spatial autoregressive model (SAR) and spatial error model (SEM), which are shown in Eqs. (3) and (4) respectively, are built to further analyze the impact of URT on urban economic agglomeration. The former mainly considers the endogenous interaction effects of the explained variables while the latter mainly focuses on the interaction effects of random interferences. Spatial autocorrelation model (SAC), which is shown in Eq. (5), is established through combining these two effects into one model. Moreover, the exogenous interaction effects of explanatory variables may also cause estimation error. As shown in Eq. (6), Spatial Durbin model (SDM) is able to take this effect into consideration on the basis of SAR.

$$y_{it} = \delta \sum_{j=1}^{40} w_{ij} y_{jt} + \alpha u_{it} + X_{it}^T \beta + \mu_i + \lambda_t + \varepsilon_{it} \tag{3}$$

$$y_{it} = \alpha u_{it} + X_{it}^T \beta + \mu_i + \lambda_t + \eta \sum_{j=1}^{40} w_{ij} v_{jt} + \varepsilon_{it} \tag{4}$$

$$y_{it} = \delta \sum_{j=1}^{40} w_{ij} y_{jt} + \alpha u_{it} + X_{it}^T \beta + \mu_i + \lambda_t + \eta \sum_{j=1}^{40} w_{ij} v_{jt} + \varepsilon_{it} \tag{5}$$

$$y_{it} = \delta \sum_{j=1}^{40} w_{ij} y_{jt} + \alpha u_{it} + X_{it}^T \beta + \sum_{j=1}^{40} w_{ij} Z_{jt} \theta + \mu_i + \lambda_t + \varepsilon_{it} \tag{6}$$

Akaike's information criterion (AIC) and Bayesian information criterion (BIC) are able to test the model. The former one can evaluate the model that is more suitable for prediction, while the latter one can select the model that fits the data better. AIC and BIC are negatively correlated with the fitting degree of the model. Furthermore, log likelihood value (Log Lik.) can also judge which model fits the data best from another perspective. Log Lik. is positively correlated with the fitting degree of the model.

4 Marginal Effect Analysis

Table 2 presents the empirical results of the relationship between URT and economic agglomeration. The explained variable is the agglomeration of GDP (GDPag) while the major explanatory variable is the operation kilometers of URT. In Table 2, the numbers in parentheses are the p values of the corresponding coefficients. The numbers in parentheses in Table 3 have the same meaning. Each column in the table corresponds to the regression results and test values of each model. Column 1 presents the estimate results of the non-spatial panel data model that do not take δ and η into consideration, the asterisk means that the corresponding coefficients are not considered in the model. The other asterisks in Table 2 and the latter tables have the same meaning. Column 2 and column 3 apply two basic spatial econometric models respectively. Column 2 reveals the results of SAR and column 3 describes the results of SEM. The results of SAC and SDM are illustrated in column 4 and column 5 respectively.

The coefficient of u_{it}*post$_{it}$ in each model is positive and significant at 1.000% level, which means the opening and operation of URT can effectively promote urban economic agglomeration. Economic agglomeration is able to expand the scale of the market and promote regional scientific and technological innovation, and then promote the development of the whole city. URT indirectly promotes the economic agglomeration of the city by improving the commuting convenience of citizens. Therefore, Cities that tend to develop economy rapidly should strengthen the construction of URT.

On the basis of the above research, this study further analyzes the heterogeneity impact of URT by investigating the relationship between URT and secondary industrial agglomeration (Sec_inag), which is presented in column 1, column 2, column 3 and column 4 in Table 3 and tertiary industrial agglomeration (Thi_inag), which is described in column 5, column 6, column 7 and column 8. Column 1 and column 5 describe the estimate results of SAR. The results estimated by SEM are represented

Table 2 Spatial relationship between URT and GDPag

Variable	GDPag				
	(1)	(2)	(3)	(4)	(5)
u_{it}*post$_{it}$	0.034 (0.011)	0.034 (0.007)	0.034 (0.006)	0.034 (0.006)	0.033 (0.004)
δ	*	−0.205 (0.400)	*	0.143 (0.425)	−0.269 (0.302)
η	*	*	−0.252 (0.437)	−0.193 (0.527)	*
AIC	2293.691	2278.698	2278.261	2279.888	2288.348
BIC	2353.563	2310.630	2310.193	2315.811	2344.228
Log Lik	*	−1131.3492	−1131.13	−1130.944	−1130.1738

Table 3 Heterogeneity analysis

Variable	Sec_inag				Thi_inag			
	(1)	(2)	(3)	(4)	(5)	(6)	(7)	(8)
u_{it}*post$_{it}$	0.066 (0.077)	0.066 (0.073)	0.065 (0.072)	0.059 (0.049)	0.244 (0.003)	0.245 (0.003)	0.244 (0.003)	0.239 (0.002)
δ	−0.027 (0.900)	*	−0.001 (0.995)	−0.133 (0.582)	−0.367 (0.157)	*	−0.266 (0.162)	−0.435 (0.109)
η	*	−0.082 (0.773)	−0.082 (0.737)	*	*	-0.391 (0.270)	−0.275 (0.456)	*
AIC	3314.527	3314.383	3316.383	3320.322	3723.558	3723.059	3723.966	3733.277
BIC	3346.459	3346.315	3352.306	3376.203	3755.490	3754.990	3759.889	3789.157
Log Lik	−1649.264	−1649.191	−1649.191	−1646.161	−1853.779	−1853.529	−1852.983	−1852.639

in column 2 and column 6. Column 3, column 7 and column 4, column 8 apply SAC and SDM respectively.

The results show positive and significant relationships between secondary industrial agglomeration, tertiary industrial agglomeration and URT. Furthermore, the effect of economic agglomeration of URT on the tertiary industry is more significant than that of the secondary industry. The knowledge-intensive tertiary industry emphasizes the benefits brought by market demand and innovation agglomeration, which means URT has spillover effect of knowledge.

5 Conclusions

This research selects forty cities in mainland of China that have opened URT by the end of 2019 as study objects and establishes spatial econometric models to discover the relationship between URT and urban economic agglomeration. The main innovation of this study is that the spatial correlation is taken into consideration because of city clusters.

The results indicate that URT is able to promote economic agglomeration of the city. In addition, the positive effect of URT on economic agglomeration is more apparent in tertiary industry. As a result, policymakers ought to take a more cautious attitude when investing in URT projects. The construction of URT should be closely combined with local industrial planning to maximize the benefits of URT. Nevertheless, the question that how does URT affect the interior of the city, such as air quality and housing prices, needs more in-depth research.

References

1. Ahn K, Jang H, Song Y (2020) Economics impacts of being close to subway networks: a case study of Korean metropolitan areas. Res Transp Econom 83, Article 100900
2. Aslund O, Blind I, Dahlberg M (2017) All aboard? Commuter train access and labor market outcomes. Reg Sci Urban Econ 67:90–107
3. Borck R (2019) Public transport and urban pollution. Reg Sci Urban Econ 77:356–366
4. Bothe K, Hansen HK, Winther L (2018) Spatial restructuring and uneven intra-urban employment growth in metro and non-metro-served areas in Copenhagen. J Transp Geogr 70:21–30
5. Gonzalez-Navarro M, Turner MA (2018) Subways and urban growth: evidence from earth. J Urban Econ 108:85–106
6. Jain P, Cullinane S, Cullinane K (2008) The impact of governance development models on urban rail efficiency. Transportation Research Part A: Policy and Practice 42:1238–1250
7. Li S, Liu Y, Purevjav A, Yang L (2019) Does subway expansion improve air quality? J Environ Econ Manag 96:213–235
8. Mayer T, Trevien C (2017) The impact of urban public transportation evidence from the Paris region. J Urban Econ 102:1–21
9. Padeiro M (2013) Transport infrastructures and employment growth in the Paris metropolitan margins. J Transp Geogr 31:44–53

10. Saxe S, Miller E, Guthrie P (2017) The net gas impact of the Sheppard Subway Line. Transp Res Part D: Transp Environ 51:261–275
11. Toble WR (1970) A computer movie simulating urban growth in the Detroit region. Econ Geogr 46:234–240
12. Vickerman R (2008) Transit investment and economic development. Res Transp Econ 23:107–115
13. Zhang H (2020) Metro and urban growth: evidence from China. J Transp Geogr 85, Article 102732

Research on Intersection Signal Control Based on WIFI Probe Vehicle Detection

Yu-xin Hou, Rong-ze Yu, Ze-nan Yu, and Wei-dong Liu

Abstract With the rapid development of China's economy and the accelerating process of urbanization, urban traffic problems are becoming more and more prominent. The purpose of this paper is to use wireless network information technology with WIFI as the core to solve related urban traffic problems. Based on WIFI probe technology can automatically collect open WIFI device's MAC address to locate the functional characteristics of the device, this article will combine WIFI probe technology with intersection traffic monitoring, design and test an intersection vehicle detection scheme based on WIFI probe technology, by determining probe placement testing area and data screening processing to the statistics, the intersection of the vehicle. And through real-time adjustment of signal timing do reduce the time of no vehicle with green light, shorten the delay time, the actual capacity maximum close to the lane on the road traffic capacity, thus achieve the purpose of improving the efficiency of traffic, which has good practical application value.

Keywords WIFI probes · Vehicle detection · Signal timing

1 Research Purpose and Significance

In recent years, with the rapid development of China's economy and the accelerating process of urbanization, the problem of urban traffic congestion is becoming more and more serious. The rapid increase in the number of urban motor vehicles has brought great pressure on the existing traditional urban road network. According to the report, as of the end of 2019, the number of motor vehicles nationwide has reached 348 million, an increase of 8.83% compared with the end of 2018. The China Business Industry Research Institute predicts that the number of motor vehicles for China will be approximately 367 million in 2020. The traditional transportation system can no longer adapt to the current doubling of traffic volume, so the creation

Y. Hou · R. Yu · Z.-n. Yu · W. Liu (✉)
Shenyang Jianzhu University, 25 Hunnan Middle Road, Hunnan District, Shenyang, China
e-mail: 8754183338@qq.com

W. Wang et al. (eds.), *Green Transportation and Low Carbon Mobility Safety*,
Lecture Notes in Electrical Engineering 944,
https://doi.org/10.1007/978-981-19-5615-7_8

of intelligent transportation system is urgently needed to improve the current urban traffic problems.

The wireless network information technology with WIFI as the core as the main development of information industry in China provides new ideas for solving the increasingly serious urban traffic problems [1]. WIFI probes technology has been widely used in many fields, such as passenger flow statistics and analysis, precision marketing, public security, intelligent attendance and intelligent transportation, etc. But the technology has not yet been applied to vehicle detection at intersections, this study is to apply the new technology on the intersection vehicle detection, to a certain extent, has carried on the innovation, design a new intelligent vehicle detection system, through the detection of intersection traffic, which can adjust the signal timing to improve traffic efficiency. The convenience of WIFI probes device for data collection is unmatched by traditional data collection methods. Therefore, it is of great significance to make reasonable use of the application potential of WIFI probe device for all walks of life, including the transportation industry.

2 Brief Account of WIFI Probe Positioning Technology

2.1 *Definition of WIFI Probe Technology*

In the 2018 white paper on WIFI probe and technology of China, it is pointed out that in a narrow sense, a WI-FI probe refers to a signal frame in the communication process of a WIFI device, this type of message is dedicated to the request of the terminal (laptop, smart phone, etc.) can be connected to the WIFI network equipment surrounding the WIFI signal, and then released by WIFI devices such as wireless router, wireless access point (AP), and other equipment in the message to give a reply. To put it simply, when a WIFI device outputs information to a mobile terminal through WIFI, the mobile terminal with WIFI technology within a certain range can also receive the information, so these data can be collected by the probe.

2.2 *Principle of WIFI Probe Positioning Technology*

WIFI probes location technology is to use WIFI probe device to collect and analyze the MAC address of the mobile device within the detection range that has opened the wireless receiver, so as to get the location of its terminal. WIFI probes itself does not have the function of data storage, and the collected data needs to be transmitted to the cloud server, so the transmission cycle needs to be configured. Due to the large amount of data in this scheme, a high concurrent processing structure is needed to cope with the large amount of data in high-frequency reports.

There are two main methods for WIFI probe positioning, namely the fingerprint positioning method and the three-point positioning method. This paper mainly applies the three-point positioning method, that is, when the same mobile terminal is detected by three or more WIFI access points at the same time in the case of multiple WIFI access points in the same area, it can be used. The distance between the mobile terminal and each WIFI access point is calculated by the signal characteristics, and then draw a circle with each WIFI access point as the center and the distance between the WIFI access point and the mobile terminal as the radius, and the intersection is the mobile terminal's position [2]. As shown in the figure, the coordinates of three WIFI access points A, B and C are set as (x_a, y_a),(x_b, y_b),(x_c, y_c) and the distances between mobile terminal O and three WIFI access points are set as, respectively d_a, d_b, d_c The following formula can be given:

$$\sqrt{(x_a - x)^2 + (y_a - y)^2} = d_a \tag{1}$$

$$\sqrt{(x_b - x)^2 + (y_b - y)^2} = d_b \tag{2}$$

$$\sqrt{(x_c - x)^2 + (y_c - y)^2} = d_c \tag{3}$$

The coordinate of point O can be solved, which is the specific address of the mobile terminal.

2.3 Research Status at Home and Abroad

Foreign WIFI probes equipment started earlier, application and development than the domestic, have more leading, in foreign, WIFI probe equipment has been officially used in the field of security, and in the domestic application of this aspect has not been fully popularized. The use of devices like WIFI probe in traffic detection has a precedent dating back to 2012, when it was used to detect floating vehicle data. However, the relevant applications in China basically started in 2016–2017, and there is a big gap in both breadth and depth compared with those in foreign countries (Table 1).

3 Design Scheme of Intersection Vehicle Detection Based on WIFI Probc

See Fig. 1.

Table 1 Research and application directions at home and abroad

Overseas research and application fields	Domestic research and application fields
Real-time passenger flows statistics and analysis, collect offline population data, and provide valuable passenger flow data for merchants	Employee attendance, employee tracking, enterprise attendance data accurate attendance
The data detected by wireless network sniffing devices are connected with relevant information saved by users to achieve offline precision marketing, such as crowd portraits, etc	Reminder to VIP customers, such as helping offline shopping guide sales
Public safety business, engineering, etc., as a supplement to the city security	Reminder to VIP customers, such as helping offline shopping guide sales
Business, public security	Floating car test data

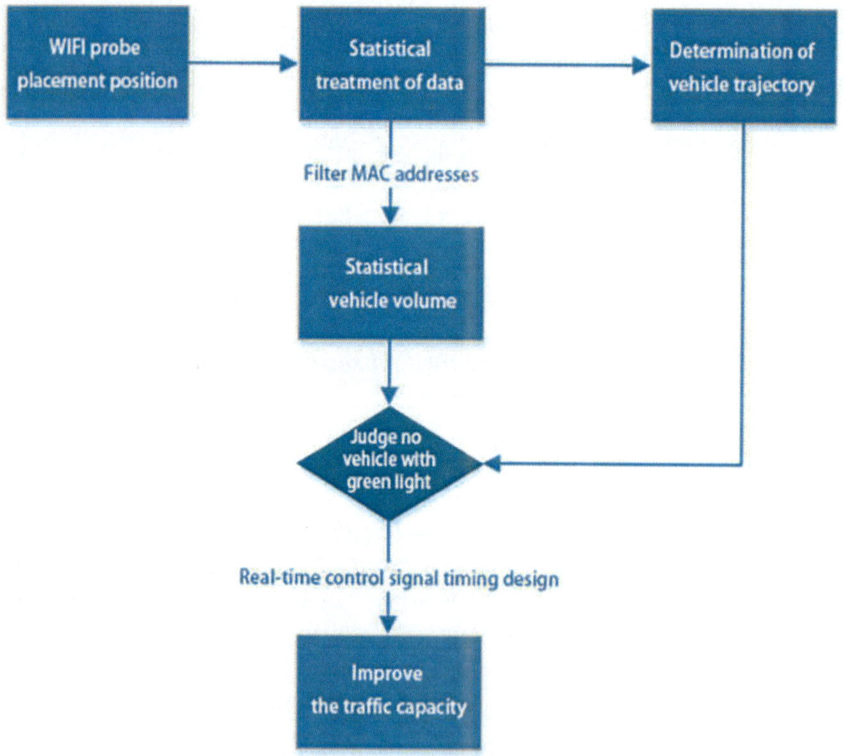

Fig. 1 Design process of traffic flow detection at intersection

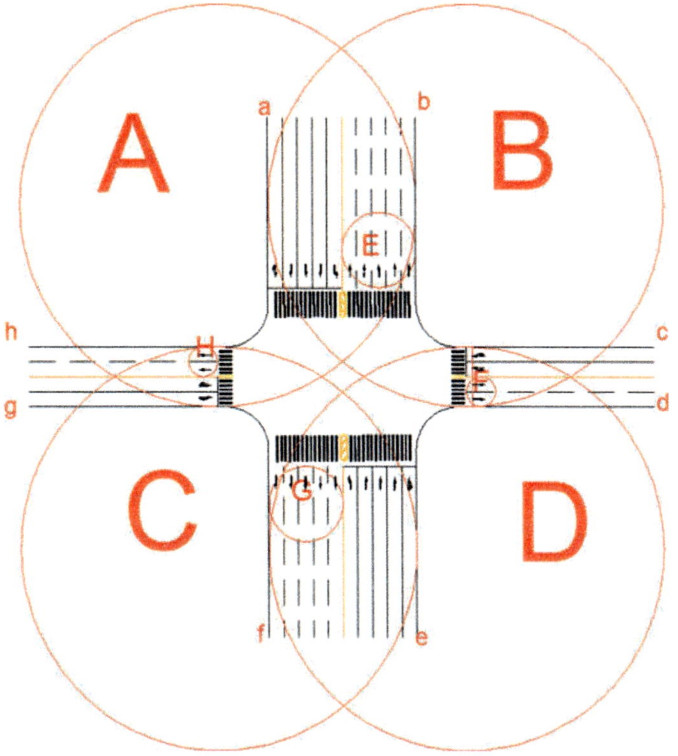

Fig. 2 WIFI probe placement and detection

3.1 Position and Detection Range of Probe

This design is based on the intersection shown in Fig. 2. The intersection has two-way ten lanes in the north–south direction, two-way four lanes in the east–west direction, and the lane width is 3.5 m. A total of 8 WIFI probes will be placed at the intersection. The range detected by probes A, B, C and D is A circle with A radius of 47 m. The circular area is tangent to the line where the corresponding edge is located and the center of the circle is on the line where the corresponding lane stop line is located. The center of the circle is the placement position of the probe.

3.2 Data Filtering and Processing Methods

MAC addresses of pedestrians and objects near intersections are removed, and MAC addresses of mobile terminals carried by vehicles running on the road are retained. Determine which lane the remaining available MAC addresses are in and track them to ensure the direction of the vehicle as it crosses the intersection. Keep track of the time the target vehicle entered the intersection and left the intersection.

As can be seen from the figure, an MAC address can be detected by AB, CD, AC and BD to determine the driving direction of the vehicle, and then the corresponding vehicle trajectory can be obtained according to the second detection by E, F, G and H probes.

By recording the filtered MAC address, the bicycle volume in and out of each direction is counted. The longest time t_1 and the shortest time t_2 are set for the north–south green light. On the basis of the shortest time t_2, if within a certain 20 s, the MAC address detected by the two probes at the same time is less than 4, the green light is identified as no vehicle with green light, and the setting time of the east–west green light is the same. The traffic efficiency can be improved by shortening the green light time [2] (Fig. 3).

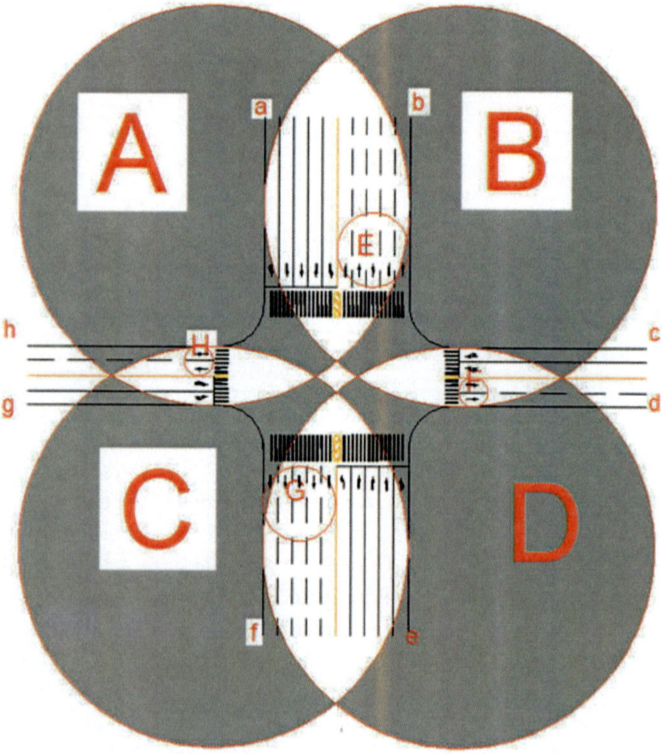

Fig. 3 The shaded area is the area detected by both probes

3.3 Research on Delay Time

The calculation formula of the delay of the entry lane at the intersection of the 2000 edition Highway Capacity Manual is

$$\mathrm{d} = d_1(PF) + d_2 + d_3 \tag{4}$$

$$d_1 = \frac{0.5C\left(1 - \frac{g}{C}\right)^2}{1 - \left[min(1, x)\frac{g}{C}\right]} \tag{5}$$

$$d_2 = 900T\left[(x-1) + \sqrt{(x-1)^2 + \frac{8klx}{cT}}\right] \tag{6}$$

$$d_3 = \frac{1800Q_b(1+u)t}{cT} \tag{7}$$

$$PF = \frac{(1-P)f_{PA}}{\left(1 - \frac{g}{C}\right)} \tag{8}$$

wherein:

d_1 is the uniform delay
PF is the correction coefficient of signal linkage obtained according to the arrival rate of vehicles and the arrival ratio of vehicles with green time
d_2 Additional delays due to random arrivals of vehicles and partial periods of absence of clearance and oversaturation
d_3 Additional delays caused by vehicles already in the system at the beginning of the analysis
g is the effective green time of the lane group analyzed at the intersection
x is the saturation of the analyzed lane group
C is the length of the analysis period, and k and l are the adjustment coefficients of the control mode and the upstream influence
c is the capacity of the analyzed vehicle group, and u is the regulation coefficient
Q_b is the initial number of vehicles in the lane group analyzed
t is the time to clear the initial vehicle within the analysis period (when t < T, , u is 0)
T is the duration of the analysis

The Time Assignment adjustment method in this study can shorten the time d_3 to a certain extent, so as to reduce the delay time on the whole.

In the Time Assignment adjustment, within the range of the shortest green time and the longest green time, if the number of mobile terminals where MAC addresses passed is less than a fixed number detected within a fixed time, it regarded as no vehicle with green light, that is, it will be converted to green light in other directions. This can shorten the time of red lights in other directions, which directly reduces the initial number of vehicles in the analysis lane group, that is, the time to clear the initial vehicles is reduced. That is, the value of Q_b and t in formula (6), (7) can be reduced to reduce the value of d_3, thereby reducing the value of d.

3.4 Analysis of Intersection Capacity

Our country often defines the road capacity for a certain point on the road a lane or a section, unit time may pass the largest traffic entity (vehicle or pedestrian) number [3]. In this paper, the stop line method proposed by Beijing Municipal Design Institute is used for calculation. The stop line of the entrance lane is taken as the Control surface. Once the vehicle passes this surface, it is regarded as passing the intersection [4]. This method is concerned with the traffic capacity of signalized intersections under the given conditions of convergence and signal control [5].

Set the capacity value of a lane as N_1, and the actual capacity as N_{actual}, and the difference N_0 between the two is the capacity within the green time.

$$N_1 - N_{actual} = N_o \tag{9}$$

This design scheme can real-time control signal timing design, shorten the time of green light, and Let N_{actual} be equal to N_1 to the maximum extent, In order to improve the traffic efficiency. Especially when the traffic flow at the intersection is very different in peak hours and off-peak hours, when the intersection is in off-peak hours, there will be a long time of empty green light, resulting in very low traffic efficiency. However, if the design scheme is adopted, it will real-time control signal timing design and improve the traffic efficiency.

3.5 Scheme Analysis

In this study, the intersection signal control system based on WIFI probes vehicle detection was designed. By determining the detection area of probe placement and data filtering, the statistical screening of vehicles at intersections can be carried out by WIFI probes. In the range of the maximum green time and the shortest green time, the green time can be reduced by real-time control signal timing design, and the HCM delay model can be used to shorten the delay time and make the actual traffic capacity of the road as close as possible to the road capacity, so as to achieve the purpose of improving traffic efficiency.

4 Conclusion

The following contents were achieved in this study:

(1) designed a vehicle detection based on WIFI probes for interring Detection signal control system, the system for mobile devices using WIFI probe the MAC address of the collection and analysis, adopt new technology, vehicle to locate and screening of intersection, the longest green light time and minimum green light time range, by adjusting the signal timing, to improve traffic efficiency.
(2) At the same time, the collected MAC address is used to accurately determine the traffic direction of each mobile terminal, so as to analyze and study the traffic behavior of the intersection. And real-time regulation of the intersection signals timing.
(3) By using HCM delay model to analyze the timing scheme of this study, it can be determined that theoretically, this design can reduce delay time and improve traffic efficiency.
(4) Analyze the capacity of intersections based on data processing. It can be concluded that the design can make the intersection through the vehicle maximum close to its capacity.

In other words, by using the design of this scheme, real-time detection and real-time adjustment of signal timing can be used to minimize the empty time of green light, so as to improve traffic efficiency, meet people's travel needs, prevent vehicle congestion and reduce air pollution.

At the same time, due to the limited time and ability, this paper has only done a preliminary research on WIFI probe and intelligent transportation, and there are still many shortcomings that can be further studied and learned.

(1) In the study, the situation that there is no mobile terminal in the vehicle is ignored, which is easy to cause detection error.
(2) Nowadays, with the gradual popularization of random MAC address algorithm in smart mobile terminal devices, the MAC addresses carried by mobile terminals with this function are randomly generated when they scan the surrounding WIFI hotspots. However, in this study, multiple WIFI probes need to record the MAC address of the same mobile terminal at the same time or at different times to determine the location of the vehicle, so it will cause errors.
(3) The security issue of WIFI probes also needs to be considered. There are some unscrupulous enterprises and businesses that illegally use WIFI technology to steal users' privacy information and then make profits. Therefore, the regulatory authorities need to strengthen the supervision of the use of WIFI probe technology and urge all parties to respect the legitimate rights and interests of users.

References

1. Wenjie D (2016) Research on traffic congestion detection technology of internet of vehicles based on data mining. Master Thesis, Nanjing University of Posts and Telecommunications
2. Shenyang Jianzhu University (2020) A signal control method for vehicle detection technology based on WIFI probe: CN202010129441.7 [P]
3. China Mobile Communications Group Corporation (2013) A traffic event detection method and system: CN201310751983.8 [P]
4. Lei H (2017) Study on the effect of illegal lane change on traffic efficiency at signalized intersections. Master Thesis, Chang 'an University, Shaanxi
5. Yuanyuan L (2016) Analysis of capacity of modern tram intersections. Master's Thesis, Southeast University, Jiangsu

Traffic Lights Recognition Based on Position Feature

Zhi-fa Yang, Xian-jun Fan, Zhuo Yu, Shi-wu Li, Ai-min liu, and Chang-an song

Abstract Due to the color and shape characteristics of traffic lights, the color model and shape detection are used to detect traffic lights in the work, the images were processed by the ROI (region of interest) extraction, image enhancement, grayscale binarization processing and morphological processing. Then the contour search and connected domain filtering algorithm were used to extract the traffic signal backlight backplane area, thus detecting and segmenting the traffic signal light backplane. Moreover, taking the traffic signal light backplane as positive sample, the other non-traffic light backplane was used as negative sample to build model library. HOG algorithm was used to extract the feature vectors of samples and exclude the false targets based on SVM classification algorithm. Finally, according to the positions of red and green signal lights on signal light, the pixel value accumulation in the area where the signal light is located was calculated as position feature to recognize the red and green signal lights.

Keywords Traffic engineering · Image recognition · Traffic lights · Position feature

1 Introduction

As important road environment information, the traffic lights control the traffic of vehicles at road intersections, and ensure the vehicles driving. However, due to driving fatigue or distraction, the drivers cannot obtain the change of signal information in time, which causes the traffic safety hazard at intersection and affects the road traffic. Therefore, the identification of traffic lights has important practical significance.

Z. Yang · X. Fan · S. Li · A. liu · C. song
School of Transportation, Jilin University, Changchun 130022, China

X. Fan · Z. Yu (✉)
School of Mathematics and Statistics, Changchun University of Technology, Changchun 130012, China
e-mail: 609611836@qq.com

W. Wang et al. (eds.), *Green Transportation and Low Carbon Mobility Safety*, Lecture Notes in Electrical Engineering 944,
https://doi.org/10.1007/978-981-19-5615-7_9

Because of obvious color and shape characteristics of traffic lights, the color model and shape detection are commonly used to detect traffic lights. Jin Tao [1] used the Gaussian model and multi-color model to segment traffic signal images, and used the correlation filtering to identify final state. Traffic lights are identified by canny operator edge extraction algorithm, Hu invariant moment and Mahalanobis distance. Ren Yong and Peng Jingyu [2] used the HSI color model to segment color threshold of image. Besides, the projection method is used to find the approximate location of traffic signal and perform the ROI segmentation. Digital Traffic Signals Indicator is identified by BP neural network. Wang Xuelin [3] combined the contrast, brightness, chrominance, saturation, and local depth information to locate the traffic signal in image. Furthermore, the stroke marking algorithm is used to mark and segment the ROI. The shape feature and wavelet decomposition are used to reduce the dimension, thus extracting ROI features. Finally, the traffic signal light arrow and digital are recognized by binary tree classification and template matching algorithm.

Shi et al. [4] used the HSV color model and traffic signal change law to mark the red, green, yellow and black colors in image, thus locking candidate area. Ji and Yang [5] integrated the VSA [6–8] (Visual Selective Attention, visual selection attention) model to obtain the candidate areas for traffic lights. Jang et al. [9] proposed the high exposure technology by integrating the low exposure technology and normal exposure technology to select candidate regions. In References [10–12], the HOG algorithm is used to extract candidate region features, and the SVM to classify traffic lights.

Most of traffic lights are identified using color models. Since the traffic signal light occupies a small area in video image, and the information such as color and brightness of light is greatly affected by ambient light, the traffic signal light in image is directly detected. On the one hand, the large search area affects the real-time performance of algorithm. On the other hand, the color threshold segmentation using color model is more sensitive to illumination. Therefore, during recognition, the red and green signal lights may be incorrectly identified.

In the work, the traffic light was identified as follows. Based on comprehensive preprocessing of image, the position of signal lamp backplane was found based on contour features, with the model library constructed. Then the feature extraction scheme was established by HOG method. Thirdly, the pseudo targets were eliminated based on SVM classification algorithm. Finally, the signal color recognition was completed according to the difference position and brightness of red and green signal lights.

Fig. 1 Determining the ROI area

2 Image Preprocessing

2.1 Extraction of ROI

In image processing, people are usually interested in an area in image instead of entire image. This area is called the ROI (region of interest). It can be processed to improve the algorithm running speed, and reduce the interference of irrelevant content in image, thus improving detection accuracy.

The traffic signal is installed at a height of 3 m (column) or 5.5–7 m (cantilever), complying with relevant regulations [13]. The traffic signal light is always located in the upper half of video image by adjusting installation position and angle of camera in vehicle. Therefore, the ROI is determined as the upper half of video image (See Fig. 1).

On the one hand, this can reduce the search time in non-target areas and improve the running speed. On the other hand, the interference of near-surface objects such as the taillights in image is avoided, with improved detection.

2.2 Image Enhancement

The image enhancement based on the white light compensation algorithm proposed by Anil K. Jain [14] is used to increase the recognition accuracy. The algorithm implementation steps are as follows.

First, calculate the gray value of pixel as reference white, and the number of pixel points as reference white. Formula (1) shows the average gray value of reference white pixel.

$$\text{avgGray}_{\text{ref}} = \text{Gray}_{\text{ref}}/\text{Num}_{\text{ref}} \tag{1}$$

Secondly, the light compensation coefficient is calculated as

$$coe = 255/avgGray_{ref} \tag{2}$$

Finally, the pixel values of all pixels in original image are multiplied by ray compensation coefficient to achieve the ray compensation of original image (See Formula (3)).

$$pi = pi \times coe \tag{3}$$

Figures 2 and 3 show the image effects using light compensation based on reference white.

Comparison chart shows that the overall brightness of image after light compensation has been improved, but the color of original image also undergone a certain change. Furthermore, the traffic signal scatters in a certain degree, resulting in easy adhesion to surrounding things. Thus, Formula (3) is transformed into Formula (4).

Fig. 2 Before compensation

Fig. 3 After compensation

Fig. 4 Improved model for ray compensated graph

Figure 4 shows the image processed by improved model, and the color of processed image maintains a good effect.

$$pi = pi/coe \tag{4}$$

2.3 *Grayscale and Binarization*

The camera is used to capture the image described by RGB color model, and the information is determined by three components of R, G, and B. Compared with color image, the gray image calculation is reduced, and the grayscale image can be retained. In the work, the image grayscale processing model (See Formula 5) in opencv is used for the grayscale image of red and green lights.

$$Gray = 0.072169*B + 0.715160*G + 0.212671*R \tag{5}$$

In order to identify traffic signal more accurately, the image is binarized based on grayscale image processing. Image binarization is to make the pixels only have 0 (black) or 255 (white) values (normalized to 0 and 1) in image, and then the white highlighted area is processed in binary image. In traffic signal detection, the key to image binarization is the threshold selection.

The traffic signal backplane area is performed under the sky with large gray value and black traffic signal light backplane. Therefore, the pixel value larger than the binarization threshold is set to 0; otherwise it is 1. This allows the traffic signal backplane area to be highlighted in the binary image. Since the traffic signal backplane area includes the traffic signal backplane and traffic light emitting area, the selected binarization threshold should ensure that the traffic signal light emitting area and signal light backplane are in the same area after binarization processing. By processing the video images of weather and illumination conditions, the binarization threshold is 120, and the obtained backlight image area in the binary image is better (See Fig. 5).

Fig. 5 Binary image

Fig. 6 Results of morphological open processing

2.4 Morphological Processing

The most basic morphology [15] treatment includes the corrosion and expansion. The work adopted the morphological opening to process image after etching and expanding. Image was corroded to eliminate the adhesion of signal lamp backplane to background object in environment. Then, the traffic signal backlight area was filled and closed to ensure the integrity and independence of traffic signal backplane area in binary image. Figure 6 shows the result of the morphological opening. After processing, the traffic signal backplane can be better separated from environmental background and remain relatively intact.

3 Traffic Signal Light Board Position Determination

3.1 Contour Extraction

After pre-processing image, the contour is extracted to identify the back panel of traffic signal from environment.

Contour finding, based on two algorithms proposed by Satoshi Suzuki [16], is used for the topological analysis of digital binary images. According to the relationship between boundary of binary image and original image region, the algorithm expresses the original image by determining the enclosing relationship of each boundary in binary image, including outer boundary, hole boundary and hierarchical relationship between them.

Formulae (6) and (7) show the method of determining outer boundary and hole boundary. Wherein, the pixel value of point in binary image is represented. The point (i,j) in Formula (6) represents the starting point of outer boundary, and that in Formula (7) represents the starting point of hole boundary.

$$f(i, j-1) = 0,\ f(i, j) = 1 \tag{6}$$

$$f(i, j) = 1,\ f(i, j+1) = 0 \tag{7}$$

When the contour search is used to detect the traffic signal backplane, the idea of finding outer boundary is utilized. When the contour is searched, the boundary pixel points are marked from the starting point of boundary. At this time, it is necessary to define an identifier NBD representing the boundary points in binary image, which is initialized to 1. NBD adds 1 every time a new boundary point is found. When $f(p, q) = 1,\ f(p, q+1) = 0$, the NBD is set to 0, and the point (p, q) represents the right end of boundary.

3.2 *Connected Domain Filtering*

The connected area in binary image is searched and marked by contour extraction, and whether the marked area is a traffic signal backplane can be judged according to geometric size. Based on the statistics the actual traffic lights and collected images, the relevant standards [17] are used to determine the geometric size range of backplane area (See Table 1).

According to the range of traffic light backplane, height, width and aspect ratio, the connected area in the image is filtered, and the possible area of traffic signal backplane is segmented.

Table 1 Geometry range of traffic signal light backplane in binary map

Area/pixel^2	height/pixel	width/pixel	pixel aspect ratio
80 ~ 1000	16 ~ 56	5 ~ 17	2.8 ~ 3.3

4 Traffic Light Recognition Based on Location Features

After detecting the possible area of traffic signal backplane, the image classification algorithm is used to eliminate the pseudo target, with the real traffic signal backplane area identified. Then, the color of signal light is identified according to the highlighted area where the traffic light is located.

4.1 Sample Library Design

After detecting and segmenting the traffic signal backplane in multiple videos, the targets mainly include traffic signal backplane, back or side of traffic signal, roadside trees, building or other facilities (See Fig. 1 in Chap. 3). In the classification training, the traffic signal backplane is used as a positive sample for classification training, and the other targets are used as negative samples. First, the positive and negative samples should be normalized. Since the intercepted target aspect ratio is between about 2.8 and 3.3, in order not to change image pixel ratio, the normalized image aspect ratio is required within this range. After statistics, the target area in the work is in the width of 5–17 pixels and the length of 16–56 pixels. Therefore, we normalize the sample into a 12×36 size image.

Traffic signal light back panel classification and recognition is a two-category problem. Its essence is to find a classification plane, so the positive and negative samples are on their own sides. Figure 7 shows the positive sample is the front of traffic signal backplane, and the negative sample should contain various non-traffic light targets. In principle, there are more negative samples in the sample library compared with positive sample. The larger diversity of negative samples can reduce the false detection rate. When the sample library is designed, the ratio of positive to negative samples is about 1:1.2, with better classification effect. There are 9,713 positive samples (with 4,526 red lights, 537 yellow lights, and 4,650 green lights), and 11,727 negative samples. Yellow light is not identified in the work.

4.2 HOG Feature Extraction

The HOG (histogram of oriented gradient) [18] feature extraction method is to characterize the image by statistically analyzing the gradient direction of local region of gray image.

First, the normalized image is grayed out, and then the gradient of image pixel is calculated. The horizontal edge gradient operator is (−1,0,1), and the vertical edge gradient operator is $(-1,0,1)^T$. Formulae (8) and (9) show the horizontal and vertical gradient values of pixels in image, respectively; $H(x, y)$ represent the pixel values of corresponding coordinate pixel points.

Fig. 7 Common target segmentation

$$G_x(x, y) = H(x+1, y) - H(x-1, y) \tag{8}$$

$$G_y(x, y) = H(x, y+1) - H(x, y-1) \tag{9}$$

The gradient magnitude and direction at the pixel (x, y) are calculated (See Formulae (10) and (11)).

$$G(x, y) = \sqrt{G_x(x, y)^2 + G_y(x, y)^2} \tag{10}$$

$$\theta(x, y) = \tan^{-1}\frac{G_y(x, y)}{G_x(x, y)} \tag{11}$$

Next, the gradient information of image is to obtain a feature vector describing the local features of image. First, the image is divided into several small cells, and the gradient direction of pixel points in each small cell is counted to generate a gradient direction histogram. In the statistics, the gradient value of pixel is used as the weight of gradient direction of pixel to participate in the gradient direction statistics. In the range of 0 to 360°, the gradient direction is divided into nine bins (See Fig. 8). For

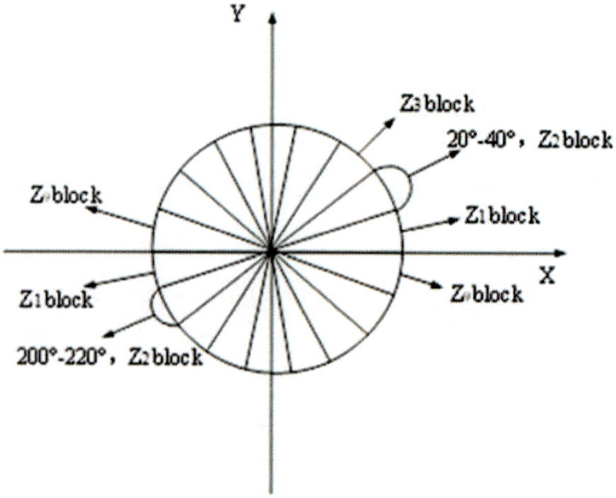

Fig. 8 Schematic diagram of HOG gradient direction area division

each small unit, the gradient direction of each pixel point is weighted to obtain a 9-dimensional feature vector describing the characteristics of small unit.

When most of small cells are combined into a block, the feature vectors of all the small cells in one block are connected in series to form a feature vector describing block. Similarly, the feature vectors of all the blocks in image are concatenated to form a feature vector describing image. Figure 9 show the calculation describing the dimensionality of image feature vector.

Since the normalized sample size is 12*36, several combinations of parameters in Table 2 is taken for HOG feature extraction.

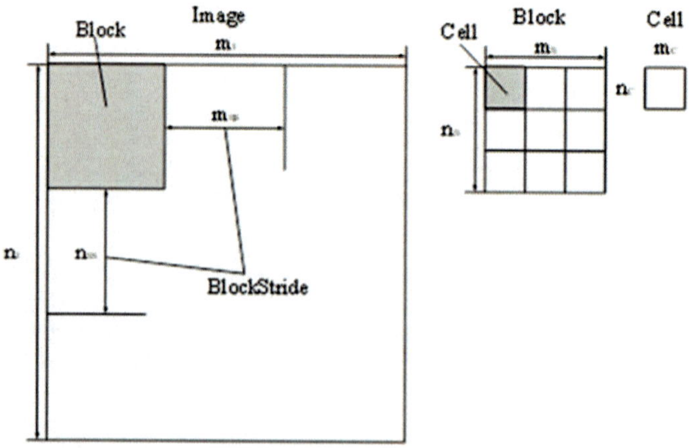

Fig. 9 Image HOG feature vector parameter calculation

Table 2 HOG feature extraction parameter settings

	Window	Block	Unit	Step size	Eigenvector dimension	Classification accuracy (%)
scheme(1)	(12,36)	(12,12)	(6,6)	(12,12)	108	56
scheme(2)	(12,36)	(12,12)	(6,6)	(6,6)	180	99
scheme(3)	(12,36)	(8,8)	(4,4)	(4,4)	576	99
scheme(4)	(12,36)	(6,6)	(6,6)	(6,6)	360	99

In Table 2, the classification accuracy rate refers to the SVM classification after feature extraction using current HOG feature extraction parameters. Moreover, the SVM classifier distinguishes the probability of positive and negative samples. In the work, the traffic signal backplane interception is carried out for specific traffic environment scenarios. Table 2 shows that in order to reduce the training time and improve the running speed of algorithm, the parameters of Scheme (2) can be used for feature extraction.

4.3 SVM Classification Algorithm

The SVM (support vector machine) is used to classify the positive and negative samples to eliminate non-traffic signal target.

Assuming that there are two sets of different data on two-dimensional plane, the two sets of data can be distinguished by a straight line, that is, the two sets of data are linearly separable. The straight line separating two sets of data is a linear classifier. In SVM, it is called hyperplane. The data point closest to the hyperplane is called the support vector. The problem is solved by SVM classification algorithm to find optimal classification hyperplane, which maximizes the distance of support vector from it.

Non-linearly separable data cannot be classified by a straight line. When the SVM classification algorithm solves the classification of nonlinear separable data, a kernel function is used to map the original data from low-dimensional space to high-dimensional space, thereby finding the optimal classification hyperplane.

In the actual use process, the original data (nonlinearly separable data) of SVM is the feature vector extracted by feature extraction algorithm, so the kernel function is needed to map the original data. OpenCV encapsulates the SVM, so it can be used with appropriate parameter selection and settings. The parameters of the work are selected as listed in Table 3. Among them, C_SVC represents the C class support vector classifier; RBF means that the kernel function type is Gaussian kernel function; parameter Gamma is the parameter corresponding to kernel function; parameter C is the penalty factor corresponding to C_SVC class support vector machine, and the iteration number represents the iteration of SVM. The SVM training function

Table 3 SVM training parameter settings

SVM type	Kernel function type	Gamma	C	iteration
C_SVC	RBF	0.01	1	1000

in OpenCV can automatically optimize the Gamma and parameter C. The optimal parameter Gamma is 0.03375, and the optimal parameter C is 12.5.

4.4 Traffic Signal Color Recognition

After removing pseudo target, the traffic signal light backplane area is obtained, and then the signal color recognition is performed. In the work, the positional features of red and green signal lights in the traffic light backplane are used for color recognition.

The general traffic signal includes three areas: red light (area 1), yellow light (area 2), and green light (area 3) (See Fig. 10). By using the difference between the gray value of traffic light emitting area and backplane area, the traffic signal light backplane area is binarized, and the binarization threshold is reasonably set. Thus, the light emitting area of traffic signal light is a white highlight area, and other areas are black. The traffic signal backlight backplane area after binarization is divided into three regions, which are labeled as Regions 1, 2 and 3.

Yellow light recognition is not performed, so only the area of highlighted Areas 1 and 3 is solved. Since the pixel value in the normalized binary image is 0 or 1 (white highlight area), it is only necessary to add the values of pixel points in area for calculation. According to the position of red light and green light in traffic light backplane, when Region 1 is larger than Region 3, the recognition result is a red light; otherwise it is a green light.

In order to test the validity of method and model, the work identified the red and green signal lights in the sample library established in previous paper. Using

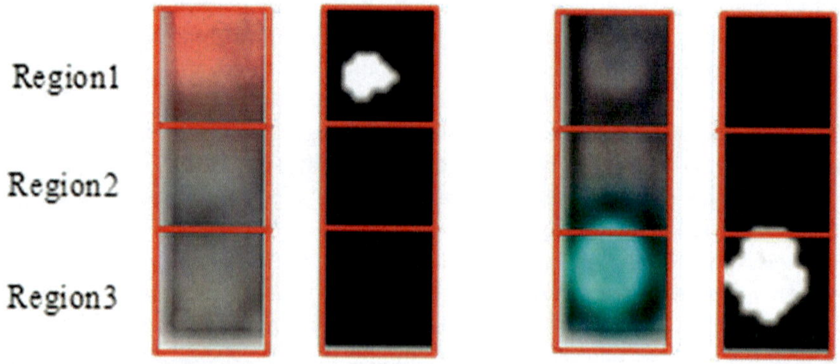

Fig. 10 Segmentation and binarized of traffic signal light backplane area

Table 4 Identification results of the proposed method in the established data set

Identification					
		Red light	Green light	Other	Sum
Fact	Red light	3962	274	290	4526
	Green light	253	4020	377	4650

Table 5 Statistical analysis results of method identification

Signal light	Accuracy rate *P*/%	Recall rate R/%	Comprehensive evaluation index F/%
Red light	94	87.54	–
Green light	93.62	86.45	–
Average	93.81	87	90.28

the accuracy rate P, the recall rate R and the comprehensive evaluation index $F = 2PR/((P + R))$, the performance of identification method was determined (Tables 4 and 5).

The average accuracy of this method is 93.81%, with the average recall rate of 87% and the comprehensive evaluation index of 90.28%, indicating that the method has better recognition of traffic lights.

5 Conclusions

The traffic signal backplane was separated from environmental background by image enhancement, morphological processing, etc. The contour search and connected domain filtering algorithm was used to extract the traffic signal backplane area, which detected and segmented the traffic signal backplane.

Taking the traffic signal backplane as positive samples and the other non-traffic light backboard as negative samples, the traffic signal light backboard classification sample library was established. Besides, the feature vector of sample was extracted by HOG algorithm, and the traffic signal light backboard classifier was generated by the SVM classification algorithm to identify traffic light backplane. The traffic signal light backplane was divided into regions. The position characteristics of red and green signal lights in traffic signal light backplane were used to calculate the difference of pixel values in the location area of red and green signal lights, thereby recognizing the traffic light colors.

Acknowledgements The work was supported by the Jilin province Science and Technology Development Project (20200201225JC), the National Natural Science Foundation of China (Grant NO. 51978310), and the Science and Technology Project in the 13th Five-Year Plan Period of Education Bureau of Jilin Province (JJKH20200686KJ, JJKH20201007KJ).

References

1. Tao J (2012) Research on methods of traffic lights detection and recognition in complex scenes[D]. Shanghai Jiao Tong University
2. Ren Y, PENG J (2013) Identification of digital traffic signals indicator based on BP neural network [J]. JISUANJI YU XIANDAIHUA (4):77–80
3. Wang X (2015) Based on dynamic scene Traffic signal detection and recognition [D]. Guizhou Minzu University
4. Shi X, Zhao N, Xia Y (2016) Detection and classification of traffic lights for automated setup of road surveillance systems[J]. Multimed Tools Appl 75(20):12547–12562
5. Ji Y, Yang M, Lu Z, et al (2015) Integrating visual selective attention model with HOG features for traffic light detection and recognition[C]// Intelligent Vehicles Symposium. IEEE 280–285
6. Zhang Q, Guochang GU, Xiao H (2009) Computational model of visual selective attention[J]. Robot 31(6):574–580
7. Yang M, Lu Z, Guo L, et al. (2014) Vision-based environmental perception and navigation of micro-intelligent vehicles[J]. 213:653–665
8. Qiong L, Shiyin Q, Li, Zhicheng (2010) The modeling of visual selective attention and its applications [J]. Sci Technol Rev 28(1001):107–115
9. Jang C, Kim C, Kim D, et al. (2014) Multiple exposure images based traffic light recognition[C], Intelligent Vehicles Symposium Proceedings. IEEE 1313–1318
10. Xie Y, Liu L F, Li C H, et al. (2009) Unifying visual saliency with HOG feature learning for traffic sign detection[C], Intelligent Vehicles Symposium. IEEE 24–29
11. Zaklouta F, Stanciulescu B (2011) Segmentation masks for real-time traffic sign recognition using weighted HOG-based trees[C], International IEEE Conference on Intelligent Transportation Systems. IEEE 1954–1959
12. Creusen IM, Wijnhoven RGJ, Herbschleb E et al. (2010) Color exploitation in hog-based traffic sign detection[C], IEEE International Conference on Image Processing. IEEE 2669–2672
13. GB 14886-2016 (2016) Specifications for road traffic signal setting and installation[S]. Standards Press of China, Beijing
14. Hsu RL, Abdel-Mottaleb M, Jain AK (2002) Face detection in color images[J]. IEEE Trans Pattern Anal Mach Intell 24(5):696–706
15. Peters RI (1995) A new algorithm for image noise reduction using mathematical morphology[J]. IEEE Trans Image Process 4(5):554–568
16. Suzuki S, Be K (1985) Topological structural analysis of digitized binary images by border following[J]. Comput Vis Graph Image Process 30(1):32–46
17. GB14887-2011 (2011) Road traffic signal lamps[S]. Standards Press of China, Beijing
18. Erazo-Aux J, Loaiza-Correa H; Restrepo-Giron AD (2019) Histograms of oriented gradients for automatic detection of defective regions in thermograms[J]. Appl Opt 58(13):3620–3629

Route Planning and Charging Navigation Strategy for Electric Vehicles Under the Mutual Assistance Trip System

Zhaohui Zeng, Jiangfeng Wang, Dongyu Luo, and Guojun Yang

Abstract Aiming at the range anxiety caused by battery capacity limitations on electric vehicles (EVs), a set of mutual assistance trip system was established, which used the mutual assistance information on EV drivers as a supplement to decision-making. The charging navigation and route selection optimization was carried out significantly, improving driver charging experience. The optimization model of the central control terminal aims at minimizing the total travel time costs of drivers, integrating real-time dynamic road conditions information, charging service information and EV mutual assistance information. It utilizes information entropy theory to quantify mutual assistance information on mutual assistance information risk factor θ_i, thus adjusts the weight of travel costs. The model is solved by the genetic algorithm based on the priority coding method. The results of the calculation example show that the mutual assistance trip system can significantly reduce the total travel costs of EV drivers. At the same time, the increase in mutual assistance information makes θ_i larger, which means the more reliable the corresponding travel cost ti, the greater the impact on the total time costs.

Keywords Mutual assistance trip · Electric vehicle · Path selection · Charging guide · Range anxiety · Information entropy · Genetic algorithm

1 Introduction

In order to reduce environmental pollution and resource dependence on petroleum energy, the large-scale popularization of electric vehicles (EVs) has become a trend. However, the limitation of battery capacity brings range anxiety to EV drivers. Therefore, effective route selection and charging navigation strategies are adopted to plan the optimal path and the charging station, which is necessary to reduce the impact

Z. Zeng · J. Wang (✉) · D. Luo · G. Yang
The MOT Key Laboratory of Transport Industry of Big Data
Application Technologies for Comprehensive Transport, Beijing Jiaotong University, Beijing 100044, China
e-mail: wangjiangfeng@bjtu.edu.cn

W. Wang et al. (eds.), *Green Transportation and Low Carbon Mobility Safety*, Lecture Notes in Electrical Engineering 944,
https://doi.org/10.1007/978-981-19-5615-7_10

on battery capacity for the travel of EV drivers. At the same time, it improves the operational efficiency of road traffic and prevents safety accidents.

The concerns on the Electric Vehicle Routing Problem (EVRP) have increased in recent years with the range anxiety and charging difficulties about EVs. Compared with traditional VRP, the more constraints, the more complex models of EVRP are. The EVRP was first studied in 2011 [1], whose model allowed charging fully or up to the 80% of the battery capacity and assume a constant recharging time. Subsequently, some papers utilize the constrained shortest path problem to find optimal routes for EVs, considering the limited driving range of EVs [2]. Given the charging behavior of EV drivers, driving direction and distance were used as choice indicators for optimal charging stations in several studies [3]. Combing driving time, charging time and queuing time, many papers discuss charging and route optimization for EVs [4–6]. Queuing theory was used to consider the effects of queuing time on charging station selection on the same time. In addition, some studies propose a multi-objective model to navigate EVs, the objectives include minimized charging costs, traveling time, and energy consumption etc. [7].

Due to the development of Connected Vehicles technology, it is feasible to utilize dynamic traffic information. In view of this, some researchers incorporated real-time traffic information in EVRP to improve the accuracy of route navigation schemes. Li et al. [8] presented a spatial–temporal distribution prediction method of charging load for electric vehicles based on dynamic traffic information and the characteristics of both transportation and mobile load were taken into consideration. Xing et al. [9] utilized the advantage of the model with time-division and proposed the "time-flow" road resistance model considering road impedance and intersection node impedance. Furthermore, obtaining charging station service information, Xu et al. [10] studied the method of dynamically adjusting charging service charge of fast charging station and analyzed the factors that influence electric vehicle users on selection behavior of fast charging station.

Overall, the current research didn't consider using the vehicle driver's perception information as supplementary information to assist the EV in charging guidance and route selection optimization when the traffic information is known. Under the above background, this paper intends to establish a mutual assistance trip system. After the EV drivers in the system send a charging request, the central control terminal will conduct charging guidance and route selection optimization, which will significantly improve the charging experience of EV drivers and relieve range anxiety. Due to the data support of the digital infrastructure, the road conditions information and charging station charging service information is updated in real time. Therefore, the system builds a charging navigation and route selection optimization model with the objective of minimizing the travel costs of EV drivers, which effectively improve the accuracy of the system's central dispatch.

2 Mutual Assistance Trip System

Mutual assistance trip refers to limited information interaction between EVs in the system, and vehicles assist each other, so as to avoid road congestion and crowded charging stations, so as to achieve a faster and better travel experience. Mutual assistance trip is not limited to the traditional collaboration between vehicles and roads. It also requires limited collaboration between vehicles and vehicles in the system to achieve a win–win situation for mutual assistance vehicles.

Mutual assistance information refers to the perception information on the current environment sent by the EV to the system, such as information like current congestion on the current road section, unobstructed current road section, crowded vehicles at a charging station and so on.

The structure of the mutual assistance trip system is shown in Fig. 1. Firstly, the EV driver sends a charging request for the central control terminal (i.e., the cloud super-computing center), which is regarded as a request for the system center of charging station recommendation and path planning. At this time, after receiving the request information about the EV driver, the central control terminal quickly invokes the latest updated road condition information, charging service information, and mutual assistance information. Through the mutual assistance information interaction between the mutual assistance trip system and the EV, the central control terminal aims at minimizing the travel costs of EV drivers, recommends the optimal charging station for them, and plans the optimal path to the charging station at the same time. The central control terminal will return the recommended information on the EV driver. The EV driver will avoid crowded charging stations and some congested road sections, significantly reducing the total costs of travel and improving the charging experience.

3 EV Charging Navigation and Route Selection Optimization Model

3.1 Objective Function

After the EV driver sends a charging request, the central control terminal allocates a charging station to the driver. At the same time, it indicates that the State of Charge (SOC) of the EV is insufficient, and the driver has a certain range anxiety. Therefore, EV drivers requesting the mutual assistance trip system to allocate charging stations need to alleviate this anxiety, and charging the EV as soon as possible, the better the driver experience.

With the objective of minimizing the total travel costs of drivers, the central control terminal constructs the following charging navigation and route selection optimization model:

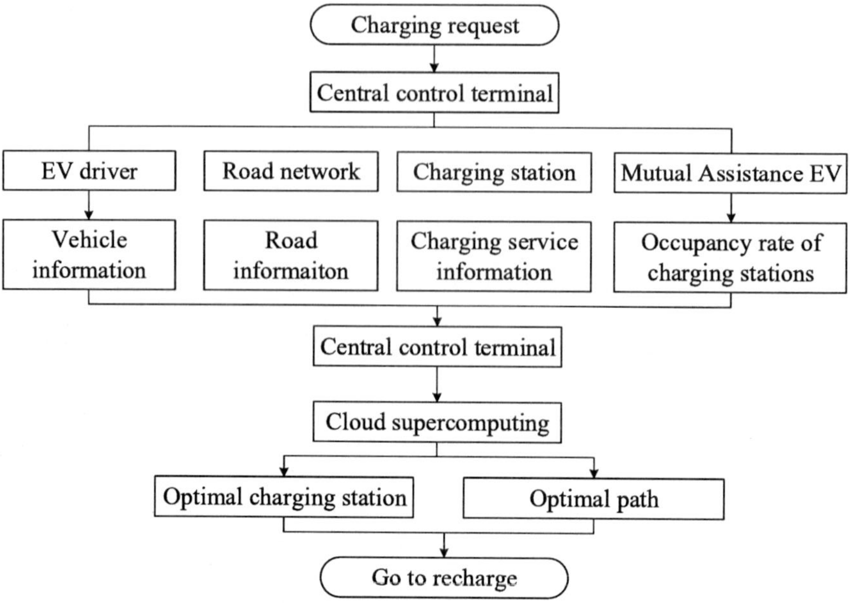

Fig. 1 Mutual assistance trip system work flow chart

$$\mathrm{minT} = \sum \theta_{\mathrm{i}} \mathrm{t_i} \tag{1}$$

where T represents the total costs of EV drivers; i represents influencing factors, ($i = 1, 2, 3, 4\ldots$) corresponding to driving time, queuing time, travel expenses, charging time etc.; t_i represents the time value cost of various factors; θ_i represents the risk factor of mutual assistance information.

3.2 *Restrictions*

3.2.1 Road Network Model

Since the mechanism of the mutual assistance trip system puts data processing before solving the optimization model, this section only needs to establish a static road network model. Nevertheless, it can overcome the shortcomings of the traditional static road network model that cannot reflect the dynamic change of the traffic network. The model is as follows:

$$\begin{cases} \boldsymbol{G} = (\boldsymbol{R}, \boldsymbol{A}, \boldsymbol{W}) \\ \boldsymbol{R} = \{r_i | i = 1, 2, 3, \dots, m\} \\ \boldsymbol{A} = \{r_{ij} | r_i \in \boldsymbol{R}, r_j \in \boldsymbol{R}, i \neq j\} \\ \boldsymbol{W} = \{w_{ij} | r_{ij} \in \boldsymbol{R}\} \end{cases} \tag{2}$$

where $\boldsymbol{G}$ represents the road network; $\boldsymbol{R}$ represents the set of all nodes in the graph $\boldsymbol{G}$; $\boldsymbol{A}$ represents the set of all directed arcs in the graph $\boldsymbol{G}$; $\boldsymbol{W}$ is the set of link weights, w_{ij} represents the length of the link from node i to node j.

3.2.2 Road Impedance Model

In order to calculate the travel time t_1, this section selects the travel time as the road impedance for modeling. In this section, the traditional BPR function is used. According to the actual situation of urban roads in my country, α and β are calibrated with historical data. The model is as follows:

$$s_{ij} = s_{ij}^0 \left[1 + \alpha \left(\frac{u_{ij}}{c_{ij}} \right)^{\beta} \right] \tag{3}$$

$$t_1 = \sum_{l=1}^{\tau-1} s_{ij}^l \tag{4}$$

where s_{ij} represents the impedance from node i to node j; s_{ij}^0 represents the travel time from node i to node j under free flow; u_{ij} represents the traffic flow from node i to node j; c_{ij} represents the actual capacity of the road section from node i to node j; α and β are calibration parameters, in this paper, the historical data is fitted, α is 0.5 and β is 1.5; τ represents the total number of nodes of the vehicle path; l represents the l-th node that the vehicle path passes.

3.2.3 Electricity Constraint Model

In the "Code for Design of Urban Roads", urban roads are divided into express roads, main roads, secondary roads, and branch roads. This paper corresponds to road grades I, II, III, IV. Reference [11] is based on the actual road driving of EV Data modeling to quantify the actual congestion level of different levels of roads and the impact on vehicle status on the power consumption per unit mileage. The power constraint model is as follows:

$$\begin{cases} E_1 = 0.247 + \frac{1.52}{v_{ij}} - 0.004v_{ij} + 2.992 \times 10^{-5} v_{ij} \\ E_2 = -0.179 + 0.004v_{ij} + \frac{5.492}{v_{ij}} \\ E_3 = 0.21 - 0.001v_{ij} + \frac{1.531}{v_{ij}} \\ E_4 = 0.208 - 0.002v_{ij} + \frac{1.553}{v_{ij}} \end{cases} \quad (5)$$

$$\sum_{l=1}^{\tau-1} w_{ij}^{l} \cdot E_g < SOC \quad (6)$$

where SOC represents the remaining power; *wk ij* represents the length of the road section corresponding to the *l*-th node of the vehicle path; E_g represents the power consumption per unit mileage of the corresponding grade g road; E_1, E_2, E_3, E_4 respectively represents the power consumption per unit mileage of road classes I, II, III, IV; v_{ij} represents the average road speed from node *i* to node *j*.

3.2.4 Charging Service Model

Combining M/M/1 queuing theory and BPR function ideas, construct a functional relationship between queuing time and charging flow at charging stations, as follows [12]:

$$t_2 = d_n^0 \left(\frac{u_n}{c_n} + \frac{u_n^2}{c_n^2} \right) \quad (7)$$

where t_2 represents the queuing time of charging station *n*;d_n^0 represents the queuing time coefficient; u_n represents the charging flow of charging station *n*; c_n represents the service capacity of charging station *n*; The queuing time coefficient d_n^0 and service capacity c_n are the calibration parameters, and this paper takes 5 min uniformly.

3.2.5 Mutual Assistance Information Risk Factor Model

The *i*-th mutual assistance information collection of EV drivers is $\left(a_1^{(i)}, a_2^{(i)}, \ldots, a_\mu^{(i)}\right) \in X^{(i)}$. Each piece of mutual assistance information corresponds to a different state of things. One piece of mutual assistance information represents a current state, and the state of things may change into the next moment. According to Markov chain theory, mutual assistance information $a_\lambda^{(i)}$ corresponding state space is recorded as $Y = \left\{ {}_1a_\lambda^{(i)}, {}_2a_\lambda^{(i)}, {}_3a_\lambda^{(i)}, \ldots, {}_qa_\lambda^{(i)} \right\}$. It's specific mapping to a certain state in ***Y*** should be recorded as${}_{k_1}a_\lambda^{(i)}$, and the next changed state ${}_{k_2}a_\lambda^{(i)}$ is relatively ideal state (Such as the state of unblocked roads, the idle state of charging stations, etc.). Using the "Evaluation Method of Road Traffic Congestion Degree" (GA/T 115–2020), the road traffic congestion degree is divided into Grade

I (severe congestion), Grade II (moderate congestion), Grade III (light congestion), and Grade IV (Unblocked). ${}_{k_2}a_\lambda^{(i)}$ may belong to one of the levels. Other types of mutual assistance information can also be divided into multiple levels. Probability of state transition $p({}_{k_1k_2}a_\lambda^{(i)})$ corresponding to mutual assistance information $a_\lambda^{(i)}$, abbreviated as $p(a_\lambda^{(i)})$, the formula is as follows:

$$p\left({}_{k_1k_2}a_\lambda^{(i)}\right) = \frac{P\left\{{}_{k_1}a_\lambda^{(i)}, {}_{k_2}a_\lambda^{(i)}\right\}}{P\left\{{}_{k_2}a_\lambda^{(i)}\right\}} \tag{8}$$

The theory of information entropy is introduced here. Equation (9) quantifies the mutual assistance information, and Eq. (10) is normalized to convert the information entropy into the mutual assistance information risk factor θ_i, as shown below:

$$H\left(X^{(i)}\right) = -\sum_{\lambda=1}^{\mu} p\left(a_\lambda^{(i)}\right) \log p\left(a_\lambda^{(i)}\right) \tag{9}$$

$$\theta_i = \frac{H\left(X^{(i)}\right)}{\sum H\left(X^{(i)}\right)} \tag{10}$$

where $\boldsymbol{H}(\boldsymbol{X}^{(i)})$ represents the information entropy of the i-th type of mutual assistance information; λ represents the λ-th piece of the i-th type of mutual assistance information; μ represents the total number of the i-th type of mutual assistance information; $p(a_\lambda^{(i)})$ represents the state transition probability corresponding to the information $a_\lambda^{(i)}$.

3.3 Solving Algorithm

EV charging navigation and route selection optimization model is a nonlinear programming problem. Genetic algorithm as a heuristic algorithm can improve the solution efficiency when solving nonlinear programming problems without losing too much accuracy. Genetic coding of individuals in the initial population is the core step of genetic algorithm solving. This paper draws on reference [13]. Using priority-based coding, this coding makes every element on the chromosome not equals to each other, and has the characteristics of permutation coding. Since the queuing time of the charging station is added to the objective function, a combined coding method combining partial arrangement coding and partial fixed coding is used when designing the chromosome coding. Except for the chromosome coding, the genetic algorithm involved is similar to the traditional method and will not be introduced.

Based on the genetic algorithm, this paper designs a fast search algorithm for the central control terminal to solve the model. The specific steps are as follows:

Step 0: The central control terminal obtains basic road network information s_{ij}^{0}, c_{ij}, w_{ij}, r_{ij}, charging service information c_n, mutual assistance information $a_{\lambda}^{(i)}$, etc. The EV driver uploads the vehicle's location information O and power information SOC. Initialization parameter setting α, β,d_n^0;

Step 1: Use the remaining mileage estimation method to estimate the maximum remaining mileage L, with O as the center and the maximum remaining mileage as the radius, the charging stations in this area are added to the set of alternative charging stations $\boldsymbol{D} = \{D_1, D_2, D_3, \cdots, D_n\}$;

Step 2: According to the area in the previous step, filter the effective mutual assistance information $\left(a_1^{(i)}, a_2^{(i)}, \cdots, a_{\mu}^{(i)}\right)$, and then obtain the corresponding probability $\left(p\left(a_1^{(i)}\right), p\left(a_2^{(i)}\right), \cdots, p\left(a_{\mu}^{(i)}\right)\right)$, information entropy$\boldsymbol{H}(\boldsymbol{X}^{(i)})$, and mutual assistance information risk factor θ_i;

Step 3: Determine the fitness function $f = \sum \theta_i t_i$, determine the traditional parameters of the genetic algorithm such as population size, selection probability, genetic probability, crossover probability, mutation probability, and number of iterations;

Step 4: Determine the elimination conditions $\sum_{l=1}^{\tau-1} w_{ij}^{l} \cdot E_g < SOC$ to eliminate the path that is not enough to reach the charging station;

Step 5: Follow the steps of traditional genetic algorithm to perform operations such as selection, crossover, mutation, etc., calculate the fitness function value, and find the best individual;

Step 6: Judge whether the convergence, if the conditions are not met, return to Step 5 to continue iterating, otherwise output the best path and the best charging station, and the algorithm ends.

4 Case Study

4.1 Case Description

The scale of the road network of the example in this paper is about 4 km × 4 km, as shown in Fig. 2. There are 20 road nodes and 31 edges in total. There are 4 charging stations located at nodes 2, 11, 13, and 20. The charging station information is shown in Table 1. The basic information of the road network includes length, free flow time, and capacity. Considering the limited space, only 5 road sections are listed as examples, as shown in Table 2. The total number of vehicles in the study area is 3350. According to the development scale of electric vehicles in Beijing, the penetration rate of electric vehicles is 20%, and the number of electric vehicles is 670.

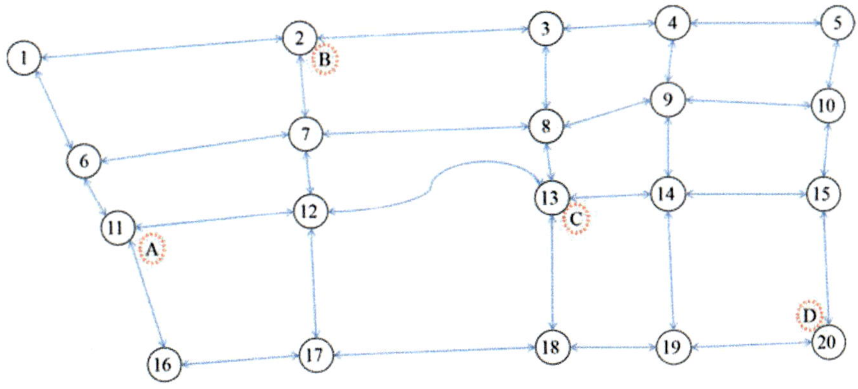

Fig. 2 Road network map

Table 1 Charging station information

Charging station number	Charging pile node	Number of charging piles	Service capabilities/(pcu/h)
A	11	14	30
B	2	20	45
C	13	10	22
D	20	4	9

Table 2 Road network basic information

Road section	Length/km	Free flow time/s	Capacity/(pcu/h)	Grade
1–6	0.6	36	3519	I
2–3	1.2	72	3519	II
6–7	1.1	66	2346	III
10–9	0.8	48	1173	IV
5–10	0.5	30	4692	II

The EV driver sends a charging request for a node in the road network. The central control terminal performs calculations based on the latest updated information, ignoring the impact on computing power on the information update time. The information calculated once is fixed information. This calculation example only considers the driving time t_1 and the queuing time t_2, so the impact of the charging pile type, brand, operator, and charging price on the driver's choice of charging station is negligible, and only the fastest charging is considered. There is enough buffer zone in the charging station for EV vehicles to queue for.

4.2 *Analysis of Results*

4.2.1 Analysis of the Effectiveness of Mutual Assistance Among EV Drivers

The node where the EV driver sends the charging request is O = 5; the SOC is 0.6 kwh. This section doesn't consider the mutual assistance risk factor θ_i. The total time costs for the EV driver to reach each charging station is calculated, and a typical representative route is selected for comparative analysis. The specific information of the route is shown in Table 3.

It can be seen from Table 3 that paths 1, 2, 3, and 4 are the paths calculated for the shortest mileage to charging stations A, B, C, and D, respectively. If the charging station is selected for charging to routes 1 to 4, route 1 has the smallest total time cost, but it is not the shortest mileage. If the EV driver chooses the charging station according to the shortest path principle without using the mutual assistance travel system, he will go to the D charging station according to route 4 for charging. The path energy consumption of path 2 is estimated to be 0.693 kwh, which exceeds the threshold limit of SOC 0.6 kwh and does not meet the mileage constraint, so the path is eliminated. Route 3 and route 5 go to the same charging station C for charging. Since route 3 passes through congested Sects. 8–13, the travel time is greatly increased, and route 5 avoids the congested section. Therefore, when choosing to go to charging station C for charging, choose Path 5 has the least total time cost.

It can be seen from Fig. 3 that choosing the charging station with the shortest driving time or the shortest queuing time is not necessarily the optimal solution. Considering Table 3 and Fig. 3, the mutual assistance travel system comprehensively considers travel time and queuing time to minimize the total time cost, which is better than considering only the mileage of a single route or the elements of travel time. The percentage of total time cost savings under different criteria in this example is shown in Fig. 4. Route 4 is the shortest mileage criterion and the total time cost is 910.81 s. Route 1 is the shortest queuing time criterion and the total time cost is 342.59 s. The total time cost of route 1 is 62.4% less than that of route 4. Comparing route 3 and route 5, the total time cost is reduced by 20% due to avoiding congested sections.

Table 3 Typical route table

Route number	Route	Route mileage/km	Total time cost/s	Route energy consumption/kwh
1	5-4-3-2	2.5	342.59	0.336
2	5-4-3-8-7-12-11	4.2	585.36	0.693
3	5-4-3-8-13	2.19	430.35	0.343
4	5-10-15-20	1.6	910.81	0.192
5	5-4-9-14-13	2.2	342.89	0.339

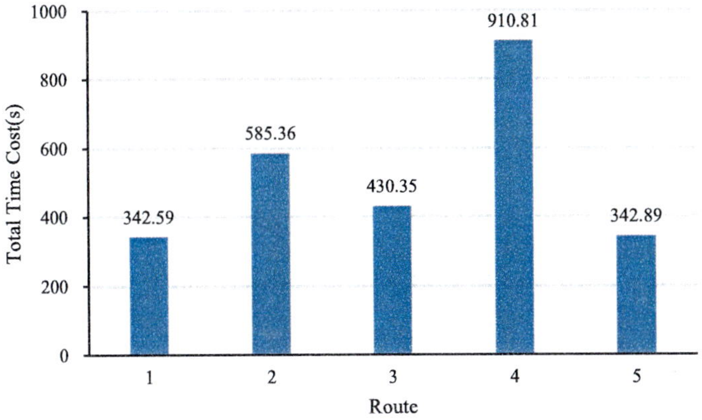

Fig. 3 Time cost of each route

Fig. 4 Total time cost fitness curve

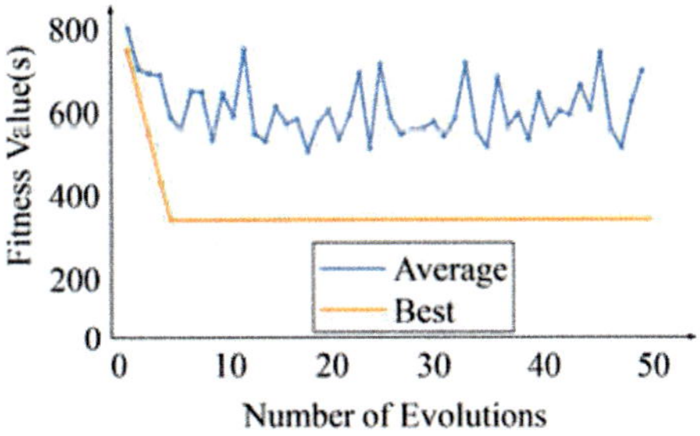

The iterative optimization process of genetic algorithm is shown in Fig. 4. The population size of genetic algorithm is set to 10, and the number of iterations is 50. The other traditional parameters of genetic algorithm, such as crossover probability, selection probability, mutation probability, etc., all adopt common values. It has been verified that the changes of these parameters will affect the convergence speed of the best fitness curve to a certain extent, but the effect is not obvious in this example, and the convergence value of the best fitness will not change. In the early stage of iterative optimization, the fitness curve gradually decreases as the number of iterations increases; in the later stage of iterative optimization, the fitness curve tends to be stable, with the best fitness being 342.59 s.

4.2.2 Sensitivity Analysis of Mutual Assistance Information Risk Factors

The higher the proportion of EVs in the total number of vehicles, the higher the penetration rate of mutual assistance EVs in the total number of EVs, which will significantly increase mutual assistance information. In order to explore the impact of the increase or decrease of mutual assistance information on information entropy, a total of no more than 10 mutual assistance vehicles are set up. Each vehicle will send at most one message in a short period of time, or may not send mutual assistance information. The relationship between the amount of mutual assistance information and information entropy is shown in Fig. 5.

It can be seen from Fig. 5 that with the increase of mutual assistance information, the information entropy generally shows an increasing trend. In this example, the information entropy is divided into two types, which are generated by mutual assistance information sent by EV drivers during driving and queuing, corresponding to road condition information entropy and charging service information entropy. Changes in the two types of information entropy will cause mutual assistance information The risk factors have changed, as shown in Fig. 6. θ_1 represents the mutual assistance information risk factor weight value corresponding to the ratio of the number of vehicles sending road condition information to the number of vehicles sending charging service information. As the number of vehicles sending road condition information increases, the mutual assistance information risk factor weight value gradually increases from 0 to 1, θ_2 on the contrary.

The larger the θ_i, the more credible the corresponding t_i, and the larger the weight corresponding to t_i, the greater the impact on T, which will ultimately have a greater impact on decision-making. The influence of the change of the mutual assistance information risk factor θ_i on T is shown in Fig. 7.

It can be seen from Fig. 7 that when θ_1 is 0, the optimal solution is route 4. If only queuing time is considered, charging station B with the shortest queuing time is selected among the four charging stations A, B, C, and D; when θ_1 is 0.1–0.5, The

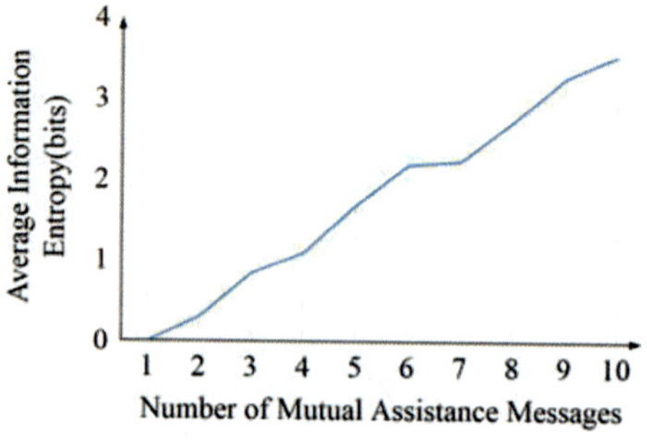

Fig. 5 The relationship between the amount of mutual assistance information and information entropy

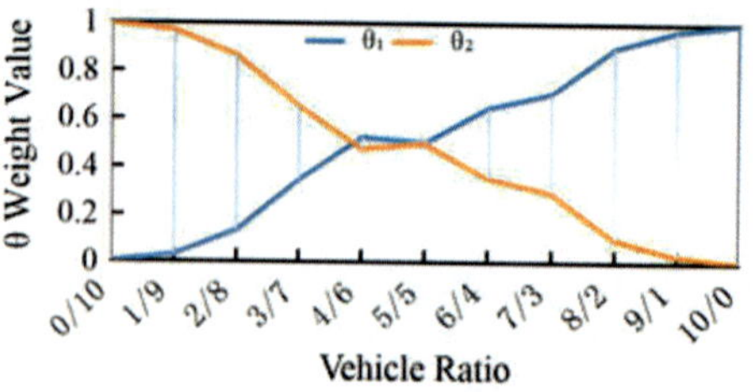

Fig. 6 Mutual assistance vehicle comparison and mutual assistance information risk factor trend change chart

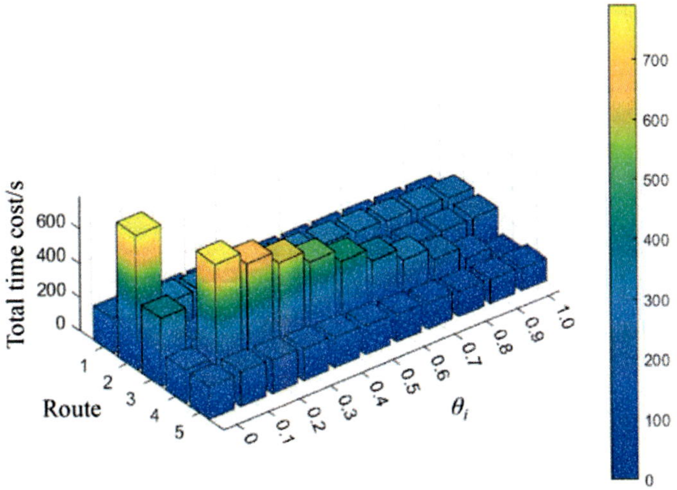

Fig. 7 Mutual assistance information risk factor θ_i change influence diagram on decision

optimal solution is route 1; when θ_1 is 0.6–0.9, the optimal solution is route 5. At this time, the weight of travel time is greater, so the driving time is mainly considered when making decisions; when θ_1 is 1, only considering the travel time, the optimal solution is route 4.

5 Conclusion

Aiming at the range anxiety of EV drivers, this paper proposes to establish a mutual assistance travel system, using mutual assistance information as a supplement to decision-making, to guide EV drivers and optimize route selection, and significantly improve the charging experience of EV drivers. The established charging navigation and route selection optimization model integrates real-time dynamic road condition information, charging service information and EV mutual assistance information, and uses information entropy theory to quantify mutual assistance information as a risk factor and adjust the weight of travel costs. The results show that the mutual assistance travel system can significantly reduce the total travel costs of EV drivers. At the same time, the mutual assistance information increases as the penetration rate of mutual assistance vehicles increases. The larger the θ_i, the more credible the corresponding t_i, and the greater the weight corresponding to t_i. The greater the impact on T, the final impact on decision-making. Due to space limitations, this paper only analyzes two factors, driving time and queuing time. In the next step, it can also consider more factors such as charging time and charging price, and analyze the effectiveness of mutual assistance information in time and space dimensions.

References

1. Conrad, RG, Figlioz Z et al (2011) The recharging vehicle routing problem. In: Proceedings of the 2011 Industrial Engineering Research Conference
2. Neaimeh M, Hill G, Hübner Y et al (2013) Routing systems to extend the driving range of electric vehicles. IET Intel Transp Syst 7:327–336
3. Wang Y, Bi J, Zhao X et al (2018) A geometry-based algorithm to provide guidance for electric vehicle charging. Transp Res Part D 63:890–906
4. Zhang Y, Aliya B, Zhou Y et al (2018) Shortest feasible paths with partial charging for battery-powered electric vehicles in smart cities. Pervasive Mob Comput 50:82–93
5. De Weerdt M, Stein S, Gerding E et al (2015) Intention-aware routing of electric vehicles. IEEE Trans Intell Transp Syst 17:1472–1482
6. Tian W, He J, Jiang J et al (2013) Electric vehicle charging load spatial allocation optimization algorithm[J]. Trans China Electrotechnical Soc 28(3):269–276
7. Wang Y, Bi J, Guan W et al (2018) Optimising route choices for the travelling and charging of battery electric vehicles by considering multiple objectives. Transp Res Part D 64:246–261
8. Li X, Li L, Liu W et al (2020) Spatial-temporal distribution prediction of charging load for electric vehicles based on dynamic traffic information. Power Syst Prot Control 48(01):117–125
9. Xing Q, Chen Z, Leng Z et al (2020) Route planning and charging navigation strategy for electric vehicles based on real-time traffic information. Proc Chin Soc Elect Eng 40(02):534–550
10. Xu S, Zhang H, Shi D et al (2020) Setting strategy of charging service fee for fast charging load of smart cities. Proc Chin Soc Elect Eng 40(02):534–550
11. Li L, Zhao X, Li X et al (2020) Electric vehicle charging demand prediction model based on dynamic traffic information and its impacts on distribution networks. Power Syst Clean Energy 36(03):107–118
12. Li H, Chen H (2020) Mixed traffic network equilibrium with battery electric vehicles considering charging queuing time. J Jilin Univ (Engineering and Technology Edition). https://doi.org/10.13229/j.cnki.jdxbgxb20200421
13. Gen M, Lin L (2009) Priority-based genetic algorithm for shortest path routing problem in OSPF. Intel Evol Syst 187:91–103

Modelling and Simulation of Speed Guidance of Multi-Intersection in a Connected Vehicle Environment

Ziwei Peng, Jiangfeng Wang, Zhijun Gao, and Haitao Huang

Abstract In this paper, we design a cooperative speed guidance algorithm for the connected vehicles going through several continuous intersections to minimize vehicles' delay or stopping time in the environment of connected vehicles. Using the vehicle's position, speed, and real-time traffic signal information (signal light's color and its' remaining time) obtained by communication of infrastructure to connected vehicle, speed advisory system can produce an optimal speed value for vehicle (to accelerate to reduce delays or decelerate to reduce stopping time) with the algorithm when the connected vehicle enters a certain range of the first intersection. When the vehicle enters a certain range of the next intersection, speed advisory system does not calculate advisory speed (to avoid frequent changes in vehicle speed) if the vehicle can pass through the intersection at the limited maximum speed without any stop, otherwise, the speed advisory system produces an optimal speed (to reduce stopping time) using information of vehicle and traffic signal like when the vehicle enters the first intersection. Speed guidance of vehicle in rest intersections is same as mentioned above. Several simulation scenarios were established to test the proposed speed guidance strategy in the connected vehicle environment. Results show that improvements in efficiency and emission are obvious, however, safety has a slight decrease. A sensitivity analysis was conducted to compare the performance of different percentages of connected vehicles and the compliance rate of advisory speed, it was found that improvements of higher penetration rate and compliance rate are more obvious.

Keywords Connected vehicle · Cooperative speed guidance · Multi-intersection · EstiNet simulation · Arterial traffic

Z. Peng · J. Wang · Z. Gao
Key Laboratory of Transport Industry of Big Data Application Technologies for Comprehensive Transport, Ministry of Transport, Beijing Jiaotong University, Beijing 100044, China

H. Huang (✉)
China Academy of Transportation Sciences, Beijing 100029, China
e-mail: huanghaitao@catsic.com

W. Wang et al. (eds.), *Green Transportation and Low Carbon Mobility Safety*, Lecture Notes in Electrical Engineering 944,
https://doi.org/10.1007/978-981-19-5615-7_11

1 Introduction

In a connected vehicle environment, vehicles can communicate with vehicles and infrastructures by using wireless communication technology (such as dedicated short-range communication). A connected vehicle can obtain other connected vehicles' real-time information (such as position and speed) and some road traffic information (such as traffic lights, and traffic accidents) in a certain range. Using this information, drivers can adopt appropriate driving strategies (such as acceleration or deceleration) in advance which may reduce congestion and delays, reduce energy consumption and emissions, and improve traffic safety at intersections. Traditional intersection management strategies focused on optimizing the signal timing scheme [1–4].

In this paper, we study how to use real-time traffic signal and vehicle's position and speed obtained by infrastructure to vehicle communication to design speed guidance strategy for the vehicle entering intersections, the following literature review focuses on the study of vehicle speed guidance.

1.1 Overview of Dynamic Speed Control

Abu-Lebdeh and Chen [5] considered the availability of wireless communication information data and proposed a dynamic speed control (DSC) system that can improve traffic efficiency and traffic safety. Chen et al. [6] took the measure of the dynamic speed limit of vehicles and optimize the signal timing scheme based on data collected by vehicles.

1.2 Overview of Dynamic Speed Guidance

Later researchers use vehicles' data and control parameters of traffic signals of intersections to design speed guidance strategies. Yang et al. [7] used the vehicle's space–time trajectory and signal timing parameters to design a speed guidance strategy based on vehicle infrastructure integration (VII). Sun and Chen [8] designed the DSG strategy based on the rolling horizon approach. He [9] used real-time platform data and current traffic control states and proposed a mixed-integer linear program (MILP) model to optimize the signal timing scheme. Cai et al. [10] proposed a no-signal cooperative optimization control method based on speed guidance and information interaction. Khondaker and Kattan [11] designed a variable speed limit algorithm in a connected vehicle environment, results showed high penetration rate of the connected vehicle can more effectively improve mobility and safety and reduce fuel consumption.

1.3 Overview of Using Connected Vehicle to Design Traffic Control Strategy

More and more studies are focused on the use of connected vehicles to design traffic control strategies. Feng and Head [12] used real-time position and velocity information of the connected vehicle to design an adaptive signal control method and optimize phase sequence and duration to reduce the delay or queue length. Gular et al. [13] proposed an optimization algorithm for one-way intersections. Argote-Cabañeroet al. [14] studied the effect of the penetration rate of the connected vehicle on arterial traffic optimization strategy. Zhao and Li [15] studied the car-following model with consideration of speed guidance in a connected vehicle environment.

1.4 Overview of Speed Guidance in Multi-Intersection

Speed guidance in multi-intersection can better improve the efficiency of vehicles traveling at several intersections than it in one intersection. Yang et al. [16] designed a speed guidance strategy for vehicles traveling several intersections without stopping, results showed that the strategy can reduce vehicle travel time. Wu, He, Yu, et al. [17] tested the proposed optimal velocity control model at arterial traffic with six intersections. Zhang, Wei and Li [18] dynamically optimized the signal control parameters and speed guidance and simulated them at multiple intersections. Wan et al. [19] proposed a driving strategy in a connected vehicle environment using the information of traffic signals with fixed timing to minimize, results showed the strategy can improve the connected vehicle's fuel economy as well as be a benefit to conventional vehicles.

However, present studies have several limitations: (i) Most studies of speed guidance focused on an isolated intersection, there are limited researchers who studied speed guidance strategy for multi-intersection. (ii) The most common intersection management strategy is traditional signal timing optimization. Few studies have taken the connected vehicle environment into consideration to design speed guidance strategies. (iii) Present simulation scenario of speed guidance in a connected vehicle environment is created with VISSIM or numerical simulation which cannot take effect of many communication factors (such as shadowed losses of communication and communication protocol) on speed guidance into consideration in infrastructure to vehicle communication. Above limitations motivate our research, in this paper, we design a cooperative speed guidance strategy for multi-intersection in a connected vehicle environment, vehicles entering intersections can obtain real-time traffic lights information and a speed advisory system in the vehicle can combine real-time information of vehicle's position, speed and signal information to produce an optimal speed for the vehicle to reduce delay or stopping time.

The rest of this paper is organized as follows: In Sect. 2, we proposed a cooperative speed guidance algorithm for vehicles traveling multi-intersection in the connected vehicle environment. In Sect. 3, a simulation scenario of speed guidance with consideration of infrastructure to vehicle communication is created, followed by simulation results and analyses in Sect. 4. Section 5 presents conclusions and future work.

2 Cooperative Speed Guidance Algorithm of Multi-Intersection

2.1 Principle of Cooperative Speed Guidance of Multi-Intersection

The simulation scenario is established with Estinet software which can realize the bidirectional coupling between the traffic simulator and communication simulator. The principle of speed guidance strategy built-in Estinet software is described as follows: Signal Agent (SA) controls traffic lights and sends its' real-time state to Roadside Unit Agent (RSUA) which can send it received information to Car Agent (CA) and CA can calculate an optimal speed with combining received traffic information and vehicle's current speed and position and drive vehicle passes through the intersection.

When the vehicle enters a certain range of the next intersection, if it can pass through the intersection with the allowed maximum speed without stopping and waiting for the green light, the speed advisory system does not guide the vehicle (avoiding a frequent change in vehicle speed), otherwise, the speed advisory system produces an optimal speed using real-time information like it in the first intersection, and vehicle passes through the intersection at the guidance speed. Figure 1 presents the principle of multi-intersection cooperative speed guidance.

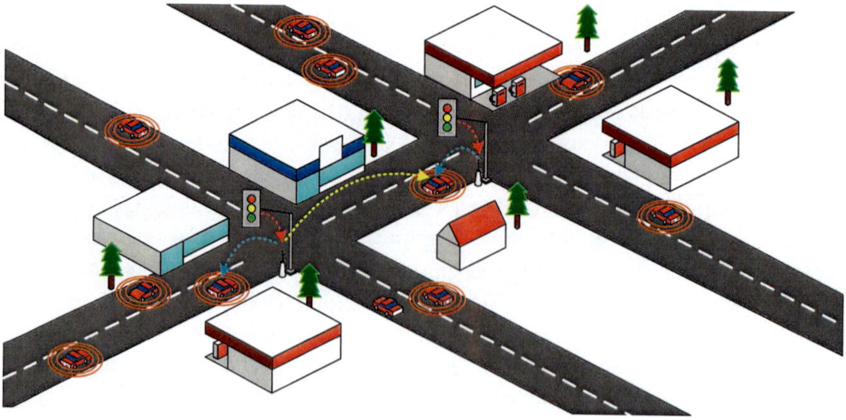

Fig. 1 Principle of multi-intersection cooperative speed guidance

2.2 Notations

t_d	driver's response time (s)
$a_{\max}$	comfortable acceleration (m/s^2)
$a_{\min}$	comfortable deceleration (m/s^2)
$v_{\max}$	maximum limit speed of road (m/s)
g	duration of green light (s)
r	total duration of red and yellow lights (s)
I	the ith intersection that vehicle goes through
v_{0i}	vehicle's speed when vehicle enters the range of speed guidance of the ith intersection (m/s)
t_{0i}	time when vehicle enters the range of speed guidance of the ith intersection (s)
$Light_i$	color of traffic light when vehicle enters the range of speed guidance of the ith intersection (m/s)
t_{gi}	rest time of green light of the ith intersection (s)
t_{ri}	rest time of red light of the ith intersection (s)
d_{0i}	distance to stop line when vehicle enters the range of speed guidance of the ith intersection (m)
$v_{\max i}$	85% of maximum limit speed of road (m/s)
$v_{\min i}$	15% of maximum limit speed of road (m/s)
v_{sugi}	suggested speed of vehicle at the ith intersection calculated by speed guidance algorithm (m/s)
$t_{\min i}$	the earliest time that vehicle can pass through the ith intersection (s)
t_{sugi}	length of time that vehicle need to pass through the ith intersection with suggested speed v_{sugi} (s)
t_{ei}	time that vehicle pass through the ith intersection with suggested speed v_{sugi} (s)
c_i	number of speed guidance at the ith intersection
a	acceleration or deceleration of vehicle (m/s^2)
x_j	x coordinate of upstream intersection j (m)
y_j	y coordinate of upstream intersection j (m)
v_j	speed of vehicle leaving upstream intersection j (m/s)
C	cycle time of signal of downstream intersection i (s)
d_i	distance that vehicle accelerates from v_j to $v_{\max}$ (m)
c_{ij}	number of cooperative speed guidance at the ith intersection
t_{ij}	length of time from the jth intersection to the ith intersection (s)
x_i	x coordinate of vehicle (m)
y_i	y coordinate of vehicle (m)
d_{ij}	distance between vehicle and upstream intersection j (m)

According to study of Yang et al. [7], the range of v_{sugi} is set between 15 and 85% of the maximum limit speed of the road to ensure the safety of the intersection, that is

$$0.15v_{\max} = v_{\min i} \leq v_{sugi} \leq v_{\max i}=0.85v_{\max} \tag{1}$$

2.3 *Cooperative Speed Guidance Algorithm of Multi-Intersection*

When the vehicle enters the range of speed guidance, the vehicle uses the received information to distinguish the color of the traffic light and develops a strategy of speed guidance. In this paper, we incorporate the time of yellow light into the time of red light for simplification. In the first intersection, schematics of speed guidance are shown in Fig. 2a and b, and the flowchart of the speed guidance algorithm is shown in Fig. 3a.

(1) State of the traffic light is the initial phase of green light or the late phase of red light when the vehicle enters the range of speed guidance, the vehicle can accelerate and pass through the intersection. A schematic of this kind of speed guidance in the first intersection is shown in Fig. 2a.

A suggested speed v_{sugi} can be calculated supposing the vehicle can pass through the intersection at the earliest time $t_{\min i}$, if v_{sugi} meets the requirement of Eq. (1) which means it is feasible that vehicle passes through the intersection at $t_{\min i}$, with the suggested speed v_{sugi}. Otherwise, add $t_{\min i}$ one minute by one minute to t_{sugi} and calculate the corresponding v_{sugi} till it meets the requirement of Eq. (1) and the vehicle passes through intersection with the new suggested speed v_{sugi}, otherwise, the speed guidance of vehicle has to be given up.

(2) State of the traffic light is the late phase of the green light or initial phase of red light when the vehicle enters the range of speed guidance, the vehicle needs to decelerate and pass through the intersection in the next green phase. A schematic of this kind of speed guidance in the first intersection is shown in Fig. 2b.

A suggested speed v_{sugi} can be calculated supposing the vehicle can pass through the intersection at the earliest time $t_{\min i}$ in the next green phase, if v_{sugi} meets the requirement of Eq. (1) which means it is feasible that the vehicle need decelerates and passes through the intersection at $t_{\min i}$, with the suggested speed v_{sugi}. Otherwise, add $t_{\min i}$ one minute by one minute to t_{sugi} and calculate the corresponding v_{sugi} till it meets the requirement of Eq. (1) and the vehicle passes through intersection with the new suggested speed v_{sugi}.

Calculation of suggested speed includes the following processes:

The driver has a response time t_d after he receives a suggested speed before he begins to accelerate or decelerate, traveling distance of the vehicle in the response time t_d is:

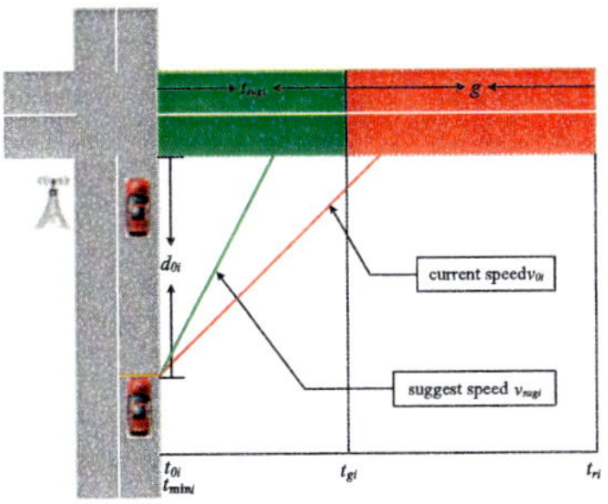

(a) State of the traffic light is the initial phase of the green light or the late phase of the red light when the vehicle enters the range of speed guidance at the first intersection

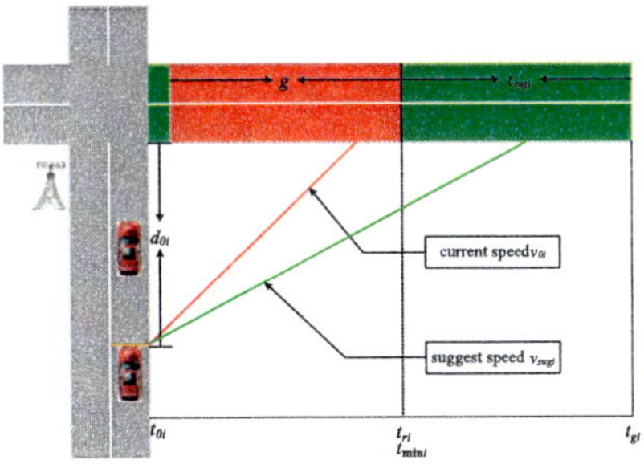

(b) State of the traffic light is the initial phase of the red light or the late phase of the green light when the vehicle enters the range of speed guidance at the first intersection

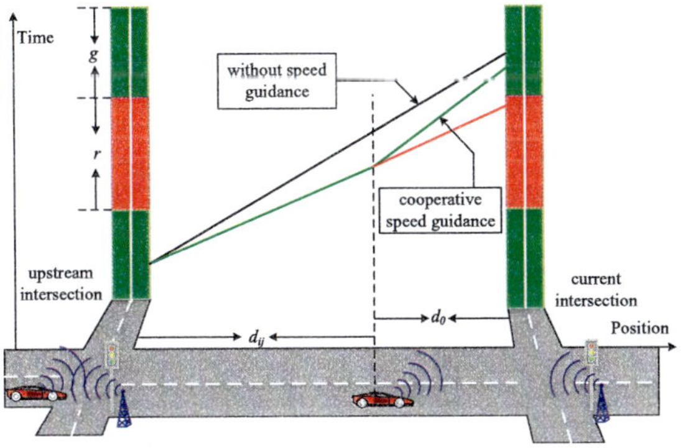

c) Cooperative speed guidance at two intersections

Fig. 2 Schematics of speed guidance

$$S_1 = v_{0i} \cdot t_d \tag{2}$$

Distance and length of time (t_1) in the process of vehicle accelerates or decelerates to suggest speed v_{sugi} meet the following equations:

$$S_2 = \frac{v_{sugi}^2 - v_{0i}^2}{2a} \tag{3}$$

$$v_{sugi} = v_{0i} + at_1 \tag{4}$$

Time (t_2) that vehicle need cost to pass through intersection with the suggested speed meets the following equation:

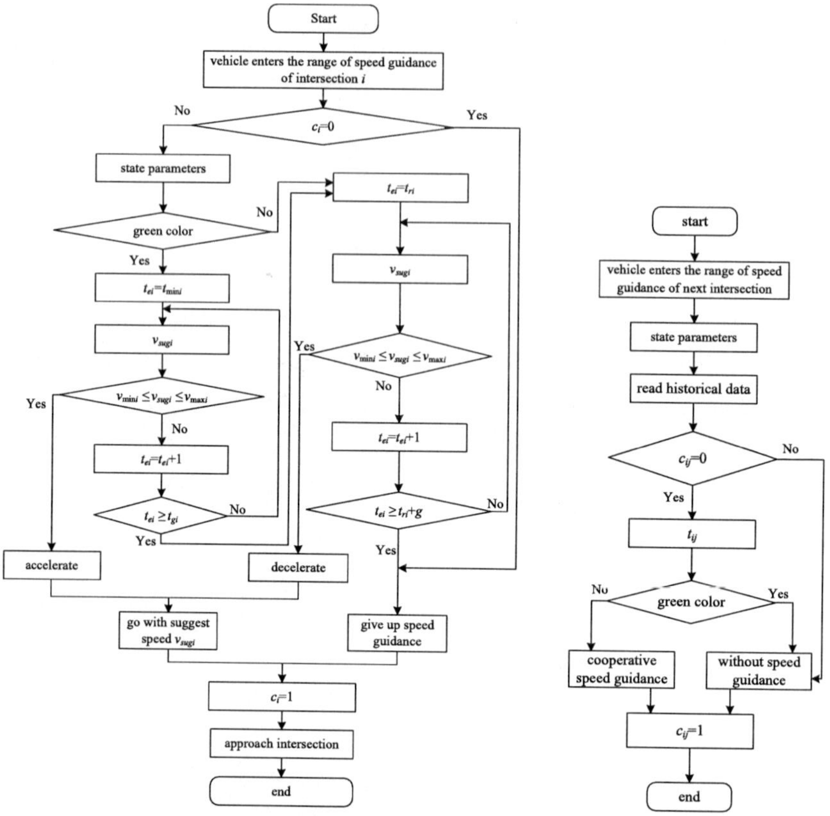

a) Speed guidance algorithm at the first intersection

b) Cooperative speed guidance algorithm at the next intersection

Fig. 3 Flowchart of speed guidance algorithm

$$d_{0i} - S_1 - S_2 = v_{sugi} \cdot t_2 \tag{5}$$

t_{ei} is time that the vehicle passes through the intersection with suggested speed, and it is sum of t_d, t_1 and t_2, and it meets the following equation:

$$t_{ei} \cdot v_{sugi} = d_{0i} + t_d\left(v_{sugi} - v_{0i}\right) + \frac{\left(v_{sugi} - v_{0i}\right)^2}{2a} \tag{6}$$

Suggest speed (need to accelerate or decelerate respectively) can be calculated by Eqs. (7) and (8):

$$v_{sugi} = [v_{0i} + a_{\max i}(t_{ei} - t_d)] - \sqrt{a_{\max i}^2 (t_{ei} - t_d)^2 + 2a_{\max i} v_{0i}(t_{ei} - t_d) - 2a_{\max i}(d_{0i} - v_{0i}t_d)} \quad (7)$$

$$v_{sugi} = [v_{0i} + a_{\min i}(t_{ei} - t_d)] + \sqrt{a_{\min i}^2 (t_{ei} - t_d)^2 + 2a_{\min i} v_{0i}(t_{ei} - t_d) - 2a_{\min i}(d_{0i} - v_{0i}t_d)} \quad (8)$$

Note that $t_{sugi} \in [t_{\min i}, t_{gi}]$ in the state of the traffic light is the initial phase of the green light or the late phase of the red light and $t_{sugi} \in [t_{\min i}, t_{gi}] \cup [t_{ri}, t_{ri} + g]$ in the state of the traffic light is the late phase of the green light or the initial phase of the red light.

In next intersection, cooperative speed guidance is conducted. Schematics of cooperative speed guidance are showed in (c) of Fig. 2, flowchart of speed guidance algorithm is showed in (b) of Fig. 3.

When vehicle enters range of next intersection, distance between vehicle and upstream intersection is:

$$d_{ij} = \sqrt{(x_i - x_j)^2 + (y_i - y_j)^2} \quad (9)$$

The distance that the vehicle needs to accelerate from the vehicle's current speed to the maximum limit speed of the road is:

$$d_i = (v_{\max i}^2 - v_j^2)/2 \cdot a_{\max} \quad (10)$$

If the value of $d_{ij} + d_O$ is less than d_i, it means the vehicle cannot accelerate with max acceleration to $v_{\max}$, so the vehicle should accelerate with comfortable acceleration. The time that the vehicle needs to arrive at the next intersection is:

$$t_{ij} = \left(\sqrt{2 \cdot a_{\max} \cdot (d_{ij} + d_{0j}) + v_j^2} - v_j\right)/a_{\max} \quad (11)$$

If value of $d_{ij} + d_O$ is greater than d_i, it means the vehicle can accelerate with max acceleration to $v_{\max}$. The time that the vehicle needs to arrive at the next intersection is

$$t_{ij} = \frac{v_{\max i} - v_j}{a_{\max}} + \frac{d_{0j} + d_{ij} - \frac{v_{\max i}^2 - v_{0i}^2}{2 \cdot a_{\max}}}{v_{\max i}} \quad (12)$$

The state of the traffic light can be judged using the time that the vehicle needs to arrive at the next intersection and the cycle time of the signal of the downstream intersection. If the color of the traffic light is green when the vehicle arrives at the next intersection, second speed guidance will not be conducted. Otherwise, if the color of the traffic light is red when the vehicle arrives at the next intersection, a suggested speed will be produced using a speed guidance algorithm like in the first intersection to guide the vehicle passing through the intersection.

3 Simulation Scenario

The simulation scenario is created with Estinet software which can realize the bidirectional coupling between the traffic simulator and communication simulator. It can simulate vehicles running in a connected vehicle environment. The simulation system of Estinet contains Car Agent (CA), Signal Agent (SA) and Roadside Unit Agent (RSUA). CA acts as an agent of the vehicle and controls vehicle mobility, it can receive information set from RSUA and produce suggest speed. SA is the agent of traffic signals and it can control the working of traffic lights and send its' information to RSUA. There are communication protocols and communication models in RSUA.

Three intersections in East Second Ring Road of Beijing are selected to build a simulation scenario of cooperative speed guidance. Figure 4 presents the road network in the simulation scenario, and Table 1 presents the parameters of the simulation scenario.

Compliance rate (CR) is the rate of vehicle running with suggest speed among all connected vehicles. Penetration rate (PR) is the rate of connected vehicles among all vehicles (including a conventional vehicle that has no equipment of vehicle to

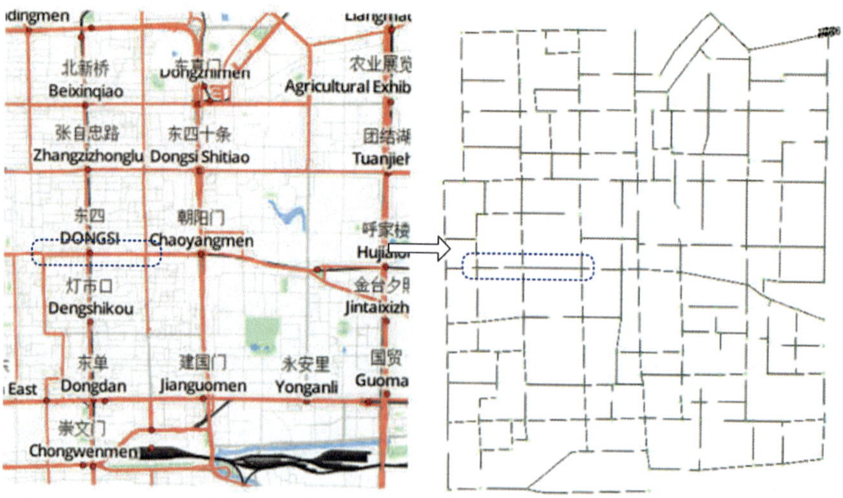

Fig. 4 Road network in simulation scenario

Table 1 Parameters of simulation scenario

Parameters	Value	Parameters	Value
Number of vehicles	100	Comfortable deceleration (m/s^2)	1
Initial headway space (m)	100	Reaction time (s)	1
Number of lanes	2	Distance between intersections (m)	736,574
Width of lane (m)	4	Timing of north–south	Red light 30 s; green light 27 s; yellow light 3 s
Maximum limit speed of road v_{max} (m/s)	18	Timing of east–west	green light 27 s; yellow light 3 s; red light 30 s
Maximum acceleration (m/s^2)	2.5	Penetration rate (PR)	0%, 25%, 50%, 75%, 100%
Minimum deceleration (m/s^2)	4	Compliance rate (CR)	25%, 50%, 75%, 100%
Comfortable acceleration (m/s^2)	2.5	Mac communication protocol	ieee802.11p

infrastructure communication). Every simulation experiment runs 1000 s, and state parameters (speed, position) of the vehicle are recorded every 0.1 s.

4 Simulation Result

4.1 Strategies of Cooperative Speed Guidance and Isolated Speed Guidance

Comparisons of different scenarios are conducted from efficiency (average speed, AS; idle period, IP; delay time, DT), safety (acceleration interference, AI) and emission (E). The idle period is the total time of vehicle stopping (speed equals 0). Delay Time can be calculated as Eq. (13):

$$DT = \sum_{i=1}^{n} \left[\left(t_i^B - t_i^A \right) - t_f \right] / n \tag{13}$$

where, DT is delay time, t_f is the time of vehicle passing intersection in state of free traffic flow (s), $t_i{}^B$ is the time that vehicle i leaving 100 m of downstream of last intersection (s), $t_i{}^A$ is the time that vehicle i entering 100 m of upstream of the first intersection (s), n is the number of vehicles.

AI can reflect frequency of the vehicle accelerating or decelerating and characterize traffic safety. Calculation of AI is presented as Eq. (14) which is in Jones and Potts's research [20].

$$AI = \sqrt{\frac{1}{T}\int_0^T (a_t - \overline{a})^2 dt} = \sqrt{\frac{1}{T}\sum_{t-0}^{T} (a_t - \overline{a})^2 dt} \tag{14}$$

where, AI is acceleration interference, T is the total travel time of the vehicle, t is the unit time (s); a_t is the vehicle's acceleration at time t (m/s^2), $\overline{a}$ is average acceleration (m/s^2). The greater of AI, the worse traffic safety.

Emission contains HC, CO and NO_x, and the calculation of emission is presented as Eq. (15) which is in Shi's research [21].

$$E = \sum_{j=1}^{n}\left(\frac{1}{3600}\sum_{i=1}^{m} T_S \times E_{Ii} + \sum_{t=0}^{T}\sum_{i=1}^{m}\frac{L_g}{1000} \times E_i\right) \tag{15}$$

where E is the total emission (g), n is the number of vehicles, T_s is the total idling time of the vehicles (s), E_{Ii} is the idling emission factor of emission i, L_g is the vehicle's driving distance in unit time (m), E_i is driving emission factor of emission i (g/veh·km), T is total time except idling time (s), m is the number of emission's kind (m equal 3 in this paper). Table 2 represents idling emission factor and the relationship between driving emission factor and speed.

Three simulation cases that are scenarios without speed guidance, with isolated speed guidance and cooperative speed guidance were conducted, and the results are presented in Table 3.

Table 3 shows improvements in cooperative speed guidance and isolated speed guidance compared to without speed guidance in aspects of efficiency, safety, and emission. Cooperative speed guidance is better than isolated speed guidance. For efficiency, AS increases about 13.36%, DT and IP reduce about 51.02% and 65.79% respectively compared without speed guidance. As for emission, a 7.07% reduction in emission, as for safety, a 6.23% increase of AI of cooperative speed guidance shows more frequent acceleration and deceleration which are bad for safety were caused by cooperative speed guidance.

Figure 5 shows the time–space trajectory of the vehicle passing three intersections with cooperative speed guidance and without speed guidance.

As shown in Fig. 5, the travel time of vehicles without speed guidance is 186.5 s, cooperative speed guidance is 127.5 s. Δt is 59 s, indicating that compared with the vehicle without speed guidance, the cooperative speed guidance for the vehicle passing three intersections can save 59 s travel time. On the other hand, the

Table 2 Relationship between emission factor and speed

Emission	Idling emission factor (g/(veh·h))	Driving emission factor E_i (g/(veh·km)) calculated by vehicle speed V
HC	18.83	$E_1 = 0.0011\ V^2 – 0.14\ V + 5.84$
COn	105.03	$E_2 = 0.0064\ V^2 – 0.63\ V + 29.72$
NO_X	9.57	$E_3 = 0.0006\ V^2 – 0.06\ V + 2.84$

Table 3 Comparison of efficiency, safety and emission under different scenarios

	Scenarios	Without speed guidance	isolated speed guidance	cooperative speed guidance
Efficiency	*AS*	8.297	9.085	9.405
	Improvement	–	9.50%↑	13.36%↑
	IP	20.254	10.740	9.921
	Improvement	–	46.97%↓	51.02%↓
	DT	33.111	12.360	11.328
	Improvement	–	62.67%↓	65.79%↓
Emission	*E*	56.518	53.012	52.520
	Improvement	–	6.20%↓	7.07%↓
Safety	*AI*	0.477	0.513	0.507
	Improvement	–	7.53%↑	6.23%↑

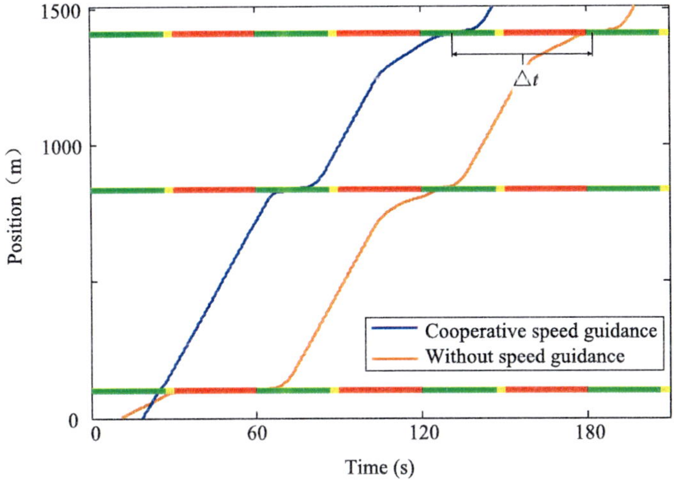

Fig. 5 Time–space trajectory of vehicle

vehicle with cooperative speed guidance has no stop, however, vehicle without speed guidance stops for a long time.

4.2 Sensitivity Analysis

Figure 6 presents the effects of different percentages of connected vehicles and compliance rates on efficiency, safety, and emission.

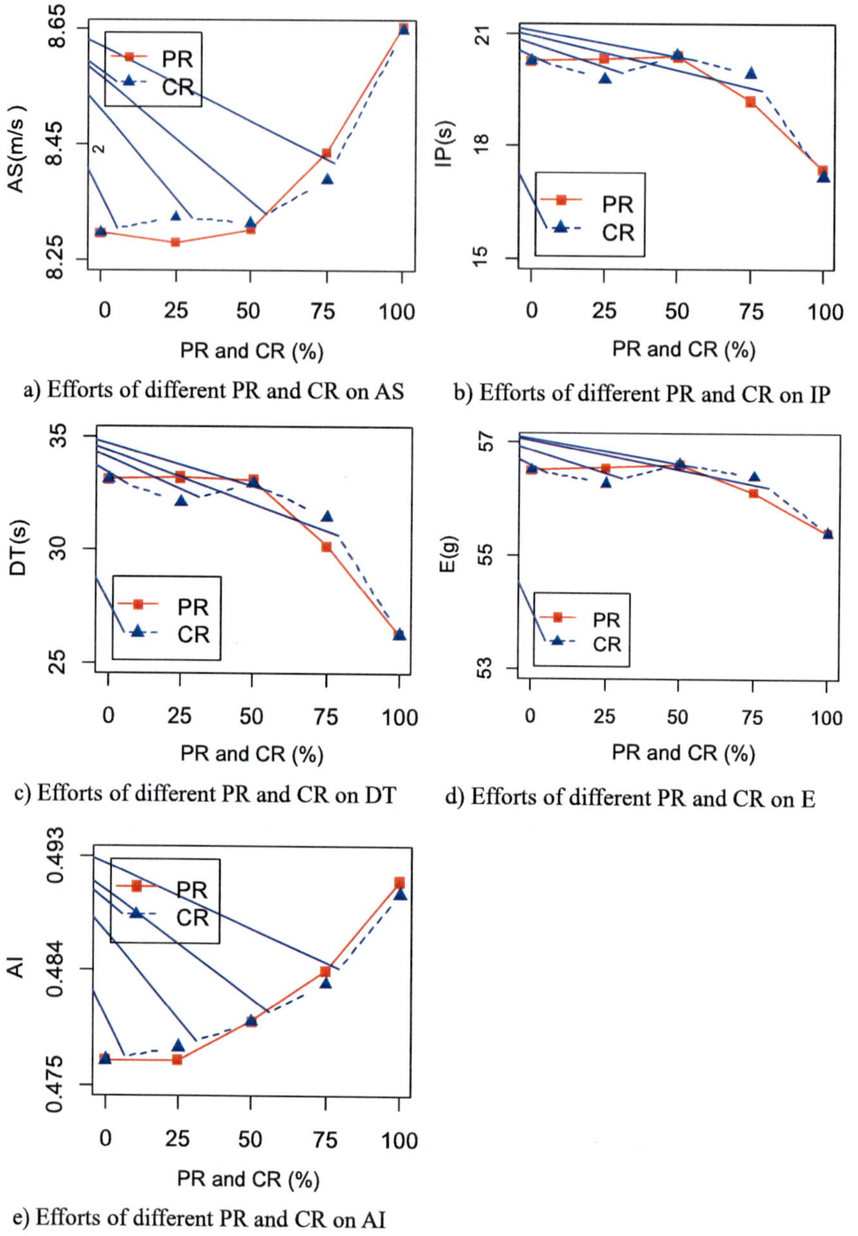

a) Efforts of different PR and CR on AS

b) Efforts of different PR and CR on IP

c) Efforts of different PR and CR on DT

d) Efforts of different PR and CR on E

e) Efforts of different PR and CR on AI

Fig. 6 Effects of different PR and CR on efficiency, safety and emission

As shown in Figs. 6a, b and c when PR and CR are less than 50%, improvements in the efficiency of cooperative speed guidance are not obvious. However, when PR and CR are greater than 50%, improvements in the efficiency of cooperative speed guidance are obvious and improvements increase with PR and CR. As shown in Fig. 6d, when PR and Cr are greater than 50%, E decreases significantly with the increase of PR and CR. But the improvement of PR and CR was not obvious when they were less than 50%. As shown in Fig. 6e, AI increases with the increase of PR and CR, which means a connected vehicle with cooperative speed guidance changes its' speed more frequently than a conventional vehicle without speed guidance. Improvement of vehicles with cooperative speed guidance is more obvious when PR and CR are greater than 50%.

5 Conclusions

In this paper, we design a cooperative speed guidance algorithm using real-time information on the traffic lights, position and speed of the vehicle to calculate the optimal suggested speed to reduce delay or stopping time. The proposed algorithm is beneficial for the vehicle cooperatively passing through multi-intersection. Compared to isolated speed guidance and without speed guidance, cooperative speed guidance can improve traffic efficiency (13.36% increase in average speed and 65.79% reduction of average delay) and reduce emission (7.09% reduction of emission). However, frequent speed change reduces vehicle safety slightly, but it is within acceptable limits. Low penetration rate (<50%) of connected vehicles and low compliance rate (<50%) of suggested speed from heterogeneous vehicles (connected vehicles are less than conventional vehicles) cause only little improvements. Improvements with a high penetration rate (<50%) and high compliance rate (<50%) are obvious. This paper did not take driver's behavior into consideration in designing the strategy of speed guidance and this is our future research.

References

1. Wong CK, Wong SC, Hong KL (2010) A spatial queuing approach to optimize coordinated signal settings to obviate gridlock in adjacent work zones[J]. J Adv Transp 44(4):231–244
2. Xuan Y, Daganzo CF, Cassidy MJ (2011) Increasing the capacity of signalized intersections with separate left turn phases[J]. Transp Res Part B Methodol 45(5):769–781
3. Skabardonis A, Geroliminis N (2008) Real-time monitoring and control on signalized arterials[J]. J Intell Transp Syst 12(2):64–74
4. Liu Y, Chang GL, Yu J (2011) An integrated control model for freeway corridor under nonrecurrent congestion[J]. IEEE Trans Veh Technol 60(4):1404–1418
5. Abulebdeh G, Chen H (2010) Exploring the potential benefits of intellidrive-enabled dynamic speed control in signalized networks[C]. Transportation Research Board 89th Annual Meeting, Washington, D.C., pp 1–19

6. Chen S, Sun J, Yao J (2011) Development and simulation application of a dynamic speed dynamic signal strategy for arterial traffic management[C]. Int IEEE Conf Intell Transp Syst. IEEE 1349–1354
7. Yang Y, Chen S, Sun J (2010) Modeling and evaluation of speed guidance strategy in VII system[C]. International IEEE Conference on Intelligent Transportation Systems. IEEE 1045–1050
8. Sun J, Chen S (2013) Dynamic speed guidance for active highway signal coordination: roadside against in-car strategies[J]. IET Intel Transport Syst 7(4):432–444
9. He Q (2010) Robust-intelligent traffic signal control within a vehicle-to-infrastructure and vehicle-to-vehicle communication environment[C]. University of Arizona, pp 101–142
10. Cai B, Zheng Z, Shangguan W, et al (2014) Unsignalized cooperative optimization control method based on vehicle speed guidance and information interaction[C]. International IEEE Conference on Intelligent Transportation Systems. IEEE, pp 57–62
11. Khondaker B, Kattan L (2015) Variable speed limit: a microscopic analysis in a connected vehicle environment[J]. Transp Res Part C: Emerg Technol 58:146–159
12. Feng Y, Head KL, Khoshmagham S et al (2015) A real-time adaptive signal control in a connected vehicle environment[J]. Transp Res Part C: Emerg Technol 55:460–473
13. Guler SI, Menendez M, Meier L (2014) Using connected vehicle technology to improve the efficiency of intersections[J]. Transp Res Part C: Emerg Technol 46(46):121–131
14. Argote-Cabañero J, Christofa E, Skabardonis A (2015) Connected vehicle penetration rate for estimation of arterial measures of effectiveness[J]. Transp Res Part C: Emerg Technol 60:298–312
15. Zhao J, Li P (2016) An extended car-following model with consideration of speed guidance at intersections[J]. Physica A 461:1–8
16. Yang B, Wang Y, Hu J, et al (2013) A traffic efficiency promotion algorithm for urban arterial roads based on speed guidance[C]. International Conference on Connected Vehicles and Expo, pp 869–873
17. Wu X, He X, Yu G et al (2015) Energy-optimal speed control for electric vehicles on signalized arterials[J]. IEEE Trans Intell Transp Syst 16(5):2786–2796
18. Zhang Y, Wei WU, Li P (2015) Modelling and simulation of vehicle speed guidance in connected vehicle environment[J]. Int J Simul Model 14(1):2015
19. Wan N, Vahidi A, Luckow A (2016) Optimal speed advisory for connected vehicles in arterial roads and the impact on mixed traffic[J]. Transp Res Part C: Emerg Technol 69:548–563
20. Jones TR, Potts RB (1962) The measurement of acceleration noise-a traffic parameter[J]. Oper Res 10(6):745–763
21. Jihong S (2012) Analysis and comprehensive evaluation of traffic pollution at urban road intersection[D]. Jilin University

Pattern Mining and Predictive Inference on Short-Term Weather and Collision Time Series Data

Yongsheng Chen and Chuanjiao Sun

Abstract Collisions are rare random events. Many traffic safety indexes with a small-sized temporal or spatial unit, e.g., daily collisions of a city or a regional highway network, are highly random and fiercely fluctuated. The descriptive and inferential analyses for this type of short-term collision time series data, abbreviated as SCTS data in this paper, are still not well-established yet. This paper is to tackle this issue by a newly emerging approach—pattern mining combined with data mining methods. Based on a collision database, calendar information, and historical weather records, the approach of descriptive statistics was employed to illustrate correlations between all data items and to identify main affecting factors for a SCTS response, with respective to single variable pattern, variable pair and multiple variable correlations. Then the structure and flow-chart of the major attributes led to different SCTS outputs were further investigated by means of decision tree method. The established decision tree structure was then utilized to predict SCTS values of future days as consequence from their calendar characters and weather forecasts. The approaches of description and inference of SCTS data developed in this paper filled in the methodological vacancy of discovering SCTS data pattern and to infer their attributes. The study of this paper also provided a viable solution to predict SCTS and therefore help to pre-schedule safety countermeasures for practitioners.

Keywords Short-term collision · Time series data · Descriptive statistics · Decision tree

1 Introduction

Collisions are rare random events. In contrast with long-range temporal and/or large-scale spatial accumulations, the measurement of collisions on a particular location during a limited time span is supposed to be short-term collision. Some examples

Y. Chen (✉) · C. Sun
Ministry of Transport, Road Safety Research Center of Research Institute of Highway, No. 8 Xitucheng Road, Haidian District, Beijing 100088, China
e-mail: yongsheng.chen@hotmail.com

W. Wang et al. (eds.), *Green Transportation and Low Carbon Mobility Safety*, Lecture Notes in Electrical Engineering 944,
https://doi.org/10.1007/978-981-19-5615-7_12

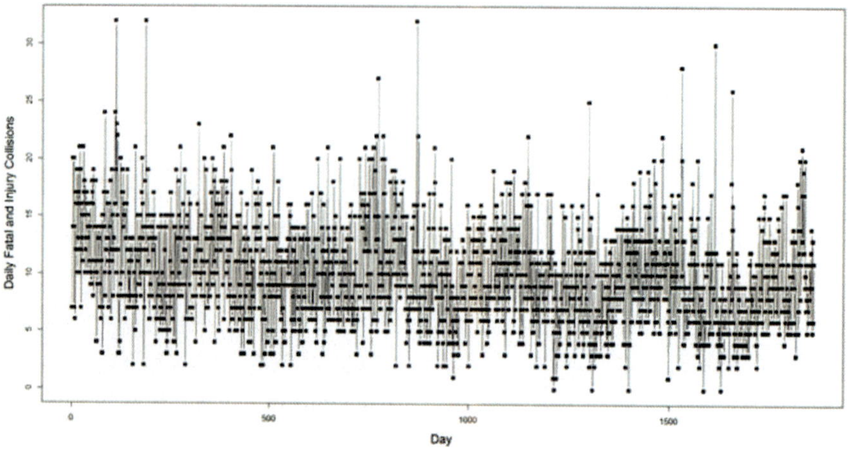

Fig. 1 Plot of daily fatal and injury collisions of the sample city

include daily fatal and injury collisions of a city, annual collisions of a neighborhood, the run-off road collisions for a certain stretch of rural two-lane highway and so on. The sequence of observations of this type of collision data, which are ordered in time (or space), forms the short-term collision time series data [10], abbreviated as SCTS data in this paper.

Typically, towards any dataset we can conduct either descriptive or inferential statistics. Descriptive statistics is used to consolidate a large amount of information [15]. This type of analysis won't be easy for the case of SCTS data. Figure 1 shows scatter plot of the "daily fatal and injury collision (FI)" of the sample city—City of Edmonton, Alberta, Canada. We can see that the observations fluctuate between 0 and 25 (with few outliers up to 30) but there is no pattern perceivable from the figure, neither character nor profile.

Meanwhile, the inferential analysis, i.e., the analysis to draw any conclusion, turns to be even tougher for SCTS data. Due to its nature of being random and small-numbered, the values of a SCTS variable are randomly fluctuating while being limited to a narrow band, causing difficulty to be attributed to particular factors.

Conventional traffic safety analysis methodology, such as the best established safety performance model form—generalized linear model (GLM)—which was originally introduced by Dr. Ezra Hauer [5, 6] and further developed by many other researchers, is good at estimating and predicting long-term aggregated collisions, other than a SCTS. Even some newly emerging methods, such as the generalized estimation equation (GEE) as an extension of the GLM [1], may be better off on aspects such as more complicated and flexible model structures, but still require the response variable has sufficient numbers in order to yield statistical significant (SS) results. That is to say, neither traditional GLM nor some newer GEE efforts are capable to deal with the small number issue.

In contrast, there are some other collision observations, have each smaller numbers. Some examples include daily fatal collision of a city, or daily run-off road collisions of a certain stretch of highway. They are either binary variable, i.e. 0/1, or categorical variable with numbers 0, 1, 2, 3, etc. Logit model is by default the conventional solution to identify, quantify the affecting factors and to estimate the probability of the occurrence of binary or categorical variable [17].

The Logit model may be transferable to estimate SCTS data however there are serious issues. First of all, SCTS is not small enough compared to the values of a typical response variable of Logit model. Over 25 values for a certain SCTS variable, for example, would cause computing difficulty for the Logit model and the outcomes might be biased; Moreover, too many variables would dampen the efforts to select optimized attributes to get a best Logit model fitting.

In summary, there is a methodological vacancy how to properly describe, analyze, and infer from the SCTS data in order to capture its pattern, affecting factors and foresee its near future trends. This paper is working on filling in this vacancy. The basic idea is to utilize some newly emerging methods in traffic domain, such as data mining to tackle the specific SCTS issues and conduct descriptive and inferential analysis in a proper and efficient way.

2 Sample Data

The daily number of fatal and injury collisions (FI) of the City of Edmonton was selected as the sample SCTS variable and calendar characters, historical weather, and weather forecasts were selected as inputs for this study.

Figure 2 demonstrates four input data sources utilized for this study. Among them, collision data came from Motor Vehicle Collision Information System (MVCIS) [11] for City of Edmonton. Historical weather data was downloaded from the Environment Canada website (Environment [4]. Qualitative weather forecasts were provided by the research team from the University of Alberta, who applied the "Weather Research & Forecasting Model" (WRF) calibrated to the City of Edmonton.

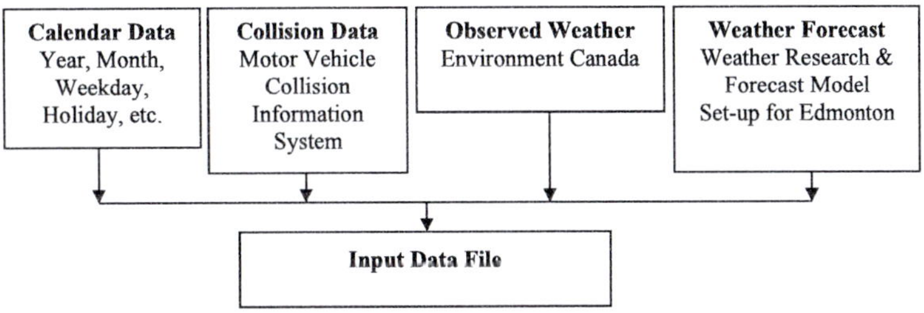

Fig. 2 Input data sources for daily collision inference and prediction

Finally, all data sources were assembled together by date and formed one unified dataset. This dataset was separated into two subsets applied respectively for model fitting and predicting.

3 Data Mining Approaches Selected to Analyze SCTS and Relevant Data

According to Ref. [2], Data mining, also called knowledge discovery in databases, in computer science, is the process of discovering interesting and useful patterns and relationships in large volumes of data. The field combines tools from statistics and artificial intelligence (such as neural networks and machine learning) with database management to analyze large digital collections, known as data sets.

The open source statistical platform R provides a wide variety of data mining tools [14]. Based on the requirement to analyze the SCTS dataset, some appropriate methods of descriptive and inferential statistics were selected in this paper's study. Within descriptive statistics category, data mining methods respectively for single variable, between pairs of variables (i.e., correlation analysis) and for multiple variables (3D analysis in this case) were utilized. For the inferential statistics, this paper introduced method of decision tree.

4 Descriptive Statistics with Data Mining Methods

4.1 Data Mining on Single Variable SCTS Data

SCTS data itself, given a specific collision feature, i.e., FI collision in Fig. 1, is categorized as single variable time series data. The data mining of this type of data is generally to figure out the pattern of this time series data, so we can also call it as pattern mining. The principal method for single variable time series data is data decomposition.

Decomposing a time series means separating it into its components which are usually a trend component and an irregular component, and if it is a seasonal time series, there is another seasonal component [3]. Figure 3 shows how the daily collision data was decomposed into the three components, among them the seasonal part signals this is a seasonal time series data.

This time series comprised of three components as:

$$y_t = T_t + S_t + R_t \tag{1}$$

where,

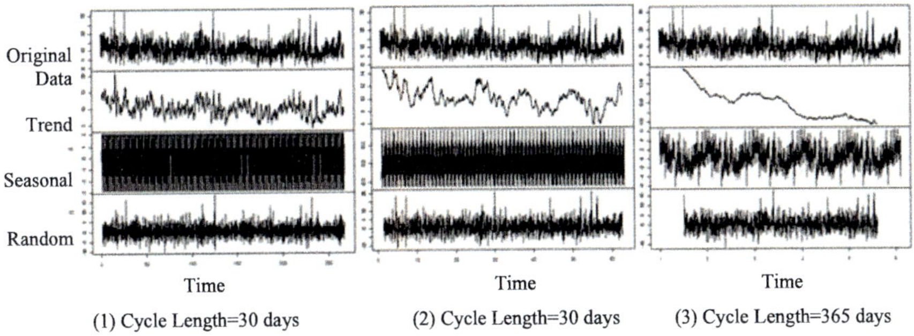

Fig. 3 Time-serial decomposition analysis on daily collision data

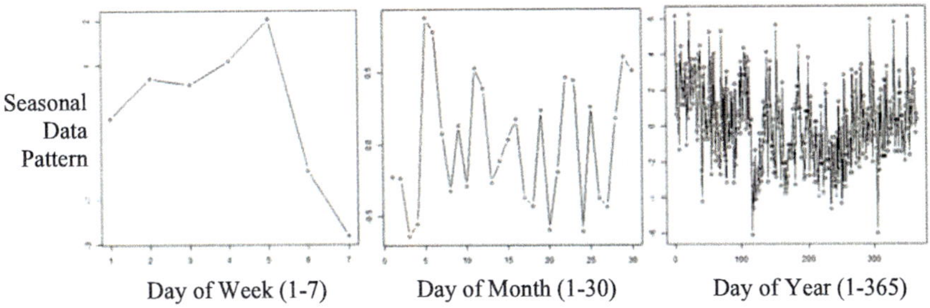

Fig. 4 Seasonality analysis of daily collision data

y_t original daily collision data,
T_t trend component,
S_t seasonal component, and
R_t random/error component.

Further decomposition analysis particularly towards the seasonality, shown as Fig. 4, reveals that the cycle length for daily collision data's seasonality is seven, i.e., it has a recurring seasonal pattern on weekdays. Figure 4 shows daily collisions increase on weekdays, peak on Friday, and then drop down on the weekend; this pattern is recurring every seven days.

4.2 Data Mining on Variable Pairs and Correlation Analysis

Data mining between variable pairs, i.e., correlation analysis, is helpful to identify correlation between outcome—the FI in this study—and its potential explanatory variables (covariates) and/or the correlations between two covariates. The principle is the covariates poised to be introduced into one model shouldn't have strong correlations between each other. That is to say, "collinearity" impact should be avoided among predictors [9]. Besides, correlation analysis may also be used to select potential covariates for next step inferential analysis given some variables show stronger correlation with the response variable.

Since the number of potential affecting factors for FI, in the sample dataset described above, is very high, only those high-dimensional data mining and visualization methods were capable to conduct correlation analysis for this study. The scatter-plot matrix with variables colored by correlation was selected, shown in Fig. 5 [13].

Scatter-plot matrices display relationship between any pair of two data items. From Fig. 5, color codes of the cells reflect the size of the correlations. "Red", "yellow" and "blue" colored cells demonstrate respectively strong, mild and weak correlation between the pair of the two data items. Figure 5 shows that the "Max Temperature", "Min Temperature" and "Mean Temperature" are correlated to each other, with almost absolute linear patterns. Besides, "Rain" and "Precipitation" showed high correlation. And also, the "Snow on Ground" and "Season" are negatively correlated to those three temperature measurements. Except for those above-mentioned covariates, the scatter-plot matrices (Fig. 5) shows that other explanatory variables display no trend between each other, providing evidence of mutual independence [7, 8, 13].

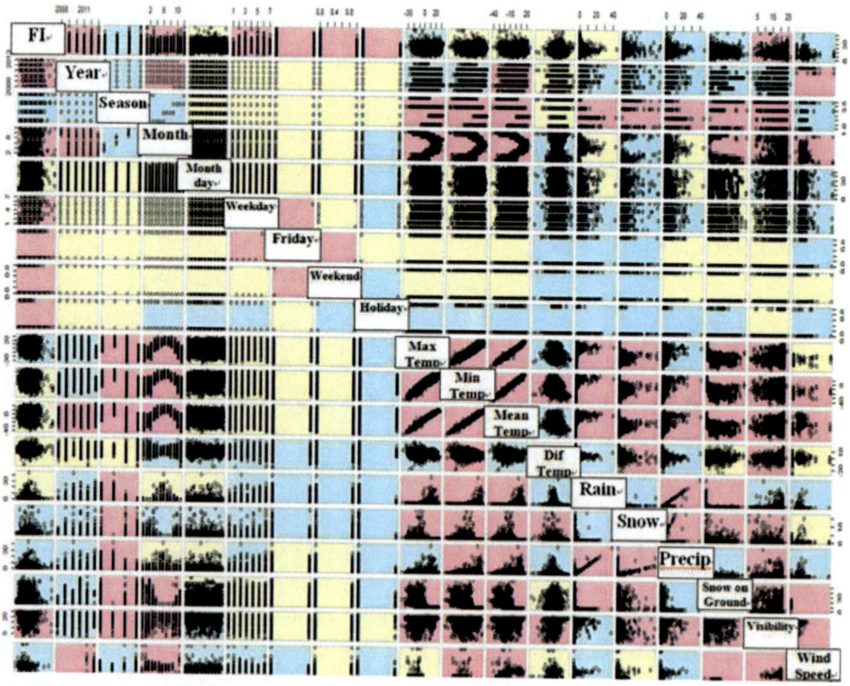

Fig. 5 Scatter-plot matrix with variables colored by correlation. *Note* red color cells—strong correlation; yellow color cells—midl correlation; blue color cells—weak correlation

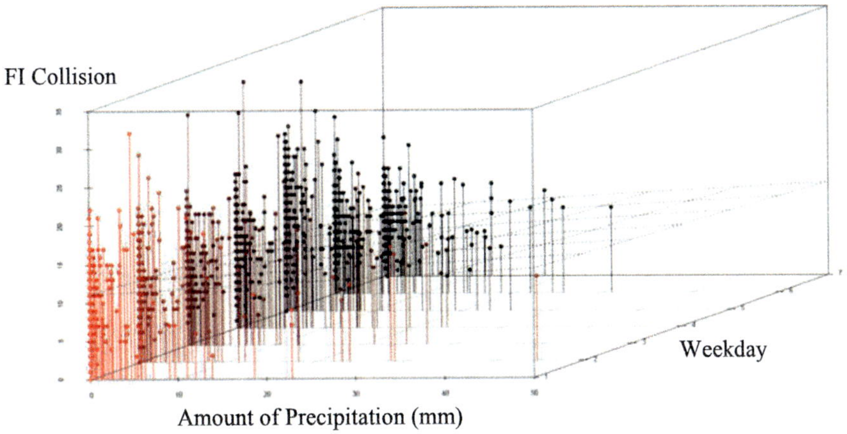

Fig. 6 3D scatter plot with coloring and vertical drop line and regression plane

4.3 Data Mining on Multiple Variables

Figure 6 is 3D scatter plot with coloring and vertical drop line and regression plane (i.e., the plane shows the regressed trend of three-dimensional variables, the extended form in 3D space from regressed curve for two variables in two-dimensional space). Figure 6 reveals relation and interactions between three variables: FI collision, amount of precipitation (mm) and weekday. The pattern depicted by scatter plot, visually augmented by the drop lines and regression plate, highlights the general trend that number of FI collision increases along with the increase of amountof precipitation, increases simultaneously from Monday to Friday, peaks on Friday and then drops during weekends.

4.4 Limitation of Descriptive Statistics

The above descriptive statistics explore patterns out of one data items and describe interactive trends among multiple variables. However there are two limitations: first, descriptive statistics is incapable to reveal the heuristic relationship among several variables; second, it displays visual pattern but cannot draw abstract conclusion. That is to say, we cannot predict by descriptive statistics. To do so, we have to utilize inferential statistics shown below.

5 Inferential Analysis with Decision Tree

5.1 Concept of Decision Tree

Moving forward from above descriptive statistics, the inferential analysis is to be facilitated through other data mining approaches so as to discover how the factors would quantitatively affect the response variable—FI in this study—and more usefully in traffic safety practices, to establish a model or a procedure to predict the values of FI in future.

In this study, decision tree method was utilized for inferential analysis. A decision tree is a decision support tool that uses a tree-like graph or model of decisions and their possible consequences, including chance event outcomes, resource costs, and utility. A decision tree is a flowchart-like structure in which internal node represents a "test" on an attribute, each branch represents the outcome of the test and each leaf node represents a class label (decision taken after computing all attributes). The paths from root to leaf represent classification rules. Decision tree is advanced to tackle the analysis of complex decisions with significant uncertainty, and with a large number of different factors that must be taken into account [12, 16].

5.2 Selection of Attributes as Inputs of the Decision Tree

Based on scatter-plot matrices and 3D scatter plot shown above, we drew some preliminary finding on which factors were correlated to each other and therefore they shouldn't be included into one model or inferential analyses like decision tree as attributes all together. Consider "Max Temperature" and "Min Temperature" were correlated to each other and "Mean Temperature" was function of them, only "Mean Temperature" was included into the decision tree analysis as the representative of the temperature measurement. For same reason, "Precipitation" was excluded if both "Rain" and "Snow" were kept. And also, "Snow on Ground" and "Season" might not be considered into the analysis due to their relatively high correlation with other attributes.

In addition, we further investigated the connections between "Dif Temp" and other temperature measurements, the coverage of "Weekday" on "Friday" and "Weekend". Eventually we eliminated attributes such as "Dif Temp", "Friday", and "Weekend". Due to missing data issue, "Visibility" and "Wind Speed" couldn't be significant in the decision tree analysis and therefore they were later eliminated.

Other attributes shown in Fig. 5, including "Month", "Weekday", "Holiday", "Mean Temp", "Rain" and "Snow" were considered as preliminary inputs of the decision tree analysis.

5.3 Results of Decision Tree Analysis

For the sample dataset described in previous section, decision tree drew a flow-chart diagram to show the paths how the factors affected the outcome and finally determined which values would be for the response variable—FI, as shown as Fig. 7.

For each internal node of Fig. 7, we saw an attribute and the branches below showed different ranges of this attribute, which led to different outcomes. Finally, each leaf of the bottom layer represented one classification consequence of the FI.

The outcome of this decision tree analysis, may not lead to a continuous numeric prediction for FI, however, it reflects the classified levels of the FI. Please be noted that as mentioned at the beginning of this paper, due to the nature of SCTS, the continuously quantitative predictions are usually unrealistic or biased. This type of classified results conveyed useful information too. We gained perception of the risk levels of FI from different outcomes of the attributes from the decision tree analysis shown in Fig. 7.

Furthermore, Fig. 7 illustrated the clear "flow-chart" how the attributes affected the final consequences of the FI, with a graphical way. And therefore, the interpretation of this analysis is very clear and straightforward.

Another function of decision tree is to predict FI classified consequences of future days with their calendar characters and weather forecasts as inputs. Figure 8 shows one prediction result.

From Fig. 8, we can see even though the values of the prediction outputs, i.e., the classifications generated by decision tree were not exactly matching the observed FI values, the profile did capture the fluctuations of real observations.

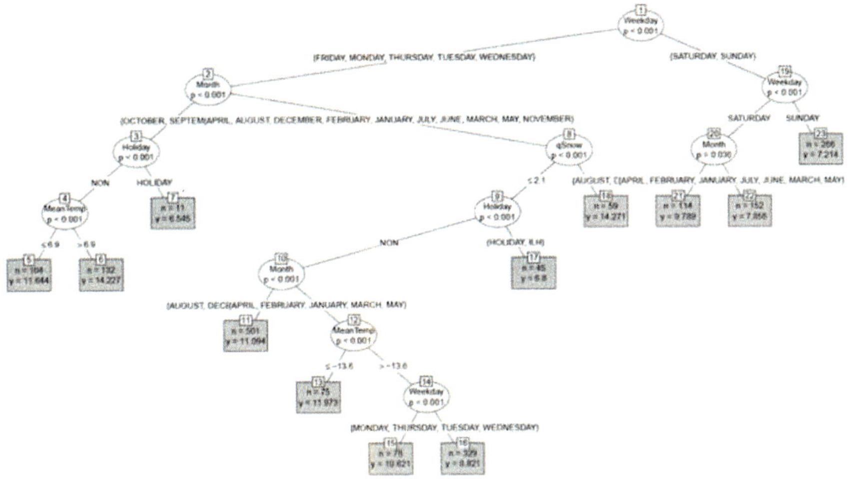

Fig. 7 Flow-chart diagram of decision tree. *Note* "ILH" in attribute "Holiday" means "day in lieu of holiday", i.e., once a public holiday falls into a weekend day, the next weekday will be non-working day

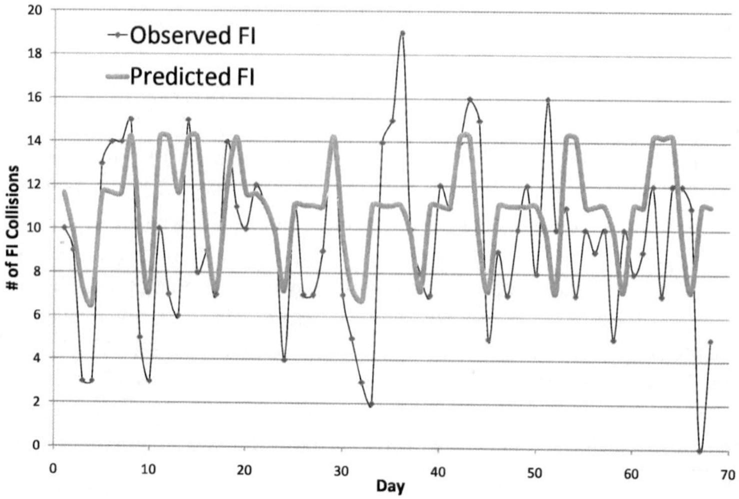

Fig. 8 Prediction of FI classifications in future days by decision tree

Probably we couldn't get exact numerical prediction from decision tree (and again, there was historically no good method to accomplish such a numerical prediction for the SCTS), we still obtained its trend: on the days we predicted higher classification levels, we also saw higher observed collisions, and vice versa. That is to say, this method can be used to identify relative trend of FI collisions: once we predict a very high classification level on a certain day, the "pre-alarm" should be triggered and we would report it as relative "high risk" day. Based on the pre-alarm, many precaution countermeasures can be pre-scheduled.

6 Conclusion

The approaches of descriptive and inferential analytics developed in this paper showed their capability to analyze SCTS datasets, filling the vacancy of traditional methods such as safety performance functions and Logit models. These newly established data mining methods have advantage to demonstrate data pattern, seasonality, hierarchical flow-chart and structure of the attributes, and the path towards particular consequences of the response variable in graphical way, and therefore the interpretation of the statistics is straightforward and informative.

The study of this paper also provided a viable solution to predict SCTS in road safety practices. It can be used to report "pre-alarms" in advance by classified prediction with the calendar characters and weather forecasts as inputs.

Considering continuously quantitative prediction of SCTS variables are usually unrealistic and biased, this series of methods of pattern mining and classified prediction by means of data mining approaches could be a realistic solution. It also provides useful analytical tools for traffic safety practitioners. The findings of this paper will support selection, pre-scheduling and evaluation of many temporally short-range safety countermeasures across multiple disciplines such as traffic operation, control, enforcement, and road maintenance.

Acknowledgements The authors acknowledge the supports of this paper by the National Key Research and Development Program of China (2018YFC0809603).

References

1. Chin H-C, Huang H (2008) Modeling multilevel data in traffic safety: a Bayesian hierarchical approach. In: Transportation accident analysis and prevention. Nova Science Publishers, Inc., New York, pp 53–106
2. Clifton C (2010) Encyclopædia Britannica: definition of data mining. http://www.britannica.com/EBchecked/topic/1056150/data-mining. Accessed 4 Aug 2014
3. Coghlan A (2012) Time Series 0.2 Documentation. http://a-little-book-of-r-for-time-series.readthedocs.org/en/latest/index.html. Accessed 3 July 2012
4. Environment Canada (2013) Daily data reports, Environment Canada, Government Canada. http://climate.weather.gc.ca/climateData/dailydata_e.html?timeframe=2&Prov=ALTA&StationID=50149&dlyRange=2012-09-01%7C2012-11-08&cmdB1=Go&Year=2013&Month=6&cmdB1=Go. Accessed 24 June 2013
5. Hauer E, Bamfo J (1997) Two tools for finding what function links the dependent variable to the explanatory variables. In: Proceedings of the ICTCT 1997 Conference
6. Hauer E (2007) Observational before-after studies in road safety. Emerald Group Publishing Limited
7. King WB (2013) R Tutorials: simple linear correlation and regression. http://ww2.coastal.edu/kingw/statistics/R-tutorials/simplelinear.html. Accessed 27 Dec 2013
8. Lund A, Lund, M (2013) Laerd statistics: pearson product-moment correlation. https://statistics.laerd.com/statistical-guides/pearson-correlation-coefficient-statistical-guide.php. Accessed on 27 Dec 2013
9. Mela C-F, Kopalle P-K (2002) The Impact of Collinearity on regression analysis: the asymmetric effect of negative and positive correlations. Appl Econ 34:667–677
10. Nagpaul P-S (2005) Time series analysis in WinIDAMS. Communication and Information, UNESCO. http://portal.unesco.org/ci/en/files/18650/11133194701TimeSeriesAnal.pdf/TimeSeriesAnal.pdf. Accessed 10 Nov 2015
11. Office of Traffic Safety (2013) Motor vehicle collision 2012. Annual Report, City of Edmonton. http://www.edmonton.ca/transportation/OTS_Motor_Vehicle_Collisions_2012_Annual_Report.pdf. Accessed 25 July 2013
12. Page S-E (2014). Chapter 1: Decision Tree. Online Course—Model Thinking, University of Michigan, Ann Arbor, Michigan, United States. http://vserver1.cscs.lsa.umich.edu/~spage/ONLINECOURSE/R4Decision.pdf. Accessed August 15 2014
13. R Development Core Team (2013) Correlation, variance and covariance (Matrices). R Documentation. http://stat.ethz.ch/R-manual/R-patched/library/stats/html/cor.html. Accessed 27 Dec 2013
14. Torgo L (2011) Data mining with R—Learning with case studies. Chapman & Hall/CRC, Taylor & Francis Group, Boca Raton, Florida, United States
15. Wallace, D. (2014). Descriptive versus inferential statistics, Lesson 1: Introduction. Lecture Note of Statistics for Psychology, Faytteville State University, North Carolina, United States. http://faculty.uncfsu.edu/dwallace/Lesson%201.pdf. Accessed 04 Aug 2014

16. Wikipedia (2014) Decision Tree. http://en.wikipedia.org/wiki/Decision_tree. Accessed 21 Aug 2014
17. Zhang H (2010) Identifying and quantifying factors affecting traffic crash severity in Louisiana. Ph.D. Dissertation, Louisiana State University, Baton Rouge, Louisiana, United States

The Residual Stress Modeling and Assessment Based on Fine Drilling

Kuikui Feng, Faping Zhang, Wuhong Wang, Zhenhe Wu, Haodong Zhang, and Qian Cheng

Abstract The diversity of fine drilling process factors results in the lack of detailed and effective modeling method for residual stress, and the sealing of the process limits the credibility assessment of the model. In order to solve these two problems, this paper studies the mechanism of stress formation and transfer in fine drilling, constructs the analytical model of residual stress on drilling surface, measures the time-varying stress of in-situ state by boundary method, and assesses the credibility of the model. The results show that the average credibility of the model is 91.30%, which verifies the efficiency of the model.

1 Introduction

In-situ state can be divided into two stages: the first one is the cutting stage, due to the role of cutting force, resulting in the workpiece temperature and stress increase. The second stage is the recovery stage, because the workpiece has been completed, the cutting force unloading leads to the gradual recovery of the temperature and stress. The temperature will return to room temperature, which will cause the redistribution of thermal stress, and the stress will change the lattice element in the workpiece due to the plastic strain, which is different from the initial value.

The cutting stress is produced by the coupling of mechanical stress and thermal stress, which results in the distribution of in-situ residual stress. At present, there are two modeling methods of machining residual stress: analytical method and finite element method (FEM).

FEM is widely used because it simulates the cutting process to represent the workpiece stress field. Its modeling methods include: The establishment and assembly of the component model, and the selection of modeling software; The mesh definition and step size division of the model; Parameter input of constitutive equation; Setting of thermal parameters in tool-workpiece contact area; Selection of finite element

K. Feng · F. Zhang (✉) · W. Wang · Z. Wu · H. Zhang · Q. Cheng
School of Mechanical Engineering, Beijing Institute of Technology, Beijing 100081, China
e-mail: zfpnew@163.com

W. Wang et al. (eds.), *Green Transportation and Low Carbon Mobility Safety*, Lecture Notes in Electrical Engineering 944,
https://doi.org/10.1007/978-981-19-5615-7_13

calculation method and numerical calculation. Shih et al. [1] took the loading and unloading of cutting force as the basis of stress change, and realized the accurate prediction of residual stress distribution by refining the grid. Nasr et al. [2] used Abaqus/Explicit program to analyze the influence of cutting edge radius and residual stress characteristics of material properties, providing a basis for the characterization of processing stress field. Wang et al. [3] proposed a modeling method based on equivalent moving load, in which the local cutting of the tool is applied to the entire surface to achieve multi-scale numerical simulation of residual stress. FEM presents the input and output of the cutting process as an implicit function and only needs to set specific model parameters, which reduces the simplified hypothesis of the model and reduces the testing cost. However, the computational efficiency and accuracy of FEM need to be greatly improved, especially in small-scale high-speed cutting simulation, where the mesh density increases exponentially, leading to the calculation cycle of several days, and prone to mesh distortion, which requires high performance of computing software.

In contrast, the analytical modeling is based on mechanical model and temperature model, combined with the thermodynamic coupling mechanism to obtain the residual stress after load release. Lin et al. [4] took shear surface load as input and used Merwin-Johnson algorithm to calculate stress increment and predict surface residual stress in orthogonal cutting. Lazoglu et al. [5] established an enhanced analytical model considering isotropy and kinematic hardening by using the superposition method of thermal stress and mechanical stress, which greatly improved the solution period. Huang et al. [6] predicted cutting residual stress based on stress influence coefficient, which is a dynamic periodic distribution of stress field. Due to the corresponding explicit relationship between independent variables and functions, analytical modeling reduces the constraint of boundary conditions by simplifying assumptions and provides a more in-depth description of the basic physical mechanism of cutting.

However, the steps of analytical modeling are complicated, especially in the field of thin-wall drilling, which contains many cutting modes and mechanisms, and lacks a unified and detailed analytical modeling method. It is difficult to measure the stress on the machined surface due to the sealing of the borehole, which poses a challenge to the credibility verification of the model. In this paper, the mechanism of stress formation and transfer in fine drilling is studied, an analytical model of residual stress on machined surface is established, and the time-varying stress of in-situ state is measured by boundary method, and the reliability of the model is evaluated.

2 Analytical Modeling of Residual Stress

Cutting stress can be divided into mechanical stress caused by cutting force and thermal stress caused by temperature. However, cutting forces are affected by the cutting mechanism and the structure of the drill tool. Therefore, with the different cutting depth, the variation of residual stress is also different.

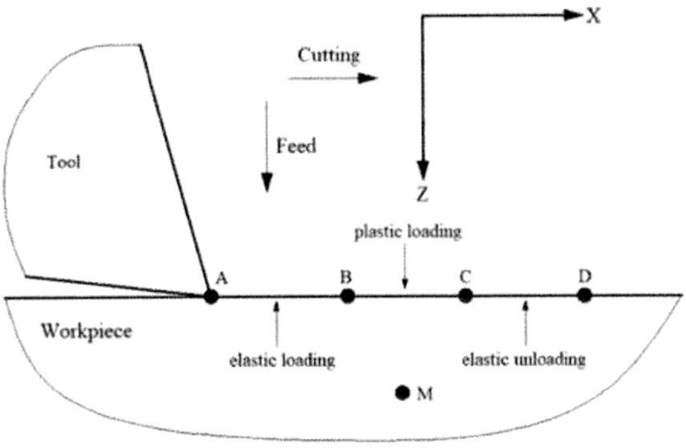

Fig. 1 The load applied to M during cutting

The stress which is lower than the yield strength is elastic stress, while the stress which is higher than the yield strength is plastic stress. The criterion is related to the spatial position of the working point. Figure 1 shows the force exerted on the target point M in the drilling based on 2D space. The cutting speed is in the X direction and the feed speed is in the Z direction. If the tool tip is far from point M (A in the Fig. 1), the loading stress of A is less than the material yield strength and elastic deformation occurs. As the tool tip moves along the X direction, the stress of M increases gradually, the loading stress of B exceeds the yield strength, and the plastic deformation begins and continues until the load unloading at C. When the tip moves to the distant D, the load applied to M can be ignored [7]. Therefore, the change of the stress of machined surface depends on the type and size of the load, and the stress change of the point at the same depth as M is similar.

Commonly used elastoplastic stress material model can be divided into ideal elastoplastic material model, bilinear elastoplastic material model and exponential hardening elastoplastic material model. Due to the different material properties, the material model is selected according to the actual engineering needs.

2.1 Loading of Stress

The mechanical stress of the tool on the workpiece can be divided into two parts: one is the stress of shear zone between the chip and the workpiece, and the other is the stress of friction zone between the flank surface and the workpiece.

Figure 2 shows the mechanical stress of M caused by the load of shear zone, while the load of shear zone is divided into normal stress $p(s)$ and tangential stress $q(s)$, which present the parabolic distribution in shear zone $[-a, a]$ [8]. The relationship between the load of shear zone and cutting force is shown in Eq. (1).

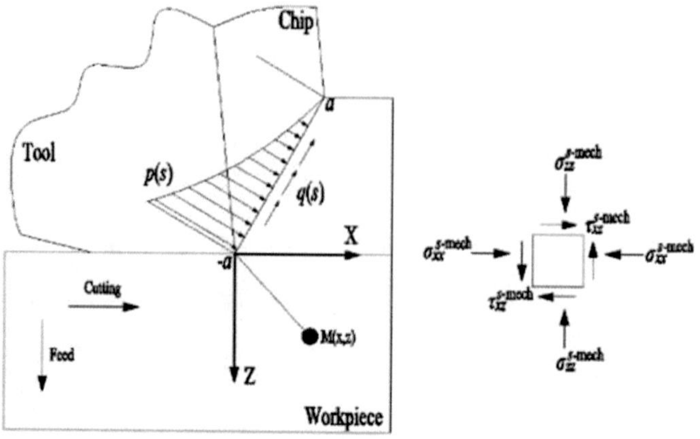

Fig. 2 The mechanical stress of shear zone

$$\begin{cases} p(s) = \dfrac{F_v \sin\varphi_s + F_f \cos\varphi_s}{2ab}\left(1 - \dfrac{s+a}{2a}\right)^2 \\ q(s) = k_s\left(1 - \dfrac{s+a}{2a}\right)^2 \end{cases} \tag{1}$$

where F_v and $F_f$are the thrust in the cutting direction and the feed direction, and the total length $2a$ of the shear zone is related to the cutting edge radius r_c.

$$a = \sqrt{\frac{4PR_c}{\pi}\left(\frac{1}{E_t} + \frac{1}{E}\right)} \tag{2}$$

where E and E_t are the elastic moduli of the workpiece and the tool. On the basis of determining the forces of the two components, the stress component of M caused by the contact force of shear zone are calculated according to Hertz contact theory [9].

$$\begin{cases} \sigma_{xx}^s = -\dfrac{2z}{\pi}\displaystyle\int_{-a}^{a} \dfrac{p(s)(x-s)^2}{[(x-s)^2+z^2]^2}ds - \dfrac{2}{\pi}\displaystyle\int_{-a}^{a} \dfrac{q(s)(x-s)^3}{[(x-s)^2+z^2]^2}ds \\ \sigma_{zz}^s = -\dfrac{2z^3}{\pi}\displaystyle\int_{-a}^{a} \dfrac{p(s)}{[(x-s)^2+z^2]^2}ds - \dfrac{2z^2}{\pi}\displaystyle\int_{-a}^{a} \dfrac{q(s)(x-s)}{[(x-s)^2+z^2]^2}ds \\ \tau_{xz}^s = -\dfrac{2z^2}{\pi}\displaystyle\int_{-a}^{a} \dfrac{p(s)(x-s)}{[(x-s)^2+z^2]^2}ds - \dfrac{2z}{\pi}\displaystyle\int_{-a}^{a} \dfrac{q(s)(x-s)^2}{[(x-s)^2+z^2]^2}ds \end{cases} \tag{3}$$

Figure 3 shows the mechanical stress of M caused by the load of friction zone. The load of friction zone is divided into normal stress $\sigma_r(s)$ in the direction of feed and

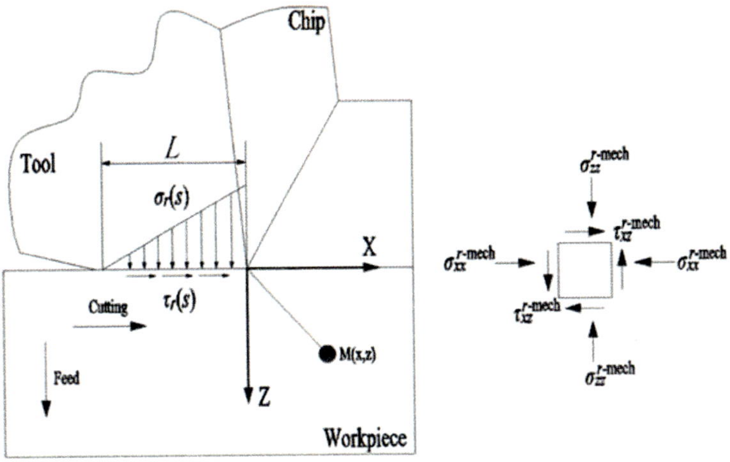

Fig. 3 The mechanical stress of friction zone

shear stress $\tau_r(s)$ in the direction of cutting, which changes linearly in the friction zone [10]. As the workpiece is a thin-walled part, the cutting time is short and the wear amount can be ignored. Therefore, the width of the friction zone approximates the contact length L between the flank surface and the workpiece.

$$\begin{cases} \sigma_r(s) = \sigma_0 \dfrac{L-s}{L}, 0 < s < L \\ \tau_r(s) = \begin{cases} \tau_0, 0 < s < L(1 - \sqrt{\dfrac{\tau_0}{\sigma_0}}) \\ \mu \cdot \sigma_r(s), L(1 - \sqrt{\dfrac{\tau_0}{\sigma_0}}) < s < L \end{cases} \end{cases} \tag{4}$$

where σ_0 and τ_0 are the normal stress and shear stress at the cutting edge respectively, μ is the coefficient of friction. The stress component of M caused by contact force of friction zone are also calculated according to Hertz contact theory.

$$\begin{cases} \sigma_{xx}^f = -\dfrac{2z}{\pi} \displaystyle\int_0^L \dfrac{\sigma_r(s) \cdot (x - s - \frac{L}{2})^2}{[(x - s - \frac{L}{2})^2 + z^2]^2} ds - \dfrac{2}{\pi} \displaystyle\int_0^L \dfrac{\tau_r(s) \cdot (x - s - \frac{L}{2})^3}{[(x - s - \frac{L}{2})^2 + z^2]^2} ds \\ \sigma_{zz}^f = \dfrac{2z^3}{\pi} \displaystyle\int_0^L \dfrac{\sigma_r(s)}{[(x - s - \frac{L}{2})^2 + z^2]^2} ds - \dfrac{2z^2}{\pi} \displaystyle\int_0^L \dfrac{\tau_r(s) \cdot (x - s - \frac{L}{2})}{[(x - s - \frac{L}{2})^2 + z^2]^2} ds \\ \tau_{xz}^f = -\dfrac{2z^2}{\pi} \displaystyle\int_0^L \dfrac{\sigma_r(s) \cdot (x - s - \frac{L}{2})}{[(x - s - \frac{L}{2})^2 + z^2]^2} ds - \dfrac{2z}{\pi} \displaystyle\int_0^L \dfrac{\tau_r(s) \cdot (x - s - \frac{L}{2})^2}{[(x - s - \frac{L}{2})^2 + z^2]^2} ds \end{cases} \tag{5}$$

The temperature distribution of workpiece is not uniform due to the influx of cutting heat, and the thermal strain caused by the workpiece affects the change of residual stress. Thermal stress is caused by three kinds of loads: The body force F_x and F_z along the cutting and feed directions at the uncut point (x_1, z_1); The hydrostatic pressure F_h of the target point (x, z); The tensile surface traction F_t of cutting boundary point $(x_2, 0)$ [11]. The three types of loads are shown in Eq. (6), and the thermal stress components are shown in Eq. (7).

$$\begin{cases} F_x = -\dfrac{\alpha E}{1-2\upsilon}\dfrac{\delta T_{(x_1,z_1)}}{\delta x} \\ F_z = -\dfrac{\alpha E}{1-2\upsilon}\dfrac{\delta T_{(x_1,z_1)}}{\delta z} \\ F_h = -\dfrac{\alpha E T_{(x,z)}}{1-2\upsilon} \\ F_t = \dfrac{\alpha E T_{(x_2,0)}}{1-2\upsilon} \end{cases} \tag{6}$$

$$\begin{cases} \sigma_{xx}^t = \int_0^\infty \int_{-\infty}^\infty (F_x G_{x1} + F_z G_{x2}) dx_1 dz_1 + \dfrac{2z}{\pi} \int_{-\infty}^\infty \dfrac{F_t \cdot (x-x_2)^2}{[(x-x_2)^2 + z^2]^2} dx_2 + F_h \\ \sigma_{zz}^t = \int_0^\infty \int_{-\infty}^\infty (F_x G_{z1} + F_z G_{z2}) dx_1 dz_1 + \dfrac{2z^3}{\pi} \int_{-\infty}^\infty \dfrac{F_t}{[(x-x_2)^2 + z^2]^2} dx_2 + F_h \\ \tau_{xz}^t = \int_0^\infty \left(\int_{-\infty}^\infty (F_x G_{xz1} + F_z G_{xz2}) dx_1\right) dz_1 + \dfrac{2z^2}{\pi} \int_{-\infty}^\infty \dfrac{F_t \cdot (x-x_2)}{[(x-x_2)^2 + z^2]^2} dx_2 \end{cases} \tag{7}$$

where υ and α are poisson ratio and linear expansion coefficient of the workpiece respectively. G_x, G_z and G_{xz} are the Green's functions under plane strain state [12].

When the tip is in the A–B of Fig. 1, the stress of M in the x and z directions can be expressed as the superposition of shear zone stress, friction zone stress and thermal stress, while the stress in the y direction is calculated by the theory of Hooke's law.

$$[\sigma]^{el} = \begin{cases} \sigma_{xx} = \sigma_{xx}^{el} = \sigma_{xx}^s + \sigma_{xx}^f + \sigma_{xx}^t \\ \sigma_{zz} = \sigma_{zz}^{el} = \sigma_{zz}^s + \sigma_{zz}^f + \sigma_{zz}^t \\ \tau_{xz} = \tau_{xz}^{el} = \tau_{xz}^s + \tau_{xz}^f + \tau_{xz}^t \\ \sigma_{yy} = \sigma_{yy}^{el} = \upsilon(\sigma_{xx}^{el} + \sigma_{zz}^{el}) - \alpha E T_{(x,z)} \end{cases} \tag{8}$$

When the tool cuts through point B, the loading stress is greater than the yield strength of the workpiece material, resulting in plastic strain. However, it is generally believed that the material overcomes the external load due to plastic strengthening, so the stress component in the loading direction is equivalent to the loading stress [5].

$$\begin{cases} \sigma_{xx} = \sigma_{xx}^{el} \\ \sigma_{zz} = \sigma_{zz}^{el} \\ \tau_{xz} = \tau_{xz}^{el} \end{cases} \tag{9}$$

Therefore, only σ_{yy} is different, which needs to be analyzed according to the linear increment theory of plastic mechanics, which could establish the relationship between the increment of stress and plastic strain [13].

Assuming that the material obeys the law of isotropic and kinematic composite hardening [14], the von Mises yield criterion is

$$f_{sur} = \frac{1}{2}(S_{ij} - \alpha_{ij})(S_{ij} - \alpha_{ij}) - k^2 = 0 \tag{10}$$

where k is shear yield strength, S_{ij} and α_{ij} are deviatoric stress and backstress.

$$\begin{cases} S_{ij} = \begin{cases} \sigma_{ij} - \dfrac{\sigma_{kk}}{3}, i = j \\ \sigma_{ij}, i \neq j \end{cases} \\ \alpha_{ij} = \langle S_{kl} n_{kl} \rangle n_{ij} \end{cases} \tag{11}$$

where $\sigma_{kk} = \sigma_{xx} + \sigma_{yy} + \sigma_{zz}$, $< >$ is MacCauley symbol, represented by $< x > = 0.5(x + |x|)$; $n_{ij} = (S_{ij} - \alpha_{ij})/(\sqrt{2}\, k)$ is the unit vector of plastic strain rate. According to the plastic flow criterion [15], the strain increment of material in the plastic loading process is

$$d\varepsilon_{ij}^{pl} = \frac{1}{h + c} \langle dS_{kl} n_{kl} \rangle n_{ij} \tag{12}$$

where h and c are coefficient of isotropic hardening and kinematic hardening. However, the y direction has no strain increment in plane strain.

$$d\varepsilon_{yy} = d\varepsilon_{yy}^{el} + d\varepsilon_{yy}^{therm} + d\varepsilon_{yy}^{pl} = 0 \tag{13}$$

The three strain increments are expressed as

$$\begin{cases} d\varepsilon_{yy}^{t} = \alpha dT_{(x,z)} \\ d\varepsilon_{yy}^{el} = \dfrac{d\sigma_{yy}}{E} - \dfrac{\upsilon}{E}(d\sigma_{xx} + d\sigma_{zz}) \\ d\varepsilon_{yy}^{pl} = \dfrac{n_{yy}(dS_{xx}n_{xx} + dS_{yy}n_{yy} + dS_{zz}n_{zz} + 2dS_{xz}n_{xz})}{h + c} \end{cases} \tag{14}$$

The stress increment in the y direction can be calculated by Eqs. (13) and (14).

$$d\sigma_{yy} = \frac{3(h+c)\upsilon - (2n_{xx}n_{yy} - n_{yy}n_{yy} - n_{zz}n_{yy})E}{3(h+c) + (2n_{yy}n_{yy} - n_{xx}n_{yy} - n_{zz}n_{yy})E}d\sigma_{xx} + \frac{3(h+c)\upsilon - (2n_{zz}n_{yy} - n_{xx}n_{yy} - n_{yy}n_{yy})E}{3(h+c) + (2n_{yy}n_{yy} - n_{xx}n_{yy} - n_{zz}n_{yy})E}d\sigma_{zz} + \frac{2d\tau_{xz}n_{xz}n_{yy} - \alpha E dT_{(x,z)}}{1 + \frac{(2n_{yy}n_{yy} - n_{xx}n_{yy} - n_{zz}n_{yy})E}{3(h+c)}} \tag{15}$$

where $d\sigma_{zz}$ and $d\tau_{xz}$ are calculated from the mechanical stress equations of Eqs. (3) and (5), while $d\sigma_{xx}$ is solved by McDowell hybrid algorithm [16].

$$\varepsilon_{xx} = \Psi\left[\frac{d\sigma_{xx} - \upsilon(d\sigma_{yy} + d\sigma_{zz})}{E} + \alpha\Delta T + \frac{n_{xx}(dS_{xx}n_{xx} + dS_{yy}n_{yy} + dS_{zz}n_{zz} + 2d\tau_{xz}n_{xz})}{h+c}\right] \tag{16}$$

where ψ is the mixing function of hardening modulus.

$$\Psi = 1 - \exp(-\kappa\frac{3h}{2G}) \tag{17}$$

where G is elastic shear modulus, κ is the algorithmic constant. Then the stress under plastic load is

$$[\sigma]^{pl} = \begin{cases} \sigma_{xx}^{el} \\ \sigma_{zz}^{el} \\ \tau_{xz}^{el} \\ \sigma_{yy}^{el} + d\sigma_{yy} \end{cases} \tag{18}$$

When the yield surface function follows the criterion $f < 0$, elastic unloading occurs, corresponding to the C-D in Fig. 1. The stress component is calculated again by Eq. (9).

2.2 *The Stress Release*

After loading, machined surface stress needs to be released (such as cutting force removal and temperature decline) to restore boundary conditions. The newly generated stress in the process of restoring boundary conditions is the residual stress, and finally the workpiece is in the stress state of self-phase equilibrium when it is not affected by any external force [17].

There are two boundary conditions for stress release: Smooth deformation surface; The cutting surface is balanced and free of traction. Therefore, the stress and strain should meet the following boundary conditions:

$$\begin{cases} \sigma_{xx}^r = f_1(z), \sigma_{yy}^r = f_2(z), \varepsilon_{zz}^r = f_3(z), \varepsilon_{xz}^r = f_4(z) \\ \sigma_{zz}^r = \sigma_{xz}^r = \sigma_{xy}^r = \sigma_{zy}^r = 0 \\ \varepsilon_{xx}^r = \varepsilon_{yy}^r = \varepsilon_{xy}^r = \varepsilon_{zy}^r = 0 \end{cases} \tag{19}$$

However, the actual rolling contact strain component ε_{xx} and stress component σ_{zz} and τ_{xz} are not zero, but reach the boundary condition after a period of time. For mechanical load and thermal load, assuming that unloading time is from t_1 to t_2 and boundary conditions are satisfied after M steps of release, then the relaxation increment is

$$\begin{cases} d\varepsilon_{xx} = \dfrac{\sigma_{xx}}{ME}(1-\upsilon^2) - \dfrac{\upsilon\sigma_{zz}}{ME}(1+\upsilon) + \dfrac{\alpha\Delta T_{(x,z)}(1+\upsilon)}{M} \\ d\sigma_{zz} = \dfrac{\sigma_{zz}^s + \sigma_{zz}^f + \sigma_{zz}^t}{M} \\ d\tau_{xz} = \dfrac{\tau_{xz}^s + \tau_{xz}^f}{M} \end{cases} \tag{20}$$

For elastic load points, the stress components $d\sigma_{xx}$ and $d\sigma_{yy}$ are calculated according to the theory of Hooke's law under plane strain condition in Eq. (21).

$$\begin{cases} d\sigma_{xx}^r = \dfrac{Ed\varepsilon_{xx}^r + (1+\upsilon)(\upsilon d\sigma_{zz}^r - E\alpha dT_k)}{1-\upsilon^2} \\ d\sigma_{yy}^r = \upsilon(d\sigma_{xx}^r + d\sigma_{zz}^r) - E\alpha dT_k \end{cases} \tag{21}$$

For the plastic load points, the relaxation increments $d\sigma_{xx}{}^r$ and $d\sigma_{yy}{}^r$ are solved by Eq. (22).

$$\begin{cases} d\sigma_{xx}^r + [\dfrac{2n_{yy}n_{yy} - n_{xx}n_{yy} - n_{yy}n_{zz}}{3(h+c)} + \dfrac{1}{E}]d\sigma_{xx}^r \\ +[\dfrac{2n_{yy}n_{zz} - n_{xx}n_{yy} - n_{yy}n_{yy}}{3h+c} - \dfrac{\upsilon}{E}]d\sigma_{zz}^r + \dfrac{2n_{yy}n_{xz}}{h+c}d\tau_{xz}^r + \alpha dT_k = d\varepsilon_{yy}^r = 0 \\ [\dfrac{2n_{xx}n_{xx} - n_{xx}n_{yy} - n_{xx}n_{zz}}{3(h+c)} + \dfrac{1}{E}]d\sigma_{xx}^r + [\dfrac{2n_{xx}n_{yy} - n_{xx}n_{xx} - n_{xx}n_{zz}}{3(h+c)} - \dfrac{\upsilon}{E}]d\sigma_{xx}^r \\ +[\dfrac{2n_{xx}n_{zz} - n_{xx}n_{yy} - n_{xx}n_{xx}}{3h+c} - \dfrac{\upsilon}{E}]d\sigma_{zz}^r + \dfrac{2n_{yy}n_{xz}}{h+c}d\tau_{xz}^r + \alpha dT_k = d\varepsilon_{xx}^r \end{cases} \tag{22}$$

The calculated result is the additional residual stress in the cyclic loading process, which is combined with the initial stress of the previous step as the initial stress of the next step, and so on until the stress and strain enter a stable state, then the stress in the workpiece is the final residual stress of the machined surface, as shown in Fig. 4.

The analytical modeling of residual stress shows that mechanical stress is negatively loaded and thermal stress is positively loaded. Since the loading degree of mechanical stress is much greater than that of thermal stress, the stress increases gradually in the negative direction and gradually offloads after reaching the peak value. At this time, the unloading degree of mechanical stress is also greater than

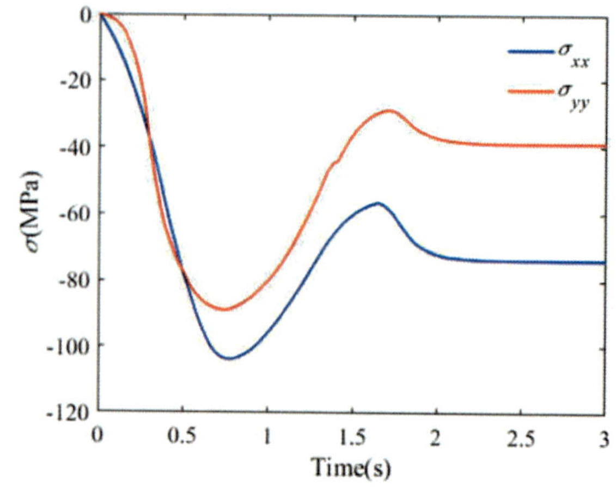

Fig. 4 The predicted of time-varying stress of in-site

that of thermal stress. The unloading time of thermal stress is longer than that of mechanical stress, so there will be a period of negative unloading of thermal stress after the completion of mechanical stress unloading.

3 Credibility Evaluation of the Model

3.1 The Experiment of Machining Stress

In order to verify the accuracy of the model, it is necessary to measure the stress change of the machined surface. Due to the sealing of drilling process, it is impossible to directly measure stress, so equivalent boundary method is adopted in this paper, as shown in Fig. 5. The $-x$ and $+y$ directions of the hole are close to the boundary of the sample (about 0.1 mm), so the stress of the hole wall on both sides is the stress of the boundary of the sample. The strain gauge is pressed on both sides of the boundary, and the stress of the measuring point is calculated by the strain at the boundary. The paste methods of strain gauge are transverse and longitudinal. The transverse strain gauge measures the stress σ_{xx} and the longitudinal strain gauge measures the stress σ_{yy}.

The samples were cuboid samples of 50 × 50 × 4 mm and made of 304 steel. The residual stress is measured by CT-5304 strain tester, which calculates residual stress by sensing the micro-strain of strain gauge. The strain tester is connected to the data acquisition card(DAQ), and the data is transferred from DAQ to PC. The experimental setup is shown in Fig. 6.

The cutting diameter of the tool is ϕ2mm, the spindle speed is set as $n = 6000$r/min, and the feed is $f = 75$ mm/min. The time-varying stress in each direction is shown in Fig. 7.

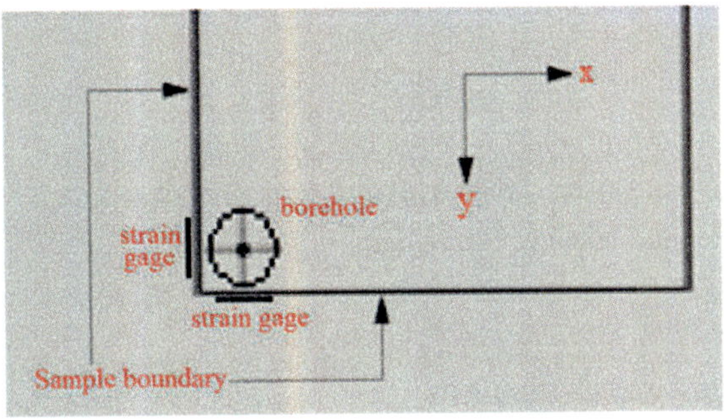

Fig. 5 The strain gauge distribution

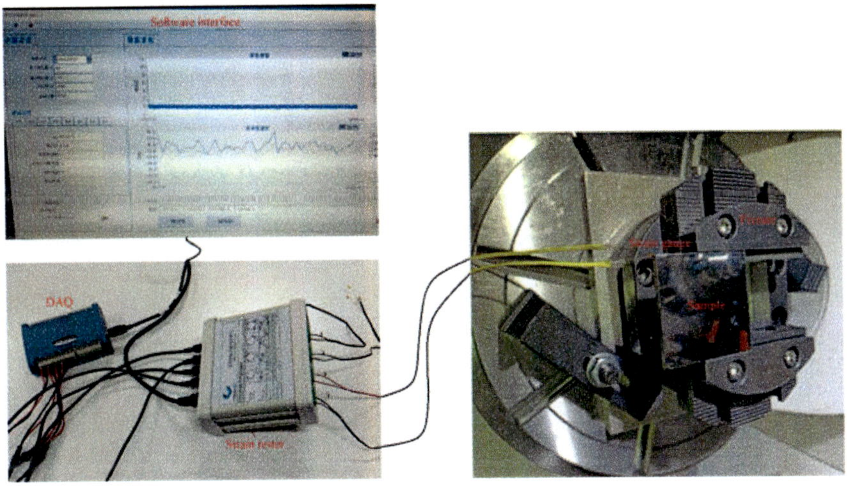

Fig. 6 The experimental setup

Fig. 7 The measured of time-varying stress of in-site

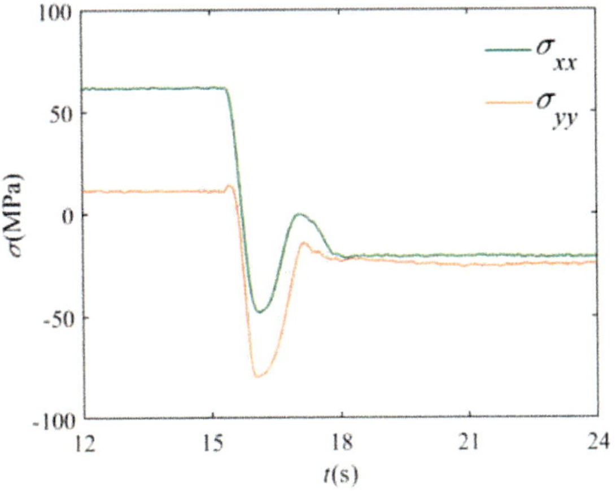

It can be seen from the figure that the change trend of the measured data is similar to that of the prediction model, which increases first and then decreases, and increases slightly to a stable state at the end, indicating that the prediction model can reflect the in-situ stress change of the processed surface. However, the model cannot completely express performance of real system, so the credibility index is used to evaluate the similarity between model and experiment.

3.2 Credibility Assessment

Firstly, the predicted data and experimental data are extracted to form samples X^{pre} and X^{test}.

$$\begin{cases} X^{pre} = \begin{cases} X_1^{pre} = \{X_{11}^{pre}, \cdots, X_{1i}^{pre}, \cdots, X_{1n}^{pre}\} & \sigma_{xx} \\ X_2^{pre} = \{X_{21}^{pre}, \cdots, X_{2i}^{pre}, \cdots, X_{2n}^{pre}\} & \sigma_{yy} \end{cases} \\ X^{test} = \begin{cases} X_1^{test} = \{X_{11}^{test}, \cdots, X_{1i}^{test}, \cdots, X_{1n}^{test}\} & \sigma_{xx} \\ X_2^{test} = \{X_{21}^{test}, \cdots, X_{2i}^{test}, \cdots, X_{2n}^{test}\} & \sigma_{yy} \end{cases} \end{cases} \tag{23}$$

Then, the weight coefficient β of each data is calculated according to the experimental data.

$$\beta = \begin{cases} \beta_{1i} = \dfrac{X_{1i}^{test}}{\sum_{i=1}^{n} X_{1i}^{test}} & \sigma_{xx} \\ \beta_{2i} = \dfrac{X_{2i}^{test}}{\sum_{i=1}^{n} X_{2i}^{test}} & \sigma_{yy} \end{cases} \tag{24}$$

After that, the relative error ε between the predicted data and the experimental data is calculated.

$$\varepsilon = \begin{cases} \varepsilon_{1i} = \left| \dfrac{X_{1i}^{test} - X_{1i}^{pre}}{X_{1i}^{test}} \right| & \sigma_{xx} \\ \varepsilon_{2i} = \left| \dfrac{X_{2i}^{test} - X_{2i}^{pre}}{X_{2i}^{test}} \right| & \sigma_{yy} \end{cases} \tag{25}$$

Finally, the model credibility Q_1 and Q_2 of x and y direction are calculated.

$$\begin{cases} Q_1 = \sum_{i=1}^{n} \beta_{1i}(1 - \varepsilon_{1i}) = 91.97\% & \sigma_{xx} \\ Q_2 = \sum_{i=1}^{n} \beta_{2i}(1 - \varepsilon_{2i}) = 90.63\% & \sigma_{yy} \end{cases} \tag{26}$$

So the average credibility Q of model is

$$Q = \frac{Q_1 + Q_2}{2} = \frac{91.97\% + 91.30\%}{2} = 91.30\% \tag{27}$$

4 Conclusions

In order to investigate the variation of residual stress field of machining surface, the representation model of residual stress of in-situ state was established. On the basis of known cutting force and cutting temperature, the credibility of the model is assessed. Finally, the average credibility of the model is 91.30%, which has a high level and can reflect the real stress change law of in-situ state. The precise characterization of residual stress lays the foundation for the evolution of workpiece geometric parameters in cutting process.

References

1. Shih AJ (1995) Finite element simulation of orthogonal metal cutting[J]. J Eng Ind 117(1):84–93
2. Nasr M, Ng EG, Elbestawi M (2007) Effects of workpiece thermal properties on machining-induced residual stresses-thermal softening and conductivity[J]. Proc Inst Mech Eng, Part B: J Eng Manuf 221(9):1387–1400
3. Wang F, Liu ZY, Guo YB et al (2017) Efficient multiscale modeling and validation of residual stress field in cutting[J]. J Manuf Sci Eng 139(9):91004
4. Lin ZC, Lee BY (1995) An investigation of the residual stress of a machined workpiece considering tool flank wear[J]. J Mater Process Technol 51(1–4):1–24
5. Lazoglu I, Ulutan D, Alaca BE et al (2008) An enhanced analytical model for residual stress prediction in machining[J]. CIRP Ann Manuf Technol 57(1):81–84
6. Huang XD, Zhang XM, Leopold J, DING H (2017) Analytical model for prediction of residual stress in dynamic orthogonal cutting process[J]. J Manuf Sci Eng 140(1):011002
7. Ulutan D, Alaca BE, Lazoglu I (2007) Analytical modelling of residual stresses in machining[J]. J Mater Process Technol 183(1):77–87
8. Agrawal S, Joshi SS (2013) Analytical modelling of residual stresses in orthogonal machining of AISI4340 steel[J]. J Manuf Process 15(1):167–179
9. Fergani O, Lazoglu I, Mkaddem A et al (2014) Analytical modeling of residual stress and the induced deflection of a milled thin plate[J]. Int J Adv Manuf Technol 75(1–4):455–463
10. Smithey DW, Kapoor SG, Devor RE (2001) A new mechanistic model for predicting worn tool cutting forces[J]. Mach Sci Technol 5(1):23–42
11. Timoshenko SP, Goodier JN (2007) Theory of Elasticity [M]. Tsinghua University Press, Beijing
12. Saif M, Hui C, Zehnder A (1993) Interface shear stresses induced by non-uniform heating of a film on a substrate[J]. Thin Solid Films 224(2):159–167
13. Chen L, El-Wardany TI, Harris WC (2004) Modeling the effects of flank wear land and chip formation on residual stresses[J]. CIRP Ann Manuf Technol 53(1):95–98

14. Yu J, Fei Q, Zhang P et al (2019) An innovative yield criterion considering strain rates based on von mises stress[J]. J Pressure Vessel Technol 142(1):014501
15. JC Su (2006) Residual stress modeling in machining processes. Georgia Institute of Technology, 2006
16. Mcdowell DL (1997) An approximate algorithm for elastic-plastic two-dimensional rolling/sliding contact[J]. Wear 211(2):237–246
17. Hoger A (1986) On the determination of residual stress in an elastic body [J]. J Elast 16(3):303–324

Research on Dynamic Equation Construction Method with Clearance

Yunhe Zhang, Faping Zhang, Wuhong Wang, Kai Wu, and Yang Li

Abstract The gap problem has always been a difficulty in dynamic analysis because it is mainly non-linear in the process of motion. Based on the L–N contact force model, the dynamic equation with clearance was built in this paper. At the same time, factors such as friction force and oil film force in the operation process were taken into account to obtain the dynamic model with comprehensive multiple factors, which laid a theoretical reference foundation for future motion analysis.

Keywords Clearance · Dynamical equation · Contact force

1 Introduction

The actual mechanism usually contains clearance, which mainly come from the clearance formed by the assembly of the mechanism, the clearance caused by errors in the design and manufacture of the mechanism and the clearance caused by wear and tear. The clearance will lead to the reduction of the movement precision of the mechanism, and the contact collision of the accessory elements in the clearance will produce the

Y. Zhang · F. Zhang (✉) · W. Wang
Laboratory of Industrial and System Engineering, Beijing Institute of Technology, Beijing, China
e-mail: zfpnew@163.com

Y. Zhang
e-mail: he005@126.com

W. Wang
e-mail: wangwuhong@bit.edu.cn

K. Wu
The Second Institute of China Aerospace Science and Industry Corporation Limited, Beijing, China
e-mail: 450747693@qq.com

Y. Li
Sinosteel Group Chifeng Jinxin Mining Co., Ltd, Beijing, China
e-mail: 37707824@qq.com

W. Wang et al. (eds.), *Green Transportation and Low Carbon Mobility Safety*,
Lecture Notes in Electrical Engineering 944,
https://doi.org/10.1007/978-981-19-5615-7_14

collision force. Therefore, it is necessary to study and predict the dynamic characteristics of the mechanism with clearance, and to establish a reasonable clearance model of the kinematic pair.

A. *Typical Model of Interstitial Mechanism*

The research on the dynamics of the mechanism with clearance began in the early 1970s, mainly represented by Dubowsky et al. Among them, Dubowsky has published one-dimensional impact pair model, one-dimensional impact rod model and two-dimensional impact ring model successively. Dubowsky has done a lot of research, achieved fruitful research results and established a relatively complete research system. A large number of domestic and foreign scholars have proposed different modeling methods, which can be roughly divided into three categories:

(1) Continuous contact model.

The continuous contact model assumes that the motion pair always maintains the contact state during the whole process of the mechanism, as shown in Fig. 1. For dynamic problems containing multiple clearance can use Lagrange method to establish dynamic model of institutions, but the model does not consider clearance collision plane elastic deformation, without considering the friction between artifacts and damping force, also can't get the collision force, therefore, this model can't reasonable describing actual collision contact clearance, the model of motion pair with clearance deviation with the actual situation of its application is restricted.

(2) Two-state model.

Considers the characteristic component separation between two-state model, the movement state of relative motion between the pair of elements is divided into "contact, separation of two states," separation "contact " two-state model is a kind of quantitative analysis model, the model than the continuous contact model is closer to the actual organization model, the kinematics and dynamics equations of the model was established based on Newton's mechanics. The two-state model can be used to approximate the contact and separation of the accessory elements by the equivalent spring damping model. The "two-state model" can well describe the dynamic characteristics of the mechanism with clearance, and the calculation process is relatively simple. However, the disadvantage of this model is the recognition of motion mode, which makes it difficult to determine whether the components are in separate state or contact state, which increases the complexity of numerical calculation and makes it difficult to obtain the steady-state solution of dynamic differential equation.

(3) Three-state model

Compared to the two state model, three state model considering the collision between the elements of motion pair properties, the relative motion state is divided into "contact, separation, collision" three states, collision state

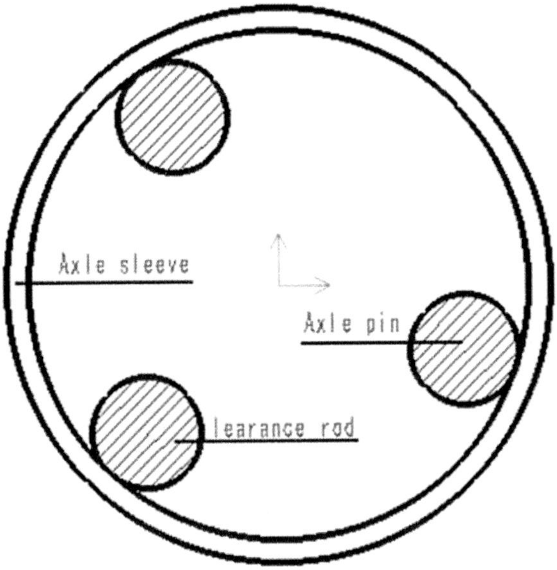

Fig. 1 Continuous contact model

is generally calculated according to component between the impact degree of conservation of momentum, "binary model" can be more real than accurately reflect the actual motion state institutions. The nonlinear dynamic differential equation of the three-state model is established based on the Lagrange equation. Due to considering the three motions, so the establishment of dynamics equations, the need to determine the time of sports belong to that kind of motion, and difficult to accurately judge the state of the conversion between the time, this creates problems for the numerical calculation, especially for more clearance problem is more difficult to get the stable solution mechanism dynamics differential equation.

(4) Four-state model

Based on the comprehensive analysis of the four modeling methods, the continuous contact model is relatively simple and can only be used to calculate the motion law of the mechanism, and it is easy to find the stable solution of the differential equation, but it is quite different from the actual clearance situation and has a low precision. "Two-state model" is relatively good in computational efficiency and accuracy, and is the most widely used and mature model. The accuracy of the three and four state models is higher than the former two, but their motion states are more numerous, which makes it difficult to determine the relative motion states among the motion accessory elements, and the numerical calculation is more difficult. as shown in Fig. 2.

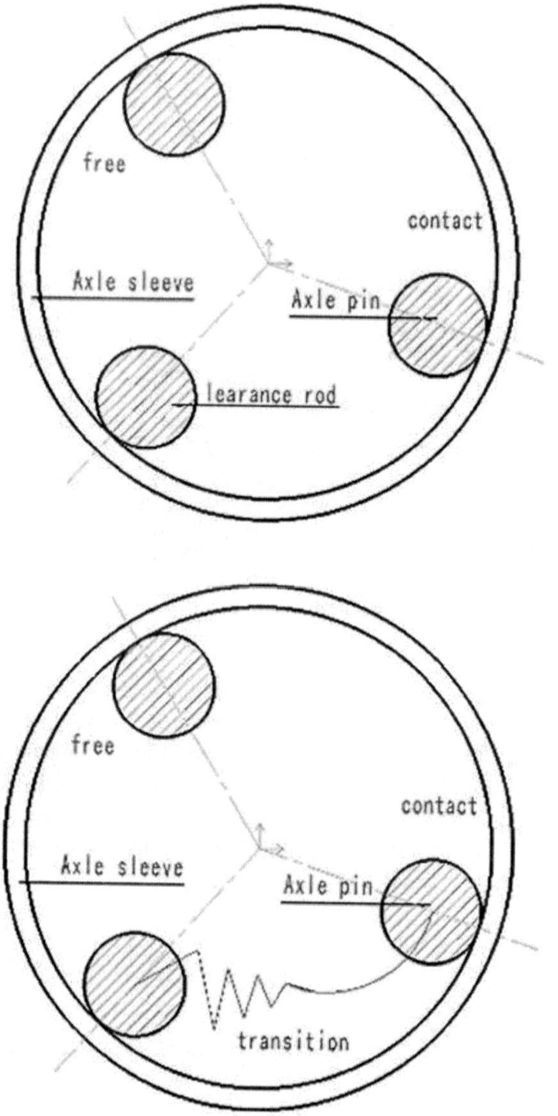

Fig. 2 The mechanical stress of shear zone

2 Construction of Clearance Dynamics Model

Finite element method are more accurate than the previous two model, considering the agency contact deformation in the process of collision and after collision contact dynamic boundary conditions, the collision force is given in the process of collision with the change of time and the spatial distribution of collision force, so for the elastic body contact collision process, finite element method is more in line with the actual physical nature, continuous contact model and the momentum balance method can overcome some of the problems, a more accurate description of contact and collision. However, the shortcomings of this method lie in its complexity and low computational efficiency.

A. *L–N Contact Force Model*

Based on the research work of Hunt and Crossley, Lankarani and Nikravesh further promoted the study of contact collision force model. Lankarani and Nikravesh hypothesized that material damping was the source of energy loss during contact and collision of interstice-containing mechanisms

$$F_N = K\delta^n\left[1 + \frac{3\left(1 - e^2\right)}{4} \cdot \frac{\dot{\delta}}{\dot{\delta}^{(-)}}\right] \quad (1)$$

$$K = \frac{4}{3\pi\left(h_i + h_j\right)}\left[\frac{R_i R_j}{R_i + R_j}\right]^{1/2} \quad (2)$$

δ: is the deformation of two spheres in contact.
n: selects 3/2 from Hertz.
K is the contact stiffness coefficient of the collision body, which is related to material properties and sphere radius.
h_i, h_j material parameters:

$$h_i = \frac{1 - v_l^2}{\pi E_l} \quad (3)$$

The parameters v_l and E_l represent poisson's ratio and young's modulus respectively.

Recovery coefficient e:

$$e = -\frac{V_i^{(+)} - V_j^{(+)}}{V_i^{(-)} - V_j^{(-)}} \quad (4)$$

$$\dot{\delta}^{(-)} = V_i^{(-)} - V_j^{(-)} \quad (5)$$

$V^{(-)}$ the velocity of two objects before they collide.
$V^{(+)}$ the velocity of two objects after collision.

Due to the relative initial velocity of the deformation:

B. *Oil Film Force Model with Clearance*

When considering lubrication, the force constraint between the shaft and the bearing changes into the oil film force constraint, or the oil film force model under the combined action of oil film force and contact force. Under Gumbel boundary condition, Hamrock obtains the oil film force model that satisfies both infinite long bearing and infinite short bearing.

For infinite long bearing, normal oil film force ***Fr*** and radial oil film force ***Ft*** are:

$$F_r = -\frac{12\mu L r_J{}^3}{\Delta R^2}\left[\left(\frac{\varepsilon^2(\omega - 2\dot{\gamma})}{(2+\varepsilon)^2(1-\varepsilon^2)} + \frac{\dot{\varepsilon}}{(1-\varepsilon^2)^{3/2}}\left(\frac{\pi}{2} - \frac{8}{\pi(2+\varepsilon^2)}\right)\right)\right] \quad (6)$$

$$F_t = -\frac{12\mu L r_J{}^3}{\Delta R^2}\left[\left(\frac{\pi\varepsilon(\omega - 2\dot{\gamma})}{2(2+\varepsilon)^2\sqrt{(1-\varepsilon^2)}} + \frac{2\varepsilon\dot{\varepsilon}}{(2+\varepsilon)^2(1-\varepsilon^2)}\right)\right] \quad (7)$$

For the infinite short bearing, the normal oil film force ***Fr*** and radial oil film force ***Ft*** are:

$$F_r = -\frac{\mu L^3 r_J}{2\Delta R^2(1-\varepsilon^2)^2}\left[\left(\frac{\pi\dot{\varepsilon}(1+2\varepsilon^2)}{\sqrt{(1-\varepsilon^2)}} + 2\varepsilon^2(\omega - 2\dot{\gamma})\right)\right] \quad (8)$$

$$F_t = -\frac{\mu L^3 r_J{}^3}{2\Delta R^2(1-\varepsilon^2)^2}\left[4\dot{\varepsilon} + \frac{\pi}{2}(\omega - 2\dot{\gamma})\sqrt{(1-\varepsilon^2)}\right] \quad (9)$$

Ratio coefficient between eccentric distance and radial clearance:

$$\varepsilon = \frac{e}{\Delta R} \quad (10)$$

Type, ***e*** for shaft, bearing, eccentric distance between ΔR for radial clearance, ε over time and the rate of change of formula:

$$\dot{\varepsilon} = \frac{\dot{e}}{\Delta R} \quad (11)$$

The minimum oil film thickness h_{min} can be expressed as:

$$h_{min} = \Delta R(1-\varepsilon) \quad (12)$$

L represents the width of the bearing.
r_J represents the radius of the bearing.
ω for relative rotation angular velocity between the shaft and bearing.
$\dot{\gamma}$ angular velocity between axis fixed along with the shaft and shaft-bearing eccentric vector.

In practical engineering design, require minimal oil film thickness h_{min} should not be less than 0.00015 times the bearing inner diameter.

According to the mechanical structure and dynamic characteristics of the body tube and shaft sleeve, the mechanical equation can be expressed as:

For the infinite short bearing, the normal oil film force ***Fr*** and radial oil film force ***Ft*** are:

$$\boldsymbol{F}_r = -\frac{\mu \boldsymbol{L}^3 \boldsymbol{r}_J}{2\Delta \boldsymbol{R}^2\left(1-\varepsilon^2\right)^2}\left[\left(\frac{\pi\dot{\varepsilon}\left(1+2\varepsilon^2\right)}{\sqrt{\left(1-\varepsilon^2\right)}}\right)\right] \tag{13}$$

$$\boldsymbol{F}_t = -\frac{\mu \boldsymbol{L}^3 \boldsymbol{r}_J{}^3}{2\Delta \boldsymbol{R}^2\left(1-\varepsilon^2\right)^2}[4\dot{\varepsilon}] \tag{14}$$

For infinite long bearing, normal oil film force ***Fr*** and radial oil film force ***Ft*** are:

$$\boldsymbol{F}_r = -\frac{\mu \boldsymbol{L}^3 \boldsymbol{r}_J}{2\Delta \boldsymbol{R}^2\left(1-\varepsilon^2\right)^2}\left[\left(\frac{\pi\dot{\varepsilon}\left(1+2\varepsilon^2\right)}{\sqrt{\left(1-\varepsilon^2\right)}}\right)\right] \tag{15}$$

$$\boldsymbol{F}_t = -\frac{12\mu \boldsymbol{L}\boldsymbol{r}_J{}^3}{\Delta \boldsymbol{R}^2}\left[\left(\frac{2\varepsilon\dot{\varepsilon}}{(2+\varepsilon)^2\left(1-\varepsilon^2\right)}\right)\right] \tag{16}$$

C. *Friction Modeling Analysis*

In the process of moving the turret, the acting force between the barrel tube and the sleeve is mainly manifested as nonlinear collision force and oil film force. In the process of working, sliding friction will occur between the barrel tube and the sleeve. In addition to the original collision force and oil film force, tangential friction force will also appear.

In the interval $v_s \leq |v_T| \leq v_d$ uses the Heaviside step function of cubic polynomial approximation, the adjusted tangential friction calculation formula

$$\boldsymbol{F}_t = -\mu(\boldsymbol{v}_T)\boldsymbol{F}_N \boldsymbol{sign}(\boldsymbol{v}_T) \tag{17}$$

$$\mu(v_T) = \begin{cases} -\mu_d sign(v_T) & |v_T| > v_d \\ -\left\{\mu_d + (\mu_s - \mu_d)\left(\frac{|v_T|-v_s}{v_d-v_s}\right)^2\left[3 - 2\left(\frac{|v_T|-v_s}{v_d-v_s}\right)\right]\right\} sign(v_T) & |v_T| > v_d \\ 2\mu_d\left(3\left(\frac{v_T-v_d}{2v_d}\right)^2 - 2\left(\frac{v_T-v_d}{2v_d}\right)^3 - \frac{1}{2}\right) & |v_T| \leq v_s \end{cases} \tag{18}$$

Parameters:

v_s large static friction critical speed.
μ_s the coefficient of friction.
v_d big slide friction.
μ_d kinetic friction coefficient.

v_T tangential sliding speed.

D. Establish the Kinetic Equation

Nonlinear overforce modeling with lubrication.

Parameter c_0 can be determined by fit tolerance between shaft and bearing, and its expression is as follows:

$$c_0 = \Delta R - \frac{(\Delta d_{max} + \Delta d_{min})}{4} \tag{19}$$

Δd_{max}, Δd_{min} are the maximum and minimum diameter clearance values of the shaft and the bearing under the determined mating grade, respectively.

Parameter e is the eccentric distance between the shaft and the bearing.

1. nWhen $e \leq \Delta R$, the nonlinear overforce manifests as oil film forceF_l

$$F_r = \frac{\mu L^3 r \pi \dot{\varepsilon}(1 + 2\dot{\varepsilon})}{2\Delta R^2 \sqrt[3]{(1 + \varepsilon^2)^2}} \tag{20}$$

$$F_t = \frac{2\mu L^3 r^3 \varepsilon \dot{\varepsilon}}{\Delta R^2 (1 + \varepsilon^2)^2} \tag{21}$$

F_r, F_t represent the normal and tangential oil film forces respectively.

In the analysis of characteristics, only normal oil film force is considered.

2. When $e \geq \Delta R + c_o$, the nonlinear excessive force Fd manifests as collision force FN and friction force FT

$$F_N = K\delta^n \left[1 + \frac{3(1 - e^2)}{4} \cdot \frac{\dot{\delta}}{\dot{\delta}^{(-)}}\right] \tag{22}$$

$$F_t = -\mu(v_T) F_N sign(v_T) \tag{23}$$

$$F_t = -\mu(v_T) K\delta^n \left[1 + \frac{3(1 - e^2)}{4} \bullet \frac{\dot{\delta}}{\dot{\delta}^{(-)}}\right] sign(v_T) \tag{24}$$

The friction factor is:

$$\mu(v_T) = \begin{cases} -\mu_d sign(v_T) & |v_T| > v_d \\ -\left\{\mu_d + (\mu_s - \mu_d)\left(\frac{|v_T| - v_s}{v_d - v_s}\right)^2 \left[3 - 2\left(\frac{|v_T| - v_s}{v_d - v_s}\right)\right]\right\} sign(v_T) & |v_T| > v_d \\ 2\mu_d \left(3\left(\frac{v_T - v_d}{2v_d}\right)^2 - 2\left(\frac{v_T - v_d}{2v_d}\right)^3 - \frac{1}{2}\right) & |v_T| \leq v_s \end{cases}$$

(25–27)

3. When $\Delta R < e \le \Delta R + c_o$, the nonlinear excessive force F_k behaves as oil film force F_l, collision force FN and friction force F_T

$$F_K = \frac{(\Delta R + c_o)}{c_o} F_l + \frac{e - \Delta R}{c_o} F_d \tag{28}$$

E. Dynamic Model of Time-varying Damping

1. nWhen $e \le \Delta R$, nonlinear excess force $F(t)$ behaves as oil film force

$$[M]\ddot{x} + [\alpha M + \beta K + i\gamma K]\dot{x} + [K]x = F(t) \tag{29}$$

$$F(t) = \frac{\mu L^3 r\pi \dot{\varepsilon}(1 + 2\dot{\varepsilon})}{2\Delta R^2 \sqrt[3]{(1 + \varepsilon^2)^2}} \tag{30}$$

2. When $e \ge \Delta R + c_o$, nonlinear excessive force $F(t)$ behaves as collision force and friction force

$$[M]\ddot{x} + [\alpha M + \beta K + i\gamma K]\dot{x} + [K]x = F(t) \tag{31}$$

$$F(t) = -\mu(v_T) K \delta^n \left[1 + \frac{3(1 - e^2)}{4} \cdot \frac{\dot{\delta}}{\dot{\delta}^{(-)}}\right] sign(v_T) \tag{32}$$

the friction factor:

$$\mu(v_T) = \begin{cases} -\mu_d sign(v_T) & |v_T| > v_d \\ -\left\{\mu_d + (\mu_s - \mu_d)\left(\frac{|v_T| - v_s}{v_d - v_s}\right)^2\left[3 - 2\left(\frac{|v_T| - v_s}{v_d - v_s}\right)\right]\right\} sign(v_T) & v_s < |v_T| < v_d \\ 2\mu_d\left(3\left(\frac{v_T - v_d}{2v_d}\right)^2 - 2\left(\frac{v_T - v_d}{2v_d}\right)^3 - \frac{1}{2}\right) & |v_T| \le v_s \end{cases} \tag{33–35}$$

3. When $\Delta R < e \le \Delta R + c_o$, the nonlinear excessive force $F(t)$ behaves as oil film force, collision force and friction force

$$[M]\ddot{x} + [\alpha M + \beta K + i\gamma K]\dot{x} + [K]x = \frac{(\Delta R + c_o)}{c_o} F_l + \frac{e - \Delta R}{c_o} F_d \tag{36}$$

$$F_l = \frac{\mu L^3 r\pi \dot{\varepsilon}(1 + 2\dot{\varepsilon})}{2\Delta R^2 \sqrt[3]{(1 + \varepsilon^2)^2}} \tag{37}$$

$$F_d = -\mu(v_T) K \delta^n \left[1 + \frac{3(1 - e^2)}{4} \bullet \frac{\dot{\delta}}{\dot{\delta}^{(-)}}\right] sign(v_T) \tag{38}$$

3 Conclusion

Based on L–N contact force model in this paper, through building the dynamic model, considering the influence of the clearance of the oil film as well as in the case of sliding friction force and the effect of building the dynamics model of sliding problem with clearance, avoid the simplifying of the general dynamics model to dynamic model is not accurate, by putting the oil film clearance and sliding factors consideration, building dynamic model can more accurately describe the mechanical changes in the process of movement.

Acknowledgements This research is from a national ministry project, the project number is JCKY2017208A001. This research is supported by the school for data and experimental assistance. The authors would appreciate for their help.

Bibliography

1. Erkaya S (2012) Prediction of vibration characteristics of a planar mechanism having imperfect joints using neural network [J]. J Mech Sci Technol 26(5):1419–1430
2. Mukras S, Kim NH, Mauntler NA (2010) Analysis of planar multibody systems with revolute joint wear [J]. Wear 268:643–652
3. Xu LX, Li YG (2014) Investigation of joint clearance effects on the dynamic performance of a planar 2-DOF pick-and-place parallel manipulator [J]. Robot Comput-Integr Manuf 30:62–73
4. Silva PC, Silva MT, Martins JM (2010) Evaluation of the contact forces developed in the lower limb/orthosis interface for comfort design [J]. Multibody SysDyn 24:367–388
5. Bai ZF, Zhao Y (2012) Dynamics modeling and quantification analysis of multibody systems including revolute clearance joints [J]. Precis Eng 36:554–567
6. Askari E, Flores P, Dabirrahmani D, Appleyard R (2014) Study of the friction-induced vibration and contact Mechanics of artificial hip joints [J]. Tribol Int 70:1–10
7. Bai ZF, Zhao Y (2013) A hybrid contact force model of revolute joint with clearance for planar mechanical systems [J]. Int J Non-Linear Mech 48:15–36
8. Wang XP, Liu G (2015) Study on impact dynamics of development for solar panel with clearance joint [J]. Procedia Engineering 99:1345–1357

A Comprehensive Evaluation Approach for Vehicle-Infrastructure Cooperation System Using AHP and Entropy Method

Wanyu Niu, Xiaofeng Liu, Dongpeng Yue, Fan Zhang, and Yonggang Yu

Abstract The vehicle-infrastructure cooperation system (VICS) has been widely used and obtained the world-wide attention. Hence, it is imperative to effectively evaluate the safety, effectiveness, and service capability of this system. Existing research mainly focus on its functional evaluation, and seldom consider the influence of human–vehicle–road–cloud cooperation, decision-making control, and system adaptability on the evaluation result. In this condition, this study selected twenty-one evaluation indicators at the aspects of environmental perception and positioning accuracy, communication and transmission capability, application scenario, decision-making control effect, and system adaptability. Then, an evaluation index system of VICS was proposed. Next, the method combining analytic hierarchy process (AHP) and entropy method was adopted, and a framework of multi-mode communication VICS was built. Finally, a field testing was implemented, and the study results show that the proposed method can evaluate the VICS effectively and comprehensively.

Keywords VICS · AHP · Entropy method · Test analysis

1 Introduction

Over the past decade, the vehicle-infrastructure cooperation system (VICS) has developed rapidly to improve traffic efficiency and traffic safety, such as vehicle active safety control, road traffic cooperation management and human–vehicle–road cooperation through the information interaction of V2V and V2I [1]. In 2003, the European

W. Niu · X. Liu (✉) · D. Yue
School of Automotive and Transportation, Tianjin University of Technology and Education, Tianjin 300222, China
e-mail: microbreeze@126.com

F. Zhang
Tianjin E-Hualu Information Technology Limited Company, Tianjin 300350, China

Y. Yu
Tianjin Highway Development Service Center, Tianjin 300170, China

W. Wang et al. (eds.), *Green Transportation and Low Carbon Mobility Safety*, Lecture Notes in Electrical Engineering 944,
https://doi.org/10.1007/978-981-19-5615-7_15

Commission proposed the e-Safety plan to make full use of advanced information and communication technologies, to accelerate the development and integrated application of safety systems to improve road traffic safety. In 2006, Japan launched the Smartway project to alleviate traffic congestion and improve traffic safety. In 2010, the United States proposed the Intelli-Drive plan to study communication and practical testing under VICS, and tested the vehicle safety assistance system. In 2020, the United States released the "Intelligent Transportation System Strategic Plan 2020–2025", which aimed at the continuous promotion of autonomous driving. In recent years, China has made great progress in terms of the VICS research, demonstration, and operation. Some VICS demonstrations have been carried out on closed sites (e.g., industrial parks, ports, and unmanned mining areas), semi-open roads, and freeways. In 2020, China released the "Intelligent Vehicle Innovation Development Strategy", which focused on the integration of VICS and demonstration applications.

As to the VICS, many studies had been conducted. Arnaout [2] and Matsumoto [3] studied the application of VICS in traffic control, and the study results showed that the use of VICS can alleviate traffic congestion and reduce traffic emissions. As to the field test and evaluation of VICS, Chai [4] built a fuzzy evaluation model, which considered the safety of the vehicle-vehicle interactive system, the safety of the vehicle-infrastructure interactive system, and the traffic signal coordinated control system. This paper does not calculate all the indicators, which may lead to incomplete evaluation results. Zhang [5] and Zhao [6] both established comprehensive evaluation system, and selected evaluation indicators at the aspects of traffic efficiency and safety, but they ignored the characteristics of VICS.

In summary, the current research mainly focus on the simulation-based functional evaluation of VICS, and seldom consider the impact of human–vehicle–road–cloud coordination, decision-making control effect, and system adaptability on the evaluation results using the field experiments. Therefore, this study proposes a VICS evaluation index system, which is consisted of environmental perception and positioning accuracy, communication and transmission capability, application scenario, decision-making control effect, and system adaptability. Then, AHP and entropy method [7] are adopted to determine the weight coefficients of different evaluation indexes, which has the advantages of fusing the subjective and objective influence. Finally, a case study using the filed experiment framework is conducted to verify the effectiveness of the proposed approach.

2 Framework of VICS Evaluation Indicator System

The VICS comprehensive evaluation should consider the system function, adaptability, control effect, and application scenario; thus the evaluation indicators should be selected scientifically, typically, comparatively, and quantitatively. The framework of VICS evaluation indicator system is presented in Table 1. As Table 1 shown, there are five second-level indicators and twenty-one third-level indicators.

Table 1 Evaluation indicator system of VICS

First-level indicators	Second-level indicators	Third-level indicators
Comprehensive evaluation of VICS (A)	Environmental perception and positioning accuracy (B_1)	Recognition accuracy of pedestrian/non-motor vehicle (C_1)
		Recognition accuracy of target vehicle (C_2)
		Recognition accuracy of vehicle driving state (C_3)
		Recognition accuracy of traffic signal and traffic sign (C_4)
		Recognition accuracy of lane line (C_5)
	Communication and transmission capability (B_2)	Communication distance (C_6)
		Communication delay (C_7)
		Communication transmission rate (C_8)
		The number of connected/covered vehicle (C_9)
	Application scenario (B_3)	The number of safety scenario (C_{10})
		The number of efficiency scenario (C_{11})
		The number of information service scenario (C_{12})
	Decision-making and control effect (B_4)	Warning accuracy rate of vehicle operating conditions (C_{13})
		Accuracy rate of traffic information reminder (C_{14})
		Accuracy rate of collision warning (C_{15})
		Improvement of traffic delay (C_{16})
	System adaptability (B_5)	Compatibility of communication equipment (C_{17})
		Coordination level of "Human–vehicle–road–cloud" system (C_{18})
		Anti-interference ability (C_{19})
		System fault tolerance and recovery processing capability (C_{20})

(continued)

Table 1 (continued)

First-level indicators	Second-level indicators	Third-level indicators
		Information security level (C_{21})

2.1 Environmental Perception and Positioning Accuracy

One of the important prerequisites for the development of VICS is environmental perception and positioning accuracy. In different traffic scenarios, there may occur human-vehicle and vehicle-vehicle conflicts. Therefore, it is necessary for VICS to accurately obtain traffic information, such as road environment information and real-time status of traffic participants, etc. So we select recognition accuracy of pedestrian/non-motor vehicle, recognition accuracy of target vehicle, recognition accuracy of vehicle driving state, recognition accuracy of traffic signal and traffic sign, and recognition accuracy of lane line as the third-level indicators.

2.2 Communication and Transmission Capability

In the vehicle scene, the relative moving speed of the vehicle is high, channel environment is complex. Therefore, it has a great influence on improving operating safety and reducing traffic accidents through a high-speed and low-latency communication technology. Currently, the commonly used communication technologies include DSRC (Dedicated Short Range Communication), LTE (Long Term Evolution)-V, 5G, etc. DSRC technology is highly reliable; LTE-V (4G) deployment is relatively easy, and its spectrum bandwidth allocation is flexible and transmission is reliable; 5G transmission rate is high and communication delay is low [8]. In the evaluation of communication technology, communication distance, communication delay and communication transmission rate are indispensable. When the number of the connected/covered vehicle (OBU) is large, it will have a great impact on the communication transmission rate and communication delay. Therefore, the number of connected/covered vehicle should be considered.

2.3 Application Scenario

For the implementation and application of VICS, it is necessary to increase the demonstration of application scenarios. At present, China has carried out the deployment, testing and application of VICS in some closed/semi-closed sites and expressways. Based on the "Cooperative Intelligent Transportation System-Vehicle Communication System Application Layer and Application Data Interaction Standard"

(T/CSAE 53-2017) released by the China Society of Automotive Engineering, this study divides the application scenarios into three types: safety, efficiency and information services. So we selected the number of safety scenarios, the number of efficiency scenarios, and the number of information service scenarios as the third-level indicators.

2.4 Decision-Making and Control Effect

VICS should have the functions of perception, planning, and decision-making, and its decision-making and control effect has important practical significance for promoting the application of VICS. VICS accurately obtains the surrounding traffic environment information, but the decision-making and control system do not make corresponding measures in time, which will also affect traffic safety and traffic efficiency. Therefore, we need to consider the warning accuracy rate of vehicle operating conditions, accuracy rate of traffic information reminder, accuracy rate of collision warning, and improvement of traffic delay as third-level indicators.

2.5 System Adaptability

There are many sensors relating to VICS, such as OBU, RSU, communication equipment, positioning equipment, mobile terminals and so on. It is necessary to consider the cooperation of "human–vehicle–road–cloud". In addition, the cloud platform of VICS requires real-time transmission, processing, and publishing information, which involves many links and may cause poor system compatibility. Weather, electromagnetic and information security also have effects on system adaptability. So we select compatibility of communication equipment, coordination level of "Human–vehicle–road–cloud" system, anti-interference ability, system fault tolerance and recovery processing capability, and information security level as the third-level indicators.

2.6 Rating Values of Evaluation Indicators

To better quantitatively assess the VICS, each indicator has five-level ranking values, that is, A [100,90], B (90,80], C (80,70) D (70,60], and E (60,0). In addition, the ranking values of these indicators are listed in Table 2.

Table 2 Classification of evaluation indicators

Evaluation index	A	B	C	D	E	Unit
C_1	[100,90]	(90,80]	(80,70]	(70,60]	(60,0]	%
C_2	[100,90]	(90,80]	(80,70]	(70,60]	(60,0]	%
C_3	[100,90]	(90,80]	(80,70]	(70,60]	(60,0]	%
C_4	[100,90]	(90,80]	(80,70]	(70,60]	(60,0]	%
C_5	[100,90]	(90,80]	(80,70]	(70,60]	(60,0]	%
C_6	>1000	[1000,500)	[500,300)	[300,100)	≤100	meter
C_7	≤20	[40,20)	[70,40)	[100,70)	>100	ms
C_8 (upload/download)	≥ 100 ≥ 300	(100,50] (300,150]	(50,30] (150,50]	(30,20] (50,30]	(20,0] (30,0]	Mbps
C_9	>30	[30,20)	[20,10)	[10,5)	[5,0]	Vehicles
C_{10}	>10	[10,8)	[8,5)	[5,2)	[2,0]	no.
C_{11}	>5	[5,4)	[4,2)	[2,1)	[1,0]	no.
C_{12}	>5	[5,4)	[4,2)	[2,1)	[1,0]	no.
C_{13}	[100,90]	(90,80]	(80,70]	(70,60]	(60,0]	%
C_{14}	>10	[10, 5)	[5, 2)	[2, 0]	<0	%
C_{15}	[100,90]	(90,80]	(80,70]	(70,60]	(60,0]	%
C_{16}	[100,90]	(90,80]	(80,70]	(70,60]	(60,0]	Scores
C_{17}	[100,90]	(90,80]	(80,70]	(70,60]	(60,0]	Scores
C_{18}	[100,90]	(90,80]	(80,70]	(70,60]	(60,0]	Scores
C_{19}	[100,90]	(90,80]	(80,70]	(70,60]	(60,0]	Scores
C_{20}	[100,90]	(90,80]	(80,70]	(70,60]	(60,0]	Scores
C_{21}	[100,90]	(90,80]	(80,70]	(70,60]	(60,0]	Scores

3 Comprehensive Evaluation Approach for VICS

This study proposed a hybrid comprehensive evaluation method, that is, AHP and entropy method. First, AHP can effectively combine both the qualitative and quantitative analysis. Second, the entropy method calculates the index weights according to the degree of variation of the index values, which is an objective weighting method to avoid the deviation caused by human factors. Finally, the comprehensive evaluation approach could make full use of the experts experiences and test data. The framework of the propose method is shown in Fig. 1.

3.1 AHP

AHP is a multi-plan or multi-objective decision-making method that combines qualitative and quantitative analysis. At present, this method has been widely used in many

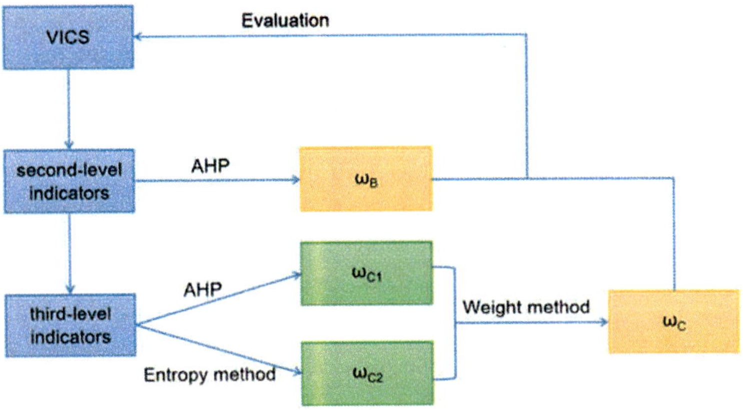

Fig. 1 Framework of the proposed method

fields, and it has also been widely used in the field of transportation [9]. The AHP steps are described as follows. First, establishing the hierarchy analysis structure. Then, constructing judgment matrix A_0 using 1–9 scale method. Finally, performing the consistency test to obtain the indicator weights, and selecting the max weight as the evaluation results.

3.2 *Entropy Method*

The entropy method is an objective assignment method, which determines the weights of different indicators according to the information provided by the observed data of each indicator. At present, this method has been widely used in the fields of agriculture, transportation [10], and its steps are stated as follows.

(1) Calculate the weight of the j-th index

$$\begin{cases} R_{ij} = x_{ij} / \sum_{i=1}^{m} x_{ij} \\ e_j = -\dfrac{1}{\ln m} \cdot \sum_{i=1}^{m} R_{ij} \cdot \ln R_{ij} \\ g_j = 1 - e_j \\ \omega_j = g_j / \sum_{j=1}^{n} g_j \end{cases} \tag{1}$$

where, x_{ij} is the detection accuracy value or the degree of membership of the j-th index in the i-th VICS; e_j is the entropy value of the j-th evaluation index; g_j is the difference coefficient of the j-th evaluation index; ω_j is the weight of the j-th secondary evaluation index.

(2) Calculate the comprehensive score of each scheme

$$S_i = \sum_{j}^{m} \omega_j \times R_{ij} \tag{2}$$

where, S_i is the comprehensive score of the i-th scheme.

3.3 Comprehensive Calculation of Coefficient Weight

To make the weights more reasonable, we used the weighting method [11] to further process the weights obtained above. The weighting method obtains a more reasonable weight by processing the index weights of different methods.

$$\begin{cases} \overline{S} = \omega_1 S_1 + \omega_2 S_2 + \cdots + \omega_n S_n \\ \sum_{i=1}^{n} \omega_i = 1 \end{cases} \tag{3}$$

where, ω is the weight set of different methods; S is the comprehensive score set of different methods. In this study, there are two methods, that is, AHP and entropy method.

4 Case Study

4.1 Establishment of VICS

A VICS experimental system was established at the campus of Tianjin University of Technology and Education in 2020, which was consisted of OBU, RSU, antenna, traffic light, and software. OBU can read the CAN information, and send warning information. With built-in WIFI, DSRC, 4G communication module, RSU communication distance is greater than 500 m. Combined antenna includes DSRC/LTE-V dual-mode communication antenna, GPS/BDS positioning antenna, and 4G cellular antenna. Mobile traffic lights support WIFI, DSRC, 4G and other communication methods, and the visible distance is greater than 300 m.

RSU is installed on the traffic lights and connects to the cloud platform through 4G communication module. Therefore, it can realize the information interaction among

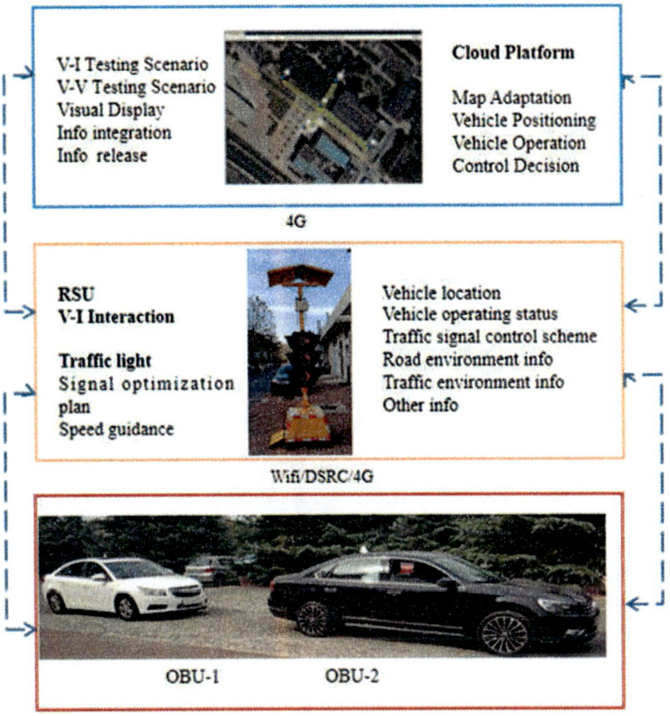

Fig. 2 The VICS experimental system

OBU, RSU, and traffic lights. The positioning accuracy of the VICS is less than 2 m, the average delay is less than 40 ms, and the data update frequency is 10 Hz. The VICS experimental system is shown in Fig. 2.

4.2 Experimental Test

The established VICS was used to conduct the communication and interaction of roads, vehicle, traffic signals, and to develop tests and evaluations of scenarios (e.g., collision warning, vehicle speed guidance, curve warning). The test scenario was shown in Fig. 3, and the test procedure was conducted as follows. The target vehicle traveled at the speed of 15 km/h, and the test vehicle followed at the same speed. When the driving was stable, the test vehicle started to accelerate, and then a forward collision warning was triggered when the safety-following threshold was reached. In addition, we used virtual information prompt sites to establish other scenarios, such as road construction, curved roads, and wet roads.

Fig. 3 Testing situation of VICS

4.3 *Evaluation and Analysis*

Due to the limited investment, this study completed some testings using evaluation indicators in Table 1. Therefore, eleven evaluation indicators were adopted, i.e., vehicle driving status recognition, communication distance, communication delay, communication transmission rate, the number of security/efficiency/information service scenario, vehicle operating condition warning, traffic information prompt, collision warning, and "human–vehicle–road–cloud" system. Other ten indicators are evaluated based on the ranking scores of experts.

(1) Index weight calculation.

1–9 scale method was applied for the second-level index layer to obtain the judgment matrix A_0, and then the sum-product method was used to calculate the weight vector of the second-level index: B = {Environmental perception and positioning accuracy, communication and transmission capability, application scenario, decision-making control effect, system adaptability}. The weights of these five indicators were obtained as follows.

$$\omega_B = \left(0.207\ 0.320\ 0.257\ 0.089\ 0.127\right)$$

Then, the consistency test was conducted, and CR equaled to 0.018, which was significantly smaller than 0.1, this demonstrated that the judgment matrix A_0 had acceptable consistency. Next, the weight values of ω_{C1} and ω_{C2} can be obtained based on Sect. 3.2 and 3.3, and $\overline{\omega}$ was the average of ω_{C1} and ω_{C2}, which are listed in Table 3.

(2) Evaluation.

Forty-five experts in the fields of traffic design, traffic control, traffic organization, wireless communication, and vehicle engineering had been invited, to score the indicators for three testing schemes. The third-level evaluation indexes of three schemes were scored, and the evaluation score using the proposed method was listed in Table 4, and the comprehensive evaluation scores using different methods were listed in Table 5.

(3) Analysis.

It can be seen that the evaluation scores of three VICS schemes are 87.82, 83.88, and 85.90 respectively. The scheme I ranks first, and these three schemes are

Table 3 Index weights

First level indicators	Second level indicators	Third level indicators			
			ω_{C1}	ω_{C2}	$\overline{\omega}$
Comprehensive evaluation of VICS (A)	B_1 0.207	C_1	0.079	0.154	0.1165
		C_2	0.263	0.252	0.2575
		C_3	0.202	0.231	0.2165
		C_4	0.175	0.194	0.1845
		C_5	0.281	0.169	0.2250
	B_2 0.320	C_6	0.243	0.213	0.2280
		C_7	0.340	0.361	0.3505
		C_8	0.243	0.175	0.2090
		C_9	0.174	0.251	0.2125
	B_3 0.257	C_{10}	0.539	0.455	0.4970
		C_{11}	0.297	0.324	0.3105
		C_{12}	0.164	0.221	0.1925
	B_4 0.089	C_{13}	0.235	0.312	0.2735
		C_{14}	0.138	0.097	0.1175
		C_{15}	0.489	0.384	0.4365
		C_{16}	0.138	0.207	0.1725
	B_5 0.127	C_{17}	0.122	0.153	0.1375
		C_{18}	0.176	0.232	0.2040
		C_{19}	0.246	0.302	0.2740
		C_{20}	0.288	0.213	0.2505
		C_{21}	0.168	0.100	0.1340

Table 4 Scores of the third-level evaluation indexes and different schemes

	B_1	B_2	B_3	B_4	B_5	Evaluation score
Scheme I	92.47	88.92	78.51	95.01	79.24	87.82
Scheme II	83.41	82.89	82.16	91.87	79.45	83.88
Scheme III	84.03	91.80	80.28	82.75	84.57	85.90

Table 5 Evaluation scores using different methods

	AHP	Entropy method	Proposed method
Scheme I	88.09	87.55	87.82
Scheme II	84.09	83.68	83.88
Scheme III	86.06	85.75	85.90

scored at the level of B. Compared with AHP, the proposed method can make full use of the test data and reduce human factors. Compared with entropy method, the proposed method can quantify the indicators which are lack of experimental data. The proposed comprehensive evaluation method can make full use of the existing experimental data and expert experience, integrate the subjective and objective factors, and reduce the errors of the single evaluation method.

5 Conclusions

For the testing and evaluation of VICS, this paper proposed an evaluation method based on AHP and entropy method, and conducted an empirical study on the three schemes of VICS. The results show that the proposed method can evaluate the VICS objectively and comprehensively. However, with the advancement of technologies, traffic application scenarios, and decreasing cost, the VICS comprehensive evaluation needs continuous and further improvement in the future.

Acknowledgements This research was supported by the Science and Technology Plan Project of Tianjin, China (22YDTPJC00120, 2021YJSO2S32, XC202028, 2021KJ018).

References

1. Zhang X, Hao S, Wenge R, Cooper D (2012) Intelligent transportation systems for smart cities: a progress review. Sci China Inf Sci 55(12):2908–2914
2. Arnaout G, Bowling S (2011) Towards reducing traffic congestion using cooperative adaptive cruise control on a freeway with a ramp. J Ind Eng Manag 4(4):699–717
3. Matsumoto S, Park T, Kawashima H (2014) A comparative study on fuel consumption reduction effects of eco-driving instructions strategies. Int J Intell Transp Syst Res 12(1):1–8

4. Chai SD (2013) Research on the function test and evaluation method of vehicle-infrastructure cooperative system. Wuhan University of Technology, Hubei (In Chinese)
5. Zhang JM (2014) Research on the test and verification method of mehicle-infrastructure cooperative simulation system. Beijing Jiaotong University, Beijing (In Chinese)
6. Zhang LS (2018) Construction of traffic simulation environment and efficiency evaluation method of VICS. Beijing Jiaotong University, Beijing (In Chinese)
7. Du D (2005) Modern comprehensive evaluation methods and selected cases. Tsinghua University Press, Beijing (In Chinese)
8. Huawei (2020) C-V2X white paper of vehicle-infrastructure integrated intelligent networking system. [R/OL] (In Chinese)
9. Abret N, Bayrak O (2021) The determination of convenient junction type utilizing from estimated traffic data with AHP method. Iran J Sci Technol, Trans Civ Eng 1–15
10. Sun XL, Jia LM, Dong HH (2010) Urban expressway traffic state forecasting based on multimode maximum entropy model. Sci China Technol Sci 53(10):2808–2816
11. Huang Z, Chai J, Bo L, Xi F (2017) Application of index weighting based on the determination method of comprehensive weight coefficient. In: Proceedings of 2017 IEEE 8th International Conference on Software Engineering and Service Science, pp 411–423

An Agent-Based Cellular Automata Model for Urban Road Traffic Flow Considering Connected and Automated Vehicles

Wang Jinghui, Lv Wei, Jiang Yajuan, Qin Shuangshuang, and Huang Guangchen

Abstract Considering the development of the vehicle to vehicle (V2V) technology and the popularisation of connected and automated vehicles (CAVs), for an extended period, urban roads will be in a mixed traffic flow scene where CAVs and human-driven vehicles (HDVs) coexist. This paper uses an agent-based cellular automata model to establish a micro-traffic simulation framework for urban roads, called the ABCA-MS model. Considering the characteristics of the intermittent flow of urban roads and signal light control, corresponding car-following and lane-changing rules are established and applied to simulate mixed traffic flow containing CAVs. The simulation results show that the traffic efficiency and the permeability of CAVs show a positive correlation; under the given traffic volume condition, the critical CAVs penetration rate for a traffic state change from congestion to unblocked is 0.4. When the penetration rate of CAVs is in the range of 0–0.4, the improvement of road traffic efficiency is the most significant, and the effect of improvement gradually slows down with the increase of CAVs penetration. Even with a low penetration rate of CAVS, the road capacity can be effectively improved, and the traffic pressure can be alleviated.

Keywords Connected and automated vehicles · Signal control traffic · Cellular automata · Agent-based model

W. Jinghui · L. Wei (✉) · J. Yajuan · Q. Shuangshuang · H. Guangchen
School of Safety Science and Emergency Management,
Wuhan University of Technology, Wuhan, China
e-mail: weil@whut.edu.cn

L. Wei
China Research Center for Emergency Management,
Wuhan University of Technology, Wuhan, China

W. Wang et al. (eds.), *Green Transportation and Low Carbon Mobility Safety*,
Lecture Notes in Electrical Engineering 944,
https://doi.org/10.1007/978-981-19-5615-7_16

1 Introduction

With the rapid development of autonomous driving technology and network communication technology, connected and automated vehicles (CAVs) have been applied to actual transportation. As an essential part of the intelligent transportation system (ITS), CAVs are also the development of future transportation direction. Researchers believe that the application of CAVs can improve traffic safety performance, reduce the frequency of accidents, and improve traffic efficiency, reduce carbon emissions, and generate significant economic benefits [1, 2]. Therefore, vehicle to vehicle (V2V) technology has received extensive attention from researchers and is also a hotspot in transportation research. Based on the development of V2V technology, in 2040, the proportion of CAVs will reach 50% [3]. Our transportation system will be a hybrid system where CAVs and human-driven vehicles (HDVs) coexist for a long time. Consequently, it is necessary to conduct extensive research on the traffic characteristics and vehicle movement in this scene.

In the 1980s, some scholars conducted research on Autonomous vehicles (AVs) [4]. In 2000, Treiber et al. [5] proposed the Intelligent Driving Model (IDM), which considers the influence of headway on vehicle speed changes, and is widely used in the car-following simulation of AVs. The PATH laboratory has carried out long-term and extensive research on V2V [6–9]. Among its proposed rules, AVs mainly adopt the car-following rules of the adaptive cruise control (ACC) system, and CAVs adopt the car-following rules of the cooperative adaptive cruise control (CACC) system. Compared with AVs, CAVs can obtain the acceleration of the preceding vehicle and maintain a more minor time headway during movement. Relevant scholars have further improved the model for the simulation of mixed traffic scenarios. Yao et al. [10–12] established the following rules of HDVs by using the full velocity difference (FVD) model [13] and the following rules of CAVs by using the CACC car-following model calibrated by the PATH laboratory [8, 9]. The traffic fundamental diagram and traffic stability under different CAVs penetration ratios are studied. Kuang et al. [14] improved the FVD [13] model and established a car-following model that considers the average headway. The simulation results show that the average headway effect plays a vital role in improving the stability of traffic flow and restraining traffic congestion. Kerner et al. [15] use the three-phase traffic theory model, investigate autonomous vehicles' driving strategy, and compare the traffic performance and parameters of AVs and HDVs. For comparison, Rios-torres et al. [16] simulated the vehicle confluence scene on the ramp based on the Gipps car-following model [17]. They studied the traffic fundamental diagram, travel time, and fuel consumption under different penetration rates of CAVs. Wang et al. [18] put forward a receding horizon control framework to model driver assistance and cooperative systems and researched CACC and ACC vehicles. Talebpour et al. [19, 20] established a traffic continuous micro-simulation framework for autonomous driving at different scales. The string stability and traffic throughput are analysed in the scenes of connected vehicles and ordinary vehicles.

On the other hand, cellular automata are widely used in transportation system simulation as a widely used simulation method. Due to the discrete characteristics of cellular automata, its car-following models are also discrete. Some scholars have adopted introduced a smaller cell size to improve the accuracy of the model, such as the MCD model [21], the KKW model [22], and Lee's model [23]. Many scholars have used cellular automata to model and analyse mixed traffic in which CAVs, AVs, and HDVs coexist in recent years. Jiang et al. [24] considered the car-following rules of the constant time headway policy and simulated the mixed traffic of ACC vehicles and HDVs. Yuan et al. [25] used the MCD model [21] and the car-following model based on constant time headway (CTH) policy to study the mixed traffic of hand motion driving and ACC and analyse different headway time and traffic breakdown probability. Lo et al. [26] established four advanced vehicle control and safety systems (AVCSS) car-following rules and compared them with traditional models. Yang et al. [27] analysed the fundamental diagrams of different CAVs ratios on single-lane and dual-lane. Wu et al. [28] studied the ant pheromone path guidance strategy in intelligent transportation systems. Zhao et al. [29–31] used the MCD model to explore the road intersections and two-lane roads on the internet of vehicles (IoV) and non-vehicle traffic. Based on the three-phase traffic flow theory, the MVDE model considering multiple factors was established by improving Gao's VDE model [32] to investigate the impact of different network penetration rates and aggressive and conservative types of drivers on traffic. Chen et al. [33] simulated the mixed traffic scenes of AVs and HDVs by improving the NaSch [34] model. They studied the optimal number of forwarding communication vehicles of AVs and the critical ratio of AVs for the maximum average flow of the system. Tanimoto et al. [35] used the cellular automata model to establish an automated vehicle control system (AVCS) and HDV mixed traffic system and studied the impact of different driving strategies on traffic through three-lane road sections. Zhou et al. [36] established a four-lane model based on the fixed headway rules of the PATH laboratory [8, 9] to study the speed changes of CACC, ACC and HVs mixed traffic. Vranken et al. [37] extended Lee's model that considers deceleration limits [23]. Each timestep is set to 0.1 s to simulate changes in traffic capacity and flux under different scenarios of CAVs penetration rate.

Many scholars have carried out a lot of exploration on the model and method of ITS, but the research scenes modelled by cellular automata are mostly single road sections or ramps. However, the application of CAVs in urban signal control traffic is more concentrated, and the difficulty of modelling is increased due to the characteristics of the intermittent flow of urban traffic and signal light control. This paper applies an agent-based cellular automata model to establish a mixed traffic simulation scenario for urban roads with two road intersections. This work is based on previous studies and the proposed model [38], which we call ABCA-MS. In the model, corresponding to different vehicle types and considering that CAVs degenerate into AVs, different car-following models are proposed to study the impact of varying penetration rates of CAVs on traffic on urban roads.

The structure of this paper is as follows. In Sect. 2, the model framework of ABCA-MS, the car-following model, and the lane changing model are introduced in detail. Then, In Sect. 3, the simulation scene is described, composed of a local road network with two road intersections, and the relevant simulation parameters are given. In Sect. 4, the results of the numerical simulation are analysed. Finally, Conclusions are provided in Sect. 5.

2 Microscopic Traffic Modelling Framework

To model the traffic scenes of urban roads, this paper uses the ABCA-MS model to establish a two-way six-lane road scene under the control of traffic lights. Considering the influence of CAVS degradation in the model, the types and ratios of each vehicle are analysed, and corresponding car-following models are established, respectively. In terms of lateral movement of vehicles, the lane-changing model is established based on the STCA model [39].

2.1 Intersection Control

Road intersections in urban traffic will divide continuous traffic flow into the intermittent traffic flow, and traffic light control is introduced to channel traffic flow in different directions to avoid traffic conflicts. In this model, the ABCA-MS model is used to establish the intersection control module by defining the dynamic and static attributes of the cell; the schematic diagram is shown in Fig. 1.

One signal cycle of the signal lamp is set to the four-phase mode, and the mode diagram is shown in Fig. 2.

2.2 Analysis of Vehicles Type

In a mixed traffic flow scene composed of CAVs and HDVs, CAVs cannot establish V2V communication in certain car-following states, which will lead to the degradation of CAV vehicles. As shown in Fig. 3, there are four car-following cases in such a mixed traffic scene.

Case 1: CAV following CAV

In this case, the follower vehicle and the leader vehicle are both CAVs; the follower vehicle can establish communication with the leader vehicle and obtain information such as the leader car's speed, position, and acceleration. The follower vehicle follows the CAVs rules when moving.

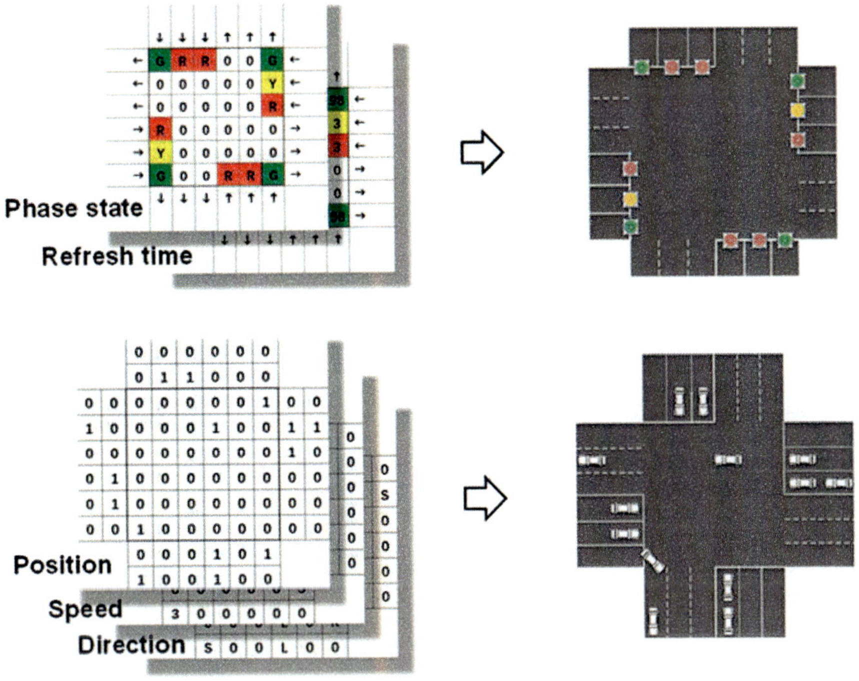

Fig. 1 Schematic diagram of ABCA-MS model

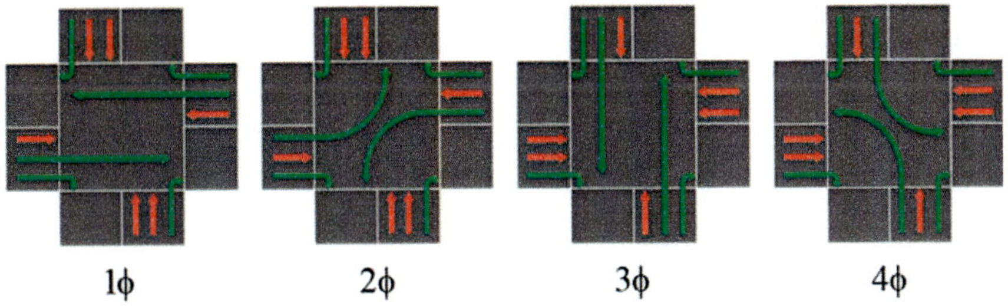

Fig. 2 Four phase release mode

Case 2: CAV following HDV

In this case, the follower vehicle is CAV, and the leader vehicle is HDV; the follower vehicle cannot establish communication with the leader vehicle and can only rely on on-board sensors to obtain the speed and position information of the leader vehicle. The follower vehicle degenerates to AV; follow the AVs car-following rules when moving.

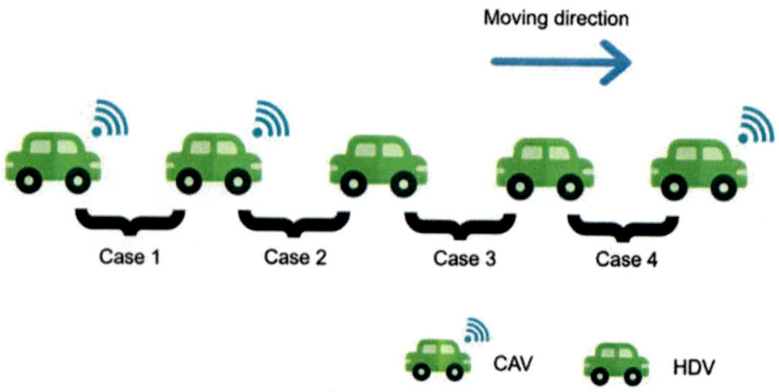

Fig. 3 Car-following cases in mixed traffic flow

Case 3: HDV following HDV

In this case, the follower vehicle and the leader vehicle are both HDVs. The follower car cannot establish communication with the leader vehicle. It can only rely on the driver to judge the position and speed of the leader vehicle, following the HDVs car-following rules when moving.

Case 4: HDV following CAV

In this case, the follower vehicle is HDV, and the leader vehicle is CAV. The follower vehicle cannot establish communication with the leader vehicle and can only rely on the driver to judge the position and speed of the leader car. The car follows the HDVs rules.

2.3 Car-Following Model

Since there are three types of vehicles, three types of car-following rules have been established accordingly. These three rules are developed according to the five steps: (1) determining the probability of deceleration, deceleration and maximum speed, (2) acceleration, (3) deceleration, (4) keep distance, and (5) position update; the model details of different vehicles are different. Compared with AVs, CAVs can maintain a minor time headway, so the space headway is smaller, and the braking is more accurate. Compared with HDVs, the AVs car-following rules do not need to consider the driver's response. Therefore, the time headway of AVs is smaller than that of HDVs, and there is no random slowdown when cruising and slowdown behavior when starting.

2.3.1 HDVs Car-Following Rules

The rules of HDVs are extended from the NaSch model [34]. The deceleration rules are added in the update step, and the vehicle speed adaptation effect and the requirements of the distance between vehicles are taken into account. When the speed of the follower vehicle is higher than the speed of the leader vehicle, its deceleration probability is positively correlated with the speed difference between the follower and leader vehicles and negatively correlated with the distance between the follower and leader vehicles. Equations (1–8) give the expression of the car-following rules (1–8).

1. Determine the probability of deceleration, deceleration and maximum speed.

$$v_{\max}=\begin{cases} 2\delta x/\tau \text{ Vehicelspassingintersections} \\ 4\delta x/\tau \text{ Vehiclescruising} \end{cases} \tag{1}$$

$$p_d=\begin{cases} p_{1-HDV} \text{ if } \mathrm{v}_{(\mathrm{i,t})}=0 \\ p_{2-HDV} \text{ if } \mathrm{v}_{(\mathrm{i,t})}\neq 0 \text{ and } \mathrm{v}_{(\mathrm{i,t})}\leq \mathrm{v}_{(\mathrm{i+1,t})} \\ \max\{\alpha(v_{(i,t)}-v_{(i+1,t)})-\beta d_{(i,t)},0\} \text{ if } v_{(i,t)}\neq 0 \text{ and } \mathrm{v}_{(\mathrm{i,t})}>\mathrm{v}_{(\mathrm{i+1,t})} \end{cases} \tag{2}$$

$$d_{(i,t)}=x_{(i+1,t)}-x_{(i,t)}-1 \tag{3}$$

$$a=\begin{cases} 1\,\delta x/\tau^2 \text{ if } v_{(i,t)}-v_{(i+1,t)}<3\delta x/\tau \\ 2\,\delta x/\tau^2 \text{ if } v_{(i,t)}-v_{(i+1,t)}\geq 3\delta x/\tau \end{cases} \tag{4}$$

2. Acceleration.

$$v_{(i,t+1/3)}=\min\{v_{(i,t)}+1, v_{\max}\} \tag{5}$$

3. Deceleration.

$$v_{(i,t+2/3)}=\max\{v_{(i,t+1/3)}-a,0\}\, if\ rand<p_d \tag{6}$$

4. Keep distance.

$$v_{(i,t+1)}=\min\{v_{(i,t+2/3)}, d_{(i,t)}+v_{(i+1,t)}-ceil(t_{HDV}\cdot v_{(i+1,t)})\} \tag{7}$$

5. Location update.

$$x_{(i,t+1)}=x_{(i,t)}+v_{(i,t+1)} \tag{8}$$

2.3.2 AVs Car-Following Rules

Compared with HDVs, AVs will not slow down due to subjective factors during driving and startup. Besides, the safety response time of AVs is shorter. The expression of the following rules is given by Eqs. (9–16).

1. Determine the probability of deceleration, deceleration and maximum speed.

$$v_{\max}=\begin{cases} 2\delta x/\tau \text{ Vehicelspassingintersections} \\ 4\delta x/\tau \text{ Vehiclescruising} \end{cases} \tag{9}$$

$$p_d=\begin{cases} p_{1-AV}\, if\; v_{(i,t)}=0 \\ p_{2-AV}\, if\; v_{(i,t)}\neq 0\, and\; v_{(i,t)}\leq v_{(i+1,t)} \\ \max\{\alpha(v_{(i,t)}-v_{(i+1,t)})-\beta d_{(i,t)},0\}\, if\; v_{(i,t)}\neq 0\, and\; v_{(i,t)}>v_{(i+1,t)} \end{cases} \tag{10}$$

$$d_{(i,t)}=x_{(i+1,t)}-x_{(i,t)}-1 \tag{11}$$

$$a=\begin{cases} 1\,\delta x/\tau^2 \text{ if } v_{(i,t)}-v_{(i+1,t)}<3\delta x/\tau \\ 2\,\delta x/\tau^2 \text{ if } v_{(i,t)}-v_{(i+1,t)}\geq 3\delta x/\tau \end{cases} \tag{12}$$

2. Acceleration.

$$v_{(i,t+1/3)}=\min\{v_{(i,t)}+1,v_{\max}\} \tag{13}$$

3. Deceleration.

$$v_{(i,t+2/3)}-a,0\}\text{ if } rand<p_d \tag{14}$$

4. Keep distance.

$$\begin{cases} v_{(i,t+1)}=\min\{v_{(i,t+2/3)},d_{(i,t)}+v_{(i+1,t)}-round(t_{ACC}\cdot v_{(i,t)})\}\text{ if } v_{(i,t)}>1 \\ v_{(i,t+1)}=\min\{v_{(i,t+2/3)},d_{(i,t)}+v_{(i+1,t)}-1\}\text{ if } v_{(i,t)}=1 \text{ and } v_{(i+1,t)}>0 \\ \text{and rand} < t_{ACC}\cdot v_{(i,t)} \\ v_{(i,t+1)}=\min\{v_{(i,t+2/3)},d_{(i,t)}+v_{(i+1,t)}\}\text{ otherwise} \end{cases} \tag{15}$$

5. Location update.

$$x_{(i,t+1)}=x_{(i,t)}+v_{(i,t+1)} \tag{16}$$

2.3.3 CAVs Car-Following Rules

Compared with AVs, CAVs can obtain the speed and acceleration information of the leader vehicle in real-time. Its safety response time is the least among the three kinds of vehicles. The expression of the following rules is given by Eqs. (17–25).

1. Determine the probability of deceleration, deceleration and maximum speed.

$$V_{\max}=\begin{cases} 2\delta x/\tau \text{ Vehicelspassingintersections} \\ 4\delta x/\tau \text{ Vehiclescruising} \end{cases} \tag{17}$$

$$p_d = \begin{cases} p_{1-CAV} \text{ if } v_{(i,t)} = 0 \\ p_{2-CAV} \text{ if } v_{(i,t)} \neq 0 \text{ and } \mathrm{v}_{(\mathrm{i,t})} \leq \mathrm{v}_{(\mathrm{i+1,t+1})} \\ \max\left\{\alpha(v_{(i,t)} - v_{(i+1,t+1)}) - \beta(d_{(i,t)} + v_{(i+1,t+1)}), 0\right\} \text{ if } v_{(i,t)} \neq 0 \\ \text{and } \mathrm{v}_{(\mathrm{i,t})} > \mathrm{v}_{(\mathrm{i+1,t+1})} \end{cases} \tag{18}$$

$$\begin{cases} v_{(i+1,t+1)} = v_{(i+1,t)} + a_{(i+1,t+1)} \\ d_{(i,t)}=x_{(i+1,t)} - x_{(i,t)} - 1 \end{cases} \tag{19}$$

$$a = \begin{cases} 1\,\delta x/\tau^2 \text{ if } v_{(i,t)} - v_{(i+1,t+1)} < 3\delta x/\tau \\ 2\,\delta x/\tau^2 \text{ if } v_{(i,t)} - v_{(i+1,t+1)} \geq 3\delta x/\tau \end{cases} \tag{20}$$

2. Acceleration.

$$v_{(i,t+1/3)} = \min\{v_{(i,t)} + 1, v_{\max}\} \tag{21}$$

3. Deceleration.

$$v_{(i,t+2/3)} = \max\{v_{(i,t+1/3)} - a, 0\} \text{ if } rand < p_d \tag{22}$$

4. Keep distance.

$$\begin{cases} v_{(i,t+1)} = \min\{v_{(i,t+2/3)}, d_{(i,t)} + v_{(i+1,t+1)} - round(t_{CACC} \cdot v_{(i,t)})\} \\ \text{if } v_{(i,t+2/3)} > 1 \\ v_{(i,t+1)} = \min\{v_{(i,t+2/3)}, d_{(i,t)} + v_{(i+1,t+1)} - 1\} \text{ if } v_{(i,t+2/3)}=1 \\ \text{and } v_{(i+1,t+1)} > 0 \text{ and } rand < t_{CACC} \cdot v_{(i,t)} \\ v_{(i,t+1)} = \min\{v_{(i,t+2/3)}, d_{(i,t)} + v_{(i+1,t+1)}\} \text{ otherwise} \end{cases} \tag{23}$$

5. Location update.

$$x_{(i,t+1)}=x_{(i,t)} + v_{(i,t+1)} \tag{24}$$

Table 1 Related variables and means of car-following model

Variable	Means
τ	Time discretization interval
δx	Space discretization length
p_d	Deceleration probability of vehicle
a	Deceleration of vehicle i
$a_{(i+1,t+1)}$	Acceleration of vehicle $i+1$ at the time $t+1$
t_{HDV}	Safety response time of HDV
t_{AV}	Safety response time of AV
t_{CAV}	Safety response time of CAV
$v_{(i,t)}$	The velocity of the vehicle i at the time t
$x_{(i,t)}$	Position of the vehicle i at the time t
$v_{\max}$	The maximum speed of the vehicle
$d_{(i,t)}$	Bumper-to-bumper space gap between vehicle i and its leader (i+1)

where, p_{1-HDV}, p_{2-HDV}, p_{1-AV}, p_{2-AV}, p_{1-CAV}, p_{2-CAV}, δ, τ, t_{HDV}, t_{AV}, t_{CAV} are both constants; the variables in Eqs. (1–24) are explained in Table 1.

2.4 Lane-Changing Model

In urban road traffic, the motivation for vehicles to change lanes is not only to obtain faster speeds. Vehicles in different directions need to change lanes to the corresponding lanes before passing the road intersection. This paper establishes a three-lane changing model applied to urban roads based on the STCA model [38]. Compared with a two-lane change, the lane-changing rules for a three-lane road are more complicated. When changing lanes, vehicles on the middle road need to consider the traffic conditions of the two adjacent lanes at the same time to make a lane-changing decision. The detailed lane-changing rules are set as follows.

The discretionary lane-changing rule is given by Eq. (25).

$$
\begin{aligned}
&Gap_{queue}(i) > 30\delta x \& Gap_{\text{int}}(i) \geq 100\delta x \& Gap_o(i) \\
&> Gap(i) \& Gap_{o,back}(i) > v_{\max} \& rand < p_{change}
\end{aligned}
\tag{25}
$$

This rule corresponds to the situation where the vehicle is far away from the queuing area and the intersection, and the motivation for changing lanes is to drive faster.

The active lane-changing rule is given by Eq. (26).

$$
\begin{aligned}
&Gap_{queue}(i) > 30\delta x \& Gap_{\text{int}}(i) < 100\delta x \& Gap_o(i) \\
&> v_{\max} \& Gap_{o,back}(i) > v_{\max} \& flag = 1 \& rand < P_{change}
\end{aligned}
\tag{26}
$$

This rule corresponds to a certain distance between vehicles, intersections, and the queuing area. The lane change requirement is to change lanes to the corresponding lane in the driving direction, but the lane change requirement is not urgent.

A mandatory lane-changing rule is given by Eq. (27).

$$Gap_{queue}(i) \leq 30\delta x \& Gap_{\text{int}}(i) \leq 30\delta x \& Gap_o(i) > v_{\max} \& Gap_{o,back}(i) > v_{back} \& flag = 1 \quad (27)$$

This lane change rule corresponds to the need for vehicles approaching the queuing area or approaching road intersections. There is an urgent need to change lanes to the corresponding lane in the driving direction. Considering the speed loss caused by the mandatory lane-changing behavior.

The variables in Eqs. (13–15) are explained in Table 2 and Fig. 4.

Table 2 Related variables and means of lane-changing model

Variable Name	Means
$Gap_{queue}(i)$	The gap between the vehicle i and the front queue area
$Gap_{\text{int}}(i)$	The gap between the vehicle i and the front intersection
$Gap(i)$	The gap between vehicle i and the nearest vehicle in front of the lane
$Gap_o(i)$	The more significant gap between the vehicle i and the vehicle ahead in the adjacent lanes
$Gap_{o,back}(i)$	The gap between the vehicle i and the rear vehicles on the adjacent lane corresponding to the $Gap_o(i)$
v_i	Speed of vehicle i
$v_{\max}$	Maximum vehicle speed
v_{back}	Speed of the nearest vehicle behind the neighboring lane of the vehicle i
p_{change}	Lane-changing probability
$flag$=1	The adjacent lane is the target lane or the middle lane determined by the direction variable of the vehicle i

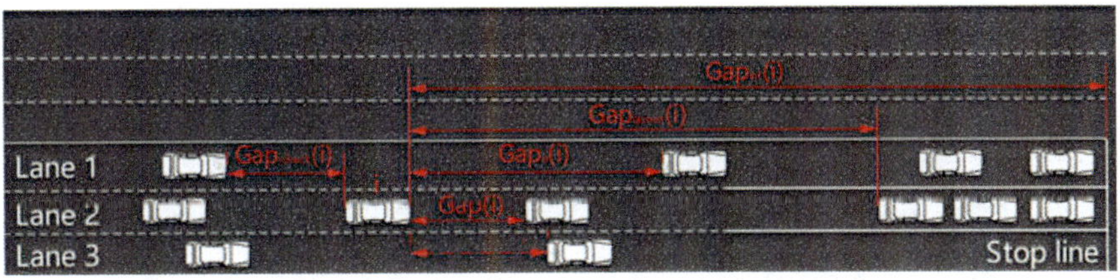

Fig. 4 The illustration for related parameters in the lane-changing model

3 Simulation

The simulation scene is an east-west road passing through two intersections (I_1, I_2) and two north-south roads. The road sections are all set to two-way six lanes, and the one-way three lanes are turn left, go straight and turn right lane. In this paper, each road segment is numbered, among which the road sections 1–6 are the vehicle arrival road segments, the road sections 7 and 8 are the connecting road segments between the intersections, and the road sections 9–14 are the vehicle departure road sections. Each road section consists of three lanes. The schematic diagram is shown in Fig. 5. The corresponding simulation user interface is shown in Fig. 6.

The length of each cell is 4 m, and the length of each road section is composed of 150 cells, corresponding to the actual distance of 600 m. The road intersection is

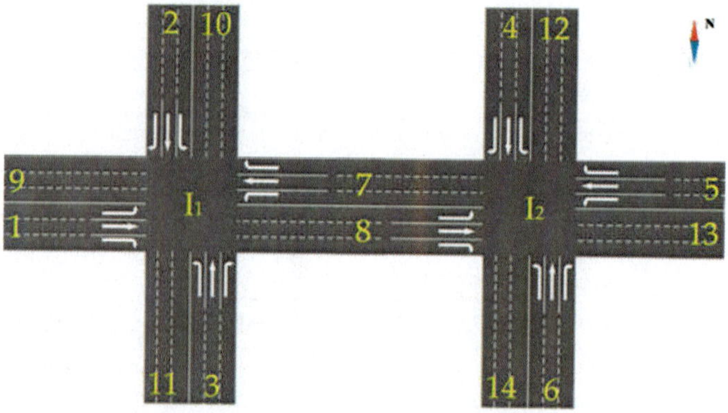

Fig. 5 Simulation scenario

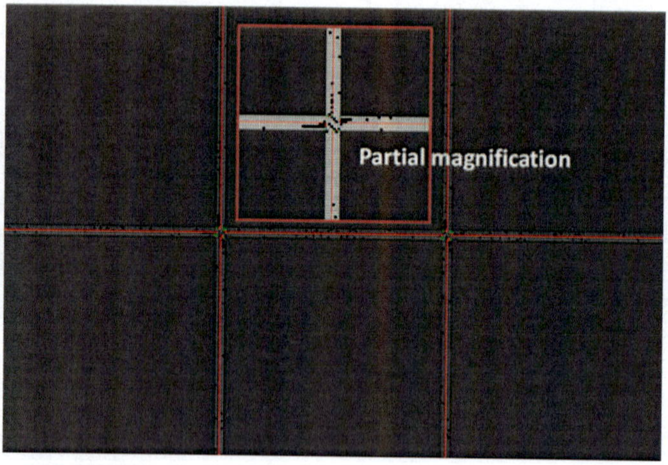

Fig. 6 Simulation interface

composed of 6 δx*6 δx, , corresponding to the actual size of 24 m × 24 m. The relevant parameter settings of the model are shown in Table 3. The vehicle's maximum speed during the cruise is 4 $\delta x/\tau$, representing the actual speed of 57.6 km/h; the maximum speed of the vehicle entering and passing the intersection is 2 $\delta x/\tau$, representing the actual speed of 28.8 km/h, meeting the speed limit requirements of urban roads.

The signal lamp's four phases of green time are set as 32 s, 24 s, 25 s and 27 s, respectively. The yellow time of each phase is 3 s, and the total duration of the signal cycle is 120 s.

4 Results and Discussion

4.1 Analysis of Traffic Fundamental Diagram

We first study the relationship between the space occupancy and flow rate and velocity of the simulation scene under different CAVs permeability conditions. With the CAVs permeability difference of 0.2 as the interval, a total of 6 groups of control simulation experiments were set up, and each group of simulations was carried out with 10 independent simulations. A total of 60 simulations were carried out. Here the first 10 00 timesteps are discarded to let the transient time out and the speed and flow rate is averaged over 4000 timesteps.

Figure 7a shows the changing relationship between space occupancy and flow in the simulation scenario under different CAVs penetration rates. It can be found that as the space occupancy rate increases, the traffic flow will gradually increase. When the space occupancy rate is higher than 0.12, it means that 10% of the cells are occupied by vehicles, and the traffic flow begins to fluctuate steadily. Figure 7b shows the relationship between space occupancy and vehicle speed under different penetration rates of CAVs in the simulation scenario. With the increase in space occupancy, the vehicle speed begins to decrease, and its downward trend increases with the rise of space occupancy. Gradually slow down, and overall, the higher the penetration of CAVs, the greater the flow and speed of traffic.

In the simulation, the road sections (9–14) are the sections where the vehicle leaves the road. The vehicle is not controlled by the signal lights when moving, and moves in the form of free-flow, which cannot reflect the characteristics of urban road traffic. Correspondingly, in the vehicle arrival road sections (1–6) and the connecting road sections (7, 8), the vehicle movement has lane-changing expectations and signal light control, which is in line with the characteristics of intermittent traffic on urban roads. Therefore, this paper conducts an independent analysis of the traffic data of road sections 1–8.

Figure 8 shows the relationship between the space occupancy, flow rate and speed of road sections 1–8 under different CAVs penetration conditions. From the figure, we can observe the changing trend of road flow rate and speed with the increase in space occupancy. The changes are similar to those in Fig. 7. In Fig. 8a, the road

Table 3 Model parameters and their values

Parameters	p_{1-HDV}	p_{2-HDV}	p_{1-AV}	p_{2-AV}	p_{1-CAV}	p_{2-CAV}	p_{change}	α	β	t_{HDV}	t_{AV}	t_{CAV} t_{CAV}
Value	0.4	0.03	0	0	0	0	0.3	0.5	0.1	1.2	0.8	0.4

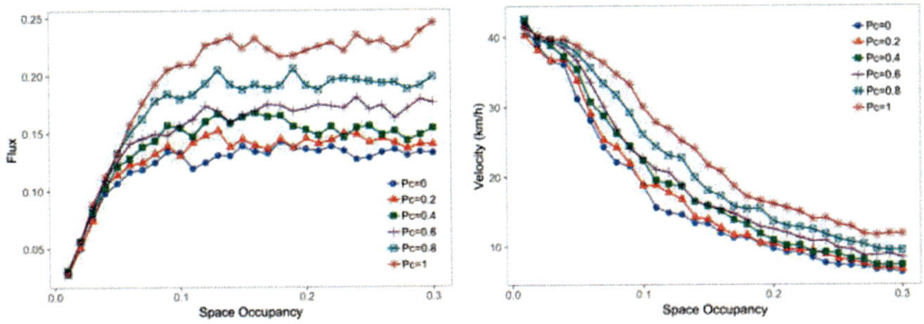

Fig. 7 Traffic flow rate and speed versus density for different CAVs penetration (data for all road sections)

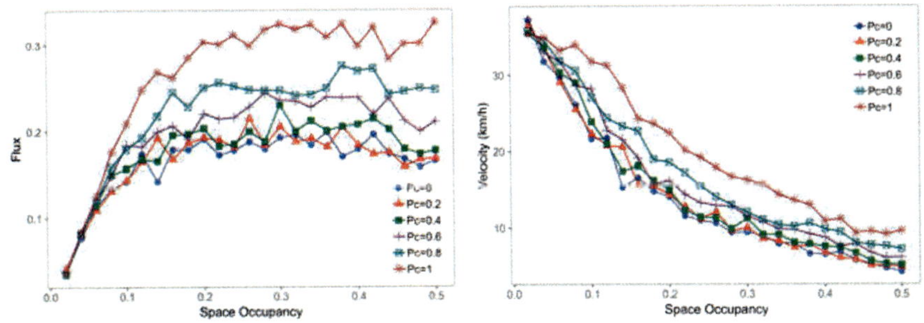

Fig. 8 Traffic flow rate and speed versus density for different CAVs penetration (data for 1–8 road sections)

flow changes from rising to a stable critical space occupancy rate of 0.2. It is worth noting that by comparing Figs. 7a and 8a, it is possible to find that the steady average flow in Fig. 8a is higher than in Fig. 7a, due to the inclusion of vehicles leaving the roadway in the count compared to the non-inclusion of the count, which reduces the spatial occupancy of the road. In Fig. 7 and Fig. 8, the corresponding traffic flow rate is completely different under the condition of the same space occupancy.

4.2 *Traffic Analysis Under Constant Traffic Volume*

By setting a constant entrance traffic flow, this model allows the traffic characteristics under different CAVs conditions to be studied. The traffic flow parameters set in this paper are shown in Table 4. by performing simulations at this constant traffic volume, we first compare the travel time and vehicle speed under different CAVs penetration conditions.

Table 4 Traffic flow data at each intersection

Vehicle arrives road		I_1				I_2			
		East	South	West	North	East	South	West	North
Input traffic volume (vehs/h)	Go straight	\	324	396	324	396	324	\	324
	Turn left	\	324	396	324	396	324	\	324
	Turn right	\	324	396	324	396	324	\	324
Simulated traffic volume (vehs/h)	Go straight	627	291	475	291	475	291	627	291
	Turn left	376	292	356	389	356	389	376	292
	Turn right	250	389	356	292	356	292	250	389

4.2.1 Speed and Travel Time of Route (1-I_1-8-I_2-13)

With the CAVs permeability difference of 0.1 as the interval, a total of 11 groups of control simulation experiments were set up, and each group of simulations was carried out with 10 independent simulations. A total of 110 simulations were carried out. Here the first 10 00 timesteps are discarded to let the transient time out and, the speed and travel time are averaged over 4000 timesteps.

Figure 9a shows the variation of average vehicle speed along the route (1-I_1-8-I_2-13) with the penetration rate of CAVs. It can be seen that with the increase of the CAVs penetration rate, the speed of vehicles will also increase; moreover, this upward trend is the most obvious in the CAVs penetration rate range of 0–0.4. When the vehicle penetration rate of CAVs is higher than 0.4, the upward trend of vehicle speed slows down. And, the fluctuating changes in the average velocity gradually converge with the increase in the penetration of CAVs.

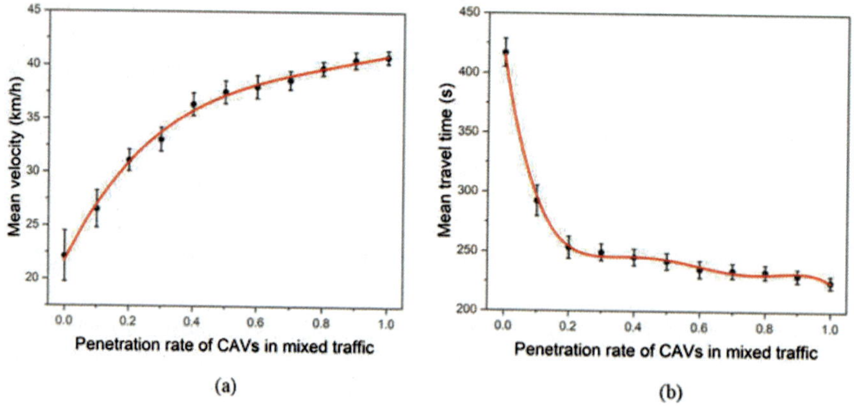

Fig. 9 Variation of traffic parameters in route (1-I_1-8-I_2-13) with the penetration of CAVs (mean ± standard deviation), where **a** shows the variation of average vehicle speed with the penetration rate of CAVs and **b** shows the variation of the average travel time of vehicles going straight from west to east with the penetration rate of CAVs

Figure 9b shows the variation of vehicle travel time along the route (1-I_1-8-I_2-13) with the penetration rate of CAVs. The total length of the road section is 1848 m. When the penetration rate of CAVs is 0, the average travel time of vehicles on this road section exceeds 400 s. When the CAVs penetration rate is 0.1, the average travel time of the road section quickly drops below 300 s. When the CAVs penetration rate is 1, all vehicles in the simulated scene are both CAVs, and the average travel time of the road section vehicles is slightly higher than 200 s. In terms of the trend of travel time, the change of travel time starts to become flat after the penetration rate of CAVs is higher than 0.3.

A separate simulation was performed for each set of CAVs permeability conditions and data was extracted for the first 5000 timesteps. Next, we studied the changes in traffic parameters over time under different CAVs penetration rates. Figure 10 shows the variation of the average vehicle speed over time in the simulation when the CAV penetration rates are 0, 0.3, and 1. As can be seen from the figure, at the beginning of the simulation, the entry lane of the road begins to generate vehicles to enter the road and quickly accelerate to the maximum speed for cruising. When the vehicles close the intersection, vehicles begin to decelerate, which corresponds to the rapid decrease of vehicle speed in the earlier period in Fig. 10.

In these three different CAVs penetration rates simulation scenarios, when the penetration rate is 0, the average speed of vehicles always shows a downward trend with the increase of simulation time, which indicates that the accumulation of blocked areas appears at the intersection sections and the traffic conditions continue to deteriorate. However, when the penetration rates are 0.3 and 1 in the simulation scenarios, the average speed of vehicles has been oscillating within a fixed speed range, indicating no deterioration of traffic. Among them, in the simulated scenario with a CAVs penetration rate of 1, the average speed of vehicles is the highest, and its average speed is close to 40 km/h.

Figure 11 shows the travel time variation of vehicles on the straight road section from west to east with time in the simulation with CAVS penetration rates of 0, 0.3

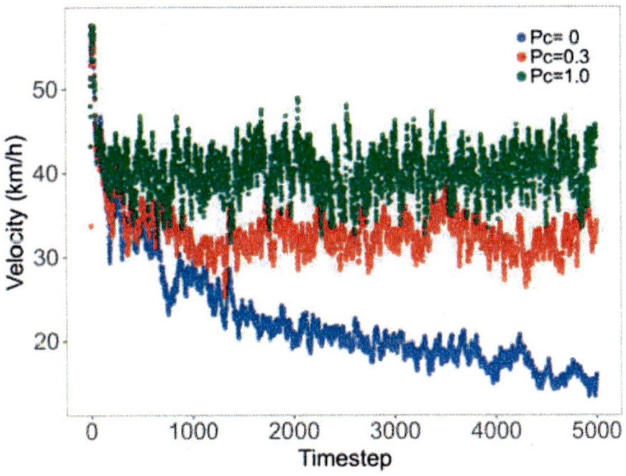

Fig. 10 The variation of vehicle velocity with time

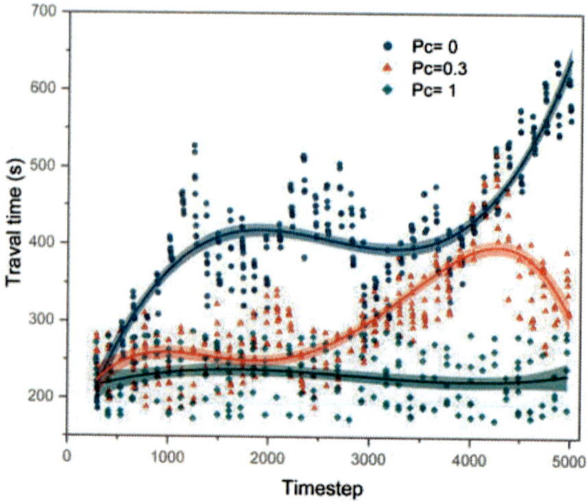

Fig. 11 The variation of vehicle travel time with timestep

and 1. At the beginning of the simulation, there are no scattered points in the scatter diagrams of the three scenarios, indicating that no vehicles have completely passed this road section in this period. After 200 timesteps, some vehicles start to arrive at the east exit of the road and leave the road, and the travel time is recorded. When the CAVs penetration rate is 0, the travel time of vehicles on the straight section from west to the east also shows a trend of oscillating increase. However, in the simulation scenario with CAVS permeability of 0.3 and 1, there is no trend of vehicle travel time increases with the increase of simulation time. The distribution of travel time is very close.

4.2.2 Queue Length at the West Entrance of Intersections I_1 and I_2

Figure 12a and b show the queuing length of vehicles in turn left and going straight lanes at the west entrance of I_1 and I_2 within 5000 timesteps when the CAVs penetration rate is 0.4. This road section also has the highest traffic volume among all road sections. When the traffic signal is red, the length of the queue of vehicles will increase rapidly, and when the traffic signal turns green, the vehicles in the queue area start to leave the intersection. The queued vehicles in the rear area are still in a queue, so the queue length will not decrease. When all the queued vehicles leave the road intersection, the figure will show a sudden decrease in the queue length (Fig. 12).

It can be seen from the figure that in turn left and go straight lanes at the west entrances of intersections I_1 and I_2, , queuing vehicles can be released entirely during the green light time. Means that when the CAVs penetration rate is 0.4, the queuing area will not accumulate under the given traffic volume in the simulation scene, and the vehicles can be released in an orderly manner. However, when the CAVs penetration rate is 0.3, queuing vehicles in the same area cannot be released entirely

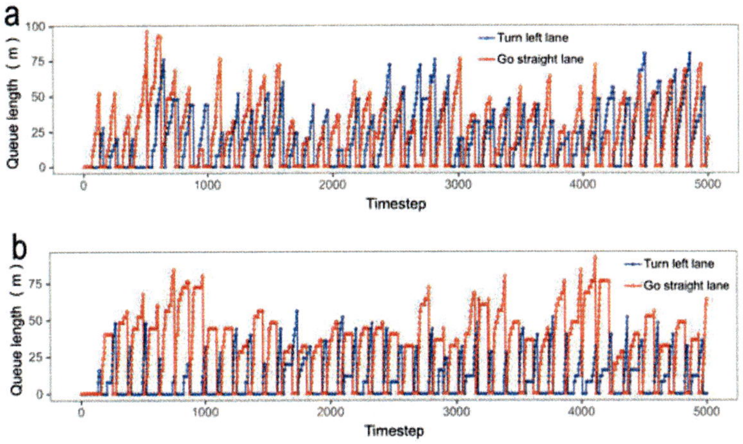

Fig. 12 When the CAVs penetration rate is 0.4, changes in the queue length of vehicles on turn left and went straight lanes over time **a** queue length at the west entrance of I_1, **b** queue length at the west entrance of I_2

within the green light time (which can be observed in Fig. 14b). So, it's a critical value where the traffic state changes from congestion to unblocked traffic.

4.2.3 Planar Distribution of Traffic Speed and Time–Space Diagram Analysis

Figure 13a–c respectively show the velocity distribution in a two-dimensional plane within 5000 timesteps in a simulated scenario with CAVS penetration rates of 0, 0.3 and 1.

As you can see from the figure that the queuing situation of vehicles in different flow directions and the variation in queuing length of vehicles in different CAVs penetration scenarios. The vehicle is generated from the road entrance, rapidly accelerates to its maximum speed and begins to cruise; when approaching the road intersection, the vehicle decelerates. As shown in Fig. 10, in the traffic scenario where CAVs penetration rate is 0, vehicles with the right-turn flow are not controlled by traffic signal lights, so there is no queuing phenomenon of vehicles, while vehicles with left-turn and straight road queuing situation are very obvious.

At the intersection, the passing speed of vehicles is about 28 km/h, and after the vehicles pass the intersection, there will be a period of acceleration. Comparing the two-dimensional distribution plane of average velocity in these three scenarios shows that, compared with the scene where the CAVs penetration rate is 0, there is no large-scale queuing phenomenon in the scene where CAVS penetration rates are 0.3 and 1. The low-speed area is concentrated near the road intersections.

This paper performs time-space diagram analysis to study more microscopic individual vehicle speed changes and vehicle queuing conditions. Figure 14a–c show the

Fig. 13 The plane distribution of vehicle velocity (The white boxes are enlarged view of the intersections)

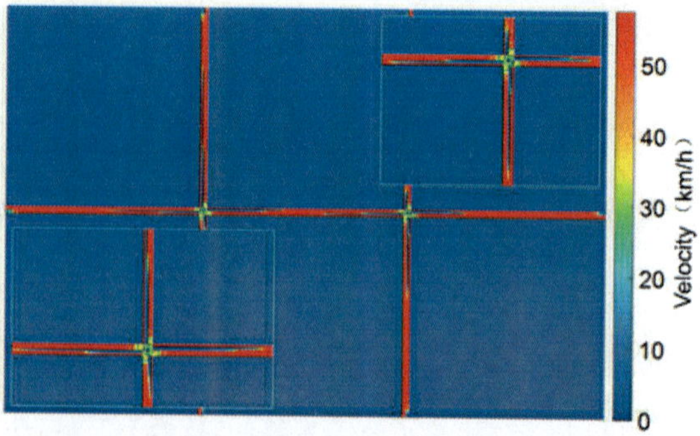

(a) Pc=0

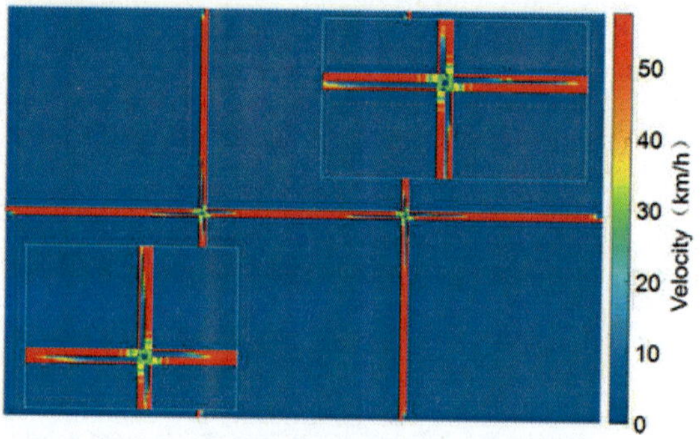

(b) Pc=0.3

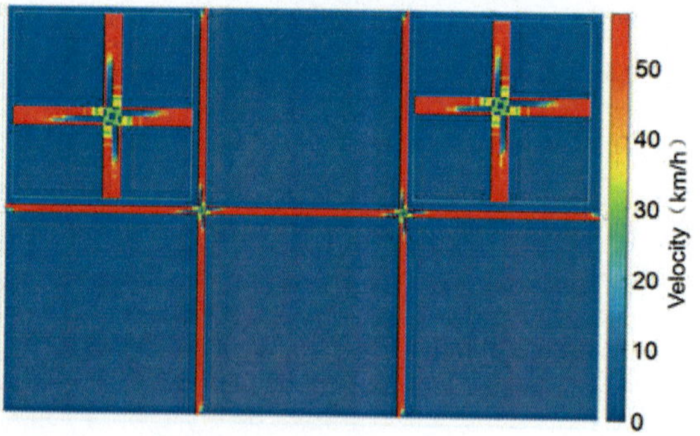

(c) Pc=1

simulation scenarios of CAVs penetration rates of 0, 0.3 and 1, respectively. The temporal and spatial distribution of vehicle speeds on a straight section from west to east. Due to the limitation of the picture ratio, only the spatial distribution of vehicle speeds at 4000–4500 timesteps are shown. The figure shows that vehicles on this section of the road will enter the road from the west entrance of the road, pass through two road intersections, and leave the road at the east exit of the road. By observing the microscopic temporal and spatial distribution of velocity, you can observe the speed

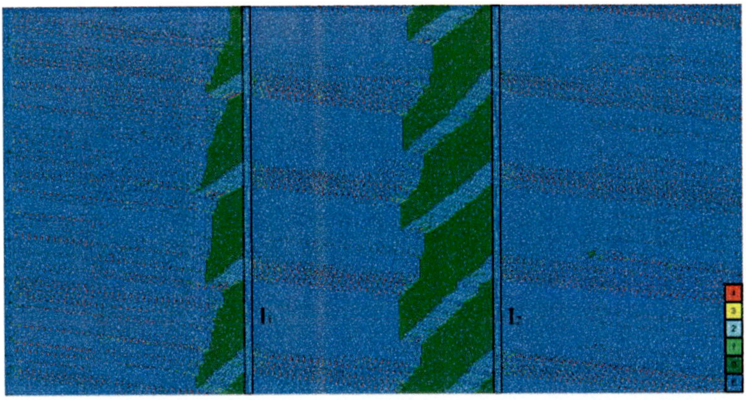

(a) Pc=0

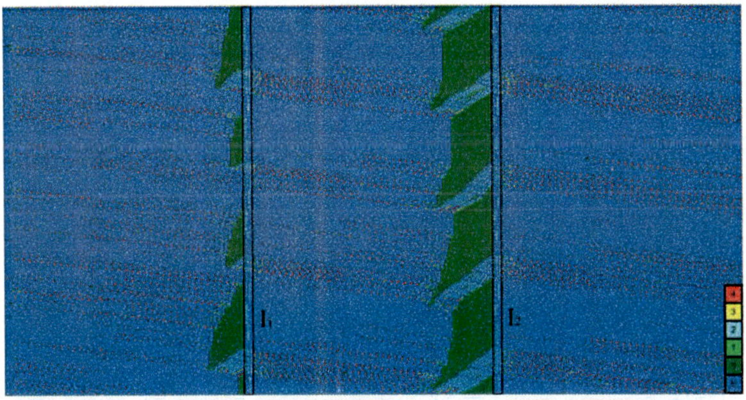

(b) Pc=0.3

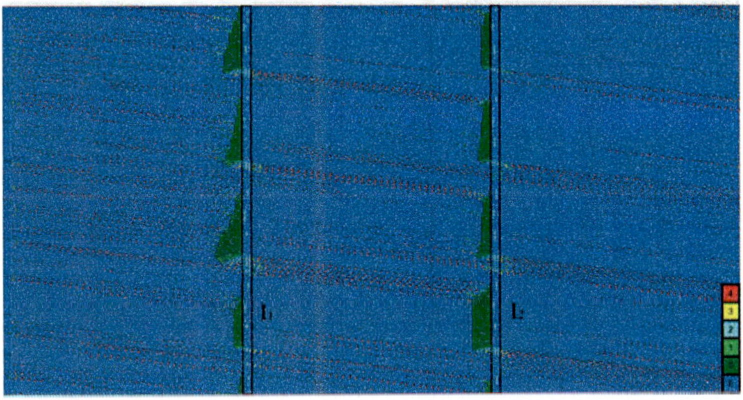

(c) Pc=1

Fig. 14 Evolution of the spatiotemporal pattern of vehicle velocity (timestep: 4001–4500) change of each vehicle, the evolution of queue length, and the vehicle release at the road intersection. In these three simulation scenarios, the vehicle will decelerate for some time before approaching the queuing area, presenting a linear change in vehicle speed. In the past large-size cell models, the vehicle deceleration is often ignored. There will be a situation where the vehicle speed decelerates from the maximum value to 0 within one timestep, which is different from the actual vehicle movement. In the car-following model proposed in this paper, the vehicle has a complete deceleration process, which can simulate the linear change of the vehicle speed when the vehicle is decelerating.

According to the comparison of Fig. 14a–c, it can be observed that when the CAVs penetration rate is 0, random slowing will occur when the vehicle is cruising; however, when the CAVs penetration rate is 0.3, random slowing will decrease to some extent; when the CAVS permeability is 1, there will be no random slowing. The same situation occurs when vehicles move from the queuing area to the intersection. Due to the slow start rule of HVDs, when the vehicle starts to accelerate from a standstill, there will be a reaction time, which causes the vehicle to not accelerate in time. In the simulation of vehicles containing CAVs, the situation has been significantly improved. The vehicles can pass quickly during the green light time, effectively avoiding the increase in the length of the line in front of the intersection. It can also be observed from the figures that when the CAVs penetration rate is 0, there will be an accumulation of queuing areas in front of the intersection I_2. When the penetration rate is 0.3, the queue length remains stable, but it cannot be released entirely during the green light time. When the penetration rate is 1, the vehicle can be released entirely within the green light time, and the queue length is concise.

Observe the image; we can find that, compared with HDVs at the same speed, CAVs can keep a more minor time headway than HDVs, so they can keep a minor space headway when moving, which means that more vehicles can be accommodated in the same length of road section without causing a speed decline. Therefore, an increase in the proportion of CAVS vehicles can improve the road's traffic capacity.

5 Conclusions

This paper uses the ABCA-MS model to build an urban road traffic simulation scenario with a complete vehicle input and output, vehicle movement, traffic control, data collection, and output modules. Because the advantages of cellular automata are easy to calculate and simulate, they can significantly improve simulation efficiency and realise the use of simple rules to simulate complex road conditions. This paper studies the influence of different CAVs penetration rates on traffic on urban roads under this model framework.

Car-following rules for three types of vehicles and lane-changing rules on urban roads are established in the simulation. The car-following model adds deceleration rules based on the NaSch model. It can express the microscopic deceleration process of the vehicle and solve the problem of substantial instantaneous deceleration caused by the forced deceleration condition of the vehicle in the previous large-size cell traffic simulation.

Through numerical simulation, it can be found that the traffic capacity of urban roads and the penetration rates of CAVs are positively correlated. Under the given traffic volume condition, the critical value of CAVs penetration rate for the traffic state change from congested to unblocked is 0.4. Road capacity improvement is most apparent when the CAVs penetration rates are between 0 and 0.4. This is because CAVs and AVs have no slow to start conditions and can quickly pass through road intersections continuously, and CAVs and AVs maintain a more minor space headway during movement; it can improve traffic capacity efficiency.

The scene set by this model is still under idealised conditions. There are heterogeneous traffic carriers such as buses and motorcycles on urban roads, and the influence of pedestrians on vehicle movement should be taken into account when passing the intersection. All these factors will bring challenges to the moving control of CAVs, which are not reflected in the simulation framework of this paper. In future research, the model will be further improved and expanded.

Acknowledgements This research was supported by National Natural Science Foundation of China (Grant No. 52072286, 72074149), and the Fundamental Research Funds for the Central Universities (Grant No. 2020VI002).

References

1. Bagloee SA, Tavana M, Asadi M, Oliver T (2016) Autonomous vehicles: challenges, opportunities, and future implications for transportation policies. J Mod Transp 24:284–303
2. Hong D, Kimmel S, Boehling R, Camoriano N, Cardwell W, Jannaman G, Purcell A, Ross D, Russel E (2008) Development of a semi-autonomous vehicle operable by the visually-impaired. In: Proceedings of IEEE International conference on multisensor fusion and integration for intelligent systems, pp 539–544
3. Litman T (2015) Autonomous vehicle implementation predictions. Victoria Transport Policy Institute, No 15-3326
4. Fenton RE, Mayhan RJ (1991) Automated highway studies at the ohio state university-an overview. IEEE Trans Veh Technol 40:100–113
5. Treiber M, Hennecke A, Helbing D (2000) Congested traffic states in empirical observations and microscopic simulation. Phys Rev E 62:1805–1824
6. Shladover SE, Desoer CA, Hedrick JK, Tomizuka M (1991) Automatic vehicle control developments in the PATH program. IEEE Trans Veh Technol 40:114–130
7. Rajamani R, Tan HS, Law BK, Zhang WB (2000) Demonstration of integrated longitudinal and lateral control for the operation of automated vehicles in platoons. IEEE Trans Control Syst Veh Technol 8:695–708
8. Milanés V, Shladover SE, Spring J, Nowakowski C, Kawazoe H, Nakamura M (2014) Cooperative adaptive cruise control in real traffic situations. IEEE Trans Intell Transp Syst 15:296–305
9. Milanés V, Shladover SE (2014) Modeling cooperative and autonomous adaptive cruise control dynamic responses using experimental data. Transp Res Pt C Emerg Technol 48:285–300
10. Yao ZH, Xu T, Jiang YS, Hu R (2021) Linear stability analysis of heterogeneous traffic flow considering degradations of connected automated vehicles and reaction time. Phys A 561:125218

11. Yao ZH, Hu R, Jiang YS, Xu TR (2020) Stability and safety evaluation of mixed traffic flow with connected automated vehicles on expressways. J Saf Res 75:262–274
12. Yao ZH, Hu R, Wang Y, Jiang YS, Ran B, Chen YR (2019) Stability analysis and the fundamental diagram for mixed connected automated and human-driven vehicles. Phys A 533:121931
13. Jiang R, Wu Q, Zhu Z (2001) Full velocity difference model for a car-following theory. Phys Rev E 64:01701
14. Kuang H, Xu ZP, Li XL, Lo SM (2017) An extended car-following model accounting for the average headway effect in intelligent transportation system. Phys A 471:778–787
15. Kerner BS (2018) Physics of automated driving in framework of three-phase traffic theory. Phys Rev E 97:042303
16. Rios-Torres J, Malikopoulos AA (2018) Impact of partial penetrations of connected and automated vehicles on fuel consumption and traffic flow. IEEE Trans Intell Transp Syst 3:1–10
17. Gipps PG (1986) A model for the structure of lane-changing decisions. Transp Res Pt B-Methodol 20:403–414
18. Wang M, Treiber M, Daamen W, Hoogendoorn SP, Arem BV (2013) Modelling supported driving as an optimal control cycle: framework and model characteristics. Transp Res Pt C-Emerg Technol 36:547–563
19. Talebpour HS, Mahmassani SH (2011) Hamdar, multiregime sequential risk-taking model of car-following behavior specification, calibration, and sensitivity analysis. Transp Res Record 2260:60–66
20. Talebpour HS (2016) Mahmassani, Influence of connected and autonomous vehicles on traffic flow stability and throughput. Transp Res Pt C-Emerg Technol 71:143–163
21. Jiang R, Wu QS (2003) Cellular automata models for synchronized traffic flow. J Phys A Math Gen 36:381–390
22. Kerner BS, Klenov SL, Wolf DE (2002) Cellular automata approach to three-phase traffic theory. J Phys A Math Gen 35:9971–10013
23. Lee HK, Barlovic R, Schreckenberg M, Kim D (2004) Mechanical restriction versus human overreaction triggering congested traffic states. Phys Rev Lett 92:1–4
24. Jiang R, Wu QS (2006) The adaptive cruise control vehicles in the cellular automata model. Phys Lett A 359:99–102
25. Yuan YM, Jiang R, Hu MB, Wu QS, Wang RL (2009) Traffic flow characteristics in a mixed traffic system consisting of ACC vehicles and manual vehicles: a hybrid modelling approach. Phys A 388:2483–2491
26. Lo SC, Hsu CH (2010) Cellular automata simulation for mixed manual and automated control traffic. Math Comput Modell 51:1000–1007
27. Yang D, Qiu XP, Ma LL, Liang HB (2017) Cellular automata–based modeling and simulation of a mixed traffic flow of manual and automated vehicles. Transp Res Record 2622:105–116
28. Wu JC, Chen BK, Zhang K, Zhou J, Miao LX (2018) Ant pheromone route guidance strategy in intelligent transportation systems. Phys A 503:591–603
29. Zhao HT, Liu XR, Chen XX, Lu JC (2018) Cellular automata model for traffic flow at intersections in internet of vehicles. Phys A 494:40–51
30. Zhao HT, Zhao X, Lu JC, Xin LY (2020) Cellular automata model for urban road traffic flow considering internet of vehicles and emergency vehicles. J Comput Sci 47:101221
31. Zhao HT, Lin L, Xu CP, Li ZX, Zhao X (2020) Cellular automata model under Kerner's framework of three-phase traffic theory considering the effect of forward–backward vehicles in internet of vehicles. Phys A 553:124213
32. Gao K, Jiang R, Hu SX, Wang BH, Wu QS (2007) Cellular-automaton model with velocity adaptation in the framework of Kerner's three-phase traffic theory. Phys Rev E 76:1–7
33. Chen BK, Sun D, Zhou J, Wong WF, Ding ZJ (2020) A future intelligent traffic system with mixed autonomous vehicles and human-driven vehicles. Inf Sci 529:59–72
34. Nagel K, Schreckenberg M (1992) A cellular automata model for freeway traffic. J Phys I(2):2221–2229

35. Tanimoto J, Futamata M, Tanaka M (2020) Automated vehicle control systems need to solve social dilemmas to be disseminated. Chaos Solitons Fractals 138:109861
36. Zhou YJ, Zhu HB, Guo MM, Zhou JL (2020) Impact of CACC vehicles' cooperative driving strategy on mixed four-lane highway traffic flow. Phys A 540:122721
37. Vranken T, Sliwa B, Wietfeld C, Schreckenberg M (2021) Adapting a cellular automata model to describe heterogeneous traffic with human-driven, automated, and communicating automated vehicles. Phys A 570:125792
38. Wang J, Lv W, Jiang Y, Qin S, Li J (2021) A multi-agent based cellular automata model for intersection traffic control simulation. Phys A 584:126356
39. Chowdhury D, Wolf DE, Schreckenberg M (1997) Particle hopping models for two-lane traffic with two kinds of vehicles: effects of lane changing rules. Phys A 235:417–439

Research on the Top-Level Design of Provincial Smart Service Areas Based on the Enterprise Architecture Approach

Rongjie Lin and Zhe Liu

Abstract To comply with the development trend of service areas from single intelligence to provincial platform construction, this paper uses the enterprise architecture method to sort out the top-level design ideas of the provincial smart service area, and proposes business architecture, application architecture, data architecture and technical architecture of intelligent service area construction from a macro level. The research results can provide theoretical reference for the provincial service area management agencies to carry out intelligent service area planning and construction, which helps promote the construction of integrated, networked, and platform-based provincial intelligent service area system.

Keywords Highway · Smart service area · Enterprise architecture · Top-level design

1 Introduction

By the end of 2021, the total mileage of highways in China reached 169,100 km and the total number of highway service areas was about 6,600 [1]. Service area as an important part of the highway, is an important window to enhance the service experience of drivers and passengers, and gradually become the core support scene of the economic development of highway road derivatives. With the development of economy and society and the upgrading of consumption, the functional demands of expressway travelers and various drivers and passengers on the service area are rapidly escalating, and the service area management and service capabilities put forward higher requirements. The national level has issued relevant policies to promote the development of intelligent service areas. For instance, the State Council issued *The "14th Five-Year Plan" for the Development of Modern Comprehensive Transportation System*, which clearly put forward "building a smart highway service area", Relevant plans such as highways and digital transportation have also deployed

R. Lin (✉) · Z. Liu
Transport Planning and Research Institute Ministry of Transport, Beijing 100029, China
e-mail: 176089100@qq.com

W. Wang et al. (eds.), *Green Transportation and Low Carbon Mobility Safety*, Lecture Notes in Electrical Engineering 944,
https://doi.org/10.1007/978-981-19-5615-7_17

the construction of smart service areas during the "14th Five-Year Plan" period, the Ministry of Transport issued an *Action Plan for New Infrastructure Construction in the Transportation Sector (2021–2025)*, which put forward the improvement of intelligent sensing facilities, the construction of a comprehensive information platform for service areas, the realization of big data in the operation and management, security and emergency, information services and other applications such as construction direction. Promoting the digital transformation of highway service areas, intelligent upgrade, and stimulating the high-quality development of service areas has gradually become the industry consensus. Jiangxi, Zhejiang, Guangdong, Shandong, and other places actively carry out intelligent service area construction practice, and from the intelligent construction of individual service areas to the construction of provincial platforms, the characteristics of networking and platform have begun to emerge. In addition, relevant scholars have proposed the general idea of smart service area construction. There are several examples involved thereof. For instance, Liu Wenhui et al. proposed eight application systems for smart service areas such as video monitoring and smart toilets [2]. Deng et al. proposed the logical architecture, physical architecture and infrastructure configuration requirements for a smart service area [3]. Cui et al. suggested the basic architecture and system functions of the provincial smart service area management service system [4].

Based on sorting out the existing provincial wisdom service area construction ideas, this study intends to introduce enterprise architecture methodology, sort out the provincial wisdom service area construction ideas, and initially propose business architecture, application architecture, data architecture and technical architecture to better adapt to the current wisdom service area development trend. The research results have a certain reference value for provincial service area management agencies to promote the construction of integrated, networked and platform-based intelligent service area systems.

2 Research Method

Through the deep integration of information technology and service area, the smart service area can realize the digital collection of core elements such as people, vehicles, merchants and consumption in the service area. possess the ability of data mining and analysis based on artificial intelligence, and build an online comprehensive application system of the whole process. Provincial smart service area construction should focus on "digitalization, platform, mobility, intelligence, networking", pay attention to the service area business pain points, enhance business efficiency, and optimize the service experience of drivers and passengers as the goal. Finally, a comprehensive application system with a horizontal network and vertical connection adapted to the characteristics of provincial service areas management will be built.

This study carries out the top-level design of provincial smart service areas based on the enterprise architecture methodology. Enterprise architecture focuses on how to map business functions and requirements to information systems and provides a balanced approach for selecting, designing, developing, and deploying information systems in line with corporate strategies and business functional requirements [5]. The use of enterprise architecture methodology to constrain and guide the construction of information technology has the core value of reducing duplication and improving collaboration and sharing capabilities [6]· Nowadays, the more popular international enterprise architecture reference models are Zachman, PERA, TOGAF, DODAF, etc. [7]. Among them, TOGAF has high openness and flexibility, and is one of the most widely used enterprise architecture standards. This study combines relevant practical experience, and refers to the TOGAF standard to carry out the top-level design of provincial smart service areas. A business, application, data, and technology architecture model with relative universality is summarized and proposed, which can effectively improve the systematic and scientific nature of intelligent service area construction.

3 Business Architecture Design

The business architecture can provide a stable and basic business requirement framework for the data and application architecture. To carry out the top-level design of provincial smart service areas, the first step is to sort out the current service area management mechanism, business features and pain point problems.

Service areas are set along the highway, with the characteristics of many points, long lines and wide areas. With the increase in the number of service areas, the provinces from the need for integrated management considerations, the establishment of provincial service area management agencies one after another. Take Jiangxi province as an example, it set up a provincial service area development and operation company in 2009, under the jurisdiction of more than 90 pairs of service areas in the province, basically realizing the unified management of the province's highway service areas. According to the "company headquarters → branch → service area" three-tier structure management model, the scope of business in the company includes service area investment, construction, property management, catering, department stores, tourism business consulting, accommodation and other categories of a wide range of business.

Combined with the features of the service management organization, business classification, management objects, etc., and fully considering the business boundary relationship, business connection and other issues, the business structure of the provincial smart service area is formed, as shown in Fig. 1, including 4 major business categories and 27 businesses module.

Focusing on the pain points such as the lack of decision-making support in the business management of the service area, the need to improve the service experience of drivers and passengers, the low efficiency of property management, and the lack of monitoring and early warning capabilities, combined with mature and feasible emerging technologies, 7 major application scenarios for the construction of provincial-level smart service areas are proposed. Including business planning and

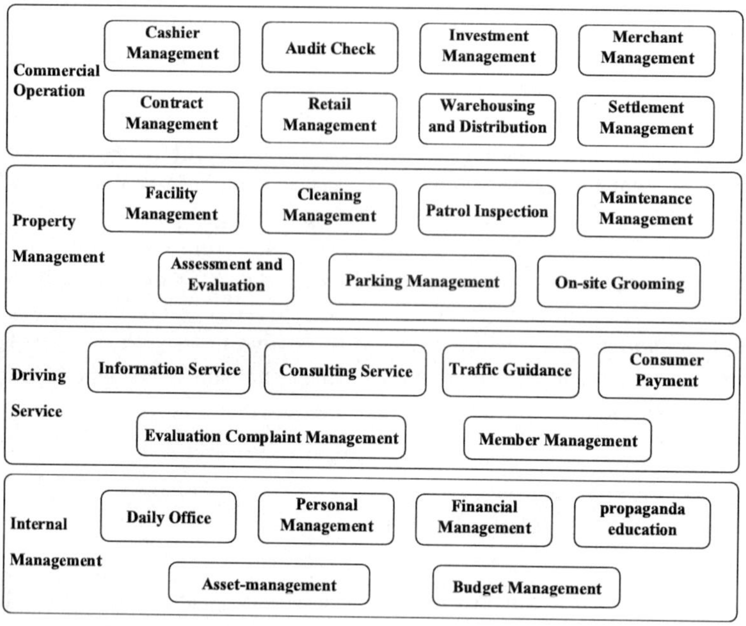

Fig. 1 Business architecture of provincial smart service areas

investment management based on big data, intelligent cashier inspection management, paperless mobile daily supervision, remote inspection and scheduling based on mobile terminals, traffic organization management based on video recognition, online power monitoring and prepayment management, cashless and diversified driver services, etc. The proposal of application scenarios helps to better connect business and application architecture, promote business and technology integration, and avoid the misunderstanding of simple technology stacking in the construction of smart service areas.

4 Application Architecture Design

By analyzing the business modules in the business framework that can apply information technology to improve efficiency and form an application framework, which can be used to guide the management agencies of various provincial service areas to carry out system development and construction. In order to build an application system with good availability, security and scalability, system development should adopt the idea of componentization, standardization, easy expansion and ease of use. According to the business characteristics of the service area, the system development to "a platform" oriented, using B/S structure, service area users directly use the browser and the system for interactive applications. Provincial smart service

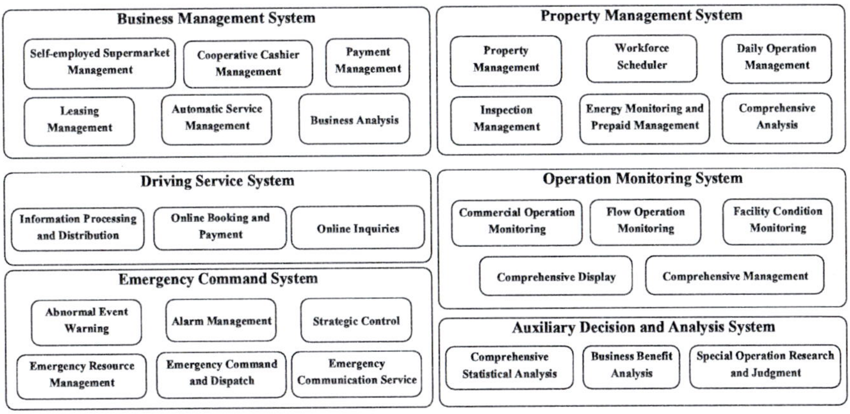

Fig. 2 Application architecture diagram of provincial smart service area

area application architecture includes 6 major application systems and 29 types of functional modules, as shown in Fig. 2.

5 Data Architecture Design

The data architecture can guide the content framework and standards for the database design of the provincial smart service area platform, and provide support for further data exchange and sharing. The data requirements are extracted through the application system, then data collection and formation program analysis is carried out. Eventually, the data architecture is sorted out and formed. The provincial smart service area data architecture includes 5 categories of basic data, 8 categories of application data and 2 categories of theme data, as shown in Fig. 3. Among them, the theme data can be further combined and expanded according to the actual application needs.

To further improve the level of comprehensive data utilization, the provincial smart service area can adopt a custom development approach to unify the construction of an interactive data middle desk. The data middle desk is forward to achieve a unified data exchange with the front-end applications, and backward to achieve docking with the back-end Internet of Things platform, third-party systems, etc., connecting all front-end collection equipment and data sources. This can achieve the aggregation, processing, analysis, sharing and presentation of key data streams in the service area.

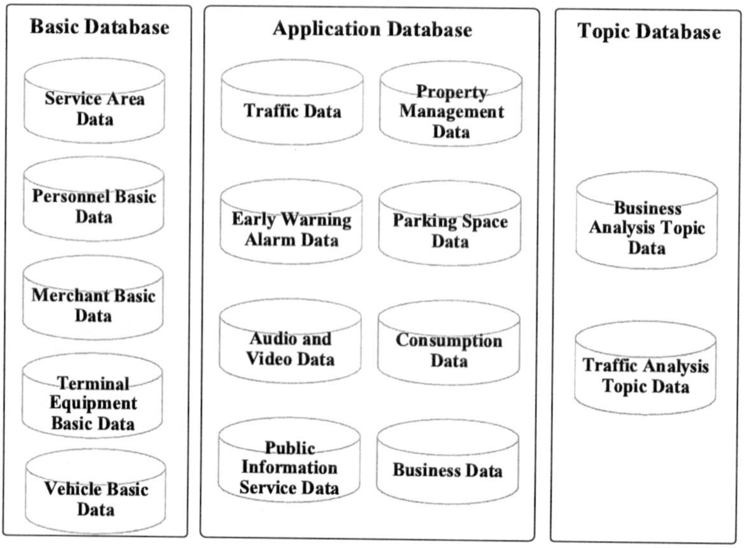

Fig. 3 Data architecture diagram of provincial smart service area

6 Technical Architecture Design

The technical architecture can provide guidance for the construction of the information infrastructure in provincial smart service area, reduce duplication of construction, and improve technical stability and system integrity. In principle, it should take a mature, common technical route, considering the advanced, and it should ensure compatibility with market mainstream technologies and standards. Considering the multi-faceted characteristics of the service area and the weak information base of the single service area, the technical architecture system based on cloud computing is adopted. The technical architecture includes three levels of application support, cloud platform facilities, and terminal systems, as shown in Fig. 4. Cloud computing technology has more significant advantages in terms of economy, security, scalability, maintainability, etc. Provincial service area management agencies can flexibly choose the technical routes of public cloud, proprietary cloud, or public and proprietary combination according to their information conditions.

7 Conclusion

This study introduces enterprise architecture methodology to form a relatively stable top-level design system for provincial smart service areas. Moreover, it discusses the design scheme in detail from four aspects: business architecture, application architecture, data architecture and technical architecture. The top-level design system is

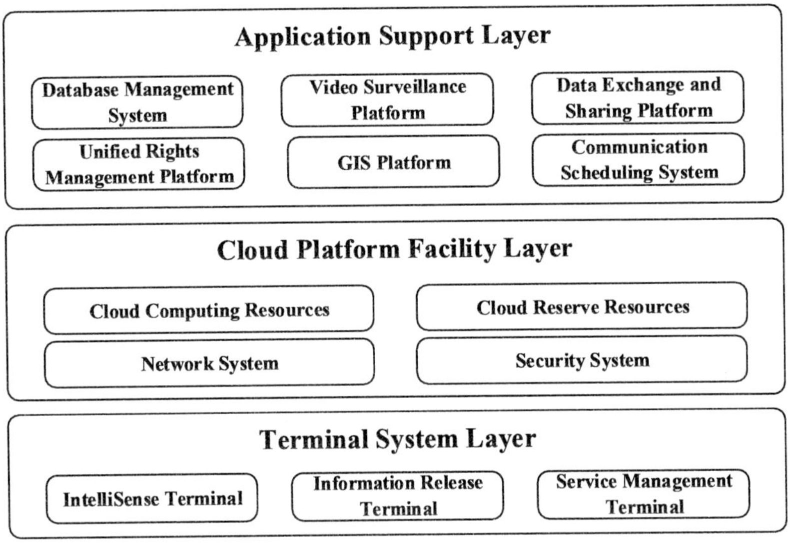

Fig. 4 Technical architecture diagram of provincial smart service area

universal and scientific, which can guide the specific construction of provincial smart service areas and improve the systematization and scientificity of smart service area construction. The research results can provide theoretical reference for the provincial service area management agencies to carry out the planning and construction of smart service areas and help promote the construction of integrated, networked, and platform-based smart service area systems. The research results of this paper have been practiced in some provinces, and the effect is relatively satisfactory. However, to confirm that the top-level design theoretical system proposed in this study has strong applicability and engineering construction effect, it is necessary to carry out deeper refinement and improvement from the level of smart service area functions, and conduct a large number of post-engineering practice evaluations to find out the deficiencies in the top-level design system and improve it continuously, so as to gradually form a universally applicable provincial smart service area construction guidelines.

References

1. Ministry of Transport of the People's Republic of China (2021) Statistical Bulletin on development of Transport industry. May 25, 2022
2. Ren-jie D, Shao-fei W, Zhi Q (2017) General conception of the intelligent expressway service area. Highway (11):140–145
3. Wen-hui L (2021) Research and application of intelligent service area information management system. Energy Conserv Environ Protect Transport 17(04):60–63

4. You-kai C, Rong W, Si-jie M (2021) Research on management system construction of smart service area in Zhejiang Province. China ITS J (11):117–120
5. Tao J, Shu-quan Z, Ming-tao L, Gen-xing Y (2009) On IT planning method driven by enterprise architecture. Comput Appl Softw (12):164–166
6. Yue H, Shuo Z, Xin-hu Z (2017) Overall framework top down design research of national trunk highway network monitoring and emergency system. Highway Eng 42(4):115–119
7. Hui CHE, Bo YANG, Hui-fen XING, Yu-qi FAN (2020) Top-level architecture design of smart city based on enterprise architecture. Technol loT&AI 3(6):16–21

Research on Investment Benefits Valuation Methods for Information Construction of Integrated Passenger Transportation Hubs

Rongjie Lin and Zhe Liu

Abstract It is very important to give full play to the powerful accelerator and multiplier of informatization through the construction of smart transport hubs to improve the integration level of regional transportation and the quality of passenger travel service. However, Our country is currently in a period of economic structural adjustment, financial constraints, limited financial support for the informatization construction of integrated passenger transportation hub, so it is urgent to build a set of scientific and effective evaluation system for the investment benefits of informatization of integrated passenger transportation hub. Given the prominent problems existing in the current construction of smart transportation hub such as the low application level of information system, waste of resources, poor investment effect, the paper selects 15 indicators to evaluate the effect of information construction from six aspects, including investment scale, perception system, operation decision, emergency command, passenger service, and benefit reflection. On this basis, the paper introduces the data envelopment analysis method to construct the investment benefit evaluation model for information construction of integrated passenger transportation hub, and finally gives the solution idea of the model. The research has guiding significance and practical application value for the informatization project planning, construction, and investment effect evaluation of our country's integrated passenger transportation hub. Meanwhile, it can effectively avoid the waste of resources in the process of information construction.

Keywords Integrated passenger transportation hub · Informatization · Investment benefits · Data envelopment analysis model

R. Lin (✉) · Z. Liu
Transport Planning and Research Institute Ministry of Transport, Beijing 100029, China
e-mail: 176089100@qq.com

W. Wang et al. (eds.), *Green Transportation and Low Carbon Mobility Safety*,
Lecture Notes in Electrical Engineering 944,
https://doi.org/10.1007/978-981-19-5615-7_18

1 Introduction

Up to now, our country's comprehensive transport system has been initially completed, various transport modes have been developing rapidly and the size of the infrastructure network has been continuously upgraded. However, the construction of an integrated passenger transportation hub is still inefficient. An integrated passenger transportation hub is a large transport station for large-scale passenger flow transfer between modes of transportation and between regions. As an important part of the integrated transportation system, it is the key to improve the efficiency of passenger transport and the core of service quality to ensure the efficient transfer and connection of large-scale passenger flow between various foreign transportation modes and cities via information technology. At the same time, in the context of new infrastructure for transportation, our country emphasizes the importance of the construction of smart hubs, and most provinces or cities of the country are also actively building smart integrated passenger transportation hubs.

Because the primeval construction of smart integrated passenger transportation hub has large initial investment, the characteristics of low capital-output ratio, high investment risk, combined with the lack of a scientific and effective investment benefits evaluation system. At the same time, the analysis of the investment cost is not sufficient, resulting in the current information construction mode is relatively extensive, the application level of the information system is low, further lead to the waste of resources, poor investment effect, and other problems. Our country is currently in a period of economic structure adjustment, financial constraints, limited financial support for the information construction of integrated passenger transportation hub, so it is urgent to build a set of scientific and effective evaluation models for the information investment benefits of the integrated passenger transportation hubs.

In view of this, the paper comprehensively analyzes and studies all the factors which are related to the informatization of integrated passenger transportation hubs. The input indexes of integrated passenger transportation hub are considered comprehensively from investment scale, perception system, operation decision, emergency command, passenger service and other aspects, the output indexes are determined from passenger flow scale, emergency decision level and service quality improvement rate of integrated passenger transportation hubs. The data envelopment analysis method (DEA) is introduced to determine the balance point of information construction benefits for integrated passenger transportation hubs, and based on the results of the model, guiding suggestions are given for the information construction of integrated passenger transportation hub. The research is helpful to provide the public with safe, convenient and comfortable transfer services in integrated passenger transportation hubs, and it is conducive to improve the efficiency of the informatization construction investment, enrich the basic theory of integrated passenger transportation hubs. It is of guiding significance and practical application value to the planning, construction and investment effect evaluation of our country's integrated passenger transportation hub project.

2 Literature Review

Informatization progresses is an important technical means to support efficient coordinated operation and convenient information service of integrated passenger transportation hubs. Its development process can be roughly divided into four stages, namely, simple integration, interconnection and interconnection, coordinated operation and intelligent operation. In the early stage, the construction of transportation hubs information system at home and abroad was in the exploratory stage, most of which were derived from the existing information based on a certain main mode of transportation, and the extension of long-distance passenger transport information was the most common and typical. Due to the lack of hub information construction experience and overall planning, the construction program was mainly to carry out simple system integration of several existing or newly built information systems. There was a serious information island phenomenon between the systems, which has limited support for the operation and management of the hubs. Then it entered the integration and intercommunication stage, which was mainly based on data standardization which relying on comprehensive data management and control technology, as well as business application and coordination based on aggregated data, forming the overall hub information architecture covering operation monitoring, emergency command, decision analysis and information service and other application systems. At the stage of coordinated operation, integrated passenger transportation hub operators and multi-mode operators are all committed to mining the hidden value behind multi-source and multi-type data, and using data for management and decision-making optimization has become the development direction of integrated passenger transportation hubs. Relying on intelligent sensing facilities and equipment and big data mining and analysis technology, the integrated passenger transportation hub has truly entered the intelligent era. For example, it can predict the situation and changes of passengers' detention and automatically generate an intelligent emergency evacuation plan for stranded passengers by combining the information of emergency evacuation vehicles and personnel. When passengers arrive at the hub, they can obtain a lot of valuable travel information through the mobile APP, including the current taxi queue length, the expected queue time, the route planning and navigation of the boarding station, etc. The future will be the wisdom stage, drived by the new generation of information technologies such as cloud computing, artificial intelligence and deep learning, the future hub information system will no longer be satisfied with the operation coordination at the level of a single node, but will form the coordinated operation of hub groups within the region to support a comprehensive transportation system with smooth regional connection and efficient transfer. At the same time, individual needs of individuation and differentiation will be satisfied to the greatest extent.

In terms of practical application, a lot of related work has been carried out at home and abroad. In 2006, the Ministry of Transport launched a demonstration project for urban integrated passenger transportation hubs. The Outline of the 12th

Five-Year Plan in 2011 pointed out that "in accordance with the requirements of zero-distance passenger transfer, we should strengthen the organic linkage of railways, highways, airports and urban public transport". The Outline of the 13th Five-Year Plan in 2016 pointed out that "improve the comprehensive service function of the hub and strengthen zero-distance transfer of passenger transport [1]. In 2020, the Ministry of Transport officially issued the Guiding Opinions on Promoting the Construction of New Infrastructure in the Field of Transport, which put forward 14 major tasks for new infrastructure in the field of transport and clearly pointed out the specific tasks and goals of smart hubs [2]. It includes intelligent upgrading of integrated passenger transport hubs and electronic tickets, and encourages the development of integrated passenger transport integrated electronic service models.

Up to now, most of the studies on the evaluation of integrated passenger transportation hubs are focused on the overall benefit evaluation. Zhu et al. proposed an intelligent concept of integrated passenger hub based on Maslow's hierarchy of needs model, with economic, ecological, cultural, safety and digital "five-type hub" as the goal, the intelligent development path is explored from the aspects of transportation, architecture, service, operation, integrated operation and management center, holographic perception system, and a multi-dimensional dynamic evaluation index system is established for the sustainable growth of the hub [3]. Chen et al. constructed the overall framework of the smart transportation hub information system of Chongqing West Railway Station based on the functional requirements of the system, and clearly proposed the comprehensive information management mechanism, in the hope of providing useful references for the construction of other hub information systems [4]. Nie et al. started from four aspects of "facilities and equipment, passenger flow organization, transfer efficiency and transfer service", analyzed the important influencing factors of passenger transfer service level of comprehensive passenger hub, and established the evaluation index system of passenger transfer service level [5]. At home and abroad, so far, has been in the field of integrated passenger transportation hubs of informatization, but informatization investment benefit evaluation of relevant research are very few, most studies mainly from the aspects such as transfer efficiency, passenger flow organization consider the information level of comprehensive passenger transport hub, the lack of scientific and comprehensive informatization investment benefit evaluation index system.

3 Comprehensive Evaluation Index System

Based on comprehensive consideration of the research status at home and abroad, the paper comprehensively considers the input indexes of integrated passenger transportation hubs from the aspects of investment scale, perception system, operation decision-making, emergency command and passenger service, and determines the output indexes from the aspects of passenger flow scale, emergency decision-making level and service quality improvement rate of integrated passenger transportation hubs. The data envelopment analysis method (DEA) is introduced to construct the evaluation model of investment benefit of integrated passenger transportation hubs informatization.

3.1 Determine Evaluation Indicators

The key to using DEA for benefit evaluation is to determine a set of scientific and reasonable evaluation index systems, which needs to meet the following requirements. (1) The data for calculating indicators is easy to obtain; (2) Arbitrariness and subjectivity should be avoided in the selection of indexes, and the attributes of each index (input or output indexes) should be correctly defined; (3) It is required that there is no obvious correlation between input indexes and there is an influencing relationship between input indexes and output indexes. A significant correlation between input indexes will not only increase the workload of managers' analysis but also affect the accuracy of evaluation results.

Taking the above principles into consideration and thoroughly investigating the actual situation of the integrated passenger transportation hub information construction, a set of evaluation indexes of the investment benefit of the integrated passenger transportation hubs information construction are determined. Indicators are shown in Table 1

3.2 Key Indicators Calculation

(1) The proportion of investment in informatization (%). Represents the proportion of total information investment in fixed-asset investment, the calculation caliber includes software, hardware, network, information training, direct costs incurred by employing IT technical personnel, communication equipment and maintenance costs, etc. 5 points if the proportion is more than 5%; 4 points if above 4%; 3 points if above 3%; 2 marks above 2%; 1 point for more than 1%.

$$R_{inv} = \frac{I_{\inf}}{I_{sum}} \times 100\% \quad (1)$$

Here R_{inv} represents the proportion of total information investment in fixed-asset investment, $I_{\inf}$ is total investment for integrated passenger transportation hub information construction, I_{sum} represents a total investment in the construction of the integrated passenger transportation hub.

(2) Integration level of information resources. Integrated passenger transportation hubs business coverage by the information system, more than 90% of the business for the integrated passenger transportation hub is covered by the information system and the effective information sharing of the business system is

Table 1 Evaluation indexes of informatization investment benefit for integrated passenger transportation hubs

Category	Indicators	Description
Investment scale	The proportion of investment in informatization	The calculation caliber includes software, hardware, network, information training, direct costs incurred by employing IT technical personnel, communication equipment and maintenance costs, etc
	Integration level of information resources	Integrated passenger transportation hub business coverage by the information system
Perception system	The monitored area ratio of passenger flow	The ratio of the area covered by passenger flow monitoring to the area of the integrated passenger transportation hub plane
	Monitoring rate of online equipment	The ratio of the equipment that can monitor to the total quantity
Operation decision-making	Real-time of passenger organization scheme evaluation system	Whether it has a real-time evaluation function
	Real-time of passenger scheduling scheme evaluation system	Whether it has a real-time evaluation function
Emergency command	The level of cross-departmental collaboration mechanism	Whether to establish coordination mechanism with police, ground bus, subway and taxi departments
	The level of dispatch emergency information in real-time	Delay in emergency information release, usually less than 5 min?
	Informatization level of emergency plan database for abnormal events	Whether the emergency plan will be automatically started for emergencies, holiday peaks and other special circumstances
Passenger service	Information guidance system composition	Whether to publish real-time information including distance, estimated time, congestion, parking guidance, etc
	The means of information release are diverse	Whether to use the display screen, WeChat public account, mobile phone SMS, voice call and other means to release travel guidance information

(continued)

Table 1 (continued)

Category	Indicators	Description
	The composition of non-inductive payment systems	The application rate of self-service ticket purchase, face recognition, WeChat, Alipay and other mobile payment methods
Benefits	Station entry and exit times increased	Represents the shortening degree of passengers' time in and out of the station before and after the information investment for the integrated passenger transportation hub
	Reduction in emergency response time	Represents the degree of shortening of emergency response time before and after the information investment for the integrated passenger transportation hub
	The intensity of transfer passenger flow On station	The number of passengers received per unit time in the integrated passenger transportation hub

scored 5 points. The score is 3 if more than 70% of the integrated passenger transportation hub business system integration. The application is scattered and the prevalence of data island score 1.

$$R_{int} = \frac{B_{\inf}}{B_{sum}} \times 100\% \quad (2)$$

Here $B_{\inf}$ is the business covered by informatization means, B_{sum} represents the total business of integrated passenger transportation hubs.

(3) The monitored area ratio of passenger flow. Represents the ratio of the area covered by passenger flow monitoring to the area of the integrated passenger transportation hub plane. The rate reaches 100% gets 5 points, reaches 90% or more gets 4 points, 70% or more gets 30 points, 50% or more gets 2 points, otherwise gets 1 point.

$$R_{\sup} = \frac{S_{\sup}}{S_{sum}} \times 100\% \quad (3)$$

Here $S_{\sup}$ is the coverage area of passenger flow monitoring and S_{sum} is the integrated passenger transportation hubs.

(4) Monitoring rate of online equipment. Represents the ratio of the equipment which can monitor to a total quantity, if the online monitoring rate of the equipment reaches 100%, 5 points will be scored; if it reaches more than 90%, 4 points will be scored; if 70%, 3 points will be scored; if 50%, 2 points will be scored; otherwise, 1 point will be scored.

$$R_{Qui} = \frac{Q_{\mathrm{sup}}}{Q_{sum}} \times 100\% \tag{4}$$

Here Q_{sup} represents the number of devices to be monitored, Q_{sum} is the total number of integrated passenger transportation hubs.

(5) For qualitative indicators, such as real-time of passenger organization scheme evaluation system, real-time of passenger scheduling scheme evaluation system, the level of cross-departmental collaboration mechanism, informatization level of emergency plan database for abnormal events, information guidance system composition, the means of information release are diverse, the composition of the non-inductive payment system. The paper plans to use the Likert 5-grade scale to grade the degree of informatization.

$$S = \frac{\sum\{(1), (2), (3), (4), (5)\}}{25} \times 100\% \tag{5}$$

For each indicator, the expert scoring method is adopted, and the values are respectively calculated according to the degree of function perfection (5 = excellent function, 4 = good function, 3 = normal function, 2 = with this service, 1 = without this service), and then the above formula is substituted for calculation.

(6) Station entry and exit times increased. Represents the shortening degree of passengers' time in and out of the station before and after the information investment for the Integrated passenger transportation hubs.

$$R_S = \frac{\sum_i^m avgT^s_{before}}{m} - \frac{\sum_j^n avgT^s_{later}}{n} \tag{6}$$

Here $avgT^s_{before}$ represents the average time of passengers entering and leaving the station before the information construction for the integrated passenger transportation hub. $avgT^s_{later}$ is the average time of passengers entering and leaving the station after the information transformation of the Integrated passenger transportation hubs.

(7) Reduction in emergency response time. Represents the degree of shortening of emergency response time before and after the information investment for integrated passenger transportation hubs.

$$R_E = \frac{\sum_i^m avgT^e_{before}}{m} - \frac{\sum_j^n avgT^e_{later}}{n} \tag{7}$$

Here $avgT^{e}_{before}$ represents the emergency response time before the informatization construction of the Integrated passenger transportation hub, $avgT^{e}_{later}$ is the emergency response time after the informatization transformation of the Integrated passenger transportation hub.

4 Building the Model

4.1 DEA Model

Data envelopment analysis is an efficiency evaluation method proposed by American operations research scientists Charnes, Cooper and Rhodes in the 1970s [6]. The original model was the CCR model and BBR model, Fig. 1 shows the production possibility sets of the CCR and BCC models. The CCR model assumes consistent returns to scale, while the BCC model does not make that assumption. Accordingly, this study conducted its analysis using the BCC model.

The model determines the relative efficiency of using decision-making units (DMUs) to compare the performance of all other DMUs in the dataset and to determine the total factor productivity ratio to measure efficiency.

In this study, we focused on the maximal movement towards the efficient frontier by the proportional reduction of inputs. By using the relative analysis in information construction of integrated passenger transportation hubs, operators can plan the extent to which inputs must be reduced to allow the stations to improve their performances.

Suppose there is a set of n DMUs denoted by $U_K(k = 1, 2, 3, \ldots, n)$ to be analyzed, each of which has m types of inputs and s types of outputs, their corresponding weight vectors are denoted as , $V = (V_1, V_2, \ldots V_m)^T$, $U = (U_1, U_2, \ldots U_m)^T$. The input and output of the j_{th} in n DMUs are denoted by vectors

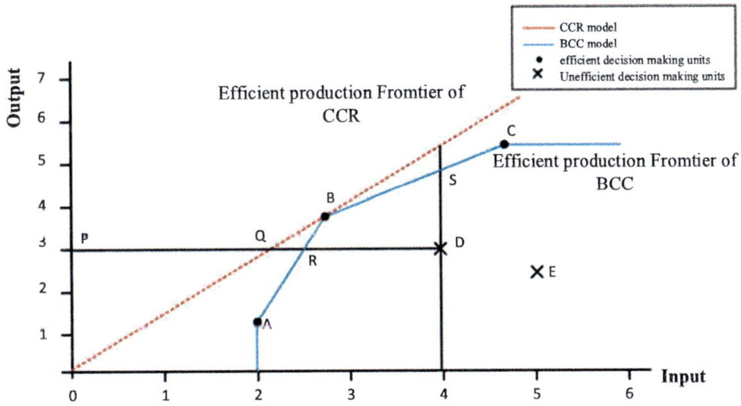

Fig. 1 The CCR model and the BCC model production possibility set

respectively: $X_j = (X_{1j}, X_{2j}, \ldots X_{ij} \ldots X_{mj})^T$, $Y_j = (Y_{1j}, Y_{2j}, \ldots Y_{rj} \ldots Y_{sj})^T$. Formula (8) shows the model formula of the CCR model.

$$Max = \frac{U^T \cdot Y_0}{V^T \cdot X_0} \tag{8}$$

$$\begin{cases} \dfrac{U^T \cdot Y_0}{V^T \cdot X_0} = h_j, \ j = 1, 2, \ldots, n \\ U \geq 0, \ V \geq 0 \end{cases} \tag{9}$$

For the convenience of calculation, we can convert it into an equivalent linear programming mathematical model, and introduce Archimedes infinitesimal quantity ε to form a CCR model with non-Archimedes infinitesimal quantity. Its dual linear programming problem is modeled as follows:

$$D(\varepsilon) = \min[\theta - \varepsilon(e^- S^- + e^+ S^+)] \tag{10}$$

$$s.t \begin{cases} \sum\limits_{j=1}^{n} X_j \lambda_j + S^- = \theta X_0 \\ \sum\limits_{j=1}^{n} Y_j \lambda_j - S^+ = Y_0 \\ \lambda_j \geq 0, j = 1, 2, 3 \ldots n \\ S^+ = (S_1^+, S_2^+, \ldots., S_s^+)^T \geq 0, S^- = (S_1^-, S_2^-, \ldots., S_s^-)^T \geq 0 \end{cases} \tag{11}$$

Here, $\theta, \lambda, j = 1, 2, \ldots n$ are dual variables, $e^- = (1, 1, \ldots 1) \in E_m$ represent dual variables of m-dimensional, $e^+ = (1, 1, \ldots 1) \in E_s$ are dual variables of s-dimensional; S^+, S^- are also slack variables.

4.2 Model Adaptability Analysis

The introduction of the DEA model to evaluate investment benefits from information construction of integrated passenger transportation hubs has the following advantages: (1) DEA model has absolute advantages in dealing with the comprehensive efficiency evaluation of multi-input and multi-output indexes. (2) There is no need to conduct dimensionless data processing before applying the DEA method. (3) There is no need to make any weight assumption for the indexes, which effectively avoids the influence of subjective factors on the evaluation results.

The BCC model corresponds to the VRS model with variable size, and the CCR model corresponds to the CRS model with constant size. In view of the fact that there is no optimal scale point for investment in information construction, with the continuous development of information technology, the benefit of investment in information technology is increasing year by year, and the difference of its scale benefit is also increasing. Therefore, this paper uses the BCC model to build an evaluation model of the investment benefit of informatization.

4.3 *Information Investment Benefit Evaluation Model*

First, the decision-making unit is determined. The model requires that the decision-making unit should have the same goal and task, as well as the same inputs and outputs. Different integrated passenger transportation hubs all undertake the important tasks of various external transportation modes and efficient connection between cities, with the same benefit objectives and tasks, and almost the same informatization input-output indexes. Therefore, this paper selects the integrated passenger transportation hub as the decision-making unit to evaluate the information investment benefit of the integrated passenger transportation hubs.

Secondly, the investment scale, the perception system, operational decisions, emergency command, passenger service in five aspects index as input indexes, The entry and exit times increased, reduction in emergency response time and intensity of transfer passenger flow on the station as the output indexes, build integrated passenger transportation hubs informatization investment benefit evaluation model based on the DEA (Fig. 2).

Then, the index is treated. Although the DEA model does not need the dimensionless processing of the indicators, it should carry out the forward processing to ensure that all the indicators in the data envelopment analysis model are positive and reasonable. According to the attribute type, it can be divided into benefit-type, cost-type and interval-type, the input and output indexes are analyzed and classified into different index attribute types [7].

(1) Benefit-index refers to the larger the attribute value, the better.

$$r_{ij} = \frac{a_{ij} - \min a_{ij}}{\max a_{ij} - \min a_{ij}} \tag{12}$$

Here r_{ij} is the forward processing result of the j evaluation index in the i decision-making unit, a_{ij} is the value of the j evaluation index of the i decision unit.

(2) Cost-type index means that the smaller the attribute value, the better.

$$r_{ij} = \frac{\min a_{ij} - a_{ij}}{\max a_{ij} - \min a_{ij}} \tag{13}$$

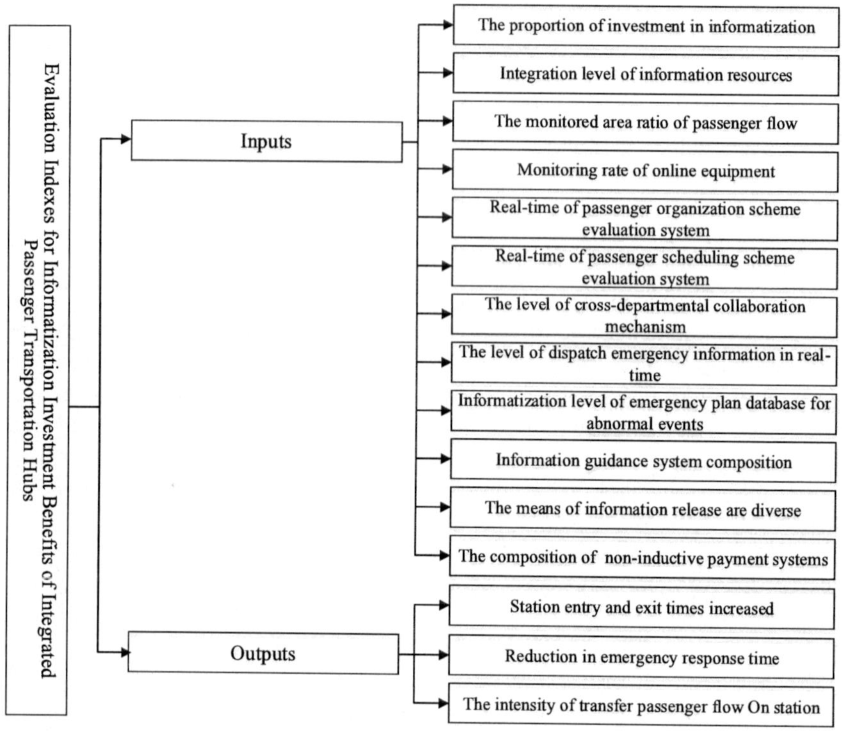

Fig. 2 The informatization investment benefit evaluation model of integrated passenger transportation hubs

(3) Interval-index means that the closer the attribute value is to a fixed interval [m, n], the better

$$r_{ij} = \begin{cases} \max(k_j^1 - \min(a_{ij}), \max(a_{ij}) - k_j^2) - (a_{ij} - k_j^2)\, if\, a_{ij} > k_j^2 \\ \max(k_j^1 - \min(a_{ij}), \max(a_{ij}) - k_j^2)\, if k_j^1 \le a_{ij} \le k_j^2 \\ \max(k_j^1 - \min(a_{ij}), \max(a_{ij}) - k_j^2) - (k_j^2 - a_{ij})\, if\, a_{ij} \le k_j^1 \end{cases} \tag{14}$$

Here k_j^1 is the upper limit of the optimal interval of the j evaluation index, k_j^2 is the lower limit of the optimal interval of the j evaluation index.

Finally, the data are collected and the model is solved. The commonly used solution software is DEAP2.1.

5 Conclusion

Because of the prominent problems existing in the current construction of smart integrated passenger transportation hub, such as the low application level of information system, waste of resources and poor investment effect, the paper establishes a set of scientific and effective evaluation index system of integrated passenger transportation hub information investment benefit, and selects 15 indicators to evaluate the effect of information construction from six aspects, including investment scale, perception system, operation decision, emergency command, passenger service and benefit reflection. On this basis, the data envelopment analysis method is introduced to define decision units and input-output indicators, and the investment benefits evaluation model of integrated passenger transportation hub informatization is constructed. Finally, the idea of solving the model is proposed.

The research has guiding significance and practical application value for the informatization project planning, construction and investment effect evaluation of Our country's integrated passenger transportation hubs. The results of the model can also provide guiding decision-making suggestions for the information construction of integrated passenger transportation hubs, which can effectively avoid the problem of resource waste in the process of information construction. However, the article still has the following major problems. Given the current difficulties in data collection, this paper did not verify the accuracy of the model results based on the measured data. In the subsequent process, the study on the applicability of the model and the accuracy of the results will be strengthened.

References

1. Xi C (2017) Study on evaluation of information management mode of urban comprehensive transportation hub. Shanghai Highways 1:72–75
2. Ministry of Transport of the People's Republic of China (2020) Guiding opinions of the ministry of transport on promoting the construction of new infrastructure in the field of transport 75. http://www.gov.cn/zhengce/zhengceku/2020-08/06/content_5532842.htm
3. Zhu J, Sun C, Lin Y et al (2020) Exploration on intelligent development path of comprehensive transportation hub under station and city integration mode. ITSAC
4. Chen C, Peng J (2020) Overall planning and design of Chongqing West railway station intelligent hub information management system. Mod Transp Technol (4)
5. Nie T, Zheng W, Sun Y et al (2019) Evaluation method of passenger transfer service level in comprehensive passenger hubs. Transp Res 5(30)(06):46–52
6. Charnes A, Cooper WW, Rhodes E (1978) Measuring the efficiency of decision making units. Eur J Oper Res 2(6):429–444
7. Liu X (2009) The necessity of data forward processing in factor analysis and its software implementation. J Chongqing Univ Technol (Nat Sci) 23(09):152–155

Learning Individual Travel Pattern by Using Large-Scale Mobile Location Data with Deep Learning

Hao-yang Yan, Yu-jie Li, Xiao-han Liu, Xi Chen, and Xiao-lei Ma

Abstract Factors found to be influencing individual travel patterns have been explored in several studies. A number of studies have suggested an association between mobile location data and individual travel patterns. This paper proposes a novel deep learning framework to extract individual travel patterns by using large-scale mobile location data. The proposed framework includes methods for extracting origin and destination points based on spatiotemporal thresholds, matching the origin and destination with traffic analysis zone, and predicting based on natural language processing methods, in which a neural network is constructed to vectorize the traffic area on the basis of spatiotemporal information. A case study is performed using mobile location data involving more than 3 million users in Beijing, and the results show that our proposed framework can effectively identify individual travel patterns. The results of this study can further provide insights into traffic demand identification, bus network optimization, and other related research.

Keywords Mobile positioning data · Origin–destination extraction · Travel pattern · Machine learning model · Data mining

1 Introduction

Travel demands in cities have significantly increased owing to rapid urban expansion and motorization. Problems, such as traffic congestion, have aggravated, and the efficiency and level of service (LOS) of transportation systems have decreased. Understanding travel patterns can provide insights into the solutions to improving transportation efficiency and LOS. Promoting the cognition of the travel demand can provide theoretical and data supports for city planning, road planning, and construction.

H. Yan · Y. Li · X. Liu · X. Chen · X. Ma (✉)
School of Transportation Science and Engineering, Beijing Key Laboratory for Cooperative Vehicle Infrastructure System and Safety Control, Beihang University, Beijing 100191, China
e-mail: xiaolei@buaa.edu.cn

W. Wang et al. (eds.), *Green Transportation and Low Carbon Mobility Safety*, Lecture Notes in Electrical Engineering 944,
https://doi.org/10.1007/978-981-19-5615-7_19

With the development of communication and computer technology, the use of smartphones is becoming prevalent. The integrated Global Positioning System (GPS) module in smartphones records the real-time location information of residents and provides large-scale mobile location data, which indicates the potential of identifying individual travel patterns. However, mobile location data usually only contain spatiotemporal information, thereby lacking travel pattern, travel purpose, and travel origin and destination. Consequently, extracting the travel activity chains from mobile positioning data and further mining and analyzing individual travel patterns are necessary.

In recent years, considerable literature has recognized the importance of individual travel patterns. The core of the extraction of travel activity chains is to identify the *origin and destination* points from the trajectory [1]. The key aspect of origin and destination extraction is to divide the continuous and stateless raw data into multiple nodes on the basis of a spatiotemporal threshold. Alexander et al. [2] clustered the trajectory points close to one another in space into several possible activity points and screened out the possible activity points again in terms of time. They analyzed the anonymous call data record (CDR) in Boston for 2 months to attain the travel OD matrix of users. Jiang et al. [3] adopted a similar idea to process the CDR of Singapore for extracting an individual travel network. The authors finally showed the visualization results of residents' travel distribution among the traffic analysis zones (TAZs) in Singapore. Ye et al. [4] developed an approach based on travel chain iteration to find the activity points that meet the space and time thresholds, thereby preventing the errors caused by uneven spatial distribution and large time intervals. They exploited the GPS data of Beijing to explore individual travel patterns on the basis of historical locations. Van et al. [5] used the random forest method to identify the origin and destination. They set multiple spatial windows in accordance with the spatial data structure to obtain local density. They concluded that the random forest algorithm could obtain higher classification accuracy in comparison with four machine learning algorithms in identifying activity and trip points.

Regarding travel pattern mining, fruitful study results have been documented by employing various methods in previous literature. Vanhoof et al. [6] extracted family–work information from French residents and studied the travel patterns by combining the mobile communication mode information and time in CDR data. Jiang et al. [3] employed the digraph model to classify individual travel into several patterns and performed a statistical analysis of Singapore residents. Ye et al. [4] used a neural network to vectorize the travel location of residents. They realized the verification of the location before or after the travel and achieved good results in the test. Yu et al. [7] proposed a travel pattern mining method combining discrete wavelet transform and neural network model. Their case study showed that this method is fast for travel pattern mining. Xiao et al. [8] developed a Bayesian network by using the K2 algorithm to identify travel patterns, and they used the maximum likelihood method to estimate the corresponding conditional probability table. Compared with the patterns obtained by telephone reminiscence survey, the recognition accuracy of the algorithm was verified.

However, the accuracy of existing travel pattern mining methods still needs to be improved. To our best knowledge, no study has combined TAZ and travel chain to identify travel patterns. Therefore, this paper proposes methods to judge the attributes of residents' trajectory points and extract the origin and destination points by using the mobile location data generated by smartphones. The main contributions of our work can be summarized as follows:

- We propose a novel two-step method to extract origin and destination from mobile positioning data, which performs well with high accuracy and low error.
- An unsupervised machine learning method is proposed to mine the travel patterns within the trip data we extracted. On the basis of the method, TAZ is vectorized for trip prediction and can be used as a foundation for further research.
- Our proposed method is evaluated on real-world mobile location data of massive users. The results of the experiment show that our method has outstanding performance.

2 Methodology

2.1 *Extraction of Origin and Destination*

Origin and destination point extraction can be transformed into the equivalent problem, in which trajectory points are divided into two types: moving and staying. The travel points between successive staying points are used to capture moving trajectories. The staying points on both sides of the trajectory represent the origin and destination points. The trajectory points in location range A can be formulated as

$$\{p_1, p_2, p_3, \ldots, p_n\} = \{(x_1, y_1), (x_2, y_2), (x_3, y_3), \ldots, (x_n, y_n)\} \in A. \quad (1)$$

For convenience, the average longitude and latitude of the trajectory points are calculated as location point representation $\overline{p}$ for location range A. This method transforms the location information from a cluster to a point. The location point representation can be formulated as

$$\overline{p} = \left(\tfrac{1}{n}\sum x_i, \tfrac{1}{n}\sum y_i\right). \quad (2)$$

In Fig. 1, $P1, P2, P3, \ldots$ represent the trajectory points in the mobile location data, and the arrows represent the temporal order of their generation. The first step is *spatiotemporal* clustering. In accordance with the given temporal threshold and spatial threshold, the trajectory points that are close enough in space and stay long enough in time are clustered as a starting point (or end point), and the other trajectory points between the starting and ending points are defined as moving trajectory points. Red points (P3-P6), purple points ($P8andP9$), and orange points (P10–P12) are

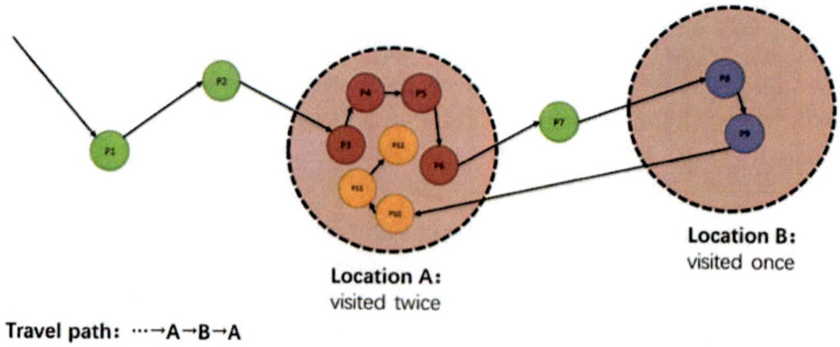

Fig. 1 Extraction of origin and destination

extracted as staying points, while green points are extracted as moving points. Then, spatial clustering is carried out. The points that are close enough in space and meet the threshold value are clustered, and a location range containing one cluster is defined as a location. The light red circle represents the extracted locations *A* and *B*, and then the number of visits to the locations is obtained in accordance with the time of the track points. The travel path from *A* to *B* and back to *A* can be extracted. The proposed approach avoids treating the same location (such as location *A* in the example) as two different locations.

2.2 *Analysis of Individual Travel Pattern*

2.2.1 Vectorization

Skip-gram is a common method of word to vector in natural language processing (NLP). It aims to predict context words via the given center word. The main idea of skip-gram is maximizing the probability of context words in the predefined word window. Let $V = \{v_1, v_2, \ldots, v_n\}$ denote all of the words and $W = \{w_1, w_2, \ldots, w_T\}$ denote the words in the training set. The objective function of neural networks of skip-gram can be formulated as

$$\max \Pr(\{w_{n-c}, \ldots, w_{n+c}\} \backslash w_n | w_n) = \frac{1}{T} \sum_{n=1}^{T} \sum_{\substack{-c \leq m \leq c \\ m \neq 0}} \log p(w_{n+m} | w_n), \quad (3)$$

where c represents the size of word window; i.e., the first and last c words are considered the context. $T = |W|$ represents the total number of words in the training set, and $p(w_{n+m}|w_n)$ represents the probability to be maximized usually using $softmax$

function, which can be formulated as

$$p(w_{n+m}|w_n) = \frac{\exp\left(u_{w_{n+m}}^T v_{w_n}\right)}{\sum_{k=1}^{T} \exp\left(u_{w_k}^T v_{w_n}\right)}, \tag{4}$$

where u_{w_k}, v_{w_n} denote the vectors of the context words and center word, respectively.

The mapping of the word to vector can be defined as

$$\phi : w \in V \to \mathbb{R}^{T \times d}, \tag{5}$$

$$\phi' : w \in U \to \mathbb{R}^{T \times d}, \tag{6}$$

where ϕ, ϕ' represent the mapping of context words and center word, respectively; and d represents the dimension of the vector.

In training, we usually build a neural network with a shallow structure and let the center word and context words be the input and output, respectively. The weight of the dense layer of the neural network can be represented by a matrix in the shape of $T \times d$. The matrix can be regarded as a matrix composed of T row vectors, and each row vector represents its corresponding word.

2.2.2 Spatiotemporal Model

Using the aforementioned skip-gram method, we regard the index of the TAZ of origin and destination points in travel as words and the travel as the sentence. Then, we transfer the skip-gram to the analysis of individual travel patterns. A trip from TAZ "origin" to TAZ "destination" can transfer to two pairs of input–output: (a) input is "origin," and output is "destination"; (b) input is "destination," and output is "origin." We transfer the trip data mentioned above and obtain the function $\phi : w \in V \to \mathbb{R}^{T \times d}$, i.e., the weights of the dense layer in the neural network, which can be formulated as $\left(v_{w_1}, v_{w_2}, \ldots, v_{w_n}\right)^T$, where v_{w_k} represents the vector of kth TAZ in wth training set.

In the above model, a neural network is trained by transforming the trip into the input and output of the neural network. After the training, TAZ can be transformed into the location vector. However, this model ignores the time characteristics of the trip. The impact of time characteristics on residents' travel patterns is obvious. For example, obvious differences should exist between the morning and evening in the association between commercial and residential areas. Therefore, the time characteristics of residents' travel should be added to the model.

Let $O \to D$ represent a trip. Consider two trips: $O \to D$ and $O' \to D$; their arriving times are t, t', respectively. If they meet the following conditions:

$$\Delta t = \left|t - t'\right| < \delta, \tag{7}$$

where δ denotes the predefined threshold, we can consider $O, O^{'}, D$ to have correlation in both spatial and temporal dimensions. D can be regarded as the center word, and $O, O^{'}$ can be regarded as the context words.

2.2.3 Trip Prediction

On the basis of the proposed spatiotemporal model, TAZs: $k \in K\{1, 2, 3 \ldots\}$ can be represented by vectors $v_{w_k} = (v_0, v_1, \ldots, v_d)$. They can be formulated as

$$p(w_{k2}|w_{k1}) = v_{w_1} v_{w_2}^T \tag{8}$$

where v_{w_1}, v_{w_2} represent the vectors of TAZ k_1, k_2, respectively; and $p(w_{k2}|w_{k1})$ represents the $softmax$ probability. Large $p(w_{k2}|w_{k1})$ indicates the high probability of a trip between k_1, k_2. Given TAZ k_m, the following equation is calculated:

$$P = \{(w_{ki}|w_{km})|k_i \in K\{1, 2, 3 \ldots\}\backslash k_m\} \tag{9}$$

The calculation results are sorted from large to small, and the top-Q values are taken. Their corresponding TAZs mean the top Q possible context TAZs.

3 Experiments and Results

3.1 Extraction of Origin and Destination

As shown in Fig. 2, the attributes of mobile location data include user ID (with anonymity generated by computer), time, longitude, and latitude. The time of location data ranges from September 1, 2018 to September 7, 2018, and the region range mainly covers Beijing, i.e., 39°26'N to 41°03'N, 115°25'E to 117°30'E. For each user, the recorded rows in a day are about tens of thousands. The total amount of data in a day is nearly 70 GB, including the travel trajectories of about three million users.

Data are normalized via two steps. The first step is to cluster data in accordance with time and user. The original data are divided into subclusters to directly study a certain day or a user to improve the speed and convenience of data processing. The second step is to remove the outliers in the data, including duplicate points, drift points, and points out of the scope of the study.

In this study, the TAZ division method is shown in Figs. 3 and 4. Several main roads in Beijing are selected as the boundary for division. The granularity of TAZ division is adaptively adjusted in accordance with urban planning and population density. The TAZ in densely populated areas (such as the city center) has a high degree of refinement and small granularity, whereas the TAZ is of a low degree of

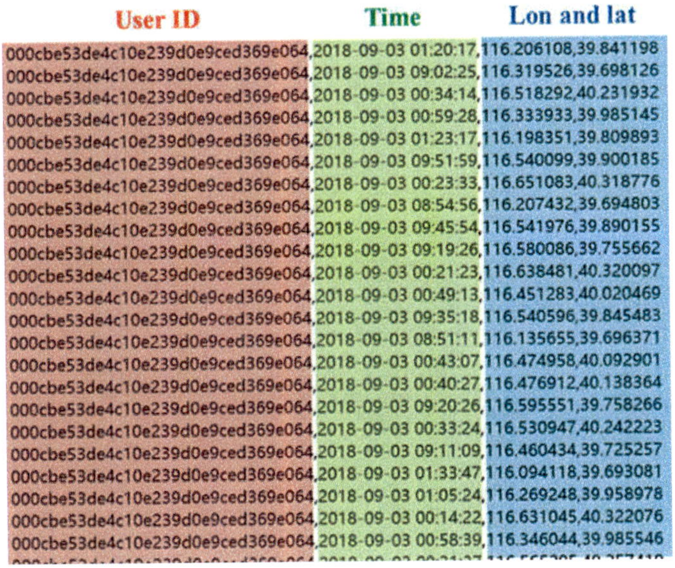
User ID Time Lon and lat

000cbe53de4c10e239d0e9ced369e064,2018-09-03 01:20:17,116.206108,39.841198
000cbe53de4c10e239d0e9ced369e064,2018-09-03 09:02:25,116.319526,39.698126
000cbe53de4c10e239d0e9ced369e064,2018-09-03 00:34:14,116.518292,40.231932
000cbe53de4c10e239d0e9ced369e064,2018-09-03 00:59:28,116.333933,39.985145
000cbe53de4c10e239d0e9ced369e064,2018-09-03 01:23:17,116.198351,39.809893
000cbe53de4c10e239d0e9ced369e064,2018-09-03 09:51:59,116.540099,39.900185
000cbe53de4c10e239d0e9ced369e064,2018-09-03 00:23:33,116.651083,40.318776
000cbe53de4c10e239d0e9ced369e064,2018-09-03 08:54:56,116.207432,39.694803
000cbe53de4c10e239d0e9ced369e064,2018-09-03 09:45:54,116.541976,39.890155
000cbe53de4c10e239d0e9ced369e064,2018-09-03 09:19:26,116.580086,39.755662
000cbe53de4c10e239d0e9ced369e064,2018-09-03 00:21:23,116.638481,40.320097
000cbe53de4c10e239d0e9ced369e064,2018-09-03 00:49:13,116.451283,40.020469
000cbe53de4c10e239d0e9ced369e064,2018-09-03 09:35:18,116.540596,39.845483
000cbe53de4c10e239d0e9ced369e064,2018-09-03 08:51:11,116.135655,39.696371
000cbe53de4c10e239d0e9ced369e064,2018-09-03 00:43:07,116.474958,40.092901
000cbe53de4c10e239d0e9ced369e064,2018-09-03 00:40:27,116.476912,40.138364
000cbe53de4c10e239d0e9ced369e064,2018-09-03 09:20:26,116.595551,39.758266
000cbe53de4c10e239d0e9ced369e064,2018-09-03 00:33:24,116.530947,40.242223
000cbe53de4c10e239d0e9ced369e064,2018-09-03 09:11:09,116.460434,39.725257
000cbe53de4c10e239d0e9ced369e064,2018-09-03 01:33:47,116.094118,39.693081
000cbe53de4c10e239d0e9ced369e064,2018-09-03 01:05:24,116.269248,39.958978
000cbe53de4c10e239d0e9ced369e064,2018-09-03 00:14:22,116.631045,40.322076
000cbe53de4c10e239d0e9ced369e064,2018-09-03 00:58:39,116.346044,39.985546

Fig. 2 Summary of mobile positioning data

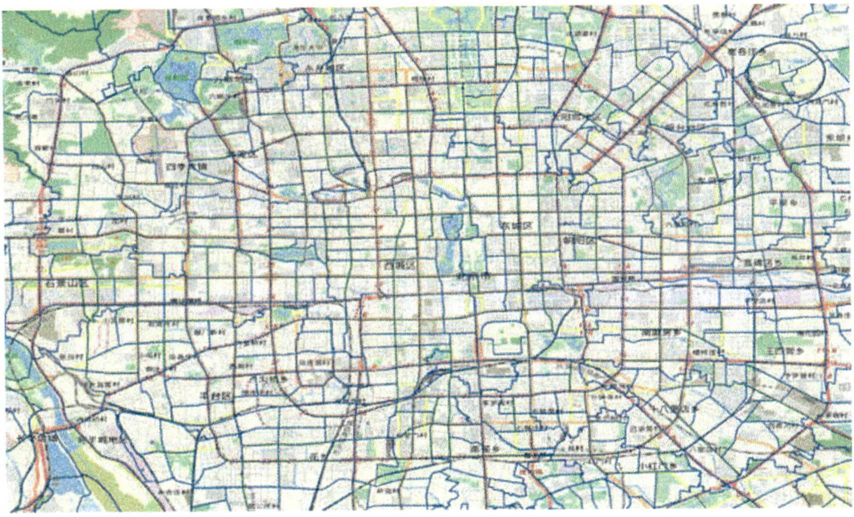

Fig. 3 TAZs in the Fifth Ring Road of Beijing

refinement and large granularity in areas with sparse population density (such as suburbs). In accordance with the above division approach, 1911 TAZs are generated. The lengths and widths of TAZs in the central area with high population density are about 500–1000 m.

Fig. 4 Overall TAZs in Beijing

This study leverages the PostgreSQL system and PostGIS plug-in to extract the origin and destination points of individual trips, match the results with TAZs, and output the trip data of all users in TAZ within 1 day. Examples of trip data are shown in Table 1. Finally, the original trajectory data are converted into trip data, which can significantly simplify the data scale. The amount of data is reduced from about 70 GB to about 1 GB in 1 day. The matched trip data can be a foundation of the subsequent travel pattern analysis.

3.2 Analysis of Individual Travel Pattern

This study chooses the extracted trip data of September 3, 2018 as the training set. For the trips on September 3, 2018, each trip can generate two pairs of input–output. One pair is the origin point as input and the destination point as output. Another pair is the destination point as input and the origin point as output.

The specific structure of the neural network is shown in Table 2. The $softmax$ function is selected as the activation function of the output layer of the neural network. The output of this layer is transformed into a vector whose dimension is equal to the number of TAZs. The value of each dimension in the output represents the $softmax$ probability of traffic TAZs corresponding to the dimension serial number, and the sum of values in all dimensions should be 1.

Figure 5 shows the changes in accuracy and loss function values during the training process. The red and blue curves represent the training and validation sets, respectively. We choose the best model of validation set and test it on the test set. The test result shows that the accuracy of the model is 0.616.

The weight matrix after training is extracted as $\left(v_{w_1}, v_{w_2}, \ldots, v_{w_n}\right)^T$, and let v_{w_k} denote the location vector corresponding to the k th TAZ in the training set w. Given TAZ k_m, calculate the values of each element in $v_{w_m} v_{w_k}^T$ on the basis of TAZ

Table 1 Examples of matched trip data

User ID	Departure time	Origin longitude	Origin latitude	Arrival time	Destination longitude	Destination latitude	Origin TAZ	Destination TAZ
004ebd2592170d84e4712609707b7c55	2018/9/3 13:55	116.412825	39.925348	2018/9/3 14:05	116.399448	39.931945	58	79
004e9f946d7663fd6c68c498bd3d82af	2018/9/3 21:25	116.4013379	39.9930654	2018/9/3 21:30	116.4013268	39.9867814	506	504
004ec242bcce062592aea608af0e084d	2018/9/3 18:25	116.269702	39.87948	2018/9/3 19:17	116.410675	40.036258	768	478
004ea27ed6c73b48a984c51fec319adf	2018/9/3 15:08	116.423678	39.899056	2018/9/3 15:49	116.421937	39.90418	92	51
004ea27ed6c73b48a984c51fec319adf	2018/9/3 15:56	116.421709	39.907361	2018/9/3 16:11	116.455258	39.906355	52	536

Table 2 Structure of the neural network. Model: "sequential_1"

Layer (type)	Output shape	Param #
dense_1 (Dense)	(None, 300)	516,300
activation_1 (Activation)	(None, 300)	0
dropout_1 (Dropout)	(None, 300)	0
dense_2 (Dense)	(None, 1720)	517,720
activation_2 (Activation)	(None, 1720)	0

Total params: 1,034,020
Trainable params: 1,034,020
Non-trainable params: 0

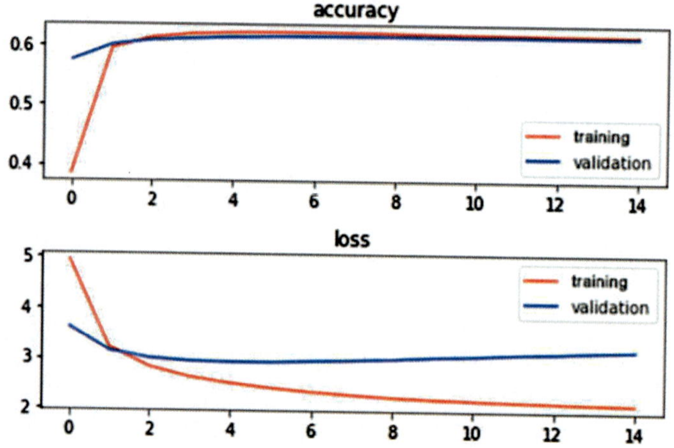

Fig. 5 Neural network training process (part one) accuracy (part two) loss function

k_m and TAZ $k \in K\{1, 2, 3 \ldots\} \backslash k_m$. Then, the probability distribution pattern of traffic assignment between this TAZ and other TAZs in the city can be obtained. The calculation results are ranked in descending order. The first q values are taken, and the corresponding TAZs are the first q TAZs to most likely come from or go to. However, in practical research, considering that the similarity of location vectors should be mainly expressed by the angle between vectors, the modulus of vectors should not represent the similarity between vectors. Calculating $v_{w_m} v_{w_k}^T$ is replaced with calculating the cosine distance of a value of $v_{w_m} v_{w_k}^T$. The calculation formulas of cosine distance are given as follows:

$$D_c = 1 - S_c(\boldsymbol{A}, \boldsymbol{B}), \tag{10}$$

$$S_c = \frac{\boldsymbol{A} \cdot \boldsymbol{B}}{\|\boldsymbol{A}\| \|\boldsymbol{B}\|} = \frac{\sum_{i=1}^{n} A_i B_i}{\sqrt{\sum_{i=1}^{n} A_i^2} \sqrt{\sum_{i=1}^{n} B_i^2}}. \tag{11}$$

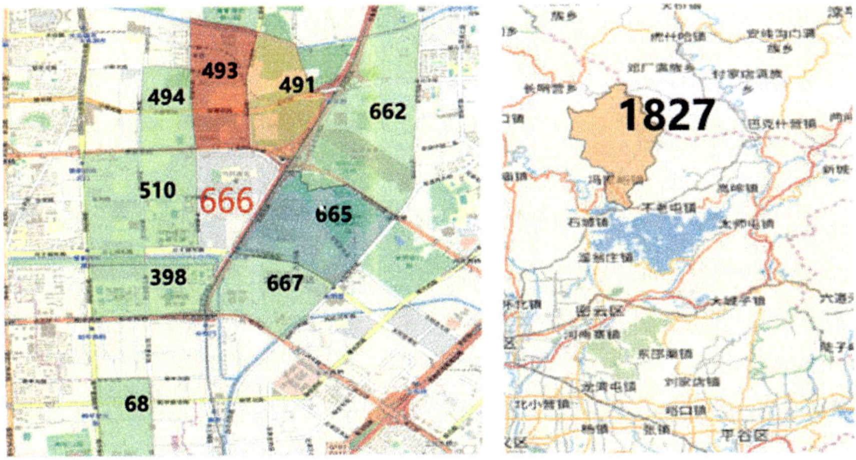

Fig. 6 Travel behavior prediction of No. 666 TAZ

Taking No. 666 TAZ as an example, the cosine distances between the corresponding vector of No. 666 TAZ and the corresponding vector of other TAZs are calculated. The results are sorted in ascending order, and the following results are obtained: the numbers of top 10 TAZs related to No. 666 most strongly are Nos. 491, 493, 667, 665, 398, 662, 494, 68, 510, and 1827. Figure 6 shows that the results are matched with the map for visualization processing. The main TAZs are located in the business districts around No. 666 TAZ. Note that No. 1827 TAZ, which ranks 10th, belongs to the boundary of Beijing. Two possible explanations for this phenomenon can be considered: (1) it is the error caused by data noise or insufficient training level; (2) the trajectory leaving or entering Beijing is regarded as the origin point or destination point in the boundary due to the removal of the points beyond the research scope of Beijing in the data preprocessing process. From the observation of the information of GIS, No. 1827 TAZ is the main road leading to Chengde, Hebei Province and has the characteristics of bus station, railway station, and other government traffic regional hubs. Therefore, the mentioned phenomenon is not an error caused by data noise or insufficient training but an acceptable result.

4 Conclusion

Motivated by the popularity of smartphones, this study focuses on the extraction of the origin and destination of residents and the analysis of individual travel patterns by using mobile location data. In the O–D point extraction process, we divide the travel origin and destination points on the basis of spatiotemporal thresholds. In analyzing individual travel patterns, the NLP method skip-gram is used to build a neural network and predict the context TAZs. In summary, we propose a complete framework for individual travel pattern mining based on mobile location data. It provides data and theoretical support for urban road network planning, bus network optimization, and traffic management and control. In future research, we can use the results of TAZ vectorization in this paper as the basis, further fuse multisource data, and establish a sophisticated individual travel pattern model.

Acknowledgements The study was supported by National Key R&D Program of China (2021YFB1600100).

References

1. Liu F, Gao Z, Janssens D, Jia B, Wets G, Yang Y (2021) Identifying business activity-travel patterns based on GPS data. Transport Res Part C Emerg Technol 128:103136. https://doi.org/10.1016/j.trc.2021.103136
2. Alexander L, Jiang S, Murga M, González MC (2015) Origin–destination trips by purpose and time of day inferred from mobile phone data. Transport Res Part C Emerg Technol 58:240–250
3. Jiang S, Ferreira J, González MC (2017) Activity-based human mobility patterns inferred from mobile phone data: a case study of Singapore. IEEE Trans Big Data 3(2):208–219
4. Ye Y, Zheng Y, Chen Y, Feng J, Xie X (2009) Mining individual life pattern based on location history. In: 2009 Tenth international conference on mobile data management: systems, services and middleware, pp 1–10
5. Van Dijk J (2018) Identifying activity-travel points from GPS-data with multiple moving windows. Comput Environ Urban Syst 70:84–101. https://doi.org/10.1016/j.compenvurbsys.2018.02.004
6. Vanhoof M, Reis F, Smoreda Z, Ploetz T (2018) Detecting home locations from CDR data: introducing spatial uncertainty to the state-of-the-art. ArXiv:1808.06398
7. Yu JJQ (2021) Travel mode identification with GPS trajectories using wavelet transform and deep learning. IEEE Trans Intell Transp Syst 22(2):1093–1103. https://doi.org/10.1109/TITS.2019.2962741
8. Xiao G, Juan Z, Zhang C (2015) Travel mode detection based on GPS track data and Bayesian networks. Comput Environ Urban Syst 54:14–22. https://doi.org/10.1016/j.compenvurbsys.2015.05.005

Short-Term Traffic Flow Forecast Based on ARIMA-SVM Combined Model

Jiaxin Peng, Yongneng Xu, and Menghui Wu

Abstract Short-term traffic flow prediction is an important basis for traffic state discrimination and traffic congestion prediction. The ARIMA-SVM combined prediction model is used to forecast the urban short-term traffic flow. Firstly, the ARIMA model and SVM model are used to forecast the urban short-term traffic flow. Then, the ARIMA-SVM combined prediction model was obtained by using the updated dynamic weight weighted fusion method to forecast the urban traffic flow, and the results were compared with the separate ARIMA model and SVM model. The empirical results show that the ARIMA-SVM combined model can more accurately predict the city's short-term traffic flow.

Keywords Short-term traffic flow prediction · ARIM model · SVM algorithm · Combined forecasting model

1 Introduction

With the rise of intelligent transportation technology, traffic flow forecasting is playing an increasingly important role as the basis for real-time traffic state discrimination and traffic flow guidance. Short-term traffic flow forecasting obtains dynamic data information in real time through various traffic road condition collection devices. Data preprocessing, model building, optimization solution and other methods obtain

J. Peng · Y. Xu (✉)
Nanjing University of Science and Technology, 200 Xiaolingwei, Nanjing 210094, China
e-mail: x780906@163.com

J. Peng
e-mail: 774027183@qq.com

M. Wu
Nanjing Normal University, 122 Ninghailu, Nanjing 210097, China
e-mail: 1657217352@qq.com

W. Wang et al. (eds.), *Green Transportation and Low Carbon Mobility Safety*,
Lecture Notes in Electrical Engineering 944,
https://doi.org/10.1007/978-981-19-5615-7_20

accurate and reliable forecast information, and release effective road condition information for road users, so as to save travel costs, reduce pollution, and save resources [1].

Juan et al. [2] pointed out in their research that the prevention of traffic congestion should be the first priority. The use of historical traffic flow data of the road section to dynamically predict the future traffic state of the road section in a short period of time and provide the prediction result to the traffic management department can be convenient for the transportation department. To induce traffic flow in time to reduce the occurrence of congestion. At the same time, travelers can also choose the appropriate travel time and route according to the predicted results, and actively avoid congested roads.

Short-term traffic flow prediction has the characteristics of short prediction time and strong randomness, so it has higher requirements on the prediction accuracy and training speed of the algorithm. At present, the widely used models in the field of traffic flow prediction mainly include models based on statistical methods, non-parametric regression models, neural network models, and support vector machine models. Based on the above model, scholars at home and abroad have carried out a lot of research. Dianxi et al. [3] optimized the neighbor matching method of the original model for the problem of neighbor state selection, and proposed a short-term traffic average speed prediction model based on non-parametric regression; Shanying [4] used fuzzy neural network for short-term Time traffic flow prediction, simulation results show that this method can accurately predict short-term traffic flow. Xiaoling et al. [5] used SVM support vector machines to predict short-term traffic conditions based on multi-dimensional space-time parameters; Li Qiaoru et al. [6] selected SVM to predict traffic flow based on the temporal and spatial characteristics of traffic flow.

Han et al. [7] improved the ARIMA model to achieve adaptive prediction of traffic flow, Manman [8] proposed a Kalman filter model based on phase space reconstruction theory, and Chengtao et al. [9] were analyzing support vector machine regression estimation. Based on the performance of the method parameters, a PSOSVM short-term traffic flow prediction model with particle swarm optimization PSO optimized parameters was proposed. Chunjiao et al. [10] used Elman neural network to achieve simultaneous multi-section traffic volume on the basis of reducing the complexity of the road network. prediction. Manchun et al. [11] divided the traffic flow sequence into a linear part and a non-linear part, and respectively used ARIMA model and artificial neural network to predict; Huili et al. [12] applied wavelet analysis method to denoise traffic data and used time series model prediction Traffic flow.

Although the existing theoretical models can predict short-term traffic flow, they still have the following shortcomings: Although the model based on statistical theory is simple to model, it is only suitable for prediction in a steady state; non-parametric regression models can deal with nonlinear problems, But the efficiency is low; the neural network model is a learning method that relies on experience. In the case of insufficient samples, it will lead to a decline in over-learning and generalization capabilities, and the sample structure will increase the complexity of the algorithm, resulting in excessive training time. The efficiency is not high; although the SVM model solves the problem of over-learning in the case of insufficient samples of the neural network model, the SVM is only limited to small cluster samples. When there are too many observation samples, its efficiency will be lower.

Due to the instability of time series data, the fluctuations in the same time period are large, and the fluctuations in another time period are small, forming a phenomenon of volatility aggregation. Therefore, it is very useful to use first-order or multiple-order differences to stabilize the original data that is not stable necessary. Based on this, the ARIMA model and the SVM model are combined to predict the city's short-term traffic flow. The short-term traffic flow data of 288 points in 18 days of the city is used for analysis, and a combined forecasting model based on ARIMA-SVM is constructed. The results show that the method proposed in this study has high forecasting accuracy.

2 Modle Approach

2.1 ARIMA Model

The Autoregressive Integrated Moving Average (ARIMA) method is one of the traditional random time series models invented by Box and Jenkins, also known as the B-J method. There are three basic types of ARIMA models: autoregressive (AR) model, moving average (MA) model and autoregressive moving average (ARMA) model. ARIMA modeling is essentially a data-oriented exploration method, which has the appropriateness of adapting to the structure of the data itself and the flexibility of the model. With the help of autocorrelation function and partial autocorrelation function, the random nature of time series can be approximated, and information such as trend, random change, periodic component, periodic pattern and sequence correlation can be found. As a result, the prediction of the future value of the sequence with a certain accuracy can be easily obtained, and the model is:

$$y_t - \varphi_0 - \varphi_1 y_{t-1} - \cdots - \varphi_p y_{t-p} = a_t - \theta_1 a_{t-1} - \cdots - \theta_q a_{t-q} \tag{1}$$

Among them, a_t, a_{t-1},..., a_{t-q} are stationary white noises with a mean value of 0 and a variance of ξ^2, and p and q are the orders of the autoregressive model and the moving average model, respectively. The model is named an autoregressive moving average sequence with model autoregressive order p and model moving average order q, abbreviated as ARMA(p, q) sequence. When $q = 0$, it becomes an AR(p) sequence; when $p = 0$, it becomes an MA(q) sequence. φ_1, φ_2,..., φ_p are autoregressive coefficients, θ_1, θ_2,..., θ_q are moving average coefficients, and they are all parameters that need to be estimated. The ARIMA model can only deal with time series of stationary processes. If you want to analyze a non-stationary time series, you must make it stationary. The most common and simplest method is to perform a difference operation on the original non-stationary time series to obtain an ARIMA (p, d, q) model, where d is the order of the difference number. In the ARIMA model, the trend term is extracted through the difference operation of the time series and transformed into a stationary term. Then, the ARMA model is estimated, and then transformed after the estimation to adapt to the original sequence model before the difference operation.

2.2 SVM Model

Support vector machine (SVM) was proposed by Vapnik in the 1990s. His basic idea is to map the nonlinear problem in the original low-dimensional input space to the high-dimensional feature space for solution. Support vector regression (SVR) introduces Vapnik's insensitive loss function into the support vector machine to solve the problem of nonlinear regression estimation. The algorithm of SVM is as follows.

For the vector X, a nonlinear mapping φ is defined to map the training (input) data to a so-called high-dimensional feature space. Then, in the high-dimensional feature space, there is theoretically a linear function f to represent the nonlinear relationship between the input data X and the output data. Such a linear function, that is, the SVM function:

$$f(X) = W^T \varphi(X) + b \tag{2}$$

Among them, f(X) represents the predicted value; the coefficients W and b are variable. As mentioned above, the purpose of the SVM method is to minimize the empirical risk.

$$R_{emp}(f) = \frac{1}{N} \sum_{i=1}^{N} \Theta_\varepsilon(y_i, W^T \varphi(x_i) + b) \tag{3}$$

Among them, Θ_ε (y, f(X)) is the insensitive loss function.

$$\Theta_\varepsilon(y, f(X)) = \begin{cases} |f(X) - y| - \varepsilon, |f(X) - y| \geq \varepsilon \\ 0, |f(X) - y| < \varepsilon \end{cases} \tag{4}$$

The research focus of SVM is to find the optimal hyperplane to minimize the training error between the training data and the insensitive loss function. Among them:

$$\min_{W,b,\xi^*,\xi} R_\varepsilon(W, \xi^*, \xi) = \frac{1}{2} W^T W + C \sum_{i=1} (\xi^* + \xi) \tag{5}$$

Among them,

$$y_i - W^T\varphi(X_i) - b \le \varepsilon + \xi_i^*, i = 1, 2, \cdots, N \tag{6}$$

$$\xi_i^* \ge 0, i = 1, 2, \cdots, N$$

$$-y_i + w^T\varphi(X_i) + b \le \varepsilon + \xi_i, i = 1, 2, \cdots, N \tag{7}$$

$$\xi_i \ge 0, i = 1, 2, \cdots, N$$

After solving the quadratic optimization problem with inequality constraints, the parameter vector W in the equation is given.

$$W = \sum_{i=1}^{N}(\beta_i^* - \beta_i)\varphi(X_i) \tag{8}$$

Among them, β_i^*, β_i are Lagrange multipliers. The final SVM prediction function is

$$f(X) = \sum_{i=1}^{N}(\beta_i^* - \beta_i)\varphi(X_i)K(X_i, X_j) + b \tag{9}$$

Among them, $K(X_i, X_j)$ is called the kernel function, and the value of the kernel is respectively equal to the inner product of two vectors X_i and X_j in the feature space $\varphi(X_i)$ and $\varphi(X_j)$, namely

$$K(X_i, X_j) = \varphi(X_i)\varphi(X_j) \tag{10}$$

There are many types of kernel functions. The most commonly used kernel functions are Gaussian radial basis functions and polynomial functions. So far, it is difficult to determine the type of kernel function for a particular data mode. However, the Gaussian radial basis function kernel is not only easy to implement, but also has the function of nonlinear mapping, which is suitable for dealing with nonlinear relationship problems. The author chooses the Gaussian kernel function as the kernel function in SVM, and its expression is:

$$K(x_i, x_j) = exp(-\|x_i - x_j\|^2/2\sigma^2) \tag{11}$$

The prediction accuracy of the SVM model depends on the size of the parameter C and the kernel parameter σ. The grid search method is used to optimize the SVM parameters. Within the specified parameter range, the optimal parameter combination is found through the traversal method to minimize the mean square error.

2.3 *Short-Term Traffic Flow Prediction Model*

(1) ARIMA forecasting model

Establishing an ARIMA model to predict traffic flow mainly includes the following steps:

1. Using the ADF unit root test to test the stationarity of the time series;
2. Difference processing is performed on the non-stationary sequence, and the original sequence is converted into a stationary sequence;
3. Use AIC and BIC criteria to determine the order of the model and estimate the parameters;
4. Eliminate some lagging terms with insignificant coefficients, revise the ARIMA model, and estimate the parameters of the revised model;
5. Carry out residual test on the revised model. Check whether the residual items conform to the white noise process. If they do not, it indicates that there is still relevant information in the residual items that has not been extracted, and the model needs to be further improved;
6. Using the determined model to predict the short-term traffic flow.

(2) SVM prediction model

Establishing an SVM prediction model to predict traffic flow mainly includes the following steps:

1. Suppose the short-term traffic flow sequence is x(t), and use the sliding window method to train SVM for x(t). The size of the sliding window is k. That is, the value of $x(t+1)$ is predicted by the SVM model from the traffic flow value of $x(t)$, $x(t-1)$,..., $x(t-k+1)$.
2. Using the grid search method to determine the parameters C and σ of the SVM, so that the trained model has the smallest mean square error between the predicted value of the historical traffic flow and the true value.
3. Using the trained SVM model to predict the traffic flow.

(3) Combination forecasting model

The combined forecasting model is a kind of forecasting model that combines the forecasting results of different forecasting models with appropriate weights. The combined forecasting model is to give the two models a certain weight, and the results of the two forecasting models are weighted and combined in order to achieve a better forecast result. Its main purpose is to maximize the synthesis of the prediction results of each model and improve the prediction accuracy. The combined model can overcome the limitations of a single model more systematically and scientifically, thereby improving the prediction accuracy.

The key to the combined forecasting model is the determination of the weights. Appropriate weight assignment will improve the prediction accuracy of the entire combined prediction model. At present, the commonly used weight determination methods include equal weight method, reciprocal variance method and advantage matrix method. This article uses the error sum of squares method to determine the weight of the combined forecasting model. The main idea is as follows.

Let x_i be the actual traffic flow at the i-th time point (i = 1, 2,..., n), and let $\tilde{x}_{ik}$ be the traffic forecast value at the i-th point in the k-th method (k = 1, 2,..., k), $e_{ik} = x_i - \tilde{x}_{ik}$ is the prediction error of the traffic flow prediction value at the i-th point of the k-th method, w_k is the estimated value of the weight coefficient of the k-th method, and $\hat{x}_i$ is the combined prediction value, then:

$$\hat{x}_i = \sum_{i=1}^{k} w_k \tilde{x}_{ik} \tag{12}$$

The weights of the combined forecasting model are established based on the minimum sum of squared forecast errors as the criterion. The specific formula is as follows:

$$minE = \sum_{i=1}^{n} \left(x_i - \hat{x}_i\right)^2 \tag{13}$$

$$s.t. \sum_{k=1}^{K} w_k = 1, w_k \geq 0, k = 1, 2, \ldots, K \tag{14}$$

3 Case Study

3.1 Data Sources

Collect traffic flow data at 288 time points in 18 days in a city as the research object, specifically: a daily sampling interval of 15 min, sampling time is 7:30–11:30 in the morning, continuous sampling for 18 days, a total of 288 time points sequence Data, unit: vehicles. The time series of traffic flow is shown in Fig. 1.

Use the first 17 days of data training to build a prediction model, and use the last day's data as test data to test the prediction effect.

3.2 ARIMA Model Prediction Results

First, use MATLAB2018 to test the stationarity of the time series. The test of time series ADF is an effective tool to test the stability of time series. If there is a unit root in the non-stationary time series, use the difference method to eliminate it, and get a stationary series. It can be seen from the ADF test that the original time series is

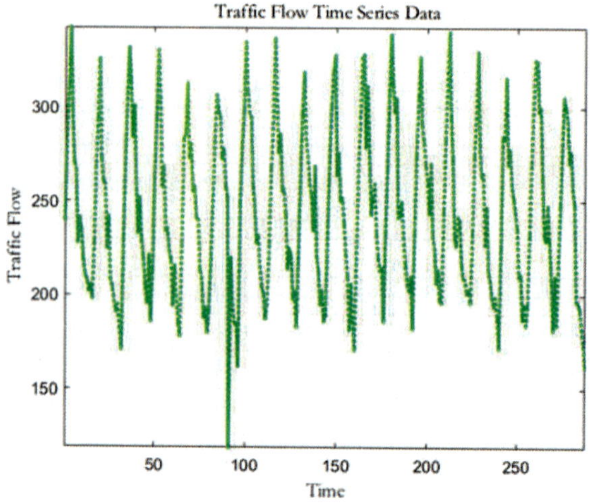

Fig. 1 Urban traffic flow time series data set

unstable. Therefore, it needs to be first-order difference, and the first-order difference data is found to be stable after testing. The sequence diagram after the first-order difference is shown in Fig. 2.

The logical value from the ADF is calculated as 0, the calculation fails, and the original time series is unstable. Perform a first-order difference on the original sequence, and the test shows that the first-order difference data is stable, as shown in Fig. 2. Therefore, it is necessary to fit a first-order difference model.

In order to determine the values of parameter p and parameter q in the ARIMA(p,1,q) model, autocorrelation and partial correlation analysis are performed on the sequence after the first-order difference.

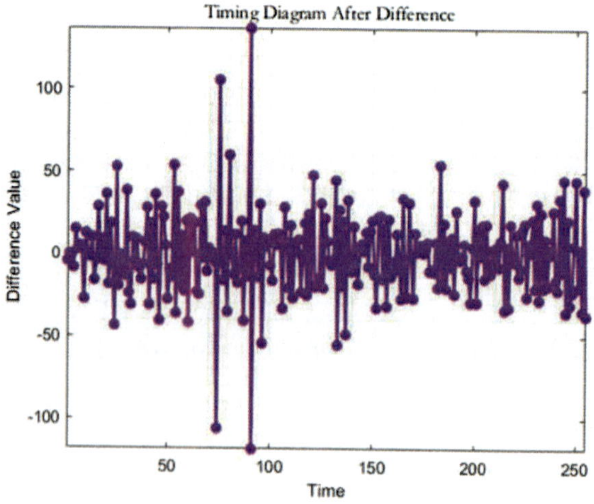

Fig. 2 Timing diagram after first-order difference

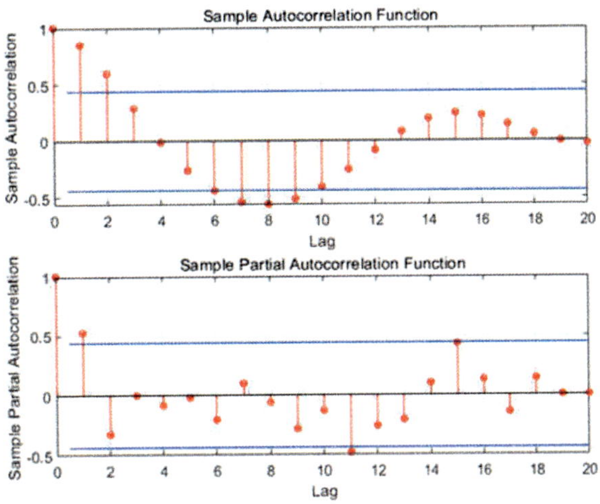

Fig. 3 ACF and PACF analysis graph

It can be seen from Fig. 3 that the autocorrelation quantity is tailing and partial correlation quantity truncation. This shows that the sequence after the first-order difference is suitable for prediction using the ARIMA model. After analyzing Fig. 3, ARIMA (4, 1, 2), ARIMA (4, 1, 3) and ARIMA (4, 1, 4) models were initially established.

The AIC and BIC criteria are used to determine the order of the model. The smaller the AIC value and the BIC value, the higher the model fit. Calculate the AIC and BIC values of ARIMA (4, 1, 2), ARIMA (4, 1, 3), and ARIMA (4, 1, 4), respectively, and the results are shown in Table 1.

It can be seen from Table 1 that the model with the smallest AIC value and BIC value is ARIMA (4, 1, 4). Using the ARIMA (4, 1, 4) model to fit the time series, the results are shown in Table 2.

It can be seen from Table 2 that the coefficient P values of the model residual estimation are all less than 0.05, indicating that the model fitting effect is better. Use the established ARIMA (4, 1, 4) model to predict urban traffic flow, as shown in Fig. 4.

Table 1 AIC and BIC values of different models

Model	AIC	BIC
ARIMA(4, 1, 2)	2256.8	2281.6
ARIMA(4, 1, 3)	2245.6	2274.0
ARIMA(4, 1, 4)	2243.2	2275.2

Table 2 Arima(4,1,4) fitted value (Gaussian distribution)

Parameter	Estimated Value	Error	T test	P
AR{1}	−1.4416	0.81971	−1.7587	0.078624
AR{2}	−1.4235	0.63636	−2.2369	0.025295
AR{3}	−0.69802	0.74001	−0.94325	0.34555
AR{4}	−0.051698	0.092777	−0.55723	0.57737
MA{1}	0.64441	0.80228	0.80323	0.42184
MA{2}	0.3945	0.082646	4.7734	1.8115e-06
MA{3}	−0.49635	0.33724	−1.4718	0.14108
MA{4}	−0.53009	0.64681	−0.81955	0.41247
Variance	346.19	20.423	16.951	1.8983e-64

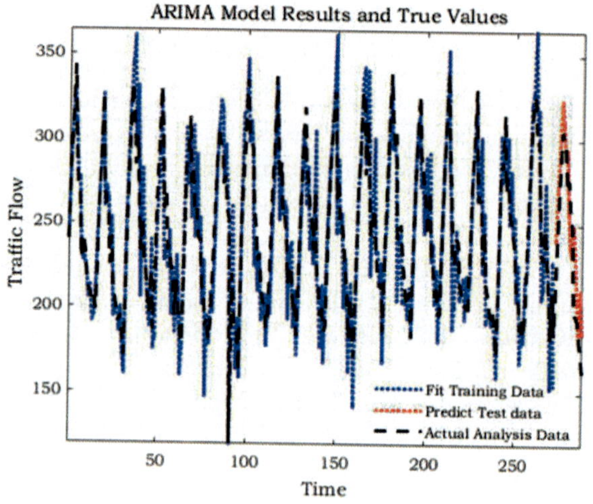

Fig. 4 ARIMA model results and true values

3.3 SVM Model Prediction Results

For the urban monthly traffic flow sequence x(t), the sliding window method is used to train the SVM. The sliding window size is set to 16. That is, the flow value at time t + 1 is determined by the prediction of the SVM model by the flow value at time t, t-1,…, t − 15.

The SVM model data needs to be normalized before input. This paper uses min-max normalization to process the original data to eliminate dimensional differences. The linear transformation of the original data is that the result falls into the interval [0, 1], and the formula is as follows:

$$X = \frac{x - min}{max - min} \tag{15}$$

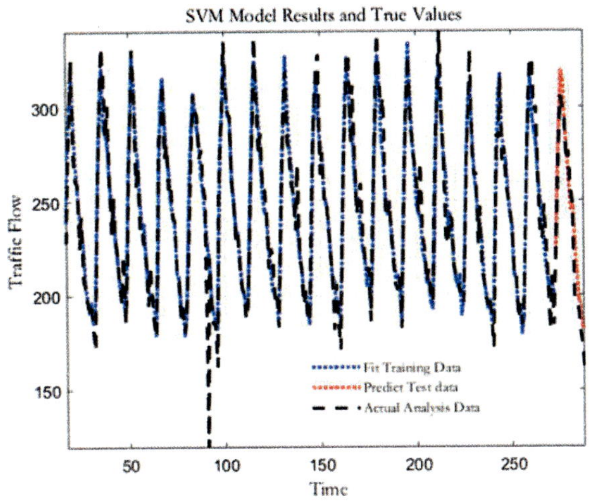

Fig. 5 SVM Model results and true values

Among them, max is the maximum value of the features in the sample data, and min is the minimum value of the features in the sample data. x is the original data, and X is the normalized data.

The SVM is trained using the sequence data of the sliding window, and the penalty coefficient C and σ the kernel parameter σ are set to optimize the grid. The sum of squared errors of the determined SVM model for the predicted value is minimized. Using the determined SVM model to predict the urban traffic flow data, the results are shown in Fig. 5.

3.4 ARIMA-SVM Combined Forecasting Model Forecast Results

In order to effectively improve the accuracy of the prediction model, fully absorb the advantages of the two prediction models, combine the two prediction models to obtain a new combined prediction model, and re-predict the data.

The principle of the smallest sum of squared errors is used to determine the respective weights of the ARIMA model and the SVM model to be 0.45 and 0.55, respectively. Calculate the prediction results of the ARIMA-SVM combined prediction model. The prediction results and prediction errors of different models are shown in Figs. 6 and 7.

Based on the above prediction results and index evaluation, it can be seen that by combining the two prediction models of ARIMA and SVM, the overall prediction effect of the model can be effectively improved, and it can contribute to the prediction of urban short-term traffic flow.

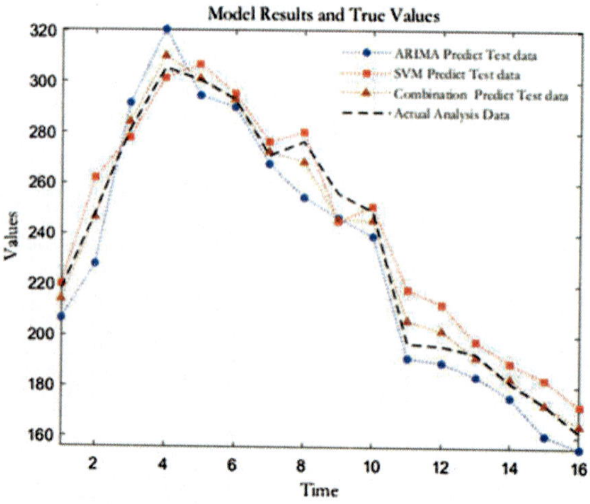

Fig. 6 Different model prediction results

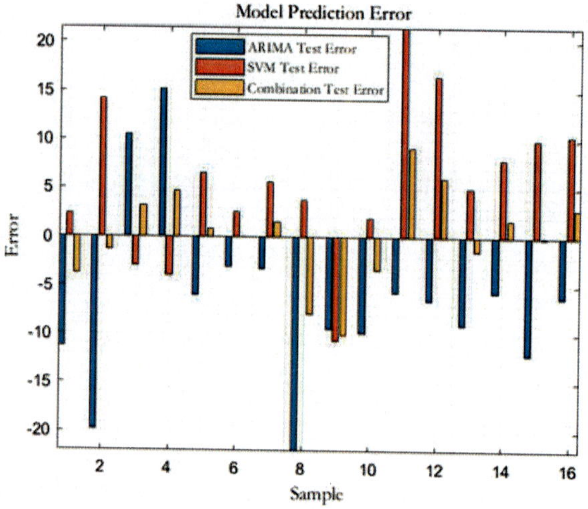

Fig. 7 Different model prediction errors

3.5 *Error Analysis*

The error evaluation index adopts MAE, MAPE, MSE, RMSE.

$$MAE = \frac{1}{N}\sum_{i=1}^{N}\left|X_{prediction,i} - X_{real,i}\right| \tag{16}$$

Table 3 Analysis of Frediction errors of different model test sets

Index	Model		
	ARIMA	SVM	Combination model
MAE	9.7386	7.8544	3.6279
MAPE	0.0417	0.0369	0.0157
MSE	122.8057	91.8170	22.5953
RMSE	11.0818	9.5821	4.7535

$$MAPE = \frac{1}{N}\sum\nolimits_{i=1}^{N}\left|\frac{\left(X_{prediction,i} - X_{real,i}\right)}{X_{real,i}}\right| \tag{17}$$

$$MSE = \frac{1}{N}\sum\nolimits_{t=1}^{N}\left(X_{predicition,t} - X_{real,t}\right)^2 \tag{18}$$

$$RMSE = \sqrt{\frac{1}{N}\sum\nolimits_{t=1}^{N}\left(X_{predicition,t} - X_{real,t}\right)^2} \tag{19}$$

where $X_{prediction}$ is the predicted value, X_{real} is the actual value, and N is the number of samples.

Table 3 shows the evaluation indicators of the prediction errors of each prediction model on the test data set. From the table, it can be seen that the ARIMA-SVM combined prediction model has obtained the best value in each evaluation index. Its MAE, MAPE, MSE And RMSE reached 3.6279, 0.0157, 22.5953 and 4.7535, respectively. Compared with ARIMA, ARIMA-SVM has increased by 62.7472%, 62.3501%, 81.6008% and 57.1053% in various indicators, respectively. Compared with SVM, these improvements are 53.8106%, 57.4526%, 75.3909% and 50.3919%, respectively. This once again proves the superiority of the ARIMA-SVM combined forecasting model in the field of traffic volume forecasting.

4 Conclusion

This paper mainly studies the application of ARIMA-SVM combined forecasting model in city short-term traffic flow forecasting. Firstly, introduce the basic principles of ARIMA time series model and SVM forecasting model, and put forward the method of combined forecasting model; secondly, elaborate the specific modeling process and method of ARIMA-SVM model; finally, through empirical research, the ARIMA-SVM model is used to predict the urban traffic flow data and performs error analysis. From the empirical results, the overall prediction accuracy of the ARIMA-SVM model is better than that of a single ARIMA prediction model and a single SVM prediction model. This shows that the ARIMA-SVM combined forecasting model can be well applied to the short-term traffic flow forecasting in Chinese cities.

Acknowledgements I would like to thank Associate Professor Yongneng Xu for his guidance on the topic and structure of my thesis. I would also like to thank the authors of the references cited in the paper, who have conducted in-depth research on the relevant content, so that I can stand on the shoulders of giants, in order to see farther, in order to find out more valuable things. Finally, I would like to thank "the National Natural Science Foundation of China" (No. 52072214) for supporting this paper.

References

1. Zhipeng G, Rui L, Jian Z, Yongkai H, Fengping Z, Binbin Y (2015) Combined forecast of short-term traffic volume based on time series and GSVMR model. J Chang'an Univ (Nat Sci Ed) 35(S1)):222–225+234
2. Juan T, Shengchun W (2015) Research on traffic congestion prediction model based on deep learning. Comput Appl Res 32(10):2951–2954
3. Dianxi S, Taojie D, Bo D, Hui L (2016) A traffic speed prediction method based on non-parametric regression. Comput Sci 43(02):224–229
4. Shanying C (2017) Research on short-term traffic flow prediction method based on fuzzy neural network. Comput Measur Control 25(08):155–158
5. Xiaoling L, Peng J, Shanhua W, Bin Y (2011) Road short-term traffic state prediction based on multi-dimensional time and space parameters. Transport Syst Eng Inf 11(04):140–146
6. Qiaoru L, Rong Z, Liang C (2015) Short-term traffic flow prediction model based on SVM and adaptive spatio-temporal data fusion. J Beijing Univ Technol 41(04):597–602
7. Chao H, Su S, Chenghong W (2004) Real-time adaptive prediction of short-term traffic flow based on ARIMA model. J Syst Simul 2004(07):1530–1532+1535
8. Manman S (2012) Research on short-term traffic flow prediction method based on Kalman filter. Southwest Jiaotong University
9. Chengtao C, Jianmin X (2007) Short-term traffic flow prediction method based on PSO-SVM. Comput Eng Appl 15:12–14
10. Chunjiao D, Chunfu S, Zhihua X, Juan L (2010) Road network short-term traffic flow prediction method based on Elman neural network. Transport Syst Eng Inf 10(01):145–151
11. Manchun T, Naobin F, Jianmin X (2007) Traffic flow prediction based on the combined model of ARIMA and artificial neural network. China J Highway Transport 04:118–121
12. Huili D, Haode L, Zhizhou W, Xiaoguang Y (2009) Traffic flow prediction method based on wavelet analysis and ARIMA model. J Tongji Univ (Nat Sci Ed) 37(04):486–489+494

A Lightweight Fine-Grained VRU Detection Model for Roadside Units

Jian Shi, Dongxian Sun, Haodong Zhang, Haiqiu Tan, Yaoguang Hu, and Wuhong Wang

Abstract Object detection of vulnerable road users (VRU) under low computing resources of roadside units is one of the key technologies to achieve vehicle-infrastructure cooperative perception. In this paper, a lightweight fine-grained VRU detection model is proposed. Analyzing the existing complex traffic environment, the traditional definition of VRU is no longer applicable. Our work includes two parts: One is to redefine the fine-grained VRU and construct a new dataset. This task makes the perceptual information obtained by detection more comprehensive and accurate. Another is to optimize YOLOv4 by using the channel pruning method in model compression. The optimized model is 60% lighter than the original model. Under the limitation of low computing resources at the roadside units, the real-time detection of VRU is realized while ensuring a certain detection accuracy.

1 Introduction

There are two kinds of technical solutions to achieve driverless, one is autonomous driverless with individual, and the other is intelligent connected vehicles with collaborative emphasis [1]. Both schemes are based on the environment perception module to provide support for subsequent control and decision-making. According to the existing research and the achievements that have been implemented, the radar-based sensing technology has become increasingly mature [2]. However, due to its poor perception of object instances, it is unable to effectively understand complex traffic scenes. Therefore, the camera-based sensing technology is the key to realize scene understanding. At present, camera-based perception mainly focuses on deep learning. The multi-class tasks for traffic participants (classification, detection, tracking, segmentation, etc.) greatly improve the recognition efficiency of complex traffic scenes.

J. Shi · D. Sun · H. Zhang · H. Tan · Y. Hu · W. Wang (✉)
School of Mechanical Engineering, Beijing Institute of Technology, Beijing 100081, China
e-mail: wangwh@bit.edu.cn

W. Wang et al. (eds.), *Green Transportation and Low Carbon Mobility Safety*,
Lecture Notes in Electrical Engineering 944,
https://doi.org/10.1007/978-981-19-5615-7_21

According to the Global Status Report on Road Safety issued by the World Health Organization [3], about 1.35 million people die in road traffic accidents worldwide every year, and more than half of the traffic accident deaths are vulnerable road users (VRU). Pedestrians and cyclists (especially bicycles and motorcycles) were defined as VRU in previous studies. As an important part of road participants, the safety of VRU is a subject that cannot be ignored in the transportation system, and the protection of VRU is urgent.

In the environmental perception module, VRU detection is a basic and important work. There are many research on it. Briefly summarized as follows: from the perspective of detection view, it is divided into ego-view for individual autonomous driverless and top-down view for intelligent connected vehicles; from the perspective of detection algorithms, it is divided into optimizing the single-stage or two-stage detection algorithm to improve the detection accuracy and speed, and integrating the detection algorithm of other information (such as point cloud information, depth information, etc.) to improve the detection effect. In more detail, Nikhil et al. [4] used YOLO algorithm to detect the triple riding and speed violation on two-wheelers. Chandan et al. [5] presented a fast and efficient approach to VRU detection and pose estimation for real-time AD applications. Kheireddine et al. [6] proposed a radar & camera fusion framework for reliable VRU detection, tracking and classification, the low detection rate of the camera in bad weather conditions is compensated by the radar detection. For a specific VRU category, like pedestrian, there are some studies for pedestrian attributes, characteristics and low computational capability framework [7–9]. Through the above research, we can find two problems: (1) Most studies focus on the research of detection algorithms, and the engineering application of algorithms is not discussed in detail. Moreover, due to the great differences between algorithm research and actual demand scenarios, the current algorithm engineering is difficult to implement; (2) Owing to the emergence of powered two wheelers and tricycles in the complex traffic scenario, and their accident rate in the transportation system is gradually increasing, the traditional definition of VRU needs to be changed. In particular, relatively fine-grained VRU detection will bring more abundant information.

Based on the above analysis, from the perspective of intelligent connected vehicles, we proposed a lightweight fine-grained VRU detection model for roadside units. The contribution of this paper can be summarized into two points: (1) According to the current actual traffic scene, we redefined the type of VRU and constructed a new object detection dataset for VRU, specifically, pedestrians, cyclists, bicycles, motorcycles, powered two-wheelers and tricycles; (2) We used the pruning strategy to optimize YOLOv4, and the VRU detection model was transplanted on the Hisilicon platform. In other words, we simulated the computing resources of roadside units, and the performance of the VRU detection model was verified.

2 Methods

As the classical tasks of computer vision, object detection is also an important research point in environmental perception. The purpose of object detection is to define the category and position information of objects in a single frame image. Object detection based on deep convolution neural network has made countless breakthroughs. From the series network of LeNet [10] to the series network of ResNet [11], feature extraction using Convolution Neural Network (CNN) has become the mainstream method for object detection. The representative work summarized from different perspectives mainly includes: (1) Algorithm structure: the two-stage object detection algorithm represented by R-CNN series [12–14] and the one-stage object detection algorithm represented by YOLO series [15–17]. (2) Using anchor: anchor-based algorithms represented by Fast R-CNN [18] and anchor-free algorithms represented by CenterNet [18]. In addition, SSD [19] for small object detection and the hottest Visual Transformer [20] can also be considered as the classic work in the field of object detection.

The commonly used model compression method is to use lightweight networks to replace the heavier networks in the one-stage or two-stage detection algorithms as the backbone network of feature extraction (for example, using MobileNets or ShuffleNet to replace Darknet in YOLO series). However, after the actual engineering model test, like the improved model obtained by MobileNetV2 instead of darknet-53, it contains a large number of 1×1 and 3×3 convolution and frequent application of exchange memory, so that the forward inference time of the network can not meet the expected speed of application in roadside units. Therefore, we use pruning method to lighten the model. Pruning [21–23] is one of the important means of model compression, pruning can be divided into structued pruning and unstructured pruning. Because unstructured pruning needs the support of customized software and hardware, we adopt structured pruning strategy.

In this paper, we use channel pruning to optimize YOLOv4. YOLOv4 integrated many excellent ideas in algorithm theory and carried out a lot of experimental demonstration. The main improvement over YOLOv3 is that on the basis of using head of YOLOv3, CSPDarknet-53 (network structure is shown in Fig. 1) was used to replace Darknet-53 as the backbone network.

Channel pruning is a commonly used pruning strategy. This paper adjusts the channel sparsity to achieve the balance of flexibility and ease of operation. Due to channel pruning, all inputs and related connections associated with the channel need to be trimmed. Therefore, it is impossible to cut the weight of the pre training model. To solve this problem, a scale factor, named γ, multiplied by the channel output is introduced into each channel, Sparse regularization of γ through joint training the network weight and γ. Then use γ to conduct channel pruning and fine tune the network after pruning. As shown in (1).

$$L = \sum_{(x,y)} l(f(x, W), y) + \lambda \sum_{\gamma \in \Gamma} g(\gamma) \tag{1}$$

Fig. 1 CSPDarknet-53 network

	Type	Filters	Size	Output
	Convolutional	32	3×3	256×256
	Convolutional	64	3×3/2	128×128
	Cross-Stage-Partial			
1	Convolutional	32	1×1	
	Convolutional	64	3×3	
	Residual			128×128
	Convolutional	128	3×3/2	64×64
	Cross-Stage-Partial			
2	Convolutional	64	1×1	
	Convolutional	64	3×3	
	Residual			64×64
	Convolutional	256	3×3/2	32×32
	Cross-Stage-Partial			
8	Convolutional	128	1×1	
	Convolutional	128	3×3	
	Residual			32×32
	Convolutional	512	3×3/2	16×16
	Cross-Stage-Partial			
8	Convolutional	256	1×1	
	Convolutional	256	3×3	
	Residual			16×16
	Convolutional	1024	3×3/2	8×8
	Cross-Stage-Partial			
4	Convolutional	512	1×1	
	Convolutional	512	3×3	
	Residual			8×8
	Avgpool		Global	
	Connected		1000	
	Softmax			

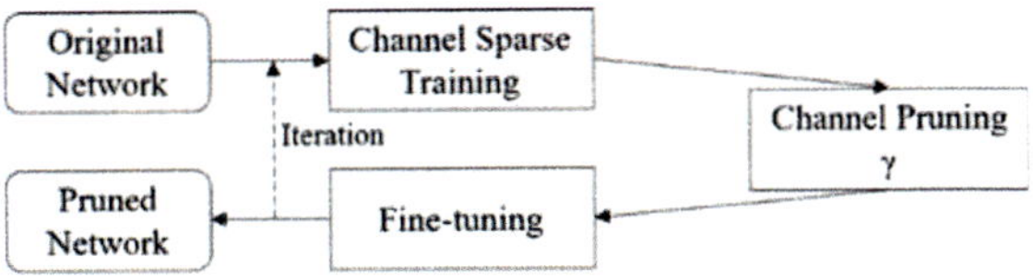

Fig. 2 Channel pruning process

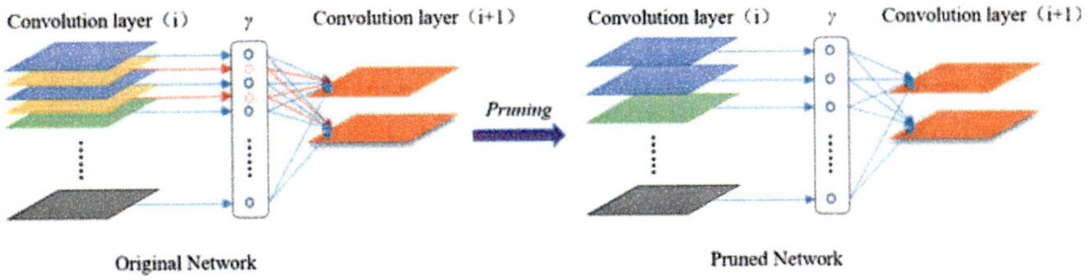

Fig. 3 Network pruning

In (1), (*x*, *y*) is the input and output of training data label, and *W* is the weight of training. It can be considered that $L = \sum_{x,y} l(f(x, W), y)$ is the loss function trained for conventional convolutional neural network. $g(y)$ is a sparse function and λ is a balance factor. The specific optimization process can be shown in Fig. 2.

The essence of channel pruning is to delete all input and output connections on "unimportant" channels. Therefore, a narrow network can be obtained through channel pruning, so as to lighten the model. γ can be used as an intermediate variable, and because it is optimized together with the network weight, the network can identify "unimportant" channels. Removing these channels will not have a great impact on the generalization performance. The specific operation is shown in Fig. 3.

Batch normalization (BN) is regarded as the current standard method to achieve rapid convergence and better generalization performance. The BN layer is used to regularize the internal activation function using mini batch data. Suppose z_{in} and z_{out} are the input and output of a BN layer, and *B* is the batch of the current mini-batch. As shown in formula (2) and (3).

$$\hat{z} = \frac{z_{in} - \mu_B}{\sqrt{{\sigma_B}^2 + \epsilon}} \tag{2}$$

$$z_{out} = \gamma \hat{z} + \beta \tag{3}$$

μ_B and σ_B are the average and standard deviation of input activation under *B*, γ and β are the scale and transformation parameters that can be trained. These two parameters theoretically provide the possibility of linearly converting the regularized activation function back to any scale. Generally speaking, the BN layer will be placed behind the convolution layer. Thus, we directly use the γ factor in the BN layer to cut the channel, so as to realize channel pruning. After network pruning optimization, we fine tune the network again.

Table 1 Number of instances of 6 classes of objects in TDVRU on training set and testing set

VRUNet-label	Pedestrian	Cyclist	PTW	Bicycle	Motorcycle	Tricycle
Training set	11,007	2065	1833	2106	549	328
Testing set	1836	319	277	315	106	87

At present, the dataset is mainly from the perspective of the main vehicle. The content of this paper is to detect VRU from the roadside perspective, and there is no sufficient data available. In order to suffice our research, we collected and labeled a new dataset for VRU in top-down view, we named it as TDVRU. It has 5000 image data (Train-4000 and Test-1000), the resolution of each image is 1920×1080, including 20,828 individual instances of 6 classes of pedestrians, cyclists, bicycles, motorcycles, powered two-wheelers and tricycles. TDVRU covers different time periods and different weather conditions to ensure integrity and rationality. Table 1 shows the details of TDVRU.

3 Results

The training equipment and environment we used mainly include: Intel(R) Xeon(R) Gold 6129 @ 2.30 GHz CPU, NVIDIA GeForce RTX 2080Ti*4 GPU, Ubuntu18.04, Pytorch1.7, Darknet, Tensorflow1.13. The model was transplanted under the Hisilicon platform. And the main equipment and environment include: Hisilicon Hi3516DV300 and caffe1.0.

We tested the accuracy and speed of VRU detection on TDVRU. The specific evaluation index is shown in (4). k is the number of object detection classes, and AP_i is the detection accuracy of each class of object. Table 2. shows the performance of our optimized VRU detection model (input: 1920*1080). In the multi-version pruned model, we selected the YOLOv4-pruned model that model size is reduced by about 60%.

$$mAP = \frac{\sum_{i=1}^{k} AP_i}{k} \tag{4}$$

Table 2 Performance of optimized VRU detection model on TDVUR

	Model size (M)	Inference time (ms)	mAP
YOLOv4-Baseline	198.4	51.0	0.820
YOLOv4-Pruned	75.3	29.2	0.822
YOLOv3	246.4	56	0.773

Table 3 Performance of Hi-model on TDVUR

	Model size (M)	FPS	mAP
Hi-model	75.9	17	0.822

Fig. 4 Performance of VRU detection model on testing set. Different classes correspond to different colors. The multi-color bounding boxes represent the detection results. Specifically, the red bounding boxes are pedestrians, green bounding boxes for cyclists, PTWs and bicycles are marked with blue boxes and inside the yellow bounding boxes are tricycles

We transplanted YOLOv4-Pruned model on Hisilicon Hi3516DV300, and the performance of Hi-Model (input: 1920*1080) as shown in Table 3.

We select roadside videos from different angles to visualize the VRU detection model. As shown in Fig. 4.

4 Conclusions

In order to achieve the real-time detection of VRU on roadside units, the situation that the category of VRU has changed greatly under the influence of emerging VRU in complex traffic scenes was fully considered. We redefined the types of VRU and further divided the cyclists and pedestrians in the traditional sense into six fine-grained categories: pedestrians, cyclists, bicycles, motorcycles, powered two-wheelers and tricycles. Based on it, a new dataset for VRU detection is constructed, which is named TDVRU, in top-down view. We used the channel pruning method to optimize the classical one-stage object detection algorithm YOLOv4. The optimized model is 60% lighter than the original model when the detection accuracy is basically unchanged. On the Hisilicon Hi3516DV300 simulating roadside units, the optimized model has the performance of real-time detection. The follow-up research will focus on the tracking of VRU. Based on the research of this paper, optimizing the two-stage target tracking algorithm will be an interesting topic.

References

1. Wei S, Yu D, Guo C et al (2019) Survey of connected automated vehicle perception mode: from autonomy to interaction. IET Intell Transp Syst 13(3)
2. Kai D, Qiren S, Fangwu M (2021) Summary of SLAM and fusion location methods based on lidar. Automot Digest 02:1–8
3. World Health Organization (2018) Global status report on road safety 2018: summary.World Health Organization (WHO/NMH/NVI/18.20)
4. Mallela NC, Volety R, N RK (2021) Detection of the triple riding and speed violation on two-wheelers using deep learning algorithms. Multimed Tools Appl 80(6):8175–8187
5. Kumar C, Ramesh J, Chakraborty B, Raman R, Weinrich C, Mundhada A (2021) VRU pose SSD: multiperson pose estimation for automated driving. Proc AAAI
6. Aziz K, De Greef E, Rykunov M, Bourdoux A, Sahli H (2020) Radar-camera fusion for road target classification. IN: 2020 IEEE radar conference (RadarConf20), Florence, Italy. IEEE, pp 1–6
7. Mordan T, Cord M, Pérez P, Alahi A (2020) Detecting 32 pedestrian attributes for autonomous vehicles. IEEE Trans Intell Transp Syst 22:1–13
8. Zhou C, Wu M, Lam SK (2019) Group cost-sensitive Boostlr with vector form decorrelated filters for pedestrian detection. IEEE Trans Intell Transp Syst 21(12):5022–5035
9. Savkin A, Lapotre T, Strauss K, Akbar U, Tombari F (2020) Adversarial appearance learning in augmented cityscapes for pedestrian recognition in autonomous driving. In: 2020 IEEE International conference on robotics and automation (ICRA), Paris, France. IEEE, pp 3305–3311
10. LeCun Y, Bottou L, Bengio Y, Haffner P (1998) Gradient-based learning applied to document recognition. Proc IEEE Inst Electr Electron Eng. 86(11):2278–2324
11. He K, Zhang X, Ren S, Sun J (2016) Deep residual learning for image recognition. In: Proceedings of the IEEE conference on computer vision and pattern recognition (CVPR). Las Vegas, NV, USA, pp 770–778
12. Girshick R, Donahue J, Darrell T, Malik J (2014) Rich feature hierarchies for accurate object detection and semantic segmentation. In: Proceedings of the IEEE conference on computer vision and pattern recognition, Columbus, OH, USA. IEEE, pp 580–587
13. Girshick R (2015) Fast r-cnn. In: Proceedings of the IEEE international conference on computer vision, Santiago, Chile. IEEE, pp 1440–1448
14. Ren S, He K, Girshick R, Sun J (2017) Faster r-cnn: towards real-time object detection with region proposal networks. IEEE T Pattern Anal 39(6):1137–1149
15. Redmon J, Divvala S, Girshick R, Farhadi A (2016) You only look once: unified, real-time object detection. In: Proceedings of the IEEE conference on computer vision and pattern recognition, Las Vegas, NV, USA. IEEE, pp 779–788
16. Redmon J, Farhadi A (2017) YOLO9000: better, faster, stronger. In: Proceedings of the IEEE conference on computer vision and pattern recognition, Honolulu, HI, USA. IEEE, pp 7263–7271
17. Redmon J, Farhadi A (2018) Yolov3: an incremental improvement. arXiv 1804.02767
18. Duan K, Bai S, Xie L, Qi H, Huang Q, Tian Q (2016) Centernet: keypoint triplets for object detection. In: Proceedings of the IEEE/CVF international conference on computer vision, Seoul, Korea (South). IEEE, pp 6569–6578. Amsterdam, pp 21–37

19. Liu W, Anguelov D, Erhan D, Szegedy C, Reed S, Fu CY, Berg AC (2016) SSD: Single shot multibox detector. In: European conference on computer vision, Amsterdam. Springer, Cham, pp 21–37
20. Carion N, Massa F, Synnaeve G, Usunier N, Kirillov A, Zagoruyko S (2020) End-to-end object detection with transformers. In: European conference on computer vision, Glasgow, UK. Springer, Cham, pp 213–229
21. Zhang X, He Y, Jian S (2017) Channel pruning for accelerating very deep neural networks. In: 2017 IEEE International conference on computer vision (ICCV). IEEE
22. Zhuang Z, Tan M et al (2018) Discrimination-aware channel pruning for deep neural networks. In: Annual conference on neural information processing systems (NIPS)
23. He Y, Liu P, Wang Z et al (2019) Filter pruning via geometric median for deep convolutional neural networks acceleration. In: 2019 IEEE/CVF conference on computer vision and pattern recognition (CVPR). IEEE

Design of Evacuation Plan for Shenyang Metro Line 9 Based on Game Passenger Flow Distribution

Weidong Liu, Quanbo Fu, Wenqi Sun, and Rongze Yu

Abstract In view of the increasingly complex road network structure and the long evacuation time of the subway, this paper studies the design of the multi-path evacuation passenger flow distribution ratio based on the game theory model and the matrix game method. The evacuation plan was designed in combination with the transfer station of the Olympic Sports Center of Shenyang Metro Line 9, and the plan was optimized from the three perspectives of facilities, passenger flow and station halls. Use Anylogic simulation software to construct the Olympic Sports Center subway station and simulate the evacuation plan. The simulation results show that the evacuation efficiency of personnel has been significantly improved after optimization; emergency treatment measures effectively guide personnel to escape from multiple evacuation exits, reducing the problem of personnel retention and improving the utilization efficiency of evacuation exits; the game path allocation method is used to clarify the number of personnel allocated in the complex subway network, and improve the utilization rate of roads and the efficiency of evacuation.

Keywords Subway · Evacuation · Game theory

1 Introduction

As the main force of urban passenger transportation, the subway needs to complete a large amount of passenger flow collection and distribution work. When an emergency occurs, due to its own structural characteristics and the nervousness of the people to be evacuated, the evacuation efficiency of personnel is not high.

Hong [1] and others studied the game characteristics between "emergency events" and passengers. The concept of relative density was put forward, and the relationship between evacuation time and passenger flow, passenger flow speed, etc. was found. Lei [2] and others studied the influence of passenger density, exit width and automatic

---

W. Liu · Q. Fu (✉) · W. Sun · R. Yu
School of Transportation Engineering, Shenyang Jianzhu University, Liaoning Shenyang 110168, China
e-mail: 44667276@qq.com

W. Wang et al. (eds.), *Green Transportation and Low Carbon Mobility Safety*, Lecture Notes in Electrical Engineering 944,
https://doi.org/10.1007/978-981-19-5615-7_22

ticket gates on evacuation time using an agent-based evacuation model. The research results will help guide the evacuation design of large underground spaces. Lo [3] and others studied the influence of multi-exit evacuation mode and time by the interaction of evacuated crowds, and based on this, proposed an evacuation exit game theory choice model. Wang [4] and others studied the revision of the social power model by considering the influence of peer interaction. An evacuation simulation was carried out in the waiting hall of Hangzhou Wulin Square subway station. Provide data support for emergency plan and emergency response.

In this paper, based on the game passenger flow distribution, combined with the actual situation of the Shenyang Olympic Sports Center subway station, the design points of the evacuation plan and the construction process of the simulation model are given. Designed an evacuation plan for the Shenyang Olympic Sports Center subway transfer station. Use AnyLogic software to simulate and analyze the established emergency evacuation plan to get the emergency evacuation plan.

2 Analysis of Evacuation Plan

2.1 Determine Impedance

Scholars at home and abroad have studied the relationship between the speed, flow rate, and density of passenger flow through investigation and other means, and used SPSS software to fit the functional relationship between the three. This section conducts follow-up work on the basis of the analysis of the relationship between speed and density in the related literature [5] (Fig. 1).

Passenger speed-density fitting function inside the corridor:

$$v(\rho) = -0.596\ln(\rho) + 1.065 \tag{1}$$

Passenger walking speed-density fitting function inside the ascending stairs:

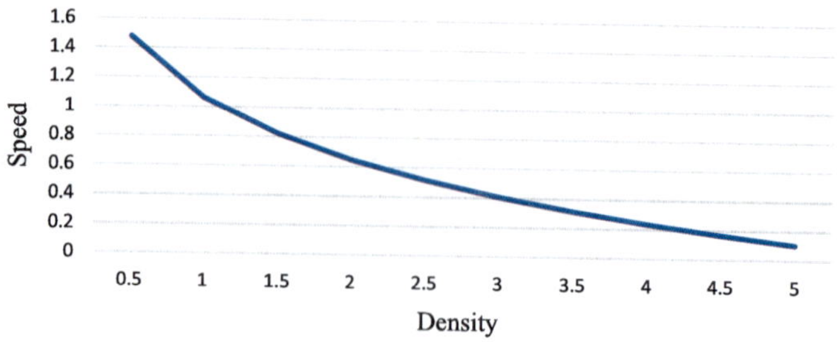

Fig. 1 Pedestrian speed-density curve in the passageway under congestion

$$v(\rho) = -0.360\ln(\rho) + 0.725 \tag{2}$$

Passenger walking speed-density fitting function inside descending stairs:

$$v(\rho) = -0.398\ln(\rho) + 0.813 \tag{3}$$

2.2 *Multi-Passageway Game Passenger Flow Distribution Model*

The development of the subway has made the internal environment more and more complicated. When distributing passenger flow in an emergency, it is necessary to take into account the passenger flow distribution among multiple OD pairs at the same time.

For any OD, define the effective evacuation route set R for passengers; there are k effective evacuation paths between OD pairs, namely $R = \{r_1, r_2 \ldots, r_k\}$; all the passenger flow between OD to be evacuated is allocated to the route r_k ($r_k \in R$), and the internal state of the station is $W = \{w_1, w_2 \ldots, w_n\}$ at this time. Under these conditions, the road network winning (evacuated passenger flow loss) matrix that can constitute a game problem [6]:

$$\begin{array}{c} \\ w_1 \\ w_2 \\ \cdots \\ w_n \end{array} \begin{array}{c} \begin{array}{cccc} r_1 & r_2 & \cdots & r_n \end{array} \\ \begin{bmatrix} \alpha_{11} & \alpha_{12} & \cdots & \alpha_{1n} \\ \alpha_{21} & \alpha_{22} & \cdots & \alpha_{2n} \\ \cdots & \cdots & \cdots & \cdots \\ \alpha_{n1} & \alpha_{n2} & \cdots & \alpha_{nn} \end{bmatrix} \end{array}$$

a_{ij} is the time required for path j to travel when allocating passenger flow to path i. Known from game theory [7], if satisfied:

$$\max_{i}\min_{j} a_{ij} = \min_{j}\max_{i} a_{ij} = a_{rs}.$$

It shows that the game problem has a pure solution strategy. At this time, the corresponding strategy combination is called the Nash equilibrium of the game, that is, all the passenger flow is allocated to the path r. On the contrary, it shows that there is a mixed strategy solution. It is necessary to divide the passenger flow to be evacuated to various roads according to the obtained optimal ratio.

In reality, there is more than one evacuation route between the platform and the exit. When escaping, passengers will comprehensively consider the distance of different routes, time spent and other external factors, in order to achieve the shortest escape time and the highest efficiency.

People's path choice behavior will cause changes in the internal network impedance, and a game relationship is formed between the two [8]. The model believes that both travelers and road networks are wise. That is, after travelers have determined their "best route", the road network will increase congestion due to travelers' choices, resulting in prolonged evacuation time and achieving the goal of reducing passenger flow. When the two compromise, the dynamic balance is achieved, and the overall evacuation efficiency is the greatest. That is, after the passenger flow distribution plan, the impedance is the smallest, and the travel time of the route chosen by all the persons to be evacuated is the shortest.

2.3 Design Ideas of Subway Evacuation Scheme

Determine the internal information of the station. Comprehensively consider the multi-path between the evacuation point and the exit, determine the OD matrix, the total evacuation flow Q, and allocate the flow rate n in stages. Suppose the initial distribution flow in each facility is 0.

Get the set K_k of effective paths in the station's internal network. According to the observation data, using the Dijsktra algorithm, using time as the impedance, retrieve the OD path with the least impedance, find the path with the shortest travel time, and place the first batch of passengers into this path using all-by-nothing allocation.

After inserting the first batch of passenger flow, update the flow and travel time of each facility, and calculate the impedance change value. Based on the calculated impedance value, the Dijsktra algorithm is used to find the second shortest path. Establish the winning (loss) matrix, use the game distribution idea to carry out the evacuation distribution of passenger flow and solve it, and determine the flow proportion of the multiple paths allocated to each path.

When there are multiple paths, it is necessary to update the travel time of the facilities and sections, use Dijsktra again to find the third shortest path, and complete the previous steps until no new shortest path is generated.

Repeat the above steps to allocate all OD pairs until the distribution ratio of each path between each OD pair is obtained, the allocation is over, and the algorithm is completed.

2.4 Evaluation System

The evacuation time is the core measure of whether the plan is an effective plan for subway evacuation. According to the requirements of 8.3.10 in the "Metro Design Code" [9]: the total time from the first person to evacuate to the last person should be less than 6 min. Calculated as follows:

$$T = 1 + \frac{Q_1 + Q_2}{0.9[A_1(N-1) + A_2 B]} \le 6\,\text{min} \tag{4}$$

where

T Evacuation time (s);
Q_1 Number of passengers on the train (p);
Q_2 Number of passengers and staff on the platform (p);
A_1 Escalator capacity (p/(min × m));
A_2 Stair passing ability (p/(min × m));
N Number of escalators;
B Stair width (m).

3 Design of Evacuation Plan for Olympic Sports Center Station of Line 9

3.1 Overview of Olympic Sports Center Station

Shenyang Metro Line 9 is 29 km long. There are 23 stations in total, all of which are underground stations. Among them, there are 4 transfer stations including Olympic Sports Center Station and Changqing South Street Station, and the remaining 19 are ordinary stations. The basic conditions such as the floor area and internal structure of the Olympic Sports Center Station match the game allocation model. At the same time, the internal passenger flow evacuation route is more complicated and typical, so this article chooses the Olympic Sports Center Station as the research object (Fig. 2; Table 1).

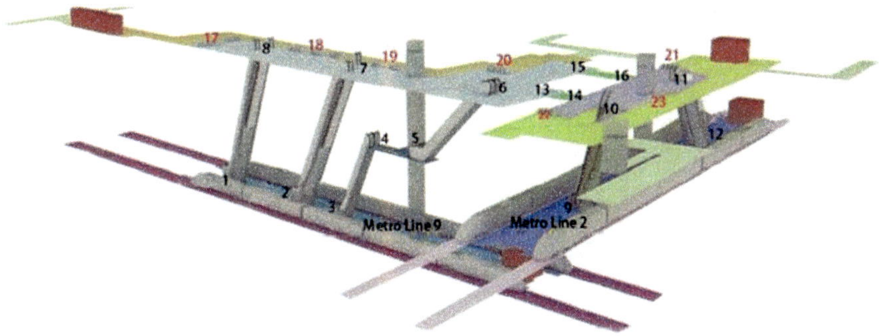

Fig. 2 Schematic diagram of the interior of the olympic sports center station

Table 1 Node division of the station

	Point
Starting point	1, 2, 3, 9 ,12
Intermediate point	4, 5, 6, 7, 8, 1, 13, 14, 15, 16
End point	17, 19, 21, 22

3.2 Passenger Flow Distribution Plan Design

3.2.1 Evacuation Plan in Case of Emergency on Line 9

Scenario simulation: A fire broke out when the train of Line 9 arrived in the Olympic Sports Center Station. Assuming that the subway is fully loaded with 1,300 passengers waiting to be evacuated, and the total number of platform workers is 200. At the same time, no trains on Line 2 arrived at the station, and trains arriving soon will not stop at the Olympic Sports Center Station (Fig. 3; Table 2).

(1) Impedance calculation: The calculation result is shown in the figure below, and the impedance unit is second.
(2) Determine the value of the number of allocations n: Here, n is set to 3, that is, the total passenger flow is divided into three times for distribution, and the number of passenger flows allocated each time is 500.
(3) The first allocation: First, an all-or-nothing allocation is performed, that is, all the first batch of 500 passengers will be allocated to the 2–7-19 route path with the least impedance. At the same time, the passenger flow of Line 2 is less, and the all-in-one evacuation is directly carried out.
(4) The second and third allocations: After each allocation is completed, the formula (1) (2) is used to determine the impedance value of each road section in the next

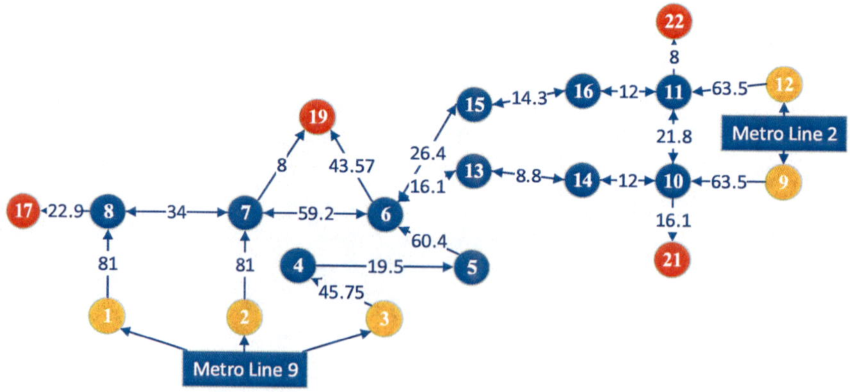

Fig. 3 The internal route impedance value of the station

Table 2 Effective evacuation path and impedance value

Number	Effective evacuation path	Impedance value
1	1-8-17	103.9
2	1-8-7-19	123
3	2-7-8-17	137.9
4	2-7-19	89
5	3-4-5-6-19	169.3
6	3-4-5-6-13-14-10-21	178.65
7	12-11-22	71.5
8	12-11-10-21	101.4
9	9-10-11-22	93.3
10	9-10-21	79.6

Table 3 Results of passenger flow distribution plan for platforms of line 9

Number	Effective evacuation path	Flow (p)	Percentage (%)
1	1-8-17	60	4
2	1-8-7-19	102	6.8
3	2-7-8-17	232	15.5
4	2-7-19	500	33.3
5	3-4-5-6-19	431	28.7
6	3-4-5-6-13-14-10-21	175	11.7

Table 4 Results of passenger flow distribution plan for platforms of line 2

Number	Effective evacuation path	Flow (p)	Percentage (%)
7	12-11-22	50	25
8	12-11-10-21	50	25
9	9-10-11-22	50	25
10	9-10-21	50	25

allocation under the congested state. According to the effective path, the winning (loss) matrix is established and then solved for distribution (Tables 3 and 4).

(5) Distribution result.

3.2.2 Evacuation Plan in Case of Emergency on Line 2

Scenario simulation: A fire broke out when the train of Line 2 arrived in the Olympic Sports Center Station. Assuming that the subway is fully loaded with 1,300 passengers waiting to be evacuated, and the total number of platform workers is 200. At

Table 5 Results of passenger flow distribution plan for platforms of line 2

Number	Effective evacuation path	Flow (p)	Percentage (%)
7	12-11-22	500	33.3
8	12-11-10-21	250	16.7
9	9-10-11-22	250	16.7
10	9-10-21	500	33.3

Table 6 Results of passenger flow distribution plan for platforms of line 9

Number	Effective evacuation path	Flow (p)	Percentage (%)
1	1-8-17	34	17
2	1-8-7-19	33	16.5
3	2-7-8-17	33	16.5
4	2-7-19	34	17
5	3-4-5-6-19	33	16.5
6	3-4-5-6-13-14-10-21	33	16.5

the same time, no trains on Line 9 arrived at the station, and trains arriving soon will not stop at the Olympic Sports Center Station.

Determine the value of the number of allocations n. Here, n is set to 3, that is, the total passenger flow is divided into three times for distribution, and the number of passenger flows allocated each time is 500 (Tables 5 and 6).

3.3 *Station Emergency Response Plan*

(1) Variable attribute equipment
According to the passenger flow distribution plan, when a subway arrives on Line 9 and no subway arrives on Line 2, the passages L13-L14 change the direction of passage to Line 2, and the influx of line 2 passengers into the passage is prohibited; the gates need to be fully open, and the opening direction is the outbound direction; the four pairs of escalators need to be stopped and used as stairs [10].

(2) Passenger separation point and concentration point
Separation point: According to the passenger flow distribution plan, guardrails are needed to divide the escalator into different escape routes at the left and right sides of the number 8, 7, and 6, so that the passenger flow at the above three points can only be passed on one side. Number 14 and 16 should adopt the principle of nearest evacuation and lead to exits 22 and 21. Prevent staying, catching up, and congestion.

Concentration point: On the one hand, the number 2 escalator needs to be stopped and used as a staircase. On the other hand, a 2 m fence needs to be installed at the entrance of the escalator to make full use of the entrance space (Fig. 4).

Reference number 13 is a one-way entrance. Compared with 15, the distance between the two is only 10 m, but the design in the evacuation plan is to give priority to the use of passage way 13. Appropriately flow the number of passengers from 13 to 15, and try to maintain a free flow state to ensure that the overall passage time of the passage is minimized (Fig. 5).

(3) Platform, station hall area

Platform: The partition method shall be adopted between the number 1, 2, and 3 escalators. When evacuating, passengers in the carriage should choose the nearest stairs to escape, which is more in line with the actual situation. This can effectively prevent the occurrence of convection.

In the station hall on the first floor, it is necessary to guide the crowd, that is, use guardrails and manual on-site command methods to restrict the flow direction of the evacuation passenger flow to prevent large-scale interweaving and convection phenomena.

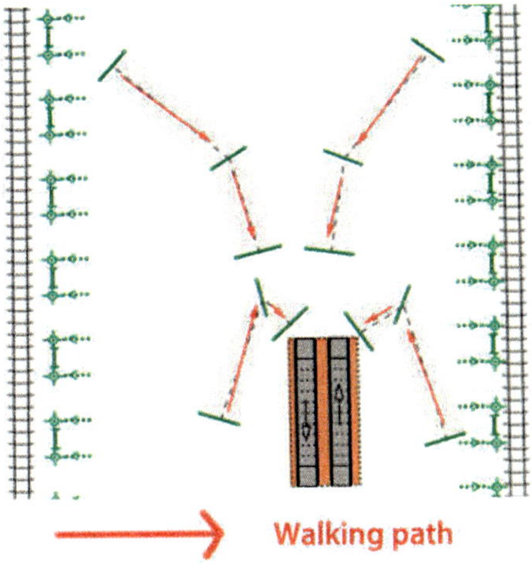

Fig. 4 Schematic diagram of optimization measures at the entrance of the stairs

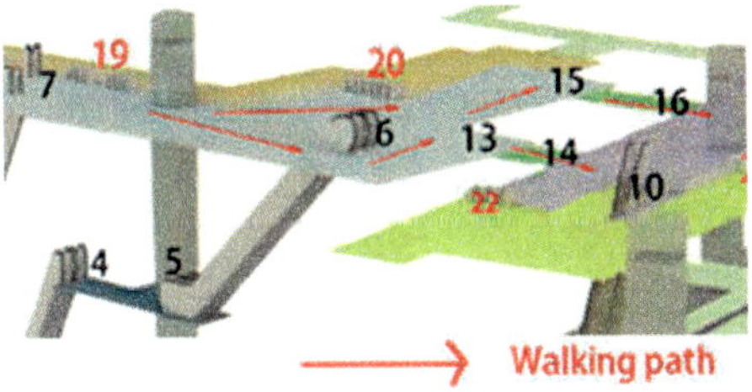

Fig. 5 Schematic diagram of crowd path

Combined with the evacuation plan, people evacuated through elevator No. 8 will use gates 17 and 18 to evacuate; people evacuated by elevator No. 7 use gates 18 and 19 to evacuate; people evacuated by elevator No. 6 use gates 20 and 22 to evacuate; Passengers evacuated from channel 13–14 use gate 22 to evacuate, and passengers evacuated from channel 15–16 use gate 21 to evacuate.

Station hall: Passengers who have reached the outside of the gate can be regarded as completed evacuation. Although these passengers have reduced their risks, they still need to be guided. Therefore, in the station hall on the first floor of Line 9, guardrails and evacuation signs are used to guide passengers passing through the turnstiles to use the nearest exit to escape to the ground.

4 Anylogic Simulation and Evaluation

4.1 Simulation Environment Construction

This simulation included the construction of the plane frame of the Olympic Sports Center Station, the layout of basic attributes (flow, passage, etc.), simulation parameter setting, allocation plan placement, running simulation, and effect evaluation (Fig. 6).

Simulation parameter setting regulations:

(1) The generation rate of passenger flow is 20p/s
(2) The free flow travel time in the emergency state is set to the maximum value in the free flow state, and the travel time of the crowded passenger flow is determined according to (1)–(3).

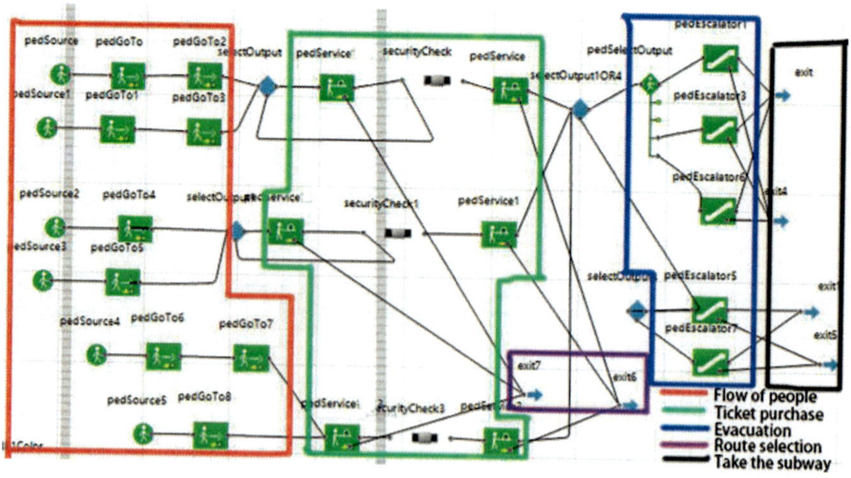

Fig. 6 Crowd flow model construction

(3) The proportion of traffic allocated for each path is determined according to the game distribution result.

4.2 Evaluation of Evacuation Plan for Olympic Sports Center Station

(1) Theoretical calculation:
1500 people to be evacuated at Olympic Sports Center Metro Station, escalator capacity 100 p/(min × m), 4 escalators, staircase passage capacity 60 p/(min × m), staircase width 3 m, calculated by 2.6 formula, theoretical evacuation time 232 s.
(2) After taking measures:
According to the game distribution plan, the direction of the flow of people is divided, and the number of results is put into the simulation software. According to the relevant requirements of emergency evacuation, measures such as path restrictions and guardrails are adopted in the software to divide the internal route of the passage and conduct simulation experiments (Fig. 7).

After the simulation, the interweaving phenomenon on the west side of the station hall has been significantly improved. The U-shaped crowd gathering area has also been relieved. The crowd density color in all passages has dropped by one level, and the purpose of crowd evacuation has become clearer. The simulation results show that the overall loss time of evacuation is reduced, the maximum density in the facility is reduced, and the final evacuation completion time is 208 s (Fig. 8).

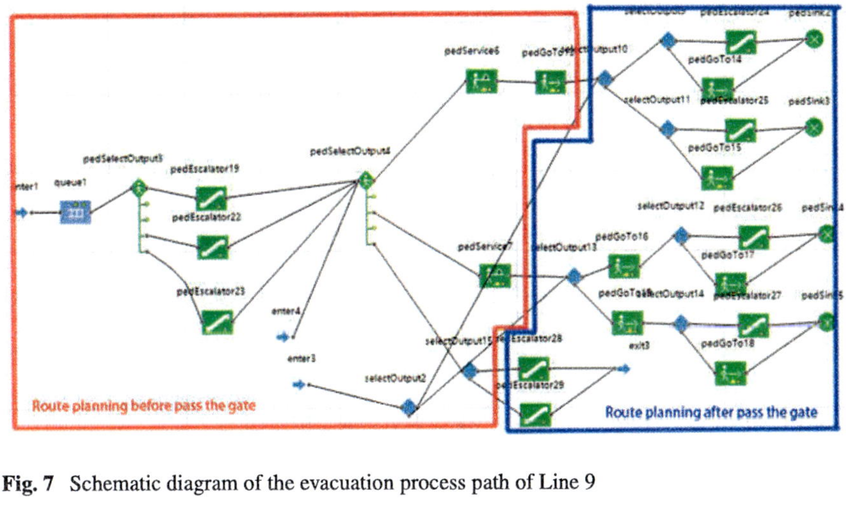

Fig. 7 Schematic diagram of the evacuation process path of Line 9

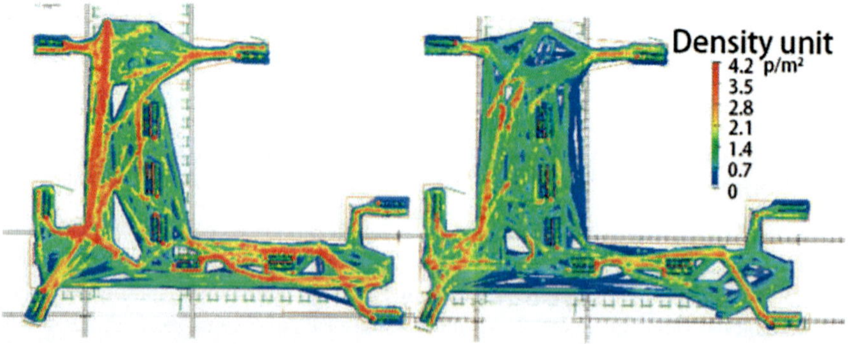

Fig. 8 Comparison diagram of measures taken and no measures taken

5 Summary

This paper takes Shenyang Metro Line 9 and Olympic Sports Center Station as the research background. Combining the established theoretical model and research data, this paper focuses on platform network construction, route retrieval, data collection, game distribution, and emergency response. This article takes Shenyang Metro Line 9 and Olympic Sports Center Station as the research background. Combining the established theoretical model and research data, this article focuses on platform network construction, route retrieval, data collection, game distribution, and emergency response. A specific evacuation plan for Line 9 has been worked out. Simulate demonstration through Anylogic simulation. Taking the evacuation time as the first priority, the route plan of Line 9 during the evacuation was optimized, and the rationality of the plan was verified. Finally, a design plan was proposed.

References

1. Hong L, Xu RH. Analysis on game behaviors of passengers in emergency evacuation in subway station. Adv Transp, Pts 1 and 2.97–98.576: 576–582
2. Lei WJ, Li AG, Gao R, Hao XP, Deng BS (2012) Simulation of pedestrian crowds' evacuation in a huge transit terminal subway station. Physica A-Stat Mech Appl 6(33):5355–5365
3. Lo SM, Huang HC, Wang P, Yuen KK (2006) A game theory based exit selection model for evacuation. Fire Saf J 2(3):364–369
4. Wang L, Zheng JH, Zhang XS, Zhang JL, Wang QZ, Zhang Q (2016) Pedestrians' behavior in emergency evacuation: modeling and simulation. Chin Phys B 25(11):689–698
5. Hui X (2013) The bottleneck identification and simulation study of the evacuation capacity of urban rail transit stations. Jiaotong University, Beijing
6. Li ZY, Tang MB, Liang D, Zhao Z (2016) Numerical simulation of evacuation in a subway station. In: 2015 international conference on performance-based fire and fire protection engineering (ICPFFPE 2015) 1.126, pp 616–621
7. Wei X (2014) Operations research, 3rd edn. Machinery Industry Press, Beijing

8. Yang XX, Dong HR, Wang QL, Chen Y, Hu XM (2014) Guided crowd dynamics via modified social force model. Physica A-Stat Mech Appl 5(68):63–73
9. Ministry of Housing and Urban-Rural Development of the People's Republic of China (2014) GB 50157-2013 metro design code. China Construction Industry Press
10. Meng YD;, Jia CQ (2017) Research and application of metro station evacuation simulation. In: 2017 4th international conference on information science and control engineering (ICISCE), vol 233, pp 1123–1125

Drivers' Visual Characteristics of Urban Expressway Based on Eye Tracker

Tianjun Feng, Ziwen Zhao, and Xiujuan Tian

Abstract In order to compare and analyze the visual characteristics of drivers in the congested and unblocked state of urban expressways, real vehicle tests were carried out on the eastern expressway in Changchun City using the German Dikablis eye tracker and its supporting D-Lab software. The test data was processed by using descriptive statistical analysis and non-parametric inspection methods to quantify the impact of congestion on the driver's visual characteristics. The results show that drivers mainly obtain traffic information by gaze when driving on the expressway, and the gaze points are mostly concentrated on the road vehicles; the driver's gaze duration and scan duration in the congested state account for the highest proportions in the 200–250 ms and 0–25 ms time periods, respectively; the average gaze duration and the average scan duration of the drivers in the congested state were higher than those in the unblocked state. The driver's gaze duration and saccade duration in the two states are significantly different, and the Mann–Whitney U test results are less than 0.05; the pupil area changes more drastically in the congested state, and the pupil area change rate is 38.67%.

Keywords Congestion state · Urban expressway · Visual characteristics · Gaze duration · Pupil area

T. Feng · Z. Zhao · X. Tian (✉)
School of Transportation Science and Engineering, Jilin Jianzhu University, Changchun 130118, China
e-mail: jidatianxj@126.com

T. Feng
e-mail: fengtianjun@jlju.edu.cn

Z. Zhao
e-mail: 1173689976@qq.com

W. Wang et al. (eds.), *Green Transportation and Low Carbon Mobility Safety*, Lecture Notes in Electrical Engineering 944,
https://doi.org/10.1007/978-981-19-5615-7_23

1 Introduction

Urban expressways are an important part of the urban road transportation network and play an important role in urban transportation. At present, most cities in China have built urban expressways, and they are of considerable scale. Taking Changchun City where the author is located as an example, a "two horizontal and three vertical" expressway system including eastern expressway, western expressway, southern expressway and northern expressway has been built, which has played a role in alleviating urban traffic congestion and shortening the travel time.

In recent years, urban areas have continued to extend, and more and more residents choose expressways when traveling. And with the rapid increase in the number of motor vehicles, the problem of expressway congestion has gradually emerged [1]. Frequent traffic congestion on urban expressways mainly occurs during rush hours. During peak hours, the traffic volume on the road is large, which exceeds its designed capacity. Moreover, at the entrance and exit of the ramps, the number of lanes changes and vehicles are frequently intertwined, causing traffic congestion on the entrance and exit of the ramps, which leads back to the main line and affects the traffic operation of the main line [2].

At present, there are many objective definitions for traffic congestion, and various indicators are used. Each country or region has established evaluation indicators according to their own road network operating conditions and the characteristics of actual urban traffic flow [3, 4]. Although the expression of each evaluation index is different, it can be judged the severity of traffic congestion through intuitive numerical values. Scholars have done research on the formation mechanism of traffic congestion, congestion status recognition, congestion charging and congestion relief strategies [5–8]. However, these studies mainly focus on objective state parameters, and there are few studies on subjective perception parameters. Moreover, studies on defining traffic congestion from the perspective of driver experience are even rarer. Different drivers have different feelings about the same road congestion, so the acceptance of congestion is also different.

In the state of traffic congestion, the speed of the vehicle cannot reach the expected speed value of the driver, causing economic loss and time loss of the traveler, and it is also easy to cause the driver's mood to change and cause traffic accidents [9]. Emotional changes are mainly reflected in visual changes. Vision is the most important way for drivers to obtain road and traffic information. Studies have shown that the amount of information obtained through vision accounts for about 80% of the total information [10].

Therefore, scholars have carried out a series of researches on the visual characteristics of drivers. The research data is mainly obtained through real-vehicle tests and driving simulation tests [11–13], mainly focusing on indicators such as gaze duration, saccade range and pupil area. These indicators can characterize the psychological and physiological characteristics of the driver, so as to judge his driving safety [14]. It has been concluded that the driver's eye movement behavior is related to its cognitive load and affects the sensitivity of its distraction [15, 16]. The driver's fatigue status can also be detected through visual features [17]. The research methods of driver's visual characteristics mainly include statistic analysis, significance test, neural network construction and Markov chain modeling [18–22]. These methods can reflect the driver's visual change rules well. According to this, this research draws on the index selection methods and data analysis methods of scholars.

Through research, it is found that under different driving environments, the driver's visual characteristics will show obvious differences. For example, compared with ordinary road sections, the average gaze time of drivers driving in the extra-long tunnel section is longer, and their gaze points are distributed more widely in the horizontal direction and tend to be concentrated in the vertical direction [23]; In the mountainous expressway section, the driver's gaze duration is shorter and the cognitive load is smaller [24].

At present, the road driving environment studied by scholars is mainly concentrated in tunnels, mountain roads and highways, but there is less research on urban roads in complex environments. As an important part of the urban road system, the urban expressway has gradually attracted the attention of researchers [25] for the impact of its driving environment on drivers. In addition, China is quite different from other countries in terms of expressway road conditions, traffic composition, and driver driving behavior. In view of this, this paper uses real-vehicle tests to collect the visual characteristic data of drivers during driving, and analyzes their multiple visual characteristics in two states of unblocked and congested urban expressways through the measured data. This study is expected to make a positive exploration for the study of drivers' psychological state and the prevention of urban road traffic accidents.

2 Test Plan Selection

2.1 Test Section Selection

The eastern urban expressway in Changchun City is selected as the test section. The starting point is the entrance of Shengtai Street and the ending point is the exit of Dongsheng Street. The whole journey is 4.9 km and the speed limit is 80 km/h. The form of the expressway is elevated. And the diversion section is two-way eight-lane, and the test section is separated by intermediate belts throughout the whole process. The plan of the test section is shown in Fig. 1.

2.2 Test Equipment and Personnel Selection

The instruments used in the test are German Dikablis eye tracker, D-lab software, positive selection inverter, camera and small car, as shown in Fig. 2. The eye tracker

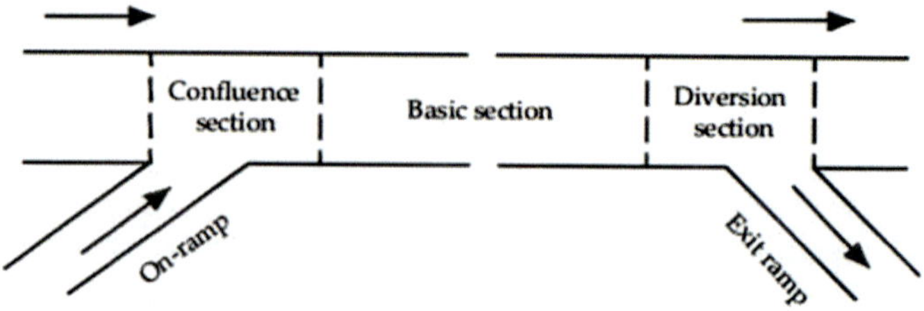

Fig. 1 Plan view of test section

Fig. 2 Test equipment

is developed by Ergoneers, Germany, with high precision and low invasiveness. The eye tracker is equipped with two cameras: eye-camera and field-camera. The former is responsible for tracking the eyes, and the latter is responsible for recording scenes. It can collect binocular eye movement data and real-world data at the same time, and can analyze eye movement data in various aspects with D-lab software. In addition, the resolution of the eye-camera is 384×288, the resolution of the field-camera is 768×576, and the sampling frequency of the eye tracker is 50 Hz.

The test personnel are skilled drivers with certain driving experience, and have normal visual functions, no physical defects or major accidents. Before the test, the driver shall be guided on the use of instruments and psychological counseling to restore the normal driving state as much as possible. The test time is from 15:00 to 17:30, with good weather conditions.

2.3 *Congestion Status Judgment*

Drivers are the main component of the road traffic system, and their mental state plays a decisive role in road traffic safety. Taking the driver as the main body, select the traffic jam state division method based on the difference between the driver's expected speed and the actual operating speed [26]. This method defines the traffic congestion pressure coefficient $\chi_{press}^{t_0-t_n}$ to characterize the driver's perception of the degree of congestion during period t_0 to t_n. The definition of variables is shown in formula (1):

$$\chi_{press}^{t_0-t_n} = \frac{\overline{V}_{qw} - \overline{V}_{yx}}{\overline{V}_{qw}} \tag{1}$$

In formula (1): $\chi_{press}^{t_0-t_n}$ is the driver's perception of traffic congestion pressure during the t_0 to t_n period on the path S; $\overline{V}_{qw}$ is the driver's expected operating speed average during the t_0 to t_n periods on the path S, in km/h, and the average expected operating speed needs to comply with the speed limit specification of expressway [27]; $\overline{V}_{yx}$ is the average value of the actual running speed of the vehicle during t_0 to t_n periods on the path S, and the unit is km/h. According to the driver's evaluation of the pressure coefficient, the 85% quantile value of the cumulative frequency distribution curve is taken as the classification threshold of the congestion state and the unblocked state. When $\chi_{press}^{t_0-t_n} < 0.65$, the traffic flow state is unblocked, and when $\chi_{press}^{t_0-t_n} \geq 0.65$, the traffic flow state is congested.

2.4 Determination of Test Procedure

a. The test personnel conduct on-site inspections on selected urban expressway sections to confirm various road conditions, weather conditions and traffic control facilities;
b. The tester selects the location where the test vehicle starts to drive, at least 3 km away from the test section, to give the driver a process of adaptation to avoid the driver from obtaining data with large deviations due to tension;
c. The driver wears the eye tracker, and the observer starts the computer and calibrates the eye tracker;
d. The driver starts the test vehicle for the test, and the observer asks the driver the expected time to pass the test section and records it;
e. During driving, observers use D-Lab software to observe the driver's point of view capture to ensure that the instrument data is valid;
f. Each time the driver completes the test section, the observer will export and store his visual characteristic data and video data. Repeat the test and obtain a total of 20 sets of test data;
g. After the test is completed, the test personnel calculate the traffic congestion pressure coefficient based on the recorded data, and classify and preprocess the eye movement data according to the traffic flow state.

3 Analysis of Gaze Characteristics

3.1 Analysis of Eye Movement Hot Spots

Eye movement heat maps, also called heat maps, represent the driver's browsing and gaze behavior within the field of view, reflecting the driver's attention distribution. The red area represents the focus area, and the yellow and green represent areas with less attention. Select the eye movement heat map of the driver in the unblocked and congested state for 4 min for analysis, as shown in Figs. 3 and 4.

It can be seen from the heat maps 3 and 4 that when the traffic flow of the urban expressway is in unblocked state, the driver is driving at a higher speed and has more concentrated gaze. The concentrated gaze area is the area directly ahead, that is, the vehicle ahead and the road ahead; In the square area, when a vehicle enters the ramp on the right, the driver's attention is drawn. When the traffic flow of the urban expressway is in a state of congestion, the area of the driver's gaze concentration is enlarged, and the gaze is more scattered. This reflects that when traffic congestion occurs on the expressway, the traffic density increases, and the driver follows the vehicle in front at a lower speed most of the time, and pays more attention to the information of vehicles in adjacent lanes and surrounding buildings, and the distribution of gaze is smoother wider.

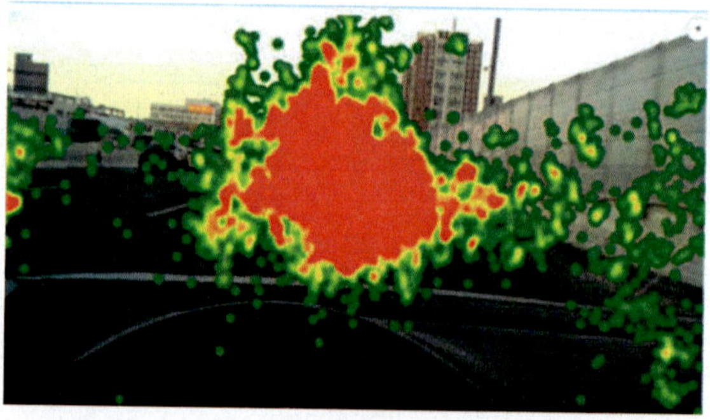

Fig. 3 Eye movement heat map in unblocked state

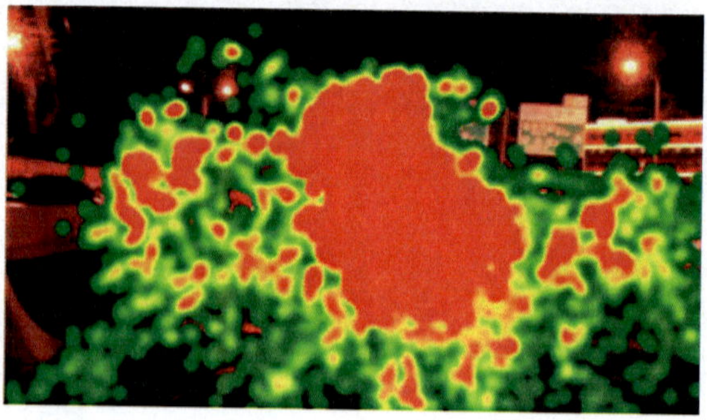

Fig. 4 Eye movement heat map in congested state

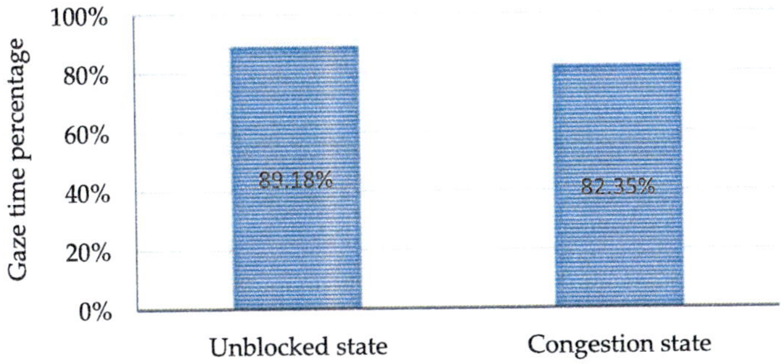

Fig. 5 Percentage of fixation time

3.2 *Analysis of Gaze Time Characteristics*

When the driver is driving on the urban express road, the visual changes are mainly manifested in the gaze and saccade behavior. Because the average speed of the express road is faster than the average speed of other types of urban roads, the driver needs a high degree of concentration, so the driver's visual behavior is mostly gaze behavior. Saccades only move quickly from one gaze point to another, and there are very few saccades without purpose. The gaze time of the actual vehicle test driver is counted, and the gaze time of the driver in the driving process is obtained as a percentage of the total time, as shown in Fig. 5.

It can be seen from Fig. 5 that when the drivers are driving in the two traffic flow states, their gaze time percentages are above 80%, and the gaze time of the drivers in the congested state is slightly less than that in the unblocked state. The main reason is that in a congested state, the driver's visual attention is more slack, the attention points of interest increase, and the saccade behavior increases, shortening the gaze time.

3.3 *Analysis of Gaze Duration Characteristics*

The gaze duration refers to the duration of time during which the center of the visual axis remains unchanged during the gaze process, in ms. The length of the gaze duration represents the difficulty of the driver to process the information. The longer the duration of a single gaze, the more difficult the information needs to be processed, but the target of the gaze duration is not necessarily longer than the gaze duration Short goals make more sense. According to statistics, more than 99% of gaze durations under unblocked and congested conditions are distributed below 1000 ms, so the sample data of gaze durations above 1000 ms is excluded. The statistical analysis results of the measured data are shown in Table 1.

Table 1 Statistical results of gaze duration

Sample	Mean (ms)	Standard Deviation (ms)	Kurtosis	Skewness
Unblocked state	309.65	234.31	0.506	0.781
Congestion state	341.93	205.49	0.742	1.049

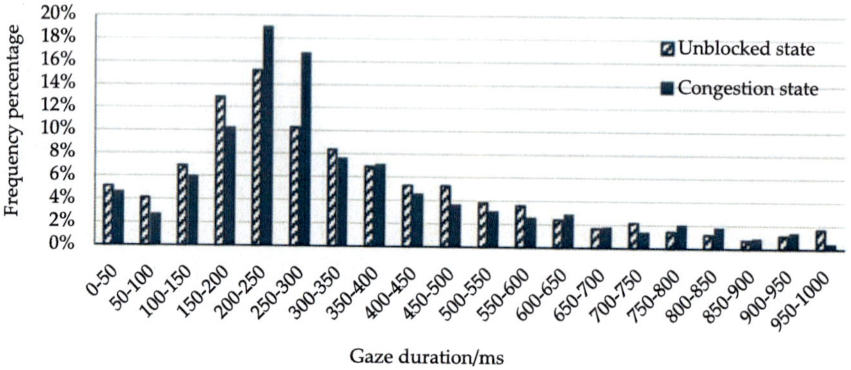

Fig. 6 Segmentation statistics of gaze duration

It can be seen from the above table that the average gaze duration of drivers in the traffic congestion state of urban expressways has increased significantly, reaching 341.93 ms, indicating that the driver's information processing intensity is greater in the congestion state, which is likely to cause fatigue.

In order to further study the characteristics of the gaze duration of the driver in the congestion state of the expressway, this paper takes the gaze duration as a time period of 50 ms, and calculates the gaze duration of the driver in the unblocked and congested states. The statistical histogram is shown in Fig. 6.

The data in Fig. 6 shows that when the driver is driving on an urban express road, whether in a smooth state or a congested state, the gaze duration is mainly concentrated between 0 and 400 ms, and the proportion in this section reaches about 70%. The frequency of the gaze duration in the unblocked state and the congested state reaches the maximum in the 200–250 ms time period, and the corresponding proportions in the two states are 15.22% and 19.01% respectively; in the 150–200 ms time period and 250–300 ms time period, the gaze duration frequency of the two states is quite different.

In order to compare the difference of the driver's gaze duration under the state of unobstructed expressway and congested traffic, two independent samples of non-parametric test method-Mann–Whitney U test method were selected to test it, and $p = 0.009 < 0.05$, indicating When the driver is driving on the urban express road, there are significant differences in the gaze duration between the unblocked state and the congested state.

4 Analysis of Saccade Characteristics

4.1 Analysis of Saccade Amplitude Characteristics

The saccade range refers to the area covered by the eye jump after completing a saccade, that is, the area where the gaze shifts from the previous gaze point to the next gaze point. The angle value of the gaze rotation is usually used to measure the saccade range. In this article, the magnitude of the saccade is the angle of the line of sight between the eye and the two fixation points. The larger the scan range, the greater the amount of information obtained during the last gaze, which helps to understand the surrounding traffic environment, reduce tension, and improve driving safety. Using 1° as an interval, draw a saccade amplitude frequency diagram in the range of 0–24° as shown in Fig. 7, and draw a saccade amplitude scatter diagram as shown in Fig. 8.

It can be seen from Fig. 7 that the frequency of the saccade amplitude distribution in the unblocked and congested states both reaches the maximum in the 2–3° interval, which is 20.17% and 14.58% respectively. In the congested state, the saccade amplitude in the 3–5° interval varies uniformly, which is different from the unblocked state. The changing law of saccade amplitude in both states showed a trend of first decreasing, then increasing and then decreasing.

After calculation, the average saccades of the drivers in the unblocked state and the congested state are 3.875° and 5.665°, and the standard deviations of the saccade angles are 3.978° and 4.674° respectively. Combined with Fig. 8, it is illustrated that the driver needs to obtain more traffic information in a congested state. In order to reduce driving tension and increase safety, and to obtain more road traffic information, the driver needs a larger scan when the urban expressway is in a congested state than in a smooth state.

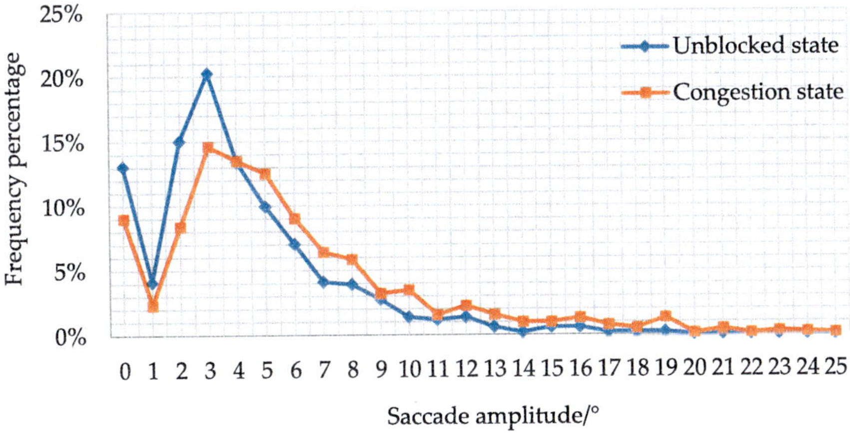

Fig. 7 Saccade amplitude frequency diagram

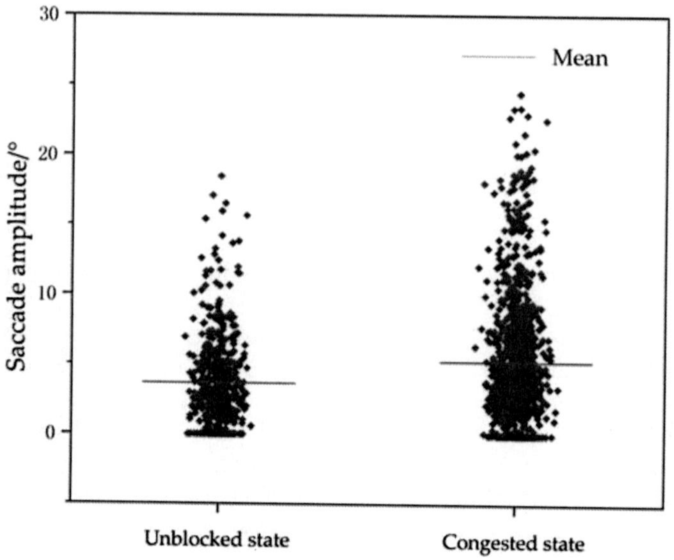

Fig. 8 Scatter plot of saccade amplitude

4.2 Analysis of Saccade Duration Characteristics

The saccade duration refers to the duration of a saccade performed by the driver, which can reflect the time it takes for the driver to search for target information during the visual search process. The shorter the saccade duration, the simpler the road condition information, the shorter the time it takes for the driver to search for the target, and the higher the efficiency. Moreover, the saccade duration is also related to the information density in the search range. The longer the saccade duration, the smaller the information density in the search range. Therefore, the duration of the scan is closely related to driving safety.

Since more than 99% of the saccade durations under unblocked and congested conditions are distributed below 100 ms, the sample data of saccades greater than 100 ms are eliminated, and the results in Table 2 are obtained by statistical analysis.

It can be seen from Table 2 that the mean value of the driver's saccade in the traffic congestion state of the urban expressway is 29.06 ms, and the standard deviation is 16.40 ms. The traffic congestion increases the driver's saccade duration. The main reason is that the urban expressway traffic situation is complex in the traffic

Table 2 Statistical results of saccade duration

Sample	Mean (ms)	Standard Deviation (ms)	Kurtosis	Skewness
Unblocked state	26.15	15.08	5.98	2.21
Congestion state	29.06	16.40	4.54	1.87

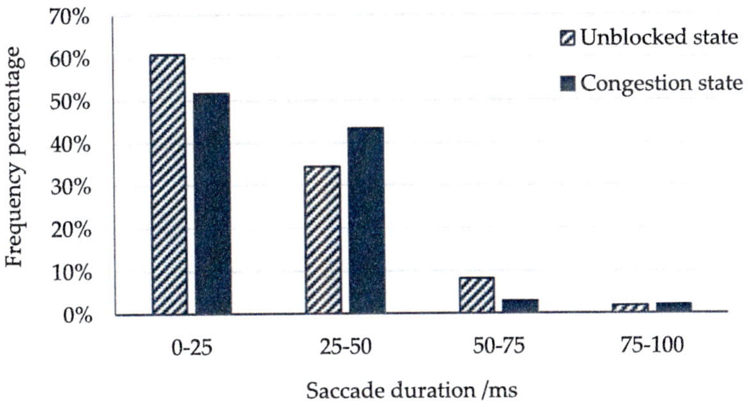

Fig. 9 Statistics of saccade duration by time period

congestion state, the information density is low, and the difficulty for the driver to obtain information increases. Therefore, searching for the target takes a long time and the search efficiency is low, which is not conducive to driving safety. In order to further study the characteristics of the saccade duration when the driver is driving in the congested state of the urban expressway, this paper takes the saccade duration as a period of 25 ms, and calculates the saccade duration of the driver in the unblocked and congested state. The statistical graph is shown in Fig. 9.

It can be seen from Fig. 9 that the distribution trend of the saccade duration segment of drivers in different traffic flow states is roughly the same, mainly concentrated in the 0–50 ms time period, and the saccade duration distribution of drivers in congested state is more uniform than in unblocked state. In the 0–25 ms time period, the saccade duration of the congested state is less than that of the unblocked state, while the congested state is slightly higher in the 25–50 ms time period.

The Mann–Whitney U non-parametric test method was used to test the saccade duration of the driver in the state of unblocked and congested traffic on urban expressways, and $p = 0.007 < 0.05$, which indicates that the driver is in unblocked state and congested state when driving on urban expressways. There are significant differences in the duration of saccades.

5 Analysis of Pupil Characteristics

When the driver is driving on an urban expressway with different traffic flow conditions, the pupil characteristics will change, and the change of pupil area is closely related to the driver's degree of psychological effort during information processing. The size of the pupil area can indicate the driver's sensitivity to changes in operating load. When the driver tries to see a target, the pupil area will increase. Based

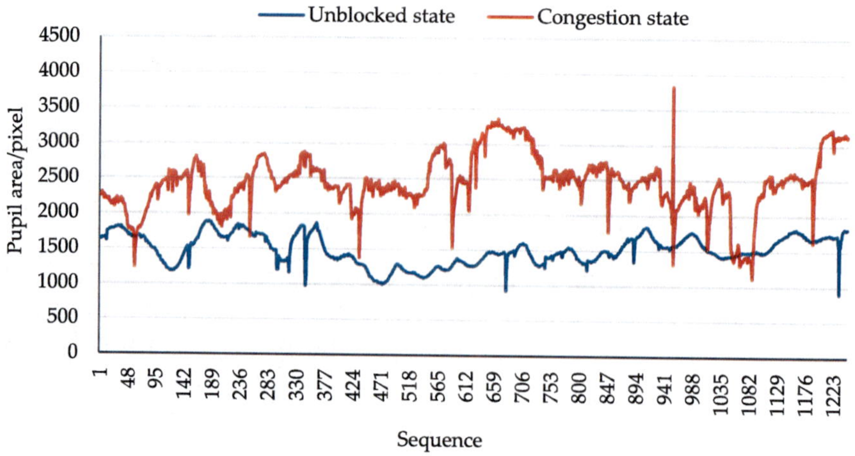

Fig. 10 Changes in pupil area section

on this characteristic, this article selects the driver's pupil area to analyze the pupil characteristics.

When the driver is driving on an urban expressway, due to different traffic flow conditions, the pupil area will also show different changes. Select the value of the pupil area of the driver during a certain period of time to draw a line graph, as shown in Fig. 10.

It can be seen from Fig. 10 that the pupil area of the driver in the congested state is larger than that in the unblocked state, mainly distributed over 2000 pixels, and the pupil area changes more strongly. In the unblocked state, the pupil area is mainly distributed around 1500 pixels, and the change is stable. This shows that when the driver is driving in a congested state, the driver's mood changes are more sensitive, which is not conducive to driving safety.

The pupil area change rate E refers to the ratio of the difference between the driver's pupil area e_2 under the operating load state and the pupil area e_1 under no operating load state to e_1, namely

$$E = (e_2 - e_1)/e_1 \tag{2}$$

Studies have shown that the pupil area change rate can reflect the driver's driving tension, thereby affecting driving safety and comfort. The definition standard of the specific threshold U [28, 29] is: when comfortable, $U < 20\%$; when more nervous, $20\% \leq U \leq 40\%$; when very nervous, $U > 40\%$. Calculating the pupil data measured by the eye tracker shows that the driver's pupil area change rate in the expressway congestion state is 38.67%, and the driver's pupil area change rate in the unblocked state is 15.28%, indicating that the driver is in a congested state Uncomfortable state.

6 Conclusions

Under the two states of unblocked and congested urban expressway traffic, the driver's visual characteristics are studied based on measured data. The results show that the two states have a significant impact on the driver's visual characteristics. The specific performance is as follows:

(1) The driver mainly obtains traffic information by gazing while driving. The average gaze time of the driver in the congested state is less than that in the unblocked state, and the gaze points are more scattered;
(2) The average gaze duration of the driver in the congested state is 341.93 ms, the average saccade duration is 29.06 ms, and the average saccade amplitude is 5.665°, both of which are higher than the unblocked state; the driver's gaze duration is mainly distributed in 0–400 ms time period, the largest proportion is in the 200–250 ms time period;
(3) Using the Mann–Whitney U non-parametric test method to test the difference of the driver's gaze duration and saccade duration in the state of unblocked and congested urban expressways, the p-values are all less than 0.05, indicating that in the two states the gaze duration and the saccade duration are significantly different.
(4) The pupil area change rate of the driver under the congestion state of the urban expressway is 38.67%, and the pupil area change degree is stronger than that of the unblocked state, indicating that the driver is in a state of driving tension.

This research establishes the correlation between the driver's visual characteristics in the two traffic states of unblocked and congested urban expressways, and provides a basis for the management department in traffic planning and traffic control. The research of this thesis is still in the exploratory stage, and many questions need to be further refined in the future research. For example, due to the constraints of time and survey area, the paper did not study the visual characteristics of drivers in different congestion states. If the follow-up conditions are available, it should be studied to enrich the research results.

Acknowledgements The authors declared no potential conflicts of interest with respect to the research, authorship, and/or publication of this article.

Funding The authors disclosed receipt of the following financial sup-port for the research, authorship, and/or publication of this article: This research has been supported by "Thirteenth Five-Year" Science and Technology Project of the Education Department of Jilin Province (JJKH20200285KJ) and Science and Technology Research and Planning Project of Education Department of Jilin Province (JJKH20210270KJ).

Declaration of Conflicting Interests The author(s) declared no potential conflicts of interest with respect to the research, authorship, and/or publication of this article.

References

1. Wang Y (2019) Exploring the problems and solutions of the urban express way. Constr Des Eng 10:61–62
2. Zhang XQ, Xie ZJ, Du ZH et al (2019) Analysis on traffic characteristics of short distance travel on urban expressway. Technol Highway Transport 35:116–122
3. Quiroga CA (2000) Performance measures and data requirements for congestion management systems. Transport Res C-Emer 8(1–6):287–306
4. Okamura H, Watanabe S, Watanabe T (2000) An empirical study on the capacity of bottlenecks on the basic suburban expressway sections in Japan. In: Proceedings of the 4th international symposium on highway capacity, vol 12, pp 120–129
5. He SL (2012) Analyze the mechanism of urban road traffic congestion with the theory of supply and demand. J People's Publ Secur Univ China (Sci Technol) 18:48–52
6. Wada T, Fukumoto J, Ohtsuki K et al (2011) A novel real-time recognition method of vehicular traffic congestion by VANET. IEICE Trans Commun 94(1):55–63
7. Wu K, Chen Y, Ma J, Bai S et al (2017) Traffic and emissions impact of congestion charging in the central Beijing urban area: a simulation analysis. Transport Res D-Tr E 51:203–215
8. Angelelli E, Arsik I, Morandi V et al (2016) Proactive route guidance to avoid congestion. Transport Res B-Meth 94:1–21
9. Li G, Lai W, Sui X et al (2020) Influence of traffic congestion on driver behavior in post-congestion driving. Accident Anal Prev 141:105508
10. Li ZM (2000) Traffic engineering, 2nd edn. China Communications Press, Beijing, p 11
11. Qin L, Dong LL, Xu WH et al (2018) Influence of vehicle speed on the characteristics of driver's eye movement at a highway tunnel entrance during day and night conditions: a pilot study. Int J Environ Res Public Health 15:656
12. Calvi A, De Blasiis MR, Guattari C (2012) An empirical study of the effects of road tunnel on driving performance. Procedia Soc Behav Sci 53:1098–1108
13. Fan J, Chen S, Liang M et al (2018) Research on visual physiological characteristics via virtual driving platform. Adv Mech Eng 10(1). https://doi.org/10.1177/1687814017717664
14. Zheng XJ, Li ZH, Zhang Y (2018) Eye movement research and development of driving safety. Technol Innov Manag 39:50–59
15. Cooper JM, Medeiros-Ward N, Strayer DL (2013) The impact of eye movements and cognitive workload on lateral position variability in driving. Hum Factors 55:1001–1014
16. Yekhshatyan L, Lee JD (2012) Changes in the correlation between eye and steering movements indicate driver distraction. IEEE T Intell Transp 14:136–145
17. Lin L, Huang C, Ni X et al (2015) Driver fatigue detection based on eye state. Technol Health Care 23:S453–S463
18. Qi WW (2014) Characteristics research for driver's physiology and lane changing in congested state of urban road. PhD Thesis, Harbin Institute of Technology, China, July
19. Li H, Wu C, Chu D et al (2016) Drivers' hazard perception analysis based on logistic regression and Cochran–Mantel–Haenszel test. Adv Mech Eng 8(9). https://doi.org/10.1177/1687814016670059
20. Wang X, Liu Y, Xu Q et al (2019) Feature extraction and dynamic identification of driving intention adapting to multi-mode emotions. Adv Mech Eng 11(4). https://doi.org/10.1177/1687814019839906
21. Li ZH (2018) Modeling analyses on driver's gazing behavior based on real vehicle eye movement experiment. Sci Technol Eng 18:312–317
22. Choi IH, Jeong CH, Kim YG (2016) Tracking a driver's face against extreme head poses and inference of drowsiness using a hidden Markov model. Appl Sci 6(5):137
23. Hu YQ, Liu HX, Zhu T et al (2017) Research on visual characteristics of drivers driving through extremely long expressway tunnel. J China Saf Sci 27:31–36
24. Ye JG (2017) Study on the driver's eye movement and psychological characteristics and its application in landscape construction in mountainous freeway. MD Thesis, Chongqing Jiaotong University, China, June

25. Zhu DC (2018) Research on the effects of expressway ramp environmental factors on driver's physiological characteristics and behaviors. MD Thesis, South China University of Technology, China, June
26. Qi W, Pei Y, Song M et al (2013) Pattern analysis of driver's "pressure-state-response" in traffic congestion. Discrete Dyn Nat Soc
27. CJJ 129:2009. Specification for design of urban expressway
28. Hu JB, Li R, Ma Y (2014) Safety threshold evaluation method for expressway tunnel lighting in entrance section. J China Highway Transport 27:92–99
29. Zhao L (2008) Two-lane highway alignment research based on driver's psychological and physiological reaction. PhD Thesis, Beijing University of Technology, China, July

Hilbert-Huang Transform in Pavement Texture and Skid-Resistance Study

Yuan-shuai Dong, Yun Hou, Jia-lei Tian, Yu-xuan Cao, Chen-wei Guo, Tuo Fang, and Jing Zhou

Abstract Insufficient skid-resistance of a pavement is a critical cause of traffic accidents. Pavement texture plays an important role in the skid performance. In this paper, the texture information was collected by a 3D laser scanner/device. The 3D texture features were restored by Gaussian filtering and 3D reconstruction. The texture signals were decomposed by Complementary Ensemble Empirical Mode Decomposition (CEEMD) algorithm. The correlation between Intrinsic Mode Function (IMF) and the original signal was analyzed. Hilbert transform on the most relevant IMF was performed to obtain the instantaneous frequency and instantaneous amplitude. The maximum correlation Hilbert transform index (Rcm) was proposed to characterize the coefficient of friction. In addition, the friction index was obtained on the same specimens. The linear fit R-square of Rcm and friction index is above 0.9. It is suggested that the finding of his study can accommodate friction management based on pavement texture collected by the 3D laser scanning system.

Keywords Pavement texture · Hilbert-Huang transform · Friction coefficient · Rcm

Y. Dong · Y. Hou · J. Tian · Y. Cao · C. Guo
China Highway Engineering Consulting Corporation, Beijing 100089, China

Research and Development Center of Transport Industry of Technologies, Materials and Equipments of Highway Construction and Maintenance Ministry of Transport, Beijing 100089, China

Research and Development Center on Highway Pavement Maintenance, CCCC, Beijing 100089, China

T. Fang
School of Civil and Environmental Engineering, UNSW Sydney, Sydney, NSW 2052, Australia

J. Zhou (✉)
Department of Civil Engineering, Tsinghua University, Beijing 100084, China
e-mail: zhou-j19@mails.tsinghua.edu.cn

W. Wang et al. (eds.), *Green Transportation and Low Carbon Mobility Safety*, Lecture Notes in Electrical Engineering 944,
https://doi.org/10.1007/978-981-19-5615-7_24

1 Introduction

One of the reasons for the high annual traffic accident rate in the world is due to the insufficient friction performance of the road surface. Pavement texture is an important factor contributing to the friction performance. By analyzing the influence of pavement texture on the friction performance of the pavement, the road safety can be improved so that the occurrence of traffic accidents could be reduced significantly. Research on texture and friction performance at home and abroad has been continuously improved. In terms of texture measurement, the traditional methods are mainly through contact with pavement surface such as sand patching. Recently, more advanced technologies such as laser scanning have been gaining more popularity. The laser scanning is a non-contact measurement approach. The laser sensor is integrated with a three-dimensional (3D) camera. Dong et al. [1] proposed a three-step de-tip method using advanced sensor technology to accurately extract micro-texture information. Ding et al. [2] used 3D laser scanning technology to collect road surface texture data, filtered the data through median filtering and Gaussian filtering, and reconstructed the processed data in three dimensions through MATLAB software. The authors used the multi-sub-composite filling algorithm of four connected regions to simulate the artificial sand patching method to calculate the Mean Texture Depth (MTD). Finally, the actual verification by the sand patching method was carried out to obtain better results. Chen et al. [3] collected the contact marks between the tire and the road surface through the pressure film system, and mainly used the pressure sensor to collect the point where the pressure of the tire and the road surface contact part is $p > 0$. And the medical image processing technology was applied to the film image processing. The pressure film image registration and fusion based on the evolutionary algorithm was used to obtain the three-dimensional contact stress distribution and the effective contact area. By analyzing the three-dimensional shape distribution of the asphalt pavement texture in the effective contact area of the tire, the results showed that the effective contact structure index of the tire could effectively characterize the friction level of the road surface.

In terms of texture evaluation, the indexes mainly include the mean profile depth (MPD) and the mean texture depth (MTD). Besides, the texture parameters commonly used in the existing studies include Ra (arithmetic mean), Rq (section standard deviation), Rz (range), Rsk (skewness), Rku (humpness), D (fractal dimension). These parameters or indexes were used to examine the correlation between texture and road friction coefficient, and establish predictive models [4, 7]. Rado and Cho [5, 6] proposed to use Hilbert-Huang transform to process the two-dimensional (2D) texture signal. They obtained the instantaneous envelope and instantaneous amplitude of the Base Intrinsic Mode Functions (BIMF) through Hilbert transform to extract the peak envelope. Li et al. [7] used PaveVision3D Ultra equipment to collect pavement texture information and verified the four evaluation indexes and slip resistance in MPD and MTD. Fractal theory has gradually been developed in new indicators in texture evaluation. Tong et al. [8] applied the multifractal theory to study the relationship between the multi-fractal dimension of pavement texture and the fractal dimension of aggregate particle size distribution. It also is feasible to use the texture fractal dimension as the friction performance evaluation index. Zhou et al. [9] studied the fractal characteristics of pavement texture under different polishing conditions by the fractal theory. The concept of cutoff wavelength was proposed as an index for evaluating ordinary polishing and differential polishing. Three-dimensional laser scanner was used to extract different polishing texture information. Yang et al. [10] used the multi-function road condition rapid detection system to collect the pavement structure depth information. They calculated the road surface wear rate, compared and analyzed the MTD and MPD values of the road wear and the worn road surface, and finally establish the relationship between the road wear and the average structural depth of the road texture. The evaluation model was intended to be applied to the texture evaluation indicator standard.

In this paper, the road surface texture information is collected by laser sensor. The relationship between road surface texture signal and friction performance is studied by nonlinear non-stationary signal time–frequency analysis technology. Hilbert-Huang transform is an effective tool to deal with nonlinear non-stationary signals. The three-dimensional texture information collected by the detection vehicle is composed of several two-dimensional nonlinear and non-stationary time–frequency signals, the algorithm can better solve the problem related to pavement texture based on its data property.

2 Hilbert-Huang Transform Principle

In 1998, Huang et al. [11, 12] proposed the Hilbert-Huang Transform for analyzing nonlinear non-stationary signals. It is composed of two parts. The first part is the empirical mode decomposition (Empirical Mode Decomposition EMD) to find the intrinsic mode function IMF. EMD can decompose complex time-series signals into a single-component signal of a series of intrinsic modal functions. Each IMF contains various frequency components from high to low in the original signal. For the local maxima and minima of any original data s(t), the upper and lower envelopes are respectively generated by cubic spline interpolation, and the mean of the corresponding data points m1 is obtained, and the difference between $s(t)$ and m is defined. The value is the first component h:

$$h = s(t) - m \tag{1}$$

Consider h as the new raw data $s(t)$. At any point, the average of the two envelopes consisting of local maxima and local minima is zero, then it is regarded as IMF component; remember $c_1 = h$ is IMF_1;

$$s(t) - c_1 = r_1(t) \tag{2}$$

Consider $r_1(t)$ as the new $s(t)$, and repeat the above process to get IMF_2, c_2, $r_2(t)$, IMF_n, c_n, $r_n(t)$, etc., and $r_n(t)$ is the remainder.

$$s(t) = \sum_{i=1}^{n} IMF_i(t) + r_n(t) \tag{3}$$

The EMD method has modal aliasing phenomena for noise signals and intermittent signals, which seriously affects the accuracy of signal decomposition.

The Ensemble Empirical Mode Decomposition (EEMD) method [13] effectively obtains multiple mean IMF components as the final result by adding Gaussian white noise decomposition multiple times in the original frequency $x(t)$ time–frequency space. Modal aliasing is suppressed.

$$x_i(t) = x(t) + z_i(t) \tag{4}$$

Repeat Eqs. (1) and (2), and finally find the average to get the i-th IMF.

$$IMF_i = \frac{1}{N} \sum_{i=1}^{N} C_{ij}(t) \tag{5}$$

Zi(t) is the added white noise signal; N is the number of added white noise; $C_{ij}(t)$ represents the ith IMF after the jth white noise processing is added.

The white noise added during the EEMD decomposition process cannot be completely eliminated, causing noise interference to the signal. Complementary Ensemble Empirical Mode Decomposition [14] Add N pairs of positive and negative auxiliary white noise to the original signal, obtain 2 N signals and perform EMD decomposition for each signal, and finally combine the calculation results. The CEEMD decomposition process is the same as EEMD, except that noise is added after adding a noise, and IMFs are obtained according to EMD decomposition. Not only the modal aliasing problem is solved, but also the original signal is accurately reconstructed [15].

The second part is to use the Hilbert transform to obtain the physical characteristics of the instantaneous frequency, instantaneous phase and Hilbert-yellow spectrum of the signal.

For any time series $x(t)$

$$\hat{x}(t) = \frac{1}{\pi} \int_{-\infty}^{\infty} \frac{x(\tau)}{t - \tau} d\tau \tag{6}$$

The parsing signal is:

$$v(t) = x(t) + j\hat{x}(t) = a(t)\exp(j\theta(t)) \tag{7}$$

The instantaneous amplitude is:

$$a(t) = \left(x^2(t) + \hat{x}^2(t)\right)^{1/2} \tag{8}$$

The instantaneous frequency is:

$$f(t) = \frac{\omega(t)}{2\pi} = \frac{1}{2\pi} \cdot \frac{d\theta(t)}{dt} \tag{9}$$

It can be seen from Eq. (7) that there is a real part $x(t)$ and an imaginary part in the Hilbert transform analysis signal, and $x(t)$ represents the original signal, indicating the Hilbert transform.

Equations (8) and (9) are the instantaneous physical characteristics of the obtained signal.

3 Data Collection

The experiment selected two types of asphalt mixtures: AC-13 and AC-16. For each type, it includes four design alternatives, as shown in Tables 1 and 2. And each design alternative has three specimens, the asphalt mixture was molded in the sill board to form the specimens with size of 300 mm × 300 mm, shown in Fig. 1.

As shown in Fig. 2, experiments were carried out on 12 sets of Hot Mix Asphalt (HMA) AC-13 and 12 sets of AC-16 with a total of 24 sets of slabs through the multi-function inspection vehicle flatness/texture module. The texture acquisition module can achieve an accuracy of 10 μm and can collect all macro textures and partial micro texture information. The exposure time of the 3D camera in the device is as low as 1 μs, which can effectively avoid the motion blur caused by the excessive speed of the vehicle. At the same time, the equipment is not affected by vehicle vibration

Table 1 AC-13 gradation scheme design

Gradation design	Pass rate of different sieve holes (mm)/%										
	16	13.2	9.5	4.75	2.36	1.18	0.6	0.3	0.15	0.075	Oil-stone ratio/%
Graded lower limit	100	100	85	68	50	38	28	20	15	8	–
Graded lower limit	100	90	68	38	24	15	10	7	5	4	–
AC-13-1	100	94	70	44	31	21	15	10	8	6	4.7
AC-13-2	100	95	75	52	35	25	18	12	9	7	4.8
AC-13-3	100	97	82	62	43	29	21	13	10	8	4.9
AC-13-4	100	98	80	64	45	33	26	16	12	7	5.0

Table 2 AC-16 gradation scheme design

Gradation design	Pass rate of different sieve holes (mm)/%											
	19	16	13.2	9.5	4.75	2.36	1.18	0.6	0.3	0.15	0.075	Oil-stone ratio/%
Graded lower limit	100	100	92	80	62	48	36	26	18	14	8	–
Graded lower limit	100	90	76	60	34	20	13	9	7	5	4	–
AC-16-1	100	96	85	70	48	33	25	18	12	10	7	4.8
AC-16-2	100	98	89	77	54	27	16	11	10	8	7	4.9
AC-16-3	100	99	91	78	62	42	22	16	12	10	6	5.0
AC-16-4	100	95	81	60	39	28	20	16	12	9	5	4.7

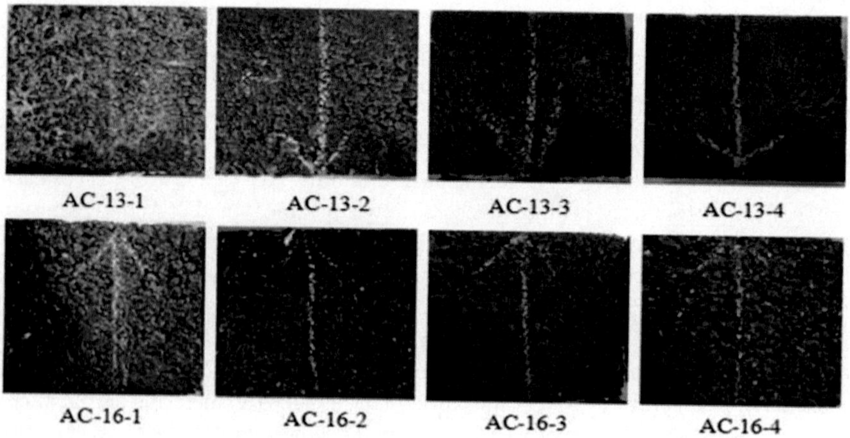

Fig. 1 The AC-13 and AC-16 rutting plate

and driving speed when detecting pavement texture [16]. In the experiment, Three-dimensional texture information was collected by pulling the rutting plate online laser sensor. In Fig. 2, on the top of the picture is the real-time 2D texture elevation information. The transverse sampling length is 270 mm. On the bottom of the Fig. 2 was consisted of 512 lines of 2D signals. Each car slab extracts 300 columns of data, and each column of 800 points constitutes a 800×300 matrix representing a rectangular road surface area of 100 mm $\times$ 100 mm. The original data is filtered by Gaussian filtering (Fig. 3), and the Gaussian filtering can filter out the singular points in the data without changing the characteristics of the original data, thereby effectively improving the purity of the data. Figure 3 shows the three-dimensional texture map after Gaussian filtering, which clearly expresses the characteristics of the pavement texture.

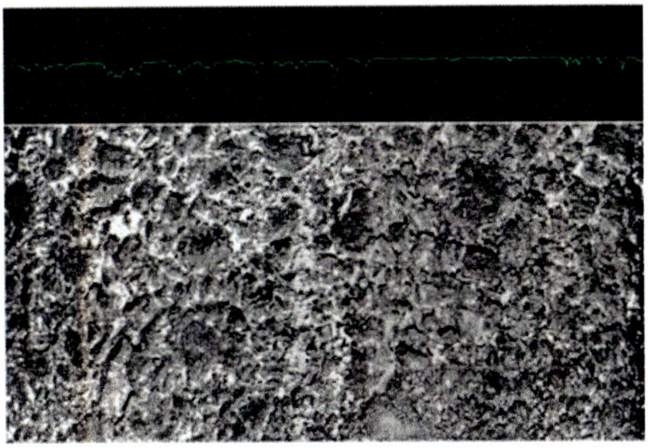

Fig. 2 Texture data acquisition

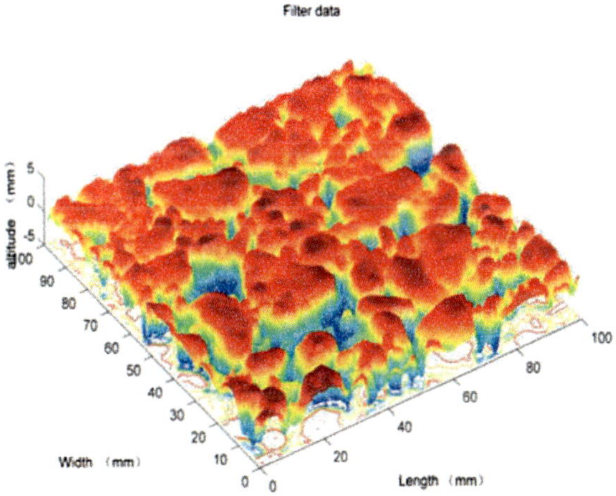

Fig. 3 3D texture after filtering

4 Data Processing and Analysis

Taking the data of the 200th column of a car slab of AC-13 as an example, CEEMD decomposition of the signal, adding the standard deviation of the auxiliary Gaussian white noise and the standard deviation of the signal is 0.2, and the number of added noise is taken 100 times, the natural mode. The function is set to 8. By decomposing, nine intrinsic mode functions IMFs and one remainder RS10 are obtained, where IMF1 is the decomposed original signal.

IMF 2-9 is a single component signal that is arranged from high frequency to low frequency after the original signal is decomposed. The correlation between the eight intrinsic mode functions and the original signal is calculated (Fig. 4). The correlation between IMF6 and the original signal is up to 0.708 as shown in Fig. 5. The highest correlation IMF6 signal is subjected to Hilbert transform to obtain the instantaneous amplitude and instantaneous frequency of the signal (Fig. 6). The frequency of the

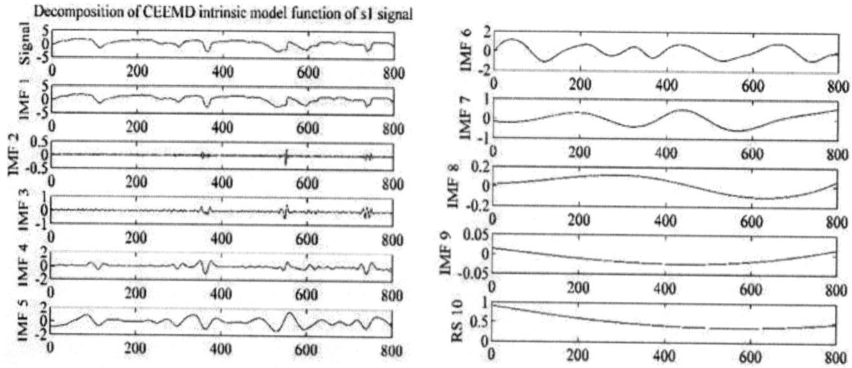

Fig. 4 Decomposition of CEEMD intrinsic modal function of s1 signal

Fig. 5 IMFs correlation index

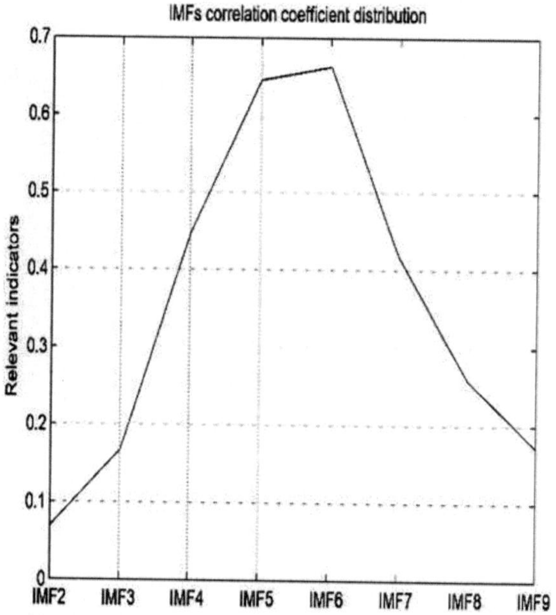

signal is mainly concentrated at about 20 Hz, and some of the frequencies are greater than 100 Hz, which is mainly the collected microscopic texture information. The amplitude varies mainly from 0 to 2 mm.

Take M as the mean of 800 points, N is the mean of 800 points, and A is the ratio of the mean M of the amplitude to the logarithm of the mean of the frequency $|lgN|$, recorded as:

$$A = M/|lgN| \tag{10}$$

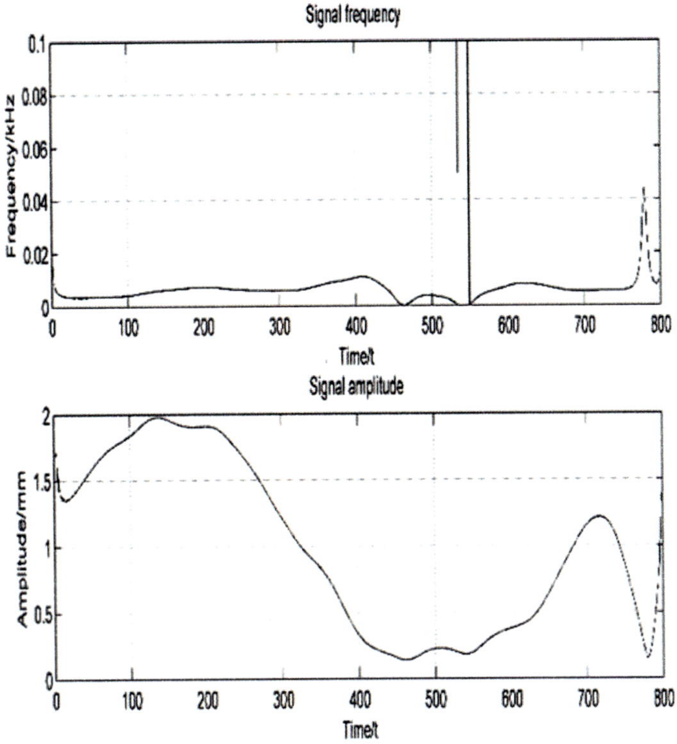

Fig. 6 Signal frequency and amplitude

The ratios of the eight groups of 300 columns of data are respectively obtained to obtain the results shown in Fig. 6. In the figure, the abscissa is the ordinal number of 300 columns of data in each group, and the ordinate is the ratio of the mean value of the four groups AC-13 and the four groups of AC-16 to the absolute value of the logarithm of the frequency mean. The data in Fig. 7 is mainly concentrated between 0 and 1, and the small part of the data is between 1 and 2. The Rcm (maximum correlation Hilbert transform coefficient) is defined as the mean of 300 A values per group.

As a logic next step, the friction data was collected on the asphalt mixture slabs. The purpose is to figure out the potential relationship between the friction and texture. There is a good correlation between MPD and skid-resistance performance. The pavement texture structure within the one-time scanning range of the laser line is evenly divided into two parts. The difference between the average value of the two peak points $(h_1 + h_2)/2$ and the median value h is taken as the average section depth MPD. In order to comprehensively characterize the MPD of rutting plate specimen, the 300 columns 2D texture signals are calculated by the Eq. 11.

$$MPD\mathrm{i} = \frac{h_1 + h_2}{2} - h \quad (11)$$

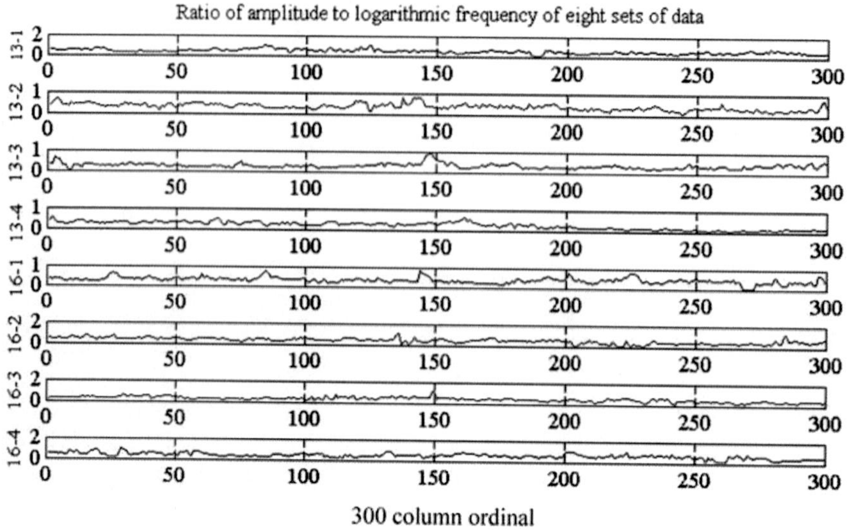

Fig. 7 Ratio of amplitude to logarithmic frequency of eight data

where h1 is the left side of the peak; h2 is the right side of the peak; h is the midline value; MPDi is the ith column MPD value;

$$MTD = \frac{1}{M \times N}\sum\nolimits_{i=1}^{N}\sum\nolimits_{j=1}^{M}\left[Z_p(x_i - y_i) - Z(x_i - y_i)\right] \quad (12)$$

where Zp(x,y) is the maximum value corresponding to the peak line of the spatial topography.

Ra is the arithmetic mean of absolute value of contour deviation within the sampling range. In a statistical sense, it takes the amplitude distribution of the surface texture morphology of the road surface as its average. In the statistical aspect, it represents the absolute first-order origin distance of the set of data points, reflecting the dispersion degree of the contour amplitude of the topography relative to the reference line. It is widely used to evaluate the roughness of the surface morphology features. The R_a is calculated by the Eq. 7 using 3D texture data.

$$R_a = \frac{1}{M \times N}\sum\nolimits_{i=1}^{N}\sum\nolimits_{j=1}^{M}|Z(x_i - y_i)| \quad (13)$$

where Z (x,y) is the elevation information of topography based on datum line; M, N are the sampling points in two perpendicular directions.

The standard deviation of contour amplitude distribution of road surface topography can be expressed by R_q. R_q has a good correlation with the anti-slip performance of asphalt pavement, and can be used to evaluate the roughness of asphalt pavement texture.

Table 3 Rcm and other Index

Correlation	Rcm	MPD	MTD	Ra	Rq
Rcm	1	0.92	0.95	0.92	0.94
MPD	0.92	1	0.96	0.95	0.93
MTD	0.96	0.96	1	0.93	0.97
Ra	0.92	0.95	0.93	1	0.97
Rq	0.94	0.93	0.97	0.97	1

$$R_q = \sqrt{\frac{1}{M \times N} \sum\nolimits_{i=1}^{N} \sum\nolimits_{j=1}^{M} (Z(x_i - y_i))^2} \tag{14}$$

where Z (x,y) is the elevation information of topography based on datum line; M, N are the sampling points in two perpendicular directions.

The 24 sets of plates data was calculated. The correlation between any two indexes is obtained by calculating Pearson simple correlation coefficient. The results on the friction data together with the texture index are presented in Table 3.

In Table 3, the correlation between Rcm and other index is above 90%, indicating that the indicator Rcm obtained by CEEMD method for 3D texture data processing has a good correlation with pavement skid-resistance performance. In particular, the correlation with MTD reached 96%. By fitting Rcm and the value of MTD, it is found that there is a linear correlation between the two (Fig. 8). The Rcm is the average of 300 maximum amplitudes. Through the linear regression analysis of the data, it is found that the data conforms to the linear equation:

$$y = 2.6x + 0.2 \tag{15}$$

The goodness of R^2 is 96%, which indicates that the two have good correlation, indicating that Rcm can be better characterize the size of the skid-resistance performance.

The experimental results show that the friction performance is closely related to the frequency and amplitude of the signal. Through the decomposition of the original signal, the complex signal information is decomposed into several single

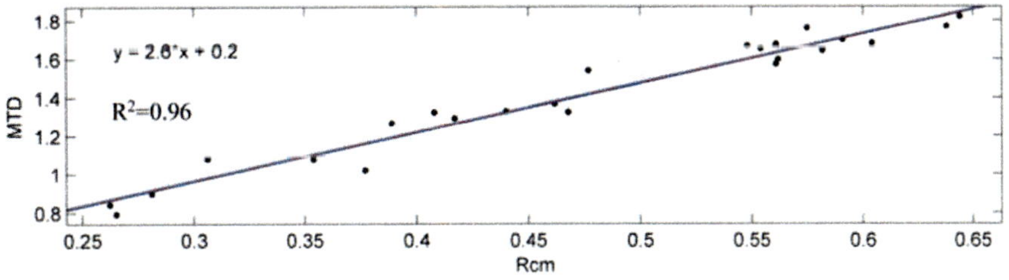

Fig. 8 The relation between Rcm and MTD

modal signals which are correlated with the original signal. The IMF of the greatest correlation is the main component of the original signal. Rcm has good applicability to AC-13 and AC-16.

5 Conclusions

Asphalt mixture surface texture was collected by a laser scanning system in the laboratory. In the meantime, the friction data index was also obtained. To correlate the texture and friction, the following analysis was conducted:

(1) Effectively extract 2D and 3D information of pavement texture from asphalt pavement by 3D laser scanning system. The 3D texture data was filtered by Gaussian filtering. The 3D texture image was reconstructed that can clearly display the road surface texture information.
(2) Decompose the 2D texture signal by CEEMD algorithm, analyze the correlation between IMFs and the original signal, and perform Hilbert transform on the most relevant IMF to obtain the instantaneous frequency and instantaneous amplitude.

As a result, an Rcm index that can characterize the road surface friction coefficient was proposed. The linear correlation analysis between the friction index MTD and the Rcm index was performed with a goodness of fit of 0.96. This implies that the proposed approach can effectively capture the relationship between the pavement texture and friction performance.

In the next steps, the correlation between the Rcm index and the friction coefficient will be verified by a larger amount of test data, covering other types of pavement surfaces. This study will be extended to the in-situ pavements, particularly at relatively high speed condition. The ultimate goal is to use the developed equipment to collect pavement texture data for network-level friction management.

Author Contributions The authors confirm contribution to the paper as follows: study conception and design: Runhua Guo, Jing Zhou and Youqiang Si; data collection: Runhua Guo, Jing Zhou and Youqiang Si; analysis and interpretation of results: Runhua Guo, Jing Zhou and Youqiang Si; draft manuscript preparation: Runhua Guo, Jing Zhou and Youqiang Si. All authors reviewed the results and approved the final version of the manuscript.

References

1. Dong N, Prozzi J, Ni F (2019) Reconstruction of 3D pavement texture on handling dropouts and spikes using multiple data processing methods. Sensors 19(2):278
2. Shihai D, Enhui Y, Chenping W, Yaying J, Kaiyun L, Xinrui Z (2019) 3 d high precision laser non-contact detection of asphalt pavement surface texture. J Southw Jiaotong Univ 1–8
3. Chen B, X Zhang, Yua J, Wang Y (2017) Impact of contact stress distribution on skid resistance of asphalt pavements. Constr Build Mater 133:330–339

4. Vladescu S-C, Medina S, Olver AV et al (2016) The transient friction response of a laser-textured, reciprocating contact to the entrainment of individual pockets. Tribol Lett 62(2):19
5. Rado Z, Kane M (2014) An initial attempt to develop an empirical relation between texture and pavement friction using the HHT approach. Wear (309):233–246
6. Choongwoo C (2010) Application of Hilbert Huang transformation to analyze pavement texture-friction relationship. The Pennsylvania State University
7. Li L, Wang KCP, Li Q (2016) "Joshua". Geometric texture indicators for safety on AC pavements with 1mm 3D laser texture data. Int J Pavement Res Technol 9(1):49–62
8. Shenjia T, Xiangbing X, Dayong Z (2016) Fractal description of asphalt pavement texture distribution and evaluation of skid resistance. China J Highw Transp 29(2):1–7
9. Xinglin Z, Yuanyuan Z, Maoping R, Shenqing X, Xiaoming H (2019) Analysis of polishing behavior of asphalt pavement surface based on segmenting variable dimension. China J Highw Transp 32(4):187–195+242
10. Guofeng Y, Haoyang W, Yuli P (2016) Wear detection and evaluation method based on multi-line texture. China J Highway Transp 29(3):35–40
11. Huang NE, Shen Z, Long SR (1998) The empiricalmode decomposition and hilbert spectrum for non-linear and non-stationary time series analysis. Proc Royal Soc London, Series A 454:903–995
12. Huang NE, Shen Z, Long SR (1999) A new view of nonlinear water waves: the Hilbertspectrum. Annual Rev Fluid Mech 31:417–457
13. Wu ZH, Huang NE (2009) Ensemble empirical mode decomposition: a noise assisted data analysis method. Adv Adapt Data Anal 1(1):1–41
14. Yeh J, Shieh J, Huang NE (2010) Complementary ensemble empirical mode d-ecomposition: a novel noise enhanced dataanalysis method. Adv Adapt Data Anal 2(2):135–156
15. Si YQ, Guo RH, Shi PC (2019) Comparative study of signal time-frequency analysis techniques based on EMD, EEMD and CEEMD. CT Theory Appl 28(4):417–426 (in Chinese)
16. Guo R, Yu Z, Zhou Y (2019) Development and preliminary evaluation of a varying-speed road profiler. J Test Eval 48. Published ahead of print, 27 March. https://doi.org/10.1520/JTE20180703

D-S Evidence Reasoning Based Transportation Project Investment Decision Model and Its Application

Ling Sui and Xiaoli Zhang

Abstract This research demonstrates five approaches of extracting degree of ignorance (DOI) coefficients of index systems based on the DOI factors of investment decision making systems. Based on Dempster -Shafer (D-S) evidence reasoning, A comprehensive decision-making model of transportation project investment (CDM-TPI) is constructed. Lastly, the case comparison studies are carried out to validate the effectiveness and practicability of the model when applying in the investment decision making systems with the DOI factors.

Keywords D-S evidence reasoning · The extraction of DOI coefficients · Data fusing · CDM-TPI

1 Introduction

Most of information for the comprehensive project investment decision is inaccurate, incomplete, fuzzy, and even contradictory, therefore presenting DOI. To effectively carry out investment analysis and decision, formal methods needed to be used for describing this DOI information to further analyze the DOI inference method. DOI inference does not necessarily involve complete logic reasoning, and it merely conduct approximate reasoning and judgment on DOI information within an allowable error range. Although such reasoning may not be able to deduce optimal decision results, it is generally capable of acquiring the decision results that equivalent to that

L. Sui
China Academy of Transportation Sciences, No. 501, The Transportation Development Research Center of China Academy of Transportation Sciences, Ministry of Transport. R. China, 240 Huixinli, Chaoyang district, Beijing 100029, China
e-mail: xulingjky@163.com

X. Zhang (✉)
China Academy of Transportation Sciences, No. 522, The Transportation Development Research Center of China Academy of Transportation Sciences, Ministry of Transport. R. China, 240 Huixinli, Chaoyang district, Beijing 100029, China
e-mail: 791867362@qq.com

W. Wang et al. (eds.), *Green Transportation and Low Carbon Mobility Safety*, Lecture Notes in Electrical Engineering 944,
https://doi.org/10.1007/978-981-19-5615-7_25

of expert-based decision, which can basically satisfy the application requirement of different fields [1]. D-S evidence reasoning, as an effective DOI reasoning approach, is considered as a decision theory. Compared to Bayes' decision theory (e.g. Bayes theory), D-S evidence reasoning can only also deal with the DOI induced by inaccurate knowledge, but also process the DOI resulting from uncertain factors, without providing prior probability [2, 3]. The key of this method is to find out the ways of extracting degree of ignorance (DOI). On this basis, this research proposes a theoretical frame based on D-S evidence reasoning; furthermore, the five approaches of extracting DOI of index system are demonstrated: lastly, a comprehensive investment decision model base on D-S evidence reasoning theory is built by integrating the comprehensive decision characteristics of investment project to perform comparison of difference cases.

2 D-S Evidence Reasoning Theory [4]

Frames of discernment Θ usually are pairwise mutually exclusive in evidence theory and consist of all targets for decision making, that is: $\Theta = \{\theta_1, \theta_2, \ldots, \theta_m\}$. We give the assignments of basic probability distribution to each of subsets θi (i = 1, …, m) contained in frames of discernment Θ, and then employ combination rule of D-S evidence reasoning theory.

Definition 1 Combination rule of D-S evidence reasoning theory [4–8]:

Assuming Bell, …, Belk are belief functions in a same frame of discernment (FOD)Θ, and m_1, …, m_k denote the assignments of basic probability distribution, while Bel, new belief function generated using D-S combination is marked as $\oplus_{i=1}^{k} Bel_i$, its corresponding basic probability assignments are determined using following equations:

For $A \subset \Theta$, $A \neq \Phi$, we obtain

$$m(A) = \left(\sum_{\bigcap_{i=1}^{k} A_i = A, A_i \subset \Theta} (m_i(A_1) \cdot \cdots \cdot m_k(A_k)) \right) \cdot \left(\frac{1}{1-K} \right) \quad (1)$$

$$K = \left(\sum_{\bigcap_{i=1}^{k} A_i = \Phi, A_i \subset \Theta} (m_i(A_1) \cdot \cdots \cdot m_k(A_k)) \right) \quad (2)$$

This rule is adopted to fuse the propositions of multiple information sources $S_1, \ldots, S_k$, where K denotes the normalization factor of $0 \leq K < 1$. If all sets of intersections are empty sets, K = 1 is present, in such case, Dempster's rule of combination is meaningless.

Definition 2 Factors of evidence conflict (con) [4, 5]

Generally, in addition to compatibility, evidences also conflict with each other. To measure conflict degrees of difference evidences, the operators provided in the literature [4] are used to obtain the equation:

$$Con(Bel_1, \cdots Bel_k) = Log(1/1 - K) \tag{3}$$

Obviously, $0 \leq \text{Con} < \infty$, when $K \to 0$, $\text{Con} \to 0$ is obtained; in the case of $K \to 1$, $\text{Con} \to \infty$, indicating the larger the Con value, the greater the conflict degrees among different evidences.

In this research, index system of the investment decision system is seen as a group of evidence information, and therefore it is fairly reasonable to use aforementioned D-S evidence reasoning rule. When D-S evidence theory is applied in the comprehensive decision of investment projects, measured features acquired from the information for each of independent indexes consist of the evidences of this theory. By constructing corresponding basic probability distribution functions, each of all evidences including Θ are endowed with certain DOI, which are employed to generate new evidences. The analysis results show that there are four key focuses in D-S evidence reasoning as follows:

(1) Solving basic probability assignments $m(i)$ of each evidence;
(2) Solving the DOI of each evidence, $m(\Theta)$;
(3) Using the combination rule of D-S evidence reasoning to generate new belief function Bel(A) and solving basic probability assignments m(A);
(4) Solving the problems relating to the combination of the evidences which are relevant and compatible with each other;

The focus of this study is to extract DOI of the decision index system and utilize D-S evidence theory to build a comprehensive decision model to carry out information fusion, which is not further illustrated in the (4) item.

According to aforementioned analysis, the premises of using the combination rule of D-S evidence reasoning included as follows:

(1) All evidences must be independent to certain degree;
(2) The conflict degree (Con) among different evidences should not exceed certain threshold Con_0: the value of K should not set to be a larger value;

The application of D-S evidence theory primarily focuses on the independence of evidences, which is difficult to be satisfied. Most of studies directly use D-S theory by supposing the evidences are approximately independent; however the performance of D-S evidence reasoning theory when applying in the cases that have high demands to fused information and require great reliability, remains unclear. Especially, in the case that evidences conflict with each other, K approximates to 1or equals to 1. The results fused using the combination rule of D-S evidence reasoning are unreasonable and conflict with each other, which even cannot be fused using the combination rule of D-S evidence reasoning. To solve this problem, existing literatures [6, 10–13] revised this combination rule. Literature [9] promotes and improves the combination rule of

D-S evidence reasoning based on the data dependence to further overcome the limits of D-S evidence reasoning. The study [14] both took the correlation and conflict of the evidences into account. In order to reduce the computation amount of multi-source information fusion, previous studies [15, 16] also put forward improved equations, In general, these approaches are not ideal: as they are complex, and therefore merely applied in limited fields. Although such contradiction in the investment decision of transportation projects is not prominent, it is still necessary to conduct in-depth study to make the D-S evidence theory have more universal significance.

3 The Establishment of Investment Decision Model

3.1 The Establishment of the Index Systems

It is essential to build the index systems to assess comprehensive decision of transportation project investment. The index system should be featured with incompleteness and relative independence: it has to reflect the principle of profit and equity and presents the requirements of overall goals for investment decision. Additionally, the concept is likely to exhibit distinct feature, easy calculation process, and accessibility to data information. Generally, the index systems do not have different dimensions, which show varying attributes. As the index systems with inconsistent dimensions make each of targets difficult to be horizontally compared, each of indexes needs to be standardized for the sake of decision analysis. There are a great many of standardizing approaches, which can be selected according to different scenarios and targets.

Assuming $\Theta = \{\theta_1, \ldots, \theta_m\}$ are m target scheme sets for comprehensively evaluating investment decision of transportation projects. Each scheme consisting of n index systems $P = \{p_1, \ldots, p_n\}$ featured with different physical meanings and varying dimensions to describe its characteristics. Hence, we obtain an evaluation index data matrix with $m \times n$ dimensions, which are normalized correspondingly to generate new index data matrix [R]:

$$[R] = \begin{matrix} & \begin{matrix} p_1 & p_2 & \cdots & p_n \end{matrix} \\ \begin{matrix} S_1 \\ S_2 \\ \cdots \\ S_m \end{matrix} & \begin{bmatrix} x_{11} & x_{12} & \cdots & x_{1n} \\ x_{21} & x_{22} & \cdots & x_{2n} \\ \cdots & \cdots & \cdots & \cdots \\ x_{m1} & x_{m2} & \cdots & x_{mn} \end{bmatrix}_{m \times n} \end{matrix}$$

where x_{ij} represents the jth index of the normalized ith scheme, [R] is supposed as evaluation environment.

3.2 Extraction of DOI Indexes

Solving the DOI of evidences is a key part in the application of D-S evidence reasoning. The selection of methods for extracting DOI should be conducted based on real cases. Theoretically, in an evaluation environment [R], the more concentrated that each of schemes in an index system distributed or the more that a index system matches its average information compared to other index systems, indicating the information containing in the index is more favorable to decision making of project investment. This outcome reflects the low DOI of the information containing in the index system according to the D-S evidence reasoning, vice versa. By conducting sufficient case studies and integrating the characteristics of the investment decisions to transportation projects, this research put forwards five approaches of extracting DOI.

For the sake of description, the definitions of indexes are made under the evaluation environment [R]:

$$\bar{X}_i = \frac{1}{N} \cdot \sum\nolimits_{j=1}^{n} x_{i,j},\ i = 1, \ldots, \mathrm{m}; \tag{4}$$

$$\delta_i = 1/\sqrt{n} \cdot \sqrt{\sum\nolimits_{j=1}^{n} (x_{i,j} - \bar{x}_i)^2},\ j = 1, \ldots, n. \tag{5}$$

(1) Correlative information entropy of KullBack [3]

Preconditions include: $\forall i, j$, when $X_{ij} = 0$, we obtain $\ln\left(X_{ij}/\overline{X_i}\right) = 0$, the DOI of the jth index system is defined as:

$$DoI\left(p_j\right) = \sum\nolimits_{i=1}^{m} x_{ij} \cdot \ln\left(\frac{x_{ij}}{\overline{X_i}}\right),\ j = 1, \ldots, n. \tag{6}$$

(2) ShanNon information entropy operator [3]

Preconditions consist of: $\forall i, j$, when $x_{ij} = 0$, we set $\ln(x_{ij}) = 0$. The DOI of the jth index system is defined as:

$$DoI\left(p_j\right) = -\sum\nolimits_{i=1}^{m} x_{ij} \cdot \ln\left(x_{ij}\right),\ j = 1, \ldots, \mathrm{n}. \tag{7}$$

(3) L-operators of DOI

Preconditions include: $\forall i,\ j$, in the case of $\delta_i = 0$, $\delta_i \prec \varepsilon$, where ε is positive infinite decimal, which should not set to be zero. The DOI of the jth index system with q orders can be defined as:

$$DoI\left(p_j\right) = \frac{1}{m}\left(\sum\nolimits_{i=1}^{m} \frac{\left|x_{i,j} - \overline{X_i}\right|^q}{\delta_i^q}\right)^{1/q},\ j = 1, \ldots, n;\ q \geq 1. \tag{8}$$

(4) Grey correlation degree operators

By combining grey correlation system theory, DOI of the jth index with q orders is defined as:

$$DoI(p_j) = \frac{1}{m}\left(\sum\nolimits_{i=1}^{m} (r_{i,j})^q\right)^{1/q} \tag{9}$$

where,

$$r_{i,j} = \frac{\min_{i=1}^{m}\left|x_{i,j} - \overline{x}_i\right| + \zeta \max_{i=1}^{m}\left|x_{i,j} - \overline{x}_i\right|}{\left|x_{i,j} - \overline{x}_i\right| + \zeta \max_{i=1}^{m}\left|x_{i,j} - \overline{x}_i\right|}, (i = 1, \ldots, m;\ j = 1, \ldots, n) \tag{10}$$

which is the mean value of grey correlation degree. The constant ζ denotes the identification coefficients of grey correlation, and its function is to adjust the range of comparable environments. In practical application, ζ is usually set to be 0.5.

(5) Fuzzy operators

Based on integrating fuzzy system theory, DOI of the jth index with q orders can be defined as:

$$DoI(p_j) = \frac{1}{m}\left(\sum\nolimits_{i=1}^{m} (\tilde{r}_{i,j})^q\right)^{1/q} \tag{11}$$

where

$$\tilde{r}_{i,j} = \frac{x_{i,j} \wedge \overline{X}_i}{x_{i,j} \vee \overline{X}_i} \tag{12}$$

The five operators aforementioned do not present high demands to sample size and small calculation with no typical distribution law being involved. Moreover their evaluating results are fairly objective and are able to extract the DOI of indexes. It is noted that: Operator 1 takes initial information of the investment schemes evaluated based on multi-attributes and multi-indexes in the decision process into account and sufficiently utilizes the information of raw data, however it pays little attention to the information of the environment [R], and it performs well in case of complete information; Operator 2 uses an overall concept to measure DOIs containing in each of indexes rather than only focusing on each of independent index, and therefore it is seen as high-level evaluation of DOI. The issues including shortage of data and unfavorable quality in the decision making process can be preferably overcome by Operator 2; the rest of three operators reflect the DOIs for each of indexes from different perspectives by integrating the information of the environment and individuals. In the case of more schemes, Operators 3 and 4 are used as Operators 2 and 5 are likely to weaken individual information.

Although these five operators can be used for the selection of schemes, their decision results show unfavorable discreteness and tend to be homogenized, which further affects the distinguishability of the decision results and utilization of existing information of the index system. For the schemes with great difficulty, high fuzzy degree and similar levels, the priority order is unlikely to be determined and therefore the DOI of evaluated results fails to be judged. Hence, the decision results are unreliable and the reasonability of selected schemes is difficult to be verified, which is generally present in the comprehensive decision of transportation projects' investment. D-S information fusing technology exhibits advantages such as the increase of confidence degree, distinguishability and spatial resolution of decision results, and decrease of DOI. The method that uses five operators to extract DOI to further be fused using D-S evidence reasoning, combines the merits of both five operators and D-S evidence reasoning and overcomes the shortfalls of five operators.

3.3 *Mathematical Model of Establishing the Functions of Basic Probability Distributions*

The key of merging acquired information (evidences) and other evidences in D-S evidence reasoning theory as an invariable method based on evidence systems is to define the functions of basic probability distribution. The processing method is described as: at firstly, the method in Sect. 3.2 is used to extract DOIs for each of indexes (evidences), and then these information are utilized to obtain basic probability assignments of each of indexes (evidences), namely,

$$m_i(j) = x_{ij} \cdot \big(1 - DoI(p_j)\big) \tag{13}$$

$$m_j(\Theta) = DoI\big(p_j\big) \tag{14}$$

$$Con(Bel_1, \ldots Bel_n) = \ln(1/1 - k) = -\ln\Big(M(\Theta) + \sum\nolimits_{i=1}^{m} M(i)\Big) \tag{15}$$

where $m_i(j)$ denotes the basic probability assignment of the *i*th scheme for the *j*th index, $m_j(\Theta)$ refers to the DOI of the *j*th index; $Con(Bel_1, \ldots Bel_n)$ is the value of conflict degree for the *n*th index (evidence); M(i), $i = 1, \ldots,$ m, is the basic probability assignment of fused *i*th scheme, and M(Θ) is the DOI of the newly fused evidence system. Hence, an investment decision model is built based on D-S evidence reasoning and then combination rule of Definition 1 is used to further solve the decision results.

3.4 Constraint Conditions

To effectively utilize the combination rule of D-S evidence reasoning to conduct decision analysis by integrating actual demands, the following constraint conditions are obtained as follows:

(1) Target schemes have to be endowed with maximum basic probability assignment;
(2) The basic probability assignments of target scheme should be greater than fused $M(\Theta)$;
(3) The difference between basic probability assignments of target schemes and other schemes should be greater than given threshold T_0;
(4) The DOI $m_j(\Theta)$ of various indexes and $M(\Theta)$ of the newly fused index should be smaller than a threshold T_1;
(5) The conflict degree of evidences—Con is below a certain threshold: K is smaller than the threshold T_2.

4 Case Study

Many schemes usually need to be selected in the investment decision of practical transportation projects when introducing comprehensive evaluation standards, among which, there are an optimal scheme, suboptimum scheme and poorest scheme. By taking urban planning issue of a city as case study in literature [17], five approaches are used to make decision and then obtained results are compared with fused results. In urban planning, there are six feasible schemes in the reconstruction of major traffic rods in the city. The schemes are denoted by $H_1, H_2, \ldots, H_6$, among them, Scheme 1 is the scheme of sub-lane; Scheme 2: the scheme of urban rapid rail transit; Scheme 3: parallel-mixed traffic stream; Scheme 4: subway schemes; Scheme 5: the rail transits of existing roads; Scheme 6: layering schemes of rail viaducts. Seven economic indexes such as technology, economy, environment and society are shown in Table 1. By feasibility analysis and experts-level investigation of the investment projects, all economic indexes are found to present profit [17].

In this research, the five approaches are used to extract the DOI coefficients of indexes, which finally satisfy the decision requirements. Table 2 shows the decision results obtained using the conventional methods based on the five operators without being fused and also displays the compared results of basic probability assignment and total DOI of different schemes fused through D-S evidence reasoning theory. As seen from the fusion results, the larger the threshold T_0, the smaller the number of chosen target schemes; T_0 can be taken a greater value if solving optimal scheme is only considered. As the purpose of investment project is to sort schemes, T_0 is set to be a small value, hence we obtain $T_0 = 0.001$; $T_1 = 0.2$ and $T_2 = 0.9$.

As seen from the case analysis aforementioned, the conclusions required by each of fusion methods are generally the same: namely, Scheme 6 is worthiest to be

Table 1 The index of technological economy

	Degree of construction easiness	Function (‰)	Traffic flow (Per thousand vehicles/h)	Vehicle speed (10 km/h)	Linear marking (%)	Safety coefficients	Overall coefficients
Scheme 1	0.85	8.8	2.2	2.5	0.51	0.33	2.25
Scheme 2	0.4	3.6	0.8	6.0	0.75	0.67	3.0
Scheme 3	0.6	6.2	2.0	3.0	0.58	0.5	2.5
Scheme 4	0.2	3.6	0.8	8.0	0.70	0.83	3.25
Scheme 5	0.4	3.6	0.8	6.0	0.75	0.5	3.0
Scheme 6	0.6	6.2	3.5	5.0	0.63	0.67	3.0

Table 2 The decision result of five different methods and the fusion result obtained using D-S evidence reasoning

	Basic probability distribution function						m (Θ)
	Scheme 1 (%)	Scheme 2 (%)	Scheme 3 (%)	Scheme 4 (%)	Scheme 5 (%)	Scheme 6 (%)	Total DOI
Information entropy operator	19.51	12.54	15.67	14.31	11.32	26.66	–
Fusion	0.08	8.15	2.64	4.51	4.60	80.02	0.00031%
Information entropy operator	11.39	18.07	12.30	20.22	16.25	21.77	–
Fusion	0.39	2.43	3.58	1.74	1.45	90.40	0.00069%
DOI of L operator	15.28	15.62	13.52	17.13	14.84	23.61	–
Fusion	0.15	5.00	3.29	3.48	2.73	85.35	0.00049%
Operator of grey correlation*	12.45	17.21	12.60	19.79	15.51	22.44	–
Fusion	0.33	3.35	3.57	2.17	1.98	88.59	0.00071%
Fuzzy operator	10.59	18.43	12.15	20.82	16.28	21.73	–
Fusion	0.44	2.04	3.62	1.47	1.25	91.18	0.00063%

invested, while Scheme 1 is the least selection to be invested. It is noted that the fusion method based on D-S evidence reasoning is more conducive to the concentration of the schemes than that of without using D-S evidence reasoning. Data fusion leads to a decrease of DOI of the index systems, and makes the distribution of basic probability functions more apparent. By comparing the schemes on the basis of same decision rules, the information of the system can be more sufficiently utilized, which increases the accuracy of the system in recognition of target schemes.

5 Conclusion

D-S evidence reasoning as a data fusion approach is able to widely process DOI information. This research attempts to apply this theory into the investment decision making and establishes a simple effective model. Large amount of case studies indicate the proposed approach intergrading D-S evidence reasoning in this research can acquire fairly satisfying results, and it specially performs well in the case that there are no big difference between schemes to be compared and calling for more precise and reliable decision results. However, duo the influence of calculation precision, distribution fusion structure is suggested to be adopted when more than ten indexes (evidences) are considered, to make the data of multiple evidences' reasoning unable to submerge some important information; or other method is used as: each of standardized indexes information are multiplied by η (The integers of $\eta > 1$), and then the data after fusing based on D-S evidence reasoning, are all divided by η.

References

1. Fan L (1992) Uncertainty of artificial intelligence. Beijing: China Meteorological Press (5):119–154
2. Buede DM, Paul G (1997) A target identification comparison of Bayesian and Dempster-Shafer multisensor fusion. IEEE Trans Syst, Man, Cybern—Part A: Syst Humans 27 (5):569–577, September
3. Mostaghimi M (1997) Bayes estimation of a decision using information theory. IEEE Trans Syst, Man, Cybern—Part A: Syst Humans 27(4):506–517, July
4. Xinsheng D (1993) Evidence theory and artificial intelligence in decision making. China Renmin University Press, Beijing
5. Murphy RR (1998) Dempster-Shafer theory for sensor fusion in autonomous mobile robots. IEEE Trans Robot Autom 14(2):197–206 April
6. Fixsen D, Mahler RPS (1997) The modified Dempster-Shafer approach to classification. IEEE Trans Syst, Man, Cybern—Part A: Syst Humans 27(1):96–104, January
7. Shengyi L, Xuezhong W, Dapeng F (1998) Multisensor information fusion theory and its application in intelligent manufacturing system. Changsha: National University of Defense Technology Press (11):119–195
8. Yanping W, Jie Y, Yuan L, Xiangfang S (2000) A method of indentifying defect targets based on information fusing technology. Beijing: J Image Graph 5(3):237–240
9. Hongzhi X, Zhizheng L, Ming Y (2000) An improved algorithm of multi-information fusion based on evidence theory. Robot 22(1):7–11
10. Yin C, Liangming L, Guozheng Y (2000) The problems concerning the application of D-S evidence reasoning in information fusion and their corrections. Syst Eng Electron 22(11):69–71
11. Quan S, Xiuqing Y, Weikang G (2000) A synthesis formula of a new challenge based evidence theory. Acta Electronica Sinca 28(8):117–119
12. Congfu X, Weidong G, Yunhe P (2000) An improved D-S algorithm. Comput Eng Appl (3):37–38
13. Yang X, Xizhi S (1999) A modification of evidence theory combination rule. J Shanghai Jiaotong Univ (Chin Edn) 33(3):354–360
14. Qi Z, Weikang G, Jilin L (1999) Vision information fusion based on Dempster Shafer evidence inference theory in ALV. Chin J Comput 22(2):193–198

15. Hongyan S, Bo Z, Kezhong H et al (1999) The identification of targets base on Dempster-Shafer fusion. J Tsinghua Univ (Sci Technol) 39(9):90–94
16. Zhongjian L (2000) An analysis of multi-source information fusion technology used in target recognition system. Syst Eng Electron 22(6):40–41
17. Mei W (1995) Research of theory and methods of comprehensive evaluation of investment projects and enterprises' financing risk and their applications. Doctoral dissertation from Hohai University, Nanjing, pp 62–78

Optimization of Brake Judder Based on Dynamic Model of Disc-Pads Spring Contact

Gongyu Pan, Yaqi Feng, Peng Liu, Qizhao Xu, and Lin Chen

Abstract In order to predict the brake judder more accurately, the mechanism of brake judder was studied. First, a dynamic model of braking system based on surface-to-surface contact between disc and pads was constructed. The correctness of the model was verified by comparing the results of bench test and simulation. Then, the key parameters of brake judder were found from the calculation expressions of brake pressure variation (BPV) and brake torque variation (BTV), and two optimization directions were proposed to improve the structure of brake caliper and brake pad backplate. Finally, a method to determine parameters of the improved disc brake was proposed, which combined the finite element analysis with theoretical calculation. The parameters of the improved disc brake were substituted into the Simulink model of the dynamic model, and the time-domain responses of BPV and BTV were obtained. The results show that the maximum BPV and the maximum BTV of the optimized braking system are reduced by 30.7% and 34.2%, respectively, thus testifying to the correctness of the optimization method of brake judder proposed in this paper. The research method proposed in this paper had a certain contribution to the study of brake judder of disc brake. It not only improved the quantization accuracy of brake judder, but also reduced the probability of brake judder greatly. Therefore, this study has a certain engineering significance.

Keywords Brake judder optimization · Surface spring contact · Parameter determination · Brake torque variation · Brake pressure variation

G. Pan · Y. Feng (✉) · P. Liu
Jiangsu University, Zhenjiang 212013, China
e-mail: fyq15751011272@163.com

G. Pan · Q. Xu · L. Chen
Jiangsu Hengli Brake Manufacturing Co., Ltd, Taizhou 214500, China

W. Wang et al. (eds.), *Green Transportation and Low Carbon Mobility Safety*,
Lecture Notes in Electrical Engineering 944,
https://doi.org/10.1007/978-981-19-5615-7_26

1 Introduction

When people pay more and more attention to drive comfort, brake judder has become an important problem affecting the comfort of drivers and passengers [1]. Brake judder is a kind of low-frequency vibration, which can be transmitted to the steering wheel, brake pedal, seat rail and body floor through the steering system, caliper system and brake hydraulic pipeline [2, 3]. The external vibration caused by it can greatly affect the driver's comfort, and at the same time, driving fatigue and misoperation caused by it have affected the driver's driving safety [4]. In addition, brake judder also causes great harm to the service life of the braking system, which increases the maintenance cost of vehicles. Therefore, it is of great significance to take measures to reduce the brake judder of automobiles.

At present, many scholars have tried to reduce brake judder by improving the structure of brake components. Sim et al. [5] obtained the structural design parameters of the brake disc sensitive to brake judder through the sensitivity analysis method, and proposed an optimization method for reducing brake judder. Bryant et al. [6] optimized the shape of the brake disc ventilation groove to reduce brake judder. According to the response surface analysis method, Jung et al. [7] optimized the design parameters of the ventilation disc to reduce the maximum temperature and thermal deformation of disc, thus reducing the occurrence of thermal judder.

However, there are few studies on brake judder from the perspective of the dynamic model of the brake system. Leslie [8] proposed a brake caliper dynamic model which could be used to predict the level of BTV at a given DTV input. In order to reduce the level of BTV, the sensitivity of brake torque variation to the stiffness of brake components was studied. Zhang et al. [9] established a single point contact brake dynamic model to predict brake pressure variation, brake torque variation and caliper vibration acceleration in Matlab/Simulink software, and verified the validity of the model by simulation and test. Most of the existing brake dynamic models are in the form of single-point contact or multi-point contact, which have low quantitative accuracy for brake judder and cannot be better helpful for brake judder analysis and control.

In this paper, a disc-pads surface contact dynamic model of disc brake which is more in line with the actual contact situation is proposed to improve the accuracy of predicting brake judder. On this basis, a method to reduce brake judder by simultaneously improving caliper and brake pads is presented. The optimization result is validated by performing the judder analysis of the original and optimized disc brake.

2 Disc-Pads Surface Contact Dynamic Model of Disc Brake

2.1 Premises of Model Construction

In this paper, the following assumptions are made when constructing the disc-pads surface contact model of disc brake.

(1) The braking system can be simplified as a multi-degree-of-freedom particle-spring system.
(2) In this model, the surface distributed spring contact mode is used between disc and pad, and other parts are point-to-point contact mode.
(3) The effect of brake pressure variation on friction characteristics between brake friction pairs is ignored, but the influence of the relative velocity between brake disc and brake block on the friction characteristics is considered.
(4) The influence of thermal factor is not considered.

Based on the above assumptions, this paper takes the initial disc thickness variation (DTV) and surface run-out (SRO) of brake disc with brake judder phenomenon as input, BPV and BTV as output for simulation analysis to study the mechanism of brake judder and put forward the optimization scheme for brake judder.

2.2 Dynamic Model of Brake

The disc-pads surface contact dynamic model of disc brake proposed in this paper is shown in Fig. 1. The basic method taken in model development was to divide the braking system into constitutive mass/stiffness elements [8]. This model includes brake disc, outer brake block, inner brake block, calipers, piston and caliper bracket, which is immovable. The components were connected by springs and dampers. In this model, the single-point contact mode in the existing model was adopted between inner brake block and piston, between piston and caliper, and between outer brake block and caliper, while the surface-to-surface contact mode proposed in this paper was adopted between brake disc and brake block. This simplified brake dynamic model for analysis of brake judder mechanism has been recognized in recent years [8–11].

When braking, the brake pedal force is transmitted to the brake wheel cylinder through the hydraulic system, and the piston under hydraulic pressure presses the inner brake pad against the brake disc surface. Due to the surface run-out on one side of the brake disc, the brake pad is forced to vibrate in the direction perpendicular to the surface of the brake disc. When the brake pad is pressed against the surface of the brake disc, the inner brake pad is subjected to friction along the rotating direction of the brake disc due to the rotation of the brake disc, so that friction is generated between brake pad and piston surface. In the tangential direction, the movement of the brake pad is limited by the caliper holder. The outer brake pad is in direct contact

Fig. 1 Schematic diagram of disc-pads surface contact dynamic model of disc brake

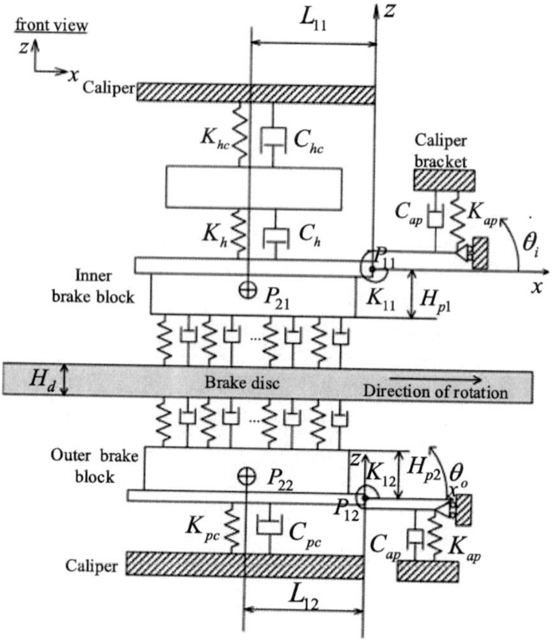

with the caliper, and the force is similar to that of the inner brake pad. According to the above, six degrees of freedom were identified in this model, including the axial freedom of inner brake pad and its rotational freedom around center of mass, the axial freedom of outer brake pad and its rotational freedom around the center of mass, and the axial freedom of piston and caliper.

According to the calculation equations of kinetic energy and bending potential energy, the expressions of kinetic energy and potential energy of brake disc, brake block and piston can be deduced. The kinetic energy of brake disc is set as T_d, and the potential energy as V_d.

$$T_d = \frac{1}{2}\rho H_d\left(\int_{r_1}^{r_2} r(w_{dr})^2 dr \int_0^{2\pi} (w_{d\varphi}(\varphi))^2 d\varphi\right)(\dot{q}_d(t))^2 \tag{1}$$

where r_1 and r_2 are the inner radius and outer radius of the brake disc, respectively; ρ is the mass density; H_d and φ are the thickness and rotation angle of brake disc, respectively; w_d is the axial displacement of brake disc surface;

$$V_d = \frac{1}{2}D\left\{\int_{r_1}^{r_2} \frac{1}{r}\left[(\dot{w}_{dr}(r))^2 + 2vr\dot{w}_{dr}(r)\ddot{w}_{dr}(r)\right.\right.$$

$$
\begin{aligned}
& + r^2(\ddot{w}_{dr}(r))^2\Big]dr \times \int\limits_0^{2\pi} (w_{d\varphi}(\varphi))^2 d\varphi + 2(1-v) \\
& \times \int\limits_{r_1}^{r_2} \frac{1}{r^3}(w_{dr}(r) - r\dot{w}_{dr}(r))^2 dr \times \int\limits_0^{2\pi} (\dot{w}_{d\varphi}(\varphi))^2 d\varphi \\
& + \int\limits_{r_1}^{r_2} \frac{1}{r^3}(w_{dr}(r))^2 dr \int\limits_0^{2\pi} (\ddot{w}_{d\varphi}(\varphi))^2 d\varphi \\
& + \int\limits_{r_1}^{r_2} \frac{2}{r^2} w_{dr}(r)(\dot{w}_{dr}(r) + vr\ddot{w}_{dr}(r))dr \\
& \times \int\limits_0^{2\pi} w_{d\varphi}(\varphi)\ddot{w}_{d\varphi}(\varphi)d\varphi \Bigg\} (q_d(t))^2
\end{aligned} \tag{2}
$$

In the above, D is the bending stiffness of the brake disc; ν is Poisson's ratio.

The kinetic energy and potential energy of inner brake block are expressed by Eqs. (3) and (4), respectively.

$$
T_{p1} = \frac{1}{2} m_{p1} \dot{w}_{p1}^2 + \frac{1}{2}(m_{p1} L_{11}^2 + I_{p1})\dot{\theta}_i^2 \tag{3}
$$

where m_{p1} is the mass of inner brake block; I_{p1} is the inertia moment; w_{p1} is the axial displacement of inner brake block; θ_i is the rotation angle of inner brake block around point P_{11}; L_{11} is the distance between the center of mass of inner brake block and the connecting point of brake block and caliper bracket.

$$
\begin{aligned}
V_{p1} = & \frac{1}{2}(K_{ap} + K_h) w_{p1}^2 - K_h L_{11} w_{p1} \theta_i \\
& + \frac{1}{2}(K_{11} + K_h L_{11}^2)\theta_i^2
\end{aligned} \tag{4}
$$

In Eq. (4), K_{ap} is stiffness of abutment shim between pad and bracket; K_h is the interface stiffness between inner brake block and piston; K_{11} is the rotational stiffness of inner brake block around point P_{11}.

Similarly, the kinetic energy and potential energy of outer brake block are expressed by Eqs. (5) and (6), respectively.

$$
T_{p2} = \frac{1}{2} m_{p2} \dot{w}_{p2}^2 + \frac{1}{2}(m_{p2} L_{12}^2 + I_{p2})\dot{\theta}_o^2 \tag{5}
$$

where m_{p2} is the mass of outer brake block; I_{p2} is the inertia moment; w_{p2} is the axial displacement of outer brake block; θ_o is the rotation angle of outer brake block around point P_{12}; L_{12} is the distance between the center of mass of outer brake block and the connecting point of brake block and caliper bracket.

$$V_{p2} = \frac{1}{2}(K_{ap} + K_{pc})w_{p2}^2 - (K_{pc}L_{12})w_{p2}\theta_o + \frac{1}{2}(K_{12} + K_{pc}L_{12}^2)\theta_o^2 \tag{6}$$

where K_{12} is rotational stiffness of outer brake block around point P_{12}; K_{pc} is contact stiffness between outer brake block and caliper.

The kinetic and potential energy of piston can be expressed as follows.

$$T_h = \frac{1}{2}m_h\dot{w}_h^2 \tag{7}$$

In the above, m_h is the mass of piston; w_h is the axial displacement of piston.

$$V_h = \frac{1}{2}K_{hc}w_h^2 \tag{8}$$

where K_{hc} is equivalent hydraulic stiffness of wheel cylinder.

The Lagrangian function is:

$$L = T_d + T_{p1} + T_{p2} + T_h - V_d - V_{p1} - V_{p2} - V_h \tag{9}$$

By substituting each value into the above equation, the following equation of motion can be obtained.

$$M\ddot{w} + C\dot{w} + Kw = 0 \tag{10}$$

In the above, M, C and K are the mass, stiffness and damping matrices, respectively; w is the displacement matrix.

According to Fig. 1, brake pressure (P_B) and brake torque (T_B) can be expressed by Eqs. (11) and (12), respectively.

$$P_B = P_0S + (w_h - w_c)K_{hc} + (\dot{w}_h - \dot{w}_c)C_{hc} \tag{11}$$

where P_0 is initial braking pressure; S is the contact area of piston; w_c is the axial displacement of caliper; C_{hc} is equivalent hydraulic damping of wheel cylinder.

$$T_B = \mu_p \cdot R_{eff} \cdot [2P_0S + (w_{ri} - w_{p1})K_p + (\dot{w}_{ri} - \dot{w}_{p1})C_p + (w_{p2} - w_{ro})K_P + (\dot{w}_{p2} - \dot{w}_{ro})C_p] \tag{12}$$

where μ_p is the coefficient of friction between pads and disc; R_{eff} is the brake effective radius; w_{ri} and w_{ro} are the displacements of inner and outer surface of disc, respectively; K_p and C_p are the dynamic compressive stiffness and damping of brake block, respectively.

Thus, BTV and BPV can be respectively expressed as:

$$P_{\mathrm{BPV}} = (w_h - w_c)K_{hc} + (\dot{w}_h - \dot{w}_c)C_{hc} \quad (13)$$

$$\begin{aligned} T_{\mathrm{BTV}} = \mu_p \cdot R_{eff} \cdot [&(w_{ri} - w_{p1})K_p \\ &+ (\dot{w}_{ri} - \dot{w}_{p1})C_p + (w_{p2} - w_{ro})K_P \\ &+ (\dot{w}_{p2} - \dot{w}_{ro})C_p] \end{aligned} \quad (14)$$

The values of all parameters involved in Fig. 1 and Eqs. (1)–(14) are listed in Appendix A.

2.3 *Input of Model*

Brake dynamo test was done using LINK model 3900. The disc brake with brake judder problem was employed in the test. The arrangement of measuring points in this test is shown in Fig. 2.

The brake disc thickness variation and surface runout measured by displacement sensors are superimposed as the system input. The initial DTV of the brake disc can be obtained by taking the thickness variation of disc during the uniform rotation of rotor at the initial braking speed, as shown in Fig. 3.

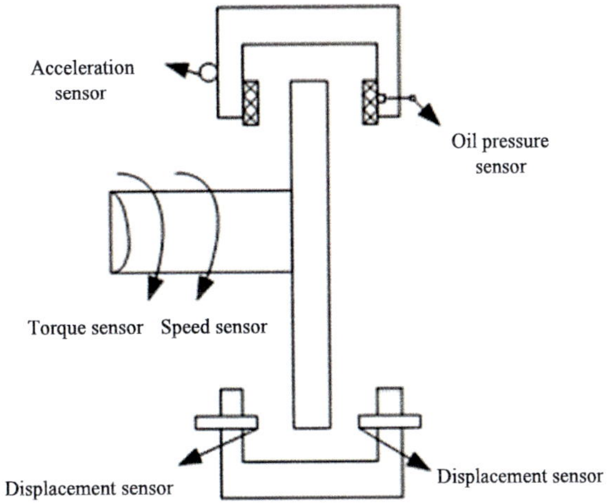

Fig. 2 Layout diagram bench test measuring points

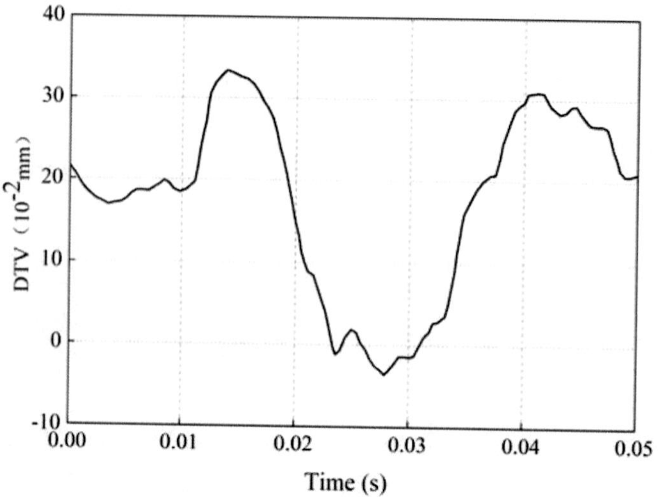

Fig. 3 The initial DTV of brake disc

The initial surface run-out of the brake disc can be measured by the displacement sensor during the uniform rotation of the rotor at the initial braking speed, as shown in Fig. 4.

It can be seen from Figs. 3 and 4 that both DTV and SRO curves have second-order characteristics. The values of DTV and SRO changing with time were converted to those changing with angle, and then the converted DTV and SRO were superimposed,

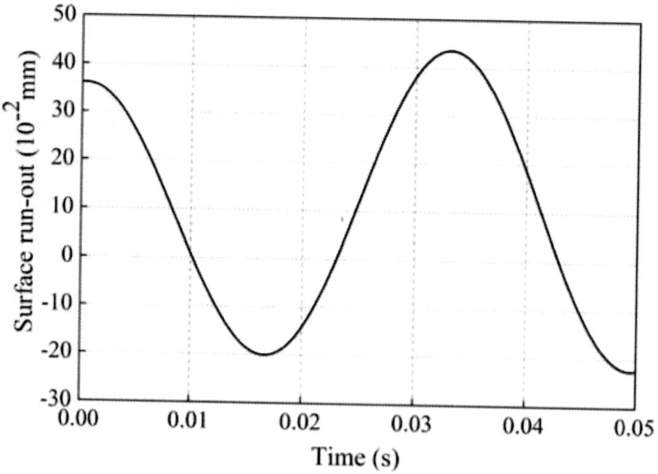

Fig. 4 The initial SRO of brake disc

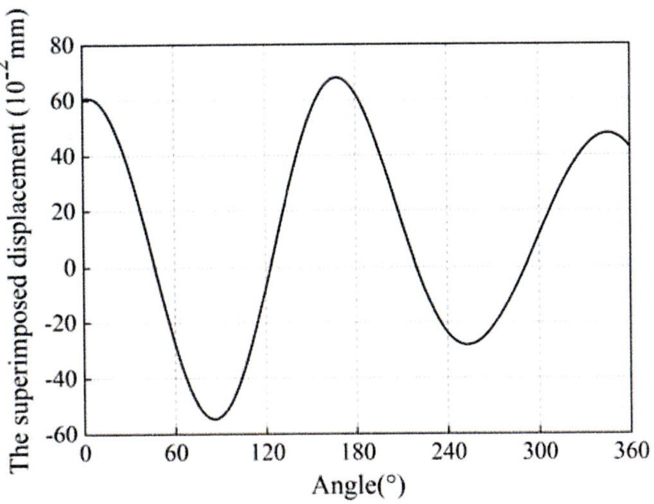

Fig. 5 Displacement curve after superposition of SRO and DTV

and the superimposed displacement curve was shown in Fig. 5. It also has the second-order characteristics and will be used as the system input of the dynamic system simulation.

2.4 Verification of Model

Vehicle brake judder is typically quantified in terms of BPV and BTV [12, 13]. So, in order to verify the correctness of disk-pads spring contact dynamics model in reflecting the phenomenon of brake judder, it is necessary to compare the simulation results of BPV and BTV with the test results.

In order to ensure the comparability between simulation results and test results, the initial conditions of simulation analysis and test should be consistent, including the initial braking pressure, initial braking speed and the input of geometric irregularities in the same brake disc surface. More specific information concerning the bench test of the brake judder can be found in the references [14].

The actual BPV and BTV of the braking system were obtained by bench test. Test process started from the initial braking speed and ended when the speed dropped to 0 km/h. The basic conditions of the test are shown in Table 1.

According to Fig. 1, Matlab/Simulink software was used to develop the Simulink model of the brake. The initial DTV, SRO and system parameters were input into Simulink model, and the BPV and BTV were obtained by simulation. Refer to Appendix A for system parameters of this model. Due to space limitation, the analysis of simulation results and test results of a braking condition (initial brake pressure

Table 1 Basic condition of bench test

Category	Value	Category	Value
Moment of inertia	83.9 kg·m^2	Brake effective radius	132.5 mm
Dynamic radius of tire	358 mm	Deceleration	3.92 km/h
Initial braking speed	120 km/h	Environment temperature	25 °C

is 2.5 MPa) is presented, and similar conclusions can be obtained under other brake pressure conditions.

It can be seen from Figs. 6 and 7 that although the amplitudes and frequencies of the simulation results are inevitably different from the test results to some extent, the overall trend of the simulation results agrees well with the test results. Obviously, the test results are larger than the simulation results, which is preliminarily thought to be due to some unrealistic assumptions in the model development and ignorance of the influence of thermal factor. In the early stage of braking, the reason for the gradual increase of the brake torque variation was mainly that the complete contact between brake block and brake disc requires a certain response time. In the later braking period, thermal expansion and thermal deformation of disc occurred due to friction heat generation, resulting in transient disc thickness variation on the brake disc surface, which aggravated the geometric irregularities of brake disc, and further increased the fluctuation of brake pressure and brake torque.

The maximum BPV of the simulation and dynamo test are 8.8 MPa and 10.4 MPa, and the simulation result is very close to the test data. The maximum BTV of the test is 54.2 N · m. The maximum BTV of the simulation is 38 N · m, which is also close to the test result. Therefore, the BTV and BPV results of the simulation follow the

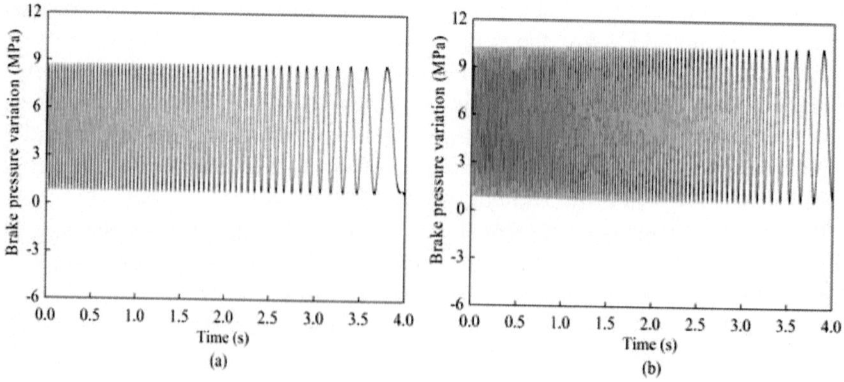

Fig. 6 Time-domain responses of BPV: **a** Simulation result; **b** Bench test result

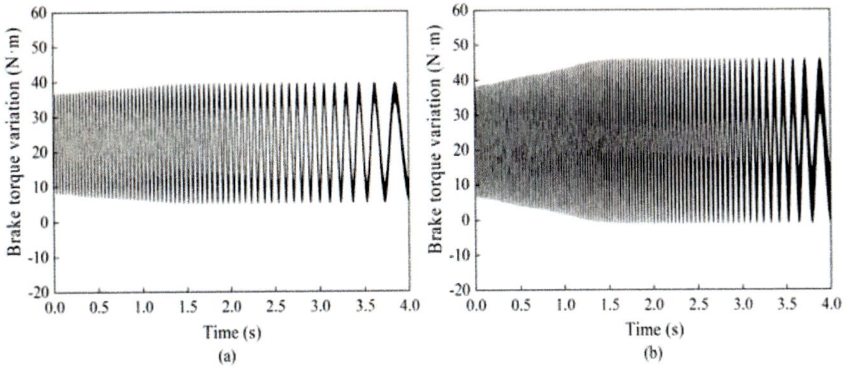

Fig. 7 Time-domain responses of BTV: **a** Simulation result; **b** Bench test result

test data successfully, which indicates that the model is an effective method to study the mechanism of brake judder.

3 Optimization of Brake Judder

3.1 The Key Influencing Parameters of Brake Judder

Because BPV and BTV are the excitation source of brake judder, the method to reduce brake judder is to decrease the magnitude of BPV and BTV. It can be seen from Eqs. (13) and (14) that the main factors affecting BPV and BTV are as follows:

(1) The geometric irregularities of the inner and outer brake disc surface, that is, DTV and SRO.
(2) The hydraulic stiffness (K_{hc}) and damping (C_{hc}) of wheel cylinder.
(3) The dynamic compressive stiffness $\left(K_p\right)$ and damping $\left(C_p\right)$ of brake block.
(4) The axial displacement of piston (w_h) and caliper (w_c), and the friction characteristic between brake pad and disc.

At present, there are few studies on reducing brake judder by changing the hydraulic stiffness and damping of wheel cylinder and by changing the dynamic compressive stiffness and damping of brake block. Therefore, this paper carries on the optimization of brake judder from these two aspects.

The evaluation equations of the optimization effect of braking judder are as follows.

$$O_1 = \frac{P_{BTV1\max} - P_{BTV0\max}}{P_{BTV0\max}} \times 100\% \tag{15}$$

$$O_2 = \frac{P_{BPV1\max} - P_{BPV0\max}}{P_{BPV0\max}} \times 100\% \tag{16}$$

where $P_{BTV1\max}$ represents the maximum BTV of improved disc brake; $P_{BTV0\max}$ represents the maximum BTV of original disc brake; $P_{BPV1\max}$ represents the maximum BPV of improved disc brake; $P_{BPV0\max}$ represents the maximum BPV of original disc brake.

Since there is no clear range of the criteria for brake judder optimization, the criterion adopted in this paper is that, when satisfying the requirements of $O_1 \geq 20\%$ and $O_2 \geq 20\%$, the optimization scheme is considered to be effective.

3.2 Improved Design of Caliper Structure

According to Eq. (13), the smaller hydraulic stiffness and damping is, the smaller BPV will be. On the premise of ensuring the safety and reliability of disc brake, in order to reduce the hydraulic stiffness and damping of wheel cylinder, this paper proposes to change the diameter of piston cylinder (the diameter of piston changes accordingly).

According to the improvement scheme shown in Fig. 8, the three-dimensional (3D) model of caliper body was constructed by using CATIA software, as shown in Fig. 9. It can be seen that the difference between original caliper and improved caliper is that the radius of piston cylinder was changed, that is, $r_0 > r_1$ or $r_0 < r_1$.

r_o is the radius of original piston cylinder with a value of 26 mm. When the inner radius of the piston cylinder meets $r_0 < r_1$, the contact area between brake fluid and piston increases. Therefore, according to Eq. (17), the hydraulic stiffness and

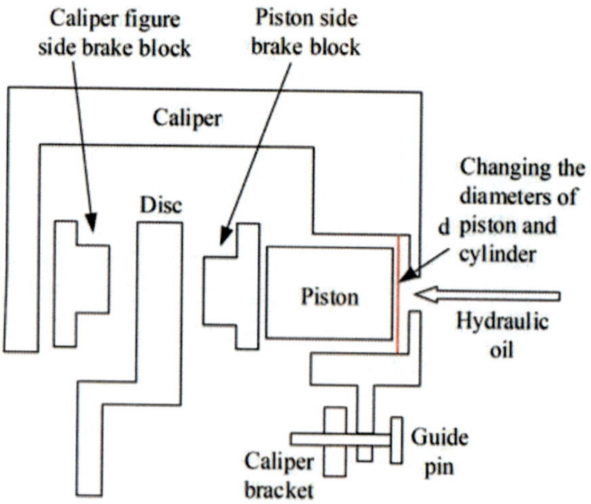

Fig. 8 Schematic diagram of improvement method of caliper structure

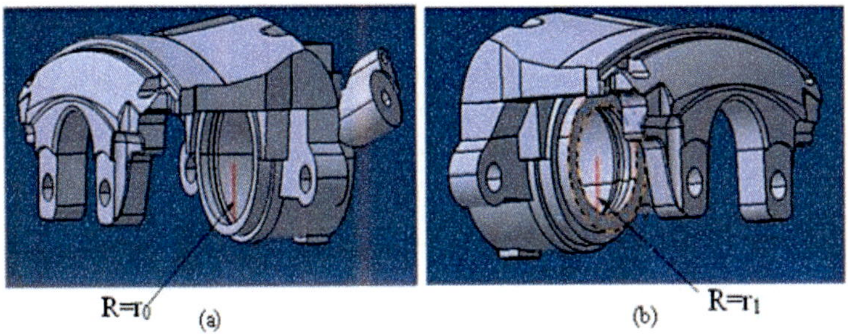

Fig. 9 Three-dimensional model of: **a** original caliper body; **b** improved caliper body

damping of wheel cylinder decreases with the decrease of brake hydraulic pressure under the condition that the same magnitude of braking force is applied. The radius (r_1) of optimized piston cylinder is 30 mm.

$$P = \frac{F}{S} \tag{17}$$

where P is hydraulic pressure, which can be measured by oil pressure sensor; F is the force exerted by piston on brake pad; S is the contact area between piston and brake fluid.

The magnitude of damping is much less than that of hydraulic stiffness, so it can be ignored. The hydraulic stiffness of wheel cylinder can be determined by the following g equations.

$$K'_{hc} = \frac{F}{h} \tag{18}$$

In the above, h is the traveling displacement of piston. By substituting Eq. (17) into Eq. (18) can be obtained:

$$K'_{hc} = \frac{P \cdot S}{h} \tag{19}$$

The traveling displacement of piston, the surface area of piston and brake pressure are known quantities. By substituting each parameter value into above equation, the hydraulic stiffness value of optimized caliper body can be obtained. The hydraulic stiffness value after optimization is 6.2×10^5 N/m.

3.3 Improved Design of Brake Pad Backplate

According to Eq. (14), the smaller dynamic compressive stiffness and damping of brake block is, the smaller BTV will be. Since the magnitude of damping is much less than that of compressive stiffness, it can be ignored. Because changing the mass of the object can change the stiffness of the object, this paper, on the premise of ensuring the safety and reliability of the disc brake, proposes to change the compressive stiffness of brake block by slotting brake pad backplate.

As shown in Fig. 10, this paper proposed three design schemes of groove layouts on brake pad backplate. The effective area of brake pad backplate is 4825 mm^2. The design size of a single transverse groove on backplate surface is 80×4 mm, and the depth of groove is 1 mm. One transverse groove, two transverse grooves and three transverse grooves were designed respectively on the backplate surface, and the percentages of slotted area in total area of the backplate were 6.6% , 13.3% and 19.9% , respectively. Excessively large groove size on the steel back surface will lead to a sharp decrease in the stiffness of brake block and affect the manufacturing process of brake block. Therefore, the percentage of slotted area in total area of the backplate shall not be more than 25% [15]. In order to reduce BTV as much as possible, the mass of backplate should be decreased as much as possible within a reasonable range, so the scheme of making three transverse grooves in the backplate was chosen. The 3D model of improved brake pad backplate is shown in Fig. 11. The mass of improved backplate is reduced by 11.1% compared to the original one, and the compressive stiffness of brake block will also be reduced accordingly.

The dynamic compressive stiffness of brake block can be determined by combining the results of finite element analysis and theoretical calculation. The

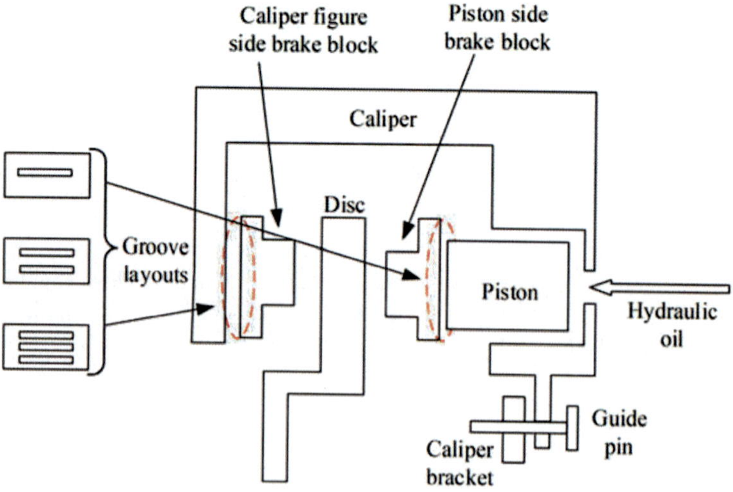

Fig. 10 Schematic diagram of improvement method of brake pad backplate

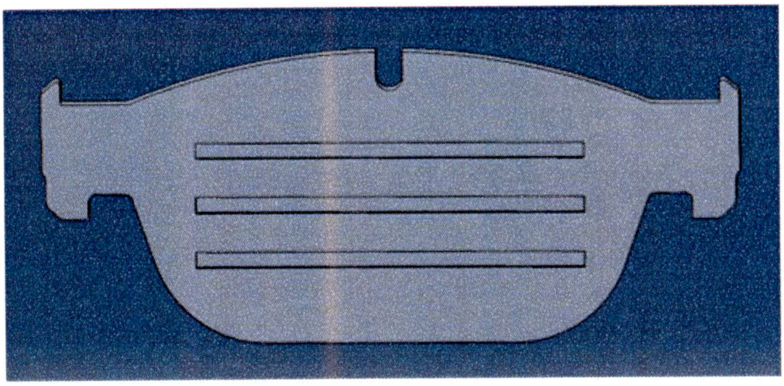

Fig. 11 Schematic diagram of improved brake pad backplate

finite element model of disc brake was established based on a simplified 3D model of an actual disc brake in ANSYS 15.0 software package. In this model, the material properties for each part of disc brake are shown in Table 2.

In the finite element analysis, the mesh type of brake disc and pad is hexahedral mesh, so each grid cell on its surface has 4 nodes, as shown in Fig. 12. It is assumed that the nodes on grid cells of the contact surface between brake disc and pad correspond one to one, and the contact stiffness of each node can be calculated according to Eq. (20).

$$K_{n1} = \frac{f}{h_1} \tag{20}$$

where f is the stress at the node; h_1 is the relative displacement at the node.

Since the four nodes of each grid cell on the contact surface are in parallel with each other, the contact stiffness of each grid cell can be calculated by the following equation.

$$\frac{1}{K_n} = \frac{1}{\sum_{i=1}^{4} K_{ni}} \tag{21}$$

Table 2 Material properties for each part of disc brake

Part	Density	Young's modulus/(MPa)	Poisson's ratio
Disc	7000	110,000	0.25
Brake caliper	7000	110,000	0.25
Piston	7200	150,000	0.28
Brake pad backplate	7800	200,000	0.28
Brake pad	2610	9600	0.30

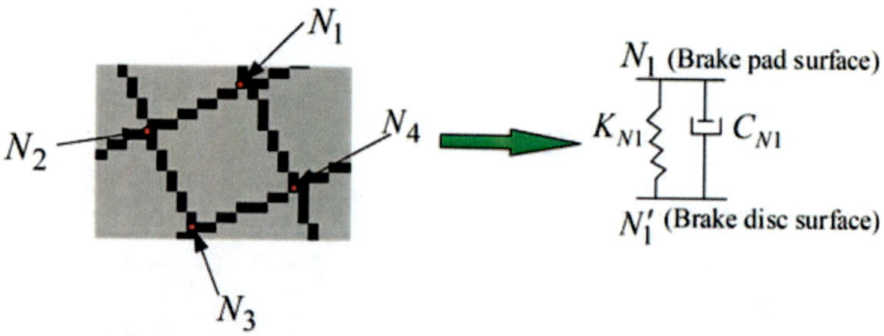

Fig. 12 Diagram of grid nodes between brake disc and pad

According to the surface spring contact dynamic model constructed in this paper, K_{ni} can be assumed to be equal, thus Eq. (21) can be simplified as:

$$K_n = 4K_{ni} \tag{22}$$

According to Eqs. (20) and (22), the calculation equation of contact stiffness between brake disc and pad can be deduced.

$$K_p' = \frac{4\mathrm{N}_1 \cdot f}{h_1} \tag{23}$$

where N_1 is the number of grid cells on the contact area between brake disc and pad.

f, h_1 and N_1 were obtained by finite element analysis. According to the above equations, the dynamic compressive stiffness of brake block can be determined. The compressive stiffness value after optimization is 3.42×10^8 N/m, which is 47.4% less than before optimization.

4 Verification of Improved Design

It is known that changing the mass of structure can change the stiffness of structure, so the stiffness values in the whole braking system change correspondingly, and the parameter values can be determined by the same method mentioned above. The parameters of the improved disc brake were all changed within a reasonable range to ensure the stability and safety of braking efficiency.

The system parameters of improved disc brake are shown in Table 3. It can be seen that the contact stiffness and damping of each part of disc brake have been reduced to a certain extent. With the decrease of the hydraulic stiffness and damping of wheel cylinder and the dynamic compressive stiffness and damping of brake block, BPV and BTV will be reduced accordingly. In order to further verify the reliability of the optimization scheme, the simulation and test of brake judder was performed.

Table 3 System parameters of optimized disc brake

Parameter	Value	Parameter	Value	Parameter	Value
m'_c	1.38 kg	K'_{pc}	1.78×10^8 N/m	C'_{ap}	5.7 N · s/m
m'_h	0.10 kg	K'_{ap}	5×10^6 N/m	C'_{ac}	3.56 N · s/m
m'_p	0.40 kg	K'_{ac}	2×10^6 N/m	μ'_h	0.075
R'_{eff}	0.145 m	C'_{hc}	0.18 N · s/m	μ'_p	0.36
K'_{hc}	6.2×10^5 N/m	C'_h	0.29 N · s/m	μ'_{ap}	0.072
K'_h	2.7×10^7 N/m	C'_p	20 N · s/m	μ'_{pc}	0.068
K'_p	3.42×10^8 N/m	C'_{pc}	0.19 N · s/m		

4.1 *Simulation Analysis of the Improved Barking System*

In order to make the results before and after optimization comparable, it is necessary to ensure that the input conditions and system parameters are consistent. The time-domain responses of BPV and BTV during braking can be obtained by Matlab/Simulink.

Figure 13 shows the BPV simulation result of the improved disc brake. Figure 14 shows the BTV simulation result of the improved disc brake. As can be seen from Figs. 13 and 14, the maximum BPV and BTV of optimized disc brake are significantly reduced compared with those before optimization. For a clearer comparison of optimization effect, the maximum BPV and the maximum BTV are listed in Table 4. The maximum BPV of optimized disc brake was reduced by 30.7% compared with that before optimization, meeting the standard of more than 20%. The maximum BTV of original disc brake and improved disc brake are 38 N · m and 25 N · m, respectively. The maximum BTV of optimized disc brake was 34.2% less than that before optimization, which also satisfied the specified condition. The simulation results show that the improved method proposed in this paper can make the maximum BPV and BTV reach the optimization standard. Thus, the possibility of brake judder occurring can be reduced, which indicates that the optimization methods of brake judder proposed in this paper are reasonable and correct.

5 Conclusion

In this study, the optimization research of a disc brake was carried out based on the established disc-pads surface contact dynamic model of disc brake. The results are as follows.

(1) The disc-pads spring contact dynamic model proposed in this paper can accurately reflect the characteristics of brake judder, and the simulation results of

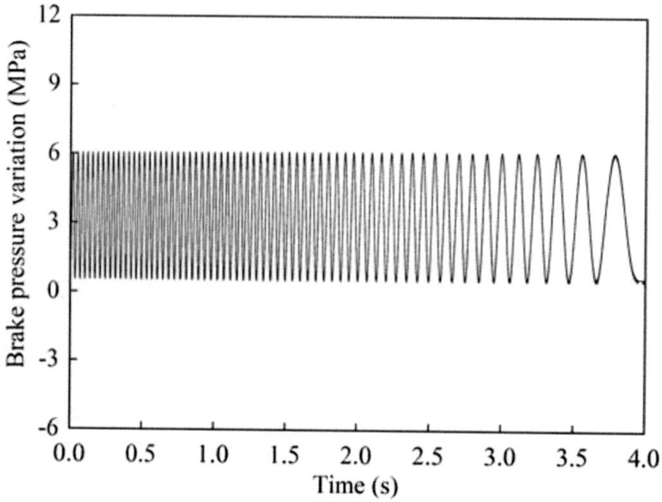

Fig. 13 Simulation result of BPV according to time after optimization

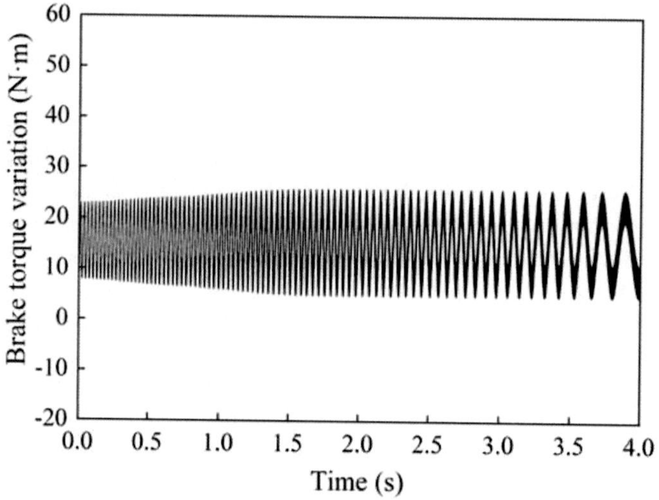

Fig. 14 Simulation result of BTV according to time after optimization

Table 4 Comparison of simulation results before and after optimization

Category	Before optimization	After optimization	Comparision (%)
BPV_{max}	8.8	6.1	−30.7
BTV_{max}	38	25	−34.2

brake pressure and brake torque are consistent with the test results, which indicates that the dynamic model is correct and effective.
(2) The theoretical derivation and simulation results show that the equivalent hydraulic stiffness of wheel cylinder and compressive stiffness of brake block have a great effect on the brake judder.
(3) The results show that the optimal scheme to reduce the maximum BTV and BPV by changing the inner diameter of piston cylinder and slotting brake pad backplate are correct and effective.
(4) The method of finite element analysis combined with theoretical calculation proposed in this paper can successfully determine the parameters of components of the optimized braking system and greatly shorten the research period.

Optimization of brake caliper and brake pad backplate is proposed to reduce brake pressure and brake torque variation in this paper, which finally successfully reduces the incidence of brake judder. By building a dynamic model of braking system, braking judder can be quantified in the form of braking pressure variation and braking torque variation, which can provide some references for reducing braking judder.

Acknowledgements This work was supported by Jiangsu Hengli Brake Manufacturing Company on the projects of "Study on Key Technology of NVH Characteristics of Braking system" (No. 201908), "High Level Innovation and Entrepreneurship Talent" (No. 2019015) of Jingjiang City and Jiangxi Province Key Laboratory of Vehicle Noise and Vibration (No. JXNVHKB-KFKT-201802).

Appendix A

Table A.1 The parameter values involved in the simulation process

Category	Symbol	Value	Category	Symbol	Value
Mass of caliper	m_c	1.5 kg	Stiffness of sliding joint between caliper and bracket	K_{ac}	3.3×10^8 N/m
Mass of piston	m_h	0.12 kg	Stiffness of abutment shim between pad and caliper bracket	K_{ap}	4×10^5 N/m
Mass of brake block	$m_{p1} = m_{p2}$	0.45 kg	Contact stiffness between brake block and caliper	K_{pc}	2×10^8 N/m

(continued)

(continued)

Category	Symbol	Value	Category	Symbol	Value
Inner radius of brake disc	r_1	0.085 m	Equivalent hydraulic stiffness	K_{hc}	7.8×10^6 N/m
Effective friction radius	R_{eff}	0.133 m	Interface stiffness between inner brake block and piston	K_h	3.5×10^8 N/m
Thickness of disc	H_d	0.024 m	Compressive stiffness of brake block	K_p	6.5×10^8 N/m
Outer radius of brake disc	r_2	0.17 m	The rotational stiffness of brake blocks around the fixed points	$K_{11} = K_{12}$	1.53×10^4 N · m/rad
Density of disc	ρ	7.80×10^3 kg/m^3	Damping of abutment shim between pad and caliper bracket	C_{ap}	6.0 N · s/m
Poisson's ratio	υ	0.3	Damping of sliding joint between caliper and bracket	C_{ac}	4.0 N · s/m
Equivalent hydraulic damping of wheel cylinder	C_{hc}	0.2 N · s/m	Damping between piston and brake block	C_h	0.3 N · s/m
Inertia moment of brake blocks	$I_{p1} = I_{p2}$	2.26×10^{-4} kg · m^2	Compressive damping of brake block	C_p	22 N · s/m
Surface area of piston	S	28.27 cm^2	Damping between brake block and caliper	C_{pc}	0.2 N · s/m
Coefficient of friction between pads and rotor	μ_p	0.4	Coefficient of friction between piston and brake block	μ_h	0.09

(continued)

(continued)

Category	Symbol	Value	Category	Symbol	Value
Distance between the center of mass of the brake block and its connection point with the caliper bracket	$L_{21} = L_{22}$	0.06 m	Coefficient of friction between caliper bracket and brake block	μ_{ap}	0.08
Coefficient of friction between brake block and caliper	μ_{pc}	0.08			

References

1. Kim J, Lee M, Kim B, Cho C (2013) A prediction of the relation between the disc brake temperature and the hot judder critical speed. Trans Korean Soc Automot Eng 21(1):61–67
2. de Vries A, Wagner M (1992) The brake judder phenomenon. SAE Technical Paper
3. Lee CF, Savitski D, Manzie C, Ivanov V (2015) Active brake judder compensation using an electro-hydraulic brake system. SAE Int J Commer 8(1)
4. Jacobsson H (2003) Aspects of disc brake judder. Proc Inst Mech Eng, Part D: J Automob Eng 217
5. Sim KS, Lee JH, Park TW, Cho MH (2013) Vibration path analysis and optimal design of the caliper for brake judder reduction. Int J Automot Technol 14(4):587–594
6. Bryant D, Fieldhouse JD, Talbot CJ (2011) Brake judder-an investigation of the thermo-elastic and thermo-plastic effects during braking. Int J Vehicle Struct Syst 3(1):58–73
7. Jung SP, Kim YG, Park TW (2012) A study on thermal characteristic analysis and shape optimization of a ventilated disc. Int J Precis Eng Manuf 13(1):57–63
8. Leslie AC (2004) Mathematical model of brake caliper to determine brake torque variation associated with disc thickness variation (DTV) input. SAE Technical Paper
9. Zhang L, Quan X (2009) Brake judder analysis based on single-point contact model. In: Proceedings of the 2009 annual meeting of China society of automotive engineering
10. Jacobsson H (2003) Disc brake judder considering instantaneous disc thickness and spatial friction variation. Proc Inst Mech Engs, Part D. J Automob Eng 217(5):325–342
11. Sen OT, Dreyer JT, Singh R (2013) Low frequency dynamics of a translating friction element in the presence of frictional guides, as motivated by a brake vibration problem. J Sound Vib 332:5766–5788
12. Sen OT, Dreyer JT, Singh R (2012) Order domain analysis of speed-dependent friction-induced torque in a brake experiment. J Sound Vib 331:5040–5053
13. Sen OT, Dreyer JT, Singh R (2013) Envelope and order domain analyses of a nonlinear torsional system decelerating under multiple order frictional torque. Mech Syst Signal Process 35:324–344

14. Pan G, Liu P, Xu Q, Chen L (2020) Brake judder analysis based on dynamic model of disc-pad spring contact. J Chongqing Univ Technol (Natural Science)
15. Massi F, Berthier Y, Baillet L (2008) Contact surface topography and system dynamics of brake squeal. Wear 265(11):1784–1792

Optimal Study of Bus Priority Signal Control Based on Service Reliability

Mengqi Wang, Rui Li, and Sulan Meng

Abstract With continuous development in society and economy, a series of problems like traffic congestion are also emerging. In solutions of alleviating these traffic problems, bus priority control is proved to be effective. However, the implementation of bus priority measures needs to meet certain conditions, for bus passengers, time is what they most concern. Therefore, to improve the time effect of passengers, bus priority control method is investigated, then real-life traffic data at the intersection of Huju Road-Beijing West Road in Nanjing City is utilized to quantitatively analyze the priority control method. The results show that, under certain conditions, the method can bring time effects to passengers both at the intersection and downstream stops, improving the service reliability of buses.

Keywords Bus priority · Signal control · Time effect · Service reliability

1 Introduction

With continuous development in traffic, a series of accompanying problem like poor road network and traffic congestion restrict its development in turn, also seriously hindered the development of urban economy. Due to limited road resources, developing public transport is considered as an efficient solution to alleviate these traffic problems. Previous research has suggested that, compared with traditional signal control strategies, traffic signal priority (TSP) control is more competitive in service level of buses, encouraging more passengers to travel by bus.

Research on TSP system abroad begin with the project of bus priority at signalized intersections conducted by Wilbur et al. in 1968 [1].

With the objective of minimum passenger delay, some studies have established bus priority control models, optimizing signal timing scheme at intersections by calculating delay before and after adopting bus priority [2–4]. Since not all impact of

---

M. Wang · R. Li (✉) · S. Meng
College of Civil and Transportation Engineering, Hohai University, Xi Kang Road 1, Nanjing 210098, China
e-mail: lirui2012@hhu.edu.cn

W. Wang et al. (eds.), *Green Transportation and Low Carbon Mobility Safety*, Lecture Notes in Electrical Engineering 944,
https://doi.org/10.1007/978-981-19-5615-7_27

bus priority control is favorable, some investigations have considered and quantified the impact on other transportation vehicles [5–7]. In order to reduce negative impact, some studies have predetermined certain control conditions to judge whether bus priority control is desirable. For instance, bus priority control is preferable only when service reliability is improved or average delay at intersections is reduced [8]. However, most of the current studies only focus on one aspect of service reliability and bus priority, rarely combining both of them.

Hence this paper combines these two aspects by establishing a bus priority control model based on service reliability, relying on selecting evaluation index of service reliability and control strategies of bus priority. Then the model is also introduced in Huju Road-Beijing West Road intersection in Nanjing to analyze the impact of bus priority control with real-life data.

The paper is organized as the following: in Sect. 2, time effect of passengers is selected as the evaluation index of service reliability, and feasibility of bus priority control is judged after comparing time effect of passengers with and without it. In Sect. 3, real-life traffic data of Nanjing Huju Road-Beijing West Road intersection is used to analyze the model and its effectiveness. The final section presents some conclusions and recommendations.

2 Methodology

2.1 Bus Priority Process

To minimize the negative impact of bus priority, which needs to meet certain conditions, the process of bus priority control is shown in Fig. 1.

2.2 Calculation the Time Effect of Passenger After Bus Priority

To judge whether the bus priority can be conducted through the time effect of the passengers after the bus priority, it needs to calculate the time effect after the bus priority. The time effect considers the passengers at the intersection and the passengers at the downstream bus stops. The intersection vehicles include bus vehicles and social vehicles, and the social vehicles are divided into social vehicles in the bus phase and social vehicles in the non-bus phase. Then, Take the priority control strategy of green extension as an example to calculate the time benefit.

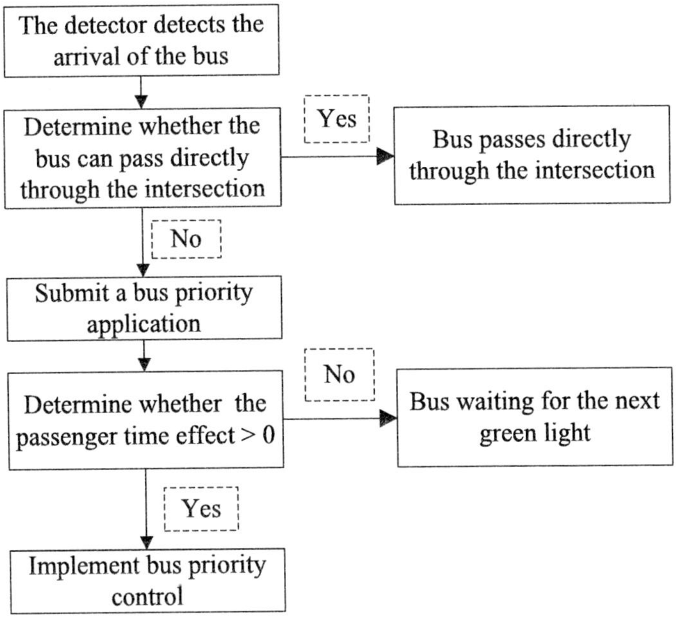

Fig. 1 Process of bus priority control

2.2.1 Time Effects for Intersection Bus Passengers

The time effect of the bus is the time saved by the bus that would otherwise need to stop and wait at the intersection due to the priority of the bus passing through the intersection. Set detector at l meters before the stop line at the intersection to detect the arrival of the bus. Assuming that the bus arrival time is t_e, the average driving speed of the bus is v_b, then the predicted time for the bus to reach the stop line at the intersection is:

$$t_a = t_e + \frac{l}{v_b} \tag{1}$$

where, t_e is the time when the detector detects the arrival of the bus (s).
l is the distance of detector from stop line at the intersection (m).
v_b is the average travel speed of the buses (m/s).

The start time of green light of the bus phase at the intersection is g_s, the end time is g_e and the maximum green light extension time is $g_{a\max}$, $g_{a\max}$ depending on the minimum green time of the adjacent phase of the bus phase, when considering the safe crossing time of the vehicles at the intersection, the minimum green time depends on the time of the vehicle safely through the intersection, which is $t_{v\min}$ (general value is 5–7 s [9]) and the time of the pedestrian safely through the intersection, which is

$t_{p\min}$. According to the U.S. transit capacity and quality of service manual, TCQSM 2nd) [10]. $t_{p\min}$ can be calculated by:

$$t_{p\min} = t_{walk} + t_{fdw} \tag{2}$$

where, $t_{p\min}$ is the time for pedestrians to pass through the crossing safely (s).
t_{walk} is the pedestrian walking time, usually takes 7 s.
t_{fdw} is the time when pedestrians all pass through the intersection (s), $t_{fdw} = \frac{L}{S_{15}^P}$.
L is the length of the pedestrian crosswalk at the intersection (m).
S_{15}^P is an average speed of 15% of pedestrians (m/s).

The minimum green time in the signal phase is calculated as:

$$g_{i\min} = \max(t_{v\min}, t_{p\min}) \tag{3}$$

The maximum extension time of the green is calculated as:

$$g_{a\max} = g_i - g_{i\min} \tag{4}$$

where, $g_{a\max}$ is the maximum green extension time for bus phase (s).
g_i is the green time adjacent to the bus phase (s).
$g_{i\min}$ is minimum green time adjacent to the bus phase (s).

If, $g_e < t_a < g_e + g_{a\max}$, it can request the bus phase green time to extend for the bus, the extended green time is:

$$\Delta t_y = t_a - g_e + T \tag{5}$$

where, T is the time for the bus to safely pass through the intersection (s).
g_e is the end time of green light for the intersection bus phase (s).

If bus priority is not given, the time bus needs to wait at the intersection is:

$$t_{wb} = C - t_a + g_s \tag{6}$$

where, C is the cycle length.

After giving the bus priority, the time effect of the bus at the intersection is the waiting time saved at the intersection, as follows:

$$t_{ib} = t_{wb} = C - t_a + g_s \tag{7}$$

where, t_{ib} is the bus time effect at the intersection after the green extension (s).

The total time effect of the passengers on the bus phase is:

$$T_{1b} = t_{1b} \cdot P_b = (C - t_a + g_s) \cdot P_b \tag{8}$$

where, T_{1b} is the total time effect for the bus after green extension (s).
P_b is the average number of passengers for buses.

2.2.2 Time Effects for Passengers in Social Vehicles

Time Effect of Passengers in the Social Vehicle for Bus Phase

If the bus phase is not given priority, the social vehicles of the bus phase also need to wait for the next green time at the intersection to pass. After giving the bus priority, the other social vehicles in this phase will also reduce the waiting time, the reduced time is the time effect of social vehicles in the bus phase.

Assume the bus phase as the first phase, and the delay reduced by the green extension is calculated as follows:

$$t_{1j} = \frac{\Delta t_y q_{1j}}{2}\left[2r_1 + \frac{\Delta t_y q_{1j}}{s_{1j}} - \Delta t_y\right] \tag{9}$$

where, t_{1j} is the reduced delay due to green extension for bus phase, which is the time effect of this part of the vehicle in the approach j.
Δt_y is the extended green time.
q_{1j} is the arrival rate of social vehicles in the approach j (pcu/s).
s_{1j} is the saturation flow rate of social vehicles in the approach j (pcu/s).
r_1 is the red time for the bus phase (s).

If there are other approaches in the bus phase, the vehicles on these approaches will also reduce the delay, the total time effect of the bus phase is:

$$T_{1v} = \sum\nolimits_{i=1}^{k} p_v t_{1j} = \sum\nolimits_{i=1}^{k} p_v \cdot \frac{\Delta t_y q_{1i}}{2}\left[2r_1 + \frac{\Delta t_y q_{1i}}{s_{1i}} - \Delta t_y\right] \tag{10}$$

where, T_{1v} is the total time effect of social vehicles after green extension for bus phase (s).
p_v is the average number of passengers in the social vehicles.
k is the total approaches for the bus phase.

Time Effects of Passengers in the Social Vehicle for Non-Bus Phase

The intersection signal cycle is certain, so when the green time is extended in the bus phase, the green time of the non-bus phase will be compressed to meet the requirements, and the start time of the green light of the non-bus phase will also

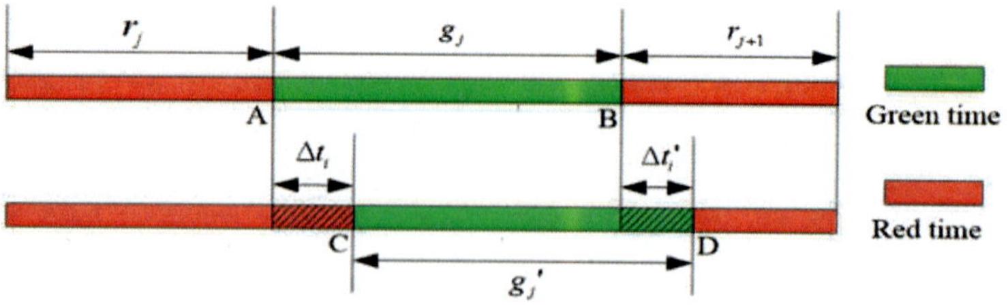

Fig. 2 Adjustment diagram of non-bus phase matching after green extension

change. The extended green time of the bus phase needs to be satisfied by squeezing the green time of the other phases.

As shown in Fig. 2, r_j, r_{j+1} are the red time of the non-bus phase before the green extension, g_j is the green time of the non-bus phase before the green extension, Δt_i is the green delay starting time of the non-bus phase after the green extension. $\Delta t_i = \Delta t_y$, where, Δt_y is green extension time for bus phase, $\Delta t_i^{'}$ is green delay ending time of non-bus phase after the green extension time, $\Delta t_i^{'} < \Delta t_i$. The AB section is the green time for the non-bus phase before the green extension, the CD section is the green time for the non-bus phase after the green extension. Vehicles arriving at the BD section reduce the waiting time at the intersection, the total time effect can be calculated as formula (10).

Vehicles arriving at the AC section will increase the waiting time at the intersection, and the time loss of this part of vehicles is calculated as follows:

$$t_{ij} = \frac{\Delta t_y q_{ij} S_{ij}}{2(S_{ij} - q_{ij})}(2r_i + \Delta t_y) \quad (11)$$

where, t_{ij} is time delay reduced in the approach j of the non-bus phase i due to the green extension, which is the increased time loss (s).

Δt_y is the green extension time (s).

q_{ij} is the arrival rate of social vehicles in the approach j of the non-bus phase i (pcu/s).

S_{ij} is the saturation flow rate of social vehicles in the approach j of the non-bus phase i (pcu/s).

r_i is the red time for the non-bus phase (s).

The time effect of vehicles in the phase i approach j is equal to the time effect of arriving vehicles in the BD section minus the time loss of arriving vehicles in the AC section, as follows:

$$T_{ij} = t_{ij}^{'} - t_{ij} = \frac{\Delta t_i^{'} q_{ij}}{2}\left[2r_i + \frac{\Delta t_i^{'} q_{ij}}{s_{ij}} - \Delta t_i^{'}\right] - \frac{\Delta t_i q_{ij} S_{ij}}{2(S_{ij} - q_{ij})}(2r_i + \Delta t_i) \quad (12)$$

where, T_{ij} is the time effect for approach j of the non-bus phase i (s).
$t_{ij}^{'}$ is the time effect for vehicles on approach j of the non-bus phase i after the green extension (s).
t_{ij} is the loss of time for vehicles on approach j of the non-bus phase i after the green extension (s).
Δt_i is the green delay starting time of the non-bus phase i after the green extension (s), $\Delta t_i = \Delta t_y$.
$\Delta t_i^{'}$ is the green delay ending time of non-bus phase i after the green extension (s).

The total time effect for all vehicles in the non-bus phase is:

$$T_i = T_{ij} P_v = \sum_{i=1}^{m_i} \sum_{j=1}^{m_j} (t_{ij}^{'} - t_{ij}) P_v$$
$$= \sum_{i=1}^{m_i} \sum_{j=1}^{m_j} \left\{ \frac{\Delta t_i^{'} q_{1j}}{2} \left[2r_1 + \frac{\Delta t_i^{'} q_{1j}}{s_{1j}} - \Delta t_i^{'} \right] - \frac{\Delta t_i q_{ij} S_{ij}}{2(S_{ij} - q_{ij})} (2r_i + \Delta t_i) \right\} P_v \tag{13}$$

where, P_v is the average number of passengers on social vehicles.
m_i is the number of non-bus phase.
m_j is the number of approach per non-bus phase.

2.2.3 Time Effect for Passengers at the Downstream Stop

If the bus needs to wait in the stop line at the intersection, the passengers at the downstream stop also need to wait. After the green extension, vehicles at the bus phase reduces the waiting time at the intersection and passengers reduce the waiting time at the downstream stop accordingly. The time effect as follows:

$$T_s = t_{ib} \cdot P_s = (C - t_a + g_s) \cdot P_s \tag{14}$$

where, T_s is the time effect of passengers at downstream stop after the green extension (s).
t_{ib} is the time effect of the bus at the intersection after the green extension (s).
t_a is the time bus reaching the intersection stop line (s).
P_s is the number of passengers waiting at the downstream stop.

3 Case Study

According to the above-mentioned bus priority control model based on service reliability, this section takes the intersection of Nanjing Huju Road-Beijing West Road for example. Meanwhile, time effect of passengers before and after introducing the model is calculated and compared to determine its feasibility.

3.1 Instance Base Data

The current signal timing (reduced to a four-phase intersection at Huju Road-Beijing West Road) is shown in Fig. 3.

The arrival rate, saturation flow rate, and number of approaches at the intersection of Huju Road-Beijing West Road are as shown in Table 1. Since the bus is given priority in phase 1, the number of buses in the other phase is converted into passenger car units.

To facilitate calculation, some data is assumed. Supposing the buried detector at 100 m before the intersection to detects the bus arrival time, the length of the intersection is 50 m, the average bus travel speed is 30 km/h, the time required for the bus to successfully pass through the intersection is 6 s, the average passenger number of buses is 25, the average passenger capacity of social vehicles is 2, the average number of passengers waiting at downstream stops is 15.

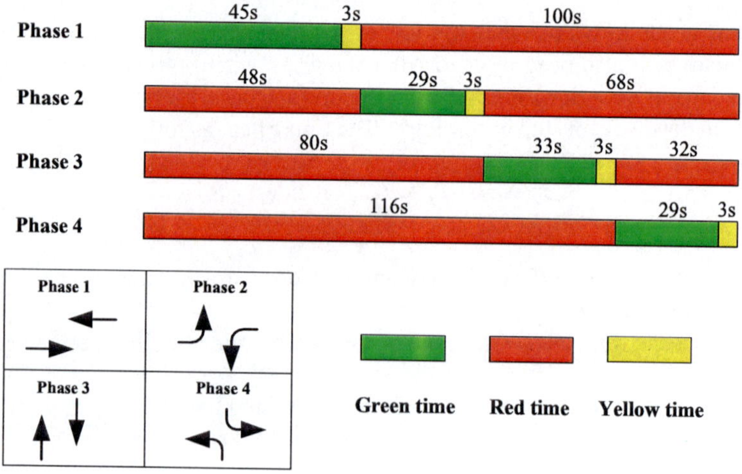

Fig. 3 Phase timing diagram of Huju Road-Beijing West Road crossing

Table 1 Traffic data composition of Huju road-beijing west road crossing

Phase	Phase 1		Phase 2		Phase 3		Phase 4	
Approach	West	East	West	East	South	North	South	North
Number of social vehicles	542	294	260	270	90	40	264	256
Number of buses	66	54	42	0	42	72	12	36
Passenger car unit (pcu)	740	456	386	270	216	256	300	364
Number of lanes	2	2	2	2	1	3	2	2
Social vehicle arrival rate (pcu/s)	0.103	0.063	0.054	0.038	0.030	0.036	0.042	0.051
Social saturation flow rate (pcu/s)	0.500	0.500	0.500	0.500	0.500	0.500	0.500	0.500

3.2 *Analysis Bus Priority Control*

Since the bus phase (i.e. phase 1) the end time of green light is 45 s, assuming the detector detects that the bus arrival time at 35 s, then the time of bus reaches the intersection stop line is 47 s. After calculation, the bus phase needs to extend the green time for 8 s. The maximum green extension in the phase is 17.5 s, therefore request green priority for the bus phase.

3.2.1 The Time Effect of Passenger on the Bus

According to Fig. 3, the current intersection cycle length is 148 s, the start time of green light of the bus phase is 0 s. The bus arrived at the intersection stop line at 47 s, so the time effect of the bus after the green extension is 101 s. After considering the number of the passengers on the bus, the total time effect of the bus is 2525 s.

3.2.2 Time Effect of Passengers in the Social Vehicle for Bus Phase

According to Table 1, the social vehicle arrival rate of the west approach in the bus phase is 0.103 pcu/s. The social vehicle arrival rate at the east approach is 0.063 pcu/s. The saturation flow rate of a single approach is 0.500 pcu/s. Bus phase approaches has 2 lanes each, and the red time of bus phase is 100 s. Then, the time effect of per lane on west approach in the bus phase after the green extension is 79.78 s. The time effect of each lane on east approach in the bus phase after the green extension is 48.64 s, since there are two lanes of each approach in the bus phase, the time effect of social vehicle in the bus phase is 256.84 s. After considering the number of passengers of the social vehicles, the total time effect of social vehicles in the bus phase is 513.68 s.

3.2.3 Time Effect of Passengers in the Social Vehicle for Non-Bus Phase

The extension of green time in the bus phase is 8 s. The compressed green time for each phase of the non-bus phase is: phase 2 is 2.55 s, phase 3 is 2.90 s, phase 4 is 2.55 s. Combining the arrival rate and saturation flow rate of social vehicles in a single lane of the non-bus phase and the number of lanes in each approach, the time effect of phase 2 on the west approach is 25.17 s. The time loss of phase 2 on the east approach is 16.29 s. The time loss of phase 3 on the south approach is 11.77 s. The time loss of phase 3 on the north approach is 14.43 s. The time loss of phase 4 on the south approach is 14.01 s. The time loss of phase 4 on the north approach is 17.35 s. So the total time loss of the social vehicle in the non-bus phase is 200.70 s. After considering the number of passengers of the social vehicles, the total time loss of social vehicles in the non-bus phase is 401.41 s.

3.2.4 Downstream Stop Passenger Time Effect

The time effect of the passengers at downstream stop is equal to the reduced delay of the bus at the intersection multiplied by the number of passengers at the downstream stop. If the average number of passengers waiting at the downstream stop is 15. After the green extension, the time effect of the downstream stop passenger is 1515 s.

According to the calculation results, the favorable impact of the implementation of the bus priority on the bus phase vehicles is far greater than the adverse impact to the non-bus phase vehicles, and the total time effect of vehicle passengers in the bus phase after the implementation of the bus priority is far greater than the total time effect of the passengers of the non-bus phase vehicles. Therefore, it can be used through the bus phase of the bus priority application.

4 Conclusion

In conclusion, this paper proposes a bus priority control model based on service reliability: after bused apply for priority control, calculate time effect of the passengers at the intersection and downstream stops. Then determine whether the application is allowable, pass it if the overall time effect is greater than 0, unless, reject it. Finally, analyze the model with real-life traffic data of Nanjing Huju Road-Beijing West Road intersection. The results show that the overall effect on passengers at the intersection and downstream stops is favorable under certain conditions.

In future work, parts of test data will be replaced by real-life data for further analysis of the model, and other bus priority control strategies will also be considered to refine the model.

Acknowledgements This work was supported by the Natural Science Foundation of Jiangsu Province (Grant No.BK20181307) and Postdoctoral Science Foundation of China (Grant No.2018M630505).

Conflict of Interests The authors declare that there is no conflict of interest regarding the publication of this paper.

References

1. Smith W (1968) Study of evolutionary urban transportation. Volumes I, II, and III, U.S. Department of Housing and Urban Development. Westinghouse Airbrake Company, and Institute of Public Administration
2. Yagar S, Han B (1999) Procedure for real-time signal control that considers transit interference and priority. Transp Res, Part B: Methodol 28(4):315–331
3. Hounsell N, Shrestha B (2012) A new approach for co-operative bus priority at traffic signals. Spec Issue Emerg Coop Technol Intell Transp Syst 13(1):6–14
4. Guo R, Zhang Y (2019) Optimization model of bus priority signalized intersection based on passenger delay. J Dalian Jiaotong Univ 40(5):1–6
5. Dion F, Hellinga B (2002) A rule-based real-time traffic responsive signal control system with transit priority: application to an isolated intersection. Transp Res, Part B: Methodol 36(4):325–343
6. Chang J, Collura J, Dion F et al (2003) Evaluation of service reliability impacts of traffic signal priority strategies for bus transit. Transp Res Rec 1841(1):23–31
7. Zhao T, Liang S, Xu H et al (2020) Limited bus priority signal control considering the reliability of station time. Intell Comput Appl 10(5):163–168
8. Bie Y, Wang D, Wei Q et al (2011) Single-point active and passive limited public transit priority strategy considering saturation constraints. J Jilin Univ: Eng Edn 41(5):1222–1227
9. Guan H (2004) Disaggregated model—A tool for traffic behavior analysis. People's Communications Press, Beijing
10. Federal Transit Administration, Transit Cooperative Research Program, Transit Development Corporation et al (2010) Transit capacity and quality of service manual, 2nd edn. National Research Council: Transport Research Board

Electric Vehicle Charging Demand Forecast Based on Residents' Travel Data

Zhule Jin, Yongneng Xu, and Zheng Li

Abstract The charging demand of electric vehicles is closely related to residents' travel behavior, this paper proposes a charging demand prediction method based on the travel behavior of electric vehicle users. By studying the probability distribution of temporal and spatial characteristics in travel behavior, constructing a charging decision model for electric vehicle users, use Monte Carlo method to simulate the travel activities of each electric car in one day, get the charging demand of the whole city. Take Nanjing as an example to verify the model, this method can effectively predict the temporal and spatial distribution of charging demand, and can provide a basis for subsequent site selection of charging facilities.

Keywords Charging demand · Travel behavior · Monte Carlo · Electric vehicle

1 Introduction

Recently, as the energy crisis has become increasingly serious, oil prices have risen, environmental pollution and calls for emission reduction have increased, electric vehicles are an effective way to solve the above problems, and are strongly supported by governments of various countries.

The large-scale application of electric vehicles will generate a large amount of charging demand. Forecasting the temporal and spatial distribution of electric vehicle charging demand can provide an important basis for the design of electric vehicle

Z. Jin (✉) · Y. Xu
Department of Transportation Engineering, Nanjing University of Science and Technology, Nanjing City 210094, Jiangsu Province, China
e-mail: 718963471@qq.com

Y. Xu
e-mail: x780906yn@163.com

Z. Li
Department of Mechanical and Electronic Engineering, Nanjing University of Science and Technology, Nanjing City 210094, Jiangsu Province, China
e-mail: 2711171685@qq.com

W. Wang et al. (eds.), *Green Transportation and Low Carbon Mobility Safety*,
Lecture Notes in Electrical Engineering 944,
https://doi.org/10.1007/978-981-19-5615-7_28

range, the location of charging facilities and the analysis of the impact of charging load on the power grid.

The charging demand of electric vehicles is closely related to the daily travel behavior of electric vehicle users Liting et al. [1] considered a number of factors affecting the charging demand of electric vehicles, combined with the statistical results of fuel vehicles, established a statistical model of charging power demand, and then obtained the total charging demand of multiple electric vehicles, and obtained the influence of different scale charging demand on the original power grid load curve through examples. Xifan [2] considered the randomness of the start time of single battery charging and the initial charging state of the battery, and proposed a modeling and analysis method for the load demand of the distribution system when the battery of electric vehicles is charging by simulating four charging scenarios for comparative study; Chuyue [3] used the combination forecasting method and grey system model to forecast the future market scale of electric vehicles, and described the electric energy supplement process of electric vehicles by establishing the travel behavior model of electric vehicles, and then predicted the charging and replacement demand of electric buses, electric taxis and the charging demand of electric private cars. Bae [4] from the perspective of power distribution system, the charging demand of electric vehicles may change with the change of space and time. Under this background, a mathematical model of charging demand of fast charging station electric vehicles is proposed based on hydrodynamics traffic model and M/M/s queuing theory, and a space–time model of charging demand of fast charging station electric vehicles near the exit of expressway is established. Yizhou [5] takes electric taxis and household electric vehicles as the research objects, analyzes their charging mode and power demand, uses gravity model for reference, and considers the utilization and path distance of different regions at the same time, establishes the travel distribution prediction model, puts forward the calculation method of charging demand, and puts forward the prediction method of spatial distribution of charging demand of electric vehicles.

With more and more in-depth research on the prediction of electric vehicle charging demand, it is not enough to consider only the time or space distribution of charging demand from the statistical results of fuel vehicles, it is necessary to comprehensively study the temporal and spatial distribution of electric vehicle charging demand based on the actual situation of electric vehicle users. In this paper, based on the spatial–temporal characteristics of electric vehicle users' travel and considering the user preferences, a charging demand forecasting method for electric vehicles is proposed, and a case study of Nanjing city is carried out to verify the analysis.

The travel behavior of electric vehicles needs to study the travel preference and driving preference of travelers, and analyze the probability relationship between the starting point and destination of travelers, combined with the actual geographic information and road traffic conditions, the specific travel starting point and destination location are determined, and the travel mileage and the required battery power are obtained. Then considering the different preferences of electric vehicle users for the lowest SOC, the charging demand is judged according to the current SOC, and the temporal and spatial distribution of charging demand is obtained. According to the data of NHTS (national household travel survey) and traffic survey in China, the average number of trips per day is 2.75, according to the statistical results of Beijing transportation development annual report, the commuting trip (commuting, going to school) accounts for 45.3% of the total trip, and the life trip (leisure, entertainment, shopping) accounts for 48.6% of the total trip, people's life patterns on weekdays are similar, and their travel activities are basically the same, so travelers' travel decisions can be simulated by probability.

The time dimension feature quantity mainly includes the starting time of the electric vehicle user's travel, driving time, arrival time, parking time, etc. The characteristic quantity of spatial dimension mainly includes starting point, parking place, destination, driving mileage,etc. The above behaviors are simulated by Monte Carlo method, and the space–time distribution of EV charging demand is obtained.

2 Electric Vehicle Charging Demand Forecast Base on Monte Carlo

This paper considers the travel activities of EV users on commuting days, that is, the travel activities between living areas, working areas and various functional areas, according to the average daily trip times of nhts, this paper studies the situation of three trips a day at most, 24 h a day.

(1) Travel start time

According to the statistical results, the starting time T_0 of EV users follows gamma distribution, and its probability density function is as follows:

$$f(T_0) = \frac{58.908^{9.881}}{\Gamma(9.881)} T_D^{8.881} e^{-58.908 T_0} \tag{1}$$

(2) Driving time

This article studies the situation of no more than 3 trips per day, and classifies the trips of electric vehicle users as follows: a. The user departs from the residential area and arrives at a certain functional area. b. The user departs from a functional area to the next functional area. c. The user departs from a functional area and arrives at the residential area. The probability density function of the three types of driving time t_i distribution is as follows:

$$f(t_i) = \frac{1}{\sigma t_i \sqrt{2\pi}} e^{-\frac{(\ln t_i - \mu_i)^2}{2\sigma_i^2}} \tag{2}$$

For a: $\mu_a = 3.02, t_a = 0.775$, for b: $\mu_b = 2.846, t_b = 0.815$, for c: $\mu_c = 3.04, t_c = 0.761$.

(3) Park time

Electric vehicles can only be charged in the parking state, so the parking time of electric vehicles directly affects the charging time of electric vehicles, the parking time t_{s_i} of electric vehicles in a certain place obeys the generalized extreme value distribution, and its probability distribution in typical functional areas (such as enterprises and governments) is as follows:

$$x = \frac{t_{s\_i} - 438.44}{164.506} \tag{3}$$

$$f(x) = \frac{1}{164.506} e^{-(1-0.234x)^{4.27}} (1 - 0.234x)^{3.27} \tag{4}$$

The probability distribution of parking time in other areas is as follows:

$$x = \frac{t_{s\_i} - 68.52}{41.761} \tag{5}$$

$$f(x) = \frac{1}{47.761} e^{-(1-0.657x)^{-1.52}} (1 + 0.657x)^{-2.52} \tag{6}$$

(4) Travel distance

Most of the electric vehicles currently sold have a cruising range of more than 200 km, and some models have a cruising range of more than 700 km. The current survey shows that the average daily mileage of private passenger cars is 64 km. According to the statistical data of nhts, the single driving distance l_i conforms to the normal distribution, and its probability density is as follows:

$$f_L(l_i) = \frac{1}{\sigma_L l_i \sqrt{2\pi}} e^{-\frac{(\ln l_i - \mu_L)^2}{2\sigma_L{}^2}} \tag{7}$$

where: $f_L(l_i)$ is the probability density of single trip mileage l_i, σ_L is the standard deviation of travel, μ_L is the average travel distance, $\sigma_L = 1.2$ km, $\mu_L = 6.9$ km.

The car departs from the starting point, and the time spent on the way includes the travel time of i trips and the stay time at the first $i - 1$ destinations. The time T_i for the car to reach the i-th destination can be obtained:

$$T_i = T_0 + \sum_{n=0}^{i} t_i + \sum_{i=0}^{i-1} t_{s_i} \tag{8}$$

Time of the car leaving the i-th destination T_{l_i}:

$$T_{l\_i} = T_i + t_{s_i} \tag{9}$$

(5) Charging decision

According to the statistical results of EV project in 2011, the initial SOC of EV follows normal distribution

$$f(x_s) = \frac{1}{\sigma\sqrt{2\pi}} e^{-\frac{(x_s-\mu)^2}{2\sigma^2}} \tag{10}$$

where $\mu = 0.46576$, $\mu = 0.17924$.

Whether electric vehicle is charged or not depends on the next trip decision. The mileage and required power of the journey expected by the user to determine whether the current battery capacity meets the next trip demand. If the current remaining battery power can meet the next trip and maintain a certain proportion of the surplus, it does not need to be charged this time. If the current battery power does not meet the above conditions, electric vehicle users need to charge for the next trip. Therefore, the charging decision of electric vehicle users can be expressed as:

When the electric vehicle reaches i, the current state of charge has fallen below the user-acceptable threshold y:

$$SOC_i < y \tag{11}$$

The current state of charge cannot meet the requirement that the remaining power after the next trip is higher than the user's acceptable threshold y. If the user does not cancel the trip plan, the electric vehicle should be charged in the current area:

$$SOC_i - \frac{l_{i+1}e_0}{B_c} < y \tag{12}$$

In the formula, SOC_i is the state of charge when the electric vehicle reaches i, and l_i is the distance traveled by the vehicle to i(Unit: km). B_c is the battery capacity (km/h), e_0 is the power consumption per kilometer during driving, according to the statistical results, y conforms to the normal distribution.

The length of parking time at the destination will determine whether the electric vehicle can be fully charged before the next trip, and the length of time will affect the user's choice of charging mode. After the user arrives at location i, he chooses the slow charging mode to charge. If the parking time is longer than the required slow charging time, the user directly selects the slow charging mode. If the parking time is not enough to get the state of charge for the next trip through slow charging, in order to meet the travel demand in a short time, the user chooses fast charging.

The user selects the slow charge when the amount of power obtained by slow charging during the parking time of location i plus the remaining power is greater than the next trip plus the minimum acceptable threshold for the user, that is:

$$SOC_i + \frac{p_j t_{s_i}}{B_c} > l_{i+1}e_0 + y \tag{13}$$

When the user's power obtained by slow charging during the parking time of place i plus the remaining power is less than the power required for the next trip plus the minimum threshold acceptable to the user, select fast charging, that is:

$$SOC_i + \frac{p_j t_{s_i}}{B_c} < l_{i+1}e_0 + y \tag{14}$$

where p_j is the power of charging pile, $j = 1$ is the slow charging power, and $j = 2$ is the fast charging power. If the battery can be fully charged during the parking time, the charging time is $(1 - SOC_i)B_c/p_j$, If it cannot be fully charged during the parking time, the charging time is t_{s_i}, the amount of electricity obtained by charging is$p_j t_{s_i}$.

According to the space–time characteristic quantity mentioned above, use Monte Carlo method to extract the user's travel start time, stop time at the destination, single travel distance, start SOC and other data to predict the charging demand of electric vehicle users, simulate the charging demand of each electric vehicle user, so as to obtain the spatiotemporal distribution of electric vehicle charging demand in the entire city. The implementation steps are shown in Fig. 1.

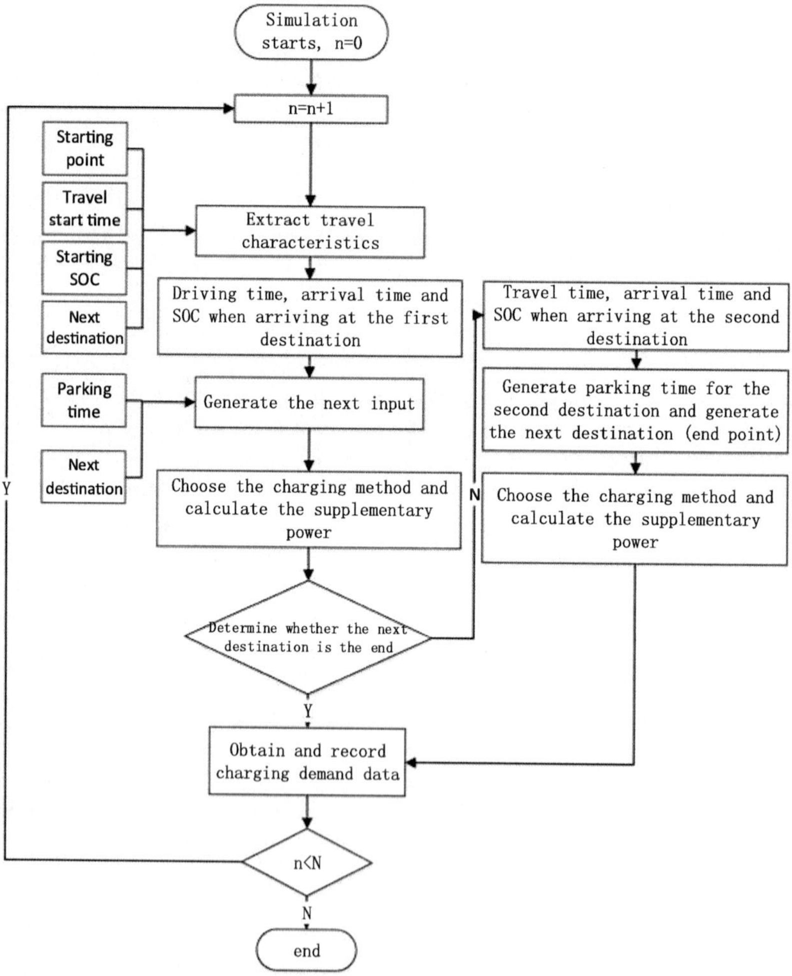

Fig. 1 Monte Carlo simulation charging demand flow chart

Simulation starts n = 0

Step 1: n = n + 1, according to the probability density function above, extract the starting time of the electric vehicle user's travel, the starting point of the travel, the starting SOC of the electric vehicle, and the destination of the next trip.

Step 2: According to the destination of step 1, generate the driving time and arrival time to the first destination; The SOC of the EV at the destination is calculated from the initial SOC and the power consumption per unit mileage.

Step 3: According to the probability density function in the space–time model above, generate the length of time that the electric vehicle user parks at the destination, and generate the next destination.

Step 4: According to the parking time of the electric vehicle user at the destination, the SOC, and the estimated power to reach the next destination, calculate whether it needs to be charged, and determine the charging method.

Step 5: Judge whether the next destination is the destination of this trip;

If the result is true: end the simulation of this electric vehicle, record its charging demand, judge whether n < N is true, if it is true, return to Step 1, if it is false, the program ends.

If the result is false: proceed to Step 6.

Step 6: Generate the travel time of the second trip and calculate the time to reach the second destination; calculate the SOC when arriving at the second destination according to the starting SOC of the second trip and the power consumption per unit mileage.

Step 7: Generate the length of time that the electric vehicle user parks at the destination, and generate the next destination.

Step 8: According to the parking time of the electric vehicle user at the destination, the SOC, and the estimated power to reach the next destination, calculate whether it needs to be charged, and select the charging method.

Step 9: Record the charging demand of the electric car.

Step 10: Judge whether n < N is true, if it is true, return to Step1 , if it is false, the program ends.

3 Simulation Analysis

In this paper, Nanjing is selected as a simulation example. More than 250,000 pieces of POI data of Nanjing in 2020 are obtained through Gaode map. Through the deletion of unqualified and repeated parts, the data suitable for this paper is selected, including the following types: government units, business units, leisure and entertainment places, shopping and consumption places, scenic spots, residential areas, a total of 168,618. The number of electric vehicles set in this simulation is 20000. According to the current performance parameters of mainstream electric vehicles, the power consumption per kilometer $e_0 = 0.15$ km.h and the battery capacity $B_c = 60$ kw.h are selected. According to the parameters of mainstream charging piles, the slow charging power $p_1 = 8$ kw.h and the fast charging power $p_2 = 45$ kw.h are selected.

This simulation environment is Matlab2016a. After setting the above parameters, read the POI data, use the Monte Carlo method to simulate the travel behavior of each car, obtain the overall travel situation of electric vehicle users, and solve and record the charging demand.

In this simulation, a total of 15,196 charging requirements were obtained, 3927 fast charging requirements, and 11,179 slow charging requirements. The spatial distribution is shown in Fig. 2:

(1) Total charging demand

Divide a day into 24 hours, count the number of charging requirements in each hour, and the time distribution of total charging requirements is shown in Fig. 3.

It can be seen that the charging demand presents a double peak shape. The demand for slow charging starts to increase significantly from 5:00 and reaches the maximum around 8:00, at this time, the demand for fast charging starts to increase, the increase in demand for fast charging continues until 16:00 in the

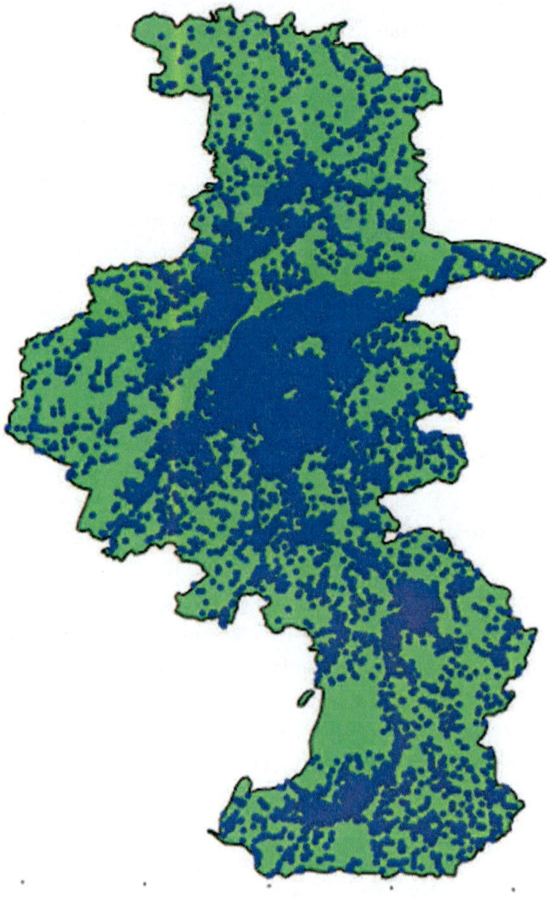

Fig. 2 Distribution of charging demand in Nanjing

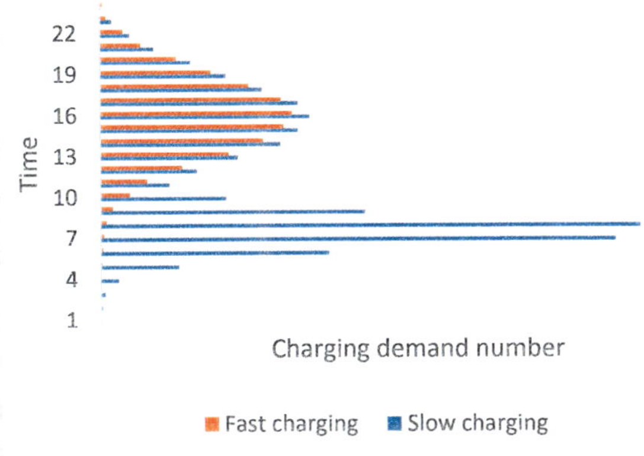

Fig. 3 Time distribution of charging demand in Nanjing

afternoon and reaches the maximum, the slow charge demand has been declining from 8:00 to 12:00, then it gradually increased until it reached a local peak at 16:00, and then continued to attenuate.

(2) Functional area 1 (corporate government departments, etc.) has a total of 5438 charging requirements, 47 fast charging requirements, and 5391 slow charging requirements, showing a single peak shape, as shown in Fig. 4. In Fig. 4a, it can be seen intuitively that there are only a few fast charging requirements in function area 1, and most of them are slow charging requirements. Among them, the slow charge demand increased significantly from 5:00, and then gradually decayed after reaching the maximum value at 8:00. After 16:00, there was little charging demand, and the change trend of fast charge demand was basically the same as that of slow charge.

(3) Functional area 2 (leisure and entertainment, etc.) has a total of 6646 charging requirements, of which 2925 are required for fast charging and 3721 are required for slow charging. The charging requirements are single-peak, as shown in Fig. 5. Figure 5a shows that the demand for fast charging and slow charging began to increase significantly at 9:00, and then continued to increase, reaching a peak at 17:00, and then the demand gradually decayed, the change trend of fast charging and slow charging in functional area 2 is basically the same.

4 Conclusion

This paper analyzes the travel behavior of electric vehicle users, uses the Monte Carlo method to simulate the travel behavior of electric vehicles based on the temporal and spatial characteristics of the travel process, and proposes a method to predict the charging demand of electric vehicles. Based on the above analysis, a simulation

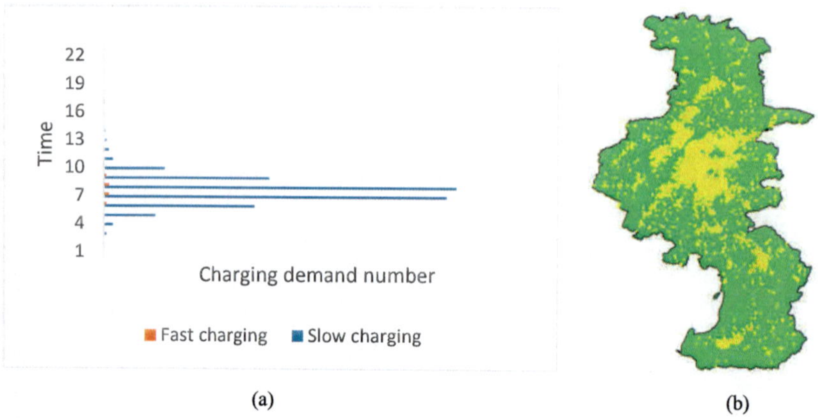

Fig. 4 Distribution of charging demand in functional area 1. **a** Time distribution, **b** spatial distribution

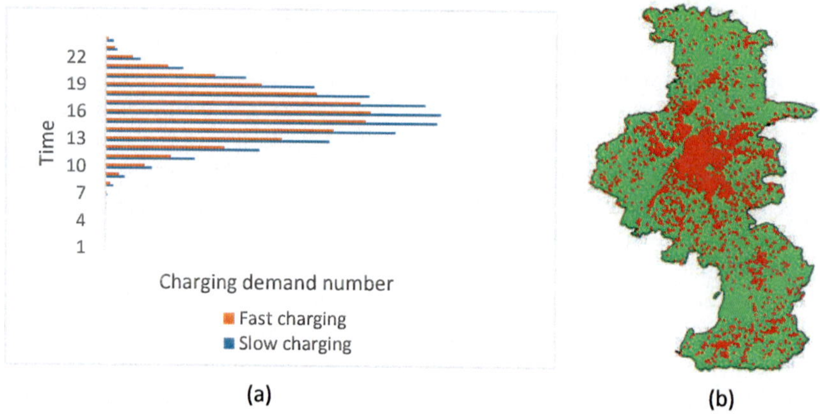

Fig. 5 Distribution of charging demand in functional area 2. **a** Time distribution, **b** spatial distribution

analysis with Nanjing City as an example shows that the method can effectively predict the temporal and spatial distribution of electric vehicle charging demand, the prediction results can provide a basis for the site selection planning of charging stations, according to the simulation results, the conclusions are as follows:

(1) The time distribution of charging demand shows obvious peaks and valleys. The charging demand of electric vehicles on weekdays is mainly concentrated from 7:00 to 9:00 and 15:00 to 18:00. This is mainly due to the rapid increase of charging demand caused by commuting and various activities after work.

(2) The population distribution, functional attributes, economic development degree and other factors in different areas of the city are closely related to the charging demand, and lead to obvious differences in the spatial distribution of the charging demand.

Acknowledgements This research is supported by National Key R&D Program of China under Grant No. 2017YFB1001801 and Fundamental Research Funds for the Central Universities (30917012102). We would like to thank the reviewers for their valuable suggestions and comments.

References

1. Liting T, Shuanglong S, Zhuo J (2010) Statistical modeling method of electric vehicle charging power demand. Power Syst Technol 34(11):126–130
2. Xifan W, Chengcheng S, Xiuli W et al (2013) Overview of electric vehicle charging load and dispatch control strategy. Proc Chinese Soc Electr Eng 33(1):1–10
3. Chuyue C (2015) Analysis and forecast of electric vehicle charging and switching demand. Beijing Jiaotong University, Beijing. https://doi.org/10.7666/d.Y2915967
4. Bae S, Kwasinski A (2012) Spatial and temporal model of electric vehicle charging demand. IEEE Trans Smart Grid 3(1):394–403
5. Yizhou J (2012) Spatial distribution forecast of electricity demand for large-scale electric vehicles. North China Electric Power University. https://doi.org/10.7666/d.y2139977
6. Zhuoran L, Jian C, Kai L et al (2015) Development status and trends of electric vehicles at home and abroad. Electr Power Constr 36(7):25–32. https://doi.org/10.3969/j.issn.1000-7229.2015.07.003
7. Rautiainen A, Repo S, Jarventausta P et al (2012) Statistical charging load modeling of PHEVs in electricity distribution networks using national travel survey data. IEEE Trans Smart Grid 3(4):1650–1659
8. Wei X, Lin L, Lixiong X et al (2017) Electric vehicle charging demand calculation based on Markov chain. J Electr Power Syst Autom 29(3):12–19.https://doi.org/10.3969/j.issn.1003-8930.2017.03.003
9. Tao Z, Xiaopu H, Xiaoyong Y et al (2013) Statistical mechanics of the temporal and spatial characteristics of human behavior. J Univ Electron Sci Technol China 4:481–540. https://doi.org/10.3969/j.issn.1001-0548.2013.04.001
10. Liu H, Dawei C, Jing L et al (2012) Modeling and simulation of resident trip times based on travel chain. J Shenzhen Univ (Sci Technol Edn) 29(3):264–269. https://doi.org/10.3724/SP.J.1249.2012.03264

Traffic Flow Prediction Based on GM-RBF

Yaxin Chen, Yongneng Xu, and Hui Cheng

Abstract In order to provide more reliable data for traffic control and guidance system, traffic flow prediction is very important. This paper proposes a traffic flow prediction method based on GM-RBF combined model. Firstly, build a GM(1, 1) prediction model to predict multiple sequences. Secondly, the GM model and RBF neural network are combined in application, and the residual feedback is established by the RBF model, so as to solve the problem of low accuracy of the grey model in the prediction. Taking the data every 15 min of the A12 highway in Suffolk, UK from January 1, 2019 to January 20, 2019 as a sample, taking into account the influence of weather factors on road traffic flow, the traffic flow per 15 min in a day on January 21, 2019 is predicted. The experimental results show that MAPE of GM-RBF model is 2.302911%, MAE is 3.625, RMSE is 5.6807. Compared with GM model and RBF prediction model, the error evaluation index of GM-RBF combination model has been significantly improved in accuracy. Therefore, the combined model has good applicability in traffic flow prediction.

Keywords Grey prediction · The neural network · Combination model · Traffic flow prediction

1 Introduction

With the continuous improvement of the urbanization, the number of vehicles and the total travel demand are also increasing, and the frequency of urban traffic congestion is also increasing. For travellers and management agencies, the accuracy of traffic flow prediction results will have a series of impacts on urban traffic. Therefore, it

Y. Chen · Y. Xu (✉) · H. Cheng
Nanjing University of Science and Technology, 200 Xiaolingwei, Nanjing 210094, China
e-mail: x780906@163.com

Y. Chen
e-mail: chenyaxin@njust.edu.cn

W. Wang et al. (eds.), *Green Transportation and Low Carbon Mobility Safety*, Lecture Notes in Electrical Engineering 944,
https://doi.org/10.1007/978-981-19-5615-7_29

is of great significance for travellers and traffic management departments to acquire urban traffic flow information in real time and accurately (2014).

Traffic flow prediction with high accuracy has a very important influence on the organization and control of urban traffic system. Traffic flow prediction is a real-time and accurate prediction of the passenger flow at a certain decision moment or even several moments in the future. At present, there are two kinds of algorithm models to predict traffic at home and abroad. One is linear evaluation model, the main models include ARIMA model (2015), MLR model (2016), Kalman Filtering model (2014 and 2011), etc., and the other is machine Learning method, such as SVM (2016), Neural Network (2016), Deep Learning (2017), etc. Wang Huiyong et al. (2016). used grey linear regression combination analysis method to predict expressway traffic volume. [1]. putted forward a prediction method based on the combination of bias correction random forest algorithm and data-driven feature selection strategy to forecast and verify the traffic flow of three urban roads. Sun Xianghai et al. (2008). putted forward the seasonal autoregressive summation moving average model on the basis of considering the daily cycle characteristics of traffic flow to predict the short-term urban traffic flow. Zhang Peilin et al. (2011), based on the judgment that the traffic flow belongs to the chaotic system, realized the short-time traffic flow prediction of expressway based on chaos theory, but its application convenience is low, so it is necessary to investigate whether the traffic flow belongs to the chaotic system before the prediction. Based on the periodic characteristics of traffic flow data, [2] realized the short-term prediction of traffic flow based on the differential integrated moving average autoregressive model. [3] introduced the spatio-temporal residual network (ST-ResNet) into the traffic flow prediction, and build the ST-ResNet model using convolution operation and residuals to improve the prediction accuracy of the model. Lu Baichuan et al. (2019) improved the prediction accuracy of short-term traffic flow by combining GA and WNN mixed model through multi-source data analysis.

The advantages of grey theory (GM) prediction are mainly reflected in the sample data demand less, simple calculation, small workload, short-term prediction accuracy is high, but in the ability to deal with complex linear problems is still poor. Artificial neural network has very powerful computing power, at the same time, its fault tolerance is strong, self-adaptability is good, but the number of sample demand is large. Combination forecasting model is to make use of the advantages of various forecasting models and combine different forecasting methods according to certain rules in order to improve the accuracy of prediction (1993). Based on the historical traffic flow data and considering the influence of weather factors on traffic flow, this paper proposes a road traffic flow prediction model, which combines the grey (GM) model with RBF neural network. The combined model not only integrates the advantages of small demand of grey model and high accuracy of short-time prediction, but also has the advantages of strong processing ability and good adaptability of RBF neural network, which improves the accuracy of short-time traffic prediction.

2 GM-RBF Neural Network Prediction Model

2.1 Grey Model

At present, the most widely used grey prediction model is GM(1,1) model. The basic idea of the model is to continuously add the original data, and then build a first-order linear differential equation model, and then calculate the predicted value after the calculated value of the model is reduced (2020). The steps are as follows.

Assuming the original non-negative data column is formula (1),

$$X^{(0)}=\left\{x^{(0)}(1), x^{(0)}(2), \ldots, x^{(0)}(n)\right\} \tag{1}$$

Processing the original data columns, the data were accumulated and listed as formula (2) and (3),

$$X^{(1)}=\left\{x^{(1)}(1), x^{(1)}(2), \ldots, x^{(1)}(n)\right\} \tag{2}$$

$$X^{(1)}k=\sum_{i=1}^{k} x^{(0)}(i), k = 1, 2, \ldots, n \tag{3}$$

The albino differential equation of grey GM(1,1) model is formula (4),

$$\frac{dX^{(1)}}{dt} + aX^{(1)} = b \tag{4}$$

Through discretization to deal with the equation, the GM(1,1) model obtained are formula (5) and (6),

$$X^{(0)}(k) + aZ^{(1)}k = b \tag{5}$$

$$Z^{(1)}(k) = \frac{\left[x^{(0)}(k-1) + x^{(0)}(k)\right]}{2} \tag{6}$$

a and b are determined by least square method are formula (7), (8), and (9),

$$\hat{a} = \left[a\ b\right]^{T} = (B^{T}B)^{-1}B^{T}Y \tag{7}$$

$$Y = \left[x^{(0)}(2)\ x^{(0)}(3)\ \cdots\ x^{(0)}(n)\right] \tag{8}$$

$$B = \begin{bmatrix} -Z^{(1)}(2) & 1 \\ -Z^{(1)}(3) & 1 \\ \vdots & \vdots \\ -Z^{(1)}(n) & 1 \end{bmatrix} \tag{9}$$

Substituting the parameters a and b obtained by formula (6) into formula (5) to get the response function of the model time is formula (10),

$$\hat{x}^{(1)}(k+1) = \left[x^{(1)}(k) - \frac{b}{a}\right]e^{-ak} + \frac{b}{a} \tag{10}$$

Using 1-IAGO to degrade the predicted sequence, the GM(1,1) model is expressed as formula (11),

$$\hat{x}^{(0)}(k+1) = \hat{x}^{(1)}(k+1) - \hat{x}^{(1)}(k), k = 1, 2, \cdots, n \tag{11}$$

The mean of the original data is formula (12),

$$\overline{x} = \frac{1}{n}\sum_{k=1}^{n} x^{(0)}(k) \tag{12}$$

The variance of the original data is formula (13),

$$S_1^2 = \frac{1}{n}\sum_{k=1}^{n} (x^{(0)}(k) - \overline{x})^2 \tag{13}$$

Residual error is formula (14),

$$e^{(0)}(k) = x^{(0)}(k) - \hat{x}^{(0)}(k) \tag{14}$$

The mean of the residual is formula (15),

$$\overline{e} = \frac{1}{n-1}\sum_{k=2}^{n} e^{(0)}(k) \tag{15}$$

The mean of the residual is formula (16),

$$S_2^2 = \frac{1}{n-1}\sum_{k=2}^{n} \left[e^{(0)}(k) - e\right]^2 \tag{16}$$

2.2 RBF Neural Network

RBF (Radial Basis Function) has simple structure, faster learning speed than other similar, and good approximation ability, which has been widely used in many fields such as nonlinear function approximation and pattern recognition (2014). RBF neural network has a three-layer network structure of N-P-M (2008, 2011 and 2012). The first layer is the input layer, which is mainly the data source node and plays the role of signal transmission. The second layer is the hidden layer, which contains the activation function of RBF neural network. The third layer is the output layer, which realizes the linear weighted sum of the results obtained from the hidden layer. Its structure is shown in Fig. 1.

The Gaussian function $H_j(x)$ is usually applied to the hidden layer as an activation function, the function is formula (17)

$$H_j(x) = \exp(-\frac{\left\| X_i - c_j \right\|^2}{2\sigma_j^2}), j = 1, 2, \cdots, n \tag{17}$$

The output value of the output layer node is formula (18),

$$y_i = \sum_{k=1}^{p} w_{ki} H_i(x), i = 1, 2, \cdots, n \tag{18}$$

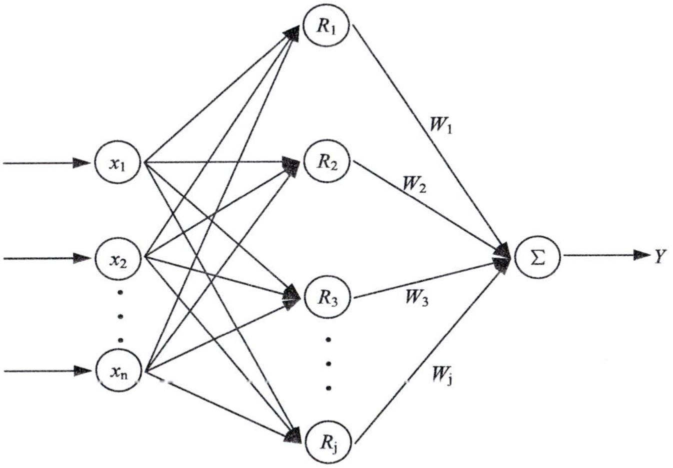

Fig. 1 The structure of RBF network

2.3 GM-RBF Model

In order to improve the accuracy of road traffic flow prediction, this paper combines the advantages of GM model and RBF neural network, and adopts the method of GM-RBF prediction model to forecast the traffic flow. In this paper, the predicted value is set as the input sample of the model, and the actual value is set as the output sample. The neural network model is adopted to predict the neural network with a certain structure, and the weights and thresholds of numerous relevant nodes can be obtained to simulate the residuals between the predicted value and the actual value, as well as the relationship between the sequences. The structure diagram of GM-RBF neural network is shown in Fig. 2.

According to the principle of GM-RBF prediction model, the specific modelling implementation method is as follows,

(1) Obtaining the sample traffic flow data and constructing the GM(1,1) model to obtain the cumulative sequence.
(2) In order to improve the prediction accuracy of GM-RBF, formula (18) is used to normalize the input and output data.

$$T = T_{\min} + \frac{T_{\max} - T_{\min}}{X_{\max} - X_{\min}}(X - X_{\min}) \tag{19}$$

(3) RBF neural network is used to learn and determine the corresponding parameters and weights to build the prediction model.
(4) Inputting the new data into the RBF neural network for prediction and output the predicted value.
(5) Formula (19) is used to process the result by inverse normalization.

$$X = X_{\min} + \frac{X_{\max} - X_{\min}}{T_{\max} - T_{\min}}(T - T_{\min}) \tag{20}$$

(6) The result of inverse normalization is then whitened to get the predicted value.

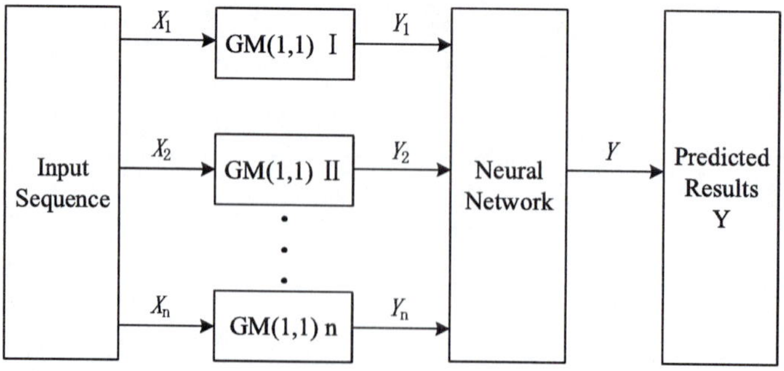

Fig. 2 The structure of GM-RBF network

Table 1 List of parameters and variables

Notation	Definition
$X^{(0)}$	Original data sequence
a	Development coefficient
b	Gray coefficient
X_i	Input variable
c_j	Centre of the j basis function
σ_j	Width of the j hidden node
n	Number of nodes in the hidden layer
y_i	Output value of the i node in the output layer
w_{ki}	Connection weight between the i unit in the hidden layer and the k unit in the output layer and the i unit in the output layer
X	Original data
$X_{\max}$	Maximum value of the original data
$X_{\min}$	Minimum value of the original data
T	Target data
$T_{\max}$	Maximum value of the target data
$T_{\min}$	Minimum value of the target data

2.4 *Definition of Parameters*

The parameters and variables involved in paper modeling are defined as shown in Table 1.

3 The Example Analysis

The data in this paper came from the British Road Open Dataset website. The road is A12 Road in Suffolk County, UK, and the longitude and latitude of the observation points are (1.094616, 52.02980), as shown in Fig. 3. The data of January 1, 2019 solstice and January 20, 2019 every 15 min were selected as the sample to forecast the traffic on January 21, 2019.

The weather data for January 1 to January 2019 and January 21, 2019 was crawled on the web using Python. For the convenience of calculation, the weather types are quantified (2016), as shown in Table 2.

In order to evaluate the prediction effect of the GM-RBF model, the prediction results were compared with the grey GM(1,1) model and the RBF neural network model. The prediction results and error curves are shown in Figs. 4, 5, 6, 7, 8 and 9.

It can be seen from Figs. 4, 5, 6, 7, 8 and 9 that the prediction curves of GM model and RBF prediction model have a low coincidence with the actual traffic flow curve,

Fig. 3 Position of observation point

Table 2 Quantification of weather types

The weather types	Quantitative coefficients
Sunny	12
Cloudy	10
Drizzle	8
Thundershower	6
Moderate rain	4
Heavy rain	2

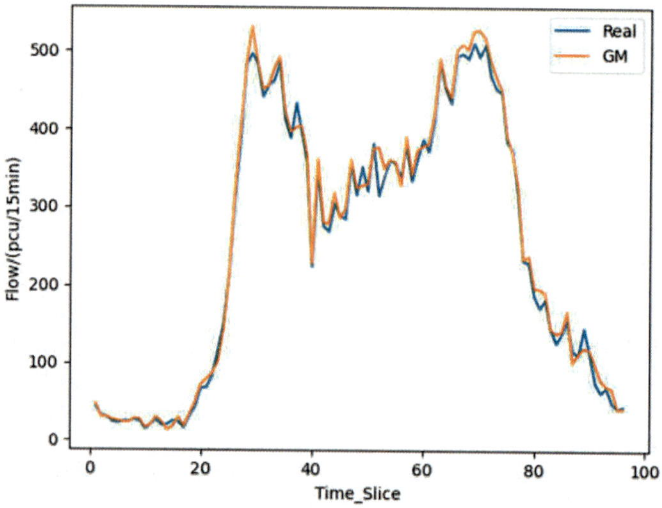

Fig. 4 The result of GM model

and most of the absolute error values are between (0, 20). However, the prediction results of the GM-RBF are highly consistent with the actual value, and the absolute error values of a few are greater than 20. In most cases, the absolute error is between

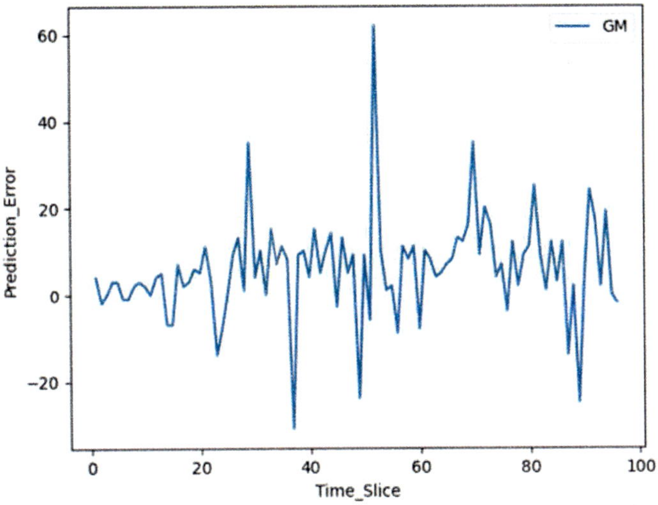

Fig. 5 The prediction error of GM

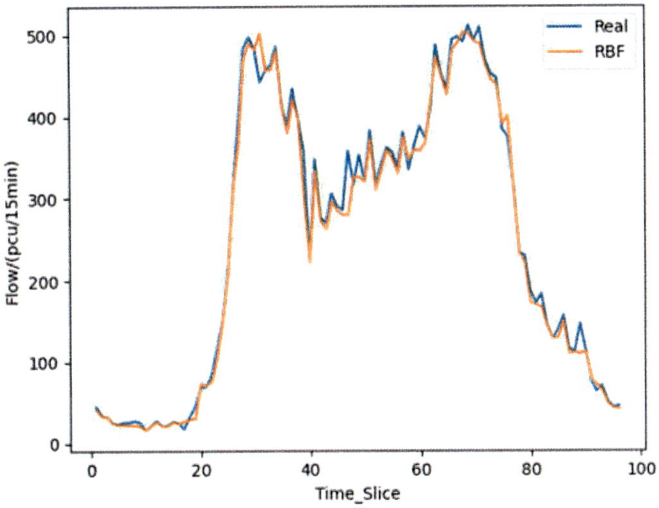

Fig. 6 The result of RBF model

(0, 10). The overall effect remains stable, and the fluctuation is less than that of the other two forecasting models.

In the process of scientific research, MAPE, RMSE and MAE are used to evaluate the performance of prediction methods and the effectiveness of prediction results (2020). The formula for calculating the three indexes are (21)–(23),

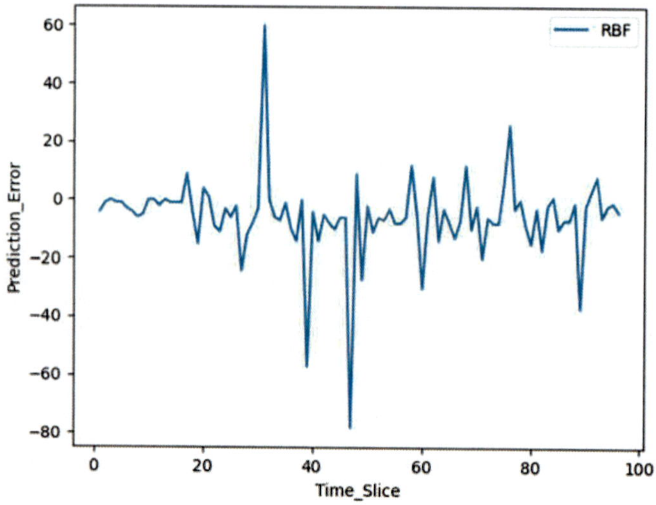

Fig. 7 The prediction error of RBF

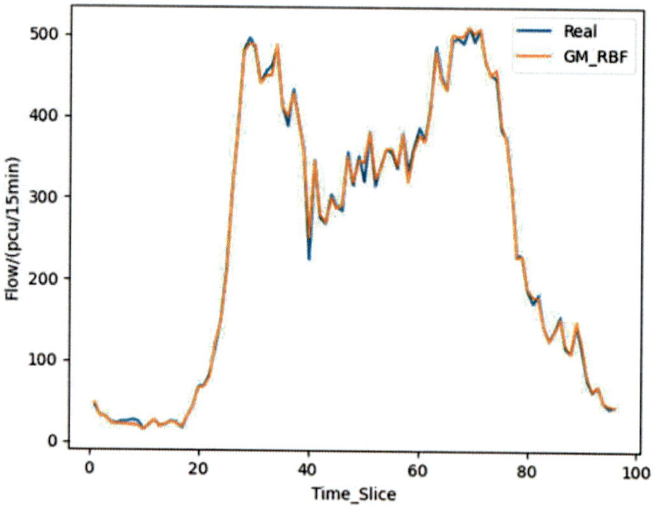

Fig. 8 The result of GM-RBF model

$$MAPE = \sum_{i=1}^{n} \left| \frac{\hat{y}_i - y_i}{y_i} \right| \quad (21)$$

$$RMSE = \sqrt{\frac{1}{n} \sum_{i=1}^{n} (\hat{y}_i - y_i)^2} \quad (22)$$

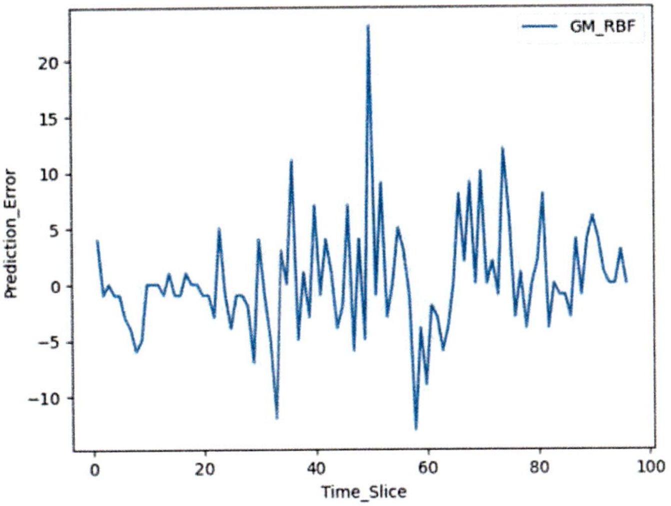

Fig. 9 The prediction error of GM-RBF

Table 3 The index analysis of three models

	MAPE	MAE	RMSE
GM	6.323928%	9.25	12.9904
RBF	5.173951%	8.90625	15.1262
GM-RBF	2.302911%	3.625	5.6807

$$MAE = \frac{1}{n}\sum_{i=1}^{n}\left|\hat{y}_i - y_i\right| \tag{23}$$

y_i is real traffic flow, $\hat{y}_i$ is forecast traffic flow.

The prediction indexes of the three models are shown in Table 3.

As can be seen from Table 3, the three error evaluation indexes of GM-RBF network are significantly smaller than those obtained by GM and RBF prediction, with higher prediction accuracy. Therefore, the GM-RBF model can achieve accurate and effective prediction of traffic flow.

4 Conclusions

Based on the analysis of GM and RBF, this paper synthesizes the advantages of the two forecasting models, and complements the disadvantages of the two forecasting models, puts forward a GM-RBF neural network forecasting model. Considering the influence factors of weather data on traffic flow, this model realizes the short-term forecasting of traffic flow. The results show that the MAPE of GM-RBF prediction model is 2.302911%, MAE is 3.625, RMSE is 5.6807, the results of the three evaluation indexes are all lower than the single GM model and RBF prediction model, which proves that this model can effectively improve the prediction accuracy. However, since this paper only considers the influence of weather factors on traffic flow, in fact, traffic flow is also affected by land use nature, holidays and other factors. These factors can be taken into account in the future to improve the accuracy of the model.

Acknowledgements This paper is supported by the National Natural Science Foundation of China (No. 52072214).

References

1. Ou J, Xia J, Wu Y, et al (2017) Short-term traffic flow forecasting for urban roads using data-driven feature selection strategy and bias-corrected random forests. Transp Res Rec 157–167
2. Kumar K, Parida M, Katiyar VK (2013) Short term traffic flow prediction for a non urban highway using artificial neural network. Proc Soc Behav Sci 104:755–764
3. Zhang J, Zheng Y, Qi D et al (2018) Predicting citywide crowd flows using deep spatio-temporal residual networks. Artif Intell 259:147–166
4. Vlahogianni EI, Karlarlaftis MG, Golias JC (2014) Short-term traffic forecasting: where we are and where we're going. Transp Rese Part C Emerg Technol 43:3–19
5. Xue R, JIAN, CHEN S K. (2015) Short-term bus passenger demand prediction based on time series model and interactive multiple model approach. Discret Dyn Nat Soc 66(1):61–78
6. Xiao-sheng L, Shen-ling W (2016) The parameter estimation of multivariate linear regression model with linear restraint. Stat Res 33(11):86–92
7. Rasyidi MA, Kim J, Ryu KR (2014) Short-term prediction of vehicle speed on main city roads using the k-nearest neighbor algorithm. J Intell Inf Syst (S095–9902) 20(1):121–131
8. Chun-hui Z, Rui S, Yang S (2011) Kalman filter-based short-term passenger flow forecasting on bus stop. J Transp Syst Eng Inf Technol 11(4):154–159
9. Xin-feng Y, Lan-fen L (2016) Short-term passenger flow forecasting on bus station based on affinity propagation and support vector machine. J Wuhan Univ Technol (Transp Sci Eng) 40(1):36–40
10. Costarelli D, Vinti G (2016) Pointwise and uniform approximation by multivariate neural network operators of the max-pro-duct type. Neural Netw 81(9):81–90
11. Vieira S, Pinaya WHL, Mechelli A (2017) Using deep learning to investigate the neuroimaging correlates of psychiatric and neurological disorders: methods and applications. Neurosci Biobehav Rev 74:58–75
12. Li-Cheng J, Jin Z, Shu-yuan Y et al (2017) Deep learning, optimization and recognition. Tsinghua University Press, Beijing, pp 100–120
13. Hui-Yong W, Qiu Y (2016) Forecast of expressway traffic volume based on the grey linear regression combined model. J Transp Eng Inf 14(1):53–57
14. Xiang-hai S, Tan-qiu L (2008) Study on urban road short-term traffic flow forecasting based on SARIMA time series model. J Highway Transp Res Dev 129–133
15. Pei-lin Z, Hong-yao Z (2011) Research on short-term traffic flow forecast of highway network based on chaos theory. Highway Eng 36(4):179–182
16. Bai-chuan L, Qin S, Guang-lu M (2019) Short-term traffic flow forecasting based on multi-source traffic data fusion. J Chongqing Jiaotong Univ (Nat Sci) 38(5):13–19, 56

17. Xiao-wo T, Chang-xiu C (1993) Study of combination forecasting method. Control Decis 8(1):7–12
18. Hong-jia L, Hong-guang Y, Si-rui L et al (2020) Passenger flow prediction of Hongqiao integrated transportation hub based on gray-neural network. Sci Technol Innov 121–122
19. Chao Z, Si-ming L, Qiao-ling X (2014) College building energy consumption prediction based on GM-RBF neural network. J Nanjing Univ Sci Technol 38(01):48–53
20. Bin L, Tao X, Ming-hua S (2008) Traffic flow combined forecast model of support vector machine. J Tiangong Univ 27(2):73–76
21. Xiao-jun L, Xiao-yan L, Jun L (2011) Forecast of railway short-term passenger flow based on RBF neural network. Railway Transport Econ 33(6):86–89
22. Min W, An-le Y (2012) A RBF neural network with optimum learning rates and its application. J Manage Sci China 15(4):50–57
23. Zhen-dong X, Zhi-ming L, Feng X et al (2016) Big data application of urban transportation card. Beijing, China Commun Press 171–188
24. He L, Zheng-min K, Cheng-gang Z (2020) Short-term load forecasting based on hybrid optimized random forest regression. Eng J Wuhan Univ 53(08):704–711

Quantitative Analysis of Characteristics and Influencing Factors' Correlation of Electric Bicycle Traffic Accidents

Yumeng Zhang and Fengchun Han

Abstract With the rapid growth of the electric bicycles ownership in P.R.C, while facilitating people's traveling, illegal behaviors of electric bicycles has become common. The frequent occurrence of traffic accidents has attracted widespread attention from all sectors of the society. In order to improve the level of electric bicycle traffic safety, based on the 10-year traffic accident data of electric bicycles in a southern city in China, the electric bicycle traffic accident characteristics was analyzed, through data screening and standardizing, on the characteristics of cyclists, temporal and spatial characteristics, accident types and causes, etc. On this basis, 16 influencing factors such as time, location, illegal behavior, accident form, and cause of accident identification of electric bicycle traffic accidents were used to construct multinomial logit model for accidents without casualties, accidents with injuries, and accidents with fatalities; and the correlation analysis of the main influencing factors was carried out for these three types of accident information of electric bicycles. Following that, electric bicycle traffic accidents shall be prevented and reduced by strengthening targeted prevention and control measures.

Keywords Electric bicycle · Accident characteristics · Multinomial logit model · Influencing factors · Correlation analysis

1 Introduction

At present, the number of electric bicycles in China is nearly 300 million, and the number of electric bicycle traffic accidents is also increasing while the number of electric bicycles increased. The traffic safety situation is severe. Therefore, making full use of existing data to explore the characteristics and influencing factors of electric bicycle traffic accidents plays an important role in the control and prevention of electric bicycle traffic risks.

Y. Zhang · F. Han (✉)
School of Traffic Management, People's Public Security University of China, Beijing, ST Huangyi, China
e-mail: hfc1966@163.com

W. Wang et al. (eds.), *Green Transportation and Low Carbon Mobility Safety*, Lecture Notes in Electrical Engineering 944,
https://doi.org/10.1007/978-981-19-5615-7_30

The research of electric bicycle traffic safety is abundant in content. Chunjiao [1] and others modeled the unsafe behavior of electric bicycles based on questionnaire survey data; Xinghui [2], Huan [3], Yiming [4] and others analyzed the injuries of cyclists in collisions with simulation methods. The majority is using electric bicycle traffic accident data to study the impact of various factors on the characteristics of the accident. Some domestic scholars chose to establish a discrete choice model to analyze electric bicycle traffic accident data. Tao [5] and others established an ordered logit model, and selected 11 significant related items from the 19 influencing factors of electric bicycle traffic accidents in Guilin; Weijie [6], Wenlong [7], Cheng [8], Dan [9], etc. used logistic regression model to analyze the main influencing factors of electric bicycle traffic accidents, and Yao [10] combined descriptive statistical analysis to find the causes of injury and death accidents through logit regression model. Some scholars also apply data mining algorithms to the analysis of electric bicycle traffic accidents. For example, Yingshuai [11] and others used random forest model to predict the degree of injury of electric bicycle traffic accident data in a certain city and rank the importance of influencing factors.

In summary, there are relatively few literatures that choose to use multinomial logit models to analyze electric bicycle traffic accidents. Based on the data of 1656 general accidents of electric bicycle in a southern prefecture-level city from 2011 to 2021, this article will establish multinomial logit models to analyze the characteristics, influencing factors and the relevance of electric bicycle traffic accidents, and to improve response measures, and governance abilities of electric bicycle traffic safety.

2 Data Description

2.1 *Data Acquisition*

Collecting accidents from 2011 to 2021 in a prefecture-level city in southern China, there are 1,656 general accidents involving electric bicycles. Data information can be divided into three categories: accident, personnel, road and facility, involving 35 dimensions, including the location and time of the accident, the injury degree of the people, the cause of the accident, education level of the people, road conditions, lighting conditions, etc.

2.2 *Data Processing*

2.2.1 Screening

16 influencing factors with high completeness, including time, location, illegal behavior, accident form, and cause of accident were selected, covering the three

categories: accident information, personnel information, road and facility information. In order to ensure the accuracy of the model, general accidents with relatively complete information records were selected as the object, and characteristic analysis and modeling are carried out.

2.2.2 Encoding

By discretizing the influencing factors, the multi-category attributes after discretized were coded. The coding results and descriptive statistical results are shown in Table 1. The percentage of each category attribute to the total number of accidents for that variable is marked in parentheses.

3 Characteristics of Electric Bicycle Traffic Accidents

3.1 Characteristics of the Driver

3.1.1 Age Characteristics

Statistics show that the age of drivers and the frequency of accidents were normally distributed, with a peak value of 35 years old. Young electric bicycle drivers had the most accidents. The highest mortality rate was for people over 69 years old, and the highest injury rate was for people aged 7–12 years old. The age groups with high casualty rates and high accident rates did not coincide. Young people were prone to accidents but not easy to be injured or killed. The main reason was that young people are prone to conduct illegal and unsafe behaviors, which lead to accidents, but their quick response and strong maneuverability can reduce accident injuries.

3.1.2 Education Level Characteristics

The data shows that the education level of drivers and the frequency of accidents were normally distributed, with junior high school education accounting for the highest proportion, up to 61.4%. The highest mortality rate was among the primary and junior high school people; the highest injury rate was among the illiterate population. The main reason was that this part of the population has a low level of education and low awareness of abiding by laws and regulations.

Table 1 Variable coding and descriptive statistics

Data category	Variable name	Classification of variable attribute and proportion (%)
Accident information	Time	1: 3:00–12:00 (37.5); 2: 12:00–21:00 (37.7); 3: 21:00–3:00 (24.8)
	Month	1: the first quarter (22.5); 2: the second quarter (23.4); 3: the third quarter (26.2); 4: The fourth quarter (28)
	B/D (drive or ride)	0: driving an electric bicycle (93.4); 1: riding an electric bicycle (6.6)
	Hurt (injury degree)	0: no-injury (20); 1: injury (59); 2: death (21)
	Accident (form)	1: collision with a moving vehicle (82.7); 2: scratching pedestrian/other vehicle and pedestrian accidents (7.4); 3: rollover / hitting a fixed object / other single object accident (4.5); 4: collision with a stopped vehicle (2.5); 5: people falling or thrown (2.1); 6: other accidents between vehicles (0.7)
	Crash (collision orientation)	1: side collision, scratching (68.7); 2: frontal collision (14.1); 3: other angle collision (10.1); 4: rear-end collision (7)
	Cause (classification of reasons)	1: other behaviors affecting safety (36.3); 2: illegal driving on the road (non-motor vehicle) (10.7); 3: other behaviors affecting safety (non-motor vehicle) (6.6); 4: violation of traffic signals (non-motor vehicle) (6.4); 5: retrograde (non-motor vehicle) (5.4)
Personnel information	Gender	0: female (30.4); 1: male (69.6)
	Age	1: 0–6 years old (0.1); 2: 7–12 years old (0.8); 3: 13–17 years old (1.9); 4: 18–30 years old (24.9); 5: 31–45 years old (38.7); 6: 46–69 years old (32.6); 7: > 69 years old (1)
	Job	0: only transportation by electric bicycle (97.6); 1: commonly used electric bicycles in working (e.g.: express delivery) (2.4)
	Edu	0: illiterate (0.1); 1: elementary school (13.7); 2: junior high school (61.4); 3: high school (15.9); 4: university (1.5)

(continued)

Table 1 (continued)

Data category	Variable name	Classification of variable attribute and proportion (%)
Road and facilities	Road (surface condition)	0: intact (96.4); 1: bumps, construction, other defects (3.6)
	Ctrl (traffic control mode)	0: none (6.1); 1: only signs (0.7); 2: only markings (1.9);3: signal light only (0.4); 4: only one other (0.7); 5: 2 control methods in the above (56.2); 6: 3 control methods in the above (30.4); 7: 4 control methods in the above (3.6)
	Site (of accident)	1: road Sect. (62); 2: intersection (33.8); 3: entrance/doorway (2.2); 4: internal road (1.7); 5: bus station (0.4)
	Light	1: daytime (54); 2: street lighting at night (41.6); 3: dusk/dawn (2.3); 4: no street lighting at night (2.1)
	Weather	0: sunny (86.4); 1: cloudy and rainy (13.6)

*Given that Cause includes many classification attributes, in order to simplify the calculation process, only attributes with a proportion greater than 5% are selected for coding and modeling, so the sum of the proportions of 5 items is less than 99.9% and only 65.4%

3.2 *Spatial Characteristics*

3.2.1 Spatial Distribution Characteristics

62% of electric bicycle accidents occurred on road sections, and 38% occurred at intersections, entrances and exits, internal roads, and bus stops. Road sections include ordinary sections, elevated sections, dangerous roadside sections and other special sections. Among them, ordinary sections accounted for 98.6%. The frequency of occurrence in other sections was low. The appearance of electric bicycles on elevated sections and other special sections is often caused by illegal behavior and speeding. Four-branch intersections had the highest frequency of traffic accidents, accounting for 53.8% of the total number of accidents apart from the road section, followed by three-branch intersections, accounting for 38.9%.

3.2.2 Road Type Characteristics

Statistics show that: 91% of electric bicycles accidents occurred on general urban roads, 5% of them occurred on self-built roads, public parking lots, squares and other roads in unit communities, and 4 of them occurred on highways and urban expressways.

3.2.3 Road Safety Characteristics

Statistics show that 8% of electric bicycle traffic accidents happened in abnormal road sections, including sections with hidden dangers that had been checked and not yet treated, sections that had been treated but still had hidden dangers, and sections that were under treatment.

3.2.4 Road Alignment

The data shows that electric bicycle traffic accidents occurred on curved sections accounted for 6% of the total, including general bends, sharp bends, steep slopes, continuous downhills, etc. The remaining 94% of accidents occurred on straight sections. Electric bicycles ridden on curved sections was usually slow, and can be braked in time in case of emergency to avoid some accidents.

3.2.5 Physical Separation of Roads

Statistics show that the proportion of electric bicycle traffic accidents on road sections without vehicle-bicycle barrier facilities was as high as 71%. Whether or not to

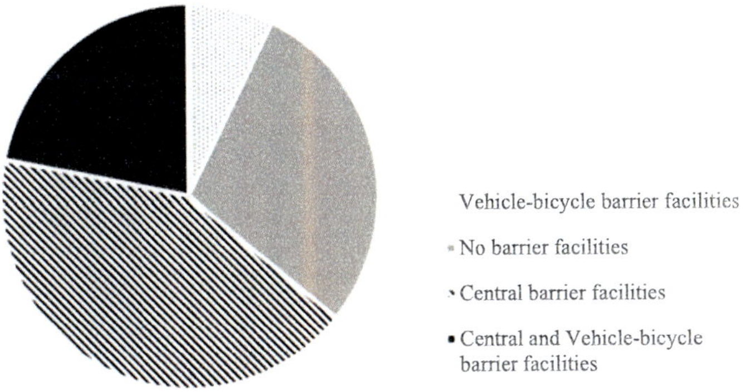

Fig. 1 Road physical isolation settings

install segregation vehicle-bicycle barrier facilities will have a greater impact on the occurrence of electric bicycle accidents, as shown in Fig. 1.

3.3 *Time Characteristics*

3.3.1 Distribution Characteristics of 24 h Per day

Statistics show that the occurrence of electric bicycle traffic accidents had obvious distribution characteristics in time, and the peak time period of accident rate and injury rate was basically consistent with the peak time of morning and evening commute. The mortality rate reached its peak at 6:00–7:00 and 10:00–11:00, which did not completely coincide with the peak time of traffic flow, as shown in Fig. 2. The reason was that 10:00–11:00 is the peak work period for takeaway and courier workers. The electric bicycles are used more, and the noon riding is tiresome and causes riders prone to slack, which is likely leading to serious accidents.

3.3.2 Distribution Characteristics of Each day of the week

Statistics show that the number of accidents occurring on seven days a week was relatively average, with the largest number of accidents on Saturday, followed by Monday, Friday, and Sunday, and it was an upward trend from Tuesday to Thursday, as shown in Fig. 3.

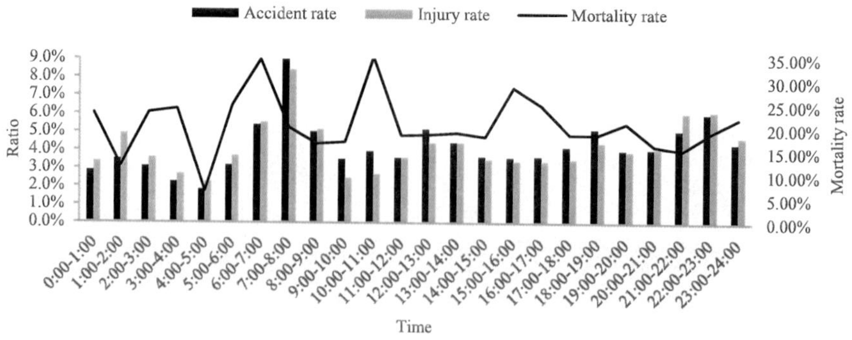

Fig. 2 Hourly trend of accident situation (accumulated 10 years)

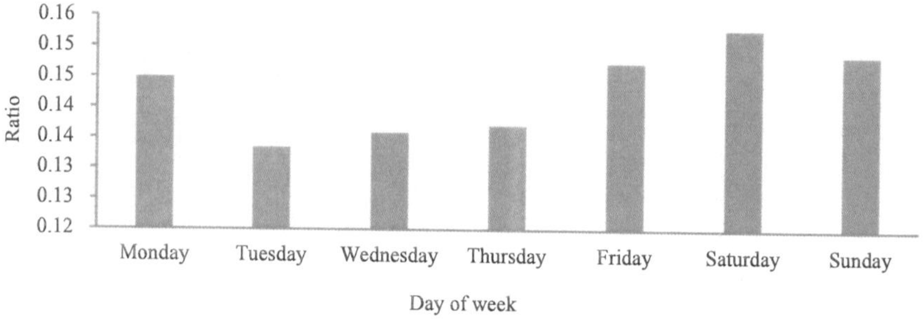

Fig. 3 The daily trend of accidents in a week

3.3.3 Distribution Characteristics of 12 months in a year

The number of traffic accidents in the first and second quarters was significantly lower than that in the third and fourth quarters, and this trend had been increasing year by year. Taking the ratio of each month's traffic accidents to the total number of accidents in ten years as the ordinate and the month as the abscissa, the monthly changes of accidents are shown in Fig. 4. It can be found that the mortality rate in December, January, March, and April was much higher than the average. The accident and injury rate showed the lowest in February, the highest in March, and a gradual increase from April to December. The reason was related to the season, climate, and holiday.

3.3.4 Ten-Year Overall Distribution Characteristics

Due to the incomplete statistics of electric bicycle traffic accidents and the small amount of data, it showed that the degree of electric bicycle traffic accidents and

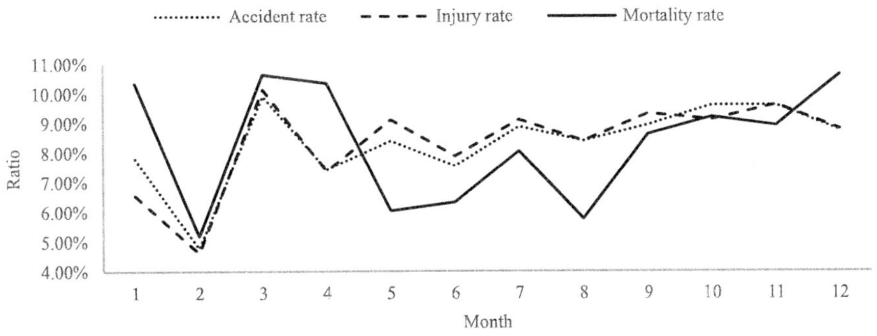

Fig. 4 Monthly trends of accidents

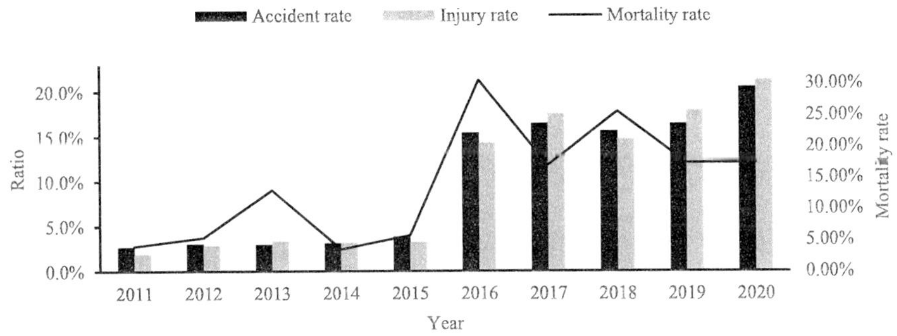

Fig. 5 Yearly trends of accidents

injuries had not changed much in 2011–2015. The situation of accidents and casualties from 2016 to 2020 changed with the year, showing a certain regularity. The number of traffic accidents and injuries was increasing year by year, whereas the mortality rate was decreasing throughout the time, as shown in Fig. 5. The reason was that the number of electric bicycles has continued to rise, and the number of accidents has also increased, but the people's safety awareness has become stronger, therefore the number of serious accidents has decreased accordingly.

3.4 Characteristics of Accident Types

The degree of injury of electric bicycle traffic accidents was relatively discrete. There were faults in the level of serious injuries (7.5%), with the highest proportion of minor injuries (51.5%), followed by no injuries (20%) and deaths (21%). Due to the small sample size of serious injury grades, this study divided the degree of injury into no-injury, injury, death, among which injury include serious and minor injuries.

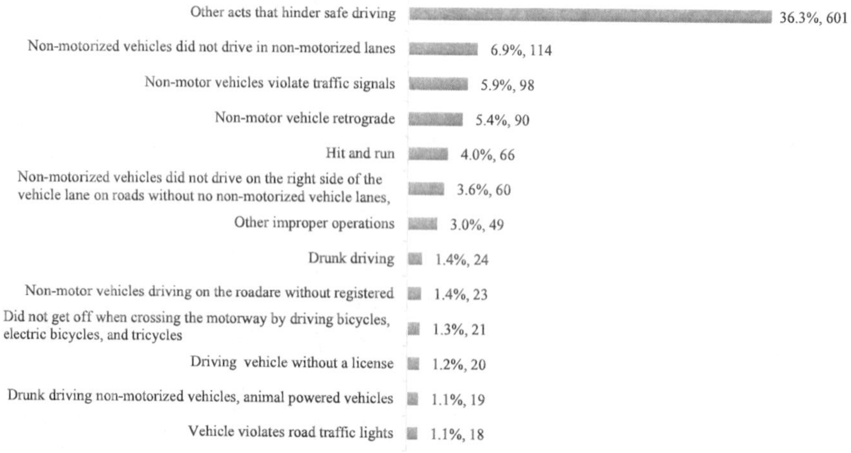

Fig. 6 Frequency of accident causes

3.5 *Cause Characteristics*

Data analysis shows that there were more than 70 identified causes of accidents, which were screened based on the standard that they account for more than 1% of the total. Among them, other behaviors that hinder safe driving accounted for the highest proportion of the identified causes of accidents, followed by illegal occupation of roads. As shown in Fig. 6, both of them were prone to conflict between the vehicles and non-motor vehicles. The proportion of sports vehicles was as high as 82.7%. Violation of signal lights and other non-motor vehicle violations is quite comment due to the low sensitivity in the mind of traffic participants,, but it has a great impact on traffic safety.

4 Correlation Analysis of Influencing Factors

4.1 *Multinomial Logit Models*

The dichotomous Logit model is the basic form of the multi-logit model, and its expression is as follows:

$$\log\left[\frac{P(y=1)}{1-P(y=1)}\right] = \sum_{k=1}^{K} \beta_k x_k \tag{1}$$

In the formula:

- K the number of independent variables;
- X_k k-th independent variable, k = 1, 2,…, K;
- β_k The regression coefficient of the k-th independent variable.

When the dependent variable has multiple values and no level or degree changes, the established regression model of independent variable and the dependent variable is multinomial logit model. When the dependent variable Y is a disordered variable of type I, the j-th category logit regression model is [12]:

$$\ln\left(\frac{P(Y=j|X)}{P(Y=I|X)}\right)=\alpha_j+\sum_{\omega=1}^{W}\beta_{jw}x_w \tag{2}$$

In the formula:

- X the collection of independent variables;
- W the number of independent variables;
- X_w w-th independent variable, w = 1, 2, …, W,
- α_j the constant term of the j-th category dependent variable, j = 1, 2, …, I − 1;
- β_{jw} The regression coefficient of the w-th independent variable in the j-th category.

In this study, a number of logit models were established based on the time series of electric bicycle traffic accident data, and the main influencing factors of the severity of electric bicycle traffic accidents were analyzed by investigating the p-values and regression coefficients of each variables.

4.2 *Model Standardization and Testing*

The 16 influencing factors of electric bicycle traffic accidents were used to construct multinomial logit regression models for accidents with no-injury, injury and fatal accidents, and standardized linear regression of the data, to obtain standardized coefficients, corresponding to the Beta value in Table 2. The Beta values of the independent variables were sorted from largest to smallest in each model. Table 2 shows the top 10 and bottom 10 factors.

The P values of the three models no-injury, injury and fatal accidents were all less than 0.05, indicating that the three models had passed the significance test. After excluding the independent variables with collinearity, there were 51 independent variables that constituted the three sets of regression models involving all 16 selected influencing factors.

Table 2 Model parameter estimation and standardized linear regression results

Dependent variable	Variable symbol	Beta	Variable symbol	Beta	Dependent variable	Variable symbol	Beta	Variable symbol	Beta
Hurt0	Edu2	0.337	Cause1	−0.137	Hurt1	Cause1	0.108	Edu2	−0.285
	Edu1	0.26	Site2	−0.073		Site2	0.078	Edu1	−0.196
	Edu3	0.247	Cause4	−0.047		B/D1	0.069	Edu3	−0.196
	Cause3	0.13	Site1	−0.041		Site1	0.064	Cause3	−0.113
	Edu4	0.092	B/D1	−0.039		Light2	0.044	Crash3	−0.091
	Gender1	0.082	Month3	−0.032		Month3	0.032	Age5	−0.089
	Age5	0.081	Weather1	−0.024		Month2	0.026	Edu4	−0.072
	Crash3	0.079	Month2	−0.02		Cause4	0.02	Age4	−0.07
	Age4	0.067	Age6	−0.019		Site3	0.016	Crash1	−0.066
	Cause5	0.067	Light2	−0.018		Age3	0.015	Ctrl5	−0.058
Hurt2	Age6	0.065	Time2	0.022	Hurt2	Light1	−0.05	Edu1	−0.019
	Ctrl5	0.033	Ctrl0	0.021		B/D1	−0.045	Light3	−0.015
	Crash3	0.032	Age4	0.019		Site1	−0.037	Gender1	−0.015
	Age5	0.028	Weather1	0.019		Light2	−0.036	Site3	−0.011
	Cause4	0.022	Crash1	0.018		Site2	−0.023	Month2	−0.011

4.3 Correlation Analysis of Main Influencing Factors

4.3.1 Correlation Analysis of Non-Injury Accidents Main Influencing Factors

The main influencing factors of no-injury accidents were education level and unsafe behavior of non-motor vehicles. In the model where the dependent variable was Hurt0, the positive value of Beta indicates that this factor was positively correlated with the occurrence of no-injury accidents. Since no-injury accidents and casualties are complementary in the accident data, a negative value of Beta means that the occurrence of no-injury accidents was complementary. To a certain extent, it could represent that the independent variable had a greater correlation with casualties.

The Beta value of elementary school, junior high school, and high school education level ranked in the top three, respectively 0.260, 0.337, 0.247. It could be seen that there was a strong correlation between the level of culture and education and the occurrence of no-injury accidents of electric bicycles. The accidents were analyzed in combination with the characteristics—the number was normally distributed in terms of literacy level. It can be seen that there were many accidents in education level of elementary school, junior high school, and high school, but they did not often cause injury or death.

The Beta value of other behaviors affecting safety of vehicles was −0.137, the absolute value was the largest in the negative values, and it had the greatest impact on casualties. For example, vehicle and carriages driven out without roadside doors being closed properly and in time could bring great safety hazards in electric bicycles. The Beta value of other behaviors affecting safety of non-motor vehicles was 0.130. Compared with vehicles, the unsafe behaviors of non-motor vehicles were less harmful and less likely to cause casualties.

4.3.2 Correlation Analysis of Injury Accidents Main Influencing Factors

The main influencing factors of injury accidents were other behaviors affecting safety and illegal behaviors of non-motor vehicles. In the model where the dependent variable was Hurt1, the Beta value of other influencing safety behaviors was 0.108, which was the maximum value among positive values. It could be seen that unsafe behavior of motor vehicles was the most important factor affecting electric bicycle injury accidents. The Beta value of cyclists was 0.069, and the violation of traffic lights by non-motor vehicles was 0.02. These non-motor vehicle behaviors were strongly correlated with the occurrence of injury accidents. The Beta value of the intersection and road section of the accident site were 0.078 and 0.064, which showed that the intersection had a greater impact on the occurrence of injury accidents than road sections.

4.3.3 Correlation Analysis of Fatal Accidents Main Influencing Factors

The main influencing factors of fatal accidents were the age of the driver and the traffic control method at the accident location. In the model with the dependent variable of Hurt2, the Beta value of 46–69 years old was 0.065, which was the most significant influencing factor for fatal accidents. People in this age group should strengthen safety education and improve safety awareness. Setting the upper age limit for the drivers could also help.

The Beta value of two or more traffic control methods in the accident site was 0.033, and the Beta value of no traffic control in the accident site was 0.021, indicating that traffic control methods should be set up reasonably according to local conditions, otherwise serious accidents would be caused.

The Beta value of other angle collisions was 0.032, and the Beta value of side collision scratches was 0.018. With the driving style of electric bicycles, side collisions were more likely to directly contact the driver's body and therefore increased the degree of injury.

The Beta value of non-motor vehicles violating the traffic signal was 0.022. Violating the traffic signal would increase the risk of conflict. When it encounters situations such as slow deceleration and distracted driving at the same time, the consequences of the accident would be more serious, causing injury or even death.

5 Conclusion

Based on the data of electric bicycle traffic accidents in an representative city in southern China, this study analyzed the characteristics of them, constructed multinomial logit models, found out and examined the correlation between the main influencing factors, leading to the conclusion as below:

(1) There were five obvious characteristics of electric bicycle traffic accidents: ①The peak of accident rate and injury rate in a day coincided with the peak of traffic flow in the morning and evening commute; ②The accident rate and injury rate were increasing year by year, whereas the mortality rate was decreasing over years; ③Crashing sports vehicles was the most common accident pattern; ④Hindering safe driving accounted for the largest proportion of the causes of accidents; ⑤The accidents mostly occurred on road sections without vehicle-bicycle barrier facilities.
(2) Through the analysis of model parameters, it was known that the cultural level had the greatest impact on non-injury accidents, unsafe and illegal behaviors had the greatest impact on injury accidents, and the driver's age and traffic control methods had the greatest impact on fatal accidents.
(3) In terms of driver characteristics, the age of 46–69 was likely to cause serious accidents. The risk could be controlled by strengthening safety education and capping the age of electric bicycle drivers.

(4) In terms of behavioral characteristics, non-motorized vehicles violating traffic signals, non-motorized vehicles retrograde, electric bicycles carrying people, and unsafe behaviors of vehicle were all major factors to cause accidents, injuries and deaths. It could improve the awareness of law-abiding by electric bicycle riders through traffic safety publicity and education, and reduce the risk of accidents by setting up complete road isolation facilities.
(5) In terms of time characteristics, for the time when electric bicycle accidents were frequent, it should strengthen road guidance, on-site and off-site enforcement to reduce accidents.
(6) In terms of environmental characteristics, bad weather such as cloudy and rainy days would increase the risk of riding electric bicycles. It was recommended to minimize riding trips and strengthen the setting of lighting facilities to improve the safety of electric bicycles traffic.

References

1. Chun-jiao D, Wei-yang X, Cheng-xiang Z (2021) Decision-making modeling for unsafe riding behaviors on electric bicycles. J Beijing Jiaotong Univ 45(01):105–110
2. Xing-hui M, Guo-qing M, Ju-hui F et al (2021) Compare analysis of collision injury between vehicle-bicycle and vehicle-electric-two-wheeler. Autom Technol (03):51–56
3. Huan L, Zhong-hao B, Wen-rui G et al (2021) Cyclist head responses in electric-bicycle-to-SUV accidents based on decision tree model. Automot Safety Energy 12(01):43–51
4. Yi-ming S, Guang-xue H, Zhao-fu D (2018) A method to calculate the vehicle collision speed based on the difference of throw distance in a car-electric bicycle crash [A]. China Acad J Electron Publ House. In: Proceedings of the 13th China intelligent transportation conference [C]. China Intelligent Transportation Association: China Intelligent Transportation Association, 8
5. Tao W, Wen-hao L, Wen-yong L (2017) Influence factors and injury severity in electric bicycles traffic crashes. J Guangxi Univ (Nat Sci Ed) 42(06):2080–2088
6. Wei-jie W, Xuan-ting S, Gui-bin W et al (2019) Analysis of factors affecting injury to electric bicycle rider in crash. China Saf Sci J 29(02):20–25
7. Wen-long J, Wan-ting Z, Fang L et al (2021) Analysis of the factors influencing traffic accidents severity involving electric bicycle—an example from Zhoushan City. J People's Public Secur Univ China (Sci Technol) 27(01):49–55
8. Cheng X, Man-quan G (2018) Analysis of the factors influencing traffic accidents severity involving electric bicycle—an example from Hangzhou [J]. J People's Public Secur Univ China (Sci Technol) 24(03):75–80
9. Dan Q, Chang-shuai W, Ao-ran D et al (2021) Research on the factors affecting injury severity of electric bicycle riders. J Wuhan Univ Technol (Transp Sci Eng) 1–12. http://www.kns.cnki.net/kcms/detail/42.1824.U.20201022.1755.010.html
10. Yao L (2017) Analysis of factors affecting personal injury of electric bicycle drivers in traffic accidents [D]. Tsinghua University
11. Ying-shuai L, Xu Z, Wei-jie W et al (2021) Factors affecting electric bicycle rider injury in accident based on random forest model. J Transp Syst Eng Inf Technol 21(01):196–200
12. Ao-ran D, Chang-shuai W, Yun-chao J et al (2021) Research on influencing factors of injury severity of electric bicycle drivers based on logit model. J Wuhan Univ Technol (Transp Sci Eng) 45(02):243–247

Study on the Operation Scheme of Standby Trains with Large Tidal Passenger Flow Under Full-Length and Short-Turn Mode

Hui Cheng, Mao Ye, and Yaxin Chen

Abstract The train operation scheme is the foundation of the urban rail transit organization. In order to improve the transportation efficiency of urban rail lines under the characteristics of tidal passenger flow, alleviate the congestion phenomenon of large passenger flow stations, and shorten the travel time of passengers, this paper focuses on the morning peak passenger flow stations on working days, and proposes a kind of urban rail backup vehicle for the characteristics of tidal passenger flow. Scheme to quickly relieve congested passenger flow. First, in the form of the upper and lower two-level full-length and short-turn mode, through the analysis and research on the conditional nature of the line site points, it is concluded that on the basis of determining the placement site of the standby train on the existing line, the heuristic algorithm particle swarm with constraints is used The optimization algorithm takes the minimum total waiting time at a large passenger flow station as the objective function, and considers the constraints of the number of trains, the cross-section full load rate, the train tracking interval and the minimum train departure frequency, and establishes the optimization model of the reserve train placement scheme. Based on the case analysis of Nanjing Metro Line 3, the results show that Liuzhoudonglu Station is a large passenger flow station during the morning peak hour on this line. There are two ready-to-deliver stations in the upstream direction. The capacity analysis shows that during the morning peak hour A total of 43 trains are operated. On the basis of the original 21 pairs of large and small roads, a standby train can be added to the No. 4 Taifunglu station to shorten the total waiting time of passengers by 721 s during the morning peak hour and alleviate the problem of tidal passenger congestion. The research results can be used for rail transit operation strategies and

H. Cheng · M. Ye (✉) · Y. Chen
Nanjing University of Science and Technology, 200 Xiaolingwei, Nanjing 210094, China
e-mail: yemao0924@163.com

H. Cheng
e-mail: 1753784137@qq.com

Y. Chen
e-mail: chenyaxin@njust.edu.cn

W. Wang et al. (eds.), *Green Transportation and Low Carbon Mobility Safety*,
Lecture Notes in Electrical Engineering 944,
https://doi.org/10.1007/978-981-19-5615-7_31

improve the efficiency of passenger flow relief under the condition of large passenger flow.

Keywords Urban rail transit · Standby train · Particle swarm algorithm · Full-length and short-turn routing

1 Introduction

Urban rail transit serves urban public transportation, and on weekdays, the rail passenger flow is mainly for commuters to work and school. It is the premise and foundation to coordinate the operation of each part and guarantee the efficiency and safety of urban rail passenger transport to formulate the corresponding train operation scheme under the specific passenger flow conditions. Therefore, under the background of large passenger flow, how to scientifically and rationally formulate the train operation scheme has become the research focus of the operation management of urban rail transit. Considering the characteristics of tidal passenger flow on weekdays under the background of large and small intersections, this paper studies the scheme of the standby train operation under this scenario.

The tidal passenger flow of urban rail transit refers to the peak passenger flow of urban rail transit system during the morning and evening commuting, school and school periods. With the expansion of urban rail transit network and the enhancement of line accessibility, more and more residents choose urban rail transit to commute, resulting in a large passenger flow in the morning and evening peak hours. How to effectively organize the operation of subway station under the condition of large passenger flow, so that the advantages of subway 'fast, convenient, comfortable and safe' can be fully reflected, has become an important research topic. On the basis of the generalized passenger flow characteristics of urban rail transit, some scholars have studied the characteristics of large passenger flow. Xiangxi [1] analysed the passenger flow characteristics of sudden large passenger flow and proposed the four-stage prediction method and process of sudden large passenger flow. On this basis, the types and characteristics of the train stop schemes of urban rail transit were analysed, and the basic optimization models of different stop schemes were established based on the operational organizational efficiency. Wang Min[2]studied the characteristics of passenger flow at entrances and exits, channels, ticket machines and landing equipment when large passenger flow occurred, and analysed the influencing factors of passenger flow organization on this basis. Jie [2] determined the influencing factors of evacuation passenger flow and evacuation organization and management based on the analysis of the characteristics of passenger flow. Jiang et al. [3]. aiming at the problem that the platform of the subway transfer station is prone to congestion in peak hours, built an optimization model of crossing stations on the basis of analysing the process of passengers' arrival and transfer.

Ruihua et al. [4]. analysed the influence of large and small intersections on the passage capacity and vehicle underbed utilization of urban rail transit lines, and established the model of passage capacity and vehicle underbed quantity under the operation mode of large and small intersections. The research shows that the loss of line passing capacity and the increase of the number of undertraining use are directly proportional to the ratio of the trains running on different routes, while the increase of the number of undertraining use varies periodically with the increase of the ratio of train travel time in different routes. Yandong [5] established an optimization model of the optimal operation scheme based on the operation mode of large and small intersections, aiming at minimizing the sum of actual passengers and train occupancy variance and the minimum number of used train bases. Junfang et al. [6]. analysed the passenger flow conditions applicable to large and small intersections, and calculated the changes of their passing capacity and the number of undertraining applications. Dandan et al. [7, 8]. and Jungang et al. [9]. established the optimization model of multi-objective train operation scheme and the optimization model of subway fast and slow train operation scheme based on flexible marshalling mode, both of which showed that the train operation scheme of flexible marshalling could minimize the operation cost. Yanxiang et al. [10]. studied the variation of the number of vehicles when different train groups were used in mixed running and the same train group were used in the operation of urban rapid rail transit lines under the condition of large and small interchanges. Kaspi et al. [11]. developed an integrated route scheme and schedule model based on the total time spent in the passenger system without considering the constraints of train capacity. Zekang et al. [12]. studied the problem of train operation adjustment under changeable train running routes in railway network, built the network flow model based on the arrival and departure time of trains and the integer programming model based on the space-time path of trains, and finally found that the train departure time could be reduced by adjusting changeable train running routes. Li et al. [13]. constructed a multi-objective nonlinear programming model for intercity train operation scheme considering regional coordination with constraints such as intercity train training capacity, departure and destination passenger flow demand and transit capacity, and solved it with simulated annealing algorithm. Yuanguang et al. [14]. proposed an optimization method of the high-speed and slow train operation scheme based on the double-layer programming model. Considering the transfer behaviour of passengers, the travel time of passengers and the uneven spatial and temporal distribution of passenger flow, it was shown that the high-speed and slow train scheme was better than the station-stop scheme, and the multi-route scheme was preferred with the train turnover time as the optimization objective.

In the operation practice of urban rail transit, there are many transportation organization and management schemes for sudden large passenger flow, but these schemes are mainly formulated using the methods and theories of daily rail transit scheming and management, and no special system has been formed [15–17]. With the rapid development of urban rail transit and the increasing sudden passenger flow, the conventional rail transit management scheme has been difficult to adapt. Through a cross road of size under the background of tidal schedule after the big standby extra train passenger flow optimization study, can make up for the deficiency of the urban rail transit operation management theory, for the operation management of urban rail transit system plays a guiding role, can lower the operating costs of operating units, improve the service level of urban rail transit system. In order to quickly relieve the passenger flow of the large passenger flow station on the urban rail line and reduce the total delay time of passengers, this paper takes the way of putting standby trains into the station and considers the station as a variable to establish the best time selection model of putting station into the station. In order to meet the needs of dynamic passenger flow changes, shorten the travel time of passengers and improve operational efficiency and safety, and enrich the technical methods of urban rail transit operation organization.

2 Problem Description

The problems studied in this paper can be summarized as follows: Known target line OD of the passenger flow distribution and related operational parameters, the upper and lower two layers of size of road form, standby extra train at appropriate site, by a large passenger flow site, the objective function is minimum total passenger waiting time, consider using the train number, section utilised, train tracking interval and the minimal frequency constraints, such as Small cross road train operation section, big and small standby the station then the frequency of the train and put on extra points, and select the relevant operational indicators, the model of the optimal train dispatching scheme is compared with the existing schemes, designed to ensure transportation efficiency, reducing the amount of all the passengers waiting time, reduce the congestion of the train at the same time, improve the passengers' travel satisfaction.

As shown in the figure, it is the schematic diagram of full-length and short-turn mode. For urban rail transit lines L, the station set is defined as S, $s = \{s_x | x = 1, 2, \ldots, n\}$, and the route directions set is D, The number of elements in the set satisfies $|D| = 2$, the route direction is d, $d \in D$, and $d = 1$ denotes the upward direction, that is the train is going from S_1 to here S_x. $d = 2$ denotes the downward direction, that is the train is going from S_x to S_1. The set of stations with the conditions of standby train is defined as N, $N = \{N_x | x = 1, 2, \ldots, n\}$. S_1 and S_n are for large interchange turnback stations, S_1 and S_a for small interchange turnback stations. At the same time, according to the train running cycle theory, it is considered that the number of trains running in the section of small interchange is m twice that of large interchange. At one time, after the station has reached the condition of putting the standby trains into the station, it is necessary to use the means of putting the standby trains into the station S_i to relieve the emergency situation of the large passenger flow station S_i in the following unit time.

In Fig. 1, the red stations are the stations with large passenger flow, and the black stations are the stations with the condition of placing standby trains. The site should be placed in the site with parking lines, and the above direction is taken as an example ($d = 1$). In order to evacuate the site with large passenger flow outbreak station S_i, the site to put the standby train must be upstream of it, that is only the station N_1

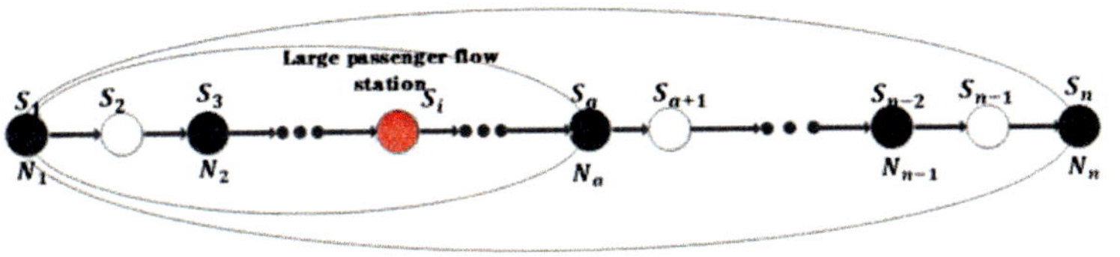

Fig. 1 Schematic diagram of large path pattern

、N_2 can be put as the option to put the standby train. The same analysis method can be used for the downward direction. This paper mainly selects the upward direction (i.e. $d = 1$) for analysis. In the next steps, we need to put the standby trains from the two stations N_1、N_2 and the subsequent schemes, which one is most suitable for relieving the emergency situation at the current station n, and get the specific departure schemes of the standby trains and the follow-up trains.

In the case of a large passenger flow in the middle station, the alternative standby is the station with the condition of placing passengers before it. Then the peak time of the large passenger flow is determined through calculation, and the train that runs in the original unit m time is changed into $m + 1$ train. If the train is put into the first station, it can be understood that the first train is the additional train. The subsequent train schedule is dynamically adjusted according to the arrival law of passenger flow in bus station and line, and a new train schedule is generated. If it is put into other stations, the impact of the newly inserted standby trains on the front and back schedule should be considered, that is, the number of passengers who are already waiting when the train arrives and the number of passengers who are stuck after the train leaves are described in the objective function.

3 Timing Selection of Tidal Large Passenger Flow Stations and Standby Vehicles

3.1 Identification of Large Passenger Flow

The occurrence of large urban rail passenger flow depends on the biggest service with all facilities in and out of the station traffic capacity, traffic station and outbound traffic can be obtained by data, usually, if the passengers personal occupy spacc arca is less than 0.4 m^2, passengers will feel particularly crowded, and the whole traffic flow rate is extremely low, at this time the station is in a state of large passenger flow obviously, In thc 'Mctro Dcsign Code', when designing the platform area, the minimum population density of the platform is also 0.4m^2. Table 1 divides passenger flow into 5 levels according to the standard of the space occupied by passengers [18].

The identification of large passenger flow needs to rely on the real-time passenger flow data in the station. Different types of rail transit systems have different methods

Table 1 Urban rail transit passenger flow identification level

Density (person/m^2)	Passenger occupied space /m^2	Passenger flow density level	Passenger flow identification level
>2.5	<0.4	1	Large passenger flow
2–2.5	0.4–0.5	2	
1.5–2	0.5–0.67	3	Medium passenger flow
1–1.5	0.67–1	4	
<1.5	>1	5	Small passenger flow

for the collection of passenger flow data. At present, the Nanjing city rail transit system in the passenger flow data through the Automatic Fare Collection (*AFC*) system analysis of system real-time statistics passengers in and out of the station number, and through the vehicle control room video monitoring system and the affair personnel on-site observation, when passengers platform and station hall layer dense and slow, can judge the station has a large passenger flow situation arises, should immediately start big traffic emergency measures at this time.

3.2 *Optimization Model of the Standby Vehicle Launch Scheme*

By adjusting the standby train delivery stations, the optimization model of the standby train delivery scheme was established with the goal of minimizing the waiting time of the passengers in the large passenger flow stations. The model considers both the number of passengers already waiting when the train arrives, $D_{i,a}$ and the number of passengers stranded after the train leaves, $L_{i,a}$. The specific calculation formula is as follows:

$$D_{i,a} = \int_{f_{i,a}}^{f_{i-1,a}} \lambda dt \tag{1}$$

$$Z = \sum_{i=1}^{n} \left[D_{i,a} \frac{(S_{i,a} - f_{i-1,a})}{2} + L_{i-1,a}(S_{i,a} - f_{i-1,a}) \right] \tag{2}$$

In the above formula, the default is that the standby train starts from the first station, and the place of the standby train is not treated as an input variable. On this basis, the 0–1 variable B_h is introduced in this paper to represent the choice scheme of the place of the standby train, where the value range of h is the number of all the possible places. If it is delivered from the first station $B_1 = 1$, the rest B_h are 0, otherwise B_1 is 0, and one of the others B_h is 1, $h \in N$. Then the selection model of the final standby vehicle placement site is established as follows:

$$\max Z = \sum_{h} B_h \cdot z_h \tag{3}$$

Considering the actual situation of train operation, the model needs to meet the following seven constraints:

(1) Stopping time of standby train:

$$f_{I,a}^{b} - S_{I,a}^{b} = b_a \tag{4}$$

(2) Stopping time of ordinary train:

$$f_{i,a} - S_{i,a} = t_a \tag{5}$$

(3) Interval travel time:

$$S_{i,a+1} - f_{i,a} = t_r \tag{6}$$

(4) Running time of the standby train:

$$S_{i,a}^{b} - f_{i,a}^{b} = t_r \tag{7}$$

(5) Tracking interval constraint:

$$S_{i+1,a} - S_{i,a} \geq h_t \tag{8}$$

$$f_{i+1,a} - f_{i,a} = h_t \tag{9}$$

$$S_{I,a}^{b} - S_{i,a}^{b} \geq h_t \tag{10}$$

$$f_{I,a}^{b} - f_{i,a}^{b} \geq h_t \tag{11}$$

$$S_{i+1,a} - S_{I,a}^{b} \geq h_t \tag{12}$$

$$f_{i+1,a} - f_{I,a}^{b} \geq h_t \tag{13}$$

(6) The waiting time of the two trains before and after the standby train stops is 1.5 times of the usual waiting time ($\bar{h} = 333$ s):

$$S_{I+1,a} - DS_{I-1,a} \leq 1.5 * \bar{h} \tag{14}$$

(7) Urban lines: small crossing section or whole line of single crossing, peak $\leq$ 6 min, flat peak $\leq$ 7 min, low peak $\leq$ 9 min; The maximum whole-day interval in the non-collinear section of the large crossing road is $\leq h_{\max} = 12$ min. Under the maximum return capacity of the signal system, the minimum tracking interval $h_{\min} = 2$ min:

$$h_{\min} < h_t < h_{\max} \tag{15}$$

3.3 Algorithm Flow

The general optimization algorithm is difficult to achieve the optimal result and the iteration speed is slow. In addition, continuous variables such as 0-1 variable and departure interval are involved in this paper. A heuristic algorithm with constraints—particle swarm optimization algorithm with penalty function is adopted to optimize the objective function by selecting different delivery sites. The specific process is as follows:

(1) Initialization
Initialization of particle swarm (there are n subway trains in total): each particle is given a random initial location and speed of the station;
(2) Calculation of fitness value
According to the fitness function, the fitness value of each particle is calculated according to the constraint condition of train operation as penalty function.
(3) Determining the best individual adaptation value
For each particle, the adaptation value of its current site placement position is compared with the adaptation value corresponding to the historical best site placement (PBEST). If the adaptation value of the current site placement position is higher, the current site placement position is used to update the historical best site placement position.
(4) Evaluate the optimal adaptive value of the group
For each particle, the adaptation value of the current site placement position is compared with the corresponding adaptation value of the global best site placement (GBEST). If the adaptation value of the current site placement position is higher, the current site placement position is used to update the global best site placement position.
(5) Update the position and speed of the particle delivery site
Update the velocity and location of each particle according to the formula;
(6) Determine whether the algorithm is over
If the end condition is not met, then return to Step 2. If the end condition is met, the algorithm will end, and the global optimal site placement position (GBest) is the global optimal solution (optimal site placement) (Fig. 2).

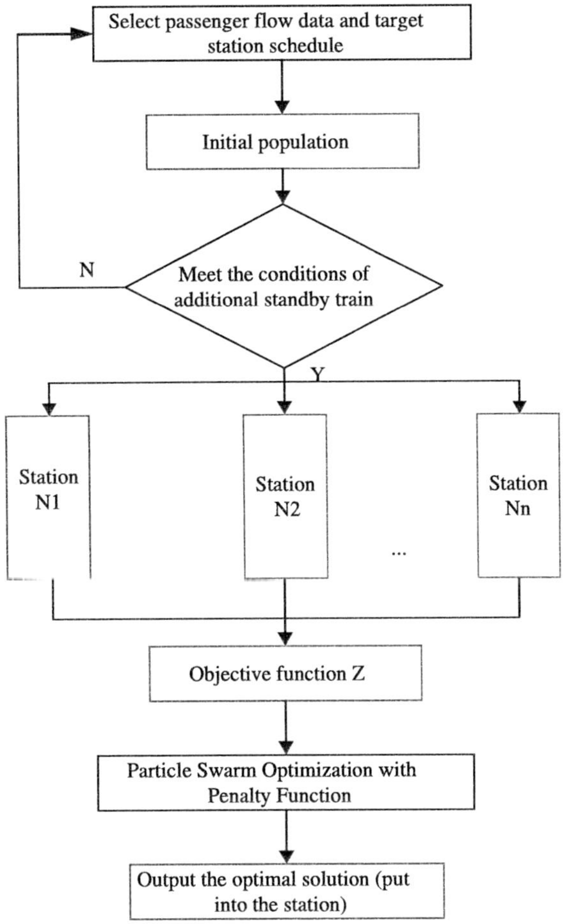

Fig. 2 Overall algorithm flow chart

3.4 Definition of Parameters

The parameters and variables involved in paper modeling are defined as shown in Table 2.

4 Case Study

4.1 Overview of Line Operation Mode

There are 29 stations in Nanjing Metro Line 3, of which only Linchang Station is an elevated station, while the other 28 are underground stations. There are 5 transfer

Table 2 List of parameters and variables

Notation	Definition
i	Train Index
n	Station index
a	Number of congestion Station
I	Number of the standby train
λ	The passenger flow arrival rate
$D_{i,a}$	Number of passengers already waiting when the train arrives at the large passenger flow station, which can be represented by the passenger flow arrival rate
$S_{i,a}$	Time when the train arrives at the station in the optimized schedule
$f_{i,a}$	Time after the train leaves the large passenger flow station in the optimized schedule
$L_{i,a}$	Number of stranded passengers after the train leaves the large passenger flow station, which can be calculated from the passenger flow arrival rate
$f_{I,a}^{b}$	Time after the number standby train leaves the large passenger flow station
$S_{I,a}^{b}$	Time when the number standby train arrives at the large passenger flow station
$S_{i,a}^{b}$	Arrival time of the number standby train in the optimized schedule
$f_{i,a}^{b}$	Time after the number standby train leaves the large passenger flow station in the optimized schedule

stations, respectively, Taifeng Road Station to transfer to Line S8, Nanjing Station to transfer to Line 1, Jimingsi Station to transfer to Line 4, Nanjing South Station to transfer to Line S1, S3 and 1. In addition to the five transfer stations, Liuzhou East Road Station is also a major passenger traffic jam station on the line.

The total length of Nanjing Metro Line 3 is 44.9 km, of which the elevated line is 2.4 km and the underground line is 42.5 km. This line adopts 6 A-type drum trains, each train can seat 310 people, A total of 1,860 people, and the maximum running speed is 80 km/h. There are 7 transfer stations on the whole line, which are the upstream and downstream lines of Taifeng Road Station, Xiaoshi Station, Fuqiao Station, Daminglu Station and Shengtaixilu Station respectively. At the same time, there are two back and back lines at both the first and the last stations. The current operation mode adopted by this line is the large and small interchange mode during peak hours. Reasonable use of this mode can relieve the congestion of trainings and improve the efficiency of passenger transport. The specific operation mode is shown in Fig. 3. However, the largest section of OD passenger flow is the round-trip section from Nanjing South Railway Station to Nanjing Railway Station. Passengers in this section can take Line 1 or Line 3 to relieve part of the passenger flow. However, passengers from Liuzhou East Road Station to Xiaoshi Station and Liuzhoudonglu Station to Nanjing Station can only choose to take Line 3, which causes the section to be extremely crowded during the morning and evening peak hours and requires extra standby trains to alleviate the large passenger flow problem.

This study is based on Nanjing Metro Line 3. Through the analysis of line station passenger flow, the passenger flow is easy to gather at East Liuzhou Road Station.

Fig. 3 Route diagram of Nanjing Metro Line 3

Therefore, this paper takes East Liuzhou Road as a typical station, and aims to relieve the congestion of passenger flow on the whole line by relieving the passenger flow at this station. The station is located at the intersection of East Liuzhou Road and Jiangshan Road, Pukou District, Nanjing City. Within the subway radiation range of 2 km, there are more than 10 large and medium-sized communities, such as Venice Water City, Mingfa Binjiang New City, Tianrun City, Xulii Upper City, and Qiaobei New Village, with a resident population of more than 300,000. The station is the first stop for Line 3 to cross the river from downtown. The passenger flow of 15 min around Tomb-sweeping Day in the past five years is shown in Fig. 4. It can be analysed that the passenger flow surged during the morning peak period from 6:00 to 9:00, with the highest daily passenger flow of 5268 person-times/15 min. The weekday passenger flow analysis of a certain week in 2019 is selected as shown in Fig. 5. It can be seen that this station is a peak-type station. During the morning peak period from 6:00 to 9:00, the peak passenger flow still continues, with the maximum daily passenger flow of 4927 person-times/15 min. In the morning peak hour, the upward train is almost full every time when it is sent out from Liuzhou East Road Station. Although some passengers get off when the train arrives at Nanjing Railway Station, a large number of commuters and students get on at the same time, which often leads to the situation that passengers "cannot get off and cannot get on". Although Nanjing Metro Station has also taken countermeasures, but the problem still cannot be fundamentally solved, we must find a way from the transport capacity. With the continuous development of urban traffic level, the passenger flow of public transportation will increase continuously, so as to ensure travel safety, improve travel efficiency and improve the waiting environment for passengers. This paper proposes a way to put extra standby trains into the crowded stations of tidal city rail line.

4.2 Data Acquisition

The actual passenger flow data of Line 3 on March 13, 2019 are selected, and the AFC data of subway is screened by Java language in 15 min. Through the statistical analysis of passenger flow, the time period for backup trains to run and the station for target distribution are obtained. Liuzhou East Road Station is finally determined to be a large passenger flow station, its station number is 6. The time conditions for satisfying the state of large passenger flow are 07:00–09:00 and 18:00–19:15. The time period from 07:00 to 09:00 was selected as the research object in this paper.

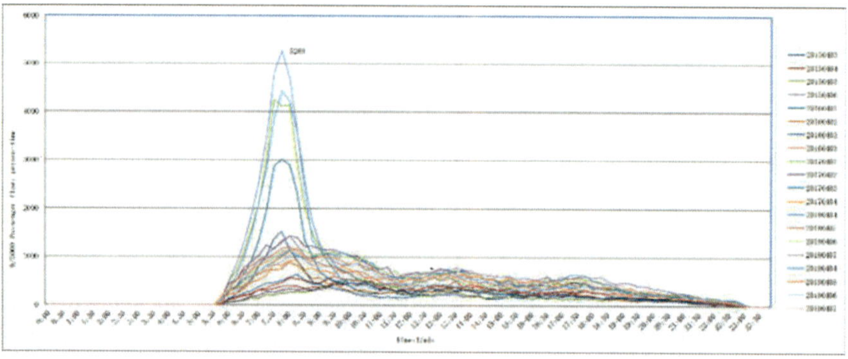

Fig. 4 Passenger flow statistics (Liuzhou East Road Station) during Tomb Sweeping Festival from 2015 to 2019

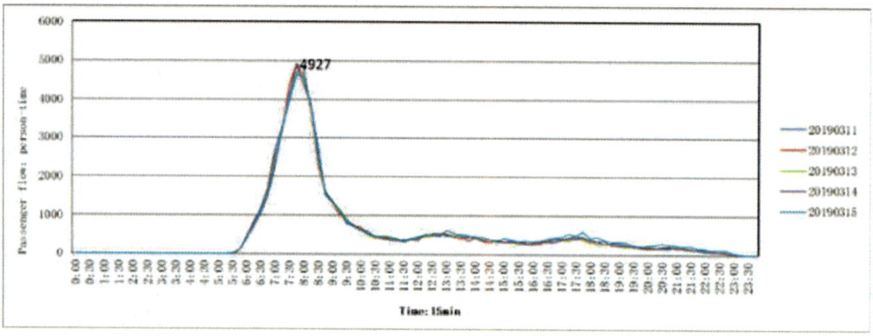

Fig. 5 Passenger flow statistics of weekday (Liuzhou East Road Station) in a certain week in 2019

The schematic diagram of the station and the time distribution diagram of passenger flow are shown in Figs. 6 and 7.

According to the image analysis, Liuzhou East Road Station No.6 is a large passenger flow station, and the stations that can put the standby trains in line No.3 are Lin Chang Station No.1, Tai Feng Lu No.4, Xiao Shi No.9, Floating Bridge

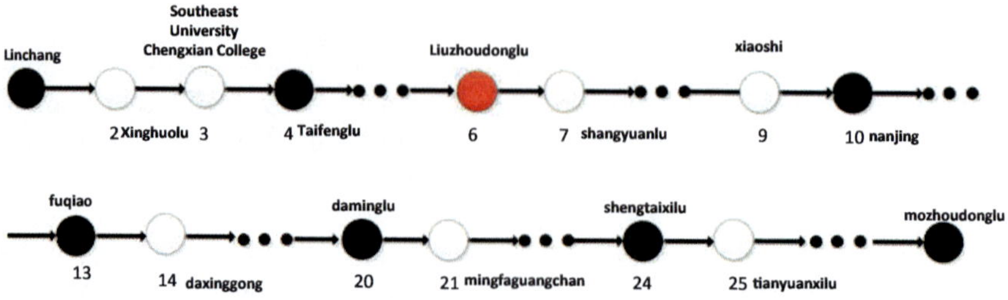

Fig. 6 Schematic diagram of large passenger flow station in the upward direction

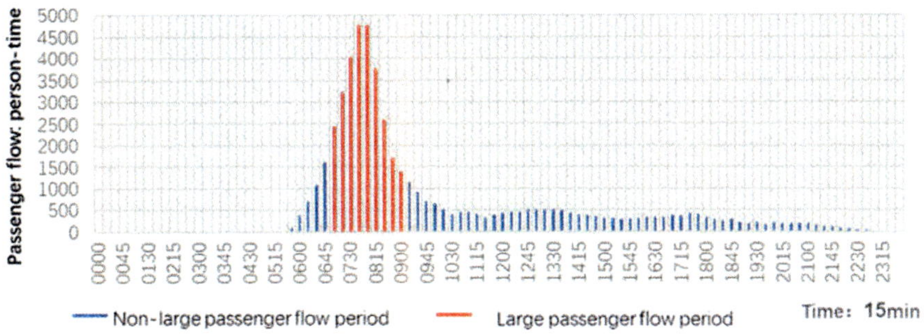

Fig. 7 Time distribution diagram of passenger flow in Liuzhou East Road Station

No.13, Da Ming Lu No.20, Shengtai West Road No.24 and Mo Zhou Dong Lu Station No.29, which meet the requirements. If you take the upward direction from Lin Chang to Mo Zhou Dong Lu, and do not stop before arriving at the big passenger flow station No. 6, only No. 1 and No. 4 can be put into the standby train; If you consider the downward direction, there are 9, 13, 20, 24, 29. Since this paper only considers the line operation during the morning peak hour, two stations, No. 1 and No. 4, are considered.

4.3 Result Analysis

MatlabR2017A programming was used to realize the selection modelling and particle swarm optimization algorithm solution of the standby vehicle loading station. Without changing the original operation scheme, we will operate the major interchange from Linchang to Mozhoudonglu, and the minor interchange from Linchang to Shengtaixilu in the middle, under the premise of 1:1 operation scheme, the corresponding schedule is calculated by adding the standby train at the No. 1 station meeting the conditions, and then the corresponding schedule is calculated by adding the standby train at the No. 4 station meeting the conditions. By putting the two results into the objective function, the results of the two schemes are compared. The fitness chart of the two subway stations is shown below. It can be seen that under the experimental background of 100 iterations and 0.97 convergence accuracy, the No. 1 station begins to converge at 45 times when it is put into operation, while the No. 4 station begins to converge after 13 times when it is put into operation, and its fitness is closer to 0 faster (Figs. 8 and 9 and Table 3).

For Nanjing Metro Line 3, the load rate of the first bus from 6 o'clock on the upward direction is more than 90%, and it reaches more than 110% from 7:30 to 8:30. From Taifenglu to Nanjing Station, it is extremely crowded, and the load rate is very high. According to the objective function, the waiting time of passengers at the station with the largest passenger flow is smaller when the standby train is put into the No. 4 station, namely Taifenglu. Through the case study, it is found that after

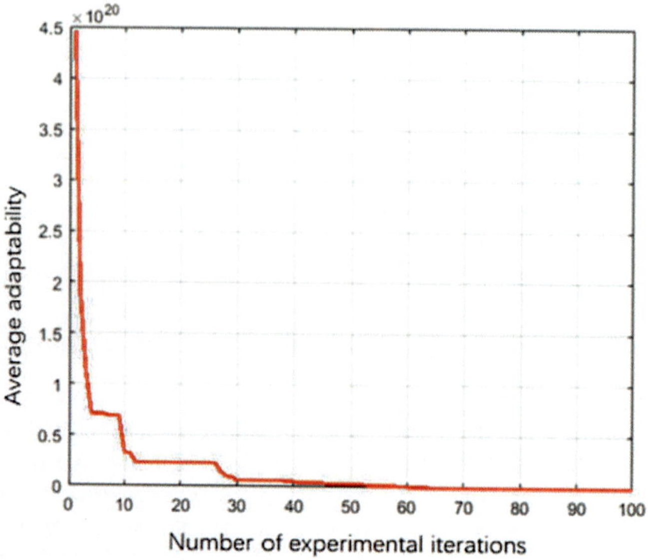

Fig. 8 Adaptability of placing standby trains in the no. 1 station

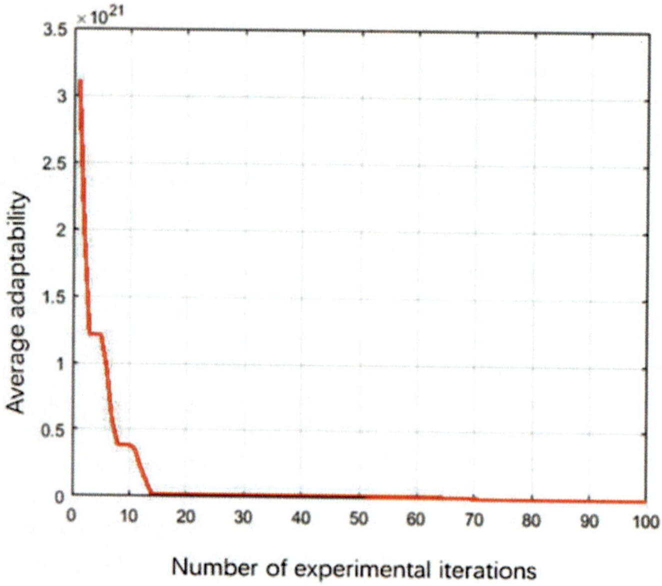

Fig. 9 Adaptability of placing standby trains in the no. 4 station

Table 3 Trains morning peak hour operation schemes

Scheme	Operating time	Operating segment	Number of trains in operation
Traditional train operation scheme	7: 00–8: 40	Linchang station to Shengtaixilu station	21
		Linchang station to Mozhoudonglu station	21
Optimized train operation scheme	7: 00–9: 00	Linchang station to Shengtaixilu station	21
		Fongtailu station to Liuzhoudonglu station	1
		Linchang station to Mozhoudonglu station	21

optimizing the departure time of the standby train and subsequent train, the waiting time of 721 s can be effectively reduced in the morning peak hour, and the passenger traffic congestion of Liuzhoudonglu Station can be alleviated.

5 Conclusion

In order to ensure travel safety, improve travel efficiency and improve passengers' waiting environment, this paper proposes an optimization model of urban rail standby train operation scheme according to the characteristics of tidal passenger flow under the background of large and small intersections. Minimize the waiting time of passengers at the large passenger flow stations by adjusting the standby bus drop-off stations. The results show that the model can alleviate the passenger traffic congestion to a certain extent. However, there are still some shortcomings in this study. For example, this study only focuses on one large passenger flow station on a single line, and it can focus on the whole road network in the future.

Acknowledgements This paper is supported by the funding from the National Key R&D Program of China (Item: 2017YFB1201202), the Jiangsu Province Graduate Student Practice Project (No. SJCX20_0126), and the Jiangsu Transportation Technology Project (Item: 2020Y17).

References

1. Xiangxi L (2012) Study on stop-schedule plan optimization for urban rail transit under outburst mass passenger flow. Beijing Jiaotong University
2. Jie L (2013) Research on metro station evacuation of large passenger flow. Southwest Jiaotong University

3. Man J, Li Hai-ying L, Xin-yue X et al (2017) Metro passenger flow control with station-to-station cooperation based on stop-skipping and boarding limiting. J Central South Univ 24(01):236–244
4. Ruihua X, Jingjing C, Shimin D (2005) Study on carrying capacity and use of rolling stock with multi-routing in urban rail transit. J China Railway Soc 04:6–10
5. Yandong W (2011) Study on optimization model of urban transit network organization and coordination with long and short routing. Logist Technol 30(21):71–74
6. Junfang L, Bingda W (2009) Analysis of typical interchange form of urban rail transit. Railway Transp Econ 31(10):54–58
7. Dandan Y, Baoming H, Qi Z et al (2015) Optimal method of train operation scheme for rail transit based on flexible grouping. J Beijing Jiaotong Univ 39(06):21–31
8. Dandan Y (2017). Research on key problems for cooperative optimization of train operation plan in urban rail transit network. Beijing Jiaotong University
9. Jungang S, Jing Y, Feng Z et al (2018) Integrate optimization model of operation schedule for metro express/local train. J Traffic Transp Eng 018(001):130–138
10. Yanxiang B, Jiahua L, Zhibin J (2006) Mixed running of different marshalling trains and its impact on URT. Urban Mass Transit 12:54–57
11. Kaspi M, Raviv T (2013) Service-oriented line planning and timetabling for passenger trains. 47(3)
12. Zekang L, Shiwei H, Haodong L (2018) Optimization model and branch-and-price algorithm for train dispatching on a railway network. J Transp Syst Eng Inf Technol
13. Li L (2019) Study on regional intercity train operation plan under emergency condition. Lanzhou Jiaotong University
14. Yuanguang S, Haiou S, Ying W et al (2018) Optimization on local/express train operation plan of urban rail transit by bi-level programming. J Transp Syst Eng Inf Technol 018(003):160–167
15. Min W (2015) Research on the organizational measures of subway station to cope with large passenger flow. Gansu Sci Technol 15:47–50
16. Guangzheng B, Jin G, Yang Y et al (2015) On operation scheme of urban rail transit in long/short routing mode. Urban Mass Transit 18(03):45–50
17. Feng S, Yanlin L, Xinlei H et al (2018) Service level-oriented optimization of high-speed railway train operation scheme. China Railway Sci 39(05):129–138
18. Lingyan L (2015) Sudden passenger flow propagation mechanism analysis and organitional optimization in urban rail transit network. Southwest Jiaotong University

Classification of Driving Tendency of Commercial Truck Drivers Based on AdaBoost Algorithm

Zhaofei Wang, Qiuping Wang, Shiqing Wang, Jianfeng Xi, and Jian Tian

Abstract In order to study the differences in the driving behavior of truck drivers, a classified management of truck drivers is implemented. Obtain the vehicle driving data of 51 commercial truck drivers in natural driving conditions through the on-board OBD device, and preprocess the original data, including detecting abnormal values, time processing, filling missing values, deleting parking data, etc. On the basis of standardizing the data of commercial cargo vehicles, the index is reduced by factor analysis to obtain the speed control behavior clusters of target vehicle drivers. By extracting variable speed factors and acceleration factors and clustering them according to factor scores, three types of driver's light, medium and heavy driving behaviors are obtained. Based on the K-means cluster analysis of the data, the AdaBoost algorithm is used to establish a classification model for the safety tendency of commercial truck drivers, and the truck drivers are divided into radical drivers and conservative drivers. First, the factor analysis method is used to extract the indicators of the two directions of speeding and acceleration, and then the K-means algorithm is used to classify from two perspectives, and finally the driver's different driving conditions can be analyzed. In addition, through further screening of all driving behavior indicators through K-means clustering, the adaboost algorithm is finally used to verify and analyze the clustering results to determine driver styles with different tendencies. Data verification classification results show that the average accuracy of the driving tendency classification model of commercial truck drivers based on the AdaBoost algorithm can reach 95.74%, which can effectively distinguish radical truck drivers from conservative truck drivers.

Z. Wang (✉) · Q. Wang
Institute of Civil Engineering, Xi'an University of Architecture and Technology, Xian 710055, China
e-mail: wzf@xauat.edu.cn

S. Wang · J. Xi
School of Transportation, Jilin University, Changchun 130022, China

J. Tian
China Academy of Transportation Sciences, Beijing 100029, China

W. Wang et al. (eds.), *Green Transportation and Low Carbon Mobility Safety*,
Lecture Notes in Electrical Engineering 944,
https://doi.org/10.1007/978-981-19-5615-7_32

Keywords Driving behavior · On-board OBD data · K-means clustering · AdaBoost algorithm

1 Introduction

In recent years, China road transport of goods surge in demand, as of 2019, the national freight vehicles up to 6.137 million, showing a steady upward trend for three consecutive years, but operating the truck road traffic accident-prone and suffered heavy losses. According to the Traffic Management Research Institute of the Ministry of Public Security, as of the end of 2017, the number of operating vehicles in my country only accounted for 7.0% of the number of motor vehicles, but the number of operating vehicle accidents and fatalities accounted for 19.87% and 31.11% of the total number of traffic accidents, respectively. The death rate per 10,000 vehicles in road traffic accidents of operating vehicles is 7.97, which is 2.9 times the national average of other types of vehicle accidents. The majority of operating vehicle accidents are freight vehicles. While realizing efficient transportation, the safety problem of the operation of trucks cannot be ignored.

According to the analysis of the causes of traffic accidents by related scholars [1, 2], traffic accidents caused by drivers' personal factors account for 88% of traffic accidents. It can be seen that the driver's personal factors have become the most important factor in traffic accidents. Therefore, studying the driving behavior of truck drivers has important practical significance for preventing and reducing the occurrence of traffic accidents.

The research on driver behavior in related literature mainly includes driving data collection and processing, driving behavior analysis model and application. Driving behavior data acquisition from the main video recognition and treatment [3, 4], vehicle sensors [5], on-board diagnostic (On-Board Diagnostic, OBD) [6, 7], driving simulator [8]. Pirdavani [8] and others developed a real-time risk model through research on collision data and real-time traffic monitoring data, which aims to predict the possibility of collisions on highways. Setting potential predictors is limited to traffic-related features. Assuming that the dependent variable (i.e. traffic safety status) is dichotomy (i.e. "no accident" or "accident"), compare the performance of rule-based classifiers with more traditional techniques (such as binary logistic regression and decision trees) and analyze the results It shows that several traffic flow characteristics such as the traffic volume, average speed and speed standard deviation of the upstream loop detector station, and the average speed difference between the upstream and downstream loop detector stations significantly contribute to the prediction of collision. The results of this study can be used in highway possibility of collision occurring within the dynamic range of the prediction.

Gershon [9] et al. conducted a study on the natural driving data of adolescents and obtained the interrelationships among risk factors. Depending on the driving environment, the variability of dangerous driving can provide information for the development of targeted interventions to reduce the risk of collisions for young novice drivers. Xiong [10] proposed a Markov chain-based driving risk state prediction algorithm, which uses clustering technology to classify the driving risk state, and determines the instantaneous risk level on the two-dimensional plane of collision time and time interval.

May [11] et al. used road driving tests to determine different driving-related risk assessment indicators, and established a driving-related risk evaluation model based on dynamic Bayesian networks.

Nengchao [12] and others to study the highway capacity expansion program with the intermediate opening portion (hereinafter simply referred OSMS) drivability, controlling respectively OSMS 9 different lengths and highway traffic flows scenario simulation. 25 participants were recruited, 225 simulated driving tests were performed using a driving simulator, and analysis of variance was used to analyze driving characteristics that can represent a safe background. The results show that when OSMS length and traffic volume is not the same, safe driving parameters are also different. When the traffic flow is low or moderate, the length of OSMS will seriously affect the maximum value of vehicle speed and collision time. The higher the traffic flow, the smaller the minimum time interval.

Shahram [13] and others summarized the existing research results and set the condition of the on-board active safety system (CASS) in the vehicle Internet environment to broadcast and send information at an interval of 100 m/s. The four communication methods in the basic network environment are periodic communication, periodic communication with built-in kinematics model, error-triggered adaptive communication, and error-triggered adaptive repetitive communication. In order to reduce the communication frequency, a motion model was established on the receiving side and the transmitting side respectively, and the corresponding joint simulation was applied to verify the differences between these four methods. Salameh [14] and others developed a high-precision positioning system for the needs of collision avoidance warning system. Japan's Masao Fukushima [15] has put forward several intelligent systems that can reduce the incidence of accidents by studying the current safety early warning applications in the current Internet of Vehicles environment. There are intersection collision avoidance support systems, accident information services caused by skidding, highway traffic safety support systems, etc. In addition, research on online monitoring of driving patterns and intentions through smartphones with built-in sensors has also shown an upward trend. Araújo [16] and others collect vehicle status information, such as speed and acceleration, through sensors embedded in smartphones, and use this information to analyze and predict their current driving safety level, and provide corresponding online alerts and prompts when necessary. Imperial College's Sun [17] also analyzed the driver's driving safety level under high-speed conditions through the acceleration and other data collected by the built-in sensors of the smartphone, and proposed intervention strategies for different driving modes.

To sum up, the driving behavior studies of domestic and foreign researchers mainly focus on fatigue driving, overspeed driving, aggressive driving and other aspects. Taking the knowledge of different disciplines as the starting point, combined with the understanding of driving behavior, three research methods, including questionnaire method, modeling method and behavior observation method, have been produced. Based on the driving data of drivers in their natural state, this paper studies the classification of truck drivers' driving tendency, which can better implement classification management for different types of truck drivers and reduce the occurrence of traffic accidents.

2 Study on Influencing Indicators of Driving Behavior

2.1 *Selection Principle*

Relevant principles should be followed and analyzed when selecting characteristic indicators, so as to lay a good foundation for the follow-up research. There are too many variables and most of them are related, which will complicate the factor analysis and cluster analysis in the future. Therefore, the following principles should be referred to in the selection of variable indicators:

(1) Scientific
Scientific is mainly reflected in the fact that the selected parameters can directly and effectively reflect the influence of changes in drivers and vehicle running conditions on driving safety. Combined with the theory and in line with the practice, it can directly evaluate the research object and has the pertinence to the research object.
(2) Operability
Feasibility and operability are the basis of evaluation indexes of automobile driving speed behavior. The collection of index data should conform to the principle of simple data and convenient calculation, but the evaluator is easy to complete the evaluation. Operability consists of the following three aspects: First, easy access to data; The second is the quantification of data; Third, we should adhere to the principle of paying more attention to quality and less quantity when selecting the index system.
(3) Effectiveness
Effectiveness means that the internal meaning of the evaluated object should be fully considered in the construction of the evaluation system, so that the internal meaning of the evaluated object is consistent with the evaluation system, so that the real level of the driver can be fully reflected.
(4) Dynamic
Driving behavior is not a constant process, but a process that changes over time. However, it can only be reflected in a long period of time. Therefore, when selecting indicators, not only static indicators should be used to evaluate driver behavior, but also dynamic indicators to evaluate the development direction of driving behavior.

(5) Adding relative value
In most cases, the analysis variables are directly extracted from the database, but the effective relative value should also be selected through calculation according to the actual research content.

2.2 Index Selection

(1) Mean speed and standard deviation of speed
The higher the mean speed is, the more accidents are likely to occur when the vehicle is driving. The standard deviation of the vehicle speed indicates the size of the vehicle speed deviation. The greater the deviation of the vehicle speed, the worse the driving stability of the driver. Therefore, the two are selected as the characteristic indexes.

$$v_a = \frac{1}{n}\sum_{m=1}^{n} v_m \tag{1}$$

$$v_s = \sqrt{\frac{1}{n}\sum_{m=1}^{n} (v_m - v_a)^2} \tag{2}$$

(2) Acceleration standard deviation, mean value of positive acceleration, positive acceleration standard deviation
Vehicle acceleration refers to the driver's operational control over the stop and operation of the vehicle. Positive and negative acceleration are the driver's reflection of the road environment in which the vehicle is running and its own driving characteristics respectively, and it is an important object in the study of driver characteristics. That is:

$$a_s = \sqrt{\frac{1}{n}\sum_{m=1}^{n} (a_m - a_a)^2} \tag{3}$$

$$a_m^+ = \frac{v_m - v_{m-1}}{t}(v_m > v_{m-1}) \tag{4}$$

$$a_a^+ = \frac{1}{n}\sum_{m=1}^{n} a_m^+ \tag{5}$$

$$a_s^+ = \sqrt{\frac{1}{n}\sum_{m=1}^{n} (a_m^+ - a_a^+)^2} \tag{6}$$

where, a_m is the MTH acceleration acquisition of the vehicle;

a_a is the mean acceleration of the vehicle;

a_s is the standard deviation of the acceleration of the car;

a_m^+ is the acceleration of the vehicle collected for the m time;

a_a^+ is the average positive acceleration of the car;

a_s^+ is the standard deviation of the positive acceleration of the vehicle.

(3) Negative acceleration mean and negative acceleration standard deviation

$$a_m^- = \frac{v_m - v_{m-1}}{t}(v_m < v_{m-1}) \tag{7}$$

$$a_a^- = \frac{1}{n}\sum_{m=1}^{n} a_m^- \tag{8}$$

$$a_s^- = \sqrt{\frac{1}{n}\sum_{m=1}^{n}(a_m^- - a_a^-)^2} \tag{9}$$

Negative acceleration represents the driver's deceleration behavior. Its influence on driving is that the larger the absolute value is, the more severe the driver's braking is, the worse the stability of driving is, and the greater the influence on driving safety is. Therefore, it is selected as the index parameter of driving characteristics.

$$a_m^- = \frac{v_m - v_{m-1}}{t}(v_m < v_{m-1}) \tag{7}$$

$$a_a^- = \frac{1}{n}\sum_{m=1}^{n} a_m^- \tag{8}$$

$$a_s^- = \sqrt{\frac{1}{n}\sum_{m=1}^{n}(a_m^- - a_a^-)^2} \tag{9}$$

3 Driver Propensity Clustering Based on Factor Analysis

3.1 Index Selection

Factor analysis and cluster analysis are the key techniques of multivariate statistical analysis. When dealing with multivariate and high-dimensional data mining systems, factor analysis is the most convenient method to combine the linear combination of characteristic parameters of driving behavior into several common variables (i.e., major components), so as to reflect that the original multiple indicators represent the majority of characteristic information of driving behavior. Then the rotation factor is obtained by factor rotation, which is clearer than the driving information in the computer. Clustering analysis is carried out for each factor to its corresponding meaning of driving behavior, and the corresponding factor score is taken as the clustering index to realize the scientific clustering of driving behavior characteristics of driving vehicles.

For the vehicle driving process, from the perspective of speed is the most intuitive, so this section selection index for the direction of the speed index: namely, average speed, average speed standard deviation, standard deviation, acceleration is the acceleration, is the average standard deviation of the acceleration, deceleration, reducing speed standard deviation the six variables of 51 freight vehicle to carry on the principal component factor analysis.

3.2 *Factor Analysis*

3.2.1 Basic Concept

In this method, a variety of indicators are summarized and grouped, and a large number of index data are condensed into several factors, which can represent the overall index data. Indicators within the same factor have a strong correlation, while data from different groups have a poor correlation.

3.2.2 Factor Model

$$\begin{cases} X_1 = a_{11}F_1 + a_{12}F_2 + \cdots + a_{1m}F_m + \varepsilon_1 \\ X_2 = a_{21}F_1 + a_{22}F_2 + \cdots + a_{2m}F_m + \varepsilon_2 \\ \vdots \\ X_p = a_{p1}F_1 + a_{p2}F_2 + \cdots + a_{pm}F_m + \varepsilon_p \end{cases} \tag{10}$$

In terms of matrix

$$\begin{pmatrix} X_1 \\ X_2 \\ \vdots \\ X_p \end{pmatrix} = \begin{pmatrix} a_{11} & a_{12} & \cdots & a_{1m} \\ a_{21} & a_{22} & \cdots & a_{2m} \\ \cdots & \cdots & \ddots & \cdots \\ a_{p1} & a_{p2} & \cdots & a_{pm} \end{pmatrix} \begin{pmatrix} F_1 \\ F_2 \\ \vdots \\ F_m \end{pmatrix} + \begin{pmatrix} \varepsilon_1 \\ \varepsilon_2 \\ \vdots \\ \varepsilon_p \end{pmatrix} \tag{11}$$

Shorthand for

$$X_{p\times 1} = A_{p\times m}F_{m\times 1} + \varepsilon_{p\times 1} \quad (12)$$

3.2.3 Specific Steps

(1) Applicability of the analytical method
Factor analysis is the process of reconstructing a few representative factor variables from many original variables. The condition is: to calculate the correlation coefficient matrix between the original variables. Before the factor analysis of the data, the applicability of the data should be judged to see whether it meets the requirements of the method. See Tables 1 and 2 for details.
(2) Factor acquisition
Principal component analysis method: this method makes the original variable change linearly through coordinate transformation, and transforms it into other groups of unrelated variables (principal component). According to the characteristic root λ_i (λ_1>λ_2>... > λp > 0) and the variance contribution rate of the common factor F_i of the eigenroot of the correlation coefficient matrix of the corresponding orthonormal eigenvector Li (equal to the sum of squares of each element in the JTH column of the factor load matrix A), and calculate the variance contribution rate and cumulative contribution rate of the common factor F_i. The number of factors is determined by the contribution rate of the cumulative variance of factors, and the factors are ranked according to their contribution rate.
(3) Factor naming
For the identified factor, according to the analysis of the specific problem, which factor has strong correlation with it, it will be named as the appropriate name based on the correlation of the original index.
(4) Factor score

Table 1 Table of phase relationship numbers

Correlation coefficient	Variable relation
r > 0	Positive correlation
r < 0	Negative correlation
|r|=1	Function relation
Is the value range of the correlation coefficient, namely $-1 < = r < = 1$	

Table 2 Correlation coefficient grade table

Grade of correlation coefficient	
|r| < 0.4	Low linear correlation
0.4 ≤ |r|<0.7	Significance correlation
0.7 ≤ |r|<1	Highly linear correlation

For well-determined factors, if the specific situation and relationship of each index data on different factors are needed, the score of factor variables can be calculated, and the following clustering indexes can be laid a foundation.

3.2.4 Verification of Possibility

For 51 freight vehicles, the following index data can be obtained through calculation. Some data processing is shown in Table 3.

(1) Adaptability test
Factor analysis can synthesize multiple indicators into several factors, so it is necessary to determine the adaptability of its method. See Table 4. for specific methods.
The test results are shown in Table 5. Thus, it can be concluded that KMO is 0.714 > 0.6, and Bartlett test $P < 0.05$, indicating that the research data are suitable for factor analysis.
(2) Factor extraction

The variance interpretation rate indicates the information of factor extraction, and determines the number of factors and the factor contribution rate. The specific analysis of the factor table is shown in Table 6.

Table 3 Data processing tables

Serial number	Mean speed	Speed standard deviation	Acceleration standard deviation	Mean of positive acceleration	Positive acceleration standard deviation	Mean of negative acceleration	Negative acceleration standard deviation
1	43.9	18.3	0.614	0.702	0.459	0.407	0.493
2	46.0	18.4	0.651	0.432	0.521	0.649	0.472
3	46.2	19.3	0.837	0.594	0.575	0.598	0.602
4	42.3	19.2	0.845	0.598	0.616	0.587	0.590
5	49.7	18.0	0.822	0.567	0.601	0.566	0.591
6	48.6	18.6	0.691	0.477	0.508	0.475	0.493

Table 4 Fitness test sheet

Methods	Requirements	Suitable for or not
Analysis of KMO value	>0.8	Very suitable for
	0.7–0.8	Is suitable for
	0.6–0.7	Can be done
	<0.6	Not suitable for
Bartlett test	P < 0.05	Suitable for

Table 5 Feasibility test sheet

KMO and Bartlett's test		
KMO value		0.714
Bartlett's test for sphericity	The approximate chi-square	454.350
	df	21
	P values	0.000

As shown in Table 6, the eigenvalues of 2 factors are greater than 1, so 2 factors have the variance interpretation rate of 51.100%, 25.169%, and the cumulative variance interpretation rate of 76.269%.

Figure 1 is the factor extraction diagram, with the number of principal components on the horizontal axis and the corresponding eigenvalues on the vertical axis. Factor extraction graph is used to assist in judging the number of factor extraction. As can be seen from Fig. 1, the horizontal axis tends to be flat after 2, so it is suitable to take 2 principal component factors for this study.

a. Factor naming
   The maximum variance rotation method was used to rotate the factors, and the load coefficients of each index on the factors were shown in Table 7 to represent different indexes.

The characteristic factors of driving behavior can be obtained by the above method, and then cluster analysis can be conducted according to the factor scores. As can be seen from Table 7, factor 2 has a higher index load coefficient corresponding to the speed, while factor 1 has a higher index load coefficient corresponding to acceleration. Acceleration represents the driver's variable speed control.

Thus, the two factors are named:

The load coefficient of factor 1 is larger in the five indexes of acceleration standard deviation, positive acceleration mean value, positive acceleration standard deviation, negative acceleration mean value and negative acceleration standard deviation. Freight vehicle because of its nature, its operation is often a long time, and often accompanied by driving at night, for inter-city freight vehicles running condition is relatively complex, often for a long time in such an environment, drivers are more likely to appear fatigue driving conditions, due to economic conditions and the reality, to improve the enterprise benefit, will often present the overload vehicles, overtaking, etc. Therefore, the acceleration change of dangerous drivers is more intense, so factor 1 is named as the variable speed driving behavior factor.

The load value of average speed and standard deviation of speed is larger on factor 2.Therefore, factor 2 was named as the overspeed tendency driving behavior factor.

Factor analysis is used to concentrate information. The relation equation between factors and research items, as shown in Tables 2, 3, 4, 5, 6, 7 and 8, can be used as the clustering basis of k-means in the following section. As follows.

Factor score 1 = −0.029* mean velocity + 0.032* variance of velocity + 0.221* variance of acceleration + 0.212* mean positive acceleration + 0.207* mean negative

Table 6 Variance interpretation rate table

Variance interpretation rate table									
Number of factors	Characteristics of the root			Rotate forward difference interpretation rate			Interpretation rate of variance after rotation		
	Characteristics of the root	Variance solution rate %	Cumulative %	Characteristics of the root	Variance interpretation rate %	Cumulative %	Characteristics of the root	Variance interpretation rate %	Cumulative %
1	4.344	54.296	54.296	4.344	54.296	54.296	4.088	51.100	51.100
2	1.758	21.973	76.269	1.758	21.973	76.269	2.013	25.169	76.269
3	0.911	11.392	87.661	–	–	–	–	–	–
4	0.597	7.458	95.119	–	–	–	–	–	–
5	0.211	2.634	97.753	–	–	–	–	–	–
6	0.086	1.078	98.831	–	–	–	–	–	–
7	0.077	0.968	99.799	–	–	–	–	–	–
8	0.016	0.201	100.000	–	–	–	–	–	–

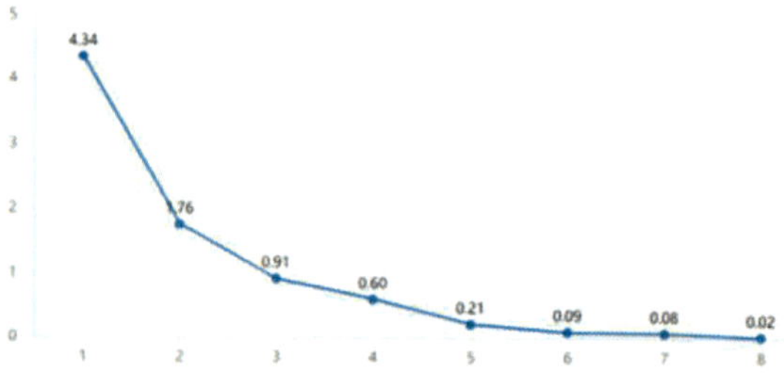

Fig. 1 Factor extraction diagram

Table 7 Table of factor load coefficient

Factor load factor		
name	Factor 1	Factor 2
Average speed	0.050	0.822
Speed standard deviation	−0.033	−0.823
Acceleration standard deviation	0.994	0.044
Mean of positive acceleration	0.957	0.054
Positive acceleration standard deviation	0.948	0.107
Negative acceleration mean	0.940	−0.069
Negative acceleration standard deviation	0.914	0.108

Table 8 Component score coefficient matrix

Component score coefficient matrix name factor		
Name	Factor	
	Factor1	Factor2
Average velocity	−0.029	0.600
The speed of variance	0.032	−0.601
Acceleration variance	0.221	−0.016
Mean of positive acceleration	0.212	−0.007
Mean of negative acceleration	0.207	0.032
Positive acceleration standard deviation	0.214	−0.096
Negative acceleration standard deviation	0.200	0.035

acceleration + 0.214* positive acceleration standard deviation + 0.200* negative acceleration standard deviation.

Factor score 2 = 0.600 * mean velocity − 0.601 * variance velocity − 0.016 * variance acceleration − 0.007 * mean positive acceleration + 0.032 * mean negative acceleration − 0.096 * positive acceleration standard deviation + 0.035 * negative acceleration standard deviation.

3.3 K-means Cluster Analysis

The K-means clustering method regards all samples as points in the n-dimensional space, and each sample can have different attributes from different perspectives, which can be expressed as a vector. The distance between samples can be represented by the distance between two points. The closer the distance is, the higher the similarity degree is, while the farther the distance is, the lower the similarity degree is.

The idea is as follows: firstly, select a certain number of samples and aggregate them to the nearest clustering point to form the initial classification; then, set the number of clustering categories by ourselves, conduct several experiments, and select the group with the best clustering result as the data classification for the last iteration.

Calculation steps of K-means clustering method:

(1) arbitrarily select n samples as the initial clustering center.
(2) For the remaining samples in the data, the distance from the initial clustering center is calculated, and it is assigned to the cluster with the smallest distance.
(3) Repeated iteration to the clustering center no longer changes.

Clustering was carried out according to the vehicle factor score, and analysis was made by SPSS software. The clustering results are shown in Tables 9 and 10.

3.3.1 Variable Speed Propensity Driving Behavior Factor Clustering

It can be seen from Tables 9 and 10 that the driving behaviors of cargo vehicles can be divided into 3 types according to the characteristics of factor 1 (driving tendency to change gears), but the severity is not too high. The proportion of drivers with severe

Table 9 Variable speed clustering results

Category	Variable speed driving behavior clustering	Change behavior
1	1, 3, 13, 14, 23, 34, 42	Heavy
2	7, 10, 11, 16, 17, 18, 20, 22, 25, 26, 32, 36, 37, 39, 41, 44, 45, 49, 50, 51	Middle
3	2, 4, 5, 6, 8, 9, 12, 15, 19, 21, 24, 27, 28, 29, 30, 31, 33, 35, 38, 40, 43, 46, 47, 48	Light

Table 10 Clustering categories

Summary of basic information of clustering categories		
Clustering categories	Frequency	Percentage (%)
1	7	13.73
2	20	39.22
3	24	47.06
Total	51	100

speed change is 13.73%.Because the truck on the volume and weight compared with other traffic participants has great advantages, so the driver is easy to produce a strong sense of security, therefore, often appear dangerous driving behavior, but as a professional driver, under the limit of specification, can inhibit its own actions, avoid sudden trample accelerator and brakes suddenly a sharp deceleration.

3.3.2 Clustering of Speeding Tendency Driving Behavior Factors

By Tables 11 and 12, the characteristics of the freight vehicles driving behavior according to the factor 2 (speeding tendency driving) clustering for three kinds of more appropriate, because of its huge size, the capacity is big, slow, thus speeding tendency is not very serious, with the tendency of speeding driving, speeding tendency degree distribution is relatively uniform.

In conclusion, through factor analysis and cluster analysis, two typical clustering of freight vehicle speed behavior characteristics are realized, and each clustering mode is reasonably explained. On this basis, the driving safety of cargo vehicles is briefly evaluated, and the clustering results provide guidance for transportation enterprises to monitor the drivers with high driving risk. In the future, more effective characteristic parameters of driving behaviors will be extracted from vehicle network data, and a prediction model of high-risk drivers will be established to improve the complete system of safety evaluation of driving behaviors.

Table 11 Clustering of speeding driving behavior

Category	Clustering speeding behavior	Over speed behavior
1	24, 26, 35, 36, 37, 38, 39, 40, 41, 46, 48, 51	Heavy
2	2, 3, 5, 7, 8, 10, 11, 12, 14, 16, 17, 18, 20, 22, 28, 42, 44, 47	Middle
3	1, 4, 6, 9, 13, 15, 19, 21, 23, 25, 27, 29, 30, 31, 32, 33, 34, 43, 45, 49, 50	Light

Table 12 Summary of basic information of cluster categories

Summary of basic information of clustering categories		
Clustering categories	Frequency	Percentage (%)
1	7	13.73
2	20	39.22
3	24	47.06
Total	51	100

4 Driver Tendency Classification Based on Adabootst

The statistical software SPSS was used to carry out K-means clustering on the data of 51 truck drivers, and the clustering results were used as the labels of the sample data of Adaboost algorithm. The Adaboost algorithm was used to establish the classification model of truck drivers' safety orientation by using Matlab software.

4.1 Index Selection and Test

4.1.1 Indicator Selection

In this paper, the transverse acceleration reflecting the safe operation of truck drivers, the longitudinal acceleration as an indicator, and the maximum speed and average speed reflecting the safety state of the vehicle are selected as characteristic parameters to identify the safety tendency of truck drivers.

4.1.2 Indicator Test

According to the independent sample T test, there were significant differences between the two groups in the maximum velocity, the mean speed, the maximum lateral acceleration and the maximum longitudinal acceleration. F values were 0.223, 2.536, 0.095, 2.246, The significance were 0.000, 0.000, 0.003 and 0.001, all less than 0.05, and the differences were statistically significant, as shown in Table 13.

It can be seen from Table 14 that all indicators of Category 2 are greater than those of Category 1. Among them, the mean maximum speed of aggressive drivers is 78.05 km/h, and the mean maximum speed of conservative drivers is 65.24 km/h. The mean value of the peak lateral acceleration of the aggressive drivers is 5.08 m/s^2, and that of the conservative drivers is 3.57 m/s^2. The mean value of the peak lateral acceleration of the aggressive drivers is 5.94 m/s^2, and that of the conservative drivers is 4.23 m/s^2.

Table 13 Independent sample tests

Parameter	F	Significant
Speed of the largest	0.223	0.000
Average speed	2.536	0.000
Maximum lateral acceleration	0.095	0.003
Maximum longitudinal acceleration	2.246	0.001

Table 14 Statistical results of identification parameters for classification results of drivers' safety propensity

Parameter	Category	The case number	Mean
Maximum speed	1	29	65.24
	2	22	78.05
Average speed	1	29	60.12
	2	22	48.72
Maximum lateral acceleration	1	29	3.57
	2	22	5.09
Maximum longitudinal acceleration	1	29	5.94241
	2	22	4.22836

4.2 K-means Clustering Algorithm

Please refer to the specific algorithm steps of K-means clustering algorithm, please refer to the above summary, and conduct the clustering analysis according to the results of the above section factor analysis as the clustering basis. Some driver cluster results using SPSS software are shown in Table 15.

Means cluster the data of 51 freight vehicle drivers, and the final clustering results are shown in Tables 16 and 17.

Table 15 Clustering results table

Number	Max. speed	Average vehicle speed value	Maximum lateral acceleration	Maximum longitudinal acceleration	Clustering
1	64	52.63	3.19	2.988	1
2	75	58.48	4.903	6.878	2
3	64	45.45	9.262	6.96	1
4	79	59.97	8.195	7.897	2
5	79	51.89	6.505	5.549	2
6	77	55.17	6.056	4.445	2
7	68	48.41	3.853	6.123	1
8	66	45.62	3.141	2.882	1
9	77	55.74	5.473	6.869	2
10	68	43.618	2.849	4.323	1
11	70	47.227	2.965	5.983	1
12	77	53.891	5.307	6.456	1
13	57	47.087	4.708	4.779	1
14	73	50.583	3.871	4.695	1
15	73	50.408	3.698	4.539	2

Table 16 Number of cases per cluster

Number of cases per cluster		
Clustering	1	29
	2	22
Effective	51	

Table 17 The final cluster center

	Clustering	
Max. speed	65	78
Average vehicle speed value	48.727	60.121
Maximum lateral acceleration	4.089	4.572
Maximum longitudinal acceleration	4.442	5.228

The cluster center represents the distance between the sample and the sample class, analyzed as follows.

Class 1 truck driver: each index value is relatively low, inclined to slow driving, driving is relatively stable, is a conservative truck driver, a total of 29 sample data.

Category 2 truck drivers: each index is relatively high value, easy to drive express, rapid acceleration and deceleration, sharp turn, for the radical truck drivers, a total of 22 sample data. Therefore, it is determined that category 1 van drivers will be conservative van drivers and category 2 vans are radical van drivers.

4.3 Adaboost Algorithm

The clustering results obtained by K-means are two types of drivers with labels of 1, 2, and then used the AdaBoost algorithm to classify truck drivers with a total of 51 freight drivers, divided into 35 training samples, and 16 samples were used to verify the accuracy of the model.

4.3.1 Introduction to Adaboost

Boosting, also known as reinforcement learning or enhancement methods, is an important comprehensive learning technique that can enhance weak learners with slightly higher prediction accuracy than random speculation to strong learners with higher prediction accuracy. This provides effective new ideas and new methods for the design of learning algorithms where it is very difficult to construct strong learners directly. The most successful one was the AdaBoost algorithm proposed by YoavFreund and RobertSchapire in 1995.

AdaBoost is the English abbreviation of Adaptive Boosting", suitable for training the following basic classifiers: the previous samples have more weight and smaller weight. Also, for each iteration, a new weak identifier is added, and the final strong identifier is not determined until a predetermined small error rate is reached or the maximum predetermined number of the iterations is reached.

4.3.2 Algorithm Process

Specified Training dataset $(x_1,y_1)\ldots(x_n,y_N)$: class tag representing training sample, $i = 1,\ldots, n$. The purpose of Adaboost is to learn a series of weak or basic classifiers from the training data, and to incorporate them into a strong classifier.

(1) initializes the weight distribution of the training data. Since each training sample is first given the same weight value, and the $w_i = 1/N$, the initial weight value in the training sample set is:

$$D_1(i) = (w_1, w_2, \cdots w_N) = (\frac{1}{N}, \cdots, \frac{1}{N}) \tag{13}$$

(2) Do Iteration, $t = 1,\ldots, T$

a. selects a weak classifier h with the lowest error rate as the t basic classifier H_t, and calculates a weak classifier $h_t : X \to \{-1, 1\}$ whose error on distributed D_t is:

$$e_t = p(H_t(x_i) \neq y_i) = \sum_{i=1}^{N} w_{ti} I(H_t(x_i) \neq y_i) \tag{14}$$

b. calculates the weight of the weak classifier in the final classifier

$$\alpha_t = \frac{1}{2} \ln(\frac{1 - e_t}{e_t}) \tag{15}$$

c. Update weight distribution of training samples $Dt + 1$

$$D_{t+1} = \frac{D_t(i) \exp(-\alpha_t y_t H_t(x_i))}{Z_t} \tag{16}$$

(3) Finally, each weak classifier is combined according to the weak classifier weight αt, namely

$$f(x) = \sum_{i=1}^{T} \alpha_t H_t(x) \tag{17}$$

A strong classifier is obtained by the action of the symbolic function sign

$$H_{final} = sign(f(x)) = sign(\sum_{t=1}^{T} \alpha_t H_t(x)) \quad (18)$$

4.3.3 Results Analysis

AdaBoost algorithm training and test iterates using adaboost, gentleboost algorithm. The number of elected iterations is 10, of which the adaboost achieves 100% accuracy after the iteration, and thus the adaboost method was selected as the truck driver safety tendency classification model algorithm. Specific model ideas are as follows

(1) Radical driver obtained by the K-means algorithm tag 1 and conservative driver tag 2 as shown in Table 18.
(2) Read 51 driver data, divide the training set into 35 samples, the test set is 16 samples, the matlab operation as shown In Fig. 2.
(3) Train the model and test the model accuracy, as shown in Fig. 3.

Table 18 Data label

Number	Max. speed	Average vehicle speed value	Maximum lateral acceleration	Maximum longitudinal acceleration	Label
1	64	52.63	3.19	2.988	1
2	75	58.48	4.903	6.878	2
3	64	45.45	9.262	6.96	1
4	79	59.97	8.195	7.897	2
5	79	51.89	6.505	5.549	2
6	77	55.17	6.056	4.445	2
7	68	48.41	3.853	6.123	1

Fig. 2 Data classification

```
temp=1:1:51;
%Training set--35 samples
P_train =result(temp(1:35), 1:4)';
T_train= result(temp(1:35), 5)';
M=size(P_train, 2);
%Testing set--16 samples
P_test = result(temp(36:end), 1:4);
T_test =result(temp(36:end), 5)';
N=size(P_test, 2);
```

```
%%Train the model
net=fitensemble(p_train, t_train, 'AdaBoost', 10, 'Tree');
%%Simulation test
T_sim1=predict(net, p_train);
T_sim2=predict(net, p_test);
```

Fig. 3 Model training

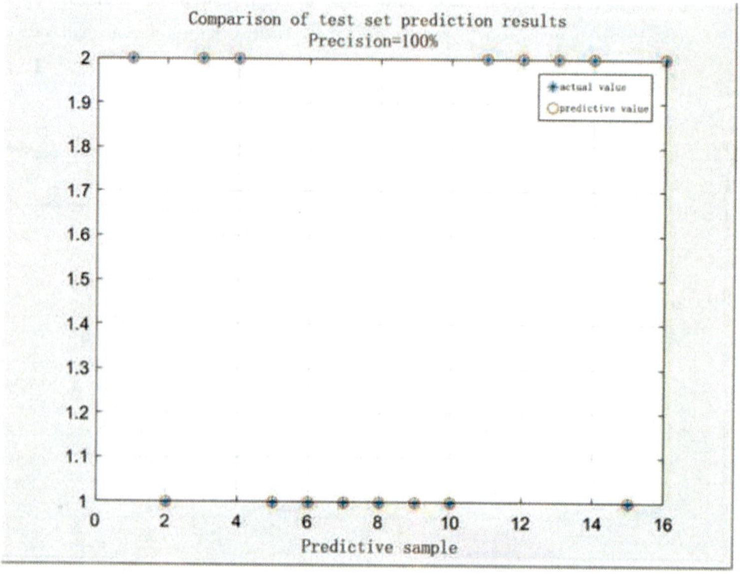

Fig. 4 Predictions of Adaboost

It is concluded that the adaboost algorithm has high prediction accuracy, so this classifier can be used to classify the driver with tendency. Vehicle supervision authorities can use this method to classify drivers, conduct targeted training for radical truck drivers, correct their operating methods, and persuade truck drivers who are not suitable for drivers, so that they can work in other industries (Figs. 4 and 5).

5 Conclusion

This paper presents a safety orientation analysis method for truck drivers based on on-board diagnostics (OBD) information and AdaBoost algorithm. The safety orientation analysis of truck drivers includes data acquisition, data processing and the safety orientation classification model based on AdaBoost algorithm.

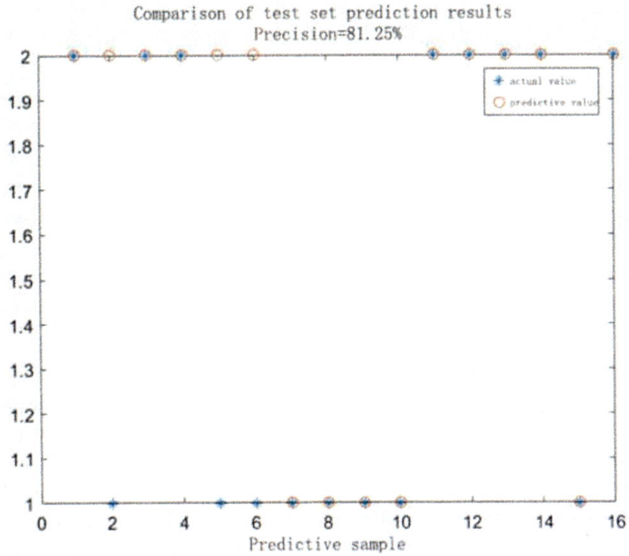

Fig. 5 Predictions of gentleboost

(1) data acquisition is to obtain the vehicle condition data of 39 truck drivers in the natural driving state through vehicle diagnosis (OBD) and smooth filtering the data to eliminate the impact of external environmental factors on the data. The maximum relative ratio of vehicle speed, lateral acceleration peak, peak acceleration of driving direction, and the maximum relative ratio of vehicle speed and engine speed of each driver are extracted as the truck driver.
(2) uses statistical software SPSS to cluster the data of 39 truck drivers, which is divided into radical truck drivers and conservative truck drivers. The classification results show that radical truck drivers tend to be easy to drive fast, easy to slow down, sharp turn, etc., conservative truck drivers drive stable and tend to drive slowly.
(3) K-means clustering results as a label of the sample data of AdaBoost algorithm, AdaBoost algorithm is applied to establish the safety tendency classification model of truck drivers. The sample data of 39 truck drivers are used as training samples and test samples of the AdaBoost algorithm, respectively, and the analysis results show that the proposed truck driver safety orientation classification method can achieve an average accuracy of 98.74%.

The truck driver safety orientation classification method can be used for vehicle management units to assess truck drivers of different personalities, targeted train radical truck drivers, correct their operation methods, and for truck drivers who are not suitable to drive, persuade them to engage in other industries. This paper only divides drivers into two categories to be improved, and the driver classification should be more detailed, not limited to two categories.

References

1. Miyaji M, Danno M, Ogurik (2008) Analysis of driver behavior based on traffic incidents for driver monitor systems. In: Intelligent vehicles symposium. Institute of Electrical and Electronics Engineers, New York, USA, pp 930–935
2. Liang J, Cheng XY, Chen XB (2008) The research of car rear-end warning model based on MAS and behavior. In: The workshop on power electronics & intelligent transportation system. Computer Society of Institute of Electrical and Electronics Engineers, New York, USA, pp 305–309
3. Juan LI, Jing ZHOU, Yuan LIN et al (2015) Research on driver behavior during yellow light based on video detection. J Transp Syst Eng Inf 15(1):69–74
4. Xiyan ZHU, Yong XU, Zhengyang PEI (2015) Application of gentle adaboost in infrared video driver fatigue detection. J Comput Measur Control 23(4):1414–1417
5. Lai SF (2017) Platform for acquiring driver behavior data: US20170185854
6. Pan YJ, Yu TC, Cheng RS (2017) Using OBD-II data to explore driving behavior model. In: International conference on applied system innovation. Institute of Electrical and Electronics Engineers, New York, USA, pp 1816–1818
7. Hwang C, Chen MS, Shih CM et al (2018) Apply scikit-learn in Python to analyze driver behavior based on OBD data. In: Proceedings of international conference on advanced information networking and applications workshops. Computer Society of Institute of Electrical and Electronics Engineers, New York, USA, pp 636–639
8. Xuedong YAN, Xiaomeng LI (2014) A review of research on hierarchical driving behavior safety model based on driving simulation experiments. J Traffic Inf Saf 32(5):1–6
9. Gershon P, Ehsani J, Zhu C et al (2018) Vehicle ownership and other predictors of teenagers risky driving behavior: evidence from a naturalistic driving study. Accident Anal Prevent 118:96–101
10. Xiaoxia X, Long C, Jun L (2018) Vehicle driving risk prediction based on Markov chain model. Discr Dyn Nat Soc 2018(8):1
11. May Q, Fan L et al (2019) Dynamic Bayesian network approach to evaluate vehicle driving risk based on on-road experiment driving data. IEEE Access 2019(7):135050
12. Lyu N, Cao Y, Wu C et al (2020) Driving behavior and safety analysis at OSMS section for merged, one-way freeway based on simulated driving safety analysis of driving behavior. PLoS ONE 15(2):e0228238
13. Rezaei S, Sengupta R, Krishnan H et al (2010) Tracking the position of neighboring vehicles using wireless communications. Transp Res Part C Emerg Technol 18(3):335–350
14. Salameh N, Challita G, Mousset S et al (2013) Collaborative positioning and embedded multi-sensors fusion cooperation in advanced driver assistance system. Transp Res Part C Emerg Technol 29:197–213
15. Fukushima M (2011) The latest trend of v2x driver assistance systems in Japan. Comput Netw 55(14):3134–3141
16. Araújo R, Igreja Â, De Castro R et al (2012) Driving coach: a smartphone application to evaluate driving efficient patterns. In: Proceedings of 2012 IEEE intelligent vehicle symposium (IV). IEEE, Madrid, Spain, pp 1005–1010
17. Sun R, Ochieng WY, Feng S (2015) An integrated solution for lane level irregular driving detection on highways. Transp Res Part C Emerg Technol 56:61–79

An Evaluation of the Effect of Urban Tunnel Lighting on Driving Comfort: A Driving Simulation Study

Yanwei Zang, Zihai Yan, Huojun Wu, Penglu Gan, Mingwei Hu, Wenlin Wu, Peng Liu, Guoqing He, and Jinghang Xiao

Abstract Urban tunnel lighting is significant for driving safety and comfort. In order to investigate the relationship between the lighting parameters and driving comfort under different situations, this paper established a tunnel lighting system model based on Shenzhen *Henglongshan* tunnel by UC-win/Road. By designing tunnel lighting driving simulation experiments, the authors changed the parameters of the tunnel lighting system model and conducted a series of tests by the driving simulator under different lighting conditions. Through analysis of the driver's eye movement data, this paper proposes that the minimum brightness of the tunnel lighting in the middle section is 0.5 cd/m^2, and the lighting brightness threshold for safe driving of the driver is 2.2 cd/m^2. The research results can provide a reference for the construction of tunnel lighting projects, and a theoretical basis for optimizing the design of tunnel lighting and improving the quality of tunnel lighting.

Keywords Transportation safety · Visual characteristics · Tunnel lighting · Driving simulation · Pupil area change rate

By the end of 2020, the total number of highway tunnels in China has reached 21,316, with a linear length of 2,199.93 million meters. Compared with 2019, it has increased by 2,249 in location and 3.0327 million meters in length. There are 1394 extra-long tunnels with 6.2355 million linear meters, and 5541 long tunnels with 9.6332 million linear meters. China has become the country with the fastest tunnel construction speed and the longest operating mileage in the world [1]. The special semi-enclosed space structure of the tunnel determines that tunnel lighting is necessary to ensure the safety and comfort of drivers driving through the tunnel

Y. Zang · Z. Yan · H. Wu · P. Gan
PowerChina HuaDong Engineering Corporation Limited, Hangzhou, China

M. Hu (✉) · W. Wu · P. Liu · G. He
College of Civil and Transportation Engineering, Shenzhen University, Shenzhen, China
e-mail: humw@szu.edu.cn

J. Xiao
School of Civil Engineering, Chongqing Jiaotong University, Chongqing, China

W. Wang et al. (eds.), *Green Transportation and Low Carbon Mobility Safety*, Lecture Notes in Electrical Engineering 944,
https://doi.org/10.1007/978-981-19-5615-7_33

section [2]. How to optimize the design improve the quality and reduce the operating cost of tunnel lighting have become popular in the field of tunnel research [3–6].

At present, the majority of scholars in the field of tunnel lighting mainly focus on the impact of tunnel entrance lighting on drivers and reducing the lighting energy consumption, In contrast, the research on the whole tunnel section lighting is less. Cantisani et al. [7] evaluated the life cycle of the tunnel lighting system. The analysis, showed that using reflective pavement materials and LED lighting lamps with good performance can effectively reduce the burden of construction, maintenance, and use for road lighting system. Liang Bo et al. [8] studied the effects of tunnel inner wall materials on tunnel lighting and driving safety. The results showed that tunnel inner wall coatings have an important influence on the tunnel lighting effect, and high-reflectivity coatings can significantly improve tunnel lighting. Using the K-MEANS clustering algorithm to divide the daily traffic volume distribution into time periods, and they proposed a strategy for dynamic control of tunnel lighting, which can significantly reduce energy consumption [9]. Based on the indoor experiments of tunnel light environment, Shiyong et al. [10] analyzed the influence of relative spectral energy and light environment color temperature on the driver's reaction time.

This paper combines the tunnel lighting theory and the driver's vision theory to establish a model of the urban tunnel lighting system, taking the change rate of the driver's pupil area as an index to evaluate driving safety and comfort. Taking Shenzhen *Henglongshan* tunnel as an example, the driving simulation of tunnel lighting is carried out based on the driving simulator. We get the minimum brightness value of driving in the tunnel and the safe and comfortable driving lighting environment through the analysis.

1 Tunnel Lighting System Modeling

1.1 Tunnel Introduction and Variable Setting of Lighting System

The *Henglongshan* tunnel is one of the longest tunnels in Shenzhen and plays an important part in Shenzhen's main road network. The *Henglongshan* Tunnel has a large traffic flow, running north-south, with an elevation of about 51 m. Each hole of the tunnel has three lanes. The left line is 2330 m and the right line is 2275 m [11]. The speed limit in the *Henglongshan* tunnel is no more than 60 km/h for trucks, and 80 km/h for cars. The Satellite map of the *Henglongshan* tunnel is shown in Fig. 1.

In order to improve the driver's visual adaptability and efficiency of tunnel lighting, it is necessary to divide the tunnel lighting into several sections. According to the relevant technical standards for road tunnel lighting design [12], the lighting zone can be taken as five sections: the approach, the entrance, the transition, the middle, and the exit, as shown in Figs. 1 and 2. Therefore, according to formula (1) and formula (2), we calculate the length of tunnel entrance section TH1, TH2,and

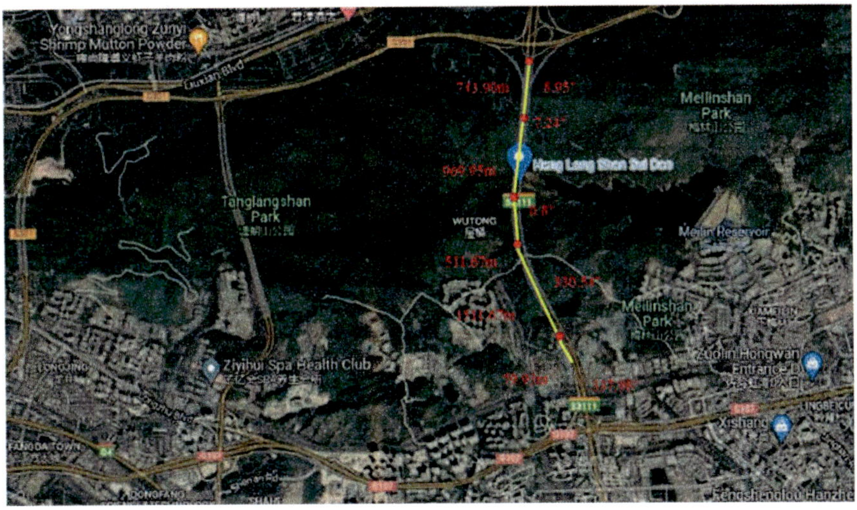

Fig. 1 Satellite map of *Henglongshan* tunnel

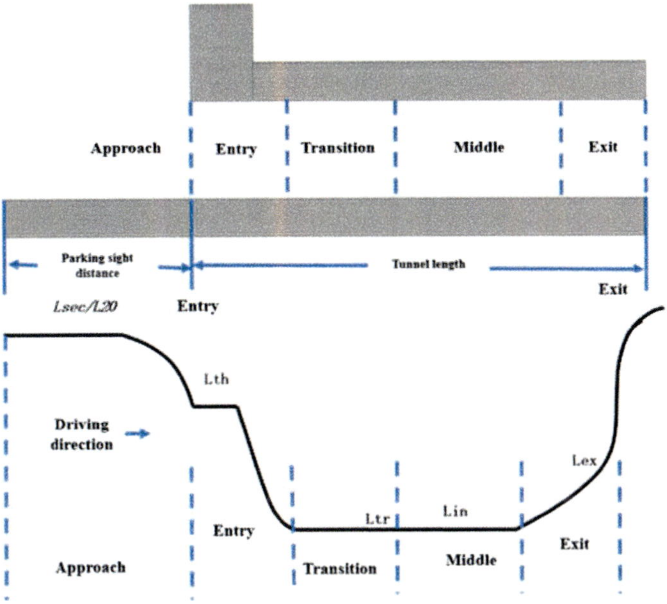

Fig. 2 Division of tunnel lighting section

transition section TR1, TR2, TR3 in turn. The "Detailed Rules for the Lighting Design of Highway Tunnels" issued by the Ministry of Transport stipulates that the length of the exit sections of the tunnels EX1 and EX2 are both 30 m, and the remaining sections are the length of the middle section.

$$D_{th1} = D_{th2} = \frac{1}{2} \times \left(1.154 \times D_s - \frac{h - 1.5}{tan10^\circ}\right) \tag{1}$$

D_{th1} indicates the length of the entrance and exit section TH1, unit: m.
D_{th2} indicates the length of the entrance and exit section TH2, unit: m;
D_s is the length of the sight distance for lighting parking, unit: m;
h is the clearance height in the tunnel, unit: m;

$$\begin{aligned} D_{tr1} &= (D_{th1} + D_{th2})/3 + V_t/1.8 \\ D_{tr2} &= 2V_t/1.8 \\ D_{tr3} &= 3V_t/1.8 \end{aligned} \tag{2}$$

V_t is the design speed, unit: km/h;
D_{tr1}, D_{tr2}, D_{tr3} represent the length of the transition section TR1, TR2, TR3, unit: m;

According to the field survey of the tunnel length and the tunnel lighting design technical rules, the tunnel is divided into four sections and the length of each section is calculated, as shown in Table 1.

The external brightness of the tunnel is the basis for determining the brightness of the entrance of the tunnel. The L20 method based on static vision theory is proposed. The brightness of the entrance of the tunnel lighting is obtained by multiplying the brightness outside the tunnel by the brightness reduction coefficient *k*. This study conducted a field investigation on the brightness outside the entrance of the *Henglongshan* tunnel. The relevant regulations and standards of domestic road tunnel lighting and the recommended values of photopic, intermediate vision, and scotopic vision by the International Commission of Illumination are taken into account. Combined with the external brightness values of the tunnel from field investigations, the five sets of intermediate lighting brightness determined in the driving simulation are 0.1, 0.5, 0.9, 1.5, and 2.5 cd/m^2. According to the survey value of the external brightness of the *Henglongshan* tunnel, combined with the relevant provisions of the tunnel

Table 1 Length of each section of *Henglongshan* tunnel

Lighting section	Entry		Transition			Middle	Exit	
	TH1	TH2	TR1	TR2	TR3	IN	EX1	EX2
Length/m	22.5	22.5	48	67	100	1965	30	30

Table 2 The length of each lighting section of the tunnel and the corresponding brightness

Lighting section	Entry		Transition			Middle	Exit
	TH1	TH2	TR1	TR2	TR3	IN	EX
Length/m	22.50	22.50	48.00	67.00	100.00	1965.00	60.00
Brightness (cd/m^2)	8.40	4.20	1.26	0.42	0.17	0.10	0.50
	35.91	17.96	5.39	1.80	0.72	0.50	2.50
	70.07	35.04	10.51	3.50	1.40	0.90	4.50
	123.73	61.86	18.56	6.19	2.47	1.50	7.50
	175.21	87.61	26.28	8.76	3.50	2.50	12.50

lighting design, the brightness value of each section of the tunnel lighting is calculated. In the study, the brightness value of the middle section of the tunnel lighting is used as the experimental data, and the brightness values of the remaining sections are calculated based on that of the middle one. The detailed values are shown in Table 2.

1.2 Tunnel Lighting System Model

The purpose of tunnel lighting is to allow drivers to approach, to drive in, and finally leave the tunnel section safely and comforta0bly within the specified speed range. The safety and comfort level in the tunnel is not lower than the section outside the tunnel. UC-win/Road is a modeling and simulation software based on modularization and graphics. Its powerful 3D modeling function and scene rendering ability can mostly simulate the driver's driving situation in the actual tunnel. The modeling process of the tunnel lighting system based on UC-win/Road mainly includes four parts: inputting terrain information, defining roads, configuring lighting fixtures, generating traffic flow, and depicting scenes.

(1) Inputting terrain information
In the process of tunnel modeling in UC-win/Road, the methods of inputting terrain include importing terrain data and customizing terrain creation.

In this study, the method of custom-created terrain project was adopted in the modeling process, as shown in Fig. 3. By setting the latitude and longitude data of the area, the appropriate terrain size and terrain elevation need to be selected according to the size of the tunnel. The created terrain data is the basis for establishing the tunnel model, obtaining the satellite image of the tunnel to be established through Google map and import the satellite image into the created terrain.

(2) Defining roads

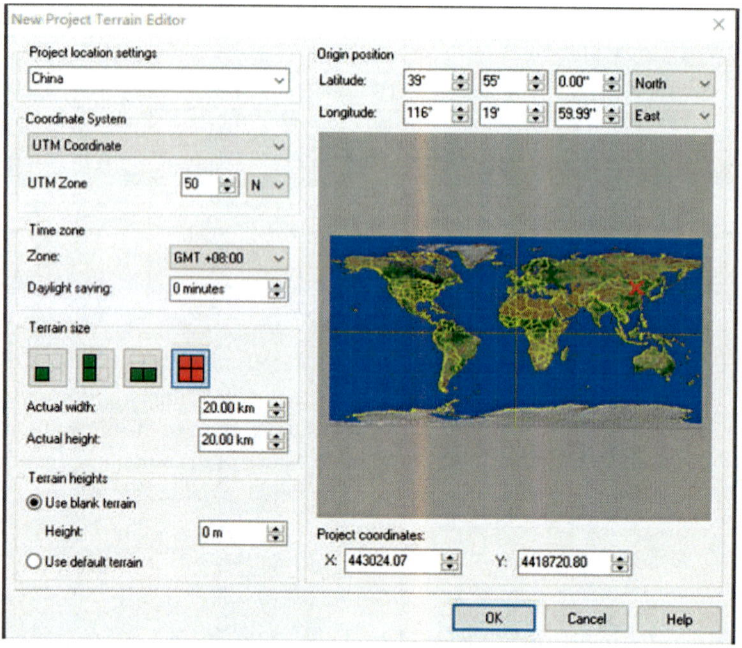

Fig. 3 Create terrain in a custom way

The establishment of three-dimensional terrain data is to define roads on it. Ordinary roads combine straight lines and curves to determine the road alignment (road path) as Fig. 4.

(3) Establishing tunnel

The section of the road crossing the mountain is the tunnel section. To make the tunnel model based of the established road, we need to select the start and end

Fig. 4 Defining roads

Fig. 5 One of the completed tunnel lighting system model

positions of the tunnel first. The start and end positions of the tunnel determine the length of the established tunnel then carry out the detailed setting of tunnel parameters setting. The shape of the tunnel opening established in this article adopts a horseshoe shape.

(4) Editing the scene outside the tunnel
The brightness outside the tunnel is the reference basis for determining the brightness of the entrance section of the tunnel lighting. Appropriately changing the scene outside the tunnel can improve the driver's light and dark adaptation speed.
After the tunnel model is established, lamps and lanterns are arranged in the tunnel to form a complete tunnel lighting system model. Then, it is necessary for us to set the parameters of the model. Finally, traffic flow needs to be generated to build a complete tunnel road traffic system. The model effect of the completed tunnel lighting system is shown in Fig. 5.

2 Driving Simulation of Tunnel Lighting Based on the Driving Simulator

2.1 Experimental Equipment and Personnel

The Logitech G923 driving simulator is used to carry out tunnel driving simulation research, as shown in Fig. 6.

In driving in the tunnel, eye movement data can reflect the driver's visual characteristics, which is the most important evaluation index of driving safety and comfort. This study uses the iView ETG 2.1 eye tracker developed by Germany Purson Technology Company to collect the driver's eye movement data. The ETG2.1 mainly consists of three modules: data acquisition module, data recording module, and data export analysis module. The data acquisition module is mainly composed of

Fig. 6 The Logitech G923 driving simulator

eye tracker glasses with camera, sensor and data transmission functions. The data recording module is the software iView ETG, in which new experiments, equipment calibration, and data recording are performed. The data export and analysis module is Be Gaze software, which is a comprehensive and powerful visual analysis software. It can provide fast, easy-to-access, and directly available visual images and data. The eye tracker uses an infrared camera to obtain the driver's eye image and sends it to computer for image data acquisition and analysis after MPEG coding, which calculates the pupil diameter, horizontal and vertical movement distance of the eyeball in real time. The sampling frequency of ETG 2.1 is 60 Hz, and the data acquisition accuracy reaches millisecond level.

2.2 *Driving Simulation Experiment and Data Acquisition*

Before carrying out tunnel lighting driving simulation, adjusting the tunnel lighting system model parameters and calibrating the eye tracker. Firstly, all drivers are trained in driving simulation operating procedures to ensure that all experimenters are familiar with the operating specification and process of the driving simulator to reduce the impact of the accidental error on the experimental results. Secondly, turn on the driving simulator and eye tracker to check whether the equipment is running normally, open the established tunnel lighting system models to set the lighting parameters. Finally, the driver is ready to drive on the driving simulator.

Before each driving simulation starts, the accuracy of eye tracker should be calibrated by a person. After driving simulation a special person is assigned to be responsible for monitoring the operation and operation status of the eye tracker. There are

Fig. 7 A driver is driving in a simulated tunnel

two scenarios here: when the driver drives the vehicle near the tunnel entrance, clicking to start the experiment in iView ETG to record the driver's eye movement data until the driver passes through the entrance section, transition section, middle section, and exit section of the tunnel. In addition, when the driver leaves the tunnel, stopping recording eye movement data and then comparing and analyzing the experimental data according to the driving video to ensure that all the collected data are scientific and effective. Based on the driving simulator, the driver carries out tunnel lighting driving simulation, as shown in Fig. 7.

3 Simulation Result Analysis

3.1 Data Pre-Processing

In the driving simulation process of tunnel lighting, the pupil area of the driver is obtained by calculating the pupil diameter.

Pr represents the pupil diameter of the right eye, the pupil diameter of the left eye is represented by *Pl*, and the average pupil diameter *Pa* is used when calculating the pupil area.

The formula for calculating the pupil area is as follows:

$$S = \pi \times \left(\frac{P_a}{2}\right)^2 \tag{3}$$

S Driver's pupil area, in mm^2;
π Pl;
P_a The driver's average pupil diameter, in mm;

This study considers the physiological differences of different drivers while avoiding large errors caused by different individual pupils. In addition, it uses the pupil area change rate to reflect the driver's pupil change. The formula is as follows:

$$P = \left| \frac{S_{t2} - S_{t1}}{S_{t1}} \right| \times 100\% \tag{4}$$

P Driver's pupil area change rate;
S_{t1} The pupil area of the driver at the previous moment (mm^2);
S_{t2} The pupil area of the driver at a later moment (mm^2);

3.2 The Influence of Lighting Brightness on Driving Safety and Comfort

To study the relationship between the driver's pupil area change rate and the brightness of tunnel lighting under different lighting environments. The color temperatures studied by the authors are orange light, white light, yellow light, and the weather are sunny and rainy. By averaging the pupil area change rates of the four drivers, we obtain the relationship curve between the driver's average pupil area change rate and the illumination brightness under six conditions, as shown in Fig. 8.We find that the average pupil area change rate of all drivers is negatively correlated with the brightness of the tunnel lighting increases, the pupil area change rate of all drivers decreases. Specifically, when the tunnel lighting brightness is lower than 0.5 cd/m^2, all driver's pupil area change rate has increased significantly. Conversely, when the tunnel lighting brightness reaches 2 cd/m^2, all drivers ' pupil area change rate has tended to 4%. Since then, as the brightness of the illumination increases, the driver's pupil change rate tends to stabilize.

By analyzing the relationship between the brightness of the tunnel lighting and the rate of change of the driver's pupil area.,we found thatwhen the tunnel illumination brightness is greater than 2.2 cd/m^2, the average pupil area change rate of all drivers tends to a stable value of 4.3%. So this study defines it as the tunnel lighting brightness threshold for the driver to drive safely and comfortably, corresponding to the pupil area change rate of the maximum pupil area change rate for the driver to drive safely and comfortably in the tunnel.

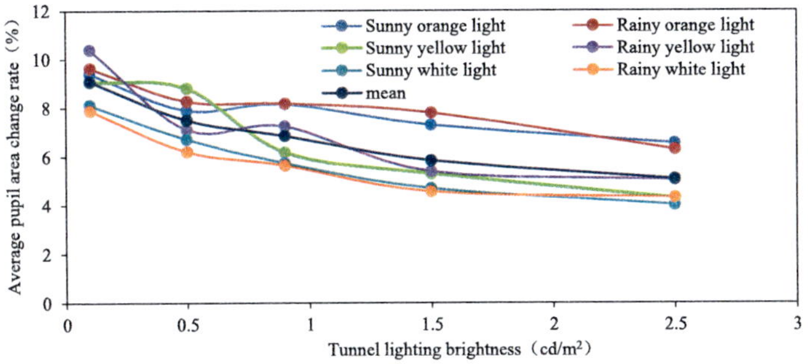

Fig. 8 Driver's average pupil area change rate under different lighting conditions

4 Conclusions

This article analyzes the influence of tunnel lighting brightness on driving safety and comfort in depth. According to the research results, there are individual differences in the driver's visual adaptability. Under the same lighting condition, the pupil area change rate of each driver is different. The driver's pupil area change rate is negatively correlated with the tunnel lighting brightness. As the tunnel lighting brightness increases, the driver's pupil area change rate gradually decreases and tends to a stable value. We obtain the quantitative relationship between the driver's pupil area change rate and the tunnel illumination brightness based on data analysis. From the relationship, the lighting brightness threshold for the driver to drive safely and comfortably in the tunnel is 2.2 cd/m^2. When the lighting brightness is greater than this value, the driver's pupil area change rate tends to a stable value of 4.3%. When the tunnel illumination brightness value is lower than 0.5 cd/m^2, the driver's pupil area change rate increases significantly. In summary, the brightness of the tunnel lighting should be greater than 2.2 cd/m^2 for safe and comfortable driving in tunnels.

References

1. Ministry of Transport of the People's Republic of China. 2020 Statistical Communiqué on the Development of China's Transportation Industry [EB/OL]. http://www.gov.cn/xinwen/2021-05/19/content_5608523.htm
2. Sun Y, Cai BF, Zheng GP et al (2020) Study on the energy-saving influence of tunnel lighting installation methods and parameters. Chin J Underground Space Eng 16(S1):389–395
3. Guo JK (2020) Research on accurate detection and intelligent control system of highway tunnel lighting. Tunnel Constr 40(S2):76–81
4. Gao XB (2020) Discussion on lighting control technology of smart highway tunnel. Highway 65(10):360–364
5. Li K, Ma HQ (2021) Research on lighting energy-saving technology for the middle section of low traffic highway tunnel. Mod Tunnel Technol 58(o1):190–196
6. Du GF, Qing HB (2021) Tunnel lighting control system based on vehicle perception and fuzzy PID. Transducer Microsyst Technol 40(02):95–97

7. Cantisani G, Di Mascio P, Moretti L (2018) Comparative life cycle assessment of lighting systems and road pavements in an italian twin-tube road tunnel. Sustainability 10(11)
8. Liang B, He SY, Pan GB et al (2015) Research on driving safety of highway tunnels based on new interior materials. Mod Tunnel Technol 52(02):16–21
9. Qin L, Shi X, Leon A S, et al. Dynamic luminance tuning method for tunnel lighting based on data mining of real-time traffic flow. Build Environ 176
10. He SY, Liang B, Zhong SM et al (2020) Research on highway tunnel lighting evaluation method based on indoor simulation of light environment and visual effect experiment. J Chongqing Jiaotong Univ (Nat Sci) 39(03):27–35
11. Kong XJ (2007) The *Henglongshan* long tunnel is planned to be completed by the end of 2007. Highway Tunnel 4:20
12. Ministry of Transport of the People's Republic of China (2014) Detailed rules for lighting design of highway tunnel

The Influence of Speed Limit Value of High-Grade Highway in High Altitude Area on Running Speed and Traffic Safety

Dawasangbu

Abstract This article collected relevant data on two high-grade highways in Tibet. Through the statistical analysis of the speed limit value, running speed and traffic accident data, the impact of the increase of the highway speed limit value on the running speed and traffic safety was studied. The regression method was used to analyze the functional relationship between the speed limit increase value and the running speed increase value and the number of traffic accidents, which provides a reference for the speed limit value setting of high-grade highways in high altitude areas. The results show that using the design speed as the speed limit cannot meet the driving expectations of most drivers, and to a certain extent reduces the capacity of high-grade highways. After the speed limit was increased, the original accident-prone road was still the key area of accidents. If no measures are taken, the number of traffic accidents may increase. In engineering practice, the determination of the speed limit value should not only consider the actual running speed of the vehicle, but also analyze the traffic accidents. The increase of the speed limit value on the accident-prone sections should be fully demonstrated and effective engineering measures should be taken.

Keywords Speed limit · High-grade highway · Traffic safety · Running speed

1 Introduction

Our country's highway speed limit generally adopts the design speed value. Such speed limit scheme takes the safe driving speed of the road section controlled by the design index as the speed limit of the vehicle, ensuring the safety of the vehicle driving on the controlled road section. However, for road sections with high design indexes, it will cause frequent speeding, which increases the difficulty of highway

Dawasangbu (✉)
Tibet Autonomous Region Research Institute of Traffic Survey and Design, No. 11, Beijing West Road, Chengguan District, Lhasa, Tibet Autonomous Region, China
e-mail: 122593163@qq.com

W. Wang et al. (eds.), *Green Transportation and Low Carbon Mobility Safety*,
Lecture Notes in Electrical Engineering 944,
https://doi.org/10.1007/978-981-19-5615-7_34

management and traffic law enforcement, and reduces the capacity and service level of highways.

In December 2020, the Tibet region increased the speed limit of highways. It was proposed to increase the speed limit value to solve the problem of low traffic capacity and low service level of high-grade highways. However, after the speed limit is increased, whether the dispersion of running speed increases, and whether traffic accidents increase has not been comprehensively studied. Through analyzing the speed limit value, traffic accidents and running speed of the two high-grade highways in the area, this article explores the impact of the increase in the speed limit value of the high-grade highway in the high-altitude area on traffic safety. This study provides a reference for the setting of speed limit values for high-grade highways in high-altitude areas in the future.

2 Related Research

Research in Australia showed that when the vehicle speed is greater than 60 km/h, the risk of traffic accidents will double for every 5 km/h increase [1].

Sweden and Finland also conducted statistical studies on the effect of speed limit implementation in developing countries. When the speed limit was greater than v_{85}, it was found that the average speed and the number and severity of accidents will increase. When the speed limit was less than v_{85}, the average speed and the number and severity of accidents will decrease. When the speed limit value was too low, only some vehicles will drive at a low speed, and the rate of overspeed violation was high [2].

In view of the speed limit on highways in developing countries, Afukaar [3] analyzed the speed limit in Ghana and other countries. It was proposed that speed control is a practical way to prevent road traffic injuries and deaths in developing countries, and a reasonable speed limit design must be adopted to make the speed limit acceptable to most drivers.

At the end of 2020, China promulgated the "Design Specifications for Highway Speed Limit Signs". The specification recommends the comprehensive evaluation method, risk factor method, and running speed method to demonstrate the speed limit value [4]. This specification has solved the problem of low capacity of our country's high-grade highways, and at the same time fully considered the factors of traffic safety. However, the research data supporting this specification were collected in low-altitude areas, and whether it is fully applicable to highways in high-altitude areas remains to be verified.

In 2004, domestic scholar Yulong Pei analyzed some highway accidents in China and believed that the accident rate increased with the increase of the speed standard deviation. That is, the more dispersed the speed of the vehicle, the higher the accident rate [5].

Sheqiang Ma and others have studied the impact of vehicle speed on traffic safety and believed that the higher the vehicle speed, the greater the dispersion, the greater the number of traffic accidents, the accident rate, and the severity of the consequences of the accident. Also, the greater the difference in the speed of vehicles on adjacent road sections, the greater the risk of road traffic [6].

At present, foreign studies on speed limit are relatively sufficient, but the characteristics of driving behavior in our country are not fully considered, so it is not appropriate to adopt them directly. Domestic research mainly focuses on the analysis of the relationship between running speed and accidents, and there is little research on the relationship between speed limit and accidents. In addition, the research data was usually collected in low-altitude areas, which is not conducive to guiding the determination of speed limit management plans in high-altitude areas.

3 Overview of the Investigated Highways

There are 5 high-grade highways in this high-altitude area, of which 2 high-grade highways have been investigated in this article. The design speed of Lagong Airport Highway is 100 km/h, and the survey mileage is 37.8 km. The design speed of Linzhi Airport Highway is 80 km/h, and the survey mileage is 45 km.

Before December 2020, the speed limit value adopted the design speed. In December 2020, the maximum speed limit was adjusted to 110 km/h.

4 Data Collection

4.1 *Collection of Traffic Accident Data*

The accident data was provided by the local traffic police department. The provided data was from January 2017 to June 2021, and there were a total of 247 traffic accidents.

In a certain statistical period, certain locations where the number of traffic accidents occurred in a road section are more prominent than other normal locations are called accident-prone points or road sections. At present, the methods for identifying the frequent occurrence points (sections) of accidents include: accident frequency method, accident rate method, cumulative frequency method, etc. Due to the difficulty in obtaining some data, this study used the accident frequency method to determine the accident-prone road sections. In the 3 years before the speed limit adjustment, more than 4 accidents within 1 km are used as the criterion for identifying accident-prone road sections. After analysis, there are 5 accident-prone sections on the Lagong Airport Highway and 3 accident-prone sections on the Linzhi Airport Highway (Table 1).

Table 1 Collected data of traffic accidents

Highway	Mileage (km)	Design speed (km/h)	Total number of accidents		Accident-prone road sections
			Before speed limit increase	After speed limit increase	Before speed limit increase
Lagong Airport Highway	37.8	100	127	11	5
Linzhi Airport Highway	45	80	91	18	3

4.2 *Running Speed*

Before and after increasing the speed limit, the project team used a handheld radar speedometer to collect the vehicle's cross-sectional running speed. For both roads, 3 accident-prone road sections and 3 general road sections were randomly selected to collect running speeds. Each collection point collects no less than 250 running speed data.

5 Safety Impact Analysis of Speed Limit Value

5.1 *Analysis of the Relationship Between Speed Limit and Running Speed*

Running speed v_{85} is affected by the driver's expected speed, traffic flow composition and vehicle performance, highway traffic environment and other factors. It is the actual speed used by a general driver when driving a car along a certain road under good weather conditions and normal traffic conditions.

The running speed data of the 6 cross sections collected this time are shown in Figs. 1 and 2. When the design speed is used as the speed limit value, on the highway with the speed limit of 80 km/h, v_{85} in most of the road sections exceed the speed limit value. On highways with a speed limit of 100 km/h, there are some sections where v_{85} exceeds the speed limit. After increasing the speed limit from the design speed to 110 km/h, the v_{85} of both highways has been increased, but the speed limit has not been exceeded.

It can be seen from the collected data that when the design speed was used as the speed limit value, it is quite different from the driver's expected speed, which increases the risk of illegal driving and cannot give full play to the efficient running efficiency of high-grade highways. After the speed limit was increased, the expectations of most drivers were met, and the operating efficiency was improved. To have

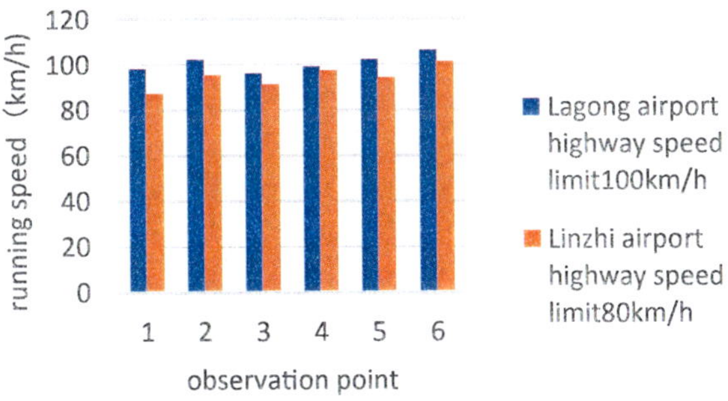

Fig. 1 The running speed collected by the observation point before the speed limit adjustment

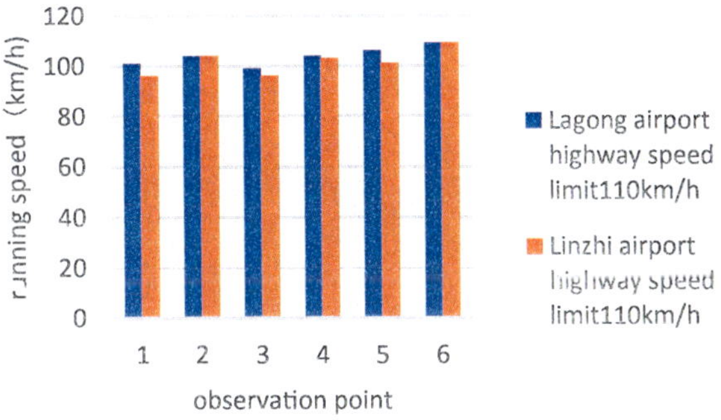

Fig. 2 Running speed of observation point collection after speed limit adjustment

a more in-depth understanding of the impact of the increase in the speed limit value of high-grade highways on the running speed, the project team performed regression analysis on the collected speed limit increase value and running speed increase value. The regression equation of speed limit increase value $\Delta \boldsymbol{v}_x$ and operating speed increase value $\Delta \upsilon_{85}$ was established:

$$\Delta \upsilon_{85} = 3.641\ln(\Delta \boldsymbol{v}_x) - 5.0503 \quad (1)$$

where $\Delta \upsilon_{85}$ (km/h) is the increase in running speed and $\Delta \boldsymbol{v}_x$ (km/h) is the increase in speed limit (Fig. 3).

The increase in speed limit has a certain functional relationship with the increase in running speed. After the speed limit was increased, the running speed also increased. Due to the influence of route alignment, driver's physical condition, vehicle characteristics, and surrounding environment, as the speed limit further increases, the increase in running speed gradually slows down. For high-speed highways with the design speed as the speed limit value, the traffic capacity of the highway can be increased by increasing the speed limit value. However, when the speed limit is increased to a certain level, the traffic capacity of the highway no longer increases.

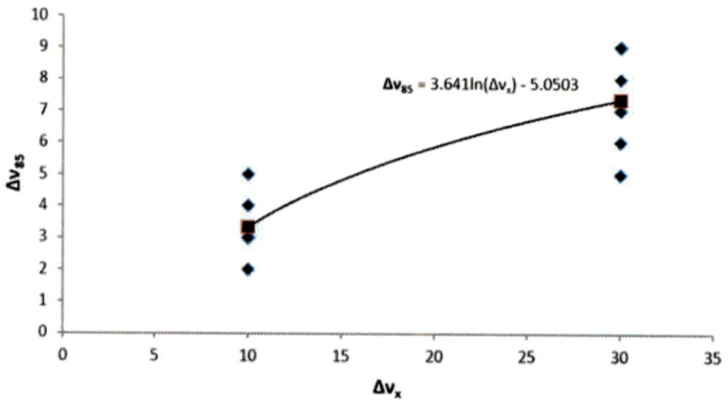

Fig. 3 The relationship between the increased value of speed limit and the increased value of running speed

5.2 *Analysis of the Relationship Between Speed Limit and Traffic Accidents*

The project team collected traffic accident data on accident-prone road sections and general road sections. It was planned to analyze the speed limit value of different dangerous road sections to clarify the impact of speed limit on traffic safety.

Through the regression analysis of the data, the regression equation of the speed limit increase value Δv_x of the accident-prone road section and the number of traffic accidents n was established:

$$n = 0.8829\ln(\Delta v_x) - 1.5364 \quad (2)$$

where n (accident number/year) is the number of traffic accidents, and Δv_x (km/h) is the increase in speed limit.

It can be seen from Fig. 4 that the number of traffic accidents on the accident-prone sections increased with the increase of the speed limit. However, as the speed limit further increased, the increase in traffic accidents gradually slowed down. After the speed limit was increased, the running speed will increase, which increased the risk of dangerous road sections. After the speed limit reached the driver's expectation, the running speed will no longer increase, and the risk of the road section will also level off.

At the same time, the project team analyzed the relationship between the number of traffic accidents on general road sections and the speed limit value. It was found that the number of traffic accidents did not change significantly after the speed limit value of general sections was increased.

Increasing the speed limit will increase the risk of accident-prone sections, but it will have little impact on general sections. When setting speed limit signs, road accident data should be collected, and road sections prone to accidents should be

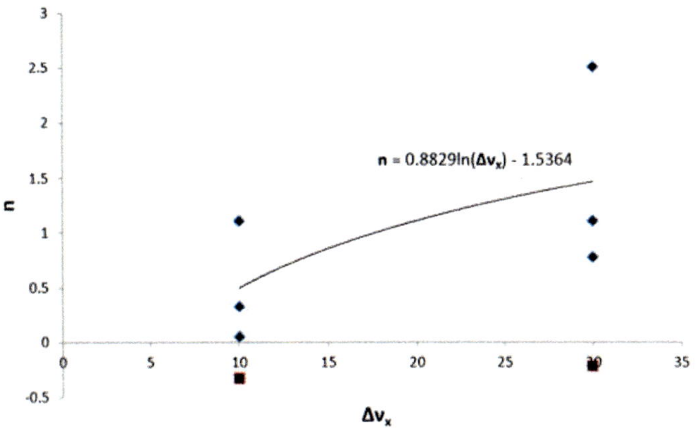

Fig. 4 The relationship between the increase of the speed limit and the number of traffic accidents on the accident-prone sections

fully demonstrated to avoid increasing the risk factor of the road section after the speed limit value is increased.

6 Conclusion

Speed limit is a key means of road safety management, which is not only related to road safety, but also related to whether the operation efficiency of high-grade roads can be fully utilized. High speed limit can improve the efficiency of highway operation, but if the speed limit value does not match the driving environment, the accident rate will increase significantly. On the other hand, low speed limit will reduce the efficiency of highway operation.

This article collected the actual vehicle speed and traffic accident data before and after the adjustment of the speed limit value on a high-grade highway in a high-altitude area in our country. The impact of the increase of the highway speed limit value on traffic safety and traffic capacity was studied, which provides a reference for the adjustment of the highway speed limit value in this area in the future. The main conclusions are as follows:

(1) The design speed limit adopted on high-grade highways cannot meet the driving expectations of most drivers, causing most vehicles to drive illegally and reducing the operating efficiency of high-grade highways. After the speed limit was increased, the actual running speed of the vehicle increased, which can increase the traffic capacity of the highway.
(2) When the speed limit value is too low, the driver will not drive according to the speed limit value without taking compulsory control conditions such as

speeding snapshots, but will determine the driving speed according to his own comprehensive judgment.

(3) After the speed limit was increased, the original accident-prone road was still the key area of accidents. If no measures were taken, the number of traffic accidents may increase. There was no significant change in the number of traffic accidents on other road sections.
(4) In engineering practice, the determination of the speed limit value should not only consider the actual running speed of the vehicle, but also analyze the traffic accident. The increase of the speed limit value on the accident-prone sections should be fully demonstrated and effective engineering measures should be taken.

Due to the limitation of data collection, the collection time of traffic accidents after speed increase in this article was short, traffic volume data cannot be obtained, and the sample size of the survey was small, which has a certain impact on actual data analysis. In the future, the corresponding results should be continuously improved through the increase of data collection. In addition, operating speed and traffic accidents are only two factors considered in the process of determining the speed limit. In actual work, the influence of road alignment, environment and other factors should be considered.

References

1. Zhang S, Li J, Li L (2006) Discussion on the speed analysis of road traffic accidents. Commun Standard 8:148–151
2. Wu J (2003) Speed control theory and application. J Highway Transp Res Dev 20(1):127–130
3. Afukaar FK (2003) Speed control in developing countries: issues, challenges and opportunities in reducing road traffic injuries. Injury Control Saf Promot 1-2, 77-81
4. JTG/T 3381-02-2020 (2020) Design specifications for highway speed limit signs. China Communications Publishing & Media Management Co, Ltd, 13–15, 30–32
5. Pei Y, Cheng G (2004) Research on the relationship between discrete character of speed and traffic accident and speed management of freeway. China J Highway Transport 17(1):74–78
6. Ma S, Liu D, Feng L (2008) The impact of vehicle speed on traffic safety and management research. J Highway Transport Res Dev 5:139–141

Research on Interference-Free Monitoring of Driver's Steering Behavior

Haodong Zhang, Haiqiu Tan, Kuikui Feng, Jian Shi, Dongxian Sun, Jie Zhang, and Wuhong Wang

Abstract The study of driver's driving behavior is not only an important part of traffic safety, but also of great significance to promoting the development of autonomous vehicles. In this paper, a Triboelectric Nanogenerators (AK-TENG) with an aluminum (Al)-Kapton friction layer structure is developed for driving behavior monitoring of drivers. AK-TENG has high sensitivity and certain flexibility. Compared with ordinary Triboelectric Nanogenerators, the structure is stronger and more stable, and less damaged. It is used to obtain the driving behavior data of the driver on the steering wheel. This paper uses AK-TENG to obtain driving behavior information data to monitor the driver's steering behavior, and also provides a new idea for the design of high-sensitivity self-powered sensors in intelligent transportation systems.

Keyword Intelligent traffic system · Driving behavior monitor · Triboelectric nanogenerator

In recent years, with the gradual increase in car ownership, the number of deaths caused by traffic accidents has surged. The misbehavior of human drivers accounts for more than 80% of all traffic accidents [1]. Autonomous vehicles and intelligent networked transportation systems are gaining more and more attention from academia and industry, and the capabilities of intelligent vehicles in highly or even fully autonomous driving are enhanced. However, challenges remain in driving highly automated vehicles on public roads. Therefore, monitoring the driver's driving behavior has always been a hotspot in the research of intelligent transportation systems, which is of great significance.

With the development trend of "intelligence" and "networking" of modern automobiles, it becomes more and more obvious. Advanced technologies such as wireless communication, advanced vehicle control, complex environment perception, and autonomous decision-making will account for an increasing proportion of the

H. Zhang · H. Tan · K. Feng · J. Shi · D. Sun · J. Zhang · W. Wang (✉)
School of Mechanical Engineering, Beijing Institute of Technology, Beijing 100081, China
e-mail: wangwh@bit.edu.cn

W. Wang et al. (eds.), *Green Transportation and Low Carbon Mobility Safety*, Lecture Notes in Electrical Engineering 944,
https://doi.org/10.1007/978-981-19-5615-7_35

core technologies of future automobiles. The information model inside the vehicle has gradually developed from a single driving and vehicle condition information model to a complex information system including vehicle information, inter-vehicle information, and information about the interaction between vehicles and other information carriers. Diversification and enrichment of information is a double-edged sword. While helping drivers better understand the operating environment of the vehicle, revolutionary changes are taking place in the interior space, human-machine interface, operation, and interaction process of the vehicle. As an independent and important research field, the human-computer interaction system faces new challenges in both theoretical research and practical application. The interaction method between driver and vehicle, the content and scene of human-vehicle interaction need to be continuously expanded. Furthermore, due to the strong uncertainty of driver behavior and cognition [2], the behavior of human drivers still needs to be better understood before transitioning to fully autonomous driving. This is necessary not only to improve vehicle safety, performance, and energy efficiency, but also to adapt to driver needs, enhance driver acceptability, and ultimately meet driver preferences within safety boundaries.

Driver status monitoring and operational behavior monitoring are the focus of driving behavior research. At present, the research methods on driver status monitoring mainly include three aspects: (1) Using physiological signals such as electroencephalogram (EEG), electromyography (EMG), electrooculography (EOG), and electrocardiogram (ECG) to monitor the driver's degree of fatigue, concentration, and emotion [3]. (2) Use data from the vehicle on the road, including the steering wheel and pedal movement information, vehicle location, and other information to detect driver fatigue and distraction [4]. (3) Using machine vision systems to detect driver fatigue and distracted behavior by extracting image features [5]. According to different data sources, the detection methods of driver operation behavior can be divided into the following: based on driver data information and based on vehicle data information. Driving behavior detection based on driver data information is usually based on the analysis of physiological data collected by wearable devices [6]. Ikenishi [7] used processed electroencephalographic (EEG) signals to classify driving intent into three categories: right-turn, left-turn, and straight-line driving. Bi et al. [8] proposed a model that uses the driver's EEG signals to complete the steering operation, which can complete three steering operations of straight, left and right turns by issuing commands. Haufe et al. [9] vehicle-based driving maneuver behavior detection uses data collected by vehicle system sensors to analyze the driver's steering behavior. Berndt et al. [10] used the driving simulator to complete the simulated driving experiment, and used the collected steering wheel motion information and vehicle lateral acceleration information to develop the recognition model. The model can complete the identification of three intentions of turning left lane change, right lane change, and normal lane keeping. [11] establishes a steering wheel motion model, detects and tracks the motion information of the steering wheel and extracts the main features through the external equipment of the vehicle, and realizes the detection of the steering wheel angle information. Tran et al. [12] proposed a new framework for driver foot tracking based on machine vision, using a HMM to judge the states of the acceleration and brake pedals, and divided the pedal states into seven categories, the accuracy rate is as high as 94%.

However, in the study of driving behavior, the above methods may have some shortcomings. Wearable devices used to acquire physiological signals can cause a great disturbance to drivers. Other data acquisition equipment may have non-negligible cost and consumption, and there are problems such as complicated installation and arrangement. Therefore, new types of sensors are needed to complement the deficiencies of existing sensors in monitoring driving behavior. In recent years, triboelectric nanogenerators have attracted increasing attention worldwide as a newly developed self-generating device. TENG based on the coupling principle of tribo-electric and electrostatic induction convert mechanical energy into electrical energy and generate characteristic signals. TENG has the advantages of simple production, low cost, high yield, lightweight, durability, and wide temperature range, and has received extensive attention from academia [13].

In this paper, a triboelectric nanogenerator (AK-TENG) with an aluminum (AI)-Kapton friction layer structure is developed for driving behavior monitoring of drivers. The AK-TENG is arranged on the steering wheel to obtain the driver's driving behavior information. A preliminary evaluation of data collected in real-time by AK-TENG shows that this self-powered high-sensitivity sensor is feasible for monitoring driver behavior. This paper lays the foundation for exploring TENG as a sensor to monitor the driver's driving state and operating behavior in intelligent transportation systems.

1 Fabrication and Testing of AK-TENG

According to different working principles, triboelectric nanogenerators (TENGs) are classified into four modes: TENGs based on independent triboelectric layers, TENGs based on vertical contact separation mode, TENGs based on single electrode, and TENGs based on lateral sliding mode. The triboelectric nanogenerator (AK-TENG) with an aluminum (AI)-Kapton tribolayer structure developed in this paper is a vertical contact separation mode based on a triboelectric nanogenerator (TENG).

1.1 Fabrication of AK-TENG

AK-TENG is simple to manufacture, low in cost, light in weight, thin in thickness, and will not interfere with the driver. It consists of three layers of AK-TENG, which are kapton layer, aluminum layer and buffer layer. At the bottom of the kapton layer, a kapton film is used as the friction layer, and the aluminum layer is located in the second layer as the back electrode of the kapton layer. The aluminum strip on top of the lower layer acts as another friction layer and electrode. The upper and lower sides of the AK-TENG are wrapped with kapton tape. The upper and lower layers of the

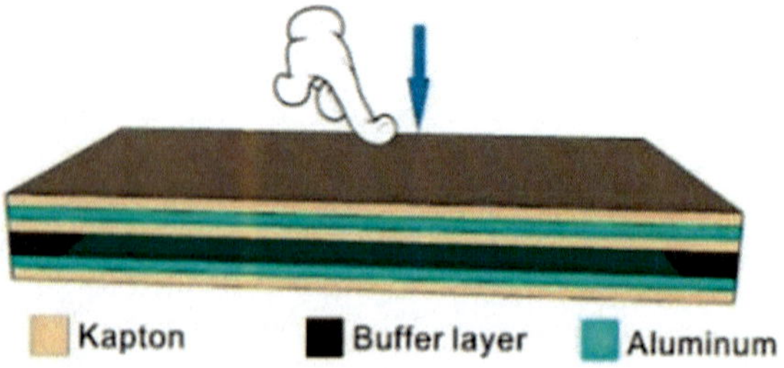

Fig. 1 The structure of AK-TENG

entire AK-TENG are separated by a buffer layer, so a gap can be formed, so that the two friction layers can be contacted and separated under the action of external force. The structure of AK-TENG is shown in Fig. 1. AK-TENG has high sensitivity and certain flexibility, the structure is firm and stable and not easy to be damaged, which fully meets the functional requirements of being installed on the steering wheel.

1.2 The Working Principle and Output Characteristics of AK-TENG

Figure 2 shows the working principle of AK-TENG. In step I, there is no distance between the two friction layers at this time, and the two are in close contact, resulting in charge separation. Due to the strong ability of aluminum to lose electrons and the strong ability of kapton to acquire electrons, the surface of the aluminum layer and the surface of the kapton layer try to have the same net positive charge and net negative charge respectively, and the output voltage of AK-TENG is 0 at this time.In step II, the two friction layers are relatively separated, so a potential difference is formed between the two electrodes. Electron flow is generated between the upper electrode and the lower electrode, forming an electrostatic balance, resulting in a positive voltage and current. In step III, the induced positive charges on the upper back electrode gradually neutralize the net negative charges on the kapton layer as the relative distance between the two friction layers increases. When the positive charge completely neutralizes the negative charge, the voltage value output by AK–TENG reaches the maximum value at this time. In step IV, the relative distance between the kapton layer and the aluminum layer gradually decreases. The voltage between the lower electrode and the upper electrode drops, and finally returns to the initial state.

An AK-TENG with a size of 3 cm $\times$ 3 cm was fabricated to verify the output characteristics. The prepared AK-TENG is mounted on a programmed linear motor capable of simulating external motion. AK-TENG output voltage and current are measured by Keithley 6514. The short-circuit current and open-circuit voltage generated by AK-TENG at different test frequencies are shown in Fig. 2b and c. When the frequency was increased from 1 to 3 Hz, the voltage peak remained basically

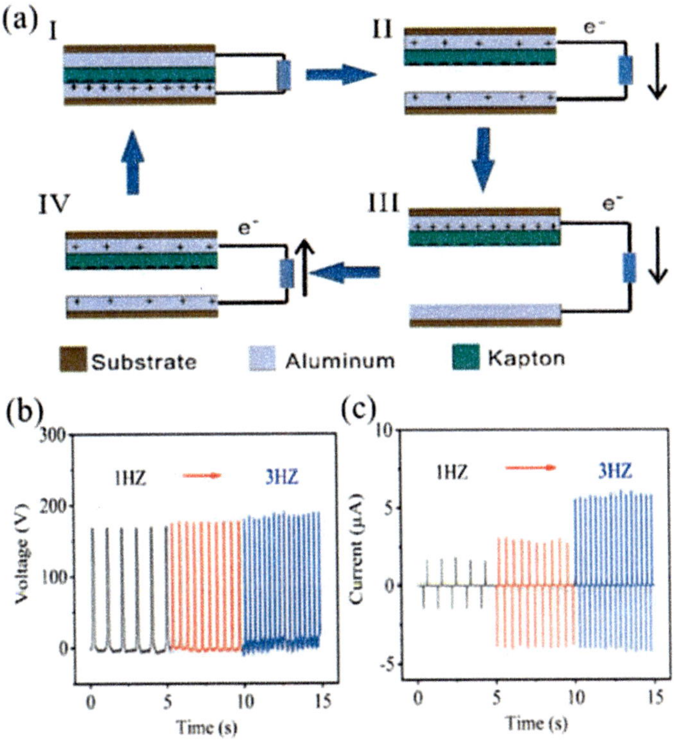

Fig. 2 The working principle of AK-TENG. **b** AK-TENG open circuit voltage. **c** Short circuit current

unchanged, as shown in Fig. 2b. The magnitude of the voltage generated by AK-TENG is independent of frequency. Figure 2c shows that the voltage generated by AK-TENG increases with frequency.

2 Installation Layout and Data Acquisition of AK-TENG

As shown in Fig. 3, two AK-TENGs with dimensions of 2.5 cm × 25 cm are arranged on both sides of the steering wheel. They are used to obtain various characteristic indicators related to the driver's operational behavior when completing the driving task. When the driver completes the steering task, the hands and the steering wheel move relative to each other, and this tiny movement contains a lot of valuable information. Arrange the AK-TENG on both sides of the steering wheel. When the driver operates the steering wheel, the force exerted by the driver's hand on the steering wheel also acts on the AK-TENG. The AK-TENG produces different contact separations according to the force applied by the driver, which in turn produces a voltage

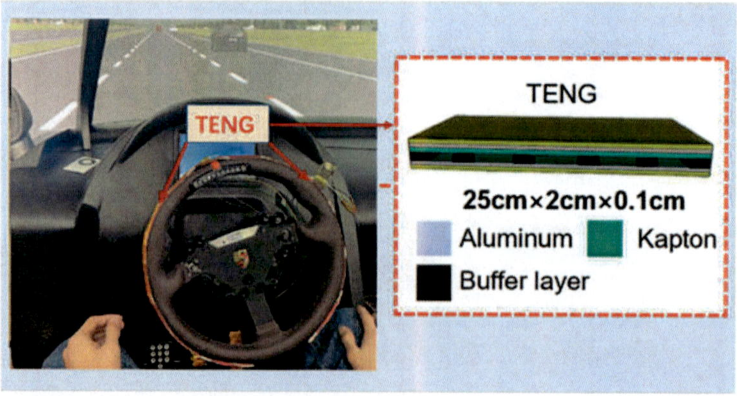

Fig. 3 Installation layout of AK-TENG

signal. Using this method, various small steering behavior information of the driver can be obtained without causing any disturbance to the driver. By analyzing the voltage signal output by AK-TENG, various characteristics and indicators of the driver's steering operation can be obtained.

The AK-TENGs are all connected to a multi-channel data acquisition device, which is connected to a laptop. During testing, drivers complete driving tasks on a driving simulation system with a 270-degree visual environment (as shown in Fig. 4). In order to capture various types of steering operations, we build a simulated driving scene in a city. The subjects were required to complete the driving task in the urban simulated driving scene, and collected the simulated driving data output by the simulated driver and the voltage signal output by the AK-TENG. When faced with complex urban simulation driving scenarios, drivers can exhibit a variety of steering and driving behaviors. By collecting and analyzing these data, the performance of the AK-TENG developed in this paper is evaluated.

Fig. 4 Driving simulator system

3 Results and Discussion

When the steering wheel is operated, the distance between the two friction layers of aluminum and Kapton becomes smaller, and the state of charge balance on the surfaces of the two friction layers is broken. But since the instrument we collect the voltage is resistive, the voltage will decay to 0 in a short time. When the steering wheel is turned to a certain position and remains stationary, the distance between the two friction layers of the AK-TENG remains unchanged, so the voltage of the AK-TENG also remains relatively stable. When the steering wheel is loosened, the newly established charge balance is broken, electrons will flow into the aluminum electrode from the outside world, and the voltage of AK-TENG decreases rapidly to a negative value. The absolute value of the voltage peak and voltage slope of AK-TENG is directly related to the angular velocity and angular acceleration of pedal motion, so the voltage waveform of AK-TENG can be used to judge the driving style of the driver when accelerating and decelerating. An AK-TENG is arranged on each side of the steering wheel, and the channels on both sides can be combined to obtain other information. Whether turning left or right, when both the left channel and the right channel fluctuate, it means that the driver operates both hands at the same time when the two steering behaviors are completed. In addition, by judging the variation law of the fluctuation range of the voltage of the right channel and the left channel, it is judged whether the driver is used to exerting force with the right hand or the left hand when completing the steering task. By comparison, it can be found that when the driver starts to implement the steering action, the voltage output by the AK-TENG changes earlier than the data output by the driving simulator, indicating that the AK-TENG has higher sensitivity and stronger ability to obtain real-time data. When the driver completes the steering operation, the voltage output by the AK-TENG returns to 0 later than the data output by the driving simulator. It means that although the driver has restored the steering wheel and pedals to their original positions, the hand still exerts a force on the steering wheel and pedals. This part of the information can be used to predict the driver's operation intention at the next moment and analyze the personalized operation habits.

In addition, the AK-TENG can be arranged on the steering wheel and pedals, and through the collected voltage data, the driver's operating behaviors such as excessive steering, rapid acceleration, and rapid deceleration can be mined. Advance prediction of current operational behavior and intent analysis at the next moment after completing this behavior. Long-term tracking of different drivers, real-time acquisition and accumulation of their subtle operating behaviors, and then obtaining the personalized operating habits of different drivers.

4 Conclusion

In this paper, a triboelectric nanogenerator (AK-TENG) with an aluminum (Al)-Kapton friction layer structure is developed for driving behavior monitoring of drivers. Firstly, the research status in the field of driver behavior monitoring is summarized and analyzed. Secondly, the working principle, preparation, electrical output characteristics, and installation arrangement of AK-TENG are studied. Use the driving simulator to build a simulated driving scene to complete the collection of experimental data. Through the analysis of the data output by the simulated driver and AK-TENG, the results show that AK-TENG has great advantages and potential in the study of driver's steering behavior.

References

1. Peden M (2005) Global collaboration on road traffic injury prevention. Int J Inj Contr Saf Promot 12(2):85–91
2. Zhang H, Wang J (2017) Active steering actuator fault detection for an automatically-steered electric ground vehicle. IEEE Trans Vehicul Technol 66(5):3685–3702
3. Murugan S, Selvaraj J, Sahayadhas A (2020) Detection and analysis: driver state with electrocardiogram (ECG). Phys Eng Sci Med 43(12)
4. Hu HY, Gao ZH (2015) Driver drives detection base on time series analysis of steering wheel angular velocity. In: Proceedings of the 2015 China society of automotive engineering annual conference, vol 2
5. Qiang J, Zhu Z, Lan P (1995) Real-time nonintrusive monitoring and prediction of driver fatigue. IEEE Trans Vehicul Technol 53(4):1052–1068
6. Gheorghe L, Chavarriaga R, Millan J (2013) Steering timing prediction in a driving simulator task. Eng Med Biol Soc
7. Ikenishi T, Kamada T, Nagai M (2008) Classification of driver steering intention at the vehicle running based on brain-computer interface using electroencephalogram. Trans Japan Soc Mech Eng C 74(741):1347–1354
8. Bi L, Lu Y, Fan X et al (2016) Queuing network modeling of driver EEG signals-based steering control. IEEE Trans Neural Syst Rehabil Eng 1
9. Haufe S, Kim JW, Kim IH et al (2014) Electrophysiology-based detection of emergency braking intention in real-world driving. J Neural Eng 11(5):56011
10. Berndt H, Emmert J, Dietmayer K (2008) Continuous driver intention recognition with hidden Markov models. In: International IEEE conference on intelligent transportation systems
11. Cucchiara R, Prati A, Vigetti F (2003) Steering wheel's angle tracking from camera-car. In: IEEE intelligent vehicles symposium
12. Tran C, Doshi A, Trivedi MM (2012) Modeling and prediction of driver behavior by foot gesture analysis. Comput Vis Image Underst 116(3):435–445
13. Zhai N, Wen Z, Chen X et al (2020) Blue energy collection toward all our self-powered chemical energy conversion. Adv Energy Mater 2001041

Freeway Design Consistency Evaluation Model Based on Alignment and Traffic Characteristics

Zhang Sufeng, Ma Yanli, Zhou Nianfa, Li Ping, and Tian Jia-jia

Abstract In order to identify the segments of China's highways that may have traffic safety hazards in the design phase, reduce the traffic accident rate and improve the operational safety of vehicles. This paper constructs a safety evaluation model based on design consistency applicable to China's highways, including the operating speed prediction model and design consistency evaluation indexes. Combined with actual cases, the effectiveness of the model is verified by comparing the constructed model, IHSDM and the evaluation method of "Specifications for Highway Safety Audit". The results show that the accuracy of the model is 42% higher compared with IHSDM, and it can effectively evaluate the design safety of China's highways. It has important theoretical and practical significance for the study of highway design safety evaluation in China.

Keywords Traffic engineering · Design phase · Design consistency evaluation model · Operating speed · Design safety

CLC Number TP374.2

1 Introduction

With the continuous construction and development of highways in China, while providing efficient and convenient services for traffic travel, it has also caused a high rate of highway traffic accidents, which is mainly due to the inconsistency of operating speed and linear design of highways. It is crucial to carry out the study of highway design safety evaluation based on design consistency.

M. Yanli (✉) · T. Jia-jia
School of Transportation Science and Engineering, 73 Huanghe Road, Harbin 150090, China
e-mail: mayanli@hit.edu.cn

Z. Sufeng · Z. Nianfa · L. Ping
Heilongjiang Highway Construction Center, Harbin 150040, China

W. Wang et al. (eds.), *Green Transportation and Low Carbon Mobility Safety*,
Lecture Notes in Electrical Engineering 944,
https://doi.org/10.1007/978-981-19-5615-7_36

Foreign research on the evaluation of highway design safety has been carried out early. Castro et al. analyzed the relationship between vehicle operating speed, operating speed difference and road design consistency and obtained a model of the relationship between vehicle operating speed and design consistency [1]. Camacho-Torregrosa et al. developed a design consistency model for road use based on Spanish two-lane operating speed data [2]. Wu et al. studied the relationship between design consistency and traffic safety, using highway data from central Pennsylvania, USA (US 322 and PA 350), and analyzed the relationship between design consistency and total accident rate using regression models, showing a significant positive correlation between design consistency and accident rate [3]. Sahili K et al. investigated the effect of geometric design consistency on road safety in the West Coast of the United States and showed that operating speed, vehicle stability, alignment index and driver workload are common factors that may affect design consistency [4]. Compared with foreign countries, the research on highway design consistency in China started late, but some results have been achieved. In 2004, the Ministry of Transportation of China, in order to reduce traffic accidents and increase highway safety, drew on advanced foreign experience and design concepts, and introduced the concept of safety design based on the actual situation in China,

and published the "Guidelines for Safety Audit of Highway" in 2004, which was 2015 revised and officially released as "Specifications for Highway Safety Audit", which started to be promoted and implemented to the whole country from April 1, 2016. In addition, Pei Yulong et al. conducted regression analysis of highway operating speed data and accident rate data and established a relationship model between the standard deviation of vehicle speed and the accident rate of billion vehicle kilometers [5]. Gao et al. established a road alignment quality evaluation model based on operating speed by using the amount of change in operating speed and acceleration between alignment units as evaluation criteria [6]. He Yulong et al. analyzed the difference of vehicle operating speed distribution between China and the United States [7]. Liu Jianbei et al. proposed road safety evaluation indexes and evaluation criteria considering highway traffic characteristics, driving behavior characteristics and road environment [8]. Yuan Zhibing established a running speed prediction model for the curved slope combination segment by studying the running speed characteristics of secondary roads in mountainous areas and applied the model results to evaluate the road running safety [9].

In summary, at present, the domestic still has not established a perfect design consistency based highway safety evaluation model, most of the research is based on the single vehicle speed difference consistency evaluation method, that is, the use of a single vehicle on the road running speed consistency to determine whether the road design is continuous, and then the road safety evaluation, and the evaluation indexes used are mostly absolute indexes, there may be some deviation in the practical application.

This paper combines the previous research results to build a highway design consistency safety evaluation model based on alignment design and traffic characteristics suitable for China, and adopts two relative indexes, namely the rate of change of operating speed of adjacent segments and the rate of deviation of operating speed

Table 1 Prediction formula for operating speed of horizontal alignment

Horizontal alignment	Operating speed prediction formula
Horizontal curve	$V_{85i,st} = 1.02R^{0.35}+88.45$
	$V_{85i,mi} = 27.85 + 0.63V_{85i,st} + 1.55\ln(L_c/2)$
	$V_{85i,en} = 1.68V_{85i,mi} - 1.58\ln R - 59.55$
Horizontal straight	$V_{85T} = \min\left(3.6 \times \sqrt{\frac{2a_{C_i,T} \times a_{T,C_{i+1}} \times L_{i,i+1} + a_{T,C_{i+1}} \times \frac{V_{85i,en}^2}{3.6^2} + a_{C_i,T} \times \frac{V_{85i+1,st}^2}{3.6^2}}{a_{C_i,T}+a_{T,C_{i+1}}}} 120\right)$

$V_{85i,st}$—the speed at the starting point of the horizontal curve (km/h);
R—the radius of the horizontal curve (m);
$V_{85i,mi}$—the speed at the midpoint of the horizontal curve (km/h);
L_c—the length of the horizontal curve (m);
$V_{85i,en}$—the speed at the end of the horizontal curve (km/h);
V_{85T}—the maximum speed at the horizontal straight (km/h),
$L_{i,i+1}$—the length of the horizontal straight segment between curve segment i and curve segment $i + 1$ (m);
$a_{C_i,T}$—the acceleration from horizontal curve I to horizontal straight (m/s^2);
a_{T,C_i}—the deceleration from horizontal straight to horizontal curve $i + 1$ (m/s^2)

relative to the design speed, as the design consistency evaluation indexes. Finally, a highway in China is used as the research object to verify the validity of the model.

2 Design Consistency Evaluation Model

2.1 *Operating Speed Prediction for Horizontal Alignment*

According to the horizontal alignment, the route is divided into horizontal straight segments and horizontal curve segments, and the operating speed prediction formulas for horizontal straight segments and horizontal curve segments are constructed respectively, as shown in Table 1.

2.2 *Operating Speed Prediction for Profile Alignment*

The profile alignment was divided into convex vertical curve and concave vertical curve. When the longitudinal gradient degree i was less than 3%, the corresponding operating speed prediction formula was constructed, as shown in Table 2.

Table 2 Prediction formula of vertical segment linear running speed

Profile alignment		Operating speed prediction formula
Vertical curve j – Longitudinal gradient – Vertical curve $j+1$, $i<3\%$	Vertical curve	$V_{85v}=\begin{cases}\min\left(\sqrt{\frac{4\times R_Z}{9.375}}+36\frac{L_Z-10}{2}100\right) convex\ vertical\ curve\\ \min\left(\sqrt{\frac{3.6\times R_Z}{1.5}},\frac{L_Z-10}{2},100\right) concave\ vertical\ curve\end{cases}$

V_{85v}—theoretical operating speed (km/h);
R_z—vertical curve radius (m);
L_c—vertical curve length (m)

Table 3 Running speed conversion model of longitudinal slope Segment [10]

Longitudinal gradient		Adjustment of operating speed
Ascending gradient	$3\% \le i \le 4\%$	Decrease 5 km/h every 1 km until minimum speed
	$i > 4\%$	Decrease 8 km/h every 1 km until minimum speed
Descending gradient	$3\% \le i \le 4\%$	Increase by 10 km/h every 0.5 km until desired speed
	$i > 4\%$	Increase by 20 km/h every 0.5 km until desired speed

When the longitudinal gradient degree $i \ge 3\%$, the operating speed conversion model of longitudinal gradient segment in "Specifications for Highway Safety Audit" can be used for calculation and adjustment, as shown in Table 3.

2.3 Operating Speed Prediction After Horizontal-Profile Combination

The road alignment of the highway is affected by the horizontal alignment and profile alignment. Therefore, the prediction formula of the combined operating speed of the highway is established by integrating the operating speed prediction of horizontal alignment and the operating speed prediction of profile alignment, as shown in Table 4.

Table 4 Prediction formula of horizontal and vertical combined running speed

Combined operating speed	$V=\begin{cases}\min\{V_{85}-(V_e-V_{85v}),90)\}V_{85}100\\ \min\{V_{85}-(V_e-V_{85v}),100)\}V_{85}100\end{cases}$

V_{85v}—Predicted operating speed of the profile alignment (km/h);
V_{85}—Predicted operating speed of the horizontal alignment (km/h)

Table 5 Calculation formula of the change rate of running speed of adjacent segments and the deviation rate of running speed relative to the design speed

Index	Calculation formula
$\Delta P_1'$	$\Delta P_1' = \frac{V_{j+1}-V_j}{V_j}$
$\Delta P_2'$	$\Delta P_2' = \frac{V_d-V}{V_d}$

V_j—the operating speed of segment J (km/h);
V_{j+1}—the operating speed of segment J + 1 (km/h);
V_d—the design speed (km/h);
V—the combined operating speed of horizontal alignment and profile alignment (km/h)

Table 6 Evaluation criteria for design consistency

Design consistency	Good	Average	Poor
$\Delta P_1'$	$\Delta P_1' \leq 0.08$	$0.08 < \Delta P_1' \leq 0.16$	$\Delta P_1' > 0.16$
$\Delta P_2'$	$\Delta P_2' \leq 0.10$	$0.10 < \Delta P_2' \leq 0.18$	$\Delta P_2' > 0.18$

3 Design Consistency Evaluation

In this paper, the rate of change of operating speed ($\Delta P_1'$) and the deviation rate of operating speed relative to design speed ($\Delta P_2'$) are adopted as the evaluation indexes of design consistency. The calculation formula is shown in Table 5.

The highway design consistency evaluation is based on the value of the change rate of the operating speed of adjacent segments and the deviation rate of the operating speed relative to the design speed, which is divided into three evaluation results: good, average and poor, as shown in Table 6.

4 Case Analysis

Choose a two-way four-lane highway in China as the research object, its design speed of 120 km/h, respectively applied design consistency model built in this paper, IHSDM and the "Specifications for Highway Safety Audit" for safety evaluation of the highway design, then the design consistency evaluation results were compared

Table 7 Comparison of evaluation results of design consistency of each model

Segment	Constructed model		IHSDM		Specifications
	Upstream	Downstream	Upstream	Downstream	
1	Average	Average	Average	Good	Good
2	Average	Average	Average	Average	Good
3	Average	Average	Average	Average	Good
4	Poor	Poor	Poor	Poor	Good
5	Average	Average	Good	Good	Good
6	Average	Average	Good	Good	Good

and analyzed with the IHSDM and specifications evaluation results. The validity of the design consistency evaluation results is verified as shown in Table 7.

It can be seen from the evaluation results in Table 7 that there are 6 segments in the upstream and downstream directions of the whole highway that are deficient in alignment design. Among them, the evaluation result of "Specifications for Highway Safety Audit" is that the design consistency of the whole alignment is good, there is no road segment with insufficient design consistency, and the "Specifications for Highway Safety Audit" is invalid. IHSDM evaluation results for the upstream direction there are 4 segments with average or poor design consistency, the downstream direction there are 3 segments with average or poor design consistency and all are duplicate segments with the upstream direction pile number, can better reflect the lack of horizontal alignment design consistency problem, but there exist certain differences with the specifications assessment results; In the upstream and downstream directions, there are 6 sections of inadequate alignment design in the constructed model, all of which are evaluated to be of average or poor design consistency, can better reflect the lack of alignment design consistency of design problems. The accuracy of the constructed model is increased by 42% compared with IHSDM, the evaluation results are highly accurate, and the constructed model can achieve effective highway safety evaluation in the design phase.

5 Conclusion

Based on the data of expressway design, a consistency evaluation model for highway design in China is established. In this paper, a design consistency evaluation model suitable for China's highways was established, and two relative indexes of design consistency evaluation were proposed, namely, the rate of change of operating speed and the deviation rate of operating speed relative to design speed. The validity of the model was verified by comparing with the evaluation results of China's relevant specifications and IHSDM.

The research results can provide a basis for quantifying the safety of highway in the design phase, provide guidance for the design and optimization of highway, and improve the driving safety of highway in the operating period.

Acknowledgements This research was supported by the Department of Transportation of Heilongjiang Province Project (MH20210247). The authors extend their gratitude to the reviewers and doctoral students of the Institute of Traffic Engineering for their valuable comments.

References

1. Castro M, Sanchez JF, Sanchez JA et al (2011) Operating speed and speed differential for highway design consistency [J]. J Transp Eng 137(11):837–840
2. Camacho-Torregrosa FJ, Pérez-Zuriaga AM, Campoy-Ungría JM et al (2012) New geometric design consistency model based on operating speed profiles for road safety evaluation [J]. Accid Anal Prev 61(8): 33–42
3. Wu KF, Eric TDPE, Himes SC et al (2013) Exploring the association between traffic safety and geometric design consistency based on vehicle speed metrics [J]. J Transp Eng 139(7):738–748
4. Al-Sahili K, Dwaikat M (2019) Modeling geometric design consistency and road safety for two-lane rural highways in the west bank, Palestine [J]. Arab J Sci Eng 44(5):4895–4909
5. Yulong P, Guozhu C (2004) Study on the relationship between speed dispersion and traffic accidents on expressway [J]. China J Highw Transp 1:125–128
6. Jianping G, Zhongyin G (2004) Evaluation of highway alignment design quality based on running speed [J]. J Tongji Univ (Nat Sci ed) 32(7):906–911
7. He Y, Wang S, Sun X et al (2010) A comparative study on the correlation between highway speed and traffic safety in China and THE United States [J]. China J Highw Transp 23(S1):73–78
8. Jianbei L, Zhongyin G, Jiangbi H et al (2010) Safety evaluation method and standard of highway route design [J]. China J Highw Transp 23(S1):28–35
9. Zhibing Y (2020) Study on prediction and application of running speed of combined segment of mountain highway with curved slope [D]. Chang 'an University
10. Huajie Engineering Consulting Co., LTD (2016) Specifications for highway safety audit: JTG B05-2015[M]. People's Communications Press Co, LTD

Driving Risk Identification Considering Coupling Coordination Degree of Primary and Secondary Tasks

Ma Yanli, Zhu Jieyu, Yining Lou, and Dong Fangqi

Abstract To study the influence of the coupling relationship between primary and secondary tasks on driving safety, four driving experiments are designed, including Bluetooth calling, conversation, screen touch operation and in-vehicle radio operation. The vehicle operation data such as standard deviation of horizontal speed, standard deviation of longitudinal speed and steering entropy are collected. The driver visual data such as entropy rate of fixation area, standard deviation of horizontal viewing angle and vertical viewing angle, and average glance speed are collected. The coupling model of primary and secondary tasks in the vehicle is established to determine the coupling coordination degree between them. The driving proportion threshold of driver's secondary tasks and the risk level of various secondary tasks are obtained. The results show that the coupling degree of normal driving and Bluetooth communication can reach high quality coordination level with high safety. Similarly, the coordination level between normal driving and talking is also high. The coupling degree of normal driving and operating radio can reach medium coordination level with general safety. The coupling degree of normal driving and touch screen operation can only achieve low coordination, which is a relatively dangerous driving state.

Keywords Traffic engineering · Driving risk identification · Primary task · Secondary task · Coupling coordination degree

M. Yanli (✉) · Z. Jieyu · D. Fangqi (✉)
School of Transportation Science and Engineering, 73 Huanghe Road, Harbin 150090, China
e-mail: mayanli@hit.edu.cn

D. Fangqi
e-mail: Fangqi_0815@126.com

Y. Lou
Department of Mathematics, University College London, WC1E 6BT London, England

W. Wang et al. (eds.), *Green Transportation and Low Carbon Mobility Safety*,
Lecture Notes in Electrical Engineering 944,
https://doi.org/10.1007/978-981-19-5615-7_37

1 Introduction

According to the FARS fatality analysis, at least 10% of fatal traffic accidents and 17% of injury traffic accidents are related to secondary task driving, which is the primary cause of distracted driving and subsequent traffic accidents [1]. Therefore, it is very important to investigate the influence of the secondary task on the primary task and determine the coupling relationship between them for improving the driving safety.

Driving secondary tasks arise from the multi-task demands of drivers in the process of driving, which are easy to compete with various resources of primary task (mainly visual and cognitive resources) [2]. As the driving process is characterized by multi-task parallel, the multi-resource theory proposed by Wickens et al. [3] is the most widely use. The multi-resource theory assumes that processing resources is a cube structure composed of channel, code and stage. The perceptual channel consists of visual and auditory levels, whose stage consists of sensory, cognitive and reactive level. The coding dimension consists of spatial coding and verbal coding level. Subsequently, Horrey et al. [4] proposed a conflict model of driving task based on the multi-resource theory. According to this model, if two tasks have competing demands for the same resource, the value of the conflict quantity is defined according to the intensity of the competition. The maximum amount of conflict occurs when a resource cannot be shared at all. A single task is accompanied by the distraction of multiple channels, which could reduce the total resource allocation as other channels are occupied [5]. The calling task is auditory input and the driving task is mainly visual input, the two tasks seem to be independent of each other and have no influence. But when the content of the call causes the driver's visual memory, the visual input will be interfered, resulting in some visual blind areas for the driver [6].

In terms of the relationship between secondary tasks and driving risk, Kidd et al. [7] pointed out in their study that the most common secondary tasks in driving are holding a mobile phone (5.1% of drivers observed), talking on a mobile phone (4.2%), eating or drinking (3.1%), talking or singing with a passenger (2.7%), operating a mobile phone (2.3%), and talking or singing without a passenger (2.1%). Primary and secondary tasks will compete within the driver's limited resources, so the driver's reaction time will slow down which would affect driving safety [8]. Dingus et al. [9] found that drivers tend to engage in at least one driving secondary task of 51.93% driving time, which would increase the driving risk by at least 2 times compared with normal driving. Jin et al. [10] the influence of cognitive secondary task on vehicle state in the process of driving through natural driving experiment. The research shows that cognitive secondary tasks requiring verbal and spatial imagination can reduce the driver's ability to scan the outside environment, which can reduce the ability to identify road information and to monitor hazards.

At present, the researches mainly study the influence of single task on driving safety and the relationship between in-vehicle secondary tasks and the environment outside the vehicle. Considering the shortcomings of previous studies, this paper introduces the coupling theory and establishes the coupling interaction model between the primary and secondary tasks in the vehicle. The coupling relationship between the primary and secondary tasks and its impact on driving safety are analyzed. It is of great significance to explore the resource allocation of primary and secondary tasks, and the study of driving safety in the driving process.

2 Experimental Data Acquisition and Processing

2.1 *Experimental Participants*

A total of 10 drivers (6 males and 4 females) with legal driving licenses are recruited for this experiment, ranging in age from 20 to 35 years old (mean of 28.5) and driving experience from 1 to 10 years old (mean of 4.3). This age group has the largest number of drivers in using mobile phones and Bluetooth devices, etc., which is representative. There is no fatigue and drunk driving of all drivers. The drivers are not influenced by drugs and can guarantee that they will not be affected by vision and hearing impairment during the driving process.

2.2 *Experimental Equipment*

This experiment adopts the driving simulation to obtain related data. The required experimental equipment includes: (1) Eye tracker (Smarteye6.0 eye-tracking system) to collect parameters of drivers' visual behavior such as fixation, scanning and blinking; (2) Driving simulator to collect vehicle operating parameters; (3) Vehicle-mounted radio equipment of simulation; (4) Bluetooth headset to make normal calls; (5) Smart phones that can be operated on a touch screen.

2.3 *The Experimental Method*

To explore the occupancy of drivers' resources by driving secondary tasks and the interaction between different in-vehicle secondary tasks and driving primary tasks, this experiment selects the most four common secondary tasks for the test, which include Bluetooth calling, in-vehicle radio operation, screen touch of smart phone and conversation. The secondary task includes the combination of different channels, such as auditory, visual, cognitive and motion, etc. The visual behavior of the driver and vehicle state parameters during driving of different secondary tasks are measured respectively.

(1) The combination of auditory and cognitive secondary task
The driver answers the phone call made by the experimental assistant through the Bluetooth headset, and then begins to talk. The assistant asks questions about the conversation and the driver answers them. There are 8 simple questions in total. If the driver feels that the current driving environment is not suitable for answering the phone, he can choose to refuse to answer.
(2) The combination of motion, cognitive and auditory secondary task
Starting from turning on the in-vehicle radio, the driver adjusts the frequency of the radio from 91.6 MHz to 102.6 MHz successively through button search, and stays for 30 ~ 60 s between every two frequencies.
(3) The combination of motion, cognitive and visual secondary task
Starting from unlocking the smart phone, the driver opens the home page of the smart phone, finds the music player in the phone, and selects the 20th music player in the music menu in the player (turn off volume).
(4) Normal driving
The driver does not perform any secondary tasks related to driving within the specified speed range.

After the start of the experiment, the driver first drives normally according to the route predetermined by the driving simulator without performing the in-vehicle tasks, and then performs different in- vehicle tasks for driving. During the experiment, the speed of each vehicle is required to maintain between 60 and 100 km.

2.4 Parameter Selection

2.4.1 Vehicle Operating Parameters

When a driver performs a more complex task in the driving process, he will reduce the speed to ensure driving safety. Steering wheel angle is an important indicator of vehicle state of lateral motion, mainly including standard deviation of steering wheel angle and steering entropy. These two indicators can not only reflect the stability of vehicle lateral movement, but also can be used to measure the driver's psychological load [11]. Therefore, three vehicle operating parameters are selected, consisting of the standard deviation of horizontal speed, the standard deviation of longitudinal speed and the steering entropy.

2.4.2 Driver Visual Parameters

The driver's visual behavior mainly includes three aspects: fixation, scanning and blinking. The driver visual parameters with high influence degree are selected to research coupling effect of the primary and secondary task.

(1) Entropy rate of fixation area, reflecting the fixation probability and average fixation duration of drivers on each fixation area in the driving process. (2) Standard deviation of horizontal viewing angle (3) Standard deviation of vertical viewing angle. The standard deviation of the view angle of horizontal and vertical directions is used as the indicator to evaluate the search breadth. The larger the standard deviation of the view angle, the wider the visual search range of the driver in this direction and the greater the external information obtained. (4) Average glance speed, the ratio of the glance distance (Angle) to the duration of the glance, which is mainly used to indicate the average speed of the driver's eye movement [11].

2.5 *Data Processing and Analysis*

2.5.1 Vehicle Operating Parameters

The significance of vehicle operating parameter can be obtained through the analysis, as shown in Table 1.

Table 1 Significance analysis of vehicle operation parameter

Vehicle operating parameters	Driving condition	Secondary tasks	Significance	95% confidence interval	
				The lower limit	The upper limit
Standard deviation of longitudinal speed	Normal driving	Bluetooth calling	0.001	1.2176	4.7104
		In-vehicle radio operation	0.016	0.4136	3.9064
		Conversation	0.000	1.8546	5.3474
		Screen touch operation	0.000	2.2176	5.7104
Standard deviation of steering wheel angle	Normal driving	Bluetooth calling	0.017	−0.5801	0.1081
		In-vehicle radio operation	0.041	−0.4581	0.2301
		Conversation	0.012	−0.7941	−0.1059
		Screen touch operation	0.005	−0.8501	−0.1619
Steering entropy	Normal driving	Bluetooth calling	0.025	−0.0784	−0.0056
		In-vehicle radio operation	0.025	−0.0784	−0.0056
		Conversation	0.022	−0.0794	−0.0066
		Screen touch operation	0.000	−0.1054	−0.0326

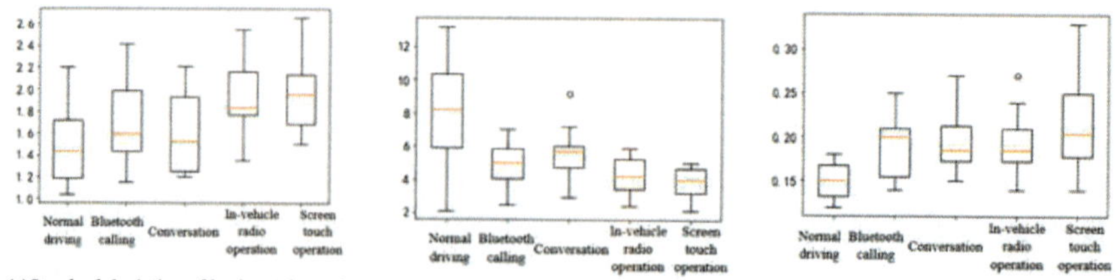

(a)Standard deviation of horizontal speed (b) Standard deviation of longitudinal speed (c) Steering entropy.

Fig. 1 Box diagram of vehicle operation parameters under different driving conditions

As can be seen from Table 1 of the significance analysis results, the differences of driving on various secondary tasks and normal driving are all lower than 0.05 (P = 0.05). Therefore, the above indicators of vehicle operating can be selected to study the coupling relationship between primary and secondary tasks during driving. Box diagram of operating parameters of each vehicle is shown in Fig. 1.

It can be seen from Fig. 1a that the mean standard deviation of steering wheel angle increases significantly in different driving conditions compared with normal driving. Compared with, the influence of steering wheel on steering stability under the driving conditions of talking is smaller than that of other secondary tasks. While the influence of screen touch operation on the steering stability is the most significant.

As shown in Fig. 1b, the mean standard deviation of the longitudinal speed in the process of secondary task driving is smaller than that of normal driving. The standard deviation of the longitudinal speed of the cognitive secondary task driving (Bluetooth calling and conversation) is larger than that of the visual secondary task driving (vehicle radio operation and touch screen operation).

As depicted in Fig. 1c, steering entropy significantly increases during the second task driving compared with normal driving. The steering entropy is the largest in the driving process of the screen touch task. As the screen touch task requires more visual resources, the driver has a large psychological load during the operation, and a worse handling stability, which is easy to affect the driving safety.

2.5.2 Driver Visual Parameters

The significance analysis of driver's visual behavior can be obtained as shown in Table 2.

As can be seen from Table 2, the differences of driving with different tasks and normal driving are all lower than 0.05 ($P = 0.05$). Therefore, four parameters including entropy rate of fixation area, standard deviation of horizontal viewing angle, standard deviation of vertical viewing angle and average glance speed could be selected to study the driver's visual behavior.

The box diagram of visual behavior parameters of drivers is shown in Fig. 2.

Table 2 Significance analysis of driver's visual behavior

Driver visual parameters	Driving condition	Secondary tasks	Significance	95% confidence interval	
				The lower limit	The upper limit
Entropy rate of fixation area	Normal driving	Bluetooth calling	0.044	−0.6144	1.1624
		In-vehicle radio operation	0.047	−0.6964	1.0804
		Conversation	0.037	−0.4924	1.2844
		Screen touch operation	0.035	−0.4754	1.3014
Standard deviation of horizontal viewing angle	Normal driving	Bluetooth calling	0.011	−0.2812	2.6952
		In-vehicle radio operation	0.019	−0.5022	2.4742
		Conversation	0.040	−0.8642	2.1122
		Screen touch operation	0.039	−0.8532	2.1232
Standard deviation of vertical viewing angle	Normal driving	Bluetooth calling	0.047	−0.3645	5.1545
		In-vehicle radio operation	0.042	−0.3195	5.1995
		Conversation	0.012	−0.5355	4.9835
		Screen touch operation	0.048	−0.2925	5.2265
Average glance speed	Normal driving	Bluetooth calling	0.011	−5.3388	53.5388
		In-vehicle radio operation	0.013	−5.8388	53.0388
		Conversation	0.045	0.6612	59.5388
		Screen touch operation	0.041	1.2612	60.1388

It can be seen from Fig. 2a that the mean entropy rate of the driver's fixation area changes significantly in the driving state of in-vehicle radio operation and screen touch operation. While the mean entropy rate of the driver's fixation area changes a little in the driving state of Bluetooth calling and conversation. The results indicate that the implementation of motion tasks has a greater impact on the entropy rate of the driver's fixation area.

As shown in Fig. 2b, the mean standard deviation of the horizontal viewing angle of drivers of secondary task driving decreases compared with normal driving. While this figure during driving of cognitive secondary task (Bluetooth calling and conversation) decreases more than that of visual secondary task driving (in-vehicle radio operation and screen touch operation).

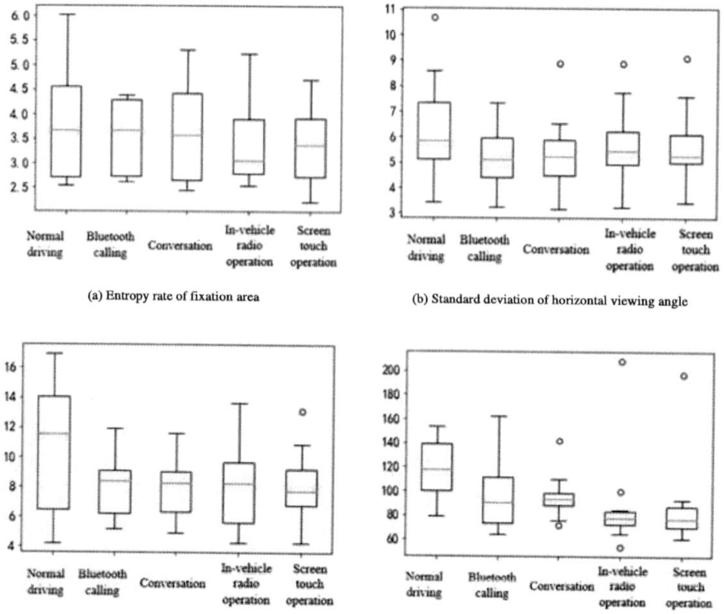

(a) Entropy rate of fixation area

(b) Standard deviation of horizontal viewing angle

(c) Standard deviation of vertical viewing angle

(d) Average glance speed

Fig. 2 Box diagram of driver's visual behavior under different driving conditions

As depicted in Fig. 2c, Compared with normal driving, the mean standard deviation of vertical viewing angle of the drivers is significantly reduced in the driving process of the second tasks.

From Fig. 2d, the average glance speed of the driver decreases significantly in the driving process of the second tasks compared with normal driving, and the average glance speed of the in-vehicle radio operation and the screen touch operation is the lowest of the second task driving. The results show that the second task of the screen touch operation is difficult because it requires more visual resources for the driver. The conversational task has the least effect on the average glance speed compared with other tasks, but there is still a significant difference compared with the normal driving.

3 The Coupling Model of Primary and Secondary Task

3.1 Coupling Degree Model of Primary and Secondary Tasks

Coupling refers to the interaction between multiple moving systems to achieve a cooperative state of a physical quantity. Considering the coupling relationship between the secondary and primary task of driving, the coupling degree model of the primary and secondary tasks of driving is presented by using the capacity coupling model, as shown in Eq. (1).

$$C = 2\left\{\frac{V_1 \times V_2}{V_1 + V_2^2}\right\}^{\frac{1}{2}} \tag{1}$$

where $V1$ is the comprehensive evaluation coefficient of driving primary task, $V2$ is the comprehensive evaluation coefficient of driving secondary task, C is the coupling degree of primary and secondary tasks.

The value of C is between 0 and 1. The closer the value of C is to 1, the higher the coupling degree of primary and secondary tasks is. The closer the value of C is to 0, the lower the coupling degree.

3.2 *The Coupling Coordination Degree of Primary and Secondary Tasks*

Coupling degree C is an important indicator of the interaction between primary and secondary driving tasks. However, it is not enough to judge the coupling relationship between the two subsystems only based on the C, because the numerical conclusions obtained by this model in some cases are different from the real coupling situation. For example, when the comprehensive evaluation coefficients of both subsystems are 0.01, compared with the case where the evaluation coefficients of the two subsystems are 0.8 and 0.9 respectively, the coupling degree calculated by the former is greater than the latter. Therefore, to make up for the deficiency of the coupling degree model, the coupling coordination degree of primary and secondary driving tasks is proposed. The coupling coordination degree can reflect the trend of the system or the internal elements of the system from disorder to order. And it is a quantitative indicator that reflects the degree of the good and bad of the coordination.

Let S represent the comprehensive coordination coefficient between primary and secondary driving tasks, reflecting the coordination level between them, and its expression is shown in Eq. (2).

$$S = \alpha V_1 + \beta V_2 \tag{2}$$

where α is the weight coefficient of the driving primary task in the whole driving process, β is the weight coefficient of driving secondary tasks in the whole driving process, $V1$ is comprehensive evaluation coefficient of driving primary task, $V2$ is comprehensive evaluation coefficient of driving secondary task, S is comprehensive coordination coefficient between primary and secondary tasks.

When considering the weight between primary and each secondary task, the influence of other secondary tasks is ignored, so for α and β, $\alpha + \beta = 1, \alpha \geq 0, \beta \geq 0$.

The coupling coordination degree model is shown in Eq. (3).

$$D = \sqrt{C \times S} \tag{3}$$

where C is coupling degree between primary and secondary tasks, S is comprehensive coordination coefficient between primary and secondary tasks, D is the coupling coordination degree between primary and secondary tasks.

Coupling coordination degree D can effectively represent the coordination situation between primary and secondary driving tasks. And it can also avoid the situation that the comprehensive evaluation coefficient between driving primary and secondary tasks is small while the coupling degree value is large. According to the theory of capacity coupling, the threshold of coupling coordination degree is defined as follows: when the coupling coordination degree D is less than 0.6, it is not coordinated. When the coupling coordination degree D is 0.6 ~ 07, it is the low coordination. When the coupling coordination degree D is 0.7 ~ 08, it is medium coordination. When the coupling coordination degree D is 0.8 ~ 09, it is a good coordination. When the coupling coordination degree D is 0.9 ~ 1.0, it is considered as high quality coordination.

3.3 Ratio of Secondary Tasks Under Different Coordination Levels

To make safety level of driving higher, the coupling degree between primary and secondary tasks will be higher and the corresponding proportion of driving secondary tasks will be reduced. The coupling model of coordination degree is transformed to study the proportion of secondary tasks under different coordination levels. Equations (2) and (3) are iterated to obtain Eqs. (4) and (5).

$$\alpha = \frac{\frac{D^2}{C} - V_2}{V_1 - V_2} \tag{4}$$

$$\beta = 1 - \alpha \tag{5}$$

The parameters in the formula are the same as above. When considering the coupling effect between the primary and secondary tasks, it is assumed that the driver does not perform other secondary tasks during the driving process. Therefore $\alpha + \beta = 1, \alpha \geq 0, \beta \geq 0$. In fact, drivers would carry out multiple tasks at the same time in real life.

4 Coupling Effect and Risk Identification

4.1 Calculation of Coupling Degree and Coupling Coordination Degree of Primary and Secondary Tasks

Before determining the comprehensive evaluation coefficient of each driving task, the weight coefficient of each parameter need to be determined first. The weight of each indicator is shown in Eq. (6).

$$U_i = \sum_{i=1}^{7}\sum_{j=1}^{7}\lambda_j x_{ij} \tag{6}$$

where λ_j is the weight coefficient of the jth index,x_{ij} is the value under the standardization of the j-th indicator of the ith task,U_i is the comprehensive evaluation coefficient of the i-th primary task or secondary task.

According to Eq. (6), the comprehensive evaluation coefficients of each driving task based on the driver's visual behavior and vehicle operating parameters are calculated as shown in Table 3.

Equations (2) and (3) can be used to calculate the coupling coordination degree between the primary and secondary tasks of drivers in the vehicle. Firstly, the value α and β of each driving task is defined. The safety level of drivers is divided into five categories, which include very high, high, medium, low and very low. The proportion of all kinds of driving tasks with high or above safety level of drivers is taken for subsequent research to ensure driving safety. Through normalization of data, the proportion of primary and secondary driving tasks is shown in Fig. 3.

Equation (2) is used to calculate the comprehensive coordination coefficient, and Eq. (3) is adopted to iterate the calculation data of coupling degree and comprehensive coordination. So the coupling coordination degree between primary and secondary tasks can be calculated. The calculation results are shown in Table 4.

The coupling degree and coupling coordination degree are shown in Figs. 4 and 5.

Table 3 Comprehensive evaluation coefficient of driving task

Driving state	Normal driving	Bluetooth calling	conversation	In-vehicle radio operation	Screen touch operation
Comprehensive evaluation coefficient	1.0000	0.3375	0.3672	0.1322	0.0461
Secondary task classification	–	Auditory + cognitive	Auditory + cognitive	Motion + auditory + cognitive	Motion + auditory + visual

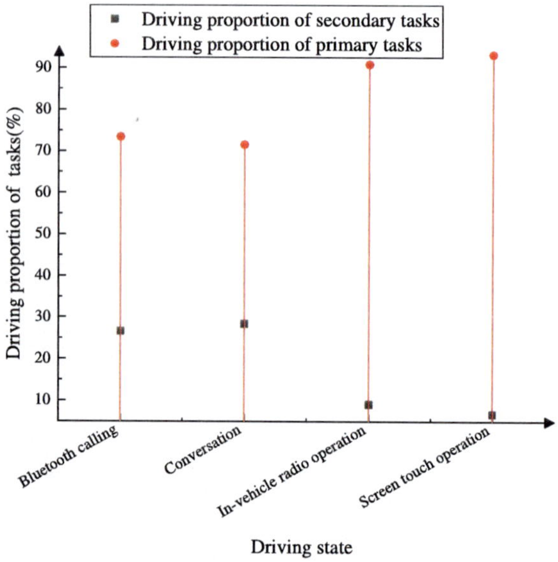

Fig. 3 Driving proportion of primary and secondary tasks in different driving states

Table 4 Coupling degree and coupling coordination of secondary task and normal driving

Driving state	Secondary task classification	Comprehensive coordination coefficient *S*	Coupling degree *C*	Coupling coordination degree *D*	Rank of coordination
Bluetooth calling	Auditory + cognitive	0.8242	0.8687	0.8462	Good
Conversation	Auditory + cognitive	0.8203	0.8864	0.8527	Good
In-vehicle radio operation	Motion + auditory + cognitive	0.9209	0.6432	0.7696	Medium
Screen touch operation	Motion + auditory + visual	0.9355	0.4105	0.6196	Low

From Fig. 4, the descending order of coupling degree between the primary and secondary tasks of driving is normal driving and talking, normal driving and Bluetooth calling, normal driving and listening to music on the radio, normal driving and screen touch operation. The result is consistent with the safety evaluation coefficient of each driving task. Combining with the actual driving, the safety of the conversation task is the highest compared with other driving tasks. The task of screen touch is the most dangerous of all tasks.

According to Table 3 and Fig. 5, normal driving and Bluetooth calling reach a good coordination level. Normal driving and conversation reach a good coordination

Fig. 4 Coupling degree of primary and secondary tasks

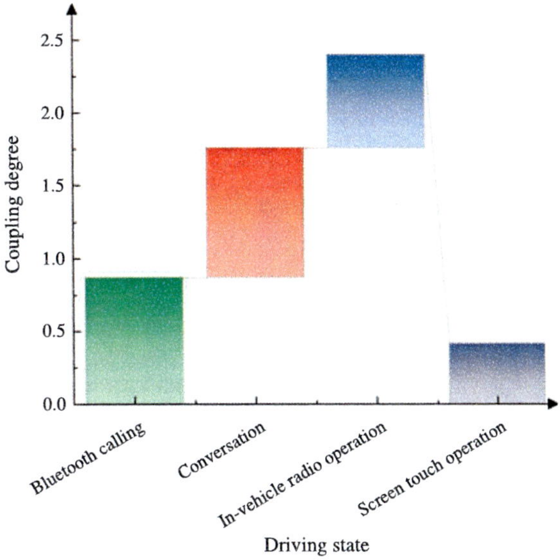

Fig. 5 Coupling coordination degree of primary and secondary tasks

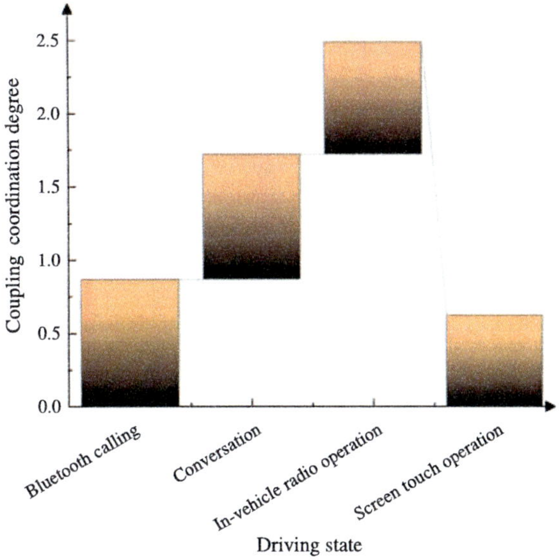

level. Normal driving and in-vehicle radio operation achieve medium coordination level. The coordination level of normal driving and screen touch operation is low.

From the perspective of multi-resource theory, screen touch operation and in-vehicle radio operation are three-channel secondary task types and they both contain motion channels. By comparing the radio operation with Bluetooth calling and conversation driving, it can be seen that the radio operation adds a motion channel, resulting in a significant difference in coupling degree. The secondary task of visual channel is far more dangerous than that of auditory channel, while the three-channel task operation is higher than that of two-channel operation.

4.2 Risk Identification of Coupling Effect of Primary and Secondary Tasks

The proportions of primary and secondary tasks at different coordination levels are calculated by Eqs. (4) and (5). D in Eqs. (4) and (5) is substituted as coordination threshold. The value of C are taken as coupling degree in Table 4. V_1 and V_2 are taken as values in Table 3. The weights of primary and secondary tasks at different coordination levels are calculated as Table 5.

The proportion of primary and secondary tasks of driving under different coordination levels can be obtained under normal condition through calculation. Combining the proportion and coordination level, the maximum coordination level can be obtained between the primary and secondary tasks. Conversation and Bluetooth calling can reach good coordination. The in-vehicle radio operation can achieve good coordination. The coordination level of screen touch operation and driving primary task is low.

To ensure the safety of driving, drivers should at least achieve good coordination in driving tasks. The weight of Bluetooth calling should be less than 39.74%. The proportion of driving during the conversation task should be lower than 43.93%. the rate of in-vehicle radio operation should be less than 0.57%. Screen touch operation during driving is not recommended and this task of driving should be minimized.

Table 5 Weight calculation results of primary and secondary tasks under different coordination levels

Driving state	$D \geq 0.9$		$D \geq 0.8$		$D \geq 0.7$		$D \geq 0.6$	
	α lower limit (%)	β upper limit (%)	α lower limit (%)	β upper limit (%)	α lower limit (%)	β upper limit (%)	α lower limit (%)	β upper limit (%)
Bluetooth calling	89.80	10.20	60.26	39.74	34.20	65.80	11.61	88.39
conversation	86.38	13.62	56.07	43.93	29.33	70.67	6.15	93.85
In-vehicle radio operation	–	–	99.43	0.57	72.55	27.45	49.26	50.74
Screen touch operation	–	–	–	–	–	–	87.10	12.90

Note "–"-means that the proportion of driving tasks at this time does not meet the condition $\beta \geq 0$, that is, the coordination degree between the primary and secondary tasks cannot reach the corresponding coordination level

Table 6 Safety level of secondary task driving coupled with normal driving

Driving state	Rank of coordination (β upper limit)	Safety level	Security attribute
Bluetooth calling	High quality (10.20)	1	Safe
Conversation	High quality (13.62)	1	Safe
In-vehicle radio operation	Good (10.57)	2	Relative safe
Screen touch operation	Low (12.90)	4	Relative risk

Whether each coordination level is safe can not be intuitively judged by Table 4, so driving safety is divided into four levels. Safety levels under different driving states can be obtained through coupling coordination degree. The results are shown in Table 6.

As indicated in Table 6, the coupling results of normal driving and Bluetooth calling can reach the high level of coordination. When the proportion of Bluetooth calling is lower than 10.20%, the whole driving state is safe. When the proportion of conversation driving is less than 13.62%, the coupling results of them can also reach the high level of coordination. The condition under which normal driving and the radio operation achieve good coordination level is that the radio operation should be lower than 0.57%. When the radio operation ratio is less than 0.57%, the driving condition is relative safe. Therefore, radio operation should be reduced. However, screen touch operation can only achieve low coordination, which is a dangerous state. Therefore, screen touch operation should be reduced as far as possible.

5 Conclusion

Previous researches show that engaging in secondary tasks in the driving process will have a certain negative impact on driving safety. However, most studies focus on the effects of different tasks on drivers' visual behavior and vehicle state. They do not show the competitive relationship between primary and secondary tasks, nor do they show the extent to which the participation ratio of secondary tasks endangers traffic safety. In this study, the coupling degree and coupling coordination degree are calculated by exploring the coupling interaction relationship between primary and secondary tasks. The driving ratio between secondary and the primary task is calculated. The driving safety conditions are discussed under different secondary tasks. The main conclusions can be drawn as follows.

(1) The models of coupling degree and coupling coordination degree of driving primary and secondary tasks are established.
(2) The relevant data are obtained by the driving simulation experiment. The parameters of vehicle operating and the driver's visual behavior are compared between the different in-vehicle secondary tasks and the normal driving. According to the coupling model of primary and secondary tasks, the coupling degree, coupling

coordination degree and coupling coordination level of primary and secondary tasks are calculated.

(3) The results show that different secondary tasks have different demands on drivers' attention and the coordination between secondary task and primary task is also different. When the driving proportion of the secondary tasks reaches a certain level, it will be in a state of coordination, which will not have a great impact on driving safety. However, when the proportion exceeds a certain level, driving safety will be endangered.

(4) In addition, the results also indicate that the coupling results between normal driving and Bluetooth calling can reach the level of high-quality coordination with high safety, so does the coupling effect between normal driving and conversation. The coupling level of normal driving and in-vehicle radio operation can reach a good coordination and the safety is medium. The low coordination of normal driving and screen touch operation is a more dangerous driving state.

Acknowledgements This research was accomplished as part of the achievements of the Natural Science Foundation of Heilongjiang Province (LH2020E056). And it was financially supported by the Department of Transportation of Heilongjiang Province Project. The authors extend their gratitude to the reviewers and doctoral students of the Institute of Traffic Engineering for their valuable comments.

References

1. Lavoie N, Lee Y, Parker J (2016) Preliminary research developing a theory of cell phone distraction and social relationships[J]. Accid Anal Prev 86:155–160
2. Tobias I, Emma J, Siri S, Anna MD (2018) An exploratory study of long-haul truck drivers' secondary tasks and reasons for performing them[J]. Accid Anal Prev 117:154–163
3. Wickens CD (1991) Processing resources and attention. Damos D. Multiple task performance. Taylor and Francis, London
4. Horrey WJ, Wickens CD (2002) Multiple rsource modeling of task interference in vehicle control, hazard awareness and in-vehicle task performance[J]. Driving assessment the second international driving symposium on human factors in driver assessment training & vehicle design
5. Naujoks F, Befelein D, Wiedemann K, Neukum A (2017) A review of non-driving-related tasks used in studies on automated driving[J]. Adv Hum Asp Transp 597:525–537
6. Ying W (2009) In-vehicle secondary task study based on human-machine interactive simulation[D]. Tshinghua University, Beijing
7. Kidd DG, Tison J, Chaudhary NK et al (2016) The influence of roadway situation, other contextual factors, and driver characteristics on the prevalence of driver secondary behaviors[J]. Transport Res F: Traffic Psychol Behav 41:1–9
8. Ma YL, Hu BY, Chan CY et al (2018) Distractions intervention strategies for in-vehicle secondary tasks: an on-road test assessment of driving task demand based on realtime traffic environment[J]. Transp Res Part D: Transp Environ 63:747–754
9. Dingus TA, Guo F, Lee S et al (2016) Driver crash risk factors and prevalence evaluation using naturalistic driving data[J]. Proc Natl Acad Sci USA 113(10):2636–2641

10. Jin L, Xian H, Yong D et al (2013) Research on secondary task safety evaluation model basrd on the vehical running state[J]. J Beijing Inst Technol 33(S1):182–186
11. Xian H (2014) Research on evaluation indexes system and evaluation model for secondary task driving[D]. Changchun: Thesis of Jilin university

Passenger Flow Organization Optimization of Xiamafang Station Based on AnyLogic

Xueyan Kong, Yongneng Xu, and Zhe Li

Abstract The article takes Xiamafang Station as an example. First, AnyLogic simulation software is used to construct the passenger flow organization model of Xiamafang Station, and the congestion points in the scene of large passenger flow in the morning peak are extracted. Then, takes the average queue length and the density of passengers as evaluation indicators to propose optimization plans. Finally, the simulation results show that the optimized scheme shortens the average queue length of passengers in the station, reduces the density, and makes trip more convenient. It is hoped that the research in the article can provide reference for the optimization of passenger flow organization in subway stations, so that rail transit can provide better services for passengers.

Keywords AnyLogic simulation · Passenger flow organization optimization · Rail transit

1 Introduction

Rail transit plays a vital role in the daily trip of residents in large cities. Compared with other urban transportation modes, its efficiency in public transportation is the highest, so it can effectively alleviate the increasing serious traffic jams. As the most common part of the urban rail transit network, subway stations are of great significance to the study of how to make them play a greater role and organize and distribute different passenger flows more efficiently.

X. Kong · Y. Xu (✉) · Z. Li
Department of Transportation Engineering, Nanjing University of Science and Technology, Nanjing City, China
e-mail: x780906yn@163.com

X. Kong
e-mail: 2251323560@qq.com

Z. Li
e-mail: 849188463@qq.com

W. Wang et al. (eds.), *Green Transportation and Low Carbon Mobility Safety*, Lecture Notes in Electrical Engineering 944,
https://doi.org/10.1007/978-981-19-5615-7_38

At present, domestic and foreign experts and scholars have conducted certain research on passenger flow organization in subway stations. Caiwen et al. [1] used AnyLogic to optimize the passenger flow at the lobby level of Dalian North Railway Station, and realized that the security check area was no longer congested and the interweaving of the security check area and the passenger flow at the exit was avoided. Shuang et al. [2] found that the installation of isolation facilities and route guidance will make the distribution of passenger flow more even, which can relieve the congestion of the station to a large extent. Zhiping [3] used AnyLogic to simulate the passenger flow organization when there was an emergency in a rail transit station, focusing on the analysis of the evacuation problem of the passenger flow in the station. Köster et al. proposed a cellular automata model, which can simply and detailedly fit the motion characteristics of pedestrians [4]. Hu [5] discussed and analyzed the passenger flow organization of urban rail transit stations. Using AnyLogic simulation software, features such as the size and frequency of subway running speed are used as the basic model parameters of the simulation model. And taking the Shenzhen Metro as an example, it shows the effectiveness and feasibility of this method, which can analyze and evaluate the performance of the passenger flow organization plan through visual and quantitative methods. Haoqing et al. [6] built a subway transfer station passenger flow organization model with the AnyLogic simulation software based on the social power model, constructed an evaluation index system and analytic hierarchy process, and proposed a comprehensive evaluation method suitable for evaluating the optimization plan of passenger flow organization, adopted AnyLogic simulation technology, and verified the feasibility of the optimization measures for passenger flow organization through a three-layer simulation experiment in a certain subway station in a certain city.

In summary, on the basis of previous research, this article fully considers the operating cost and the feasibility of the plan, and combines the causes of congestion points, and proposes a plan for optimizing passenger flow organization at Xiamafang Station, hoping to provide the facilities of the subway station. Layout planning and daily operations provide some references.

2 Model Selection

A. *Social Force Model*

AnyLogic is used in subway station passenger flow organization simulation mainly through the pedestrian library. The pedestrian library includes global parameter setting objects PedComguration, environmental objects, pedestrian objects, etc. The pedestrian flow model mainly uses the social force model. The social force model was first proposed by German traffic flow expert Helbing et al. The principle is to regard the internal and external factors in pedestrian movement behavior as "social forces" imposed on individual pedestrians, through the interaction of different "social forces" The resulting force affects the movement of a single pedestrian or a large number of pedestrians, and uses a mechanical equation set based on Boltzmann's equation of motion to describe the relationship between "social forces" [7–10]. Among the simulation software constructed with the social force model as the algorithm, the AnyLogic software is more in line with the modeling requirements of the subway station, the simulation effect is more realistic, and can be combined with the track library for comprehensive simulation of pedestrians and trains. At the same time, the software has a more convenient operation interface and data analysis tools, and supports secondary development [11]. In summary, the article intends to build a subway station simulation model based on AnyLogic 8.5.0.

In the social force model, pedestrians are mainly affected by the driving force, the force between people, and the force between people and the boundary.

Driving force refers to the influence of subjective consciousness on an individual, which is reflected in how fast pedestrians hope to reach their destination. It can be expressed by the following formula:

$$\frac{d\vec{r_\alpha}}{dt} = \overline{v}_\alpha(t) \tag{1}$$

where: $\overline{r}_\alpha$ is the position vector of pedestrianα, and $\overline{v}_\alpha$(t) is the pedestrian's walking speed.

The force between people is embodied as the "force" exerted by pedestrians in order to maintain a safe distance from other pedestrians. The equation for pedestrians to increase or decrease walking speed and walking direction is:

$$\frac{d\vec{r_\alpha}}{dt} = \overline{f}_\alpha(t) + \overline{\varepsilon}_\alpha \tag{2}$$

where: $\overline{f}_\alpha$(t) is the social force, and $\overline{\varepsilon}_\alpha$(t)is the disturbance term, which is embodied in the random behavior error of pedestrians.

The force between the person and the boundary, that is, the "force" exerted by pedestrians avoiding the wall, can be described by the following formula:

$$\overline{f_\alpha}(t) = (\overline{v_\alpha}) + \overline{f_{\alpha\beta}}(\overline{r_\alpha}) + \sum_{\alpha \neq \beta} \overline{f_{\alpha\beta}}(\overline{r_\alpha}, \overline{v_\alpha}, \overline{r_\beta}, \overline{v_\beta}) + \sum_i \overline{f_{\alpha i}}(\overline{r_\alpha}, \overline{r_i}, t) \tag{3}$$

where: $\overline{f}^0_\alpha(\overline{v}_\alpha)$ is the acceleration of pedestrian walking, $\overline{f}_{\alpha\beta}(\overline{r}_\alpha)$ is the force between the person and the wall, $\sum_{\alpha\neq\beta} \overline{f}_{\alpha\beta}(\overline{r}_\alpha, \overline{v}_\alpha, \overline{r}_\beta, \overline{v}_\beta)$ is the interaction force between pedestrian α and pedestrian β, $\overline{f}_{\alpha i}(\overline{r}_\alpha, \overline{r}_i, t)$ is the attraction effect.

B. ***Model parameter selection***

Considering the simulation function of AnyLogic software, the article selects the following three quantifiable evaluation parameters.

(1) The average queue length of passengers

Definition: The average length of passengers' queuing at various facilities in a subway station within a certain period of time, which can be used as a patency index. The longer the average queuing length, the more serious the queuing phenomenon at the facility and the lower its patency.

$$\bar{t} = \frac{\sum_{i=1}^{n} t_i}{n} \tag{4}$$

where: $\overline{L}$ is the average queue length of passengers, l_i is the queue length of passengers at the i-th facility, and n is the number of facilities in the subway.

(2) Average queue time of passengers

Definition: The average of the time that passengers queue up at each facility in a subway station within a certain period of time, which can be used as an efficiency indicator. The longer the average queuing time, the more serious the queuing phenomenon at the facility and the lower the utilization efficiency of the facility.

$$\bar{t} = \frac{\sum_{i=1}^{n} t_i}{n} \tag{5}$$

where: $\bar{t}$ is the average queue length of passengers, t_i is the queuing time of passengers at the i-th facility, and n is the number of facilities in the subway station.

(3) Passenger density

Definition: Refers to the number of passengers per unit area in different areas of a subway station, which can be used as a comfort index. The more passengers in a unit area, and the more crowded passengers feel subjectively, the worse the comfort of passengers.

$$\rho_i = \frac{n_i}{S_i} \tag{6}$$

where: ρ_i is the passenger flow density of a certain area in the subway station, n_i is the number of passengers in a certain area, and s_i is the area of a certain area.

3 Case analysis

A. Overview of Xiamafang Station

(1) Introduction to Xiamafang Station

Xiamafang Station is a station on Nanjing Metro Line 2. It is located on the north side of the intersection of Zhongshanmen Street, Bo'ai Road and Longgong Road in Xuanwu District, Nanjing. It is named after the site of Xiamafang in the Ming Xiaoling Mausoleum. Two-story island station. Moving from east to west, the station is 172.4 m long, 19 m wide, and 14.11 m deep, with a total construction area of 9586.7 m^2 (Fig. 1).

The underground floor of Xiamafang Station is the station hall floor, with 4 entrances and exits on the east and west sides. The station hall floor is mainly used for security inspection, ticket sales and ticket checking services. The second underground level is the platform level. According to field research, the number of facilities in Xiamafang Station is shown in Table 1.

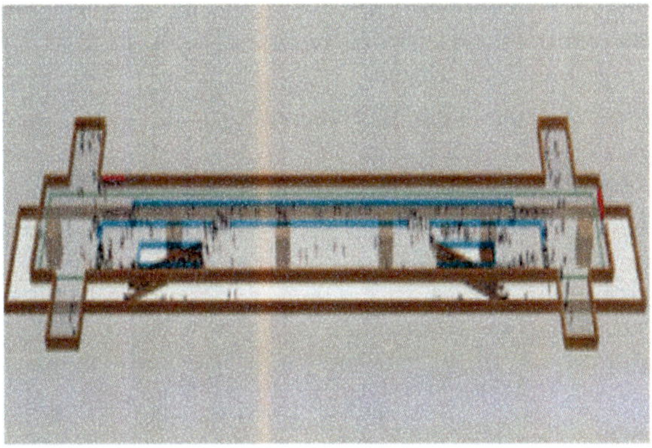

Fig. 1 3D Environmental modeling of Xiamafang station

Table 1 Service facility

Service facility	Quantity
Automatic ticket vending machine	11
Security check facility	2
Customer Service	2
Inbound ticket gate	9
Outbound ticket gate	10
Escalator	2

Table 2 Simulation parameters of facilities in the station

Facility	Parameter	Assignment parameters	Parameter unit
Automatic ticket vending machine	Service Hours	Uniform(10.0,15.0)	s
Security check	Service Hours	Uniform(3.0,5.0)	
Entrance ticket gate	Service Hours	Uniform(2.0,3.0)	
Exit ticket gate	Service Hours	Uniform(2.0,3.0)	
Escalator	Speed	0.65	m/s
23 entrance passenger flow	Arrival Rate	1250	person/h
14 entrance passenger flow	Arrival Rate	1250	

The attribute parameters of each facility are shown in Table 2, and the data are all from a field survey in the morning and peak hours of a working day.

(2) Passenger Flow Line of Xiamafang Station.

Passenger flow lines in the station are divided into two categories: inbound flow lines and outbound flow lines. Among them, the behavior of passengers in the entrance and exit flow line: passengers first arrive at the station hall level on the first floor from the entrance and exit, purchase tickets through the automatic ticket vending machine or directly go through the security check, arrive at the ticket gate area of the subway, and enter the payment area of the subway. Finally, take the escalator to the platform-level waiting area and board the train. Passenger behavior in the outbound flow line: Passengers get off the subway, reach the hall floor through the escalator, and then exit the station (Figs. 2 and 3).

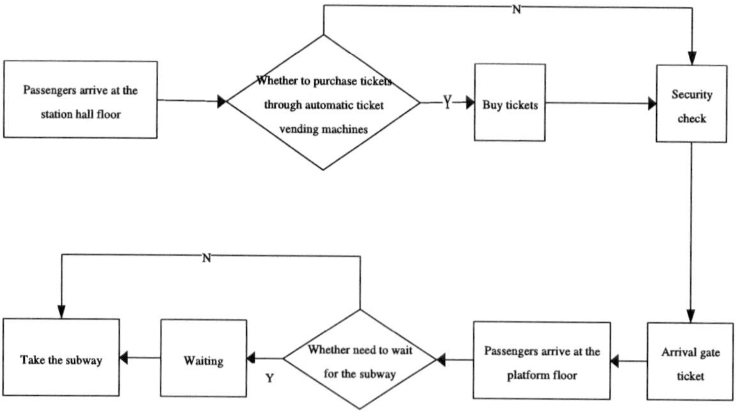

Fig. 2 Passenger arrival flow

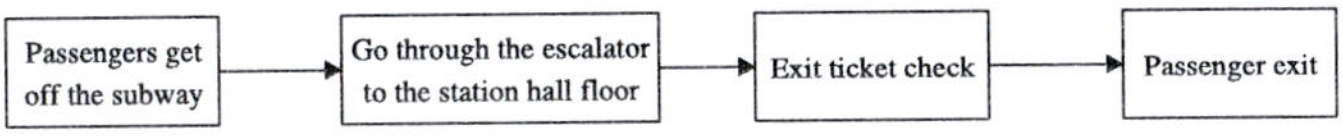

Fig. 3 Passenger departure flow

B. Case Simulation

The model is imported into the morning peak survey data of a working day for simulation, and the location of the congestion point under the condition of large passenger flow is obtained. The simulation result of the station hall is shown (Fig. 4).

(1) Station hall floor
(2) Platform floor

The simulation result of the platform is shown in Fig. 5.

C. Simulation Results

The simulation results can be seen from Figs. 4 and 5, and the analysis is shown in Table 3.

D. Optimization Plan and Comparative Analysis

Based on the above simulation results and the analysis of the causes of congestion points, from the perspective of controlling operating costs, we can focus on alleviating congestion problems in terms of standardizing and guiding passengers to walk through the flow line and adding isolation fences. The specific optimization plan is as follows:

Fig. 4 Passenger density map at station hall floor

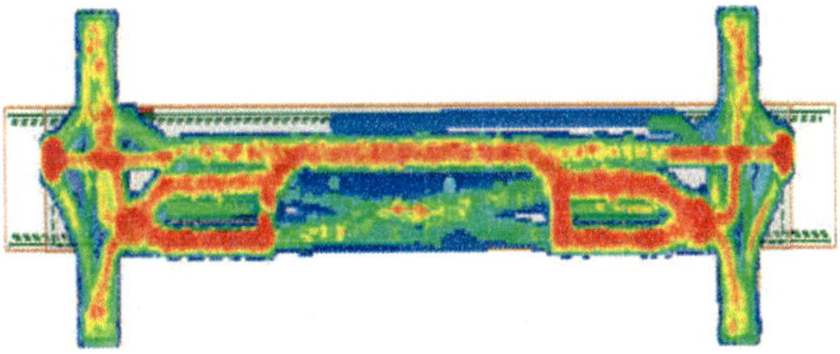

Fig. 5 Passenger density map at the platform floor

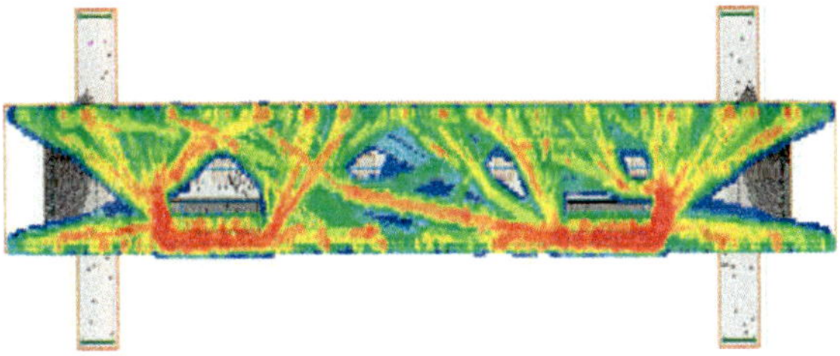

Table 3 Passenger congestion analysis

Congestion point	Station hall floor	Station hall floor	Platform floor
Location	Security checkpoint	Outbound brake	Near the escalator
Cause Analysis	The security check area is not closed, and it is easy to cross with the outbound crowd; the number of security check equipment is small, and the passenger queue is long	The exit route is not specified, and it is easy to cross with the incoming crowd	Passengers are easy to wait for the subway near the escalator

(1) Optimized plan
   (1) Closing the security check area, adding a set of security check equipment and increasing security check personnel can effectively alleviate the security check pressure in the case of large passenger flow in the morning peak, avoid passenger congestion, and speed up circulation;
   (2) Add route guidance near the exit gates to prevent exit passengers from crossing with incoming passengers;
   (3) Since there are many passengers waiting for the train near the escalator at the platform level, some passengers can be guided to go far away from the escalator and wait for the train by setting up signs, broadcasts, etc., so as to avoid stampede incidents caused by excessive congestion near the escalator;
   (4) It is recommended to reduce the number of automatic ticket vending machines, because there are very few passengers using automatic ticket vending machines to buy tickets in both research and simulation situations, so this can avoid waste of resources.
(2) Comparative analysis.

According to the above optimization scheme, the optimized passenger flow organization is simulated, and the result is shown in the figure below (Fig. 6).

Through comparison, it is found that the optimized passenger density map of the station hall floor is lighter in color, and the passenger density in the same simulation time is lower than before the optimization, which shows that the optimized solution reduces congestion. Similarly, comparing the

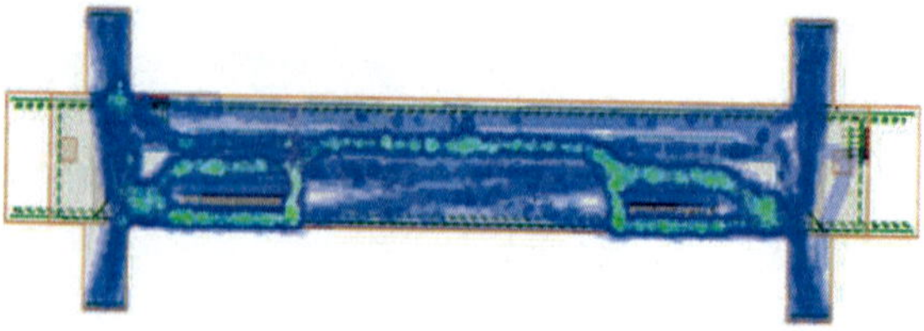

Fig. 6 Passenger density map of the station hall after optimization

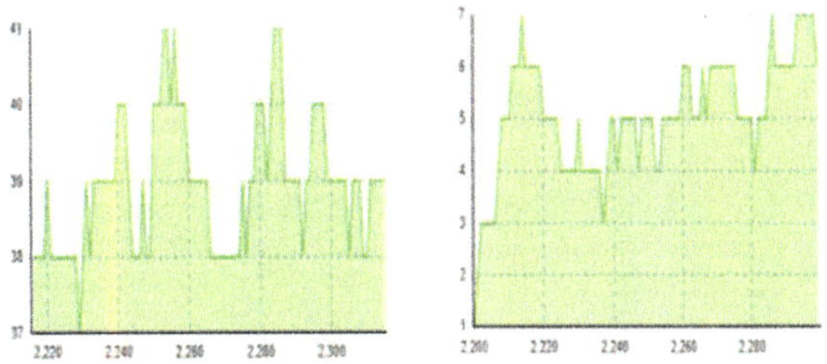

Fig. 7 The length of passenger queues at security checkpoints before and after optimization

passenger queue length at the security checkpoint before and after optimization, it is found that the passenger queue length at the security checkpoint after the optimization is significantly shorter than before the optimization (Fig. 7).

4 Conclusion

The article uses AnyLogic software to simulate the Nanjing Metro Xiamafang Station, and finds the location of the station that is prone to congestion in the morning peak passenger flow on working days, then analyzes the reasons, proposes an optimization plan, and verifies that the security inspection area is closed and the security inspection equipment is added. The effectiveness of such measures has obtained relatively good simulation results, which verified the feasibility of the optimization scheme, and also provided a solution for the Nanjing subway station to respond to the frequent large passenger flow after the response.

References

1. Caiwen M, Yongwei S (2018) Research on the simulation and optimization of passenger flow organization at the hall level of Metro Dalian North Station based on AnyLogic [J]. Transp Technol Econ 20(6):4–8
2. Shuang F, Zhili L (2018) Simulation analysis and optimization research on congestion points of metro transfer system [J]. Urban Rail Transit Res 21(7):100–105
3. Zhiping L (2018) Simulation research on emergency evacuation of urban rail transit stations[J]. Compr Transp 2018(02):64–69+72
4. Köster G, Hartmann D, Klein W (2010) Microscopic pedestrian simulations: from passenger exchange times to regional evacuation. In: Proceedings of the International Conference of Operations Research
5. Hu, M (2011) a survey and simulation of passenger flow organization of the Shenzhen Urban Rail Transit Station[C]. In: 11th international conference of Chinese Transportation Professionals (ICCTP)
6. Haoqing T, Xi Z (2018) Research on optimization evaluation of passenger flow organization in subway transfer station based on Anylogic simulation technology [J]. Electron Des Eng 6(19):90–94

7. Lihong C (2015) Simulation research on Urban Rail transit transfer station based on Anylogic[D]. Chang'an University, Xi'an
8. Jing Z, Haiyun H, Junping Z (2013) Research on fluid dynamics traffic flow model considering the influence of overtaking and changing lanes [J]. J Guangzhou Univ: Nat Sci Ed 12(5):45–49
9. Guo Q, Rui S, Shiwei H et al (2016) Traffic characteristics of pedestrian flow in Y-shaped channel of subway station [J]. J Beijing Univ Technol 42(10):1526–1531
10. Schadschneider A, Klingsch W, Klüpfel H, Kretz T, Rogsch C, Seyfried A (2009) Evacuation dynamics:empirical results, modeling and applications [J]. Encycl Complex Syst Sci 3142–3176
11. Lumin Z, Yu Z, Jinxin X (2016) Research on the simulation application of urban rail transit station based on Anylogic[J]. Railw Comput Appl 25(3): 62–66

A New Car Following Model Considering the Multi-headway Variation Forecast Effect

Yi-rong Kang and Shu-hong Yang

Abstract An extended car following model is presented by considering the effect of Multi-Headway Variation Forecast (MHVF) effect in the real world. The model's linear stability criterion was obtained by employing the linear stability theory. Theoretical analysis result shows that the new consideration leads to the stabilization of traffic systems. By means of nonlinear analysis method, the modified Korteweg-deVries (mKdV) equation near the critical point was derived, thus the propagation behavior of traffic jam can be characterized by the kink-antikink soliton solution for the mKdV equation. Numerical simulation is carried out and its results is in good agreement with the aforementioned theoretical analysis. Both of them show that the MHVF effect can suppress the emergence of traffic jamming and stabilize the vehicular system.

Keywords Multi-headway variation forecast · Car-following model · Anticipation driving behavior · Stability

1 Introduction

In the past decades, the problem of traffic congestion has been widely concerned by scholars at home and abroad. A large number of traffic flow models [1–5] have been proposed to reveal the complex mechanism behind traffic congestion. These traffic flow models can be roughly divided into macroscopic models [6–8], microscopic models [9, 10], and mesoscopic models [11, 12] according to the modeling scale. The microscopic traffic flow model takes the individual vehicle as the research object. Generally, it can describe a single vehicle dynamics process in a very simple way, and can capture rich detailed characteristics of vehicle flow movement, so it

Y. Kang (✉)
School of Transportation Engineering, Guizhou Institute of Technology, Guiyang 550003, China
e-mail: kyr6700256@126.com

S. Yang
School of Computer Science and Communication, Guangxi University of Science and Technology, Liuzhou 545006, China

W. Wang et al. (eds.), *Green Transportation and Low Carbon Mobility Safety*,
Lecture Notes in Electrical Engineering 944,
https://doi.org/10.1007/978-981-19-5615-7_39

plays an important basic role in many traffic flow models. Among a large number of microscopic traffic flow models, the most well known one is the optimal velocity (OV) model [10]. Hereafter, many researchers have attempted to improve the OV model. Since then, by introducing different factors, a series of extended ov models have been put forward to be closer to the reality of the transportation system. These expansion factors mainly include: negative speed difference information [13], positive speed difference information [14], multiple information of headway [15] and relative velocity [16], backward-looking effect [17], driver's anticipation [18], curved road condition [19], interruption factors [20] and so on.

All of the above models can reproduce many complex phenomena in actual traffic. However, the variables of these models are generally dependent on current information and rarely involve the driver's prediction information about the future traffic situation. Recently, the study on the driver's anticipation effect has attracted considerable attention of scholars since it is a universal psychological phenomenon in drivers' behavior. Tang et al. [6, 18] explored how the Driver's Forecast Effect (DFE) affects the stability of traffic flow. By introducing the factor of anticipation driving behavior into the car following model, Zheng et al. [21] examined the influence of anticipation effect upon the traffic flow via linear stability analysis and nonlinear analysis. In 2019, Wang et al. [22]developed a new car following model that takes into account the effect of headway variation tendency (HVT), the results show that HVT can improve the stability of traffic stream. By considering the effect of expected traffic variation tendencies on traffic flow, Zhang et al. [23] examined the predictive effect within a microscopic traffic model. The predictive effect on dynamics of traffic flow is also analyzed by employing the lattice hydrodynamic model [24]· Further, Daljeet Kaur and Sapna Sharma [25] investigated the predictive effect in the two lane framework on the basis of lattice model in 2020. Although many forms of the anticipation effect are incorporated into the traffic flow model to stabilize the traffic flow system, the effect of multi-headway variation forecast (MHVF), i.e. the multiple headway variation tendencies of local traffic conditions at the future moment resulting from the driver's forecast behaviour of the preceding vehicles group, has not been explored in the traffic flow model up to now.

In fact, with the help of intelligent transportation system platform, drivers can effectively perceive and estimate the information of headway variation tendency of preceding vehicles group in the next moment (multi-headway variation forecast, MHVF effect). Depending on this on-line traffic data, drivers take the appropriate measures (accelerate, maintain or slow down) to adjust their driving behavior in advance to adapt to downstream traffic conditions. And one can fully expect that the MHVF effect will have a noticeable impact on the dynamic characteristics of the vehicle following system. However, how does this effect affect the traffic dynamics under the ITS environment? This is an interesting but still open problem.

In view of the above reason, in this paper, we build a new model based on the car following theory, trying to reveal the effect of MHVF on the traffic flow. The stability and nonlinear characteristics of the new model are studied analytically, and the theoretical analysis results are verified by numerical simulation.

2 Models

For the single lane vehicle following system, Bando et al. proposed the famous OV model [10] as follows:

$$\frac{dv_j(t)}{dt} = a[V(\Delta x_j(t)) - v_j(t)] \tag{1}$$

where the subscript j is the vehicle label, t represents the time variable. a represents the driver's sensitivity coefficient, and its value is the reciprocal of the delay parameter τ. $\Delta x_j(t) = x_{j+1} - x_j$ is the headway between the leading car $j+1$ and following car j at time t. variables $v_j(t)$ and $x_j(t)$ are the velocity and position of the jth car respectively. $V(\cdot)$ represents the optimal speed function. Through the comparative study with the traffic field data, it is found that the acceleration and deceleration parameters of the OV model are not within the normal range.

In order to improve the OV model's ability to simulate actual traffic, Helbing and Tilch [13] introduced the negative speed difference effect into the OV model and proposed the GF model. Further, Jiang et al. [14] investigated the positive speed difference effect and proposed the full velocity difference (FVD) model as follows:

$$\frac{dv_j(t)}{dt} = a[V(\Delta x_j(t)) - v_j(t)] + \lambda \Delta v_j(t) \tag{2}$$

where $\Delta v_j(t) = v_{j+1}(t) - v_j(t)$ is the relative speed of two consecutive car at time t.

$\lambda = k/\tau$ is the responding factor of the relative speed. The results show that FVD model can better describe the dynamic characteristics of the traffic system than OV and GF models. However, the defect of unrealistically high deceleration in the FVD model has not been eliminated [14].

In 2012, by considering the anticipation driving behavior, Zheng et al. [21] established a extended anticipation driving car-following (AD-CF) model. The model's dynamics equation is as follows:

$$\frac{dv_j(t)}{dt} = a\{V[\Delta x_j(t) + T\Delta v_j(t)] - v_j(t)\} + \lambda \Delta v_j(t) \tag{3}$$

where T represents the driver's forecast time, $T\Delta v_j(t)$ denotes the estimation of space headway in the next moment.

The AD-CF model can effectively stabilize traffic flow by introducing anticipation driving behavior. But the AD-CF model only considers the predictive information of nearest front car, and does not considers the effect of Multi-Headway Variation Forecast (MHVF effect), i.e. the prior variation tendency of space headway of multiple vehicles ahead in the next moment. In fact, the MHVF effect of preceding vehicles group reflects the overall traffic change trend of the downstream area, i.e., whether the traffic flow of preceding vehicles on a segment will cluster, dissipate, or simply maintain a constant headway. With the help of MHVF traffic data shared by preceding vehicles, the following car can sense the downstream traffic situation, then make a decision and adjust its velocity to the optimal state in advance. To capture the effect of MHVF on the stability of traffic flow, in this paper, we developed an extended car-following model (named MHVF model) under ITS environment, the dynamics equation reads:

$$\frac{dv_j(t)}{dt} = a\{V[\Delta x_j(t) + T\Delta\overline{v}_j(t)] - v_j(t)\} + \lambda\Delta v_j(t) \tag{4}$$

where $\Delta\overline{v}_j(t) = \sum_{l=1}^{n} \beta_l \Delta v_{j+l-1}(t)$ is the weighted sum of velocity difference of the preceding vehicles group, β_l is the corresponding coefficient, n denotes the number of the vehicles ahead considered. Term $T\Delta\overline{v}_j(t)$ denotes the predictive variation tendency of space headway of downstream traffic flow (consisting of car j and its leading car $j+1, j+2, \ldots, j+n$) in the future time, i.e. the key parameter $T\Delta\overline{v}_j(t)$ represents the MHVF effect. The modeling idea of the new model is that the acceleration output of the current *jth* vehicle at time t in the following system is not only affected by the velocity $v_j(t)$ and relative velocity $\Delta v_j(t)$, but also determined by the driver's estimation of multi-headway variation tendency $T\Delta\overline{v}_j(t)$. Therefore, the proposed model can be used to explore the dynamic property resulting from the MHVF effect under ITS environment.

It is well known that the effect of the vehicles ahead on vehicle motion decreases gradually as the distance between the considered vehicle and that ahead increases. So in this paper, we select the weighted function tentatively as $\beta_l = (1/3)^{l-1}$.

For the convenience of subsequent analysis, the following formula can be obtained by first-order Taylor expansion of variable $V[\Delta x_j(t) + T\Delta\overline{v}_j(t)]$ in Eq. (4).

$$\begin{aligned} V[\Delta x_j(t) + T\Delta\overline{v}_j(t)] &= V[\Delta x_j(t) + T\sum_{l=1}^{n} \beta_l \Delta v_{j+l-1}] \\ &= V(\Delta x_j(t)) + TV'(\Delta x_j(t))\sum_{l=1}^{n} \beta_l \Delta v_{j+l-1}(t) \end{aligned} \tag{5}$$

Thus, Eq. (4) can be rewritten as follows:

$$\frac{dv_j(t)}{dt} = a\{V(\Delta x_j(t)) + TV'(\Delta x_j(t))\sum_{l=1}^{n} \beta_l \Delta v_{j+l-1}(t) - v_j(t)\} + \lambda\Delta v_j(t) \tag{6}$$

When $T = 0$, the new model is reduced to the FVD model [14]. When $T > 0$, $n = 1$, the new model degenerates into the AD-CF model [21].

In this paper, we adopt the following optimal velocity function [10]:

$$V(\Delta x_j(t)) = \frac{1}{2}v_{\max}[\tanh(\Delta x_j(t) - h_c) + \tanh(h_c)] \quad (7)$$

where $v_{\max} = 2$ is the maximum velocity and $h_c = 4$ is the safe distance.

In order to facilitate the follow-up computer simulation and nonlinear analysis, Eq. (6) is rewritten into the form of headway variable:

$$\begin{aligned}\Delta x_j(t + 2\tau) = {} & \Delta x_j(t + \tau) + \tau[V(\Delta x_{j+1}(t)) - V(\Delta x_j(t))] \\ & + T[V'(\Delta x_{j+1}(t)) - V'(\Delta x_j(t))]\sum_{l=1}^{n} \beta_l[\Delta x_{j+l}(t + \tau) - \Delta x_{j+l}(t)] \\ & + TV'(\Delta x_j(t))\sum_{l=1}^{n} \beta_l[\Delta x_{j+l}(t + \tau) - \Delta x_{j+l}(t) - \Delta x_{j+l-1}(t + \tau) + \Delta x_{j+l-1}(t)] \\ & + k[\Delta x_{j+1}(t + \tau) - \Delta x_{j+1}(t) - \Delta x_j(t + \tau) + \Delta x_j(t)]\end{aligned} \quad (8)$$

3 Linear Stability Analysis

In order to investigate the effect of MHVF on jamming transition in traffic flow, the linear stability analysis is carried out below. At the initial time, it is assumed that all vehicles move on the circular road at a uniform speed with headway b and optimal speed $V(b)$. Obviously, at this time, the traffic flow is in an equilibrium state, and the coordinates of its steady-state solution can be expressed as:

$$x_j^0(t) = bj + V(b)t,\ b = L/N, \quad (9)$$

where, N represents the total number of vehicles, and the parameter L is the length of the ring road. In order to study the stability of the traffic flow under the small disturbance condition, the small disturbance signal $y_j(t)$ is applied to make the traffic flow produce motion deviation:

$$x_j(t) = x_j^0(t) + y_j(t) \quad (10)$$

Substituting Eq. (10) into Eq. (8) and linearizing them yield the following equation:

$$\begin{aligned}\Delta y_j(t + 2\tau) = {} & \Delta y_j(t + \tau) + \tau V'(b)[\Delta y_{j+1}(t) - \Delta y_j(t)] \\ & + TV'(b)\sum_{l=1}^{n} \beta_l[\Delta y_{j+l}(t + \tau) - \Delta y_{j+l}(t) - \Delta y_{j+l-1}(t + \tau) + \Delta y_{j+l-1}(t)] \\ & + k[\Delta y_{j+1}(t + \tau) - \Delta y_{j+1}(t) - \Delta y_j(t + \tau) + \Delta y_j(t)]\end{aligned} \quad (11)$$

where $V' = dV(\Delta x_j)/d\Delta x_j|_{\Delta x_j=b}$ and $\Delta y_j(t) = y_{j+1}(t) - y_j(t)$.

By expanding $\Delta y_j(t) = Ae^{ikj+zt}$, we obtain the following equation for z:

$$\begin{aligned} &e^{2z\tau} - e^{z\tau} - \tau V'(e^{ik} - 1) - TV'(b)(e^{z\tau} - 1)\sum_{l=1}^{n}\beta_l[e^{ikl} - e^{ik(l-1)}] \\ &-k(e^{ik} - 1)(e^{z\tau} - 1) = 0 \end{aligned} \tag{12}$$

Inserting $z = z_1 ik + z_2(ik)^2 + \cdots$ into Eq. (13) and neglecting the higher order terms, one has the first order and second order terms of ik respectively:

$$z_1 = V' \tag{13}$$

$$z_2 = \frac{V'}{2}(1 - 3V'\tau + 2TV'\sum_{l=1}^{n}\beta_l + 2k) \tag{14}$$

If $z_2 < 0$, the uniformly steady-state flow becomes unstable for long-wavelength models, while the uniform flow is stable when $z_2 > 0$. Thus the neutral stable criteria for this steady state is given by

$$\tau = \frac{1 + 2k + 2TV'\sum_{l=1}^{n}\beta_l}{3V'} \tag{15}$$

For small disturbances with long wavelengths, the homogeneous traffic flow is stable in a condition where

$$\tau < \frac{1 + 2k + 2TV'\sum_{l=1}^{n}\beta_l}{3V'} \tag{16}$$

As $T = 0$, the result of stable condition is the same as that of the FVD model [14].

$$\tau < \frac{1 + 2k}{3V'} \tag{17}$$

As $T > 0, n = 1$, the stable condition is in accordance with the result of AD-CF model [21].

$$\tau < \frac{1 + 2k + 2TV'}{3V'} \tag{18}$$

The result of stable condition for MHVF model is influenced by the parameter T and n. This indicates that the stability of traffic is closely related to the prediction time step and the number of leading vehicles considered.

Figure 1 shows the neutral stable curves in the headway-sensitivity space $(\Delta x, a)$ for the MHVF model with $T = 0.3, k = 0.1$ under different values of n. From Fig. 1, it can be seen that each neutral stability curve has a vertex (h_c, a_c), which is called the critical stability point. The neutral stability curve divides the phase space into two different regions. Above the neutral stability curve is the stable region, where the small interference signals in the traffic flow will evolve and disappear with the development of time t. Below the neutral stable is the unstable region, where the small disturbance signals will gradually diverge with the movement of traffic flow, Finally, it evolves into traffic jam spreading along the upstream of traffic flow. As $n = 1$, the corresponding neutral stability curve is consistent with AD-CF model [21]. In addition, Fig. 1 also clearly shows that with the increase of the number of vehicles n considered ahead, the corresponding stability area gradually expands, which indicates that by considering the MHVF effect of preceding vehicles group, the new model has greatly improved the stability of vehicular system. Meanwhile, it is easy to find that in Fig. 1 the more vehicles we consider in the effect of MHVF, the more stable the traffic will be. However, it should be noted that when $n = 3, 4$, the corresponding curves almost coincides, which shows that just considering the information of three vehicles in front ($n = 3$), the traffic jam can be suppressed effectively. That is to say $n = 3$ is the optimal state for MHVF model.

Figure 2 is obtained by giving the different values of T when $n = 3$, $k = 0.1$. It is easy to find that the neutral stability curves move down with the increase of forecast time T, which reveals that the estimated duration T in MHVF effect has an important impact on traffic flow. What's more, the stability is gradually improved with the increase of forecast time T. Especially, as $T = 0$, the neutral stability line is the same as that in the FVD model [14].

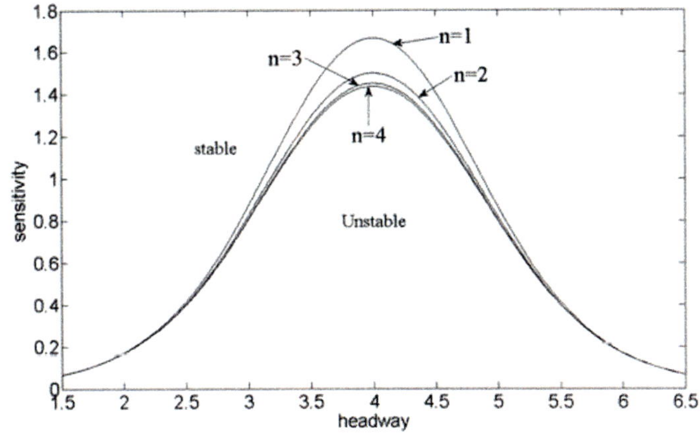

Fig. 1 Phase diagram in headway-sensitivity space $(\Delta x, a)$ for MHVF model ($T = 0.3, k = 0.1$) under different values of n

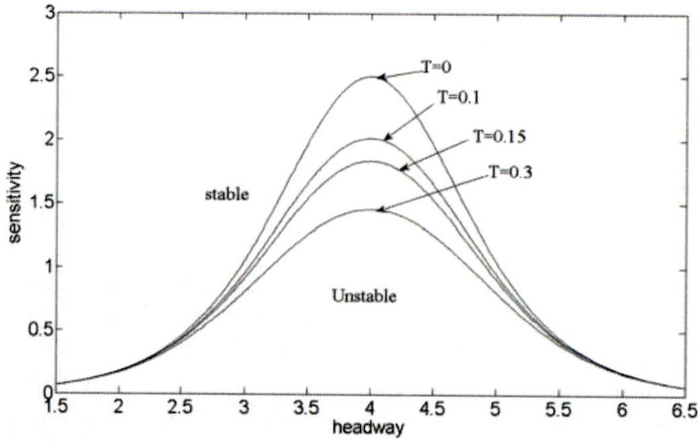

Fig. 2 Phase diagram in headway-sensitivity space ($\Delta x, a$) for MHVF model ($n = 3, k = 0.1$) under different values of T

4 Nonlinear Analysis and mKdV Equation

To investigate the effect of MHVF on traffic flow, nonlinear analysis is conducted to study the slowly varying behavior near the critical point (h_c, a_c). For extracting slow scales with the space variable j and the time variable t, the slow variable X and T are defined as follows:

$$X = \varepsilon(j + bt) \text{ and } T = \varepsilon^3 t,\ 0 < \varepsilon \leq 11 \tag{19}$$

where b is a constant to be determined. Given

$$\Delta x_j(t) = h_c + \varepsilon R(X, T) \tag{20}$$

Bring formulas (19) and (20) into Eq. (8), then expand each item to the fifth order of ε by using Taylor expansion method, and sort out the following formula:

$$\begin{aligned} &\varepsilon^2[b - V']\partial_X R + \varepsilon^3 f_1 \partial_X^2 R + \varepsilon^4[\partial_T R + f_2 \partial_X^3 R - f_3 \partial_X R^3] \\ &+\varepsilon^5 f_4 \partial_T \partial_x R + f_5 \partial_X^4 R - f_6 \partial_X^2 R^3] = 0 \end{aligned} \tag{21}$$

where

$$V' = dV(\Delta x_j)/d\Delta x_j \big| \Delta x_j = h_c V''' = [d^3 V(\Delta x_j)/d\Delta x_j^3] \big| \Delta x_j = h_c,$$

$$f_1 = \frac{3b^2\tau - V\prime - (2TV' \sum_{l=1}^{n} \beta_l + 2k)b}{2},$$

$$f_2 = \frac{7b^3\tau^2 - V\prime - 3bk(b\tau + 1) - 3b\tau TV' \sum_{l=1}^{n} \beta_l (b\tau + 2l - 1)}{6},$$

$$f_3 = \frac{V\prime\prime\prime}{6},$$

$$f_4 = 3b\tau - k - TV' \sum_{l=1}^{n} \beta_l,$$

$$f_5 = \frac{5b^4\tau^3}{8} - \frac{V'}{24} - \frac{k(4b + 4b^2\tau + 4b^3\tau^2)}{24} - \frac{TV' \sum_{l=1}^{n} \beta_l [4b^3\tau^2 + 6b^2\tau(2l - 1) + 4b(3l^2 - 3l + 1)]}{24},$$

$$f_6 = \frac{V'''}{12}(1 + 2bT \sum_{l=1}^{n} \beta_l).$$

Near the critical point (h_c, a_c), $\tau = (1 + \varepsilon^2)\tau_c$, taking $b = V'$ and eliminating the second order and third order terms of ε from Eq. (21) result in the simplified equation:

$$\varepsilon^4[\partial_T R - g_1 \partial_X^3 R + g_2 \partial_X R^3] + \varepsilon^5[g_3 \partial_X^2 R + g_4 \partial_X^4 R + g_5 \partial_X^2 R^3] = 00 \quad (22)$$

where

$$g_1 = -\frac{7b^3\tau_c^2 - V' - 3bk(b\tau_c + 1) - 3b\tau_c TV' \sum_{l=1}^{n} \beta_l (b\tau_c + 2l - 1)}{6} \quad (23)$$

$$g_2 = -\frac{V'''}{6}, \ldots g_3 = \frac{3}{2} b^2 \tau_c \quad (24)$$

$$g_{41} = -\frac{1}{6}(3b\tau_c - k - TV' \sum_{l=1}^{n} \beta_l)[7b^3\tau_c^2 - V' - 3bk(b\tau_c + 1) - 3b\tau_c TV' \sum_{l=1}^{n} \beta_l (b\tau_c + 2l - 1)] \quad (25)$$

$$g_4 = g_{41} + \frac{5b^4\tau_c^3}{8} - \frac{V'}{24} - \frac{k(4b + 4b^2\tau_c + 4b^3\tau_c^2)}{24} - \frac{TV' \sum_{l=1}^{n} \beta_l [4b^3\tau_c^2 + 6b^2\tau_c(2l - 1) + 4b(3l^2 - 3l + 1)]}{24} \quad (26)$$

$$g_5 = \frac{1}{12} V'''[6b\tau_c - 2k - 2T(V' + b) \sum_{l=1}^{n} \beta_l - 1] \quad (27)$$

To derive the regularized equation, the following transformations are performed on Eq. (22):

$$T' = g_1 T \qquad R = \sqrt{\frac{g_1}{g_2}} R' \tag{28}$$

The standard mKdV equation with a $O(\varepsilon)$ correction term is given as follows:

$$\partial_{T'} R' - \partial_X^3 R' + \partial_X R'^3 + \varepsilon M[R'] = 00 \tag{29}$$

where

$$M[R'] = \sqrt{\frac{1}{g_1}}[g_3 \partial_X^2 R' + g_4 \partial_X^4 R' + \frac{g_1 g_5}{g_2} \partial_X^2 R'^3] \tag{30}$$

Ignoring the term $O(\varepsilon)$ in Eq. (30), we obtain the standard mKdV equation with the kink–antikink wave solution.

$$R'_0(X, T') = \sqrt{c} \tanh \sqrt{\frac{c}{2}}(X - cT') \tag{31}$$

With the method described in Ref. [26], we obtain the selected velocity C.

$$C = \frac{5 g_2 g_3}{2 g_2 g_4 - 3 g_1 g_5} \tag{32}$$

Hence, we obtain the kink-antikink soliton solution as follows:

$$\Delta x_j(t) = h_c + \sqrt{\frac{g_1 C}{g_2}(\frac{\tau}{\tau_c} - 1)} \tanh \sqrt{\frac{C}{2}(\frac{\tau}{\tau_c} - 1)}[j + (1 - C g_1(\frac{\tau}{\tau_c} - 1))t] \tag{33}$$

Then, amplitude A of the kink-antikink soliton is given by

$$A = \sqrt{\frac{g_1 c}{g_2}(\frac{\tau}{\tau_c} - 1)} \tag{34}$$

The kink–antikink soliton solution shows that for the MHVF car following model, the traffic congestion is a kind of density wave near the critical point, which can be characterized by the free flow phase (low-density traffic flow) and the blocking phase (high-density traffic flow), and verifies that the occurrence of traffic jams(density wave) are related with both the forecast time duration T and the number n of the preceding vehicles considered in MHVF effect. Actually, in Figs. 3, 4, 5 and 6, the propagating backward kink–antikink density wave appears, which is in good agreement the analytical ones.

Through the linear stability criterion Eq. (16), we get the value of the critical sensitivity a_c. According to the nonlinear analysis, we obtain the propagation velocity C of the kink–antikink soliton solution by using Eq. (32). The computational values

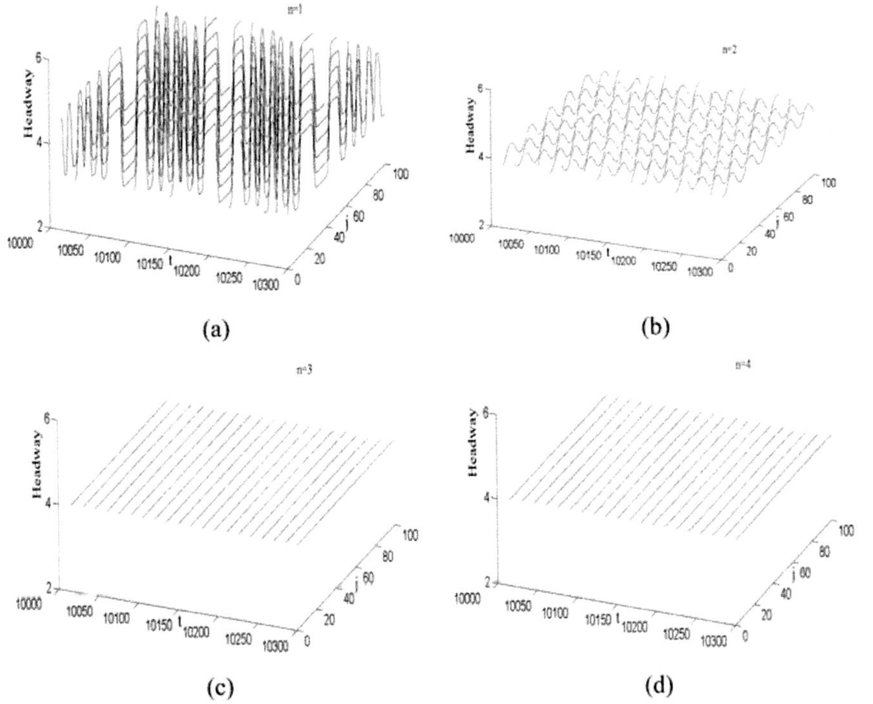

Fig. 3 Space–time evolution of the headways after $t = 10{,}000$ for the MHVF model ($a = 1.48$, $k = 0.1$, $T = 0.3$) under the different value of n, where (a) $n = 1$, (b) $n = 2$, (c) $n = 3$ and (d) $n = 4$

of a_c and C for MHVF car following model corresponding to the case of different parameters are listed in Tables 1 and 2, respectively. Table 1 is obtained by giving the different values of n under $k = 0.1$ and $T = 0.3$. Table 2 shows the a_c and C for various values of T when $k = 0.1$ and $n = 3$. It can be found clearly that with the increase of n or T, the corresponding absolute values of both a_c and C decrease gradually, which means that the performance of stabilizing traffic flow has been improved.

5 Numerical Simulation

In this section, the computer simulation is conducted to investigate the effect of MHVF on suppressing traffic jams, as well as, to check the validity of the above theoretical results. Under the periodic boundary condition, the following initial conditions are adopted:

$\Delta x_j(0) = \Delta x_j(1) = 4.0$, for $j \neq 50, 51$, $\Delta x_j(1) = 4.0 + 0.1$, for $j = 50$, and $\Delta x_j(1) = 4.0 - 0.1$, for $j = 51$. The total number of cars is $N = 100$.

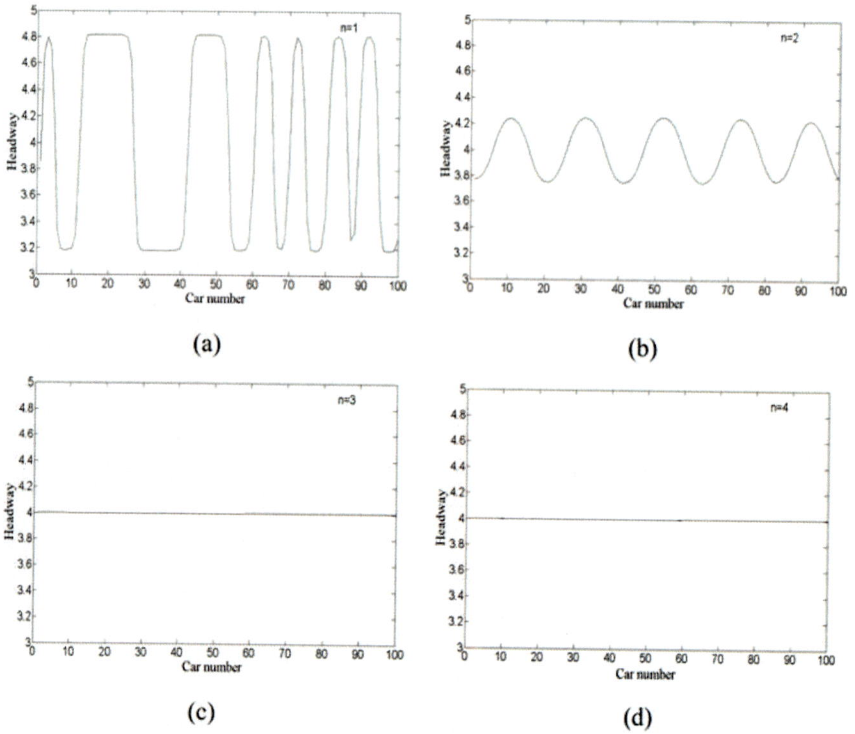

Fig. 4 The headway profile at $t = 10{,}300$ for the MHVF model correspond to the panels in Fig. 3

Figure 3 displays the typical traffic patterns after a sufficiently long time $t = 10^4$ for the MHVF model under the condition $a = 1.48$, $k = 0.1$, $T = 0.3$. The patterns (a), (b), (c) and (d) are the space–time evolution of the headway corresponding to the cases of $n = 1, 2, 3$ and 4 respectively. Pattern (a) with $n = 1$ shows the results obtained from AD-CF model [21]. In patterns (a) and (b), due to the linear stability condition is not satisfied according to Eq. (16), when the equilibrium traffic is disturbed by small disturbance, the traffic flow will fluctuate, and the propagating backward kink-antikink density wave appears as traffic jams. However, in comparing pattern (a) with (b) under the same sensitivity, traffic congestion is found to be much less serious in pattern (b), indicating that the number of preceding cars in the MHVF positively affects the stabilization of traffic flow. Besides that, from patterns (b)-(d) one can find that as the value of n increases further, small disturbances are quickly absorbed. Especially, in patterns (c) and (d), where $n = 3{,}4$, the traffic jams disappear and the inhomogeneous traffic flow recovers to the uniform state under the same sensitivity, which means that just considering the signal of three vehicles ahead is enough for suppressing the traffic jams quickly and efficiently. Hence, for the MHVF model it means $m = 3$ is the optimal state.

Figure 4 indicates that the headway profiles obtained at $t = 10{,}300$ correspond to the panels in Fig. 3. Similar results can be concluded in Fig. 4. Therefore, the

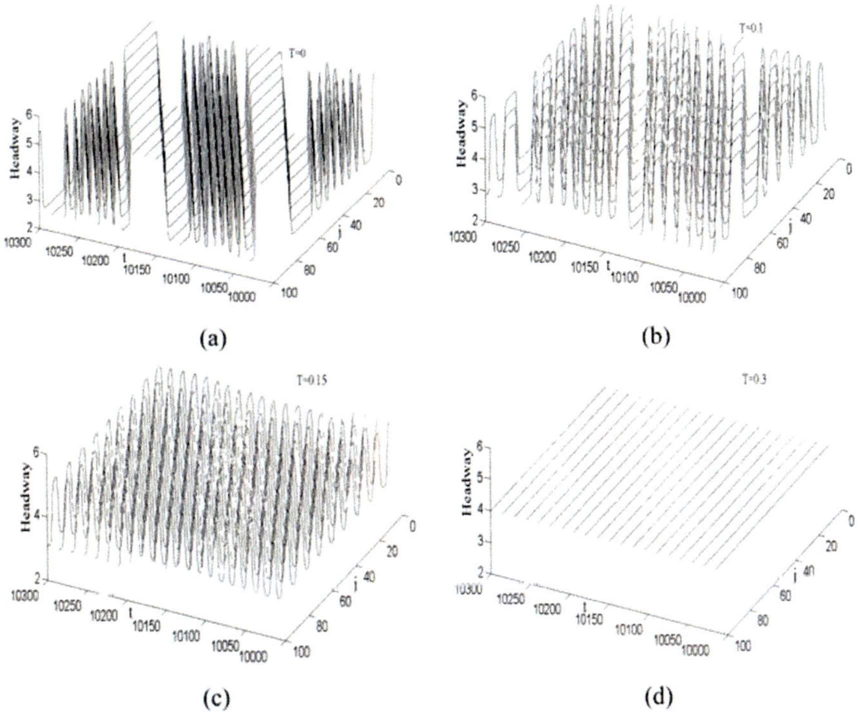

Fig. 5 Space–time evolution of the headways after $t = 10{,}000$ for the MHVF model ($a = 1.6$, $k = 0.1$, $n = 3$) under the different value of T, **a** $T = 0$, **b** $T = 0.1$, **c** $T = 0.15$ and **d** $T = 0.3$

results of the simulation are in good agreement with those of the theoretical analysis for various value of *n*, and verifies that considering the effect of multi-headway variation forecast (MHVF) in traffic flow system is necessary.

When the model is in the optimal state ($m = 3$), the influence of prediction time T on the stability of traffic flow is further studied. The simulation inputs are shown in Figs. 5 and 6.

In Fig. 5, the headway of spatiotemporal evolution pattern at time $t = 10^4$ for different $T(a = 1.6, k = 0.1, n = 3)$ is given. The headway profiles of density waves corresponding to Fig. 5 were shown in Fig. 6 at time t = 10,300 s. The patterns (a) of Figs. 5 and 6 with $T = 0$ corresponds to the solution of FVD model [14]. The patterns (b), (c), and (d) of Figs. 5 and 6 correspond to $T = 0.1$, 0.15 and 0.3, respectively. Comparing patterns (a) with patterns (b), (c), and (d) In Fig. 6 under the same sensitivity coefficient, it can be found that due to the forecast effect are considered, the amplitude of the stop-and-go wave in the patterns (a) is wider than that of patterns (b–d). The significant differences between pattern (a) and patterns (b–d) in Fig. 6 indicate that drivers' forecast information of has an important effect on the traffic stability. In addition, one can find that with the increase of the forecast time coefficient T, the amplitudes variation trend of headway is decreased, which means

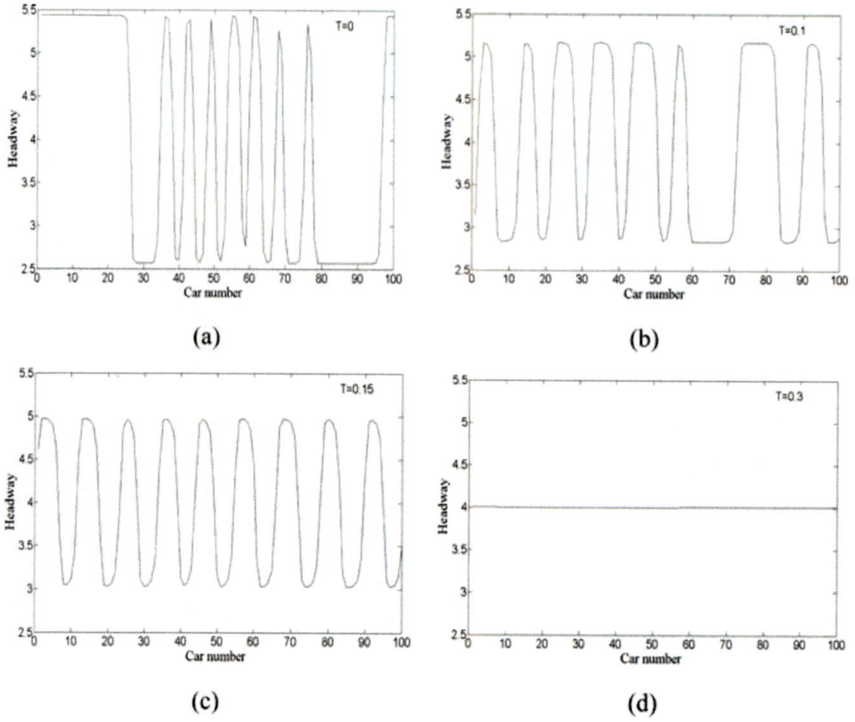

Fig. 6 The headway profile at $t = 10{,}300$ for the MHVF model correspond to the panels in Fig. 5

Table 1 The critical sensitivity a_c and the propagation velocity C for various n when with $k = 0.1$ and $T = 0.3$

n	1	2	3	4
a_c	1.6667	1.5000	1.4516	1.4362
C	−18.7813	−17.9760	−16.6285	−14.7859

Table 2 The critical sensitivity a_c and the propagation velocity C under different values of T with $k = 0.1$ and $n = 3$

T	0	0.1	0.15	0.3
a_c	2.50	2.0149	1.8367	1.4516
C	20.7373	−39.6528	−21.7354	−16.6285

the traffic flow system becomes more stable. Especially, setting $T = 0.3$, because the stability conditions are satisfied, the stop-and-go phenomenon disappears and traffic flow finally become stable in Fig. 6d as time progresses, further demonstrating that there is a positive correlation between the duration of forecast time in MHVF effect and the traffic flow stability. The similar results can also be concluded in Fig. 5.

Combined with the above theoretical analysis and numerical simulation, we can draw the following conclusions:

(1) The results of numerical analysis are in good agreement with those of theoretical analysis, and show that the MHVF effect is conducive to enhance traffic stability and suppress traffic jams.
(2) When MHVF effect is considered in the car following process, the stability of traffic flow relates with not only the number of leading vehicles n whose information is involved, but also the forecast time duration T.
(3) The stability of traffic system improves with the increase of the number n of vehicles considered or the increase of the prediction time T in MHVF effect.
(4) $n = 3$ is the optimal state for the MHVF car following model, which means just collecting the information of three preceding vehicle is enough to smooth the traffic fluctuation and suppress traffic jams effectively, and illustrates there is no need to consider more preceding vehicles than three in the proposed model, so as to avoid collecting information more than needed and increasing the information burden of drivers.

6 Summary

In recent years, wireless communication and information technologies have been widely applied in ITS environment, and thus much more information is available for drivers than ever before. Many traffic flow models have been proposed to study the complex traffic phenomena by incorporating other vehicles' traffic information provided by ITS, but few existing car-following models directly studied the multi-headway variation forecast (MHVF) effect. In this paper, we constructed a new car following model that considers the effect of MHVF to effectively curb the traffic congestion. Then, the stability judgment conditions of the new model are obtained by linear stability analysis. And the results show that the MHVF effect can effectively stabilize the car following system. In addition, through nonlinear analysis, the mKdV equation describing the propagation law of traffic density wave near the critical point of the system was derived, and its kink-antikink soliton solution was obtained. The analytical results were in good agreement with the simulation results.

Acknowledgements This work is supported by Science and Technology Plan Projects of Guizhou Province: (No.[2018]1059) and Natural Science Foundation of Guangxi (No.2018GXNSFAA050020).

References

1. Tang TQ, Wang YP, Yang XB, Wu YH (2012) A new car-following model accounting for varying road condition. Nonlinear Dynam 70:1397–1405
2. Kaur R, Sharma S (2017) Analysis of driver's characteristics on a curved road in a lattice model. Phys A 471:59–67
3. Peng GH, Kuang H, Qing L (2018) Feedback control method in lattice hydrodynamic model under honk environment. Phys A 509:651–656
4. Peng GH, Yang SH, Zhao HZ (2018) New feedback control model in the lattice hydrodynamic model considering the historic optimal velocity difference effect. Commun Theor Phys 70:803–807
5. Sun DH, Kang YR, Yang SH (2015) A novel car following model considering average speed of preceding vehicles group. Phys A 436:103–109
6. Tang TQ, Huang HJ, Shang HY (2010) A new macro model for traffic flow with the consideration of the driver's forecast effect. Phys Lett A 374:1668–1672
7. Jiang R, Wu QS, Zhu ZJ (2002) A new continuum model for traffic flow and numerical tests. Transp Res B 36:405–419
8. Zhou J, Shi ZK (2016) Lattice hydrodynamic model for traffic flow on curved road. Nonlinear Dyn 83:1217–1236
9. Watanabe MS (2006) Dynamics of group motions controlled by signal processing: a cellular-automaton model and its applications. Commun Nonlinear Sci Numer Simul 11:624–634
10. Bando M, Hasebe K, Shibata A, Sugiyama Y (1995) Dynamical model of traffic congestion and numerical simulation. Phys Rev E 51:1035–1042
11. Nagatani T (1996) Gas kinetic approach to two-dimensional traffic flow. J Phys Soc Jpn 65:3150–3152
12. Felipe S, Omer V, Joshua A (2019) Mesoscopic traffic flow model for agent-based simulation. Procedia Comput Sci 151:858–863
13. Helbing D, Tilch B (1998) Generalized force model of traffic dynamics. Phys Rev E 58:133–138
14. Jiang R, Wu QS, Zhu ZJ (2001) Full velocity difference model for a car-following theory. Phys Rev E 64:017101–017104
15. Ge HX, Dai SQ, Xue Y, Dong LY (2005) Stabilization analysis and modified Korteweg–de Vries equation in a cooperative driving system. Phys Rev E 71:066119
16. Ge HX, Cheng RJ, Li ZP (2008) Two velocity difference model for a car following theory. Phys A 387:5239–5245
17. Ma GY, Ma MH, Liang SD, Wang SY, Zhang YZ (2020) An improved car-following model accounting for the time-delayed velocity difference and backward looking effect. Commun Nonlinear Sci Numer Simul 85:105221
18. Tang TQ, Li CY, Huang HJ (2010) A new car-following model with the consideration of the driver's forecast effect. Phys Lett A 374:3951–3956
19. Zhang LD, Jia L, Zhu WX (2012) Curved road traffic flow car-following model and stability analysis. Acta Phys Sin 61(7):074501
20. Tang TQ, Huang HJ, Wong SC, Jiang R (2009) A new car-following model with consideration of the traffic interruption probability. Chin Phys B 18(3):975–983
21. Zheng LJ, Tian C, Sun DH, Liu WN (2012) A new car-following model with consideration of anticipation driving behavior. Nonlinear Dynam 70:1205–1211
22. Wang T, Li GY, Zhang J, Li SB, Sun T (2019) The effect of Headway Variation Tendency on traffic flow: modeling and stabilization. Phys A 525:566–575
23. Zhang J, Wang B, Li SB, Sun T, Wang T (2020) Modeling and application analysis of car-following model with predictive headway variation. Phys A 540:123171
24. Wang T, Zang RD, Xu KY, Zhang J (2019) Analysis of predictive effect on lattice hydrodynamic traffic flow model. Phys A 526:120711

25. Kaur D, Sharma S (2020) A new two-lane lattice model by considering predictive effect in traffic flow. Phys A 539:122913
26. Ge HX, Cheng RJ, Dai SQ (2005) KdV and kink-antikink solitons in car-following models. Phys A 357:466–476

Optimum Design of Urban Road Intersection Signal Timing Based on VISSIM Simulation

Zhe Li, Congyong Cao, and Chaoqun Kong

Abstract In the urban road network, road intersections have always been an important node of the entire urban traffic system, and they are also the areas with the densest traffic volume on urban roads. At present, there are still many problems in the traffic of some urban road intersections. On the one hand, it leads to lower traffic efficiency of vehicles, and on the other hand, it also poses a threat to the safe driving of vehicles. Based on this, in order to improve the traffic capacity, operation efficiency and safety of urban road intersections, this paper takes the Xinghai Road-Shuangqi Road intersection in Nanjing as an example, and uses VISSIM micro-simulation software to analyze. The average queuing length, maximum queuing length, and number of stops for the entrance, south entrance, and west entrance have been improved, and the index corresponding to the north entrance has become worse. If joint optimization is used in the future, the channelization of the south entrance will be changed to a straight lane and left and right shared lanes. And design the signal timing through the Webster method, and its simulation effect will be greatly improved. It can provide an effective reference for the optimal design of signal timing at intersections of the same type of urban roads.

Keywords Urban roads · Intersections · Average queue length · Maximum queue length · Number of stops

Z. Li · C. Cao (✉) · C. Kong
Department of Transportation Engineering, Nanjing University of Science and Technology, Nanjing City, China
e-mail: jy108@njust.edu.cn

Z. Li
e-mail: 849188463@qq.com

C. Kong
e-mail: 1354660862@qq.com

W. Wang et al. (eds.), *Green Transportation and Low Carbon Mobility Safety*, Lecture Notes in Electrical Engineering 944,
https://doi.org/10.1007/978-981-19-5615-7_40

1 Introduction

Since the reform and opening up, my country's economy has developed rapidly and the number of domestic cars has continued to increase. Although people's travel has become more convenient, road traffic is also under increasing pressure, traffic conflicts have increased, and the efficiency of motor vehicles has been significantly reduced [1, 2]. Urban road intersections have the function of turning traffic flow. A reasonable signal timing scheme can make full use of limited road resources and provide traffic participants with a high-quality travel experience [3, 4].

Shuang and Tongmei [5] proposed to use delay and saturation as the main evaluation indicators for the service level of signalized intersections, and at the same time use the satisfaction degree of the driver and passengers when passing the intersection as the auxiliary evaluation, and determine the signal control intersection by weighting analysis and weighted average. Intersection service level grade, and the index weight coefficient under each score is given to calculate the intersection service score and determine the grade. Miaomiao et al. [6] studied the signal timing optimization plan of existing intersections, and compared and analyzed the specific impact of the number of stops and vehicle delays on blocked intersections. Junqing et al. [7] conducted simulation analysis and evaluation on urban arterial roads. Through the analysis of vehicle delays and finding the optimal vehicle operating speed, Jia et al. [8] added a signal timing system to the roundabout, and conducted a comparative analysis of vehicle delays through VISSIM traffic simulation software, with initial results.

Based on the previous research results, this paper uses the VISSIM traffic micro-simulation software to simulate and evaluate the dynamically optimized signal timing program with the average number of vehicle stops, vehicle queue length and maximum queue length as output parameters.

2 Intersection Status Survey

A. *Overview of the basic situation of the intersection*

At the intersection of Xinghai Road and Shuangqi Road, the north–south direction is Shuangqi Road, and the east–west direction is Xinghai Road. Shuangqi Road is a two-way four-lane road with a driveway width of 3.5 m. It is connected to Ziyun Avenue and Shiyang Road in the south, and connects many corporate areas, residential communities, and shops and hospitals. At the same time, it is connected to the Shanghai-Chengdu Expressway, which bears a considerable amount of traffic on and off the expressway. Xinghai Road is a two-way four-lane road with a lane width of 3.25 m. It is connected to the ring road to the east, mainly in some urban areas under construction or to be developed, and some industrial bases; to the west are mostly homestay communities, and because of the railway barrier, it is impossible to travel Connected to the north. The channelization form of the north entrance is a straight left lane and a straight right lane, the south entrance channelization form is a straight left lane and a straight right lane, the west entrance channelization form is a left turn lane and a straight right lane, and the east inlet channelization form is a straight left lane. Lane and right turn lane. Compared with the adjacent intersection of Shuangqi Road, this intersection has fewer lanes, and there are more straight left and straight right shared lanes, and there is a lack of dedicated lanes such as straight and left. The geographic location of the Xinghai Road-Shuangqi Road intersection is shown in Fig. 1.

B. ***Current status of traffic flow***

This article conducts a field survey to obtain traffic flow data. According to the "Urban Road Traffic Planning and Design Code" (GB 50,220–95), through the sorting of the equivalent car conversion coefficients, the statistics of the peak and peak traffic flow at the Xinghai Road-Shuangqi Road intersection are obtained, as shown in Tables 1 and 2. Show. This survey is for the traffic flow during the evening peak period.

It can be seen from Table 1 that during the peak period, the traffic flow in the north–south direction is relatively large, while the traffic flow in the east–west direction is relatively small. During the peak period, the north and south inlets have a large number of straight-way traffic, while the left-turn and right-turn traffic volume is small. Compared with other entrances, the north entrance has the largest traffic volume, so the north entrance is the most prone to congestion.

Fig. 1 Geographical location of Xinghai Road-Shuangqi Road intersection

Table 1 Statistics of hourly traffic flow at the intersection of Xinghai Road and Shuangqi Road

Flat peak hour	Straight			Turn left			Turn right		
Traffic	Large car	Mid-size car	Small car	Large car	Mid-size car	Small car	Large car	Mid-size car	Small car
East Import	0	1	6	12	48	141	6	18	147
West Import	0	9	18	21	2	93	30	15	27
South Import	21	15	492	3	1	42	0	9	81
North Import	39	48	350	27	18	84	28	2	119
Total	62	72	866	63	69	360	64	44	374

Traffic volume increases during peak periods and the signal timing of intersections changes. Table 2 shows the peak traffic flow statistics at the intersection of Xinghai Road and Shuangqi Road.

It can be seen from Table 2 that during the peak period, the traffic flow in the north–south direction is relatively large, while the traffic flow in the east–west direction is relatively small. During the peak period, the traffic volume of straight, left-turn, and right-turn traffic at the north entrance is relatively large, while the traffic volume of the south entrance through-line traffic is relatively large, and the traffic volume of left-turn and right-turn traffic is relatively small. The traffic volume of straight traffic at the east entrance is small, and the traffic volume of left-turn and right-turn traffic is relatively large. The traffic volume of left-turning traffic at the west entrance is relatively large, while the traffic volume of straight-going and right-turning traffic is relatively small. It can be seen that the traffic volume of the north and south imports is relatively large, which can easily cause traffic congestion.

3 Road Intersection Improvement Plan

Through the investigation and analysis of the current situation at the Xinghai Road-Shuangqi Road intersection, to further improve the traffic capacity of the intersection, it is necessary to start with the three aspects of intersection geometric conditions, signal timing and traffic flow guidance. This chapter discusses the improvement plan The design is elaborated.

A. ***Traffic status***

According to the road geometric conditions of the field survey, the current channelization map of the Xinghai Road-Shuangqi Road intersection was drawn using Aidaroe software, as shown in Fig. 2.

Table 2 Peak hour traffic flow statistics at the intersection of Xinghai Road and Shuangqi Road

Peak hour	Straight			Turn left			Turn right		
Traffic	Large car	Mid-size car	Small car	Large car	Mid-size car	Small car	Large car	Mid-size car	Small car
East Import	0	2	18	9	39	165	9	6	174
West Import	0	3	27	22	2	116	6	12	37
South Import	20	18	460	8	0	34	4	20	54
North Import	21	33	468	26	24	96	24	15	141
Total	41	56	973	61	65	411	43	53	412

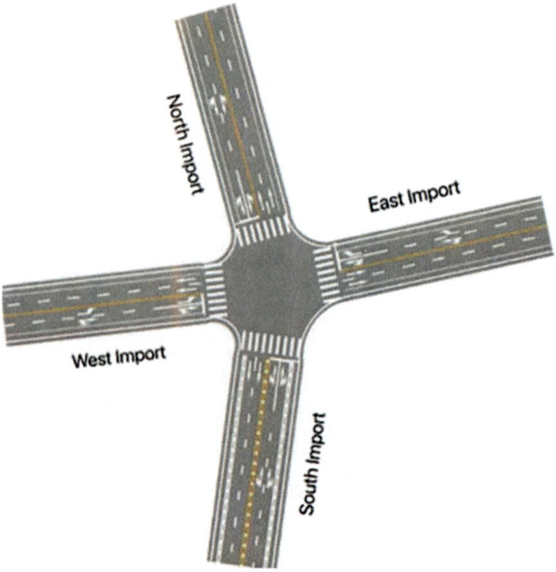

Fig. 2 Current state of channelization at the intersection of Xinghai Road and Shuangqi Road

Through field surveys, traffic flow data is obtained. According to the "Urban Road Traffic Planning and Design Code" (GB50220-95), the traffic flow statistics at the Xinghai Road-Shuangqi Road intersection during the peak period are obtained through the sorting of the equivalent car conversion coefficients.

During the peak period, the north–south traffic volume is relatively large, while the east–west traffic volume is small. During the peak period, the traffic volume of straight, left-turn, and right-turn traffic at the north entrance is relatively large, while the traffic volume of the south entrance through-line traffic is relatively large, and the traffic volume of left-turn and right-turn traffic is relatively small. The traffic volume of straight traffic at the east entrance is small, and the traffic volume of left-turn and right-turn traffic is relatively large. The traffic volume of left-turning traffic at the west entrance is relatively large, while the traffic volume of straight-going and right-turning traffic is relatively small. It can be seen that the traffic volume of the north and south imports is relatively large, which can easily cause traffic congestion.

According to the results of the traffic survey, the Aidaroe software was used to draw the flow distribution map of the Xinghai Road-Shuangqi Road intersection, as shown in Fig. 3.

It can be seen from Fig. 7 that the flow of straight traffic at the north entrance is relatively large, and the flow of left-turning and right-turning traffic is relatively large, so the channelization design of straight left and right is more reasonable; the flow of straight traffic at the south entrance is large, and the flow of left-turning traffic, The flow of right-turning traffic is small, so the channelization design of straight left and straight-right is unreasonable; the west entrance has a large left-turn traffic flow, and the flow of straight and right-turning traffic is small, so the left-turning and straight-right channelization design is more Reasonable; the east entrance has

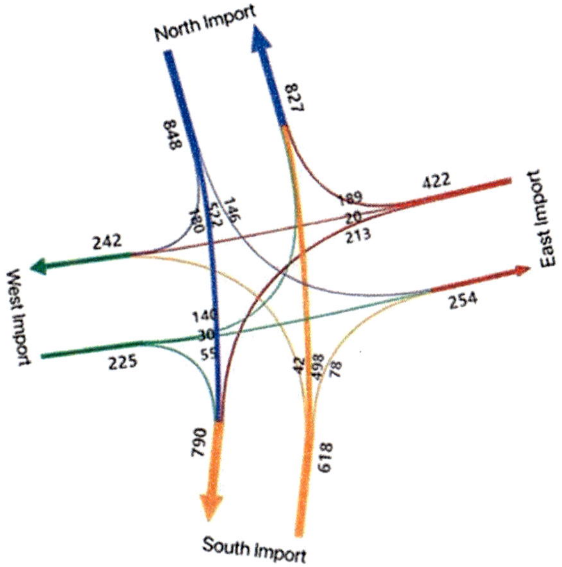

Fig. 3 Traffic flow diagram at the intersection of Xinghai Road and Shuangqi Road

a large left–right turn traffic volume and a small straight-way traffic volume, so the channelization design of straight left and right turns is more reasonable.

B. ***Signal timing plan improvement***

(1) SYNCHRO signal timing optimization

Construction of road network (Figs. 4, 5 and 6).

Due to the SYNCHRO software, the left and right turn shared lanes cannot be set, so it can only be used to improve the signal timing optimization program under the original channelization conditions. The Webster method will be used to design signal timing optimization schemes under improved channelization conditions, and the simulation effects of the two optimization schemes will be compared.

(2) Signal timing optimization under original channelization conditions.

Timing optimization results (Fig. 7).

According to the timing optimization results obtained by SYNCHRO, the improved signal timing diagram and phase diagram are drawn (Fig. 8).

4 Intersection Optimization Simulation

A. ***Intersection signal timing optimization***

According to the current channelization method of the intersection, the signal timing optimization result is obtained using SYNCHRO software, and the simulation is performed again according to the optimized signal timing.

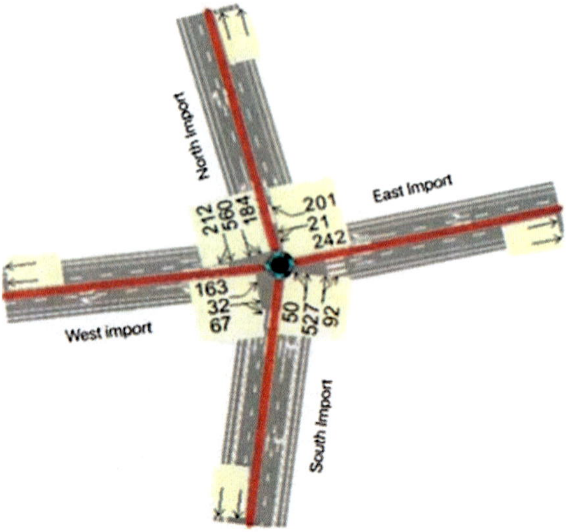

Fig. 4 Xinghai Road-Shuangqi Road intersection road network

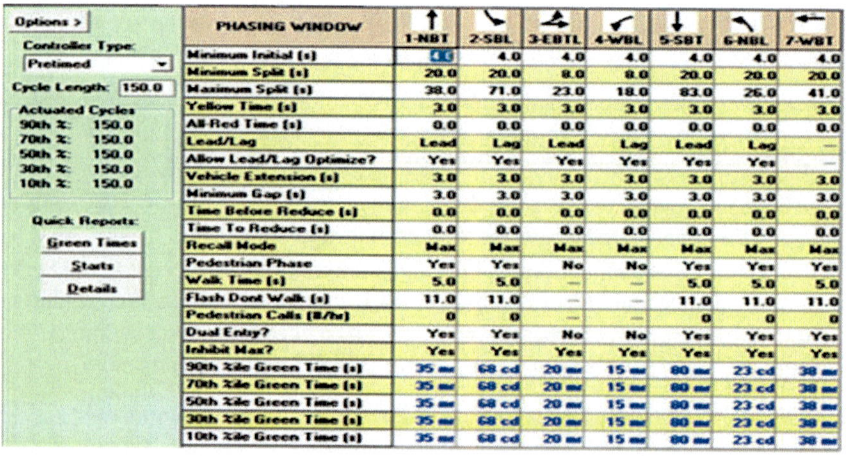

PHASING WINDOW	1-NBT	2-SBL	3-EBTL	4-WBL	5-SBT	6-NBL	7-WBT
Minimum Initial (s)	4.0	4.0	4.0	4.0	4.0	4.0	4.0
Minimum Split (s)	20.0	20.0	8.0	8.0	20.0	20.0	20.0
Maximum Split (s)	38.0	71.0	23.0	18.0	83.0	26.0	41.0
Yellow Time (s)	3.0	3.0	3.0	3.0	3.0	3.0	3.0
All-Red Time (s)	0.0	0.0	0.0	0.0	0.0	0.0	0.0
Lead/Lag	Lead	Lag	Lead	Lag	Lead	Lag	—
Allow Lead/Lag Optimize?	Yes	Yes	Yes	Yes	Yes	Yes	—
Vehicle Extension (s)	3.0	3.0	3.0	3.0	3.0	3.0	3.0
Minimum Gap (s)	3.0	3.0	3.0	3.0	3.0	3.0	3.0
Time Before Reduce (s)	0.0	0.0	0.0	0.0	0.0	0.0	0.0
Time To Reduce (s)	0.0	0.0	0.0	0.0	0.0	0.0	0.0
Recall Mode	Max	Max	Max	Max	Max	Max	Max
Pedestrian Phase	Yes	Yes	No	No	Yes	Yes	Yes
Walk Time (s)	5.0	5.0	—	—	5.0	5.0	5.0
Flash Dont Walk (s)	11.0	11.0	—	—	11.0	11.0	11.0
Pedestrian Calls (#/hr)	0	0	—	—	0	0	0
Dual Entry?	Yes	Yes	No	No	Yes	Yes	Yes
Inhibit Max?	Yes	Yes	Yes	Yes	Yes	Yes	Yes
90th %ile Green Time (s)	35 mr	68 cd	20 mr	15 mr	80 mr	23 cd	38 mr
70th %ile Green Time (s)	35 mr	68 cd	20 mr	15 mr	80 mr	23 cd	38 mr
50th %ile Green Time (s)	35 mr	68 cd	20 mr	15 mr	80 mr	23 cd	38 mr
30th %ile Green Time (s)	35 mr	68 cd	20 mr	15 mr	80 mr	23 cd	38 mr
10th %ile Green Time (s)	35 mr	68 cd	20 mr	15 mr	80 mr	23 cd	38 mr

Fig. 5 Time setting parameter setting window original timing plan

The queuing results of SYNCHRO signal timing optimization are shown in Table 3.

B. *Comparison of various plans*

(1) Comparison of average queue length

Comparing the current situation with the simulation results of SYNCHRO optimization, the average queue length comparison is shown in Table 4.

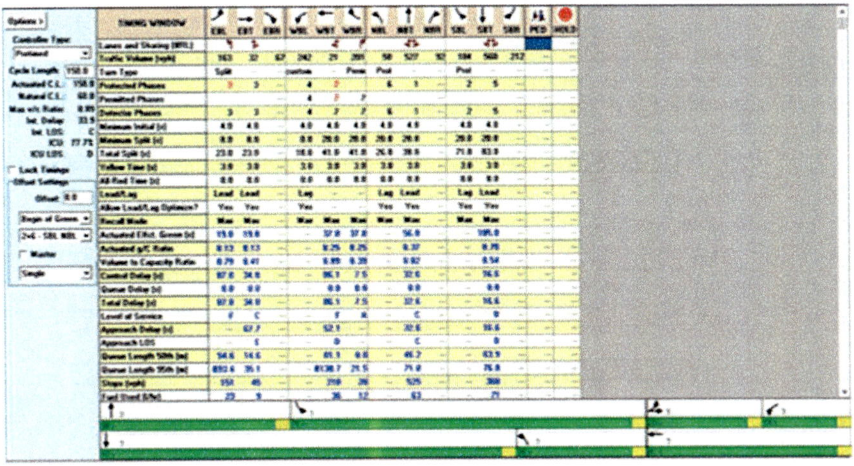

Fig. 6 Original timing scheme

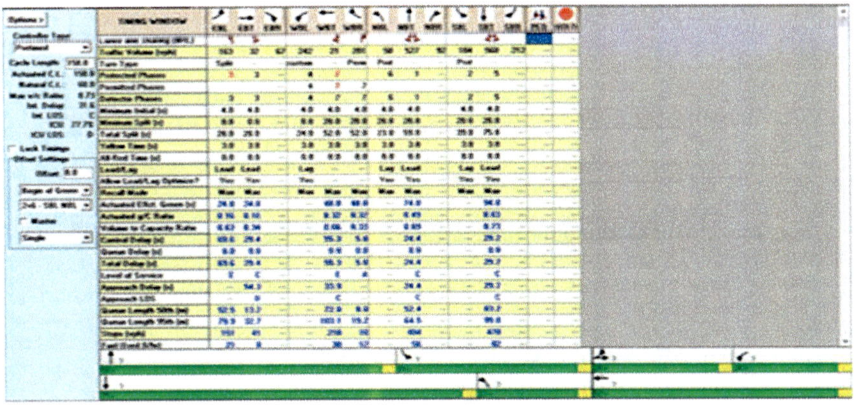

Fig. 7 Timing optimization results

Draw a histogram of the average queue length, as shown in Fig. 9.

It can be seen from Fig. 9 that compared to the current situation, the average queue length of the east, south, and west inlets optimized by SYNCHRO has been improved, but the average queue length of the north inlet has increased instead.

(2) Comparison of maximum queue length

Compare the current situation with the simulation results of SYNCHRO optimization, and the comparison of the maximum queue length is shown in Table 5.

Draw a histogram of the maximum queue length, as shown in Fig. 10.

It can be seen from Fig. 10 that compared with the current situation, the improvement effect of the maximum queue length after SYNCHRO optimization is not

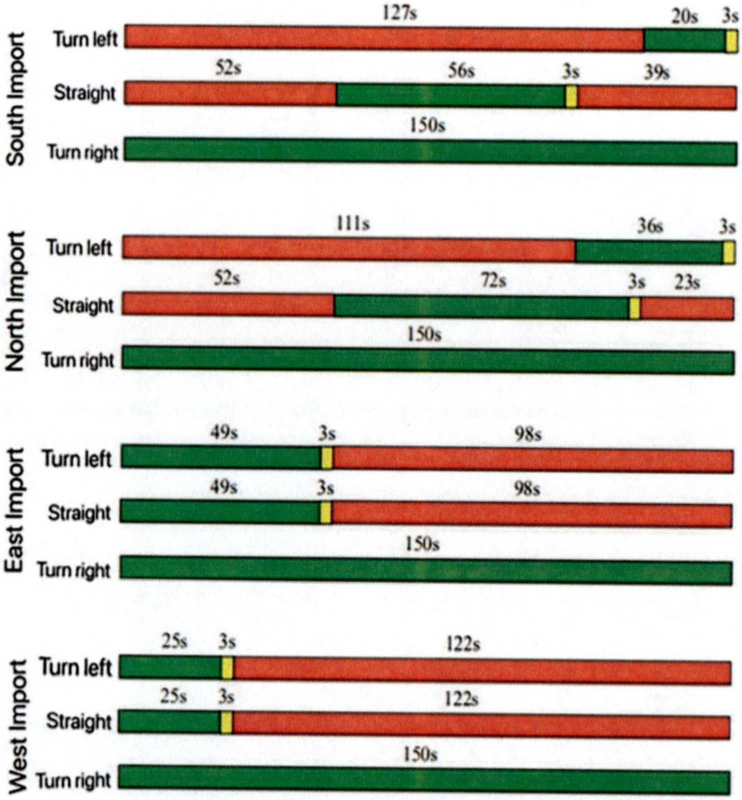

Fig. 8 Improved timing diagram for Xinghai Road-Shuangqi Road intersection

Table 3 SYNCHRO queuing results

SYNCHRO optimization	Average queue length	Maximum queue length	Number of stops
East Import	16.14	67.63	41
South Import	21.08	80.51	84
West Import	11.49	46.27	23
North Import	30.74	76.64	138

Table 4 Average queue length

Average queue length	Status quo	SYNCHRO optimization
East Import	18	16.14
South Import	32.26	21.08
West Import	13.72	11.49
North Import	18.87	30.74

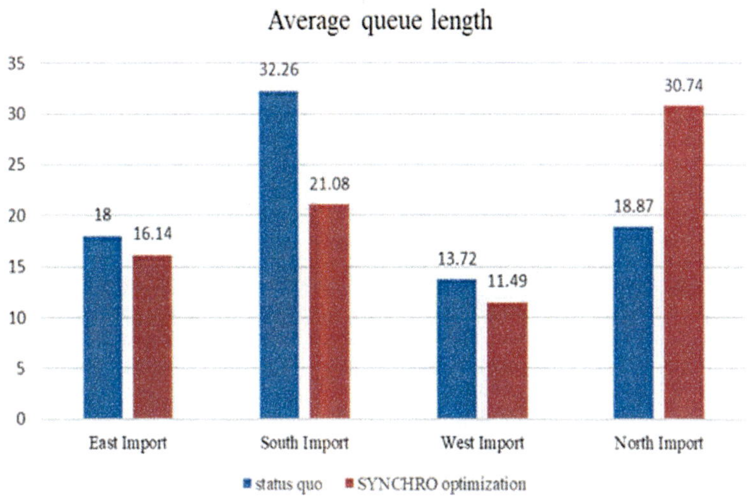

Fig. 9 Average queue length

Table 5 Maximum queue length

Maximum queue length	Status quo	SYNCHRO optimization
East Import	67.7	67.63
South Import	75	80.51
West Import	45.73	46.27
North Import	76.43	76.64

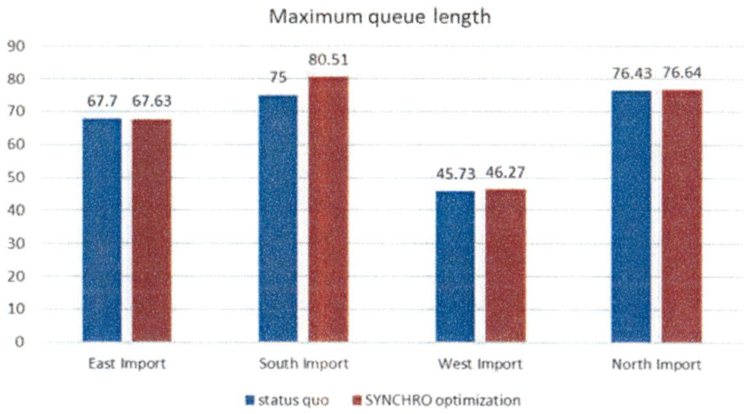

Fig. 10 Maximum queue length

Table 6 Comparison of parking times

Number of stops	Status quo	SYNCHRO optimization
East Import	51	41
South Import	105	84
West Import	27	23
North Import	97	138

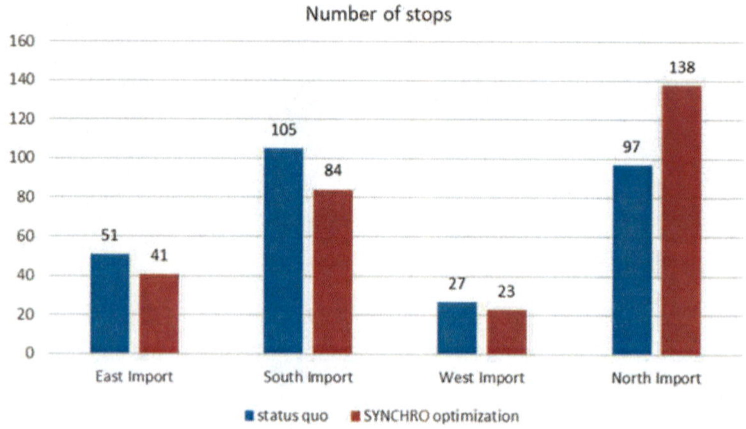

Fig. 11 Number of stops

obvious, and the maximum queue length of the south entrance has increased instead; due to the current serious congestion at the north and south entrances, the traffic of the east entrance can be sacrificed to a certain extent. Capacity to improve the traffic capacity of north–south imports.

(3) Comparison of parking times

The simulation results of the current situation, SYNCHRO optimization, and joint optimization are compared, and the comparison of the number of stops is shown in Table 6.

Draw a histogram of the number of stops, as shown in Fig. 11.

It can be seen from Fig. 11 that compared to the current situation, the number of stops at the east, south, and west entrances optimized by SYNCHRO has been improved, but the number of stops at the north entrance has increased.

5 In Conclusion

In summary, it can be seen that after Synchro optimization, the average queue length, maximum queue length, and number of stops for the east, south, and west entrances have been improved, and the corresponding indicators for the north entrance have become worse. If joint optimization is used in the future, That is, the south entrance canalization is changed to a straight lane, left and right to the shared lane, and the signal timing is designed through the Webster method, and the simulation effect will be greatly improved.

References

1. Jing L (2015) Research on open signal control method of urban traffic [D]. Southwest Jiaotong University, Chengdu
2. Li Songlin. Research on optimization of short-distance intersection control scheme [D]. Xi'an: Chang'an University, 2018.
3. Xiaohong C (2012) Research on optimization model of urban road traffic signal control in mixed traffic environment [D]. Beijing Jiaotong University, Beijing
4. Li C (2019) Research and simulation of urban road traffic system signal control strategy based on deep reinforcement learning [D]. Chengdu: Southwest Jiaotong University
5. Shuang W, Tongmei Z (2007) Research on the evaluation method of service level at signalized intersections[J]. J Chin People's Public Secur Univ (Nat Sci Ed) 53(03):77–79
6. Miaomiao L, Yuna W, Ye S (2016) Optimal analysis of signal timing at urban intersections based on VISSIM simulation [J]. J Luoyang Inst Technol (Nat Sci Ed) 26(4):22–26
7. Junqing S, Jueqiang T, Jufang Z et al (2015) Simulation and evaluation of two-way green wave traffic on urban arterial roads based on VISSIM[J]. J Zhejiang Norm Univ (Nat Sci Ed) 38(2):220–225
8. Jia J, Yuanxing P, Xinming Q (2016) Realization of signal control simulation of five-pronged roundabout based on VISSIM[J]. J Sch Manag Cadres Minist Transp 26(2):41–44

Speech-Based Driver Emotion Recognition

Haiqiu Tan, Haodong Zhang, Jian Shi, Dongxian Sun, Jie Zhang, Xiaobei Jiang, and Wuhong Wang

Abstract The premise that vehicles bring convenience to human life is to ensure the safety of people in vehicles. However, driver's negative emotions are an important cause of risky driving, road rage, and traffic crashes, which seriously endangers traffic safety. In this paper, we proposed a driver emotion recognition method based on driver's speech using audio features. Firstly, we extracted 6 features for speech gender recognition. After gender recognition, a combination of gender and MFCCs features were used for negative emotion recognition. Finally, a driver emotion recognition application was developed for function display.

Keywords Driver emotion · Speech emotion recognition · Traffic safety · MFCCs · Psychological state

1 Introduction

Safety has been widely identified as one of the most important issues in vehicles [1]. According to 2020 annual traffic crash data released by National Highway Traffic Safety Administration (NHTSA), 38,824 lives were killed by traffic crashes in U.S, showing that traffic safety issues are still very serious.

Compared to other causes of traffic crash such as vehicle, road, human factor is the most primary cause of traffic crash. Hence, in order to prevent traffic crash, it is of great importance to monitor driver's state [2]. Most existing studies recognized driver's state by utilizing physiological acquisition devices such as the electroencephalogram (EEG) [3], electrooculography (EOG) [4], electrocardiogram (ECG) [5], eye-tracking [6], etc. Compared to driver's physiological state, driver's psychological state referring to emotions is as important as driver's physiological state, among which negative emotions are the main culprit leading to risky driving. Therefore, it is necessary to recognize driver's emotion, especially negative emotions.

H. Tan · H. Zhang · J. Shi · D. Sun · J. Zhang · X. Jiang · W. Wang (✉)
Beijing Institute of Technology, Beijing 100081, China
e-mail: wangwh@bit.edu.cn

W. Wang et al. (eds.), *Green Transportation and Low Carbon Mobility Safety*,
Lecture Notes in Electrical Engineering 944,
https://doi.org/10.1007/978-981-19-5615-7_41

According to Oxford dictionary [7], emotion is a strong feeling deriving from one's circumstances, mood, or relationships with others.

There are a lot of emotion classifications in previous studies. James proposed four basic emotions, which were fear, love, rage, grief in 1890 [8].

After, the most generally accepted one, also as known as "The Big Six", including happiness, sadness, anger, surprise, fear, disgust, was firstly used by Ekman's group in their emotion recognition experiment [9]. Subsequent research in 1971 Ekman and Friesen confirmed that "The Big Six" was accurately recognized regardless of cultures, language, and who expressed them [10]. Later, scholars more or less extended or cut down the numbers of basic emotions in "The Big Six". Camras and Plutchik extended the list to 8 emotions by adding acceptance and anticipation [11]. In 1995, a book called "Passion and Reason"proposed a list of 15 emotions, and there were aesthetic experience, anxiety, compassion, depression, envy, fright, gratitude, guilt, happiness, hope, jealousy, love, pride, relief, sadness, and shame [12].

In this paper, we adopted the most generally accepted one, Ekman's 6 basic emotions, also, "The Big Six", and replaced "disgust" in it with another widely recognized emotion "neutral".

Previous studies have shown that emotional states have a significant effect on speech generalization.

It is obviously that when people talk, the speech not only conveys the linguistic meaning (words, sentences, text), but also contains emotional information with non-verbal features [13].

There are a variety of acoustic features can affect a person's speech, such as pitch, timbre, tempo, prosody, intensity and human's emotion differ with respect to these acoustic features. For instance, a person's voice containing high voice intensity level, fast tempo implies that he or she may be angry. A person with high mean fundamental frequency (F0), fast tempo, moderate mean voice intensity implies that he or she may be happy. A person's voice shows low mean fundamental frequency (F0), decreasing F0 contour, low mean voice intensity and slow temporal rate indicates that he or she may be sad [14]. Speech emotional recognition extracts these acoustic features from people's speech to make inferences about people's emotion.

Another commonly used feature is MFCC (Mel Frequency Cepstral Coefficients), which is considered as the most popular feature for speech emotional recognition [15].

2 Proposed Work

2.1 Data Collection

Speech were gathered by smart phone from 10 volunteers aged 18–26, 5 males, 5 females who performed normal driving task created by Scanner Studio and simulated 6 emotions ("Angry", "Fear", "Sad", "Happy", "Neutral", "Surprised") in the driving

Fig. 1 Driving simulator

simulator shown in Fig. 1. Each emotion was simulated three times through role playing, 180 speech clips were collected. In order to get high quality audio sample, audio samples were collected with 44100 Hz sample rate, basic noise reduction system (low frequency noise reduction) and ACG (Automatic Gain Control) amplifier to compensate the loss of sound through the air, as well as the loss due to distance between mic and mouth.

2.2 Data Pre-processing

Speech collected from volunteers cannot be used directly, because there are long silences and unvoiced parts, which can influence the accuracy of analysis. In order to get more reliable results, this paper removed unvoiced parts and long silences (over 0.2 s).

Figures 2 and 3 show present the original speech signal before audio pre-processing (including unvoiced part and silence) and the speech signal after audio pre-processing (excluding unvoiced part and silence), respectively.

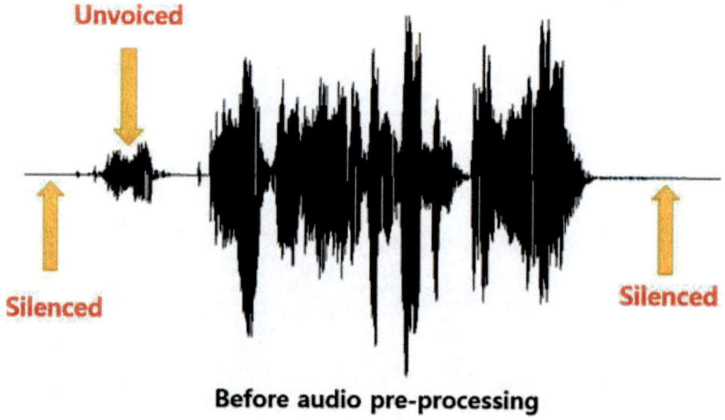

Fig. 2 Audio signal before audio pre-processing

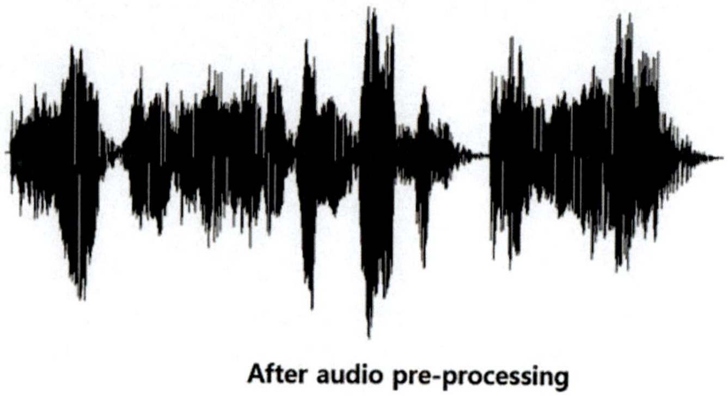

Fig. 3 Audio signal before audio pre-processing

2.3 *Gender Recognition*

Gender recognition is an important component for the application embedding speech recognition as it reduces the computational complexity for the further processing in these applications [16]. Some studies in the literature show that speech recognition and speaker identification would be simpler, if we could automatically recognize a speaker's gender (sex) [17]. Sex identification was used primarily as means to improve recognition performance and to reduce the needed computation [17]. Differences in features for male and female speakers are a well-known problem and it is established that gender-dependent emotion recognizers perform better than gender-independent ones [18].

Gender classification not only reduces the computational complexity, but also increases the performance for further emotion recognition.

Hence, 6 features (maximum of pitch, minimum of pitch, standard deviation of pitch, mean of pitch, median of pitch, mean of intensity) were extracted for gender classification. There are many existing classifiers working on gender classification, in order to find the optimal classifier for our cases, we tested main stream algorithms including KNN (K-Nearest Neighbors), SVM (Support Vector Machine), Gaussian Process, Gradient Boosting, Decision Tree, Extra Tree, Random Forest, ANN (Artificial Neutral Network), Naïve Bayes, Ada Boost, SGD (Stochastic Gradient Decent), 13 algorithms in total. We applied these 13 algorithms mentioned above to our audio database, and the results for gender recognition are shown in Fig. 4.

By comparing these algorithms mentioned above, the optimal algorithm is Random Forest whose recognition accuracy is around 95%. Therefore, this paper adopted the Random Forest as classifier for gender classification.

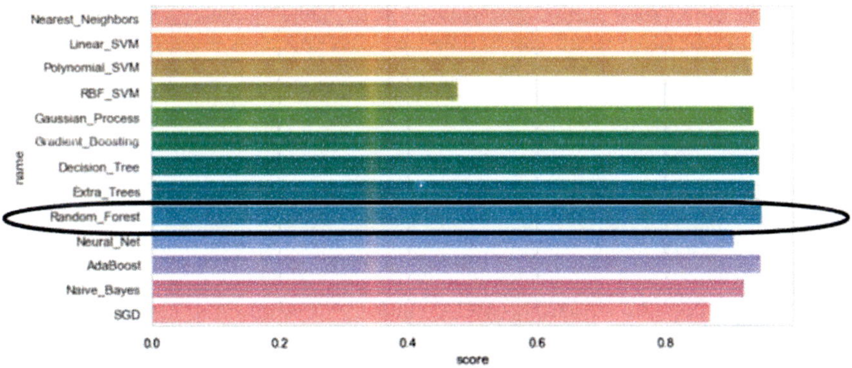

	name	score
0	Nearest_Neighbors	0.947917
1	Linear_SVM	0.933333
2	Polynomial_SVM	0.935417
3	RBF_SVM	0.477083
4	Gaussian_Process	0.937500
5	Gradient_Boosting	0.945833
6	Decision_Tree	0.945833
7	Extra_Trees	0.939583
8	Random_Forest	0.950000
9	Neural_Net	0.904167
10	AdaBoost	0.947917
11	Naive_Bayes	0.920833
12	SGD	0.866667

Fig. 4 Gender recognition results of 13 algorithms

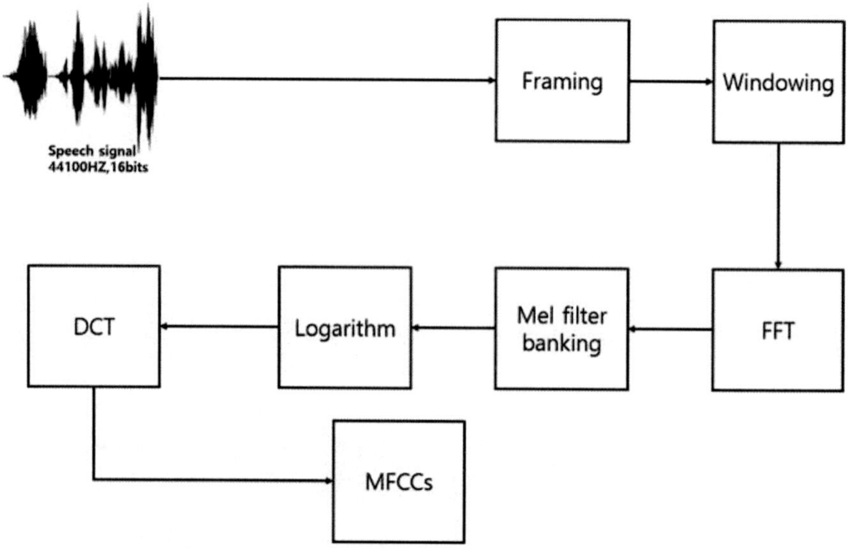

Fig. 5 The flowchart of MFCCs extraction

2.4 MFCCs Features Extraction

After gender recognition, MFCCs (Mel Frequency Cepstral Coefficients) which are widely used in analyzing audio signals, were introduced for emotion recognition.

Many existing researches reveals that MFCCs and its 12 first derivatives and 12 s derivatives are the most powerful parameters in audio recognition. In this paper, we chose to extract 12 MFCCs, 12 first derivatives and 12 s derivatives for further emotion classification. Figure 5 depicted the flowchart of MFCCs extraction.

2.4.1 Framing

In the process of the framing, speech signal will be sliced into small chunks, in this paper we used a Python's library called librosa, which is one of the popular libraries in audio processing, to frame the speech signal. Generally, the duration of each frame ranges from 20 to 30 ms, in this step we segmented the audio signal to 20 ms frames, and frame shift taken was 10 ms to make some overlap to frames so as to avoid excessive changes in two adjacent frames.

2.4.2 Windowing

Next step in the flow is windowing. After the step 'framing', audio signal was segmented into 20 ms frames and then we multiply each frame by a window function. The purpose of windowing is to eliminate to a certain extent the discontinuity between frames that appear after framing. Commonly used window functions are Hanning window, Hamming window, rectangular window, triangle window and so on. In this step, we adopted Hanning window to enhance continuity between the left and right ends of each frame. And the formula of Hanning window is shown as below, where N is the number of frames.

$$w(n) = 0.5 - 0.5\cos\left(\frac{2\pi n}{N-1}\right), 0 \le n \le N-1$$

2.4.3 FFT (Fast Fourier Transform)

In order to observe the more features of audio signal, audio signal was transferred from time-domain to frequency-domain. Audio signal on frequency-domain showed energy distribution of the signal, and different distribution of the energy represented different features for audio signal. In order to get energy distribution of audio signal, original audio signal processed after framing and windowing and then each frame went through FFT (Fast Fourier Transform) to get energy distribution of audio signal on frequency domain. The formula of FFT (Fast Fourier Transform) is shown as below, where $S_i(k)$ is the signal in time domain, $S_i(n)$ is the signal in frequency domain.

$$S_i(k) = \sum_{1}^{N} S_i(n)e^{\frac{-2\pi kn}{N}}, 0 \le k \le N-1$$

2.4.4 Mel Filter Banking

Pass the power spectrum which is got from previous step through a set of Mel-frequency triangular filters, which are set to 128 in this article, to extract frequency bands. We can use the following formula to convert between Hertz and Mel.

$$f_{mel} = 2595 \times log_{10}^{(1+\frac{f}{700})}$$

$$f = 700(10^{f_{mel}^{2595}} - 1)$$

The triangular filter can smooth the frequency spectrum and eliminate the effect of harmonics, highlight the formant of the original audio signal, and also reduce the amount of calculation.

2.4.5 Logarithm

The human ear's perception of the loudness of sound is not a linear relationship. For a small sound, the human ear can feel it as long as the loudness increases slightly, but when the sound loudness has reached a certain level, even if there is a greater increase, there is no obvious change in the human ear perception. According to research, the human ear's perception of sound loudness is logarithmic. Therefore, the reason for taking the logarithm of the Mel spectrogram is to simulate the "logarithmic" characteristic of the human ear.

2.4.6 DCT (Discrete Cosine Transform)

The next step is Discrete Cosine Transform also known as DCT, which is aim to change the distribution of data and separate unnecessary data. The formula of DCT is shown below, where k is the number of cepstral coefficients. After this step, most of the signal data will be concentrated in the low frequency area, usually we only need to take the first part of the transformed data. In this paper we only took first 12 values.

$$C(k) = 2\sum_{0}^{N-1} x(n)\cos\frac{(2n+1)k\pi}{2N}$$

2.4.7 Delta and Delta-Delta Coefficient

According to previous research, the first and the second derivatives of MFCCs are very useful for audio analysis, so after last step DCT, we got first 12 MFCC values and then calculated their first and second derivatives of MFCCs. Finally, we extracted first 12 MFCC values and their first 12 derivatives and second 12 derivative, 36 values in total. Finally, 17 features (mfcc1, mfcc2, mfcc3, mfcc4, mfcc5, mfcc6, mfcc7, mfcc8, mfcc9, mfcc10, mfcc11, mfcc12, d1, d2, dd1, dd2, and gender) were retained.

3 Result

This step we used KNN model to train our database, which is one of the widely used instance-based algorithms. Since the purpose is to recognize negative emotion and non-negative emotion, we separated all 6 emotions into two sets, one is negative emotions and the other is non-negative emotions. We classified neutral, happy, surprised as non-negative emotions and sad, angry, fearful as negative emotions. After all previous processing, 17 features were used in KNN model training, with training size 0.7. The training result of negative and non-negative emotions recognition is 85.5% (K equals to 5). The result of test accuracy is shown in Fig. 6. Since the purpose is to distinguish between negative emotions and non-negative emotions, the six emotions were further classified. "Angry", "Fear", "Sad" were classified as negative emotions, and "Happy", "Neutral", "Surprise" were classified as non-negative emotions. After classification, Table 1 described the result of emotion recognition with audio features, and the overall accuracy achieved 76.5%.

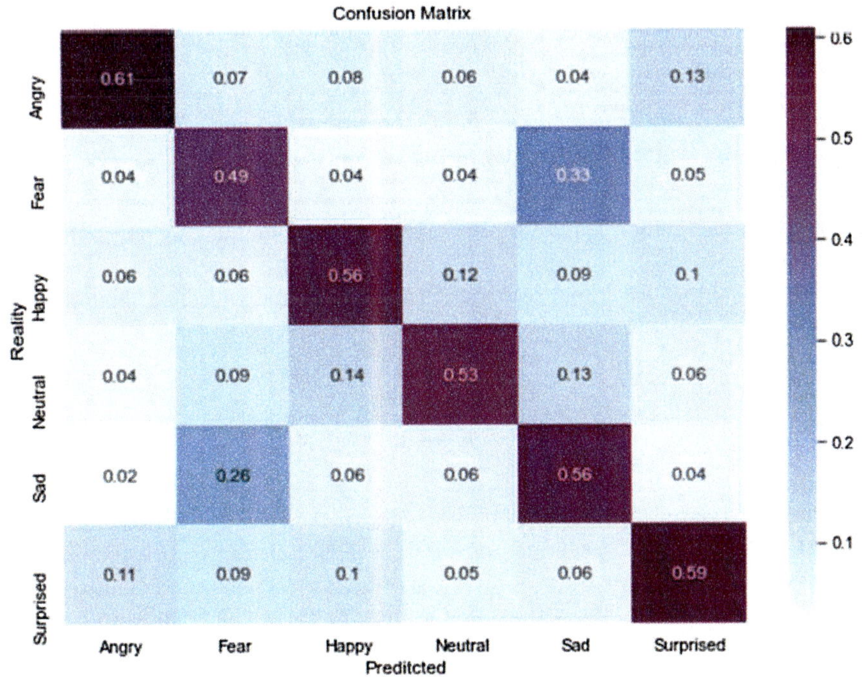

Fig. 6 Confusion matrix of emotion recognition

Table 1 Confusion matrix of emotion recognition with audio features

Classified as	Negative	Non-negative
Negative	0.73	0.27
Non-negative	0.21	0.79

Fig. 7 A show case of emotion recognition application

4 Emotion Recognition Application

An emotion recognition application was developed for function display, which is shown in Fig. 7.

Firstly, loading speech file, secondly clicking 'Emotion Detection' button, then the result of emotion recognition will pop up on the right side.

5 Conclusion

This article proposed a driver emotion recognition method using audio features. First, gender recognition was performed on the speech, which can reduce the amount of calculation and improve the accuracy of subsequent emotion recognition, and then extracted Mel Frequency Cepstral Coefficients (MFCCs) features to classify emotions, achieving the accuracy of 76.5%. Moreover, a driver emotion recognition application was developed, which may decrease risky driving in the future.

References

1. Ma J, Gong Z, Tan J et al (2020) Assessing the driving distraction effect of vehicle HMI displays using data mining techniques[J]. Transport Res F: Traffic Psychol Behav 69(1):235–250
2. Leng H, Lin Y, Zanzi LA (2007) an experimental study on physiological parameters toward driver emotion recognition [C]. In: International conference on ergonomics & health aspects of work with computers. Springer
3. Kim I-H et al (2014) Detection of braking intention in diverse situations during simulated driving based on EEG feature combination. J Neural Eng 12(1):016001
4. Zhang C, Wang H, Fu R (2014) Automated detection of driver fatigue based on entropy and complexity measures. IEEE Trans Intell Transp Syst 15(1):168–177
5. Murugan S, Selvaraj J, Sahayadhas A (2020) Detection and analysis: driver state with electrocardiogram (ECG). Phys Eng Sci Med 43:525–537. https://doi.org/10.1007/s13246-020-00853-8
6. Nguyen TP, Chew MT, Demidenko S (2015) Eye tracking system to detect driver drowsiness. In: 2015 6th International conference on automation, robotics and applications (ICARA), pp 472–477. https://doi.org/10.1109/ICARA.2015.7081194
7. Oxford University Press (2019) Oxford dictionaries. Oxford University Press
8. James W (2018) The principles of psychology. In: Personality culture society. https://doi.org/10.30936/1606-951x-2018-20-3/4-27-64
9. Ekman P, Friesen WV (1971) Constants across cultures in the face and emotion. J Pers Soc Psychol. https://doi.org/10.1037/h0030377
10. Ekman P, Sorenson ER, Friesen WV (1969) Pan-cultural elements in facial displays of emotion. Science. https://doi.org/10.1126/science.164.3875.86
11. Camras L, Plutchik R (1980) Emotion: a psychoevolutionary synthesis. Am J Psychol. https://doi.org/10.2307/1422394
12. Passion and reason: making sense of our emotions (1995) Choice reviews online. https://doi.org/10.5860/choice.32-4774
13. Laukka P (2017) Vocal communication of emotion. In: Encyclopedia of personality and individual differences. https://doi.org/10.1007/978-3-319-28099-8_562-1
14. Laukka P (2020) Encyclopedia of personality and individual differences. encyclopedia of personality and individual differences, pp 1–6. https://doi.org/10.1007/978-3-319-28099-8
15. Godino-Llorente JI, Gómez-Vilda P, Blanco-Velasco M (2006) Dimensionality reduction of a pathological voice quality assessment system based on gaussian mixture models and short-term cepstral parameters. IEEE Trans Biomed Eng 53(10). https://doi.org/10.1109/TBME.2006.871883
16. Pahwa A, Aggarwal G (2016) Speech feature extraction for gender recognition. Int J Image, Graph Signal Process 8(9):17–25. https://doi.org/10.5815/ijigsp.2016.09.03
17. Sigmund M (2008) Gender distinction using short segments of speech signal. Int J Comput Sci Netw Secur 8(10)
18. Vogt T, André E (2006) Improving automatic emotion recognition from speech via gender differentiation. In: Proceedings of the 5th international conference on language resources and evaluation, LREC 2006

Research on OD Estimation of Public Transit Passenger Flow Based on Multi-source Data

Chaoqun Kong, Tangyi Guo, and Liu He

Abstract OD calculations based on big data of public transport passenger flow can effectively improve the service quality of public transport and increase its attractiveness. In this paper, the bus GPS data and station GIS data are processed and time–space correlation is established to obtain the time interval of bus station, clustering analysis of bus IC card transaction data, and matching to get bus passengers' boarding station and time. Then, based on the probability of passenger travel behavior, the station's attraction power and transfer capacity are used to calculate the passenger's alighting station. Use actual cases to verify the algorithm, and finally calculate the OD matrix and passenger flow statistics of the bus line, so that the public transportation department can accurately obtain the bus travel information of urban residents, and provide data support for traditional public transportation planning.

Keywords Passenger flow big data · Time–space correlation · Attractiveness model · OD matrix calculation

1 Introduction

Improving the management and service level of the urban public transport system depends on the comprehensive mastery, analysis and application of public transport passenger outflow data, and the collection of the origin–destination matrix (OD) data of public transport travel is the most important and basic step. Public transport OD reflects the spatial distribution of public transport trips, and is a prerequisite

C. Kong · T. Guo (✉) · L. He
Department of Transportation Engineering, Nanjing University of Science and Technology, Nanjing City, China
e-mail: transtor@njust.edu.cn

C. Kong
e-mail: 1354660862@qq.com

L. He
e-mail: 124675282@qq.com

W. Wang et al. (eds.), *Green Transportation and Low Carbon Mobility Safety*, Lecture Notes in Electrical Engineering 944,
https://doi.org/10.1007/978-981-19-5615-7_42

for public transport planning and operation management. As an important mode of transportation for urban passenger flow transportation, urban public transport has become the consensus of local governments at all levels to improve urban traffic congestion by vigorously developing a comprehensive public transport system. At present, the OD survey of bus passenger flow needs to be obtained through residents' travel surveys. According to the source of the obtained data, the conventional survey methods mainly include: the residents questionnaire survey method, the surveyor on-board observation survey method, and the bus swipe statistical method. These survey methods require the active cooperation of citizens, and the survey methods have limitations, and they cannot fully and accurately grasp the travel needs of residents in the city. The use of big data mining technology can obtain statistics of passenger flow, passenger flow direction and residence time in time, realize a comprehensive and accurate grasp of bus passenger flow, which can be used as basic data for urban comprehensive transportation system planning and evaluation, and reduce the manpower for OD survey of bus passenger flow the input of material resources has greatly improved the accuracy.

Hu et al. [1] combined the passenger travel chain, considered the influence of passenger travel characteristics on and cited weights, and established a model for judging passenger alighting stations. Li et al. [2] comprehensively identified bus transfer behavior from the perspective of space and time on the basis of judging the passenger boarding station. Qin [3] established a public transport OD model based on station attraction, taking into account the influence of station transfer capacity. Munizaga [4] and others used three databases (automatic ticketing system, automatic car positioning system, and geocoding to define the public transportation network) according to the assumed time and distance between adjacent transfer stations, and combined with the personal travel chain to determine the drop-off location. Martin [5] proposed that for passengers with different travel patterns, the drop-off location should be determined based on the shortest transfer distance. Wang [6] et al. used the passenger flow survey data of the London Department of Transport to compare the passenger flow obtained through the personal travel chain method, which proved the feasibility of mining OD through IC card data, and analyzed the impact of transfer time. Through the above research, it is found that the boarding position of bus passengers can be judged by clustering analysis of bus IC card and gradually transitioning to the method of combining IC card data and GPS data. However, the correction factor established by the getting off model is too single. Constrained only by the attraction strength of the site.

This paper firstly carries out the process of data screening and sorting on the bus IC card consumption records and GPS data records, and then defines the concept of GPS feature points and station identification intervals to identify passengers' boarding stations. Secondly, the theory of travel connection is introduced, and an integrated recognition method for passenger disembarkation and transfer based on space and time dimensions is proposed to determine whether adjacent bus rides are connected in space and time, and then based on "time The principle of "cost minimization" is to determine whether the "get off" action at this time is the get off behavior or the first half of the transfer behavior, and identify the station corresponding to the passenger's getting on and off behavior and the station corresponding to the transfer behavior, and finally get The true origin and destination of the passenger's bus trip.

2 Multi-source data analysis

A method of fusing and matching bus IC card consumption records, GPS data records, and GIS information data is proposed, especially the method of selecting feature points of GPS data records, and the method of setting station identification intervals based on feature points, which can be more direct and rapid Filter out the data used for matching. Before data matching, the data needs to be processed and analyzed.

A. ***Basic data analysis of bus IC card consumption records.***

In this paper, the data structure of the one-invoice bus IC card charging system is used as the related object, and the key fields are selected as the data sample table of the bus IC card, as shown in Table 1.

B. ***Analysis of Basic GPS Data of Public Transport.***

The basic data collected comes from the vehicle GPS system. The main function of GPS is to provide the vehicle's positioning function, position information, direction angle, driving speed, etc. When the bus is driving on different routes, GPS will collect the vehicle's latitude and longitude coordinates and the current running time in real time to form the position, time, speed, direction angle, etc. The data sample table of the bus GPS system is shown in Table 2.

C. ***Analysis of GIS Basic Data of Public Transport Stations.***

The GIS data of the bus station is a long-term unchanging data. It records information such as station name, station longitude, station latitude, line number, line number, line name, station number, station direction angle, etc. The value of the latitude and longitude field is the same as GPS data. The values of the latitude and longitude fields are in the same coordinate system, so you can match the two. The Station_Azimuth column in the table is a new column. The specific methods are as follows:

STEP 01: Convert a physical station to a line station (regardless of direction);

Table 1 Bus IC card data sample sheet

DateTime	Card_ID	Price	Card_Type	Line_ID	Bus_ID
2019/9/1 5:11	3,104,830,115,000,550,000	2.4	205	6220	480
2019/9/1 5:13	9,616,900,730,122,490	1	6	2280	434
2019/9/1 5:15	3,104,830,115,001,280,000	1.5	207	6220	480
2019/9/1 5:17	3,104,830,115,000,440,000	1.6	205	2180	274
...	...	...	...	...	...

Table 2 Bus GPS system data sample sheet

DateTime	Line_ID	Bus_ID	Lng	Lat	Speed	Azimuth
2019/9/1 6:34	1003	1398	119.417347	32.477616	7.43	240
2019/9/1 6:34	1103	1392	119.434245	32.444883	8.85	184
2019/9/1 6:34	3010	2230	119.505548	32.392786	0	341
2019/9/1 6:34	1099	1915	119.550605	32.430517	7.86	69
...	...	...	...	...	...	...

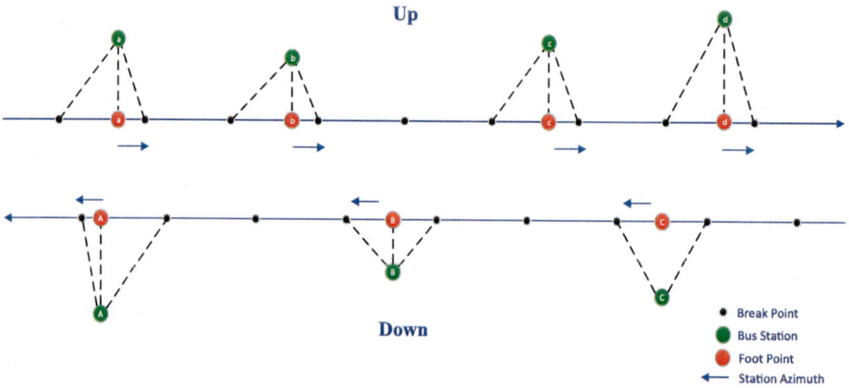

Fig. 1 The practice of foot point at bus station

STEP 02: Clean and create Stop table according to "night, night shift, interval, temporary, outage" and Line index table;

STEP 03: Import Route line position and attribute table from TC, clean and create Route table;

STEP 04: Traverse all stations, draw vertical lines (up to 50 m) section by section on the line trajectory in the bisecting direction of the line station, draw vertical lines to the line to find the position of the vertical foot (the red vertical foot point shown in Fig. 1), Determine the Route_ID, Order and Station_Azimuth of the station (the direction angle between the vertical foot point and the next interruption point in the same direction), and add the vertical foot coordinates to the route trajectory list to update the geographic information data table of the bus station (Table 3).

3 Data Association and Fusion Method

A. *Data processing flow*

The basic data of this article is provided by the big data and digital R&D center of Yangzhou Bus Company and Huashe Design Group, including Yangzhou bus

Table 3 GIS data sample form of bus station

Station_Name	Lng	Lat	Route _ID	Line _ID	Line_ Name	Order	Station_ Azimuth
Yangnong Group	119.426691	32.372699	1	69	2 Way	1	71
Wenfeng Tower	119.42825	32.37296	1	69	2 Way	2	60
Pharmaceutical factory	119.432293	32.373743	1	69	2 Way	3	88
West Garden	119.434356	32.377721	1	69	2 Way	4	351
...	...	...	...	...	...	...	...

network (510 bus lines), bus station GIS data (3779 physical stations), and bus IC cards in September 2019 Data (3,218,976 items) and bus GPS operation data (176,518,776 items), etc.

In order to facilitate the management of various data tables, the basic data storage adopts SQL Server, the algorithm processing adopts Python language programming (Pandas, Geopandas, Numpy data modules, etc.), the traffic OD recognition adopts the Transcad shortest path algorithm (time cost is minimized), and the data processing flow As shown in Fig. 2 [7].

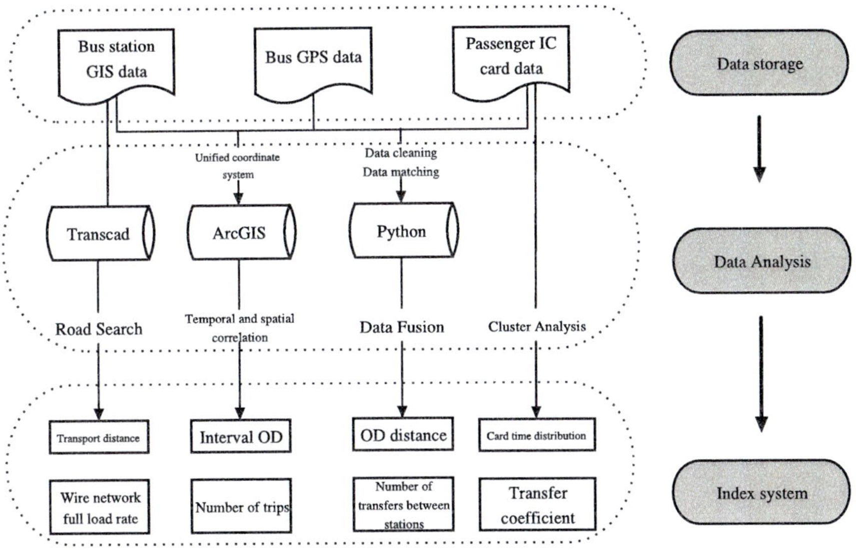

Fig. 2 Flow chart of data processing

B. ***Correlation and matching between public transit GPS data and public transit station GIS data***

According to the longitude and latitude values of GPS and GIS to determine the spatial distance between the bus and the bus stop [8], the calculation formula is as follows:

$$d = R * arccos[sin(Lat_{GPS}) * sin(Lat_{GIS}) + cos(Lat_{GPS}) * cos(Lat_{GIS}) * cos(Lon_{GPS} - Lon_{GIS})]\frac{\pi}{180} \quad (1)$$

In the formula, R is the radius of the earth, the average radius of the earth is 6371004 m, d is the space distance between the station and the vehicle, Lon_{GPS} is the value of Lng in the bus GPS trajectory information table, Lat_{GPS} is Lat in the bus GPS trajectory information table The value of Lon_{GIS} is the value of the longitude field of the bus line station in the GIS, and Lat_{GIS} is the value of the latitude field of the bus line station in the GIS.

Note: The recorded time of GPS data is not necessarily the actual bus stop time. Therefore, it is necessary to set a spatial error range according to the size and shape of the bus stop and the system error of the on-board GPS. The latitude and longitude position of the station is the center of the circle, and the circular recognition area (buffer) with the error radius L_Z is set. Generally, it can be set as the maximum length of the bus station along the road, generally 30 m), if there is GPS data falling in the buffer Record point, and the speed value of the record point is less than the set speed threshold V_Z (generally set to 3 km/h), and the absolute value of the difference between the direction angle of the vehicle and the direction angle of the station is $\leq$60, the record can be determined The point is the characteristic point, that is, the effective entry point.

If there is no feature point in the identification area of the bus station, look for the two recording points closest to the station on both sides of the station on the same line. It is reasonable to assume that the bus runs at a constant speed between the two points, and then combine the bus lines in the figure, the bus line length d_z between these two points and the line lengths d_1 and d_2 between them and the station are obtained from the vehicle position coordinates of the two records, and the recording time t_1 and t_2 of the two points, we can know the assumption between the two points The uniform running speed $\overline{\overline{V}}$ is:

$$\overline{\overline{V}} = \frac{d_z}{|t_1 - t_2|} \quad (2)$$

Thus, the approximate time $\overline{\overline{t}}_z$ when the bus enters the station is obtained as:

$$\overline{\overline{t_z}} = t_1 + \frac{d_1}{\overline{\overline{\overline{V}}}} \tag{3}$$

Or the opposite direction is:

$$\overline{\overline{t}}_z = t_2 + \frac{d_2}{\overline{\overline{\overline{V}}}} \tag{4}$$

After obtaining the characteristic point of the location near the bus station, since the speed of the characteristic point is the smallest, it can be considered that the time recorded by the characteristic point is either very close to the time of inbound or out of the station. In the case that the time of entry and exit of the vehicle cannot be accurately identified, the recording time of the characteristic point is taken as the center of the section, and the maximum parking time of the vehicle is the radius of the section, to construct a station identification section (buffer). When the feature point recording time is t_z and the maximum parking time of the vehicle is t_{smax} (generally set to 30 s), the parking identification interval is $(t_z - t_{smax}, t_z + t_{smax})$. Similarly, in the case of station hopping, the station identification interval of the hopped station is $(\overline{t}_z - t_{smax}, \overline{t}_z + t_{smax})$.

The overall idea of association and matching between GPS data and GIS data is shown in Fig. 3.

C. ***Match the identification section of the bus entering the station with the IC card data***

In the morning and afternoon rush hours, the time when passengers swipe their cards is not strictly between the inbound and outbound bus stops. Therefore, when making the correlation, the specific exit time of the vehicle is not considered, and the station identification section of the current station is regarded as the identification section. In addition, there will always be some abnormal situations in the actual operation of public transportation vehicles. For example, the IC card swiping time is earlier than the first station identification interval recorded by the GPS, and the IC card swiping time is later than the last station identification interval recorded by the GPS. It is necessary to classify the discussion on the premise of trying to restore the actual situation. The entire association process is shown in Fig. 4.

D. ***Recognition of passenger disembarkation and transfer stations based on travel chain.***

The travel chain is a closed chain that connects all travel activities within a day in the order of travel time, centered on the place of residence of the traveler. The characteristics of the bus travel chain are: travelers have at least two urban public transportation trips a day, and the time of getting off the bus before taking the bus must be earlier than the time of getting on the bus after taking the bus. Into a chain. This article only considers the bus travel chain formed by the bus travel mode, and does not consider the bus travel chain mixed with other travel modes. The overall algorithm flow is shown in Fig. 5.

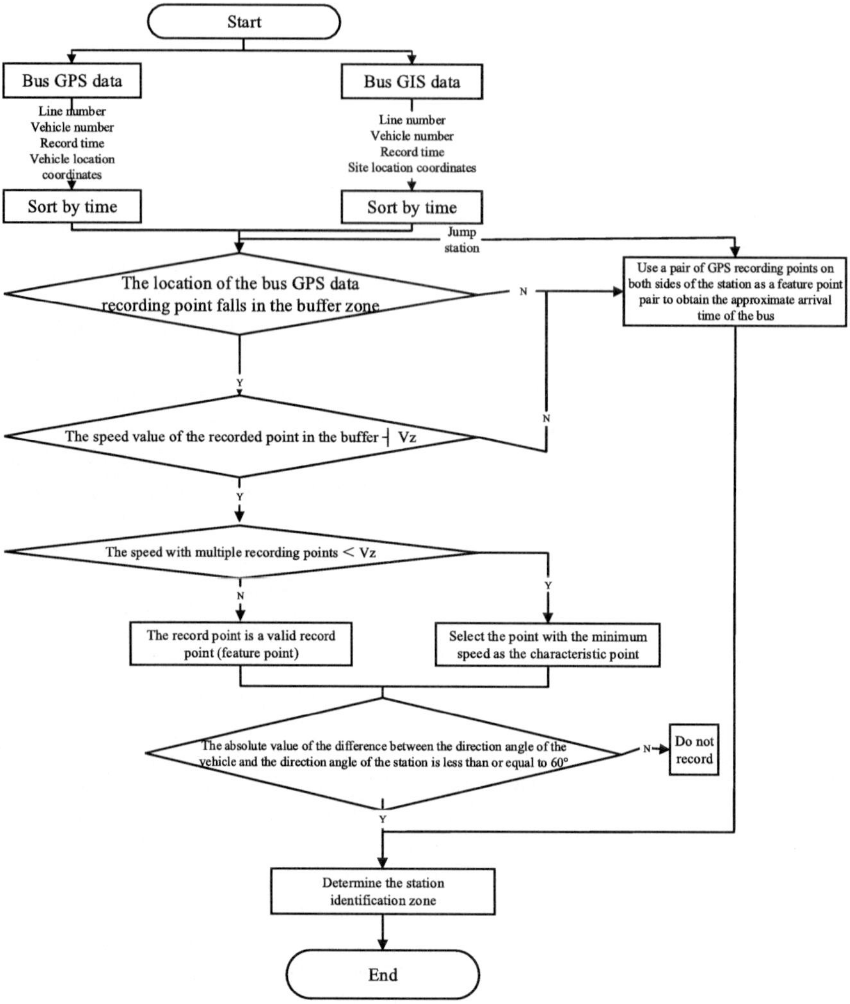

Fig. 3 The idea of association between GPS data and GIS data

SETP 01: Cluster travel time for passenger card swiping data. Passengers with travel days ≤ 9 in 30 days are defined as low-frequency passengers, and passengers with travel times > 9 are defined as high-frequency passengers.

STEP 02: Use the above method C to match the passenger boarding station, and count the travel distance and travel time of the passenger on the day, and whether it meets the threshold range of the travel chain in space and time.

STEP 03: If the passenger has a travel chain on the day, use the time difference of the passenger's card swiping or the transfer distance as the constraint parameter to determine the transfer to determine the true starting point, transfer, and end point; if the passenger does not have a travel chain on the day, the high-frequency

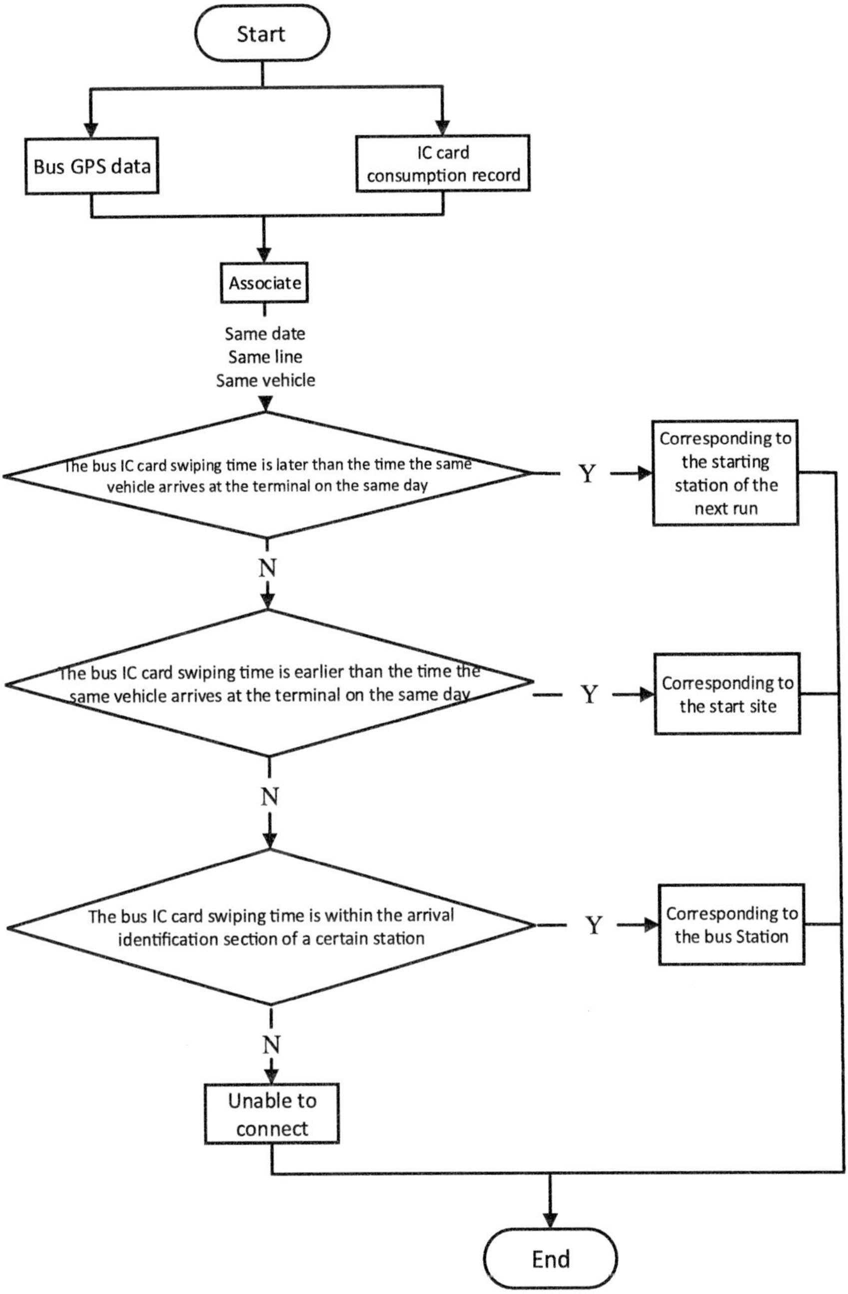

Fig. 4 GPS-based bus IC card consumption data matching flowchart

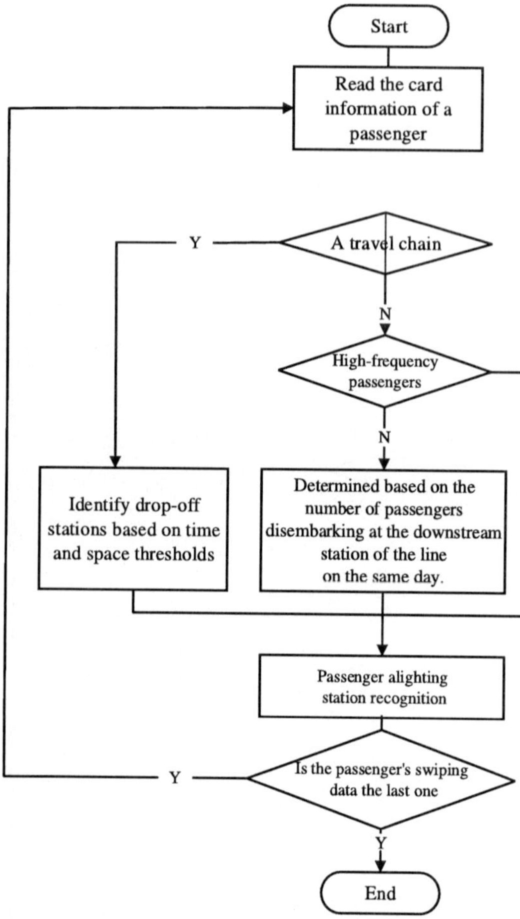

Fig. 5 Flowchart of recognition algorithm for passenger alighting stations

passenger gets off the bus The stop is determined according to the number of passengers alighting at the downstream station of the line for multiple days, and the low-frequency passenger alighting station is determined according to the number of passengers alighting at the downstream station of the line on the same day.

STEP 04: Cluster the 24 h according to a 15-min time slice (96 time slices in total), and perform statistical analysis on the number of passengers on, off, and transfers on each line in each time slice.

4 Case Study

The basic data of this article is provided by the big data and digital R&D center of Yangzhou Bus Company and Huashe Design Group, including Yangzhou bus network (510 bus lines), bus station GIS data (3779 physical stations), and bus IC cards in September 2019 Data (3,218,976 items) and bus GPS operation data (176,518,776 items), etc. (Fig. 6).

Add grid information to the site. The accuracy is low in the west and high in the east, and the latitude is low in the south and high in the north. Therefore, the coordinate axis direction is consistent with the plane coordinate system. Generally, the east–west direction of a city is at the level of 100 km, because it can be rasterized numbering according to 6 bits. Grid size, unit: m, default 500*500 m.

According to the boarding station, the getting off station, the 15-min boarding time slice, and the date, the OD including the transfer and not including the transfer is counted, and then the OD is rasterized (Fig. 7).

Statistics are based on OD with transfer between stations, including multiple time and space dimensions such as network, OD pair, route, station, section, etc. (Fig. 8)

Bus passenger flow index statistics: the distribution of passengers on and off the 500-m grid (Figs. 9, 10, 11).

The areas with a large amount of distribution in the main urban area of Yangzhou are concentrated in the city center and Yangzhou Railway Station. The monthly distribution volume of the top ten bus routes is more than 30,000 passengers. Among them, the largest passenger flow distribution route is 26, with a monthly distribution

Fig. 6 Yangzhou bus line network and Station Layer

Fig. 7 Raster image that has been matched to OD

Fig. 8 Top 10 bus lines in passenger traffic(2019.09.01–30)

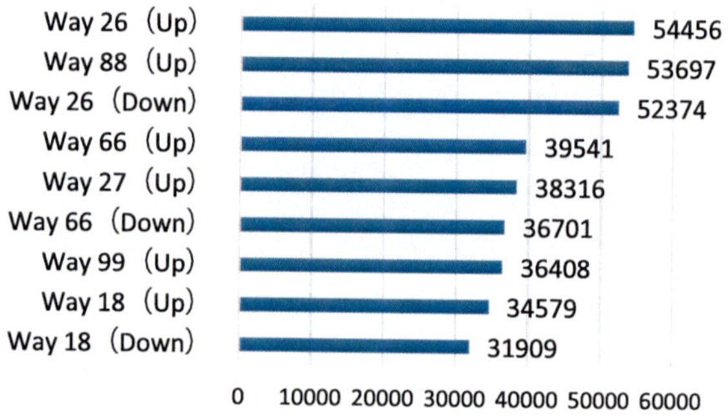

volume of 54,000; followed by 88 roads, with a monthly distribution volume of 53,000 passengers. The top ten points are shown in Fig. 8.

Based on the traffic distribution sub-software Transcad, the traffic flow distribution is performed on the bus OD, and the cross-sectional flow distribution map of the bus is obtained (Fig. 12). The overall passenger flow distribution is in a central radial pattern.

Combining the line network and station layout, it is found that the important role of public transportation in urban transportation has not yet been well brought into play. With the rapid increase in the level of motorization, to make urban traffic sustainable development, it is necessary to further strengthen the guidance to other areas, strengthen the concept of residents' public transportation, and enhance the attractiveness of public transportation to residents' travel.

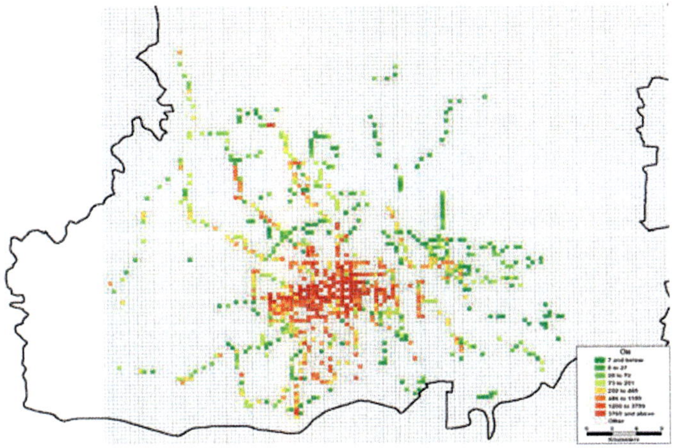

Fig. 9 Heat map of bus boarding(2019.09.01)

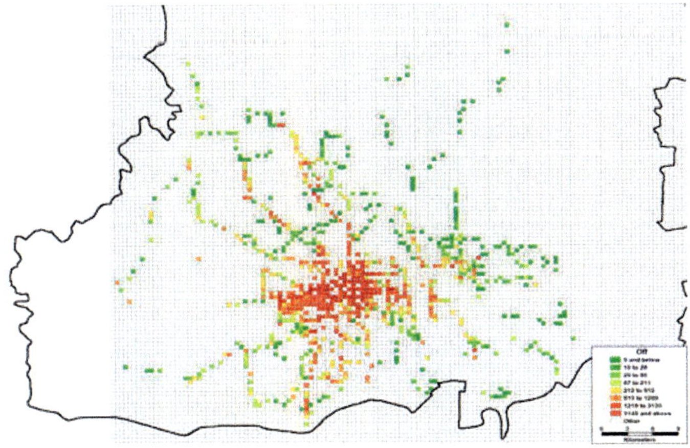

Fig. 10 Heat map of bus disembarkation(2019.09.01)

5 Conclusion

Taking Yangzhou City as an example, this paper introduces the method of data fusion analysis of bus IC card, and verifies and displays the results. Use the bus IC card and GPS data to match the time and space to determine the passenger's boarding location. By introducing the idea of travel chain, starting from the connectivity of adjacent bus riding processes in time and space, the constraints of space and time are set. Condition, use the attraction strength of the station and the transfer coefficient of the station bus line to modify it, judge the alighting behavior and the transfer behavior, and clarify the complete travel process OD of the bus traveler.

Fig. 11 500-meter grid bus travel expectation line. (2019.09.01)

Fig. 12 Bus line passenger flow

On the basis of in-depth analysis of public transportation demand, by integrating the public transportation system's road network, routes, stations and various operating information, the public transportation system supply and demand relationship is modeled and simulated, and the weak links of public transportation services in time and space are found. It can provide decision support for actual work such as network planning and optimization.

References

1. Hu J, Deng J, Huang Z (2014) Probabilistic model for judging the bus IC card passenger alighting station combined with the travel chain [J]. Transp Syst Eng Inf 14(2):62–67
2. Li Y, Weng X (2017) Transfer recognition method based on public transport IC card and GPS data[J]. J Guangxi Univ (Nat Sci Ed) 42(2):579–586
3. Qin Z (2017) Research on passenger boarding and disembarking stations based on public transport IC card and GPS data [J]. West Transp Sci Technol (8):115–119

4. Munizaga M, Palma C (2012) Estimation of a disaggregate multimodal public transport Origin-Destination matrix from passive smartcard data from Santiago, Chile [J]. Transp Res Part C 24(9):9–18
5. Trepanier M, Chapleau NTR (2007) Individual trip destination estimation in a transit smart card automated fare collection system [J]. J Intell Transp Syst Technol Plan Oper 11(1):1–14
6. Wang W, Attanucci JP, Wilson NHM (2011) Bus passenger start-end estimation and related analysis using automated data collection system[J]. J Public Transp 14(4):131–150
7. Zhang T, Gao H, Ju B, Hu Q (2019) Derivation and application of bus OD based on IC card data [J]. Urban Public Transp 2019(06):27–32
8. Qian Y (2017) Research on bus OD estimation method based on multi-day bus IC card data and AVL data [D]. Southwest Jiaotong University

Research on Metro Vehicles Allocation Based on Capacity and Maintenance Model

Fang Gao, Yongneng Xu, Jue Zhang, and Zhonglin Tan

Abstract With the continuous acceleration of urban rail transit construction mileage and the opening of existing extension lines, the demand for metro vehicles on the extended lines has also increased accordingly. However, in the actual vehicle configuration, there is room for optimization of the number of vehicles allocation on each line. By analyzing the relevant factors affecting the assignment of metro vehicles, including signal system, passenger flow conditions, maintenance mode, passenger flow organization and passenger flow development theory, comprehensively consider the length of the line, the speed of the train, the turn-back time of the train, and the interval between trains, construct a metro vehicles allocation model based on capacity demand and vehicle maintenance. Taking the extension of subway line M in a certain city as the research object, through the "four-stage method", predict the passenger flow, and analysis the key factors including the maximum cross-section passenger flow in the peak hour of metro and the mixed running of long and short routing in the case city, the allocation plan is obtained according to the allocation model. Compared with the original plan allocation plan, the train line rate is increased and the vehicle cost is reduced. Verifies the effectiveness and practicability of the vehicle model based on capacity and maintenance.

Keywords Rail transit · Capacity demand · Vehicle maintenance · Metro vehicle allocation

F. Gao (✉) · Y. Xu · J. Zhang · Z. Tan
Nanjing University of Science and Technology, No. 200 Xiaoling Wei Street, Nanjing 210094, Jiangsu, China
e-mail: gaofang1919@163.com

Y. Xu
e-mail: x780906yn@163.com

J. Zhang
e-mail: 1415920196@qq.com

Z. Tan
e-mail: 1685683699@qq.com

W. Wang et al. (eds.), *Green Transportation and Low Carbon Mobility Safety*, Lecture Notes in Electrical Engineering 944,
https://doi.org/10.1007/978-981-19-5615-7_43

1 Introduction

Before the opening of each line of urban rail transit, relevant experts will be invited to conduct research on the conditions around the line and predict the passenger flow of the line, and formulate a project feasibility study report based on the results of the investigation and prediction, so as to determine the number of vehicles allocated on the line [1]. However, with the continuous acceleration of urban metro construction mileage and the continuous increase of existing cable extension lines, the corresponding train demand is also increasing. However, in the actual vehicle configuration, due to the deviation of passenger flow forecast and actual situation, the transportation capacity does not match the transportation volume, so that there is room for optimization of the number of vehicles assigned on each line. Therefore, it is necessary to rationally optimize the number of metro vehicles allocation to save costs and improve train utilization. Secondly, in the existing train operation plan, it is necessary to further analyze the factors affecting the train on-line rate. From the perspective of analyzing the capacity demand and optimizing the maintenance mode, preliminary research and formulation of management measures and related coordination requirements are required to help increase the om-line rate of trains. In addition, due to the increase in the number of trains, operating costs such as procurement costs, maintenance costs, and staffing costs have increased. It is necessary to further optimize the allocation of trains to reduce costs and improve the level of subway operation and management.

Therefore, it is necessary to analyze the factors affecting the train on-line rate, and to reasonably optimize the number of metro assigned from the perspective of analyzing the capacity demand and optimizing the maintenance mode, so as to achieve the purpose of reducing the cost and improving the on-line rate of trains.

The number of subway trains affects the normal operation of the subway department, it will also affect related operating costs. At present, some domestic scholars have conducted a series of studies on the issue of subway trains.

Zhao [2] proposed some new understanding of the vehicle allocation formula of the Urban Light Rail Transit Engineering Design Guide based on many years of metro design experience and exchange experience with experts.

Luo et al. [3] established a model for the number of train allocations for metro lines based on passenger flow demand, especially in the case of peak passenger flow.

Zhao [4] took the calculation formula for the assigned trains of urban rail transit as the starting point, deeply analyzed the five major factors affecting the accuracy of the number of assigned trains, and gave specific optimization schemes and setting standards.

Gong et al. [5] tested the goodness of fit between the predicted passenger flow data and the actual AFC (automatic fare collection) system data, and established a vehicle allocation model.

Peeters and Kroon [6] proposed a model of vehicle turnover plan based on multiple routes and a solution method using branch pricing. According to the change of the passenger flow density of the section, the number of EMUs was changed.

Hong et al. [7] proposed three application modes: path coverage, cycle coverage, and periodic coverage. Path coverage means that each operating line on the train operation diagram is covered by an EMU turnover line. The cyclic coverage does not require empty trains to return. It is only for symmetrical operation diagrams. Periodic coverage means that the EMU performs long-distance running and go through a suitable maintenance base when needed. It provided new research ideas for the application mode and application plan of EMUs.

Maróti and Kroon [8] have established integer programming models for various problems of short-term EMU application planning, used commercial mathematical software to solve them, and designed various techniques to speed up the efficiency of the solution.

At this stage, there are not many direct research results on the issue of metro vehicles allocation. The above studies all started from a certain relevant factor and carry out relevant research on the number of vehicles, but they have not modeled the problem of vehicle allocation from a macro perspective. Based on relevant specifications and professional books, this article will analyze the factors affecting the train on-line rate. From the perspectives of analyzing capacity demand and maintenance mode, a more complete model will be constructed to rationally optimize the number of metro vehicles to reduce costs and increase the on-line rate of trains.

2 Analysis of the Factors Influencing Metro Vehicle Allocation

2.1 Signal System

Under different signal systems, the interval between trains is also different. The driving interval is inversely proportional to the transportation capacity of the track system. The shorter the driving interval, the more trains can pass per hour, the more vehicles are needed, and the greater the transportation capacity of the track [9]. Reasonably arrange the online vehicles by using large and small roads, adding trains, which can increase the utilization rate of the online trains.

2.2 Passenger Flow Conditions

In the calculation process of vehicle allocation, passenger flow occupies an important position, which is related to the design standards of railway lines, carrying capacity, vehicle marshalling and station design. How to accurately predict passenger flow is the key to obtaining the number of vehicles allocation. The passenger flow forecast data is generally the result of passenger flow forecast in the preliminary design stage. Therefore, it is recommended that the operating unit reorganize the scientific research unit and conduct a pre-study on the initial passenger flow of the new line before the new line is opened for operation [10]. On the one hand, it can better provide a basis for setting new line operation service indicators; on the other hand, it can also provide reference for car purchase arrangements of other affected lines.

2.3 Maintenance Mode

The current main factors affecting the on-line rate of electric buses are investigated and analyzed in this paper. There are two main aspects [11]:

One is the vehicle repair process factor deduction. For example, overhaul deduction and full effect repair deduction car. Among them, the overhaul operation cycle of the frame is longer, and the method of lap start is often adopted.

The second is accidental car detention. There are several situations such as rectification deduction, wheel deduction, fault detention, first-line multi-field detention and bad weather.

2.4 Passenger Flow Organization

When the number of vehicles is insufficient and the traffic adjustment cannot meet the demand of passenger flow, the station will face a backlog of passenger flow. In the case of insufficient capacity, in order to ensure the orderly progress of the safe production of passenger transportation, the station needs to control the passenger flow of the station through passenger flow restriction and passenger flow guidance.

2.5 Passenger Flow Development

In order to facilitate the future metro operation management and the construction of the extension of each metro line, it is necessary to predict the future passenger flow of each line. The passenger flow forecast of rail transit at home and abroad usually adopts a four-stage prediction method [12]. The four-stage prediction method analyzes the current situation and future traffic conditions of the city according to the four stages of traffic generation prediction, traffic distribution prediction, traffic mode division and traffic distribution.

3 The Establishment of Vehicle Allocation Model

3.1 Model Based on Capacity Demand

3.1.1 Passenger Flow Analysis

By studying the passenger flow growth pattern of metro in some mature cities, it is found that the trend of passenger flow change is very similar everywhere, and the expansion of the line network has a more significant impact on the backbone line. With the opening of adjacent new lines, the general backbone line has a large increase in daily passenger traffic during the initial period of opening and in the near future, while the long-term growth is relatively slow.

Considering the line length, train speed, train turn-back time and train interval, according to the idea that the time required to operate a round-trip cycle divided by the interval time is equal to the number of trains in operation [13], the calculation model for the number of trains can be established as follows:

During peak hours, the number of train pairs required is:

$$\mathrm{n} = \frac{Q}{AW2 \times \beta} \tag{1}$$

n The number of train pairs that should be operated during peak hours.
Q Cross-sectional passenger flow during peak hours (passengers).
$AW2$ The designed passenger capacity of the train (person).
β Full load rate of train.

Train interval:

$$\mathrm{T} = \frac{3600}{n} \tag{2}$$

T Train interval (s).
n The number of train pairs that should be operated during peak hours.

Operating cycle time of train:

$$\theta_{\text{cycle}} = \sum t_{\text{run}} + \sum t_{\text{stop}} + \sum t_{\text{turn-back}} \tag{3}$$

θ_{cycle} Running cycle time of train (s).
t_{run} Line running time of train (s) $\left(\text{i.e.: } t_{\text{run}} = \frac{totallinelength}{averagetrainspeed}\right)$.
t_{stop} Stop time of train (s).
$t_{\text{turn-back}}$ Turn-back time of train (s).

Number of trains required:

$$N_{\text{required}} = \frac{\theta_{\text{cycle}}}{T} \tag{4}$$

N_{required} Number of trains required.

3.1.2 Signal System

The calculation method of the minimum train interval of the main line section under a fixed block (11) is:

$$T_{min} = \frac{L + S}{v} \tag{5}$$

T_{min} Minimum interval time of the line (s).
S Minimum distance between trains in operation (m).
L Maximum length of train (m).
v The normal running speed of the train (m/s).

The minimum distance between trains also needs to consider the operating margin and safety interval, therefore:

$$S = S_{\text{safe}} + S_{\text{error}} + S_{\text{margin}} \tag{6}$$

S_{safe} Safe braking distance under common braking conditions (m).
S_{error} Detection error distance of train (m), which reflects the length factor of the block area under the fixed block conditions.
S_{margin} Operation margin (m).

In the CBTC mode, since the passenger carrying capacity of each train, which is the product of t the passenger carrying capacity of each train and the number of trains in operation, needs to meet the needs of the maximum passenger flow in one-way section during peak hours, the calculation model for the train interval is established as follows:

$$\beta = \frac{Q_{max}}{Q \times n} \tag{7}$$

β Full load rate of train.
Q_{max} Maximum passenger flow in one-way section during peak hours (passengers).
Q Passenger capacity of train (passengers).
n The number of train pairs that should run during peak hours, where n = 3600/T and T is the interval time (s).

When β > 1, passenger detention will occur. Therefore, the following needs to be met:

$$Q \times n = \frac{\text{Q} \times 3600}{T} \geq \text{Q}_{max} \tag{8}$$

Which means:

$$T \leq \frac{\text{Q} \times 3600}{\text{Q}_{max}} \tag{9}$$

3.1.3 Long/Short Routing

On rail transit lines, the distribution of passenger flow in various sections is uneven, especially in long-distance lines, the uneven distribution of passenger flow in each section is particularly obvious. In this case, adopting a reasonable arrangement of multiple traffic routes can improve operating efficiency, save operating costs, and improve economic benefits for operating companies. The road setting principle [14] is as follows:

(1) Based on the forecast of passenger flow, combined consideration of initial, short-term and long-term;
(2) The wiring arrangement of the line should have multi-path organization conditions;
(3) Fully consider the difficulty of passenger transport organization at the station;
(4) People-oriented and convenient for passengers. Each section, especially the section with long-traffic trains, needs to ensure a certain level of service, and the interval between trains should not be too large;
(5) Reduce the number of passenger transfers or waiting time;
(6) Reduce the number of vehicles in use and reduce operating costs;
(7) Facilitate operation management and ensure the safety;
(8) Combining the planning of the rail transit network, consider facilitating passenger transfers.

After determining the long/short routing, the number of vehicles allocated to be determined according to the passenger flow of the road, so the model is constructed as follows:

During peak hours, the number of train pairs required to operate:

$$n = \frac{Q_{long} - Q_{short}}{\text{AW2} \times \beta} + \frac{Q'}{\text{AW2} \times \beta} \tag{10}$$

n The number of train pairs that should be operated at peak hour.
Q_{long} Passenger flow during peak hours on long routing (passengers).

Q_{short} Passenger flow during peak hours on short routing (passengers).
Q' The passenger flow of the cross-section at peak hour after removing the influence of long routing minus short routing (passengers).
AW2 The designed passenger capacity of the train.
B Full load rate of train.

3.2 Model Based on Vehicle Maintenance

Take Nanjing Metro as an example. The main repair procedures of Nanjing Metro include major repair, shelf repair, full-effect repair and daily inspection. This section considers three maintenance operations: major repair, shelf repair, full-effect repair. According to the "Five-Year Plan for Overhaul of Electric Passenger Car Frames in Nanjing Metro Lines" and "Metro Design Specifications" [15], in addition to major repair, shelf repair and full-effect repair deduction, there are also deductions due to fault rectification, so the number of daily maintenance deduction in each line is converted as follows:

$$N_{\text{maintenance}} = \mathrm{N}_1 + \mathrm{N}_2 + \mathrm{N}_3 \tag{11}$$

$N_{\text{maintenance}}$ The number of trains deducted for maintenance in the warehouse per day.
N_1 The number of trains deducted from frame overhaul per day.
N_2 The number of trains deducted from Full-effect repair per day.
N_3 The number of vehicles deducted for fault rectification per day.

The maintenance model can be modified and calculated according to the specific maintenance process of the specific line, and the number of deductions for maintenance in the warehouse per day for the requested line can be obtained.

3.3 Overall Metro Vehicle Allocation Model

The overall train allocation model can be constructed by integrating the capacity-based train allocation model of 3.1 and the maintenance-based metro allocation model of 3.2, an overall vehicle dispatching model can be constructed. Because the metro allocation is divided into two parts, one is the number of trains running on the main line; the other is the number of trains in the depot for maintenance and standby. Taking into account the line length, train travel speed, train turn-back time, and interval, according to the idea that the time required to operate a round-trip cycle divided by the interval time is equal to the number of trains in operation, the calculation model for the number of trains is established as follows:

$$N = N_{used} + N_{maintenance} + N_{standby} \tag{12}$$

Integrating the passenger flow density and the minimum interval of the signal system, the number of vehicles used on the line is calculated by the following formula:

$$N_{used} = \left(\frac{2L}{\overline{v}} + T_{stop} + T_1 + T_2\right) \times \frac{n}{3600} \tag{13}$$

L Total length of the line (m).
$\overline{v}$ The average speed of the train (m/s).
T_{stop} Train stopping time (s).
T_1, T_2 Return time of terminal stations at both ends (s).
N The number of train pairs that should be operated during peak hours, depending on whether there is a route to determine the calculation method.

4 Analysis of Metro Vehicle Distribution Cost

For the problem of metro vehicle allocation, in addition to establishing an allocation model, it is also necessary to carry out relevant analysis on the cost of allocation. The cost incurred by each vehicle allocation has the following aspects [16]:

(1) Labor cost: the cost of salary and other expenses for employees involved in related vehicle operations;
(2) Material cost: the cost of purchasing trains and storing purchased trains;
(3) Maintenance cost: the related cost of repairing the train;
(4) Depreciation cost: the relevant depreciation cost after the train reaches its service life.

In this paper, after calculating the number of allocated vehicles based on the model, we will derive the number of surplus vehicles and calculate the cost of surplus vehicles over the lifetime of the surplus trains.

5 Case Analysis

5.1 Train Organization Scheme

This section will take the extension line of a certain metro line in a certain city as an example (hereinafter referred to as line M) for case application. The train organization scheme for the case is shown in Fig. 1. After the opening of the extension line, according to the current actual passenger flow and the predicted passenger flow after

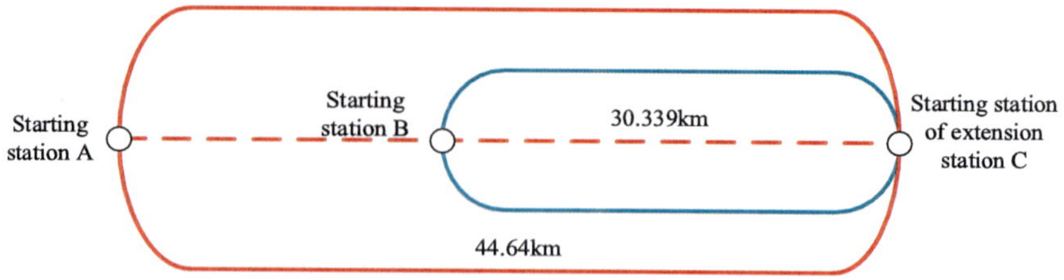

Fig. 1 Train organization scheme of route M

the extension line is opened, both mixed running of long and short routing and peak period bidirectional encryption operation will be used.

5.2 *Research on the Number of Metro Vehicle Allocation*

Line M is the extension of the original line, the vehicles composition is the same as the marshalling of the original line. The 6-carriage composition of B-type vehicles is adopted, and the passenger capacity of the train is AW2. Passenger flow forecasting is based on the changing trend of urban population and employment in various districts of the city. In the city' s road network, public transport network and rail transit network, the "four-stage method" [17] commonly used internationally is used to establish the city's public transport prediction model, and based on this model, the passenger flow prediction analysis of the extension line opening year and the next five years is carried out, as shown in Table 1.

After the opening of the extension of line M, the maximum section passenger flow is 29,800 in the expected peak hour in the opening year, and the maximum section passenger flow is 33,500 in the next five years. The number of vehicles allocated by the operation company of this line is shown in Table 2.

According to the overall vehicle allocation model established in Chap. 3, substituting the data into the calculation, the number of vehicles allocated for the extension line's opening year and the next five years can be obtained, as shown in Table 3.

It can be seen from Table 3 that compared the number of vehicles allocated by the model and the number of vehicles allocated to the original planned, the number of vehicles allocated by the model is less, and the online rate is higher, which is a big improvement compared to the original plan. Comparing the predicted number of vehicles allocated and the original number of vehicles allocated, and calculating the

Table 1 Passenger flow prediction of line M

Years	2015	2016	2017	2018	2019	2022	2027
Passenger flow prediction (unit: 10,000 person-times)	76.22	77.45	85.59	90.17	91.12	98.92	104.62

Table 2 The number of vehicles allocated to the original planned extension of line M

Years	Passenger flow prediction (unit: 10,000 person-times)	Maximum cross-section in peak hour (unit: 10,000 person-times)	Number of vehicles used (rows)	Number of maintenance vehicles (rows)	Number of vehicles allocated (rows)	On-line rate (%)
2022	98.92	2.98	57	7	67	89
2027	104.62	3.35	57	7	67	89

Table 3 The number of vehicles allocated to the extension of line M in the year of opening and the next 5 years

Years	Passenger flow prediction (unit: 10,000 person-times)	Maximum cross-section in peak hour (unit: 10,000 person-times)	Number of vehicles used in the model (rows)	Number of maintenance vehicles in the model (rows)	Number of spare vehicles (rows)	Number of vehicles allocated in the model (column)	On-line rate (%)
2022	98.92	2.98	55	5	2	62	91.94
2027	104.62	3.35	58	6	2	66	92.31

Table 4 Comparison of the surplus cost between the original number of vehicles and the model number of vehicles in extension of line M

Years	Number of vehicles allocated in the model (column)	Original number of vehicles allocated (column)	Number of surplus vehicles (column)	Cost of surplus vehicles (ten thousand yuan)
2022	62	67	5	50991.6
2027	66	67	1	10198.32

relevant costs based on the analysis in Chap. 4, the relevant surplus costs are shown in Table 4.

6 Conclusions

1. Firstly analyze the factors that affect the allocation of vehicles, including signal system, passenger flow conditions, maintenance mode, passenger flow organization and passenger flow development theory.
2. Analyze the model based on capacity and maintenance, comprehensively consider the length of the line, the speed of the train, the turn-back time of the train, and the interval between trains to establish the allocation model.

3. Analyze the cost of vehicle allocation, the cost of each vehicle allocation includes labor costs, material costs, maintenance costs, and depreciation costs.
4. By studying the key factors such as the maximum cross-section passenger flow in the peak hour of metro in the case city, the mixed running of long and short routing, the maximum nuclear passenger capacity of the vehicle, the minimum interval and the maintenance of vehicles, the appropriate model of the number of metro vehicles allocation is established. The number of vehicle allocation in the opening year and the next five years of the extension line in the case city is obtained. Compared with the original vehicle assignments, the number of vehicles allocation is reduced, the train on-line rate is improved, and the related cost is reduced. The validity and practicability of the model based on the capacity demand and vehicle maintenance are verified.

With the acceleration of urbanization, the scale of urban rail transit construction has grown rapidly, and interconnection has begun to be widely used. For urban rail transit, its interconnection refers to the safe operation of trains from different manufacturers in multiple lines or networks. Under the interconnected and networked operation mode, trains can not only travel on the specified line, but also can run across the line adjacent to the line. The operation of the train has become more diversified and organized. It is also more flexible. In the interconnection mode, the allocation of vehicles on each line can be more flexible. During peak times, according to the passenger flow, the cross-line call of trains between different lines can be studied to improve the on-line rate of trains and relieve the passenger flow pressure.

Acknowledgements This research is supported by National Key R&D Program of China under Grant No. 2017YFB1001801, "the Fundamental Research Funds for the Central Universities" (No. 30917012102), the Key Projects of International Scientific and Technological Innovation Cooperation between Governments (Grant No. 2016YFE0108000), the Natural Science Foundation of Jiangsu Province, China (Grant No. BK20171426) and the Opening Fund of Key Laboratory of Urban ITS Technology Optimization and Integration Ministry of Public Security, China (Grant No. 2017KFKT03). We would like to thank the reviewers for their valuable suggestions and comments.

References

1. Sergelen B (2019) Factors affecting rail reform model choice and cases comparison around the world. Beijing Jiaotong University
2. Zhao J (2000) Discussion on the calculation of the number of vehicles attached to the urban subway engineering. In: Metro and light rail, pp 45–46
3. Luo X, Wang Z, Lei Z (2012) Research on the calculation method of the number of metro train configurations and its application. Railw Tech Superv 40:51–53
4. Zhao W (2018) Calculation of the number of trains assigned to urban rail transit lines at the beginning and the short term. Urban Mass Transit Res 21:129–130, 135
5. Gong Y, Ji T, Xu S (2018) Study on the number of trains assigned to the extension of existing metro lines. Heilongjiang Transp Sci Technol 41(10):190–192
6. Peeters M, Kroon L (2006) Circulation of railway rolling stock: a branch-and-price approach. Comput Oper Res 35(2):538–556

7. Hong S-P et al (2008) A pragmatic algorithm for the train-set routing: the case of Korea high-speed railway. Omega 37(3):637–645
8. Maróti G, Kroon L (2005) Maintenance routing for train units: the transition model. Transp Sci 39(4)
9. Jani M (2016) A multidimensional examination of performances of HSR (high-speed rail) systems. J Mod Transp 24:1–21
10. Li H, Wang Y, Xu X, Qin L, Zhang H (2019) Short-term passenger flow prediction under passenger flow control using a dynamic radial basis function network. Appl Soft Comput J 83
11. Yang S, Wang Y, Xu Y, Wang M, Sun D, Li W (2018) A condition-based maintenance strategy considering opportunistic maintenance in multiple failure modes. IOP Conf Ser: Earth Environ Sci 186
12. Zhang N, Wang Z, Chen F, Song J, Wang J, Li Y (2020) Low-carbon impact of urban rail transit based on passenger demand forecast in Baoji. Energies 13(4)
13. Qian Q, Lin P (2016) Safety risk management of underground engineering in China: progress, challenges and strategies. J Rock Mech Geotech Eng 8(4):423–442
14. Xu Y, Jia B, Li M, Li X (2016) An improved discrete-time model for heterogeneous high-speed train traffic flow. Commun Theor Phys 65(3):381–392
15. Zhou M, Dong H, Ioannou PA, Zhao Y, Wang F-Y (2019) Guided crowd evacuation: approaches and challenges. IEEE/CAA J Autom Sin 6(5):1081–1094
16. Metro design specifications (with instructions):GB 50157-2013 (2013)
17. Jingwen X, Bingfeng S, Hongmeng C, Shiguang Z (2021) Research on hierarchical clustering method of urban rail transit passengers based on individual portrait. J Phys: Conf Ser 1883

Research on Traffic Flow Model Based on Lattice Hydrodynamics

Jie Yang and Tangyi Guo

Abstract The operating state of road traffic flow directly affects the stability of the entire urban transportation system. Once the urban transportation system becomes unstable, it will cause problems such as aggravation of traffic congestion. In the intelligent transportation environment, this paper establishes a lattice hydrodynamics model considering the density difference and the flow difference at the same time, adopts the linear stability theory, and conducts theoretical analysis and simulation research on the traffic flow characteristics through numerical simulation. The results show that the stability of traffic flow can be improved by considering both the density difference and the flow difference in the two-lane system.

Keywords Traffic flow model · Lattice hydrodynamics · Two-line · Numerical simulation

1 Introduction

Many problems such as frequent accidents with urban traffic congestion, increased vehicle exhaust emissions, smog and other issues have become an important bottleneck restricting urban economic development and improving people's living standards [1, 2]. Therefore, exploring the intrinsic characteristics of traffic flow and the causes of traffic congestion, and analyzing the mechanism of traffic congestion can help alleviate traffic congestion and provide a reliable scientific basis for actual traffic planning and traffic management [3, 4].

With the continuous development of autonomous driving technology, macroscopic traffic flow modeling in complex traffic scenes containing autonomous vehicles and artificially driven vehicles has become a research hotspot. Many scholars

J. Yang (✉) · T. Guo
School of Automation, Nanjing University of Science and Technology, 200 Xiaolingwei, Nanjing 210094, China
e-mail: yangjie930103@njust.edu.cn

T. Guo
e-mail: transtor@njust.edu.cn

W. Wang et al. (eds.), *Green Transportation and Low Carbon Mobility Safety*, Lecture Notes in Electrical Engineering 944,
https://doi.org/10.1007/978-981-19-5615-7_44

have proposed different traffic flow models for this purpose. Among them, the micro models include the car-following model [5, 6], the cellular automata model [7–9], and the mesoscopic model includes the Boltzmann model [10], etc., and the macroscopic lattice hydrodynamic model [11–13]. Kuang et al. [14] considered a mixed traffic flow model that considers the lateral expansion and contraction effect of the velocity space, and Ramanpreet and Sapna [15] proposed a lattice hydrodynamic model that considers a delayed feedback control method. In the macroscopic traffic flow dynamics modeling, the research focuses on the average speed information and average flow information feedback [16], but there are relatively few studies on the information feedback strategy of the combined effect of density difference and flow difference.

In this paper, based on the intelligent transportation environment, considering the density difference and flow difference, a lattice hydrodynamics model is established. The linear stability theory is used for analysis, and the traffic flow characteristics are simulated and analysed by numerical simulation. The results show that considering the density difference and the flow difference can improve the stability of the two-lane traffic flow.

2 Related Work

The lattice hydrodynamics model simplifies the motion equation of Payne's high-order continuum model, referring to the optimized speed function idea of the Bando optimized speed-following model, and derives the motion equation [17, 18]:

$$\partial_t(\rho v) = \alpha\rho_0 V(\rho(x+\delta)) - \alpha\rho v \tag{1}$$

where ρ_0 refers to the average density, α represents the driver's sensitivity coefficient, and δ is the average headway. Dimensionlessly transform the spatial variable x to obtain the basic lattice hydrodynamic model:

$$\begin{cases} \partial_t\rho_j + \rho_0(\rho_j v_j - \rho_{j-1}v_{j-1}) = 0 \\ \partial_t(\rho_j v_j) = \alpha\rho_0 V(\rho_{j+1}) - \alpha\rho_j v_j \end{cases} \tag{2}$$

where $V(\rho_{j+1})$ is the local density at $x+\delta$ at time t, j is the j-th grid on the grid point, ρ_j is the local density at j grid point at time t, and v_j is the local velocity at j grid point at time t.

In Fig. 1, take the grid point j on lane 1 as an example, the grid point j on lane 1 has both the flow from the grid point to the grid point $j+1$ on lane 2 and the flow from the grid point $j-1$ on lane 2 flow. In the two-lane system, the density of the grid point j on lane 1 at each time after the lane change is completed is $\gamma|\rho_0^2 V'(\rho_0)|(\rho_{2,j+1}(t) - 2\rho_{1,j}(t) + \rho_{2,j-1}(t))$.

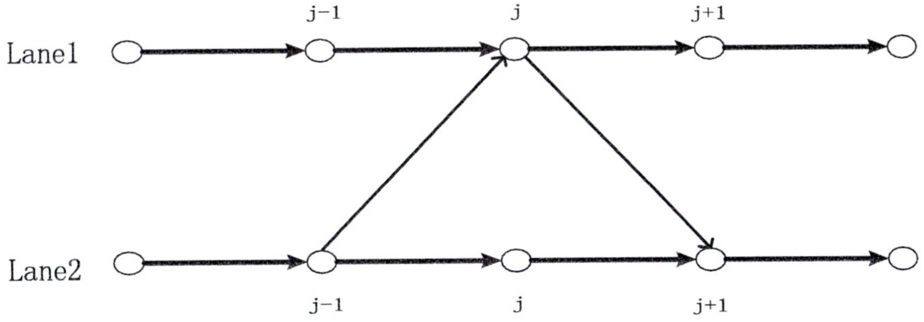

Fig. 1 Schematic diagram of two-lane traffic flow model

Conservation equation on lane 1:

$$\begin{aligned}&\partial_t\rho_{1,j}(t)+\rho_0(\rho_{1,j}(t)v_{1,j}(t)-\rho_{1,j-1}(t)v_{1,j-1}(t))\\&=\gamma|\rho_0^2V'(\rho_0)|(\rho_{2,j+1}(t)-2\rho_{1,j}(t)+\rho_{2,j-1}(t))\end{aligned}\quad(3)$$

Conservation equation on lane 2:

$$\begin{aligned}&\partial_t\rho_{2,j}(t)+\rho_0(\rho_{2,j}(t)v_{2,j}(t)-\rho_{2,j-1}(t)v_{2,j-1}(t))\\&=\gamma|\rho_0^2V'(\rho_0)|(\rho_{1,j+1}(t)-2\rho_{2,j}(t)+\rho_{1,j-1}(t))\end{aligned}\quad(4)$$

Add formulas (3) and (4) to get the continuity equation of the two-lane:

$$\partial_t\rho_j+\rho_0(\rho_jv_j-\rho_{j-1}v_{j-1})=\gamma|\rho_0^2V'(\rho_0)|(\rho_{j+1}-2\rho_j+\rho_{j-1})\quad(5)$$

Assuming that the evolution equation has nothing to do with lane changing, the evolution equation of the two-lane traffic flow model can be obtained:

$$\partial_t(\rho_jv_j)=\alpha\rho_0V_e(\rho_{j+1})-\alpha\rho_jv_j\quad(6)$$

where:

$$\begin{aligned}&\rho_j=(\rho_{1,j}+\rho_{2,j})/2\\&\rho_jv_j=(\rho_{1,j}v_{,j}+\rho_{2,j}v_{2,j})/2\\&V_e(\rho_j)=(V(\rho_{1,j})+V(\rho_{2,j}))/2\end{aligned}\quad(7)$$

3 Methodology

Based on Nagatani's two-lane grid hydrodynamics model, considering the effects of the density difference and flow difference between the current grid and the previous neighboring grid, a new evolution equation is constructed:

$$\partial_t(q_j) = \alpha\rho_0 V_e(\rho_{j+1}) - \alpha q_j + \lambda(\rho_j - \rho_{j+1})/\rho_0 + \beta(q_{j+1} - q_j) \tag{8}$$

$$\partial_t\rho_j + \rho_0(q_j - q_{j-1}) = \gamma|\rho_0^2 V'(\rho_0)|(\rho_{j+1} - 2\rho_j + \rho_{j-1}) \tag{9}$$

where $q_j = \rho_j v_j$ represents the flow of grid point j. λ and β respectively represent the reaction coefficients of the density difference and flow difference between the current grid and the previous grid, and the value interval is [0, 1].

Eliminate the flow term, and get the following equation:

$$\begin{aligned}&\partial_t^2(\rho_j) + \alpha\rho_0^2[V_e(\rho_{j+1}) - V_e(\rho_j)] - [(\alpha+\beta)\gamma|\rho_0^2 V'(\rho_0)| + \lambda]\\&(\rho_{j+1} - 2\rho_j + \rho_{j-1}) - \beta\partial_t(\rho_{j+1}) + (\alpha+\beta)\partial_t(\rho_j) + \beta\gamma(\rho_{j+2} - 2\rho_{j+1} + \rho_j)\\&- \gamma|\rho_0^2 V'(\rho_0)|(\partial_t(\rho_{j+1}) - 2\partial_t(\rho_j) + \partial_t(\rho_{j-1})) = 0\end{aligned} \tag{10}$$

where $V(\rho)$ is the optimal velocity function, the equation is as follows:

$$V(\rho) = (v_{\max}/2)[\tanh(2/\rho_0 - \rho/\rho_0^2 - h_c) + \tanh(h_c)] \tag{11}$$

where $v_{\max} = 2$ is the maximum speed, $h_c = 1/\rho_c$, $\rho_c = 0.25$ is the inflection point of the optimal speed function.

4 Linear Stability Analysis

Assuming that under the initial conditions, the system is in a stable state. The initial density is constant ρ_0, and the optimal speed is $V(\rho_0)$, that is, the stable state solution is: $\rho_j(t) = \rho_0$, $V_j(t) = V(\rho_0)$. Add a small perturbation $y_j(t)$ after the steady-state solution to make the density of the grid point j be $\rho_j(t) = \rho_0 + y_j(t)$. Substituting it into Eq. (10), we get after linearization:

$$\begin{aligned}&\partial_t^2(y_j) + \alpha\rho_0^2[V'(\rho_0)(y_{j+1} - y_j)] - [(\alpha+\beta)\gamma|\rho_0^2 V'(\rho_0)| + \lambda]\\&(y_{j+1} - 2y_j + y_{j-1}) - \beta\partial_t(y_{j+1}) + (\alpha+\beta)\partial_t(y_j) + \beta\gamma(y_{j+2} - 2y_{j+1} + y_j)\\&- \gamma|\rho_0^2 V'(\rho_0)|(\partial_t(y_{j+1}) - 2\partial_t(y_j) + \partial_t(y_{j-1})) = 0\end{aligned} \tag{12}$$

where: $V'(\rho_0) = dV(\rho)/d\rho|_{\rho=\rho_0}$, $V(\rho_j) = V(\rho_0) + V'(\rho_0)y_j(t)$, $V(\rho_{j+1}) = V(\rho_0) + V'(\rho_0)y_{j+1}(t)$.

Expand $y_j(t)$ according to Fourier series, where $y_j(t)=e^{ikj+zt}$, $y_{j+1}=e^{ik(j+1)+zt}$:

$$\partial_t(y_j) = ze^{ikj+zt},\ \partial_t^2(y_j) = z^2e^{ikj+zt} \tag{13}$$

$$\partial_t(y_{j+2}) = ze^{ik(j+2)+zt},\ \partial_t(y_{j+1}) = ze^{ik(j+1)+zt},\ \partial_t(y_{j-1}) = z^2e^{ik(j-1)+zt} \tag{14}$$

Substitute formulas (17) and (18) into formula (16):

$$\begin{aligned} &z^2 + \alpha\rho_0^2V'(\rho_0)(e^{ik}-1) - \left[(\alpha+\beta)\gamma|\rho_0^2V'(\rho_0)|+\lambda\right](e^{ik}+e^{-ik}-2) - \beta ze^{ik} \\ &+(\alpha+\beta)z + \beta\gamma(e^{i2k}-2e^{ik}+1) - \gamma|\rho_0^2V'(\rho_0)|z(e^{ik}+e^{-ik}-2) = 0 \end{aligned} \tag{15}$$

When $ik \to 0, z \to 0$, the expansion of z in Eq. (17) can be obtained: $z = ikz_1 + (ik)^2z_2 + \cdots$, and the higher-order terms above the second order are ignored, and they are substituted into Eq. (17). The real and imaginary parts in Eq. (17) respectively equal to zero, the results are as follows:

$$\begin{aligned} &k^4z_2^2 - k^2z_1^2 + \left[(\alpha+\beta)\gamma|\rho_0^2V'(\rho_0)|+\lambda\right](\cos k - 1) - k^2z_2(\cos k + \sin k) \\ &-k^2(\alpha+\beta)z_2 + \beta\gamma(\cos 2k - 2\cos k + 1) + 2k^2\gamma|\rho_0^2V'(\rho_0)|z_2(\cos k - 1) = 0 \end{aligned} \tag{16}$$

$$\begin{aligned} &-2k^3z_1z_2 + \left[(\alpha+\beta)\gamma|\rho_0^2V'(\rho_0)|+\lambda\right]\sin k + \beta kz_1(\cos k + \sin k) \\ &k(\alpha+\beta)z_1 + \beta\gamma(\sin 2k - 2\sin k) - 2k\gamma|\rho_0^2V'(\rho_0)|(\cos k - 1)z_1 = 0 \end{aligned} \tag{17}$$

When λ=0, obtain the first-order term coefficient and the second-order term coefficient in the expression:

$$z_1 = -\rho_0^2V'(\rho_0) \tag{18}$$

$$z_2 = 1/\alpha(-\rho_0^4V'(\rho_0))^2) - 0.5\rho_0^2V'(\rho_0) + \alpha\gamma|\rho_0^2V'(\rho_0)| + \lambda \tag{19}$$

If z_2 is a negative value, the uniform traffic flow will become an unstable traffic flow. The neutral and stable condition can be given by z_2=0:

$$\alpha > -\frac{2\rho_0^2V'(\rho_0)}{1+2\gamma} + \frac{1}{\rho_0^2V'(\rho_0)}\frac{2\lambda}{1+2\lambda} \tag{20}$$

When β=0, obtain the first-order term coefficient and the second-order term coefficient in the expression:

$$z_1 = -\rho_0^2V'(\rho_0) \tag{21}$$

$$z_2 = 1/2(1+2\gamma) + 1/2\alpha\left(\lambda - 2z_1^2\right) \tag{22}$$

According to the stability analysis theory, if z_2 is a negative value, the steady-state solution to the uniform distribution of the long-wave mode will tend to be unstable. On the contrary, it maintains a stable uniform flow state. Get the neutral stability condition of the new model:

$$\alpha > -\frac{2\rho_0^2 V'(\rho_0) + 2\lambda}{1 + 2\gamma} \tag{23}$$

From the above stability conditions, it can be seen that studying the effects of flow difference and density difference can improve the stability of traffic flow.

5 Numerical Simulation

For the convenience of simulation, formula (10) is rewritten into a differential form:

$$\begin{aligned}
&\rho_j(t + 2\tau) = 2\rho_j(t + \tau) - \rho_j(t) - \left[(\alpha+\beta)\gamma\rho_0^2 V'(\rho_0)+\lambda\right] \\
&\tau^2(\rho_{j+1}(t) - 2\rho_j(t) + \rho_{j-1}(t)) - (\alpha + \beta)\tau(\rho_j(t + \tau) - \rho_j(t)) \\
&- \beta\tau^2(\rho_{j+2}(t) - 2\rho_{j+1}(t) + \rho_j(t)) - \beta\tau(\rho_{j+1}(t + \tau) - \rho_{j+1}(t)) \\
&- \tau\gamma\rho_0^2 V'(\rho_0)(\rho_{j+1}(t + \tau) - \rho_{j+1}(t) \\
&- 2\rho_j(t + \tau) + 2\rho_j(t) + \rho_{j-1}(t + \tau) - \rho_{j-1}(t)) \\
&- \alpha\tau^2\rho_0^2(V(\rho_{j+1}(t)) - V(\rho_j(t)))
\end{aligned} \tag{24}$$

Under periodic boundary conditions, the initial conditions of the numerical simulation are set as follows: the number of grids $N = 100$, the initial density distribution is a piecewise function:

$$\rho_j(0) = \rho_0 = 0.25,\ j \in [1, N] \tag{25}$$

$$\rho_j(1) = \begin{cases} \rho_0\,;\ 1 \le j < 49,\ 50 < j \le N \\ \rho_0 + 0.05\,;\ j = 49 \\ \rho_0 - 0.05\,;\ j = 50 \end{cases} \tag{26}$$

Given the sensitivity coefficient value α=1.6 during the simulation process, the modes a, b, and c are all in the unstable region. Disturbance in the three modes will evolve more and more with time, and eventually the system will evolve into blocking traffic flow. Figure 2 shows that, unlike the first three modes, mode d is in a stable state. In this mode, the disturbance will gradually disappear over time, and eventually the traffic flow system will return to the original stable state.

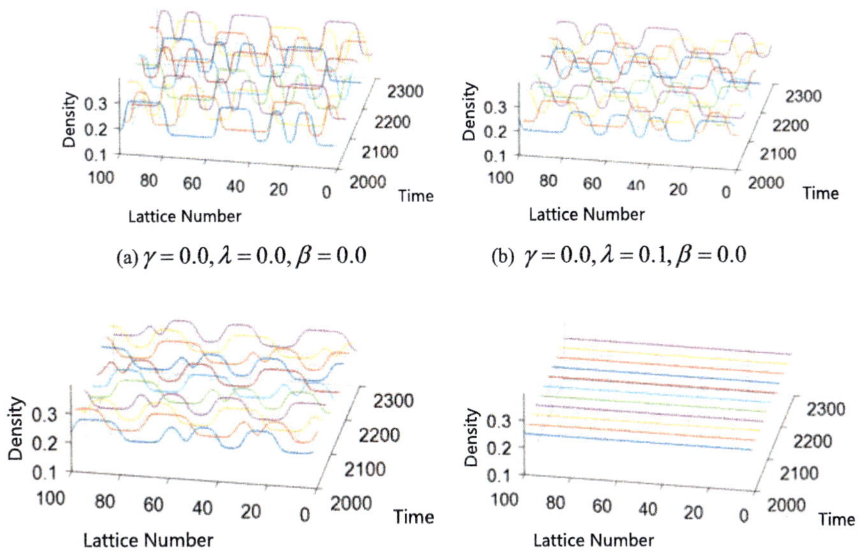

(a) $\gamma = 0.0, \lambda = 0.0, \beta = 0.0$

(b) $\gamma = 0.0, \lambda = 0.1, \beta = 0.0$

(c) $\gamma = 0.1, \lambda = 0.0, \beta = 0.0$

(d) $\gamma = 0.1, \lambda = 0.1, \beta = 0.0$

Fig. 2 Spatiotemporal evolution of density wave with given $\alpha = 1.6$ after $t = 2000$ s

Figure 3 shows the density distribution diagram corresponding to Fig. 2 when $t = 2000$ s. Based on the above theoretical analysis and simulation research, considering the effect of density difference and lane change rate can improve the stability of traffic flow.

Figure 4 shows the temporal and spatial evolution of the density wave after a long enough time without changing lanes ($\gamma = 0$). The four traffic flow patterns in the figure are all unstable traffic flows. Although the four modes are all unstable modes, with the increase of the flow difference response intensity coefficient, the amplitude of the density wave fluctuation gradually decreases. When the flow difference response intensity coefficient β increases to 0.5, the fluctuation amplitude of the disturbance decreases significantly. It shows that even if the system is in an unstable state, considering the effect of poor flow can improve the stability of traffic flow and help alleviate traffic congestion.

Figure 5 shows the density distribution diagram of the system when the flow difference response intensity coefficients take different values without changing lanes at $t = 2000$ s.

Figure 6 corresponds to the density distribution diagram when $\gamma = 0.1$ and $t = 2000$ s. Through the comparative analysis of Figs. 5 and 6, it can be clearly seen that the amplitude of the density wave is significantly reduced when a certain lane change rate is considered, and it is easier to reach a stable state than when the lane change rate is not considered.

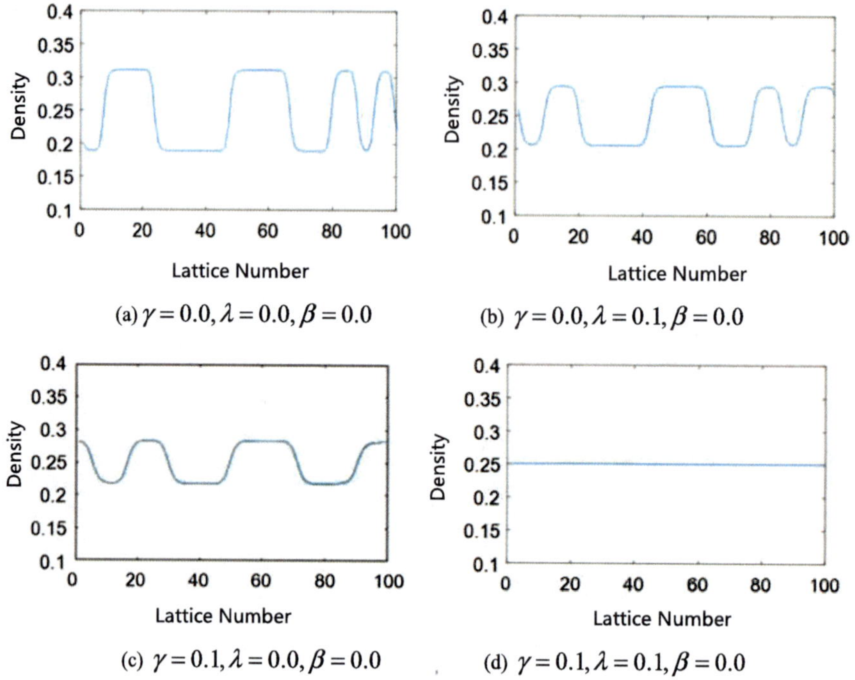

(a) $\gamma = 0.0, \lambda = 0.0, \beta = 0.0$ (b) $\gamma = 0.0, \lambda = 0.1, \beta = 0.0$

(c) $\gamma = 0.1, \lambda = 0.0, \beta = 0.0$ (d) $\gamma = 0.1, \lambda = 0.1, \beta = 0.0$

Fig. 3 Density profile with given α = 1.6 after t = 2000 s

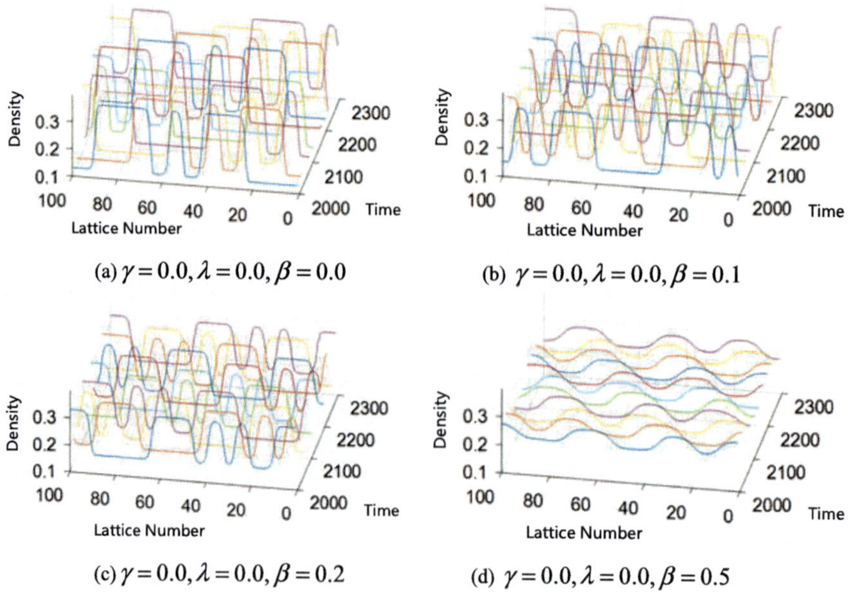

(a) $\gamma = 0.0, \lambda = 0.0, \beta = 0.0$ (b) $\gamma = 0.0, \lambda = 0.0, \beta = 0.1$

(c) $\gamma = 0.0, \lambda = 0.0, \beta = 0.2$ (d) $\gamma = 0.0, \lambda = 0.0, \beta = 0.5$

Fig. 4 Spatiotemporal evolution of density wave with given α = 1.6 after t = 2000 s

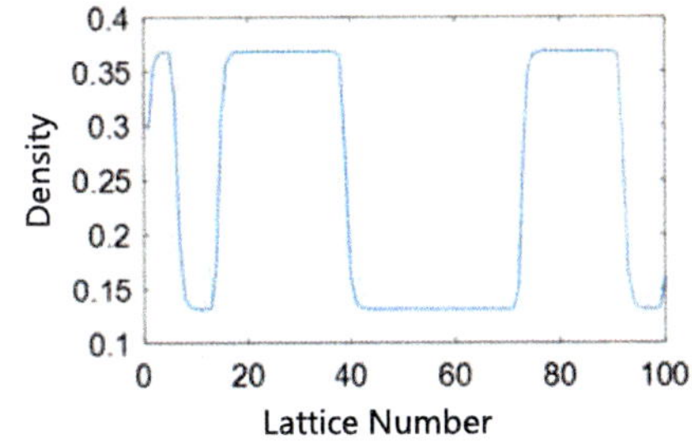

(a) $\gamma = 0.0, \lambda = 0.0, \beta = 0.0$

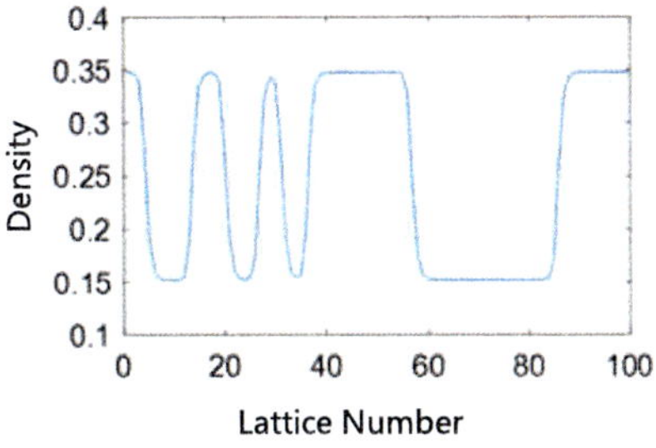

(b) $\gamma = 0.0, \lambda = 0.0, \beta = 0.1$

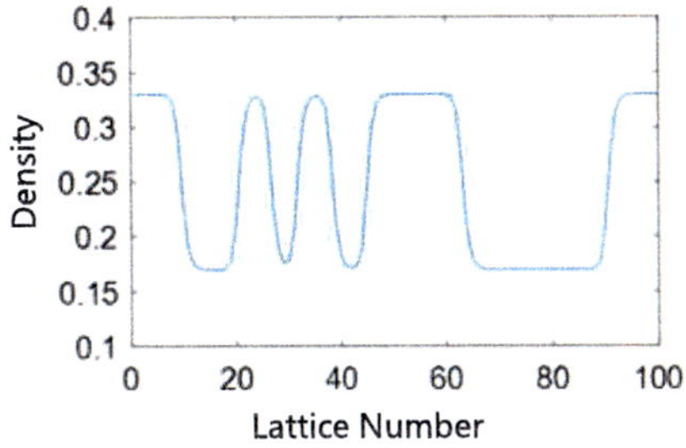

(c) $\gamma = 0.0, \lambda = 0.0, \beta = 0.2$

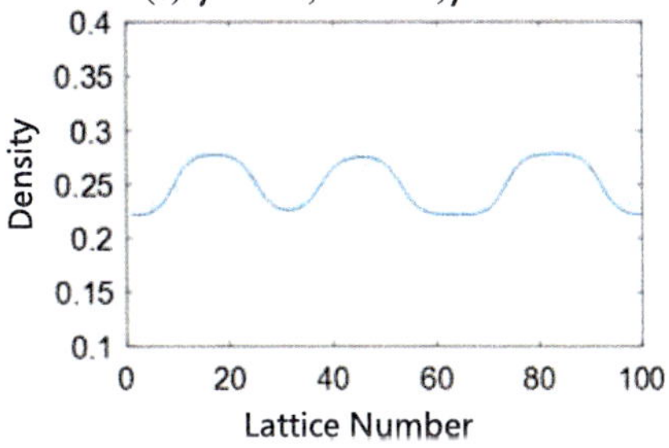

(d) $\gamma = 0.0, \lambda = 0.0, \beta = 0.5$

Fig. 5 Density profile with given $\alpha = 1.6$ after $t = 2000$ s

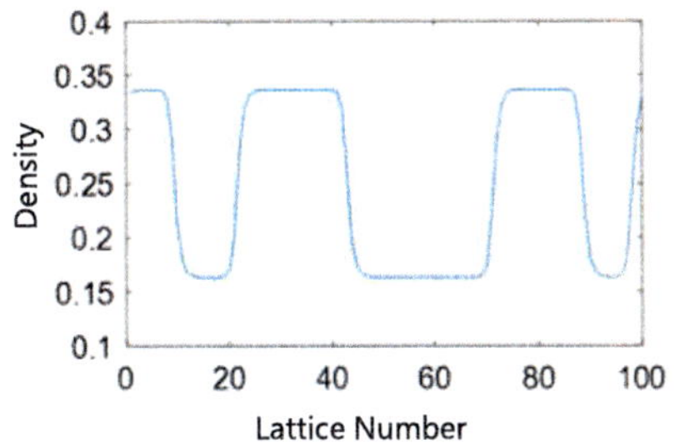

(a) $\gamma = 0.1, \lambda = 0.0, \beta = 0.0$

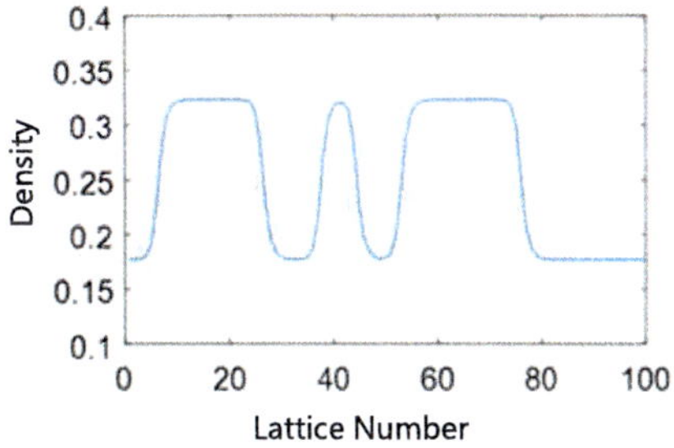

(b) $\gamma = 0.1, \lambda = 0.0, \beta = 0.1$

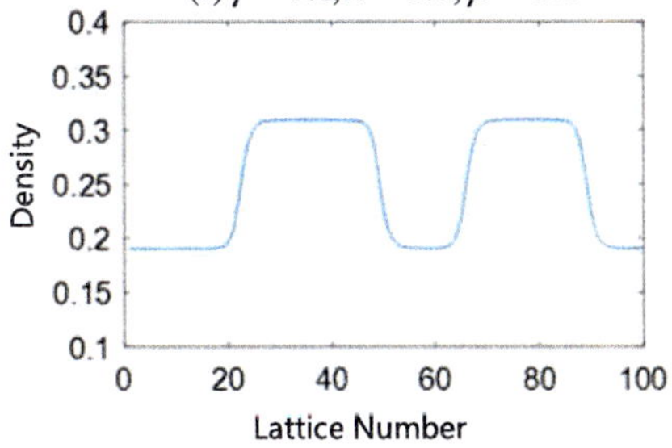

(c) $\gamma = 0.1, \lambda = 0.0, \beta = 0.2$

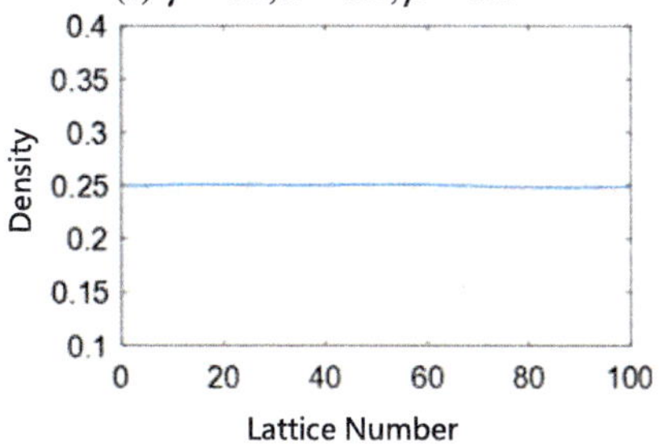

(d) $\gamma = 0.1, \lambda = 0.0, \beta = 0.5$

Fig. 6 Density profile with given $\alpha = 1.6$ after $t = 2000$ s

6 Conclusions

Based on linear stability analysis theory, analyze traffic flow conditions. The linear stability condition shows that considering the effects of density difference and flow difference can improve the stability of two-lane traffic flow. The numerical simulation further verified the theoretical analysis results, and the simulation results are consistent with the theoretical analysis. In the model that simultaneously considers the density difference, the flow difference and the lane change, it is found that the system eventually evolves into a uniform and stable traffic flow state. Based on the widespread application of intelligent transportation systems, drivers can obtain effective information (such as average flow information) of the road ahead more accurately in real time. This research can provide corresponding theoretical support for the feedback and control of average flow information of intelligent transportation.

Since this thesis is mainly a theoretical research, it is necessary to carry out actual survey research on typical road sections in the future.

References

1. Qin Y, Wang H, Wang W (2018) LWR model for mixed traffic flow in connected and autonomous vehicular environments. China J Highw Transp 31(11):151
2. Tian JF, Yuan ZZ, Jia B, Wang T (2013) Dynamic congested traffic states of density difference lattice hydrodynamic model with on—ramp. Discrete Dyn Nat Soc 941238:1–9
3. Chowdhury D et al (2000) Statistical physics of vehicular traffic and some related systems. Phys Rep
4. Nagatani T (2002) The physics of traffic jams. Rep Prog Phys 65:1331
5. Tang TQ, Shi WF, Shang HY, Wang YP (2014) An extended car-following model with consideration of the reliability of inter-vehicle communication. Measurement 58:286–293
6. Cao BG (2015) A new car-following model considering driver's sensory memory. Phys A 427:218–225
7. Zeng J, Qian Y, Mi P et al (2021) Freeway traffic flow cellular automata model based on mean velocity feedback. Phys A 562(15):125387
8. Deng JH, Feng HH (2019) A multilane cellular automaton multi-attribute lane-changing decision model. Phys A 529:121545
9. Krzysztof M (2018) A computer simulation of traffic flow with on-street parking and drivers' behaviour based on cellular automata and a multi-agent system. J Comput Sci 28:32–42
10. Kemer BS (2016) Failure of classical traffic flow theories: stochastic highway capacity and automatic driving. Phys A 450:700–747
11. Gupta AK, Sharma S, Redhu P (2015) Effect of multi-phase optimal velocity function on jamming transition in a lattice hydrodynamic model with passing. Nonlinear Dyn 80:1091–1108
12. Sharma S (2015) Lattice hydrodynamic modeling of two-lane traffic flow with timid and aggressive driving behavior. Phys A 421(1):401–411
13. Wang Q, Cheng R, Ge H (2019) A new lattice hydrodynamic model accounting for the traffic interruption probability on a gradient highway. Phys Lett A
14. Kuang C, Wu Y, Liu P, Zhang J, Chen Y A hybrid traffic flow model considering lateral expansion effect in velocity space. J Highw Traffic Sci 38(7):114–123
15. Ramanpreet K, Sapna S (2018) Analyses of lattice hydrodynamic model using delayed feedback control with passing. Phys A 510:446–455

16. Ge HX, Zheng PJ, Lo SM et al (2014) TDGL equation in lattice hydrodynamic model considering driver's physical delay. Nonlinear Dyn 76:441–445
17. Zhu HB, Zhang NX, Wu WJ (2015) A modified two-lane traffic model considering drivers' personality. Phys A 428:359–367
18. Zhang M, Sun DH, Tian C (2014) An extended two-lane traffic flow lattice model with driver's delay time. Nonlinear Dyn 77:839–847

Regionally Differentiated Real-Time Energy Consumption Prediction of Electric Vehicles Oriented to Travel Characteristics

Cheng Wang, Ya-nan Wang, Ji-yuan Tan, Fu-yu Liu, Yuan-yuan Jiang, and Zhen-po Wang

Abstract Real-time prediction of electric vehicle energy consumption is of great significance to users' travel planning and charging decisions. This paper analyzed the influence of travel characteristics and regional differences on the power consumption of electric vehicles, and built a regional electric vehicle energy consumption model based on travel characteristics prediction: In this paper, a large number of travel samples are obtained by preprocessing the real-time operation data of electric vehicles, and the influencing factors of power consumption in the travel samples are analyzed to determine that the most relevant characteristic parameters are travel mileage and time, which are used as the main characteristic indicators of energy consumption prediction. On this basis, a single-region BP neural network energy consumption prediction model was built, and the optimal network model structure was adjusted and determined through error feedback, which achieved a prediction accuracy of 93.2%; then, the travel samples of different cities are modeled and cross predicted, and established a multi-regional energy consumption prediction model; finally, the prediction results of different models are compared. The results show that this model has the highest accuracy in the energy consumption prediction of the actual operation of urban electric vehicles, which can reach 92% and above. Combining the existing electricity with the predicted energy consumption results can provide effective support for users to make reasonable charging decisions before travel.

C. Wang · Z. Wang
National Engineering Laboratory for Electric Vehicles, School of Machinery and Vehicles, Beijing Institute of Technology, Beijing 100144, China

C. Wang
Beijing Catarc Technology Center Co., Ltd., Beijing 100070, China

Y.-n. Wang · J. Tan (✉) · F. Liu
Beijing Key Lab of Urban Intelligent Traffic Control Technology, North China University of Technology, Beijing 100144, China
e-mail: asderytqaqwqer@foxmail.com

Y. Jiang
Beijing Catarc Automotive Testing Technology Center Co., Ltd., Beijing 100176, China

W. Wang et al. (eds.), *Green Transportation and Low Carbon Mobility Safety*, Lecture Notes in Electrical Engineering 944,
https://doi.org/10.1007/978-981-19-5615-7_45

Keywords Road transportation · Energy consumption prediction · BP neural network · Electric vehicles · Travel characteristics · Regional differences

1 Introduction

Electric vehicles have attracted much attention due to the advantages of low carbon, environmental protection, low noise and low cost of use. However, due to the constraints of battery capacity, the limited driving range restricts their own development [1]. If the energy consumption of the journey can be obtained in advance before the trip, it will help to relieve the driver's mileage anxiety and improve the travel experience. Therefore, how to solve the problem of energy consumption prediction before the trip of electric vehicles is worth exploring. Considering that most modern users rely on GPS positioning and map navigation to plan travel routes, and the characteristics of travel time and mileage can be obtained before travel, this paper takes this as an entry point to study the real-time prediction of electric vehicle travel energy consumption, which can provide users with the energy consumption prediction results of the trip before departure, so that users can make reasonable charging decisions and travel planning.

According to the basic data sources, the existing research on electric vehicle energy consumption prediction can be divided into two types of prediction methods based on experimental simulation and based on actual operating data.

(1) Research on energy consumption prediction based on simulation data

Due to the limitation of research data sources, some scholars start from the perspective of simulation to predict and calculate the energy consumption of electric vehicles, which has better data controllability and calculation accuracy.

Reference [2] developed an artificial neural network-based battery model to predict SOC based on the measured current and voltage. Reference [3] used the short-stroke method to construct the actual operating conditions of electric commercial vehicles, and conducted energy consumption simulations that fit the actual operating scenarios of commercial vehicles. Since the constructed driving conditions only include vehicle speed parameters, the simulation accuracy of the real road scene needs to be further improved. Reference [4] explored the influence of transmission efficiency optimization on vehicle energy consumption by means of simulation methods, and provided theoretical guidance for effectively reducing the energy consumption of electric vehicles. Reference [5] considered the influence of intersection signal lights and energy recovery on energy consumption, and proposed an improved algorithm for the optimal energy consumption of electric vehicles, which can reasonably estimate the energy consumption at the intersection according to the traffic density at the intersection. The headway in the traffic flow distribution in different time periods is quite different, which will affect the estimation accuracy. Reference [6] made further detailed research, considering the energy consumption caused by the horizontal and vertical slippage of electric vehicles, built a vehicle energy consumption model of four-wheel drive electric vehicles, and used the simulink simulation platform for simulation verification. Reference [7] used the bench method to simulate the driving conditions of electric vehicles, and collected data such as the motor speed and torque of the vehicle, the current and voltage of the battery pack through experiments, and calculated the energy consumption of electric vehicles per 100 km based on the dynamic principle. On this basis, some scholars have supplemented the influence of operation route, terrain level, traffic conditions and passenger capacity on the energy consumption of electric passenger vehicles to establish an energy consumption model [8]. The energy consumption prediction model with the parameters of vehicle driving conditions as input features mainly involves speed, acceleration and power parameters to estimate energy consumption. Reference [9] comprehensively analyzed the influence of weather, social and road network characteristics on vehicle speed, and used long short-term memory (S-LSTM) neural network to construct an electric vehicle energy consumption model based on average speed prediction. Reference [10] supplemented the influence of acceleration on energy consumption, and proposed a model framework for estimating energy consumption using speed and acceleration by analyzing the laws from floating car data. Reference [11] processed the speed of vehicle operating conditions with time series to obtain acceleration time series, calculated the corresponding probability of speed and acceleration state, and established an electric vehicle energy consumption model based on VA (speed-acceleration) probability distribution. Literature [12] proposed an electric vehicle power estimation model by analyzing the relationship between vehicle power, speed, acceleration and road gradient to estimate energy consumption. Reference [13] proposed a power-based energy consumption model for electric vehicles based on previous research results on electric passenger vehicles, taking into account the actual road conditions and driving parameters of electric trucks. Such methods of estimating energy consumption based on vehicle speed, acceleration and power models will generate secondary calculation errors, and there are many uncertain factors affecting the prediction accuracy.

(2) Research on energy consumption prediction based on actual travel data

Energy consumption prediction based on actual travel data includes two types of methods: using post-trip data to build empirical models and using real-time data during travel to build models. The research on the former is relatively mature, but the latter is relatively insufficient. The former collects the actual running data of the vehicle, analyzes the influencing factors and mechanism of driving energy consumption, combines the vehicle dynamics principle and energy consumption characteristics, and integrates engine parameters (speed, torque, etc.), operating condition parameters (speed, acceleration, power, etc.) and external environmental characteristics (road slope, elevation and weather temperature, etc.) to establish energy consumption prediction models. The latter obtains real-time data during vehicle driving according to the on-board equipment and inputs it into a corresponding model or algorithm for calculation to obtain the real-time energy consumption of the vehicle.

Reference [14] built a temperature-based energy consumption model by analyzing the collected electric taxi operation data and using machine learning methods, and the prediction accuracy was higher than that of the traditional prediction method. Reference [15] analyzed the battery operation data stored in the battery database of a battery monitoring cloud platform, and proposed to use the support vector machine and Bayesian optimization method to predict the SOC of electric buses autonomously. Reference [16] used the test data of pure electric vehicles to identify vehicle energy consumption parameters based on the recursive least squares algorithm, and predicted vehicle energy consumption in combination with driving conditions. Reference [17] used on-board sensors to return three parameters of vehicle speed, traction and road elevation during vehicle operation, and used a deep convolutional neural network to establish an energy consumption prediction model to achieve real-time prediction of electric vehicle energy consumption. Reference [18] analyzed the relationship between different driving characteristics and power consumption, used principal component analysis to reduce the dimensionality of the features, and established a real-time energy consumption model for electric vehicles. It is difficult to obtain data during driving, and it is impossible to provide users with the required energy consumption prediction results before traveling. Although the above research has achieved real-time energy consumption prediction during vehicle driving, it is difficult to obtain data during vehicle driving and cannot provide users with the required energy consumption before travel. Modern travelers usually use navigation software to complete route planning and selection before and during travel, so as to complete travel tasks quickly and efficiently. For electric vehicle users, they should pay more attention to the energy consumption required for the trip before traveling.

Considering the characteristics of the close combination of modern travel and map navigation technology, this study uses the two main travel characteristics of travel mileage and time, which are easily obtained by users before travel, as input to establish a BP neural network-based energy consumption prediction model. This realizes the real-time prediction of the energy consumption required for future travel of electric vehicles, and provides a reference for users to make reasonable charging decisions and travel planning in combination with the existing power and energy consumption predictions before travel. The innovative significance of this method is mainly reflected in several aspects:

(1) It is closely integrated with the travel mode of modern users, and the travel mileage and time obtained before travel can be used to predict the energy consumption of the trip, and provide users with a basis for charging decision-making before travel;
(2) When the historical travel samples are known, the energy consumption prediction model can be trained, and the travel mileage and time parameters can be input to realize the real-time prediction of electric vehicle energy consumption;
(3) Considering the impact of urban differences in different regions on the energy consumption of electric vehicles, establish electric vehicle energy consumption prediction models for different regions.

Table 1 Data information table

Field	Representing meaning
daq_time	Data collection time
mileage	The accumulated mileage of the vehicle at the current moment, km
speed	The current speed of the vehicle, km/h
soc	The ratio of remaining power to battery capacity (0–100%)
status	1: startup state; 2: flameout state
c_stat	1: parking and charging; 2: driving and charging; 3: not charged; 4: finished charging

2 Data Introduction and Processing

2.1 *Data Introduction*

The data comes from the real-time data of electric vehicle operation, including running time, mileage, vehicle speed, state of charge (SOC) and other information, as shown in Table 1. The sampling frequency varies between 1 and 0.1 Hz. This paper screened the data of 75 pure electric vehicles of the same model, and the travel range covered 6 cities of different scales.

2.2 *Data Processing*

2.2.1 Data Preprocessing

There are time redundancy and frequency inconsistency in the data transmission process. First, sort the data by time and delete data with duplicate timestamps. Resampling is carried out for the two sampling frequency data of 1 and 0.1 Hz, and the unified frequency is 0.1 Hz, and the travel itinerary is divided on the basis of the above data preprocessing. Take the actual running data of a vehicle for one month as an example for preprocessing, and the statistical results of the original and preprocessed data are shown in Table 2.

Table 2 Data statistics table before and after preprocessing

Data	Sampling frequency (Hz)	Number of rows of data
Original data	1	751,450
Preprocessed data	0.1	118,420

2.2.2 Travel Itinerary Division

The vehicle operation data includes two states: start-up operation and stop and flameout. A trip may include a short-term shutdown state, such as urban road signal control, short-term parking caused by taxi pick-up and drop-off, etc. should be regarded as the same trip, so this paper sets the maximum flameout and parking time during travel as 30 min. The specific division process is shown in Fig. 1.

When the vehicle is parked and turned off, it includes charging and non-charging. When the vehicle is in the flame-off state, if it is in the flame-off charging state, the two trips before and after the flame-off charging are divided into two separate trips; If it is in the flameout and uncharged state, it needs to be judged according to the time interval between the start time of flameout and the time of the next start state. If the parking interval is within 30 min, it will be regarded as the same trip, otherwise, it will be regarded as two trips before and after. Figure 2 shows the results of trip division of a certain vehicle within a period of time. As shown in the figure, the first trip includes a parking time of about three minutes in the middle, and the driving after half an hour of parking is divided into the second trip.

Taking the actual running data of a certain vehicle as an example, the data of 57 trips, 42 parking times and 25 charging times are obtained through trip division. The distribution of travel time and travel mileage is shown in Fig. 3. After the above processing of all vehicle data, a total of 3775 trip samples were obtained, and all samples were marked and stored by city. The processed data adds information such as travel start time, travel end time, accumulated mileage, and soc. See Table 3 for details.

3 Analysis of Influencing Factors of Electric Vehicle Energy Consumption

The energy consumption of electric vehicles is affected by travel time, distance, working conditions and temperature. Since the research data is the actual operation data of the vehicle, this paper focuses on the impact of travel characteristics on energy consumption. The travel characteristics of users operating in urban areas with different scales, structural layouts, and road network congestion are also different. Therefore, regional differences are also an important factor affecting energy consumption. Based on the above analysis, the influence of travel characteristics and regional differences on the travel energy consumption of electric vehicles was studied.

Fig. 1 Travel division process. (*Note* trip num indicates the travel itinerary number)

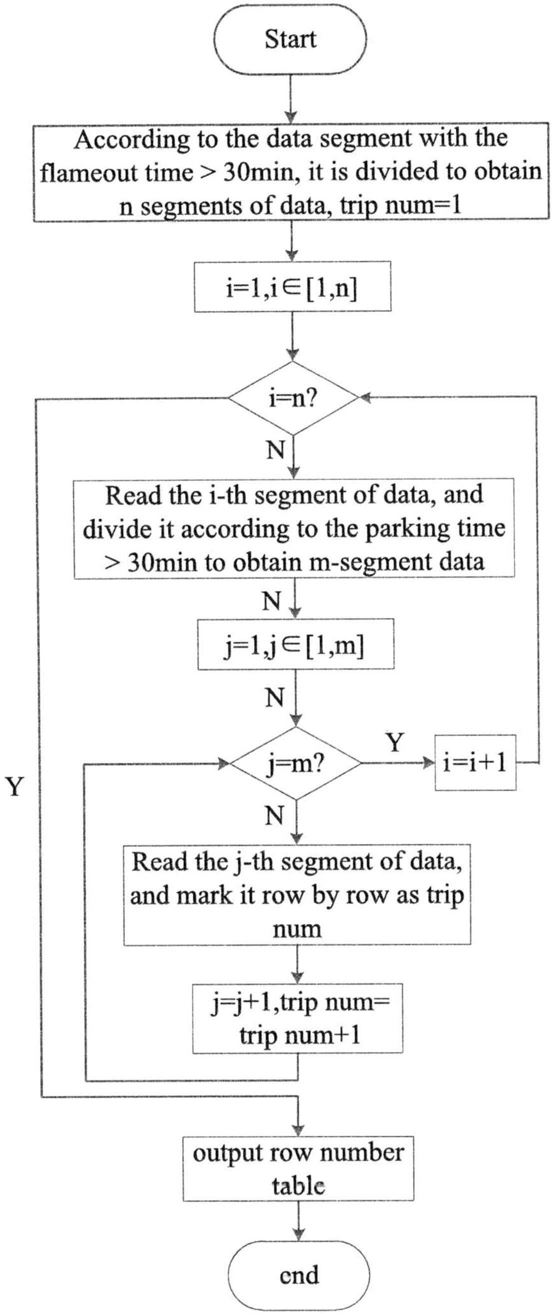

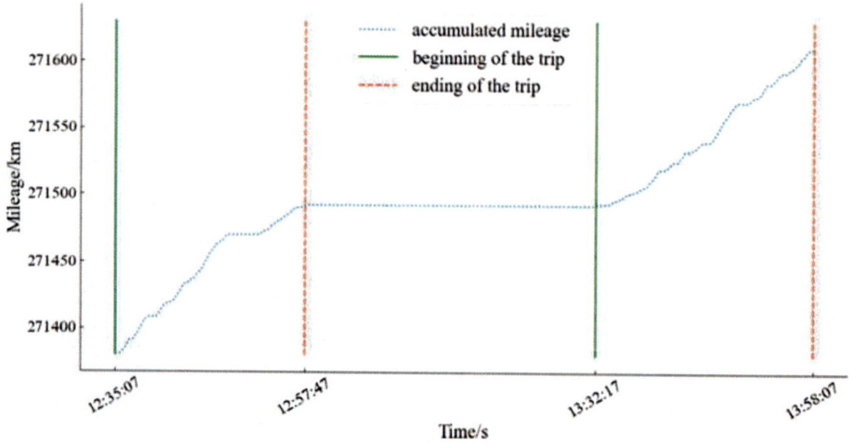

Fig. 2 Results of trip division

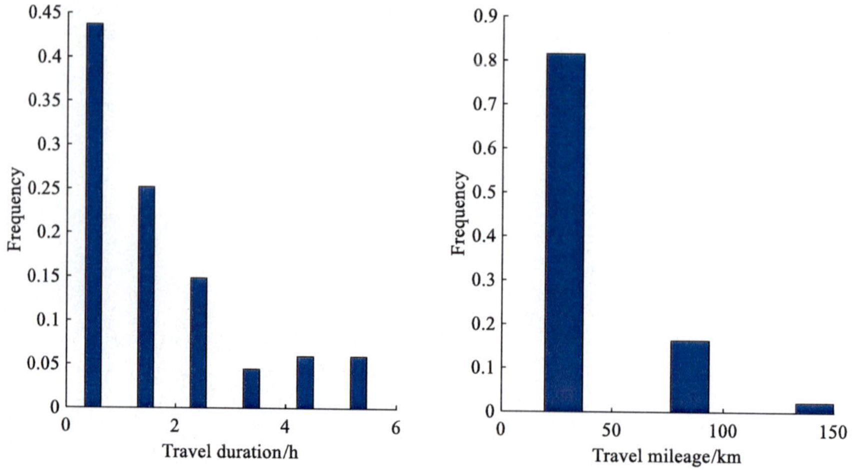

Fig. 3 Travel time and mileage distribution

Table 3 New data information table after processing

Field	Representing meaning
start_time	Start time of the travel
end_time	End time of the travel
start_m	Accumulated mileage at the beginning of the travel
end_m	Accumulated mileage at the ending of the travel
start_soc	Soc of charge at the beginning of the trip
end_soc	Soc of charge at the ending of the trip

3.1 The Influence of Travel Characteristics on the Energy Consumption of Electric Vehicles

A trip can be described as a series of processes from a starting point to a destination. Travel characteristics reflect the daily use of vehicles and have a greater impact on power consumption. Travel characteristics mainly include travel time, travel mileage, etc., which reflect travel intensity and determine travel energy consumption. The energy consumption studied in this paper is the electricity consumption of a single trip of an electric vehicle, which is represented by the change in the state of charge (State of Charge) before and after the trip. Figure 4 shows the relationship between energy consumption and time of multiple trips of a vehicle, and Fig. 5 shows the relationship between energy consumption and mileage for multiple trips of a vehicle.

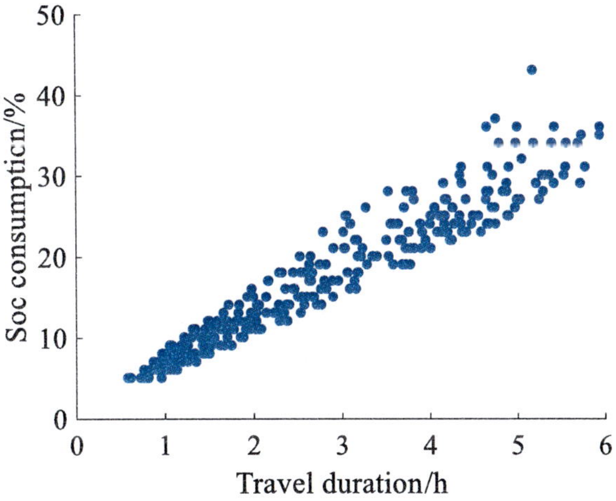

Fig. 4 Travel time and energy consumption

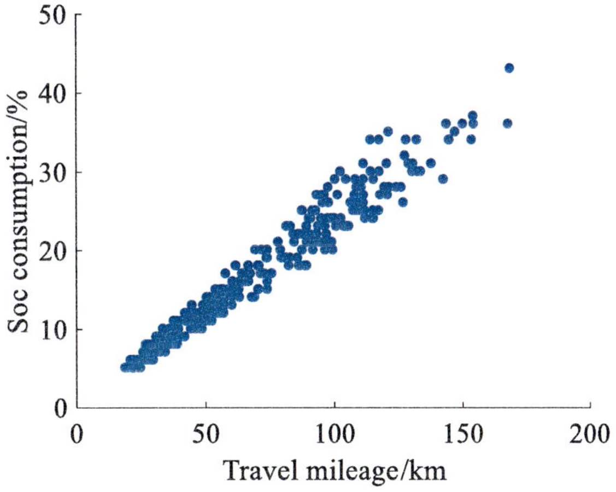

Fig. 5 Travel mileage and travel energy consumption

Table 4 Correlation analysis results

Parameter	Correlation
Accumulated mileage before travel/km	0.0059
Remaining battery before travel/soc	0.1361
Travel mileage/km	0.9849
Travel time/h	0.9415
Average speed/km*h^{-1}	0.3467

It can be seen from the figure that with the increase of travel time and travel mileage, the power consumption increases, and the two are basically linearly related to energy consumption. This law has also been verified in the literature [19]. Due to differences in parking delays, operating conditions, and driving maneuvers during travel, the same travel time and mileage will result in different power consumption.

Therefore, travel characteristics significantly affect energy consumption, and statistical methods are used to analyze the correlation between multiple travel characteristic parameters and energy consumption, as shown in Table 4. Among them, travel time and travel mileage have the highest correlation with energy consumption, which can be used as the main basis for energy consumption prediction.

3.2 *The Impact of Regional Differences on the Energy Consumption of Electric Vehicles*

There are differences in the population, administrative area, terrain structure and climate temperature of different cities, which directly or indirectly affect the energy consumption of electric vehicles. The data of travel energy consumption and travel mileage in the travel samples are counted by city, and the fitting results are shown in Table 5. The visualization of the data and fitting effect of two cities is shown in Fig. 6, and the specific relationship is shown in Eqs. (1) and (2).

$$y_1 = 0.2462x_1 + 0.2021 \tag{1}$$

Table 5 Multi-region travel energy consumption and travel mileage fitting result table

City	Coefficient	Intercept
A	0.2856	0.2105
B	0.2462	0.2021
C	0.2972	0.3236
D	0.2732	0.1332
E	0.2678	0.1220
F	0.2587	0.1123

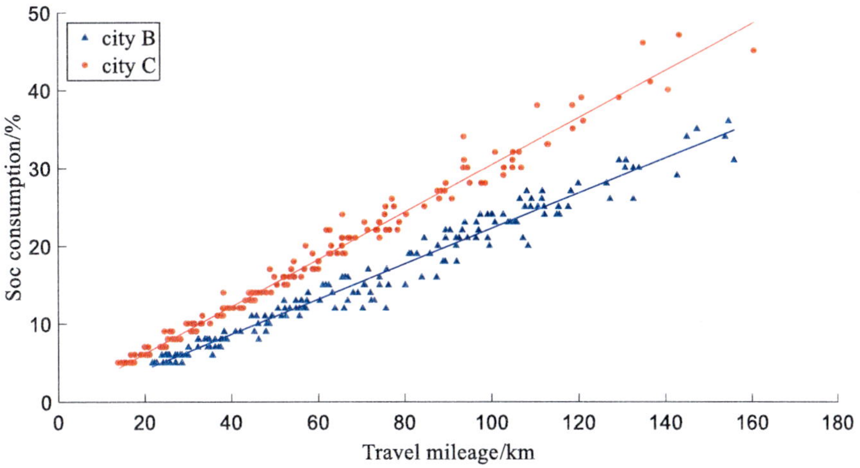

Fig. 6 Regional differences in energy consumption and mileage of vehicle travel

$$y_2 = 0.2972x_2 + 0.3236 \quad (2)$$

Among them, y_1, y_2 means travel energy consumption/soc; x_1, x_2 means travel mileage/km.

It can be seen that the fitting results of the travel mileage and travel energy consumption of vehicles in different cities are different, which verifies the influence of regional differences on the energy consumption of electric vehicles. Therefore, the energy consumption model is separately trained and established for the travel samples of different cities in the future.

4 Real-Time Prediction Model of Electric Vehicle Energy Consumption Based on BP Neural Network

In the previous paper, a large number of travel feature samples were obtained through data analysis, and it was determined that travel features and regional differences had a significant impact on energy consumption. Correlation analysis was used to screen travel feature parameters that are strongly related to travel energy consumption, and based on this, a BP neural network was constructed. Correlation analysis is used to screen out the travel characteristic parameters that are strongly related to travel energy consumption, and based on this, a BP neural network energy consumption prediction model is built. This model only needs to input the travel time and mileage to realize the energy consumption prediction before travel and obtain the energy consumption required for travel in advance. The optimal network model parameters are determined by error comparison, so that the network performance can be optimized, and a higher accuracy can be achieved in single-region energy consumption prediction.

4.1 Construction of Energy Consumption Model of BP Neural Network

Network parameter setting steps:

(1) The number of network layers: The structure of the BP neural network affects the accuracy of the network prediction. The Komogorov theory shows that any continuous system can be accurately fitted by a three-layer neural network [20]. The more network layers are set, the more complicated the calculation is, so the BP neural network prediction model established in this paper includes an input layer, a hidden layer and an output layer.

(2) The number of neurons in each layer: After the previous analysis, the input variables are travel mileage and time, and the output variables are travel energy consumption. Therefore, the number of neurons in the input layer is 2, the number of neurons in the output layer is 1, and the number of neurons in the hidden layer is usually determined according to the empirical formula:

$m = \sqrt{n+k} + a$, where m is the number of neurons in the hidden layer, n is the number of input nodes, k is the number of output nodes, a is a constant from 1 to 10, so the number of hidden layer nodes is set from small to large, and the optimal number of neurons is set by observing the error situation. Use the error evaluation index Mean Absolute Error (MAE: Mean Absolute Error) and Mean Absolute Percentage Error (MAPE: Mean Absolute Percentage Error) to analyze the prediction accuracy of the network, see Eqs. (3) and (4).

$$\mathrm{MAE} = \frac{1}{n}\sum_{i=1}^{n}\left|y_i - y_i^{'}\right| \tag{3}$$

$$\mathrm{MAPE} = \sum_{i=1}^{n}\left|\frac{y_i - y_i^{'}}{y_i}\right| \times \frac{100\%}{n} \tag{4}$$

where y_i means actual value; $y_i^{'}$ means predictive value; n means number of samples.

4.2 Energy Consumption Model Tuning and Prediction

Taking the actual operation data samples of electric vehicles in city A as an example, the network prediction errors when different numbers of hidden layer nodes are set are shown in Table 6.

Table 6 Prediction results of different network structures

Network structure	MAE	MAPE (%)
2-3-1	1.24	7.73
2-4-1	1.08	7.34
2-5-1	1.00	6.70
2-6-1	1.10	7.42
2-7-1	1.21	7.91
2-8-1	1.22	7.82
2-9-1	1.41	8.23
2-10-1	1.13	7.53
2-11-1	1.27	8.07
2-12-1	1.37	8.24

According to the principle of minimum prediction error, obtained from the above table, the "2-5-1" BP neural network structure is used to predict the travel energy consumption of electric vehicles. The regression results of the network training process are shown in Fig. 7. The ratio of training set and test set in the sample data of city A is set to be 4:1, and the prediction result of travel energy consumption is shown in Fig. 8.

It can be seen from Fig. 8 that when the overall prediction error is 6.70%, the prediction result of each trip energy consumption is relatively close to the actual value, and the accuracy of the network prediction is good. Calculate the absolute

Fig. 7 Network regression analysis

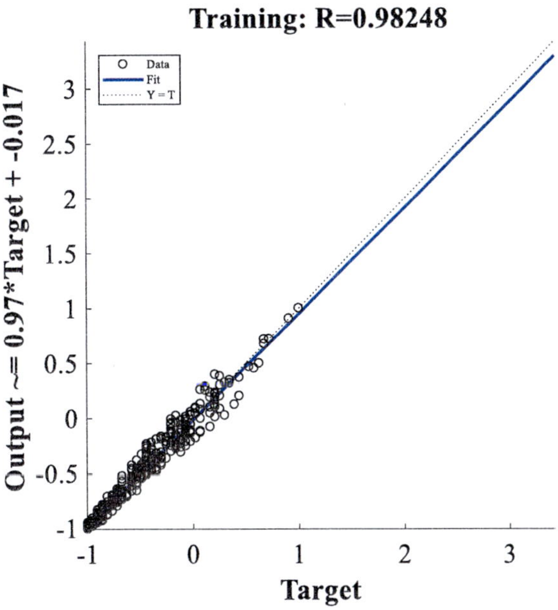

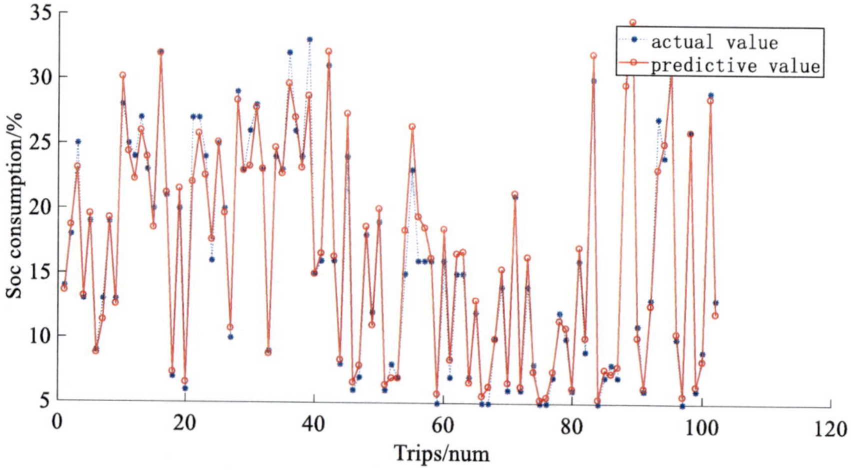

Fig. 8 Predicted and actual energy consumption

error predicted by the network with the mileage change, and take the average value of the absolute error at the corresponding travel mileage to obtain the absolute error change per unit mileage. In order to describe the error change trend, it is drawn according to the 95% error distribution. The envelopes of absolute error and absolute error per unit mileage are obtained, and the results are shown in Figs. 9 and 10:

It can be seen from the relationship between the above prediction error results and travel mileage that the absolute error of energy consumption prediction will increase with the increase of the input variable of travel mileage, and the change trend with mileage is basically distributed in (y: absolute error, x: travel mileage) range, and the

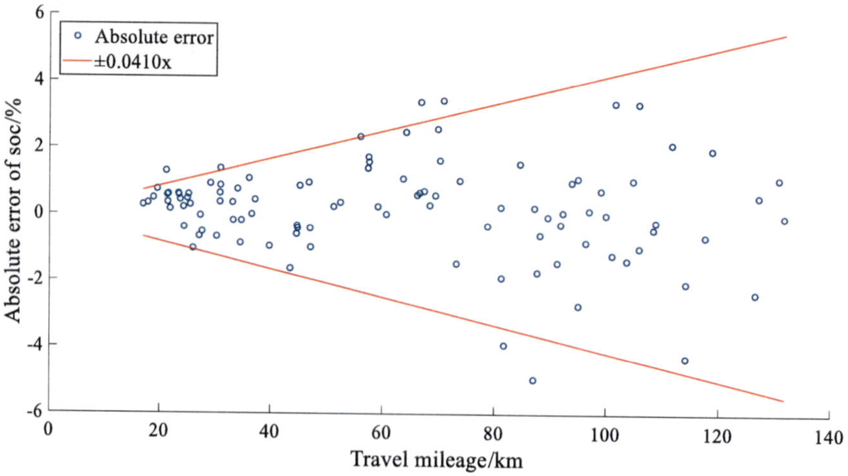

Fig. 9 Trip mileage and absolute error

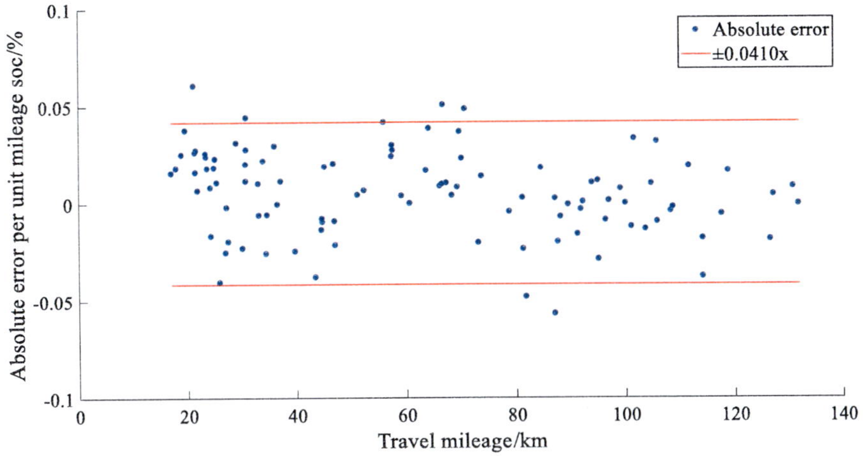

Fig. 10 Absolute error of unit trip mileage

absolute error of unit mileage is basically not affected by the input value of travel mileage, most of which are smoothly distributed in the range of [−0.04, 0.04]. By predicting the actual travel energy consumption of electric vehicles in city A, the validity of the energy consumption prediction model is verified.

Examples of prediction application scenarios are as follows:

When there is a travel demand with a travel time of t and a travel mileage of d, these two features are input into the energy consumption prediction model to obtain the required energy consumption of x. If the current soc of the vehicle is much greater than x, the user can successfully complete the next time Travel; if the current soc of the vehicle is less than or close to x, the user will have mileage anxiety, and there is a high probability of choosing to charge the vehicle before traveling. The real-time prediction of energy consumption is realized based on the real-time acquisition of the two parameters of mileage and duration in the travel task. Inputting them into the prediction model can obtain the energy consumption required for the corresponding travel, which enables electric vehicle users to understand before travel. The power required for this time provides a certain reference for travel planning and charging decision-making, and when combined with the travel distribution law, it can provide data support for the research on the spatial demand distribution prediction of charging loads.

5 Construction and Prediction of Multi-Region Energy Consumption Model

The above model has high accuracy in single-region energy consumption prediction, but because regional differences will affect energy consumption, the energy consumption prediction model obtained by single-region training may not be suitable for energy consumption prediction in different regions. Carry out research and comparison and verification, and build and predict the energy consumption model of multiple regions.

5.1 *Construction of Multi-Region Energy Consumption Model*

Considering the influence of regional differences on energy consumption, samples from six cities were trained to establish energy consumption prediction models, and the average absolute errors of single prediction and cross prediction results of the models were compared, as shown in Table 7.

As can be seen from the table, due to the existence of regional differences in energy consumption, it is necessary to build a real-time prediction model of energy consumption suitable for specific cities. The travel samples in different cities are trained separately, and the best network structure and prediction results of electric vehicles in each city are obtained as shown in Table 8.

According to Table 8, the forecast mean absolute error (MAE) of the five cities is within 2, and the mean absolute error percentage (MAPE) is within 8%.

Table 7 Multi-region single prediction and cross prediction results

Training	Prediction					
	A	B	C	D	E	F
A	6.70	9.00	8.50	12.00	9.80	11.00
B	7.60	7.15	9.50	10.00	9.80	8.90
C	8.00	8.90	6.74	11.00	9.00	8.90
D	8.00	8.50	8.90	8.00	9.00	8.90
E	7.70	8.50	8.00	10.00	7.50	8.70
F	7.40	8.00	8.00	8.80	9.00	7.93

Table 8 Network structure and prediction results of different cities

City	Network structure	Input layer-hidden layer weights	Hidden layer-output layer weights	MAE	MAPE (%)
B	2-6-1	−9.40, 23.22, 1.80, −0.22, −2.80, 0.71; 1.21, −14.58, 8.10, −3.25, 1.03, 2.67	−0.74, 0.20, −0.09, −0.19, −0.32, 0.30	1.18	7.15
C	2-9-1	31.30, −28.94, −0.80, −227.60, 0.19, −3.08, −32.36, 1.54, −1.78; −10.00, 9.30, 1.09, −34.14, −0.90, −33.62, −4.35, 0.72, −1.05	6.12, 6.19, −1.74, −0.12, −2.33, 0.03, −0.03, −6.61, −3.10	1.01	6.74
D	2-7-1	0.88, −35.14, −2.47, −4.93, −3.65, −3.66, 0.28; −3.05, 14.13, 0.79, 5.24, 2.46, 2.53, −2.25	−4.92, −0.29, −0.53, 0.23, −13.94, 13.82, 24.22	1.32	8.00
E	2-8-1	−4.54, 1.62, 4.40, −2.38, −24.67, −12.97, −0.86, 1.63; −6.42, 56.41, 5.70, 22.71, −1.18, 17.88, −0.28, 54.31	16.38, 16.42, 16.62, 0.07, −0.06, −0.05, −0.94, −16.42	1.32	7.50
F	2-6-1	0.27, −4.06, −27.95, 1.45, −0.16, 2.83; 3.06, 2.76, −117.85, 0.75, −3.14, 0.27	2.98, −0.27, 0.04, 0.48, 2.87, 0.28	1.31	7.93

Table 9 Prediction results of different models

Model	Error					
	A		B		C	
	MAE	MAPE (%)	MAE	MAPE (%)	MAE	MAPE (%)
Multiple linear regression	1.23	7.23	1.23	7.83	1.06	7.55
RBF	1.34	8.41	1.25	8.20	1.12	8.50
BP	1.00	6.70	1.17	7.15	1.01	6.74
Model	Error					
	D		E		F	
	MAE	MAPE (%)	MAE	MAPE (%)	MAE	MAPE (%)
Multiple linear regression	1.36	8.51	1.41	8.92	1.39	8.66
RBF	1.23	8.70	1.52	9.12	1.43	8.95
BP	1.32	8.00	1.32	7.50	1.30	7.93

5.2 *Comparison of Prediction Effects of Different Energy Consumption Models*

BP neural network, multiple linear regression, and radial basis (RBF) neural network are used to predict the energy consumption of electric vehicles in cities of different scales. The tuning process of radial basis (RBF) neural network is similar to that of neural network. The learning rate and threshold of the network are adjusted by using the error feedback, the network parameters are set according to the principle of minimum error, and the optimal network structure is determined for energy consumption prediction. The prediction results of the three methods are shown in Table 9.

6 Conclusions

(1) By analyzing the actual travel energy consumption of electric vehicles in different regions, it is proved that the difference in energy consumption between different regions is real;
(2) A prediction model of electric vehicle travel energy consumption is established. Given historical travel data and input of the required time and mileage for future travel tasks, the proposed method for electric vehicle energy consumption prediction can realize real-time prediction of pre-trip energy consumption.

Due to the characteristics of the research data sources, this paper does not consider the influence of the external environment during vehicle driving, such as road network conditions, temperature, driving style and accessory energy consumption, on energy consumption. The degree of dependence is high, and it is difficult to ensure the prediction accuracy when the input features are inaccurate.

Acknowledgements This research is supported by the following projects:

1. National Key Research and Development Program of China (No. 2018YFB1601100).
2. R&D Program of Beijing Municipal Education Commission (No. KM202010009007).
3. Youth Top Talent Training Program of Beijing Municipal Education Commission (No. CIT&TCD201904013).

References

1. Chen L-L, Zhang H, Ni F (2011) Discussion on the current situation and development of electric vehicle energy supply facilities. Power Syst Autom 35(14):11–17
2. He W, Williard N, Chen C et al (2014) State of charge estimation for Li-ion batteries using neural network modeling and unscented Kalman filter-based error cancellation. Int J Electr Power Energy Syst 62:783–791
3. Zheng Y-J, Sun J, Nian G-Y (2022) Energy consumption simulation and parameter optimization of electric commercial vehicles based on real-world driving cycle. China J Highw Transp 1 21
4. Wang J-N, Liu J, Chu L et al (2016) Optimal design of driving motor structural parameters for electric vehicle. J Traffic Transp Eng 16(06):72–81
5. Hu L, Zhou D-H, Huang J et al (2021) Optimal path planning for electric vehicle with consideration of traffic light and energy consumption. Automot Eng 43(05):641–649, 666
6. Xu X-Y, Li G-Y, Tao S-Y et al (2021) Simulation and analysis on longitudinal and lateral slipping energy consumption of four-wheel independently driven electric vehicle tire. J Mech Eng 57(04):92–102
7. Liu Z-T, Wu Q-L, Zong Z-J (2011) Study on the energy consumption economy of electric vehicle based on test bench simulation. J Sun Yat-sen Univ (Nat Sci Ed) 50(01):44–48, 52
8. Olaf C, Ulrich J (2019) Design and control of electric bus vehicle model for estimation of energy consumption. IFAC PapersOnLine 52(24):59–64
9. Cheng J-Z, Yu Z-R, Cheng S et al (2020) Energy consumption prediction of electric vehicle considering multiple influences in urban road network. Electr Meas Instrum 57(20):90–97
10. Antonello IC, Giuseppe M, Corrado R et al (2020) Energy consumption of electric vehicles: models' estimation using big data (FCD). Transp Res Procedia 47(03):211–218
11. Wei H-L, Lai X-X, Huang C-S et al (2014) EV calculation model of energy consumption based on velocity and acceleration distribution. J Jilin Univ (Eng Technol Ed) 44(06):1591–1595
12. Wu X, Freese D, Cabrera A, Kitch WA (2015) Electric vehicles' energy consumption measurement and estimation. Transp Res Part D 34:52–67
13. Chiara F, Vittorio M (2018) Modelling energy consumption of electric freight vehicles in urban pickup/delivery operations: analysis and estimation on a real-world dataset. Transp Res Part D 65(09):658–673
14. Hu J, Gao Z W (2021) A data driven SOC prediction scheme for traction battery in electric vehicles. Automot Eng 43(01):1–9, 18
15. Bao W, Ge J-J (2020) Study on battery SOC prediction method for electric bus based on sparsely sampled data. Automot Eng 42(3):367–374
16. Liu G-M, Ouyang M-G, Lu L-G (2014) Driving range estimation for electric vehicles based on battery energy state estimation and vehicle energy consumption prediction. Automot Eng 36(11):1302–1309, 1301
17. Shatrughan M, Jhilik B, Prasenjit B (2020) Estimation of energy consumption of electric vehicles using deep convolutional neural network to reduce driver's range anxiety. ISA Trans 98(08):454–470

18. Su S, Yang T-T, Li Y-J et al (2019) Electric vehicle charging path planning considering real-time dynamic energy consumption. Power Syst Autom 43(7):136–143
19. Bi J, Zhang J-W, Zhang D et al (2015) A correlation analysis and modeling for battery SOC and driving mileage of electric vehicle. Transp Syst Eng Inf 15(01):49–54
20. Harald HO, Helge H (2010) The Kolmogorov-Riesz compactness theorem. Expo Math 28(4):385–394

Road Traffic Accident Prediction Based on BP Neural Network

Yan Xing, Wen-hao Song, Wei-dong Liu, and Shu-shida Gao

Abstract In order to predict road traffic accident indicators scientifically and accurately, this paper established a road traffic accident prediction model based on the basic theory of BP neural network. The 14 main influencing factors of traffic accidents were selected by using correlation analysis theory as the input variable of the prediction model and traffic accident deaths was taken as the output variable of the prediction model. Data related to road traffic accidents in China from 2000 to 2017 were selected as training samples of the model, MATLAB nntool was used to train the prediction model by using traingdm, traingda, trainlm and traingd training functions, and predict the deaths of road traffic accidents in 2018 in China. It is verified that the relative error of the constructed road traffic accident prediction model is within 1%, which can be used for road traffic accident prediction.

Keywords Traffic safety · Traffic accident prediction · BP neural network · Correlation analysis

1 Introduction

By predicting the traffic accident indicators in the future, it can provide a basis for traffic safety evaluation and decision-making, which is of great significance for exploring traffic safety countermeasures. At present, the prediction methods of road traffic accidents mainly include: regression analysis method, time series method, grey model method and neural network model method. Scholars at home and abroad have carried out a lot of research on traffic accident prediction: Milton and Mannering [1] the influence of road conditions and traffic conditions on traffic accidents was analyzed, and the number of traffic accidents was predicted by using a negative binomial regression model; Ma et al. [2] a nonlinear negative binomial regression model is constructed to predict the number of highway traffic accidents. The results show that the prediction accuracy of the model is better than the traditional negative

Y. Xing · W. Song · W. Liu (✉) · S. Gao
School of Transportation Engineering, Shenyang Jianzhu University, Shenyang 110168, China
e-mail: 8754183338@qq.com

W. Wang et al. (eds.), *Green Transportation and Low Carbon Mobility Safety*,
Lecture Notes in Electrical Engineering 944,
https://doi.org/10.1007/978-981-19-5615-7_46

binomial regression model. Shen [3] the multiple linear regression method is used to establish and optimize the prediction model of road traffic accident fatalities and traffic accident influencing factors. Wang et al. [4] based on the RVM model theory, the traditional traffic accident time series prediction model is improved; Sun et al. [5] from the perspective of mining the evolution information of the time series itself, the trend prediction of road traffic accident indicators is carried out based on the ARIMA prediction model, and the prediction model is revised according to the SVR residual prediction value. Yang and Hu [6] using the method of grey system theory, a dynamic grey prediction model is established to predict the property loss of traffic accidents in my country. Limitations of a single grey forecasting model, Li and Hou [7] based on GM(1,1) model and Verhulst model, a grey combined accident prediction model is established; Wang and Liu [8] by combining the grey model and the Markov chain model, the grey transition state Markov model is established.

The occurrence of road traffic accidents is affected by many factors such as people, vehicles, roads, and the environment, showing random and nonlinear characteristics. Most of the above studies are based on the evaluation indicators of traffic accidents in historical years. The indicators are used to predict, and the influencing factors of traffic accidents are not involved in the model, which lacks scientific rationality. The neural network has strong robustness and generalization ability, and has good fitting in nonlinear problems. By mining the historical data of road traffic accidents, the influencing factors of road traffic accidents and road traffic accident evaluation indicators can be established. The complex mapping relationship between them can achieve a better traffic accident prediction effect in a short training time. Liu and Deng [9] a traffic accident micro-prediction method based on fuzzy neural network is proposed; Yi et al. [10] construction of RBF neural network based on genetic algorithm optimization to predict traffic accidents in highway tunnel groups; Zhang et al. [11] the traffic accident safety level is predicted based on the LSTM neural network model. Based on the above analysis, this paper establishes a road traffic accident prediction model based on BP neural network, selects the data related to road traffic accidents in my country from 2000 to 2017 as training samples, and uses different training functions to train the prediction model. Traffic accident fatalities are forecasted to test the accuracy and applicability of the constructed forecasting model.

2 Research Ideas

BP neural network is a kind of multi-layer feedforward neural network, which consists of an input layer, multiple hidden layers, and an output layer, transformation ability. The occurrence of road traffic accidents is affected by many factors such as people, vehicles, roads and the environment, and has the characteristics of randomness and nonlinearity. Therefore, the prediction of road traffic accidents is a typical nonlinear problem. BP neural network has good robustness and fitting for such nonlinear problems. Therefore, this paper takes the influencing factors of road traffic accidents as the sample data of the input layer and the road accident evaluation index as the sample data of the output layer. The road traffic accident prediction model of BP neural network can establish the complex mapping relationship between the influencing factors of road traffic accident and the evaluation index of road traffic accident by training the model, so as to achieve the purpose of predicting the evaluation index of road traffic accident.

3 Influencing Factors of Traffic Accidents and Selection of Evaluation Indicators

3.1 Selection of Influencing Factors of Traffic Accidents

The occurrence of road traffic accidents has strong randomness and uncertainty, and the factors affecting road traffic safety are complex and diverse. In the field of macro forecasting, the occurrence of road traffic accidents is closely related to human social activities, the number of motor vehicles, the mileage of graded highways, the traffic operation environment, and the level of social and economic development. In order to ensure the accuracy of traffic accident prediction, 17 influencing factors of road traffic accidents are comprehensively selected, namely: GDP, population, urban population, rural population, civilian vehicle ownership, private vehicle ownership, road operation Vehicle ownership, number of motor vehicle drivers, number of vehicle drivers, road mileage, expressway mileage, road passenger volume, road passenger turnover, average distance of road passenger transport, road freight volume, road freight turnover, road freight transport average distance. The original sample data for road traffic accident prediction comes from the National Bureau of Statistics from 2000 to 2018《China Statistical Yearbook》[12], as shown in Tables 1, 2, and 3.

Table 1 Raw sample data for road traffic accident prediction (1)

Years	GDP (billion)	Population (10,000 people)	Urban population (10,000 people)	Rural population (10,000 people)	Civil vehicle ownership (ten thousand)	Private car ownership (ten thousand)
2000	100280.1	126,743	45,906	80,837	1608.91	625.33
2001	110863.1	127,627	48,064	79,563	1802.04	770.78
2002	121717.4	128,453	50,212	78,241	2053.17	968.98
⋮	⋮	⋮	⋮	⋮	⋮	⋮
2016	740060.8	138,271	79,298	58,973	18574.54	16330.22
2017	820754.3	139,008	81,347	57,661	20906.67	18515.11
2018	900309.5	139,538	83,137	56,401	23231.23	20574.93

Table 2 Raw sample data for road traffic accident prediction (2)

Years	Ownership of road vehicles (ten thousand)	Number of motor vehicle drivers (10,000 people)	Number of car drivers (10,000 people)	Road mileage (10,000 km)	Highway mileage (10,000 km)	Road passenger traffic (10,000 people)
2000	702.82	7655.56	3746.51	167.98	1.63	1,347,392
2001	764.39	8455.04	4462.68	169.80	1.94	1,402,798
2002	826.34	9362.03	4827.08	176.52	2.51	1,475,257
⋮	⋮	⋮	⋮	⋮	⋮	⋮
2016	1435.77	35876.98	30328.77	469.63	13.10	1,542,759
2017	1450.22	36016.94	31658.20	477.35	13.64	1,456,784
2018	1435.48	41030.16	36923.42	484.65	14.26	1,367,170

Table 3 Raw sample data for road traffic accident prediction (3)

Years	Road passenger turnover (million person kilometers)	Average distance of road passenger transport (km)	Road freight (tons)	Road freight turnover (billion ton kilometers)	Road freight average distance (km)	Death toll (people)
2000	6657.4	49	1,038,813	6129.4	59	93,853
2001	7207.1	51	1,056,312	6330.4	60	105,930
2002	7805.8	53	1,116,324	6782.5	61	109,381
⋮	⋮	⋮	⋮	⋮	⋮	⋮
2016	10228.7	66	3,341,259	61080.1	183	63,093
2017	9765.2	67	3,686,858	66771.5	181	63,772
2018	9279.7	68	3,956,871	71249.2	180	63,194

3.2 *The Selection of Traffic Accident Evaluation*

Traffic accident evaluation indicators can be divided into two categories: absolute indicators and relative indicators. Because some parameters of relative indicators cannot be counted scientifically and accurately, and there is no uniform standard for the selection of correction coefficients, such as death intensity indicators, equivalent comprehensive fatality rates, and operating accident rates, four absolute indicators are generally used in my country to evaluate traffic accidents. As for the number of accidents, traffic accidents that do not involve casualties are omitted from the statistics; the number of injured and economic losses are relatively subjective, and the comparability is poor. The number of deaths in traffic accidents is an important indicator to measure the level of national and regional traffic safety. Countries pay more attention to traffic accidents involving fatalities, and the accident files archived by the traffic control department have more detailed records of fatalities. Therefore, this paper selects the number of deaths in traffic accidents as Road traffic accident evaluation index.

3.3 *Selection of Influencing Factors of Traffic Accidents*

Considering the differences in the correlation between the selected 17 traffic accident influencing factors and the traffic accident evaluation index, if the correlation between the selected traffic accident influencing factors and the traffic accident evaluation index is low, it will affect the prediction accuracy. Therefore, the correlation analysis was carried out with the help of the statistical analysis software SPSS, and the influencing factors of traffic accidents with low correlation were eliminated. The output results of SPSS are shown in Table 4. According to the correlation analysis theory, there is a strong correlation when the significance coefficient (sig) is less than 0.05 and the absolute value of the Pearson coefficient is greater than 0.7. From the SPSS output results, it can be seen that the significant coefficients of highway passenger volume and the average distance of highway passenger transportation are all greater than 0.05, and there is no correlation; the correlation between highway passenger turnover and traffic accident fatalities is weak, and the above three influencing factors are considered rounding. Therefore, we select: GDP, population, urban population, rural population, civilian vehicle ownership, private vehicle ownership, road operation vehicle ownership, motor vehicle drivers, car drivers, road mileage, and highway mileage, Road freight volume, road freight turnover, and average road freight distance, a total of 14 highly correlated road traffic accident influencing factors are used as the input variables of the road traffic accident prediction model.

4 Establishment of Traffic Accident Prediction Model

4.1 *Predictive Model Structure*

Hetht-Nielson [13] It is proved that a single hidden layer BP neural network can approximate any continuous function in a closed interval. In theory, increasing the number of hidden layers can enhance the fitting ability of the traffic accident prediction model, but at the same time, it will increase the difficulty of training, making the model difficult to converge, and prone to over-fitting problems, reducing the generalization ability of the model. Therefore, a three-layer structure road traffic accident prediction model is selected in this paper, that is, a single hidden layer structure.

Table 4 Correlation coefficient of influencing factors of traffic accidents

Influencing factors	Pearson coefficient	Significance coefficient
GDP	−0.857	0.000
Population	−0.887	0.000
Urban population	−0.893	0.000
Rural population	0.895	0.000
Civil vehicle ownership	−0.784	0.000
Private car ownership	−0.771	0.000
Ownership of road vehicles	−0.832	0.000
Number of motor vehicle drivers	−0.838	0.000
Number of car drivers	−0.826	0.000
Road mileage	−0.928	0.000
Highway mileage	−0.872	0.000
Road passenger traffic	−0.442	0.058
Road passenger turnover	−0.651	0.003
Average travel distance of road passengers	−0.451	0.052
Road freight	−0.901	0.000
Road freight turnover	−0.924	0.000
Average distance of road freight	−0.948	0.000

The input samples of the model are the 14 influencing factors of road traffic accidents selected in Sect. 3.3, and the output samples are the number of deaths in road traffic accidents selected in Sect. 3.2. The basic structure of the road traffic accident prediction model based on BP neural network is shown in Fig. 1.

4.2 *Number of Hidden Layer Nodes*

The selection of the number of hidden layer nodes has a great influence on the fitting ability and generalization ability of the road traffic accident prediction model. Too many nodes can easily lead to overfitting of the model, and too few nodes can easily

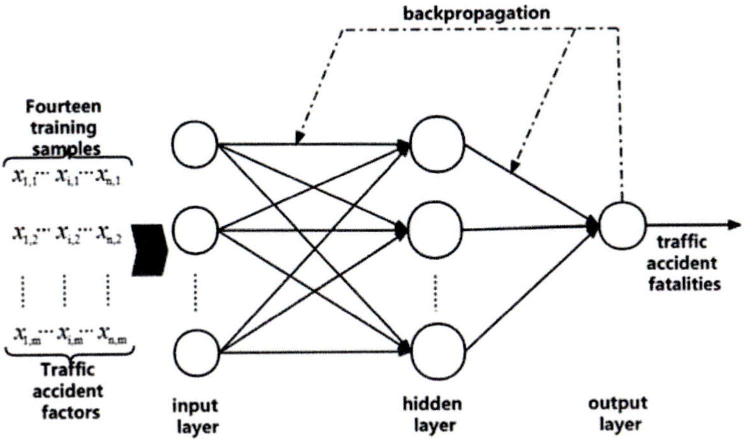

Fig. 1 Structure diagram of traffic accident prediction model

lead to underfitting of the model. Therefore, it is more critical to reasonably select the number of hidden layer nodes. At present, there are few theoretical studies on the selection of the number of hidden layer nodes, so this paper refers to the empirical formula to determine the value range of the number of hidden layer nodes, and uses the trial algorithm to select the number of hidden layer nodes when the prediction model accuracy is optimal. The empirical formula is shown in (1):

$$Z = \sqrt{X + Y} + C \tag{1}$$

in the formula: Z—Number of hidden layer nodes; X—Number of input layer nodes; Y—The number of output layer nodes; C—Take a constant from 1 to 10.

In this paper, the number of nodes in the input layer X is 14, and the node Y in the output layer is 1. It can be seen from the above formula that the number of nodes in the hidden layer Z of the traffic accident prediction model is 5–14.

4.3 Normalization of Training Samples

Since the data related to road traffic accidents are not in the same order of magnitude, it needs to be limited to an appropriate range to ensure that different types of data can be cross-input, and the training speed of the traffic accident prediction model can be improved. Using dispersion normalization to linearly transform the traffic accident-related data into a dimensionless constant between [0, 1]. The conversion formula is shown in (2), and the data after sample normalization are shown in Tables 5 and 6.

$$N = \frac{S - S_{\min}}{S_{\max} - S_{\min}} \tag{2}$$

Table 5 Normalized data (1)

Years	GDP	Population	Urban population	Rural population	Civil vehicle ownership	Private car ownership	Ownership of road vehicles
2000	0.00000	0.00000	0.00000	1.00000	0.00000	0.00000	0.00000
2001	0.01323	0.06909	0.05796	0.94786	0.00893	0.00729	0.07373
2002	0.04643	0.19414	0.11566	0.89376	0.02054	0.01723	0.14791
⋮	⋮	⋮	⋮	⋮	⋮	⋮	⋮
2016	0.79970	0.90098	0.89689	0.10525	0.78464	0.78723	0.87767
2017	0.90056	0.95858	0.95192	0.05156	0.89249	0.89675	0.89497
2018	1.00000	1.00000	1.00000	0.00000	1.00000	1.00000	0.87732

Table 6 Normalized data (2)

Years	Motor vehicle driver	Car driver	Road mileage	Highway road mileage	Road cargo volume	Road freight turnover	Average distance of freight transport	Death toll
2000	0.00000	0.00000	0.00000	0.00000	0.00000	0.00000	0.00000	0.69766
2001	0.02395	0.02159	0.00575	0.02455	0.00600	0.00309	0.00781	0.93281
2002	0.05113	0.03257	0.02697	0.06968	0.02656	0.01003	0.01563	1.00000
⋮	⋮	⋮	⋮	⋮	⋮	⋮	⋮	⋮
2016	0.84560	0.80123	0.95257	0.90816	0.78903	0.84384	0.96875	0.09874
2017	0.84979	0.84130	0.97695	0.95091	0.90747	0.93124	0.95313	0.11196
2018	1.00000	1.00000	1.00000	1.00000	1.00000	1.00000	0.94531	0.10070

in the formula: N—normalized data; S—Raw data; S_{min}—sample minimum; S_{max}—sample maximum.

5 Traffic Accident Prediction Model Test

Selecting the influencing factors and fatalities of road traffic accidents in my country from 2000 to 2017 as the training samples of the model, using different training functions and the number of hidden layer nodes, and using the MATLAB neural network nntool toolbox to build and train the road traffic accident prediction model. The optimal combination is selected according to the model training error. Finally, the model is used to predict the number of road traffic accident deaths in my country in 2018, and the prediction results are denormalized and error analyzed, as shown in Table 7.

Table 7 Traffic accident death toll prediction results and error analysis

Training function	traingdm	traingda	trainlm	traingd
Number of nodes	5	7	8	11
Output result	0.10546	0.10067	0.10653	0.098297
Denormalization	63,438	63,192	63,493	63,070
Absolute error	244	2	299	124
Relative error	0.00386	3.17E−05	0.00473	0.00196

From the error analysis results, for different training functions and the number of hidden layer nodes, the relative errors of the prediction model are all within 1%, indicating that the prediction model has high accuracy, meets the requirements of practical applications, and can be used for road traffic accident deaths. Population forecast.

6 Conclusion

Based on the relevant data of traffic accidents from 2000 to 2018 of the National Bureau of Statistics, this paper determines the input variables of the prediction model through the correlation analysis of the influencing factors of road traffic accidents, selects the number of traffic accident deaths as the output variable, and establishes a BP neural network based on BP neural network. The road traffic accident prediction model finally predicts the number of traffic accident deaths in my country in 2018. The verification results show that the traffic accident prediction model has high accuracy, and the relative error is within 1%, which meets the requirements of practical applications and can be used to predict the number of deaths in road traffic accidents, which is of great significance for exploring road traffic safety countermeasures. The model established in this paper also has certain shortcomings: it only predicts road traffic accidents from the macro level, and lacks the prediction and analysis of specific problems. Subsequent research will predict various indicators of traffic accidents from the perspective of the mechanism of traffic accidents.

References

1. Milton J, Mannering F (1998) The relationship among highway geometrics, traffic-related elements and motor-vehicle accident frequencies. Transportation 25(4):395–413
2. Ma C, Zhang S-R, Ma Z-L et al (2018) Nonlinear negative binomial model of expressway traffic accident frequency prediction. China J Highw Transp 31(11):176–185
3. Shen K (2017) Casualty toll prediction and management countermeasures for road traffic accident in China. Saf Environ Eng 24(05):138–144

4. Wang W-B, Chen H, Wei L-X (2016) Study on method for prediction of traffic accident time series. China Saf Sci J 26(06):52–56
5. Sun Y-X, Shao C-F, Ji X et al (2014) Urban traffic accident time series prediction model based on combination of ARIMA and information granulation SVR. J Tsinghua Univ (Sci Technol) 54(03):348–353, 359
6. Yang S-X, Hu Y (2017) Traffic accident loss prediction based on dynamic grey prediction. J Wuhan Univ Technol (Inf Manag Eng) 39(06):674–678
7. Li Q, Hou F-Z (2017) Grey combined forecasting model for road traffic accidents based on coefficient of variation method. Saf Environ Eng 24(06):130–133
8. Wang X, Liu X-Y (2017) Research of traffic accident prediction based on gray Markov model. Technol Econ Areas Commun 19(04):9–13
9. Liu J, Deng W (2011) Research on microcosmic forecast methods of road accident based on fuzzy neural network. J Transp Eng Inf 9(04):69–75
10. Yi F-J, Han Z, Deng W (2012) Application of RBF neural network optimized by genetic algorithm for traffic accident microcosmic forecast of highway tunnel group. J Transp Eng Inf 10(01):64–72
11. Zhang Z-H, Yang W-Z, Yuan T-T et al (2019) Traffic accident prediction based on LSTM neural network model. Comput Eng Appl 55(14):249–253
12. Nationl Bureau of Statistics (2000–2018) China statistical yearbook. China Statistics Press, Beijing
13. Hecht-Nielsen (2002) Theory of the backpropagation neural network. In: International 1989 joint conferencc on ncural networks. IEEE

Autonomous Vehicle Path Planning Based on Improved Ant Colony Algorithm

Yan Xing, Xin Rao, Weidong Liu, and Wenhao Song

Abstract Path planning is one of the key technologies for autonomous vehicles. Ant Colony Algorithm can effectively achieve the goal of path planning for autonomous vehicles, but the algorithm has the problems of low search efficiency and local optimal solution in path planning. Therefore, this paper improves the classical ant colony algorithm, using adaptive initial pheromone distribution range build initial pheromone distribution, at the same time improve stimulating factor enhanced heuristic search efficiency, and introduces the rollback strategy self-locking and deadlock problem and adopt preferential set limit to update pheromone strategy, help reduce blind ant search path, and reduce the redundancy of map information. The simulation results show that the improved ant colony algorithm can greatly improve the global search ability and convergence speed, and can help the autonomous vehicle to find the optimal path quickly.

Keywords Ant colony algorithm · Path planning · Pheromone · Autonomous vehicle

1 The Introduction

The path planning problem of autonomous driving vehicles has become a research hotspot. The feasibility of the path is the key link to complete autonomous movement. Choosing a good route can not only reduce the use cost, but also save people's time. There are many algorithms in traditional path planning, including breadth-first

About the author: Yan Xing, 1985, male, doctor, associate professor, traffic management and control, 13332425491.

Foundation item: Natural Science Foundation of Liaoning Province (2019-ZD-0658); Liaoning Federation of Social Sciences Foundation, No. 2018LSLKTQN 058.

Y. Xing (✉) · X. Rao · W. Liu · W. Song
Shenyang Jianzhu University, Shenyang 110000, China
e-mail: 770859389@qq.com

W. Wang et al. (eds.), *Green Transportation and Low Carbon Mobility Safety*,
Lecture Notes in Electrical Engineering 944,
https://doi.org/10.1007/978-981-19-5615-7_47

search (BFS), A* algorithm, Artificial potential field method [1], Dijkstra algorithm, intelligent optimization algorithms including genetic algorithm, particle swarm optimization algorithm, ant colony algorithm, Cuckoo algorithm [2], BUG algorithms, etc. As compared with other algorithms of ant colony algorithm has good robustness, parallel computing, positive feedback, so in the path planning problem has a very wide range of research and application, a relatively good results have been achieved at the same time, but also the algorithm search space is large, large amount of calculation, low search efficiency, the most superior local problems. Zhang et al. [3] proposed an improved multi-step ant colony algorithm. The optimal path generated by each iteration of ant colony was taken as the guiding path, and the multi-step moving path was determined by the path-guided search strategy. Zhang et al. [4] planned the local path from the starting point to the intermediate point through force field and boundary conditions, and then improved the traditional artificial potential field algorithm by iterative method, which had deadlocks and local path underoptimization problems. The quality of the improved solution is greatly improved, but the time of the whole search process is still long. Wang et al. [5] proposed an improved pSO path optimization algorithm by using the individual weighted average of PSO and adding inertia weight, but the algorithm had more iterations and longer search time. Jiang et al. [6] proposed the establishment of favorable matrices for starting point and target point. The initial pheromone in the matrix is larger than that in other regions, but the search scope is expanded and the search time is increased. Li et al. [7] proposed the strategies of ant detection and ant optimization combined with artificial bee colony algorithm, assigning different weights to pheromones respectively. However, this method could not supplement the number of dead ants caused by deadlocks, so it was easy to fail in complex path planning. Zhou and Li [8] proposed that the traditional artificial potential field method is relatively smooth and has good real-time performance. However, when there are obstacles near the target point, the repulsive force is far greater than the gravitational force, and the vehicle will be difficult to reach the destination. When the gravity and repulsion forces of an autonomous vehicle are exactly equal at a certain point, the autonomous vehicle will fall into local optimum.

Therefore, in order to improve the speed and accuracy of autonomous vehicle path planning, an improved ant colony algorithm was proposed. The distribution of the initial pheromone was improved, and the uneven distribution of the initial pheromone was set to reduce the blindness of ant search in the early stage. For deadlocks and self-locks, the rollback strategy was adopted in this paper. Tabu tables were divided into global tabu tables and local tabu tables to record the information of deadlock location, ant walking path node and self-lock location node respectively, reduce ant mortality and increase the number of ants that make it to the finish line. For the pheromone update, only the ant pheromone of the better path is updated, which can reduce the redundancy of map pheromone and improve the convergence speed. Simulation experiments are carried out in different mobile environments, and the experimental results show that the improved ant colony algorithm significantly improves the performance index.

2 Raster Environment Modeling

The construction of environment map is an indispensable part in the path planning of autonomous vehicles. The real working environment of autonomous vehicles is irregular physical space. The process of environment modeling is actually mapping the real working environment into a virtual environment.

The methods of environment modeling mainly include grid method, configuration space method, free space method and so on. Raster method is most used in environment modeling [9]. The raster method is used to build the environment model. When grid method is used to describe environmental information, the area where obstacles are located is shown as black in the grid map, and the map matrix is marked as 1. The free passage area is shown as white in the grid map, and the map matrix is marked as 0. Autonomous vehicles can only move in the white free grid and need to avoid the black obstacle grid. According to the location of grid, grid can be divided into intermediate grid and boundary grid. For the intermediate grid, the vehicle can choose 8 directions for the next driving direction. When moving to the boundary grid, its direction of motion should be removed from the unreachable direction. The purpose of path planning is to find an optimal passable path in the established virtual environment map, so when using the grid method to build environment map, the reasonable setting of grid is very important. The raster environment map constructed by the raster method is shown in Fig. 1.

When constructing a barrier map, first construct a Cartesian coordinate system with the first barrier at the bottom left of the map as the coordinate origin. Therefore, each grid can be represented in the coordinate format of (x, y) so that the first grid in the left corner is (1, 1). Then, each barrier is assigned a number N starting from 0 from the lower left. The conversion formula for numbers and coordinates is:

$$x = \text{int}(N/G_{\text{size}}) + 1 \tag{1}$$

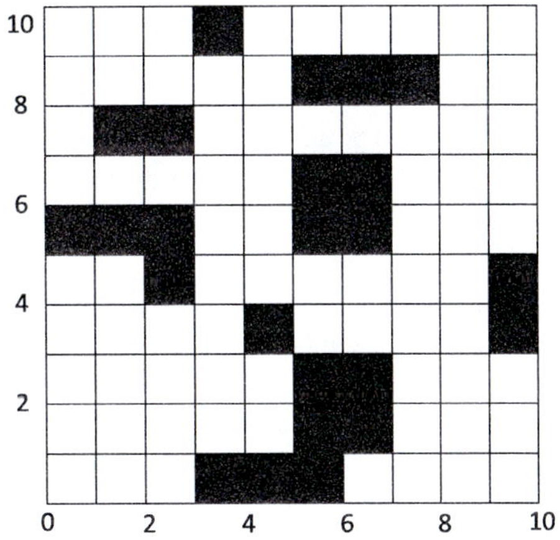

Fig. 1 Raster environment map

$$y = N\%G_{\text{size}} + 1 \tag{2}$$

Type: G_{size} is the number of grids per row; int is the integral function; % is the remainder.

3 Improvement of Ant Colony Algorithm

The pheromone, pheromone update rule, heuristic pheromone, state transition rule, deadlock and self-lock of ant colony algorithm are the factors that directly affect the performance of the algorithm. However, in the classical ant colony algorithm, these factors are insufficient, so this paper aimed at the existing shortcomings of the classical ant colony algorithm to improve.

3.1 *Adaptive Initial Pheromone Distribution*

In classical ant colony algorithm, the initial pheromone distribution is uniform. In this way, in the initial search process, ants will be aimless and unable to reach the destination directly and quickly, which will increase the time of path search and the length of path planning. Therefore, an adaptive distribution method of initial pheromone is proposed in this paper. The path distance of the start–end line is the shortest, and the optimal path is very likely to appear near the start–end line. So adjust the distribution of the initial pheromone to decrease from the middle line and the obstacle to both sides. The formula is as follows:

$$\begin{cases} \tau = \tau_0 + \vartheta C \\ \vartheta = \mu\varepsilon \end{cases} \tag{3}$$

In the formula τ is the pheromone of ant colony algorithm, τ_0 is the basic pheromone, ϑ is the adaptive parameter, μ is the distance between the node and the line, the closer the distance is, the greater the value is, and the range of its value is (0, 1). ε The smaller the proportion of obstacles is, the larger the value is, and its value range is (0, 1). Figure 2 is the schematic diagram of the initial pheromone when there are no obstacles on the line from the starting point to the end point. In the figure, the closer the grid is to the line, the higher the pheromone is, and the number of pheromone is represented by different colors. The darker the color, the higher the pheromone is, and the lighter the color, the lower the pheromone is. Figure 3 is the schematic diagram of the initial pheromone when there are obstacles on the starting point and ending point line. The black block is the obstacle, and the existence of the obstacle will decrease the pheromone of the whole. The starting and ending line paths are the global most planned paths, but the planning of the most planned path

Fig. 2 Accessibility initial pheromone distribution

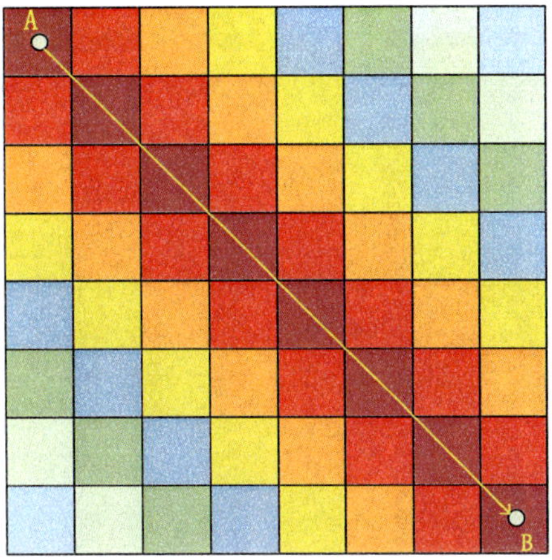

Fig. 3 Initial information distribution with obstacles

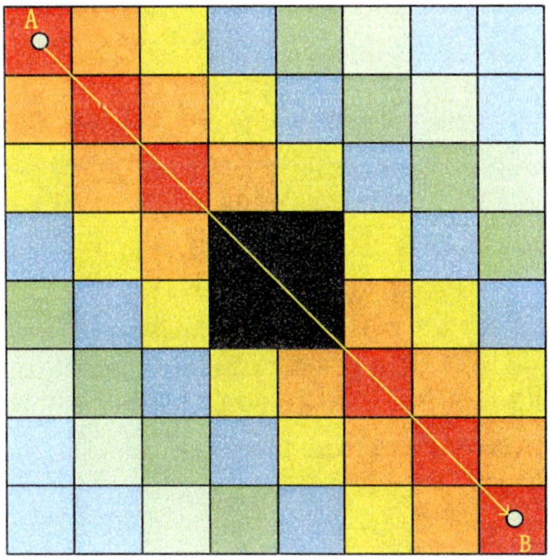

will be affected by obstacles in the map, but the optimal path is most likely to appear near the starting and ending line paths. Therefore, uneven distribution of pheromones can effectively improve the early search efficiency.

3.2 *Improvement of Heuristic Function*

In the initial stage of the classical ant colony algorithm, there is no pheromone left on the path, so the ant cannot choose the path according to the pheromone concentration, and the search is blind. In the original algorithm, the heuristic factor is the reciprocal of the node I and the lower node J, which makes the ant's search efficiency low and enlightening, and increases the operation time. Therefore, the inspiration factor is improved by taking the reciprocal of the distance between the next node j and the target point B as the inspiration factor. The formula is as follows.

$$\eta ij = 1/djb \tag{4}$$

In the formula, j is the next node and B is the target node, djb is the distance between the next node j and the target node.

3.3 Deadlocks and Self-Locks

In the initial route search of ants, there is a possibility that you will get stuck in front of the goal due to the restrictions of the map environment and contraindication table. There are two types of deadlocks and self-locks, as shown in Figs. 4 and 5, when the search ends, the ants enter a dead state, and the ant's elemental search path becomes invalid. In the main text, the key to death and the solution of the problem of the sword are taboo table, general taboo is divided into table and local taboo, general taboo table record death lock position information, local taboo table inheritance Record the general taboo table, information on the ant walking route node and the position node of the death, and when selecting the ant route, select the temporary taboo table. If the ant has a deadlock problem due to environmental issues, the ant will fall into the same deadlock again by moving the ant back two steps and adding the place where the ant encountered the deadlock to the global contraindication table. To reduce. If an ant causes a deadlock due to his path, move the ant back two steps. Unlike deadlocks, ants put a local contraindication table where they have self-locked to prevent them from repeatedly entering the same self-locked node.

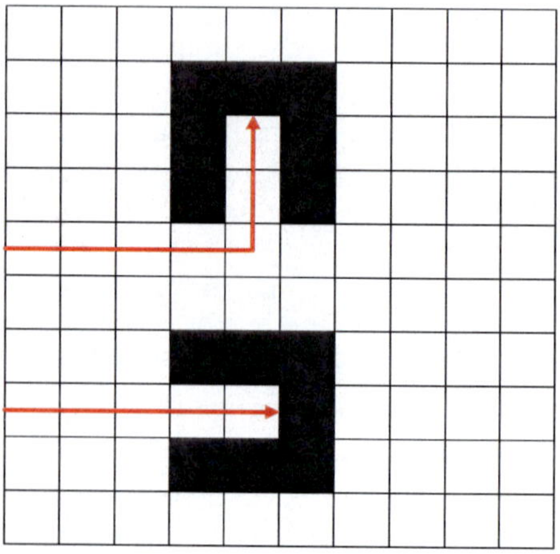

Fig. 4 Ant deadlock

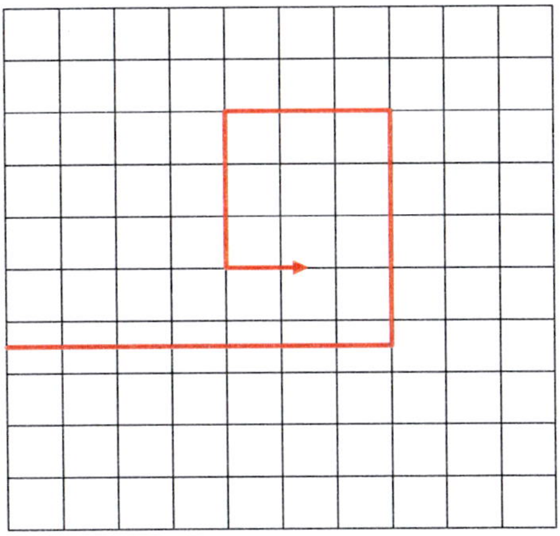

Fig. 5 Self-locking of ants

3.4 Pheromone Update Rules

Pheromone renewal is primarily used to simulate the temporal accumulation and natural volatilization of natural ant pheromones. By introducing rollback, all ants succeed in finding the goal and the search ability of the algorithm is improved, but by increasing the number of ants that succeed in the goal, the pheromone is updated on all routes and redundancy. Therefore, by updating only the optimal ant pheromone with the shortest route and the shortest route, the redundancy of the map pheromone can be reduced and the convergence of the algorithm can be accelerated. The number of ant pheromones to be updated depends on the size of the map, and the larger the map, the more ants will update the pheromones.

3.5 Improvement of Ant Colony Algorithm Flow

The application flow of the improved ant colony algorithm in the path planning of autonomous vehicles is as follows. The specific flow chart is shown in 6.

(1) Based on the known static two-dimensional space, virtual raster map is established to model the environment.
(2) Initialize the parameters and determine the starting and ending positions.
(3) Put the ant in the starting position, calculate the heuristic information, and start the search.
(4) Calculate the probability of an executable node from the transfer probability Eq. (3), search for the next selectable node using the roulette method, add it to the taboo table, and then update the route length.
(5) Determine if the ant has reached target point B, and if so, record the barrier number and route length that the ant followed. If the target point cannot be

reached, repeat step 4 until the target point B is found. Roll back if the search for an ant is delayed, and perform step 6 when all ants reach the target point.

(6) Determine if the algorithm has reached the maximum number of iterations. When the algorithm reaches the maximum number of iterations, it outputs the shortest path that the current ant colony is searching for. If the maximum number of iterations is not reached, the taboo table is emptied, the maximum number of iterations is reached, the path search continues until the algorithm finishes executing, and the shortest currently searched. The route is output.

4 Experimental Simulation and Analysis

The grid method is used to model the road in an area, and the autonomous vehicle needs to start from the upper left of the road in the area and reach the target point in the lower right. At the same time, because the vehicle itself has a certain volume, some relatively narrow roads cannot be passed, so the volume of the vehicle needs to be considered in the grid method modeling, and the road that the vehicle cannot pass is obstructed.

Simulation of route planning by MATLAB software, environment modeling by raster method. The scale of the grid environment map was 20 × 20 m. In 20 × 20 m experimental environment, the path planning method of classical ant colony algorithm is compared with the path planning method of improved ant colony algorithm proposed in this paper and the path planning method of literature [10] algorithm. The parameter settings of this experiment are shown in Table 1.

As the simulation results are shown in Figs. 6 and 7, the classical ant colony algorithm has a slow convergence rate and a long search for the optimal path. Reference [10] Although the convergence speed was improved by improving the algorithm of the ant colony, it was not possible to identify the search direction quickly in the early

Table 1 Parameter settings

Parameter	Classical ant colony algorithm	In this paper, algorithm	Literature [10] algorithm
The number of ant M	30	30	30
Pheromone elicitation factor α	1	1	1
Expectation heuristic factor β	7	7	7
Maximum iteration N max	50	50	50
Pheromones increase the intensity factor Q	1	1	1
Pheromone volatile factor ρ	0.7	0.7	$\frac{1}{\sqrt{2\pi}\sigma}e^{\frac{(n-\mu)^2}{2\sigma^2}}$

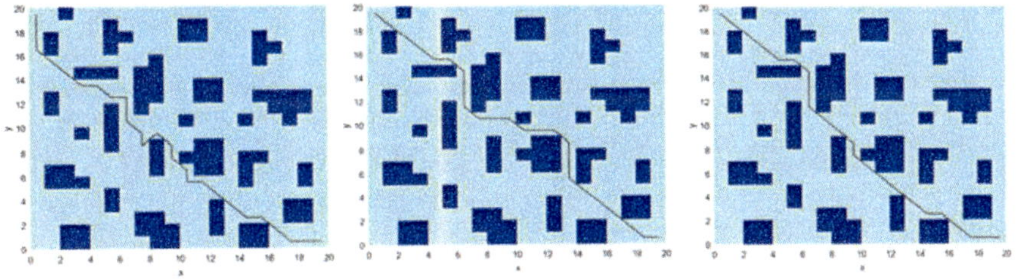

(a) Classical ant colony algorithm path planning diagram (b) Path planning diagram of improved algorithm in literature [10]

(c) This paper improves the ant colony path planning diagram

Fig. 6 Path planning

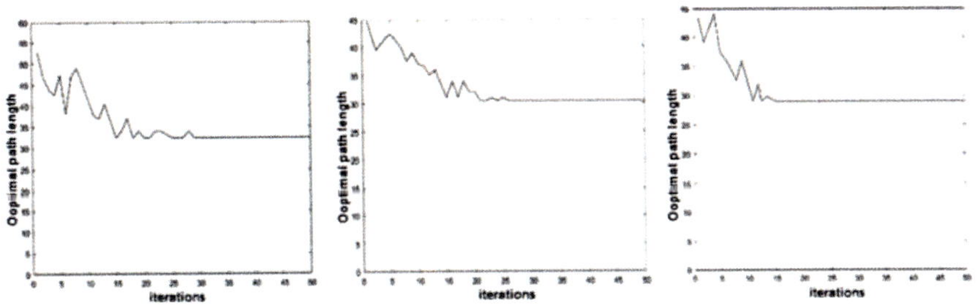

(a) Classical ant colony algorithm iteration curve (b) Literature [10] improved algorithm iteration curve

(c) In this paper, the iterative curve of ant colony algorithm is improved

Fig. 7 Iteration curves of path planning for three algorithms

stage. As shown in Table 2, the improved ant colony algorithm in this paper reduced the optimal path length searched by 2 m and the number of iterations by 14 times in the same barrier environment compared to the classical ant colony algorithm... Compared with the literature [10], the optimum path length searched by the algorithm is reduced by 1.1 m, the number of iterations is reduced by 10, the optimum path length is reduced, and the convergence performance of the algorithm is greatly improved. In summary, the route planning method proposed herein has significant improvement and better search performance in this experimental environment.

Table 2 Comparison of simulation results

Algorithm	Optimal path length	Iterations
Classical ant colony algorithm	32.4	29
Literature [10] algorithm	30.4	25
In this paper, algorithm	29.3	15

5 Conclusion

In this paper, we propose an improved ant colony algorithm for the route planning of self-driving vehicles. The path planning is pre-planned in the early stage, and the initial pheromone adaptive uneven distribution is established to avoid blind search in the early stage. Then the heuristic function is improved to improve heuristic and search efficiency, and a rollback strategy is proposed to prevent deadlock and self-lock in search process. Finally, the pheromone updating rules are improved and optimized to reduce pheromone redundancy and accelerate convergence. Through comparison experiments of different algorithms, it is found that the improved ant colony algorithm has fewer iterations and faster convergence speed, so it has better path optimization ability when applying the algorithm to autonomous vehicle path planning.

References

1. Fan X, Luo X, Yi S, Yang S, Zhang H (2003) Optimal path planning for mobile robots based on intensified ant colony optimization algorithm. In: Proceedings of the IEEE international conference on robotics, intelligent systems and signal processing, 2003, Changsha, Hunan, China, vol 1, pp 131–136
2. Mohanty PK, Parhi DR (2016) Optimal path planning for a mobile robot using cuckoo search algorithm. J Exp Theor Artif Intell 28(1–2):35–52
3. Zhang Y-Y, Zhang Z, Wang Q (2018) Robot path planning based on improved multi-step ant colony algorithm. Comput Eng Des 39(12):3829–3834, 3866
4. Zhang Q, Chen B-K, Liu X-Y, Liu X-Y, Yang H (2019) Optimal path planning for mobile robot based on improved potential field ant colony algorithm. Trans Chin Soc Agric Mach 50(05):23–32, 42
5. Wang H, Wang G, Pan D (2017) Path planning for mobile robot based on improved particle swarm optimization algorithm. Sens MicroSyst 36(05):77–79
6. Jiang M, Wang F, Ge H, Sun L (2019) Research on path planning of mobile robot based on improved ant colony algorithm. Chin J Sci Instrum 40(02):113–121
7. Li S, You X, Liu S (2020) Double ant-state ant colony algorithm combined with ABC algorithm for dynamic classification. Comput Eng Appl 56(12):37–46
8. Zhou W, Li J (2018) Review on obstacle avoidance path planning for autonomous vehicle. Automot Eng 2018(05):55–58
9. Du Y, Zhang Y, Zhao H (2020) Research on robot path planning based on parameter optimization ant colony algorithm. Mod Manuf Eng 2020(09):7–14
10. Niu L, Ji Y (2020) Robot path planning based on improved ant colony algorithm. Microprocessors 41(01):37–40

Optimized Component Learners Diversity of Traffic State Forecasting Model with Multimode Perturbation

Qingchao Liu, Tianyu Xu, Chun Li, and Shiqi Nie

Abstract Based on optimizing the diversity of component learners, this paper puts forward a method of traffic state prediction NNPDAP. In this paper, the perturbation training data set, perturbation input attribute and perturbation learning parameter are used to construct eight perturbation modes for optimizing the diversity of component learners. There have built three groups of experiments respectively for comparing the accuracy of traffic state prediction, error distribution, and time efficiency. The experimental results show that, by enhancing the diversity of component learners can improve the accuracy of prediction, so this method has a stronger competitiveness compared with no perturbation method.

Keywords Traffic state forecasting · Perturbation mode · Diversity · Component learner

1 Introduction

The ability to timely and accurately forecast the evolution of traffic is very important in traffic management and control applications [1]. In recent years, traffic forecasting has become a crucial task in the area of intelligent transportation systems (ITS), playing a fundamental role in the planning and development of traffic management and control systems. The goal is to predict traffic conditions in a transportation network based on its past behavior. Improving predictive accuracy within this context would be of extreme importance, not only to inform travelers about traffic states

Q. Liu (✉) · T. Xu
Automotive Engineering Research Institute, Jiangsu University, Zhenjiang, China
e-mail: lqc@ujs.edu.cn

Q. Liu
State Key Laboratory of Advanced Design and Manufacturing for Vehicle Body, Hunan University, Changsha, China

Q. Liu · C. Li · S. Nie
Higer Bus Company Limited, Suzhou, China

W. Wang et al. (eds.), *Green Transportation and Low Carbon Mobility Safety*,
Lecture Notes in Electrical Engineering 944,
https://doi.org/10.1007/978-981-19-5615-7_48

but also to design and realize infrastructures and mobility services and schedule interventions [2].

Traffic flow forecasting aims at estimating the number of vehicles given a specific region and a time interval, which is an important problem to address in transportation management [3–5]. Reliable, accurate, and consistent real-time traffic flow prediction should support: (1) real-time route guidance in advanced traveler information systems for saving time and money; (2) reliable traffic control strategies in advanced traffic management systems for reducing traffic congestion and accidents; and (3) the evaluation of these dynamic guidance and control strategies [6].

Many traffic flow prediction methods have been proposed in the past. These include ARIMA [7–12], Kalman Filter [13, 14], Nonparametric Regression [15–17], Support Vector Regression [18–20], and Neural Networks (NNs) [21–23], among others Time Series models. However, these models have poor prediction performance under some conditions. For example, Kalman filtering theory has been proposed for predicting short-term traffic states [24], it needs stable traffic flow timely and accurately but the real traffic data are unstable and nonlinear. The NNs model can accurately describe the nonlinear relationship among the influencing factors in traffic flow and provide easy to understand results [25]. In addition, some neural network models need to set appropriate parameters to achieve better prediction results. The Support Vector Regression model is based on statistical learning theory and structural risk minimization principle. When there are many prediction variables, the model will take a long time to train, and appropriate parameters need to be set to obtain better prediction results [26]. The K Nearest Neighbor (KNN) model is an effective nonparametric regression model applied in short-term traffic forecasting that can provide reliable data to guide travelers, which is easy to implement because the process of training data and estimating parameters is simple. It has more portability, higher accuracy, and a simpler structure than the parametric models do [27]. Nonetheless, the search algorithm and method of optimizing the diversity in this model should be improved.

Smith et al. [28] pointed out in the study that traffic flow has complex spatiotemporal behaviors, which are characterized by irregular randomness. No model can accurately predict traffic flow under all conditions, so it is difficult for a single model to capture all disturbance modes. Single model is unable to meet the diverse features of historical traffic flow. Recently, many scholars have used the generalization ability of multiple forecasting methods to establish combined models. For example, Li et al. [29] proposed an ensemble learning framework to appropriately combine estimation results from multiple macroscopic traffic flow models. Moretti et al. [21] composed of a neural networks ensemble with a simple statistical model and compare the results over the one hour forecast, then we improved ensemble model with BAGGING (Bootstrap Aggregation). When bagging is working, it will bootstrap sampling and then interfere with the training data. This method can generate accurate but diverse component learners on unstable basic learners, but it can not be applied on stable basic learners (such as KNN).

It can be found from the above literatures that there are still two defects in the existing traffic state prediction method, Firstly, the traffic flow parameter prediction has important significance for traffic managers, but for travelers, they prefer to pay more attentions to the evolution of traffic state, and require timely and accurate traffic state information. As the traffic flow parameters been predicted by the traditional method, then the parameters are transformed into traffic states and released to travelers, therefore the timeliness of the use has been reduced. Secondly, the existing traffic state prediction methods mainly research the potential law in history traffic flow time series, including the change rule of three parameters of traffic flow, space–time laws in the morning and evening peak and so on, but a lot of internal correlations existing in traffic phenomenon rendered in history traffic flow data have yet to be discovered, and cannot be quantitative description through the formulation method. For example, whether there has correlation in upstream traffic flow data of this Monday and last Monday incidents. Therefore, by building the diversity of component learners to carry on the traffic state prediction will be conductive to finding the inner evolution law of the macroscopic traffic flow, and helping travelers to obtain real-time traffic state information.

Based on the above analysis, this paper proposes an approach combining the component learners (in this study we choose KNN as component learner) with multimode perturbation strategy for short-term traffic state forecasting. Adopting multimode perturbation can identify the most important features to further improve the prediction performance, and it also helps to reduce the forecasting bias. Although these methods and tools have been applied and developed, they are rarely applied to traffic state prediction, and their integration into the proposed framework of this approach is new.

2 Methodology

2.1 Traffic Flow Forecasting

The overall framework of the proposed approach NNPDAP (Nearest Neighbor by Perturbing Data-Attribute-Parameter) is shown in Fig. 1. As we all know, when component learners have high accuracy and diversity, they can form a strong ensemble [30]. This paper from the perspective of enhancing the diversity of member learners, optimize the ensemble learning algorithm to improve the accuracy of traffic state forecasting. The most classic ensemble learning method is bagging algorithm, which can generate multiple samples from the original training set, and then train a component learner for each sample. Its prediction results are determined by the combination of votes. In addition, other types of perturbation can also be used for ensemble construction [31]. Including input attribute of perturbation traffic flow time series, parameters of perturbation traffic flow time series model, the diversity of component learners will be enhanced through multimode perturbation, the complex

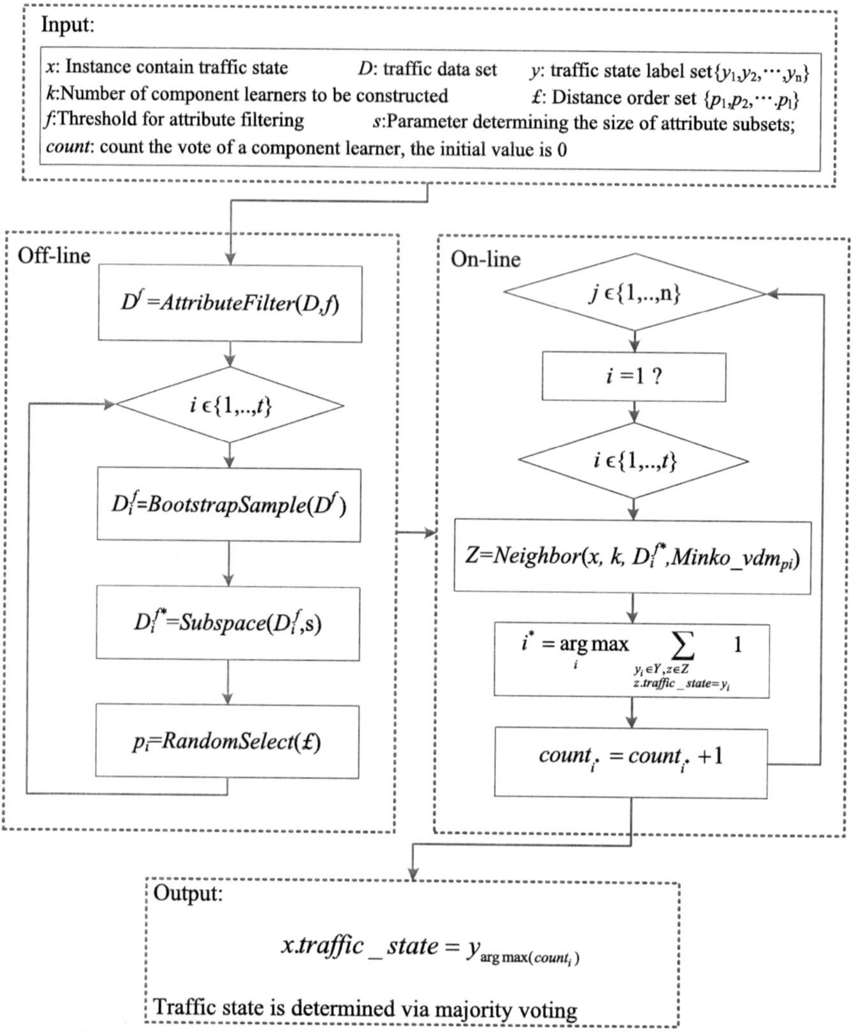

Fig. 1 The overall framework of the proposed approach NNPDAP

space–time characteristics of traffic flow time series will be grasped. All these are respectively described as follows:

(1) perturbation traffic flow training data: introduction of Bootstrap sampling of Bagging algorithm, given a learning algorithm Learner abbreviated as L and a training set T, let the learning algorithm training several rounds, the training set of each round is composed by n training samples taken from the initial training sets randomly, the initial training examples can be concentrated appear multiple times in one round of training or not.

(2) perturbation input attribute of traffic flow time series, with the passage of time, the space–time characteristics of traffic flow data regularity are presented, so with multiple section history traffic flow data to construct the training set, thus become the multidimensional attribute data, by using random subspace strategy to conduct attribute perturbation, enhance the diversity of component learners, and reduce the time cost. Random subspace strategy to ensure the performance of component learners, irrelevant attributes should be filtered. A simple attribute filtering process based on information gain criterion will be used here. Its working principle is to evaluate the information gain of all original input attributes and remove the attributes whose information gain is less than the threshold f.

(3) perturbation of traffic flow time series model parameters: the KNN classification using Euclidean distance measure, Euclidean distance is actually special minkowski distance, namely when $p = 2$, minkowski distance would translate into Euclidean distance, as shown in formula 1.

$$Minkowsky_p(x_1, x_2) = \left(\sum_{n=1}^{d} |x_{1,n} - x_{2,n}|^p\right)^{1/p} \tag{1}$$

Here x_1 and x_2 are two instances descryibed by d-dimensional continuous attribute vectors.

In order to allow Minkovsky distance to deal with the categorical attributes, the introduction of VDM (Value Difference Metric) and calculate the distance between u and v by Eq. 2. The number of training examples holding u on a is $N_{a,u}$, the number of training examples belonging to the c-*th* traffic state and holding u on a is $N_{a,u,c}$, and the number of traffic states is n.

$$VDM_P(u, v) = \sum_{c=1}^{n} \left|\frac{N_{a,u,c}}{N_{a,u}} - \frac{N_{a,v,c}}{N_{a,v}}\right|^p \tag{2}$$

The new distance metric (Minko_vdm) will be developed using the combination of Minkowsky distance and VDM. In Eq. 3, the first j attribute is the classification attribute, and the remaining (d–j) ones are continuous attributes normalized to [0, 1]. We perturbs the learning parameter of KNN through randomly choosing a p value to instantiate Eq. 3 for each component KNN.

$$Minko\_vdm(x_1, x_2) = \left(\sum_{h=1}^{j} VDM_p(x_{1,h}, x_{2,h}) + \sum_{h-j+1}^{d} |x_{1,h} - x_{2,h}|^p\right)^{1/p} \tag{3}$$

The NNPDAP algorithm works as follows.

First, filter the original training set D and eliminate irrelevant attributes. The result data set containing the remaining attributes is represented in Df. Next, t bootstrap

Table 1 Multimode perturbation and level of diversity

Perturbation mode	Perturb data with bootstrap sampling	Perturb input attributes with		Perturb learning parameters with distance configuration	Diversity level
		Attribute filtering	Attribute subspace		
Mode 0	X	X	X	X	–
Mode 1	√	X	X	X	A_1
Mode 2	X	X	√	X	A_2
Mode 3	X	X	X	√	A_3
Mode 4	X	√	√	X	B_1
Mode 5	√	X	X	√	B_2
Mode 6	√	X	√	√	C_1
Mode 7	X	√	√	√	C_2
Mode 8	√	√	√	√	D

samples will be generated from Df using NNPDAP algorithm, and the attribute subset of each sample will be randomly described.

Parameter *s* controls the size of the attribute subset. The *p* value of the Minko_vdm distance will be randomly selected from the $\{p_1, p_2,\ldots, p_l\}$ distance sequence set. When the new instance is marked, the *p* value will be used to identify its adjacent training examples on different samples. On each sample, the identified neighbors vote for the label of the new instance, while the result is regarded as the prediction made by the component KNN of the sample. Finally, the predictions of the component KNN are combined via majority voting.

In the process of traffic state prediction, will the above way of disturbance are combined, can generate eight kinds of perturbation method, the specific combination as shown in Table 1. The Mode 0 indicated there was no disturbance, Mode 8 indicated NNPDAP algorithm. An empirical analysis of the follow-up will contrast analysis of the eight kinds of perturbation method.

2.2 Data Source

In this study, Traffic flow data collected by 8 loop detectors for 31 days, from Dec. 1st to Dec. 31st in the year 2016 is used in the experiments. These loop detectors were installed around the U.S. Route 26 (US-26) in Portland, Oregon, USA. US-26 is a major cross-state state highway in the U.S. state of Oregon, connecting U.S. Route 101 on the Oregon Coast near Seaside with the Idaho state line east of Nyssa. The Portland Oregon Regional Transportation Archive Listing (PORTAL) has been developed by Portland State University in cooperation with the Oregon Department of Transportation (ODOT) and other regional transportation agency partners. The

Table 2 Traffic data collection location

Detector ID	Station ID	Lane number	Direction	Highway	Milepost	Data integrity
100857	1085	3	Eastbound	US 26	65.9	Yes
100858	1085	2	Eastbound	US 26	65.9	Yes
100859	1085	1	Eastbound	US 26	65.9	No
100860	1086	3	Eastbound	US 26	67.4	Yes
100861	1086	2	Eastbound	US 26	67.4	Yes
100862	1086	1	Eastbound	US 26	67.4	No
100863	1087	3	Eastbound	US 26	68.55	Yes
100864	1087	2	Eastbound	US 26	68.55	Yes

identification numbers (IDs) for these loop detectors are 100857, 100858, 100859, 100860, 100861, 100862, 100863 and 100864. As shown in Table 2, 100859 and 100862 collected traffic flow data are seriously missing, so they did not use. The aggregation period is 5 min, thus leading to 288 sample points per day. The total number of sample points is $288 \times 6 \times 31 = 53{,}568$. The proposed method was further evaluated under different traffic states that were simply defined by different speed ranges, i.e. 0–20, 20–40, 40–60, 60–80, and $\geq$80 km/h, according to the measured speeds by the loop detectors.

3 Empirical Study

3.1 *Forecasting Performances Comparison*

Prediction performance of perturbation mode is evaluated from the perspective of error, MAE, RMSE and MAPE are chosen as the evaluation index [1]. As shown in Table 3, Mode 1 predicts in six sections and takes 3 times of optimal value, Mode 2, 4, 5, 6, 7 take four times of optimal values, compared with Mode 0 prediction performance has improved. Mode 8 takes 5 times of optimal value, forecasting is the best. It indicates various perturbation modes can get desirable prediction performance in parallel. Mode 3 only gets one time of optimal, indicates that the disturbance model has no improvement in prediction. Thus, through the perturbation, enhanced the diversity between the members can improve the efficiency of prediction to a certain extent.

The diversity level can be divided into four levels, that is, A, B, C, D, by level A to level D, diversity increased gradually. According to different perturbation mode of each level, can be divided into eight levels, namely $A_1, A_2, A_3, B_1, B_2, C_1, C_2, D$. Diversity of Level A has obtained the optimal value eight times, including A1, A2, A3 each 3 times, 4 times and 1 times; Level B and Level C diversity have all eight times, including B_1, B_2, C_1, C_2 each 4 times; Level D obtains 5 times of optimal value. From

Table 3 Forecasting performances of different perturbation modes (%)

Perturbation modes (diversity level)		Mode 0	Mode 1 (A1)	Mode 2 (A2)	Mode 3 (A3)	Mode 4 (B1)	Mode 5 (B2)	Mode 6 (C1)	Mode 7 (C2)	Mode 8 (D)
100857	MAE	23.07	23.07	25.17	23.77	**22.72**	24.12	25.17	**22.72**	24.47
	RMSE	52.88	56.71	58.63	53.54	**52.55**	57.63	58.53	**52.55**	57.93
	MAPE	11.97	11.62	11.62	11.62	13.46	12.32	**11.27**	13.46	11.62
100858	MAE	10.8	10.45	10.45	**10.1**	10.45	9.75	**10.1**	10.45	10.45
	RMSE	38.7	36.38	36.38	37.79	**35.41**	**35.41**	35.9	**35.41**	36.38
	MAPE	4.93	4.99	4.61	4.64	5.08	**4.52**	**4.52**	5.08	4.61
100860	MAE	11.88	11.88	12.23	12.23	**10.48**	11.88	11.88	**10.48**	10.83
	RMSE	50.86	49.47	49.11	51.88	48.76	49.47	49.47	48.76	**47.67**
	MAPE	6.35	6.78	**6.26**	6.52	6.35	6.78	6.61	6.35	**6.26**
100861	MAE	14.63	15.67	**14.28**	14.63	15.33	14.98	14.63	15.33	**14.28**
	RMSE	47.95	53.12	47.59	47.95	50.08	50.43	47.95	50.08	**47.59**
	MAPE	5.95	6.67	**5.6**	5.95	7.17	6.44	5.95	7.17	**5.6**
100863	MAE	38.81	37.76	37.06	37.06	36.36	**36.01**	37.06	36.36	36.71
	RMSE	65.04	63.68	63.13	63.13	63.68	**62.29**	63.13	63.68	62.85
	MAPE	20.07	19.43	**18.24**	18.73	19.17	18.73	**18.24**	19.17	18.47
100864	MAE	25.78	**23.69**	25.43	25.43	26.13	24.39	25.08	26.13	24.04
	RMSE	57.22	**52.79**	55.05	56.92	56.92	53.45	54.74	56.92	53.12
	MAPE	11.49	**10.74**	10.91	11.46	13.55	10.97	10.91	13.55	10.91

the perspective of diversity level, the diversity of Level D is the highest, and Level D perturbation model has the minimum error and better prediction precision, to some extent, it indicates that the higher the level of diversity, the more accurate prediction it will be. From the perspective of perturbation model, 8 kinds of perturbation model corresponding to 8 kinds of diversity level, perturbation model between A_1, A_2, A_3, is relatively independent, diversity between B_1 and B_2 is global complementary, diversity between C_1 and C_2 is local complementary, Level D belongs to the parallel perturbation model. Independent perturbation obtains 8 times of diversity optimal value, and complementary perturbation obtains 12 times of diversity optimal value, the parallel perturbation obtains the least diversity optimal value, that is to say, when making traffic state prediction, using complementary way to enhance the diversity of the component learners, to a certain extent, enhanced the robustness and fault tolerance of the proposed method for traffic flow forecasting model provides a new idea.

3.2 Error Distribution

In order to further study the error distribution of traffic state prediction, this paper adopts the Formulas (4)–(6) to do comparison on the error distribution of the multi-mode perturbation method. Dev-0 indicates the deviation of forecasting the level of traffic state is zero, also the accurate of traffic state prediction. By analogy, Dev-1represents the deviation of traffic state prediction is Level 1, which is the actual traffic state and predicted traffic state differs a level. Dev-H indicates the predicted traffic state differ with the actual traffic state by multiple levels, including Level 2, 3 and 4. As shown in Table 4, Mode 0 didn't get the optimal value of Dev-0, it reflects that multimode perturbation to increase the diversity of the component learners and improve the prediction precision. Compared with no perturbation method, prediction accuracy of Mode 1, 3, 4, 5, 7 and 8 have increased about 1%, the prediction accuracy of 100863 section has increased more than 3%. Additionally, among eight kinds of perturbation method, NNPDAP obtains optimal value twice, but it can be found that NNPDAP obtained optimal in Dev-0 at the same time, the ratio of Dev-H is higher, easy to affect NNPDAP error evaluation results. Accordingly, if you want to reduce prediction error, must reduce the proportion of Dev-H, in other words, the proportion of multi-level traffic state deviation needs to be reduced. In the stage of training, how to introduce Dev-H index optimization model will be the focus of the next step.

Table 4 Error distribution of different perturbation modes (%)

Error distribution		Mode 0	Mode 1	Mode 2	Mode 3	Mode 4	Mode 5	Mode 6	Mode 7	Mode 8
100857	Dev-0	79.02	**80.07**	77.97	78.32	79.37	79.02	77.97	79.37	79.02
	Dev-1	19.23	17.83	19.93	19.93	18.88	18.88	19.93	18.88	18.88
	Dev-H	1.75	2.10	2.10	1.75	1.75	2.10	2.10	1.75	2.10
100858	Dev-0	90.94	90.94	90.94	**91.99**	90.59	90.94	91.28	90.59	90.94
	Dev-1	7.67	7.67	7.67	6.62	8.36	7.67	7.32	8.36	7.67
	Dev-H	1.39	1.39	1.39	1.39	1.05	1.39	1.39	1.05	1.39
100860	Dev-0	92.66	92.66	91.96	92.66	**93.71**	91.96	91.96	**93.71**	93.35
	Dev-1	4.89	4.54	5.59	4.54	4.19	4.89	5.24	4.19	4.19
	Dev-H	2.45	2.80	2.45	2.80	2.10	3.15	2.79	2.10	2.45
100861	Dev-0	88.15	88.15	88.50	88.50	87.80	87.80	88.15	87.80	**88.85**
	Dev-1	10.10	9.75	9.76	9.76	10.45	10.80	10.10	10.45	9.41
	Dev-H	1.74	2.09	1.74	1.74	1.74	1.39	1.74	1.74	1.74
100863	Dev-0	62.94	64.68	64.34	63.64	65.73	**66.08**	63.98	65.73	64.34
	Dev-1	35.31	33.57	34.27	34.61	32.17	32.52	34.61	32.17	34.27
	Dev-H	1.74	1.75	1.39	1.75	2.10	1.39	1.39	2.10	1.39
100864	Dev-0	77.35	77.70	77.00	77.70	77.00	77.00	77.00	77.00	**78.39**
	Dev-1	19.86	20.21	20.56	19.51	19.86	19.86	20.91	19.86	18.47
	Dev-H	2.79	2.10	2.43	2.78	3.14	3.14	2.10	3.14	3.14

$$\text{Zero-Level-Deviation} = \frac{n_{right}}{n_{total}} \times 100\% \tag{4}$$

$$\text{One-Level-Deviation} = \frac{n_{one-deviation}}{n_{total}} \times 100\% \tag{5}$$

$$\text{High-Level-Deviation} = \frac{n_{total} - n_{right} - n_{one-deviation}}{n_{total}} \times 100\% \tag{6}$$

3.3 *Efficiency Comparisons*

Not only the accuracy but also the efficiency should be considered in the prediction model. Table 5 shows the training time (TT) and the forecasting time (FT) of each of the twelve competing models. The time required to use the training data set to estimate the model is the training time, and the time required to output the predicted values of all test samples is the prediction time. The computation time of each model was calculated and recorded on a computer equipped with a 2.5 GHz 4-core CPU and 4G RAM. In the experiment, JAVA language was used to compile all prediction models.

Table 5 Computation time of different perturbation modes (in millisecond)

Mode	Time	100857	100858	100860	100861	100863	100864	Average
Mode 0	TT	32	16	31	16	32	31	26.33
	FT	15	15	16	15	15	16	15.33
Mode 1	TT	31	30	32	15	31	31	28.33
	FT	31	12	15	32	16	15	20.16
Mode 2	TT	47	63	46	47	31	47	46.83
	FT	16	15	16	15	16	15	15.5
Mode 3	TT	31	30	32	31	31	31	31
	FT	32	13	15	16	31	16	20.5
Mode 4	TT	265	218	296	219	203	265	244.33
	FT	140	109	156	109	127	156	132.83
Mode 5	TT	31	31	31	32	15	31	28.5
	FT	16	32	15	15	30	31	23.16
Mode 6	TT	47	46	47	47	46	62	49.16
	FT	16	17	31	16	32	16	21.33
Mode 7	TT	390	203	280	351	219	234	279.5
	FT	250	172	219	188	156	187	195.33
Mode 8	TT	265	297	265	205	265	202	249.83
	FT	203	218	250	187	234	172	210.67

As shown in Table 5, in terms of training time, Mode 0, 1, 2, 3, 5, 6 consumed less than 50 ms in training, and the average time was about 35 ms. However, the training time of Mode 4, 7, 8 is more than 200 ms, compared with Mode 0 1, 2, 3, 5, 6, training time of Mode 4, 7, 8 has extended more than 150 ms, that is because Mode 4, 7, 8 adopted the integrated perturbation mode, this perturbation method needs to build 100 component learners to improve the level of diversity, each component learner requires prediction and consumes too much time, in the time consumption of prediction, Mode 4, 7, 8 are more time-consuming. Although, as integrated perturbation model enhancing the level of diversity of component learners, at the same time, it consumes more time resources, so you can transfer the time-consuming modeling process to offline by offline modeling way without affecting the real time traffic state prediction.

4 Conclusions

Based on the diversity of optimized component learners, this paper proposes an NNPDAP method to forecast traffic states. The proposed approach can be divided into three parts. First, in order to improve the diversity and remove some irrelevant attributes, NNPDAP performs attribute filtering on the original training set. Second, Perturbation training data with bootstrap sampling strategy are used, after that, the p value will then be randomly selected from Minko_vdm to improve diversity. Third, the predictions of the component KNN are combined via majority voting.

In the empirical analysis stage, with six sections of the real traffic flow data taken from the United States US26 highway as the basis, three sets of experiments are constructed for comparison of prediction accuracy, error distribution, time efficiency respectively. Experimental results show that the proposed method NNPDAP in MAE, RMSE and MAPE index achieve optimal value most times, its prediction performance is superior to Mode 0, 1, 2, 3, 4, 5, 6, 7, enhancing the diversity of component learner can help to improve the accuracy of traffic state prediction, it can be found in terms of the diversity level analysis that the complementary perturbation method can obtain better prediction effect, provide a new way of thinking for the future further research; in aspect of error distribution, NNPDAP obtained optimal value twice, but still need to reduce the proportion of Dev-H, and in the future study, it needs to consider introducing Dev-H index; In terms of time efficiency, the time costs of NNPDAP is higher than that of other disturbance forms in the stage of training, can undertake offline model training, in order to reduce the time cost of overall prediction. Based on the above experimental results, NNPDAP method can better extract the underlying information in history traffic flow data, and help to dig out inner evolution law of the unknown macroscopic traffic flow.

Acknowledgements We want to thank the anonymous reviewers for their helpful comments and suggestions. This work was supported by the National Natural Science Foundation of China (U20A20331, U20A20333, 51905223, 52072160, 51875255); Natural Science Foundation of Jiangsu Province (BK20190845, BK20180100); China Postdoctoral Science Foundation

(2021M690069); Key Research and Development Program of Jiangsu Province (BE2020083-3, BE2019010-2, BE2021011-3); Transportation Science and Technology Project of Jiangsu Province (2021G05); Six Talent Peaks Project of Jiangsu Province (2018-TD-GDZB-022); State Key Laboratory of Advanced Design and Manufacturing for Vehicle Body (31915005). Jiangsu Province Research and Practice Innovation Program Project (SJCX20_1425). The authors would like to make a grateful acknowledgment to PORTAL (the official transportation data archive for the Portland-Vancouver Metropolitan region), for providing traffic flow data.

References

1. Habtemichael FG, Cetin M (2016) Short-term traffic flow rate forecasting based on identifying similar traffic patterns. Transp Res Part C-Emerg Technol 66:61–78
2. Lippi M, Bertini M, Frasconi P (2013) Short-term traffic flow forecasting: an experimental comparison of time-series analysis and supervised learning. IEEE Trans Intell Transp Syst 14(2):871–882
3. Pan TL, Sumalee A, Zhong RX, Indra-payoong N (2013) Short-term traffic state prediction based on temporal-spatial correlation. IEEE Trans Intell Transp Syst 14(3):1242–1254
4. Ye Q, Szeto WY, Wong SC (2012) Short-term traffic speed forecasting based on data recorded at irregular intervals. IEEE Trans Intell Transp Syst 13(4):1727–1737
5. Zhang J, Wang F-Y, Wang K, Lin W-H, Xu X, Chen C (2011) Data-driven intelligent transportation systems: a survey. IEEE Trans Intell Transp Syst 12(4):1624–1639
6. Zheng W, Lee D-H, Shi Q (2006) Short-term freeway traffic flow prediction: Bayesian combined neural network approach. J Transp Eng 132(2):114–121
7. Ahmed MS, Cook AR (1979) Analysis of freeway traffic time-series data by using Box-Jenkins techniques. Transp Res Rec: J Transp Res Board 722:1–9
8. Davis GA, Nihan NL, Hamed MM, Jacobson LN (1990) Adaptive forecasting of freeway traffic congestion. Transp Res Rec: J Transp Res Board 1287:29–33
9. Hamed MM, Al-Masaeid HR, Said ZMB (1995) Short-term prediction of traffic volume in urban arterials. J Transp Eng 121(3):249–254
10. Williams B, Durvasula P, Brown D (1998) Urban freeway traffic flow prediction: application of seasonal autoregressive integrated moving average and exponential smoothing models. Transp Res Rec: J Transp Res Board 1644:132–141
11. Williams BM, Hoel LA (2003) Modeling and forecasting vehicular traffic flow as a seasonal ARIMA process: theoretical basis and empirical results. J Transp Eng 129(6):664–672
12. Stephanedes YJ, Michalopoulos PG, Plum RA (1981) Improved estimation of traffic flow for real-time control. Transp Res Rec 795:28–39
13. Yao B, Chen C, Cao Q, Jin L, Zhang M, Zhu H, Yu B (2017) Short-term traffic speed prediction for an urban corridor. Comput-Aided Civ Infrastruct Eng 32(2):154–169
14. Wu Y-J, Chen F, Lu C-T, Yang S (2016) Urban traffic flow prediction using a spatio-temporal random effects model. J Intell Transp Syst 20(3):282–293
15. Huang W, Song G, Hong H, Xie K (2014) Deep architecture for traffic flow prediction: deep belief networks with multitask learning. IEEE Trans Intell Transp Syst 15(5):2191–2201
16. Lv Y, Duan Y, Kang W, Li Z, Wang F-Y (2015) Traffic flow prediction with big data: a deep learning approach. IEEE Trans Intell Transp Syst 16(2):865–873
17. Yoon B, Chang H (2014) Potentialities of data-driven nonparametric regression in urban signalized traffic flow forecasting. J Transp Eng 140(7):04014027
18. Hu J, Gao P, Yao Y, Xie X (2014) Traffic flow forecasting with particle swarm optimization and support vector regression. In: IEEE 17th international conference on intelligent transportation systems, Qingdao, China. IEEE, pp 2267–2268

19. Wang J, Shi Q (2013) Short-term traffic speed forecasting hybrid model based on chaos–wavelet analysis-support vector machine theory. Transp Res Part C: Emerg Technol 27:219–232
20. Wei D, Liu H (2013) An adaptive-margin support vector regression for short-term traffic flow forecast. J Intell Transp Syst 17(4):317–327
21. Moretti F, Pizzuti S, Panzieri S, Annunziato M (2015) Urban traffic flow forecasting through statistical and neural network bagging ensemble hybrid modeling. Neurocomputing 167:3–7
22. Zhu JZ, Cao JX, Zhu Y (2014) Traffic volume forecasting based on radial basis function neural network with the consideration of traffic flows at the adjacent intersections. Transp Res Part C: Emerg Technol 47:139–154
23. Zhang Y, Zhang Y, Haghani A (2014) A hybrid short-term traffic flow forecasting method based on spectral analysis and statistical volatility model. Transp Res Part C: Emerg Technol 43:65–78
24. Wang Y, Papageorgiou M (2005) Real-time freeway traffic state estimation based on extended Kalman filter: a general approach. Transp Res Part B: Methodol 39(2):141–167
25. Ma T, Zhou Z, Abdulhai B (2015) Nonlinear multivariate time–space threshold vector error correction model for short term traffic state prediction. Transp Res Part B: Methodol 76:27–47
26. Caruana R, Karampatziakis N, Yessenalina A (2008) An empirical evaluation of supervised learning in high dimensions. In: Proceedings of the 25th international 13 conference on machine learning, Helsinki, Finland. ACM, pp 96–103
27. Smith BL (1997) Forecasting freeway traffic flow for intelligent transportation systems application. Transp Res Part A 1(31):61
28. Smith BL, Williams BM, Oswald RK (2002) Comparison of parametric and nonparametric models for traffic flow forecasting. Transp Res Part C: Emerg Technol 10(4):303–321
29. Li L, Chen X, Zhang L (2014) Multimodel ensemble for freeway traffic state estimations. IEEE Trans Intell Transp Syst 15(3):1323–1336
30. Krogh A, Vedelsby J (1994) Neural network ensembles, cross validation, and active learning. In: Proceedings of the 7th international conference on neural information processing systems, Denver, Colorado. pp 231–238
31. Dietterich TG (2002) Ensemble learning. In: The handbook of brain theory and neural networks. The MIT Press, Cambridge, MA

Snow Depth Inversion and Analysis of Temporal-Spatial Snow Distribution Along Tianshan Highway Based on MODIS Data

Bo Yin

Abstract Most of Tianshan Highway is located in the high and cold mountain areas over 2,000 m above the sea level in Xinjiang. Highway distress in late stage is exacerbated and frequently occurs especially avalanche, wind blowing snow, geohazard, flood damage, and other disasters along the highway, which seriously affects the traffic. Restricted by the climate and topography and other factors, conventional ground monitoring is hard to be applied into dealing with the emergency situation. In contrast, remote sensing technology can monitor snows in large-scale and real-time. Based on the remote sensing data sources and studied areas, MODIS data is selected to reversely derive snow depth and analyze the temporal-spatial snow distribution in Dushanzi-Kuqa region of Tianshan Highway of northern Xinjiang in recent years. Analysis revealed that the snowfall was mainly concentrated on late October to the early April in the next year. The snow area in July is the least in summer, and reaches largest in winter, accounting for about 35.7% of the entire studied area. Most of the snow depth is at 12 to 15 cm, and the maximum snow cover depth is about 23 cm in the Duzishan region of Wusu County. The study conclusion is of great theoretical and practical significance for distress monitoring and controlling.

Keywords MODIS inversion · Snow cover · Tianshan Highway

The Tianshan Highway, also known as the Doku Highway (Dushanzi-Kuqa County in northern Xinjiang), is a major traffic artery connecting the northern Xinjiang and southern Xinjiang. Most of Tianshan Highway is located in high and cold mountain areas above 2,000 m above the sea level, along which the terrain is steep, the geological structure is complex with much fold and faul. Affected by topography, geological conditions, climate, and limited capital and technology in construction, the Highway distress of Tianshan in the later-stage happens constantly, especially the frequent avalanches, snowstorms, geological disasters, floods and other disasters, seriously affecting road traffic. Avalanches and wind blowing snow often cause road

B. Yin (✉)
Shandong Police College, No. 54 Wenhuadong Road, Lixia District, Jinan, Shandong, China
e-mail: 114136171@qq.com

W. Wang et al. (eds.), *Green Transportation and Low Carbon Mobility Safety*,
Lecture Notes in Electrical Engineering 944,
https://doi.org/10.1007/978-981-19-5615-7_49

obstruction and damage. Some sections of the road can only be maintained about four months for traffic each year, which causes great harm to the national economy. Snow cover, as the most active natural factor with multi attributes on the earth's surface, has attracted many geoscientists' attention.

In order to prevent and control snow damage of the present roads, it is necessary to master the snow cover's range, type, characteristics, natural conditions, movement law and the degree of snow damage [1]. Restricted by climate and topography, the routine ground station monitoring is difficult to be used to meet the emergency needs. With the development of space technology, snow information extraction from remote sensing has become an important subject of the snow research. Remote sensing technology can be used to monitor snow cover on a large scale. Most of the snow information is extracted by remote sensing based on the high and low reflectivity of snow in the visible and near-infrared bands respectively, and the snow area and depth are retrieved by regression models. The basic research works on snow cover monitoring based on remote sensing data have been done a lot by predecessors [2–8]. For example, Chen and Chen made good use of the snow depth data and the reflectivity RCHI of 127 meteorological stations in the northern Qinghai-Tibet Plateau during the winter of 1985–1987, to obtain the snow depth inversion formula of Qilian Mountains [4]; Wang conducted a snow information extraction and monitoring study on the snowfall in Nagchu, Tibet, from February to March, 1995 by using NOAA/AVHRR and other remote sensing data. And the inversion result is basically consistent with the disaster situation reported by the local government [5]; Zhou et al. used the weather satellite to monitor the snow cover in winter and spring in the pastoral area of Qingnan plateau in southern Qinghai Province, and put forward a method to estimate the snow cover depth and area through AVHRR data, and then tested the accuracy of this method [6]; Based on the MODIS data from July 2003 to March 2005, Wang et al. proposed to combine the normalized difference snow index (NDSI) with the Moderate-Resolution Imaging Spectroradiometer for snow identification, and proved that method can effectively identify the snow cover in Qilian Mountain area [8]. Previous studies show that, thanks to the improved sensors, MODIS satellite data are superior to AVHRR data in spatial resolution and snow retrieval algorithm [9]. In view of the characteristics of The Tianshan Highway, MODIS remote sensing data is applied to study the spatial-temporal distribution and analyze the depth of snow cover along the highway, which is of great theoretical and practical significance for road disease monitoring and controlling.

1 Survey of the Study Area

Construction of the Tianshan Highway began in 1969 and was completed and opened to traffic in 1983 after reconstruction. It is a major artery that connects northern Xinjiang and southern Xinjiang, passing through four Bingdaban namely Tielimaiti, Laerdun, Yuximolegai and Haxilegen, and three great Tianshan prairies, namely bayanbulak, kongnaiz and Qiaoerma, with a total distance of 532 km (in Fig. 1).

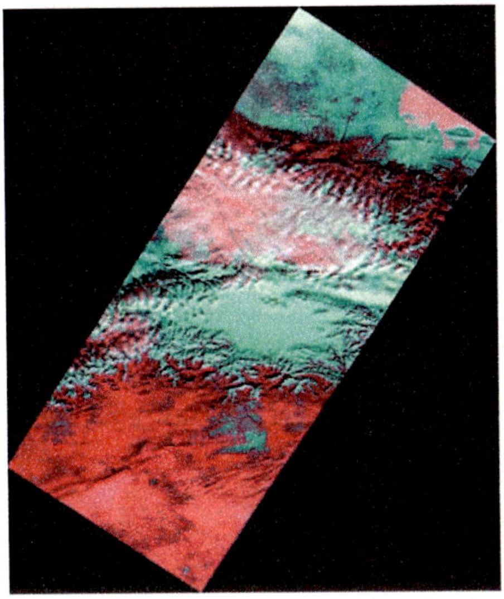

Fig. 1 Mosaic image of study area on day 337, 2010

Most of the Tianshan Highway is located in the Alpine Mountain area above 2,000 m above the sea level. The terrain along the Highway is steep, the geological structure is complex, folds and faults contained, the bedrock is exposed, and the physical and chemical weathering is strong. Vegetation in the region is sparse, mainly coniferous forest and grassland and other alpine cold-resistant vegetation. The climate here is the mountain climate in the temperate inland arid area, and the temperature difference between day and night is big in four seasons. The highest temperature is ≥30 °C, and the lowest temperature is ≤−40 °C. The rainy season is mainly concentrated in May to September, most of which are continuous rainstorm; The snowfall season is mainly concentrated in October to May in the next year. The area is passed through by Tianshan Highway where the annual average temperature is below zero, and precipitation is solid with occasionally sleet [1]. The heavy snowfall and deep accumulated snow cover leave material conditions for avalanche in this area. With limited finance and techniques during the construction, as well as complicated topography, geological conditions and harsh climatic environment in Tianshan mountains, the Highway distress in the later stage continued to be worsen, especially along the road frequently occurred avalanche, wind blowing snow, geological disasters, floods and other disasters, seriously interfering the smooth traffic of the road.

2 Introduction to MODIS

From analysis of spectral curves, snow has high reflectance in visible light band and low reflectance in short-wave infrared band. Accordingly, the snow information can be extracted on the basis of spectral-curves difference. AVHRR and MODIS are the major methods to extract snow cover information by optical remote sensing. AVHRR is an improved very high resolution scanning radiometer on board the TIROS-N/NOAA series satellites of the third generation operational meteorological observation launched by the National Oceanic and Atmospheric Administration. MODIS is Moderate-resolution Imaging Spectroradiometer. It was launched in 1999 on the Earth Observing System (EOS) on the morning star TERRA and the afternoon star AQUA. It has 36 spectral bands ranging from 0.4 to 14.4 μm and its orbital scan width is about 2,300 km. Twice daily exploratory data are available worldwide except in the equator. The data of 36 bands can reveal the numerous reflection information of terrain at the same time, and can be used for long-term observation for land surface, biosphere, atmosphere and ocean. The reflection band of MODIS is from 0.4 to 2.5 μm. The bands related to snow cover are 1, 2, 4, 6, 7, 18, 26 and 31.

Compared with AVHRR, MODIS is with higher temporal and spatial resolution. In particular, MODIS can provide daily and composite snow products with the advantages of wide spectrum and narrow channel, which solves the AVHRR sensor saturation problem caused by the high reflectivity of fresh snow [10]. MODIS snow product is the improved snow monitoring algorithm, which enhanced its sensitivity to snow, less influenced by the atmosphere, solar angle, observation angle, cloud and so on, and provided continuous temporal-spatial comparable global snow information [11]. The snow products include daily product MOD10A1 with 500 m resolution, 8-day composite product MOD10A2 with 500 m resolution and daily product MOD10C1 with 5 km resolution, 8-day composite product MOD10C2 with 5 km resolution and monthly composite product MOD10CM with 5 km resolution [12]. The 8-day snow composite product can effectively remove the cloud influence. This paper is mainly about the eight-day snow composite product of MOD10A2 with 500 m resolution (Maximum Snow Cover). MODIS data of MOD10A2 (MSC) from 2010 to 2011, and 8-day composite product MOD10A2 are also selected.

3 Study on Inversion of Snow Depth

3.1 Principle of Snow Depth Inversion

There is a good correlation between the depth of snow cover and the reflectivity of snow cover in visible light. For example, Giddings et al. carried out related experiments on the relationship between snow depth and reflectivity by using special instruments. According to the experimental results, when the snow depth is less than 30 cm and the underlying surface is black, that is, the visible light can be fully absorbed, the reflectivity of snow surface increases with the increasing of snow depth. There is a good linear relationship between them [13]. When the depth of the snow is over 30 cm, as the snow thickness increases, the surface reflectivity tends to change slowly and becomes saturated [3]. Regardless of the influence of the atmosphere influence and other factors, in fine weather, the relationship between the incident radiation of the Sun at the snow surface can be shown as the following [14, 15]:

$$Rsun = Rsa + Rsr + Rua + Rur + Rga + Rgr \tag{1}$$

$Rsun$ is the total radiation of the sun to the pixel. Rsa is the absorption radiation of the snow particles. Rsr is the reflection radiation of the snow particles. Rua is the absorption radiation of the underlying surface of the snow particles. Rur is the reflection radiation of the underlying surface of the snow particles. Rga is the absorption radiation of the objects not covered by the snow surface, Rgr is the reflection radiation of the objects not covered by the snow surface. Meanwhile, the relationship among the reflections received by the satellite can be shown as the following:

$$Rsat = Rsr + Rur + Rgr \tag{2}$$

$Rsat$ is the reflected radiation from the satellite. Under the premise that the total radiation is the same, as the snow depth increases, Rua and Rga decrease and the total radiation loss is less. What's more, Rur and Rgr also reduced. As a result, in the visible and near-infrared wavelengths, the total radiation mainly comes from Rsr reflected by snow particles. Therefore, the deeper the snow, the higher the reflectivity of near-infrared and visible light. When the snow depth reaches a certain degree, the satellite's reflection in these bands tends to be stable. In short infrared band, snow particles absorb Rsa mainly. The deeper the snow, the stronger the absorption and the lower the reflectivity. A negative correlation exists between snow depth and the reflectivity in short infrared band [16].

3.2 *Inversion of Snow Depth from MODIS Data*

It is found that MODIS band 1 and band 3 are suitable for snow depth retrieval. A bilinear regression equation for snow depth inversion is established based on the reflection spectra of band 1 and band 3 and the regression analysis of snow depth data from 80 samples in 35 stations. On the basis of the research, this paper establishes the snow depth inversion equation:

$$S = -0.18 + 15 \times ch1 + 24 \times ch3 \tag{3}$$

S refers to snow depth, $ch1$ and $ch3$ refer to MODIS band 1 and band 3 respectively.

The study area of Tianshan Highway is long and narrow, extending from the north to the south, spanning 80–85°E, 40–45°N. There are two data blocks, namely h24v04 and h23v04. After transforming the data format and projection, the data map projection is transformed to UTM. Using the snow depth retrieval model, the MODIS data in the winter of 2010 were processed to generate the snow depth retrieval image. According to the analysis, most areas of snow depth along the road is 12–15 cm deep, the deepest part locates in Wusu's Duzi Mountain, as deep as 23 cm around.

4 Snow Coverage Extraction

4.1 *Spatio-Temporal Distribution of Snow Cover*

In this study, the MODIS image is preprocessed by remote sensing software, and the band offset in the image is eliminated. The temporal distribution of snow cover was analyzed based on the data of snow cover during 2002 to 2011, and the detailed information of each period in the ten years was obtained.

According to the analysis, the seasonal variation of snow cover in the study area is significant. It mainly snows from the end of October to the beginning of April of the next year, with a maximum snow cover in winter. The snow begins to cover the land from the middle of October, and comes to the peak from late November to early March of the following year. The snow cover decreases sharply from April to June, because of the rising temperature, and reaches the minimum during early June to middle October, especially in summer, namely in July. According to the analysis of the average annual snow area, the maximum appeared in 2006, reaching 27,413.2 km^2, accounting for 17.2% of the total study area. The minimum snow cover in 2007 was 24,221.3 km^2, accounting for 15.1% of the total area. The study found that:

Spring: snow cover is distributed in a relatively wide range, accounting for 12.29% of the total area. The snow mainly covers the north of the Irena, Habierga Mountain, locating in the north and centre of the study area, but less at high-altitude area. In the centre Ulatus basin and southern Xinjiang of the study area, there's almost no snow.

Summer: the snow cover distribution is the smallest, accounting for 1.86% of the total study area. It is mainly distributed in the high parts of the Irena, Habierga Mountain, and also a little in the high mountains of Kirkteik Mountain.

Autumn: snow cover is distributed in a relatively small range, accounting for 6.29% of the total study area. Snow cover is mainly distributed in mid-high and the highest parts of the mountains, including the high-altitude area of the Irena, Habierga Mountain, and Kirkteik Mountain. However, the snow cover is less in the Ganyuladus basin or even none, due to the climate.

Winter: snow cover is distributed the most widely, mainly in the northern and central parts of the study area, accounting for 35.7% of the total. Almost most of the north of the Irena, Habierga Mountain is covered with snow, but the south slope at a higher altitude and the Ulatus basin are less covered. Meanwhile, in the south of Xinjiang, the distribution of snow cover is less in winter, because the water vapor from the north is blocked by the mountain.

5 Conclusion

Remote sensing technology can be used for large-scale and real-time monitoring of snow cover, especially in areas with bad climate and complex terrain. Tianshan Highway is located in the high and cold mountainous area of Xinjiang. Avalanche, wind blowing snow and other disasters occurs frequently along the way, which seriously affect the traffic [17]. In this paper, the MODIS remote sensing data are used to inverse the depth and distribution of snow cover from Dushanzi District to Kuqa County in the north of Tianshan Highway. It is found that, among recent years, 2006 witnessed the maximum snow cover area, reaching 27,413.2 km^2, accounting for 17.2% of the total study area. 2007 witnessed the minimum snow cover area, as low as 24,221.3 km^2, accounting for 15.1% of the total research area. Based on the analysis of snow cover in 2011, the snow cover area in the four seasons is different, and from the largest to the smallest one can be ranked as: winter, spring, autumn and summer, with a snow cover area accounting for 35.7%, 12.29%, 6.29% and 1.86% of the total area in study respectively. The snow cover in winter is the widest one. In most areas, the snow cover is 12–15 cm deep, and the deepest part is in the Mount Duzi in Wusu, reaching about 23 cm deep.

References

1. Hu R (1978) Study on snow cover, snow damage and its prevention and control in Ten-zan Mountain area of China. Xinjiang Geogr 4(2):84–94
2. Hall DK, Riggs GA (2007) Hydrol Process 21:1534
3. Xu XL, Liang D, Tsang L et al (2010) IEEE J Sel Top Appl Earth Obs Remote Sens 3(4):689
4. Chen G, Chen T (1991) Climatic analysis of seasonal snow cover resources in Qilian Mountain area. Study Geogr (01)
5. Wang L (1998) Application of remote sensing technology in monitoring snow disaster in pastoral area. Remote Sens Technol Appl 13(2):32–36
6. Zhou Y, Jia S, Liu P (2001) Estimation of snow parameters using NOAA/AVHRR data. Meteorol Sci 21(1):117–121
7. Guo Y, Zhai P, Li W (2004) A comparative study of snow cover in China between NOAA satellite remote sensing and conventional observation. Glacial Permafr 26(6):755–760
8. Wang X, Zhang Q, Guo N (2007) Identification of snow cover in Qilian Mountains using MODIS data. Dry Weather 25(2):29–34
9. Wu Y, Zhang J, Xu H et al (2007) Research progress of snow information retrieval by satellite. Meteorology 33(6):3–10
10. Zhang J, Xu X, Yan X et al (2003) Progress and direction of remote sensing information application in multi-layer land surface process parameterization research. Appl Meteorol 14(6):745–755
11. Xu H (2007) Application of Modis data in monitoring snow cover in Yunnan Province. Study Geogr Environ Yunnan Prov 19(1):60–62
12. Wu Y, Zhang J et al (2007) Research progress of snow cover retrieval by satellite. Meteorology 33(6):3–10
13. Jiang H, Qin Q et al (2011) Effects of different snow depth and area on snow cover retrieval using remote sensing data. Spectrosc Spectr Anal 31(12):3342–3346
14. Lei H, Li H, Wang J (2020) Preparation of MODIS snow area scale product on Tibetan plateau based on environmental information and regression model. Remote Sens Technol Appl 35(06):25–30

15. Lucas RM, Harrison AR (1990) Snow observation by satellite: a review. Remote Sens Rev 4(2):285–348
16. Shi J, Dozier J (1997) Mapping seasonal snow with SIR2C/X2SAR in mountainous areas. Remote Sens Environ 59(2):294–307
17. Cai H, Han D, Yang L (2020) Thickness evolution of permafrost active layer in pan-arctic region. J Remote Sens 24(08):23–29

Study on Echo Model of Saturated Traffic Flow in Subsea Tunnel

Chuanjiao Sun

Abstract In order to solve the traffic flow changes caused by accidents in subsea tunnel, the theoretical echo model is established. The characters of the subsea is considered such as the different reaction time and the broadcast process. The speed of the echo model in the subsea is given after the incident. The study focused on the influence of traffic flow changes caused by accidents on subsequent vehicles in the saturated state of subsea tunnel. The result will provide technical support for the operation of the subsea tunnel.

Keywords Echo model · Saturated traffic flow · Subsea tunnel

1 Introduction

There are more and more subsea tunnels such as the Shenzhong highway are building in China. The subsea tunnels connect the two important locations. The whole project is made of tunnels, islands, bridges and underwater interchanges. The subsea tunnels have the characteristics of long bridge tunnel transition, large longitudinal slope, and separation and confluence in the underground interchanges. Due to the important location, the subsea tunnel has large traffic volume. In case of traffic accidents, it will lead to disastrous consequences, such as huge casualties, great social losses and environmental hazards. It is significant to study the influence of traffic flow changes caused by accidents on subsequent vehicles in the saturated state of subsea tunnel. It is important to take good measures after knowing the traffic flow changes which will give great support to the operation and management of the highway.

C. Sun (✉)
Road Safety Research Center, RIOH, Beijing 100088, China
e-mail: cj.sun@rioh.cn

W. Wang et al. (eds.), *Green Transportation and Low Carbon Mobility Safety*,
Lecture Notes in Electrical Engineering 944,
https://doi.org/10.1007/978-981-19-5615-7_50

2 Reviews

Generally the research on traffic flow regards traffic flow as a fluid composed of traffic individuals and is assumed as a continuous flow. The macro indexes describing the basic characteristics of traffic flow are traffic volume, speed and density.

There are various theories which try to explain the mechanism of the traffic flow. Echo transmission is one of them. The changes of traffic flow density are similar to the wave propagation. In normal traffic flow, the traffic flow can be regarded as stable. Once there is interference such as traffic blocking occurred in front of the road, the traffic flow that first arrives at the interference point will take a change. Then the following traffic flow density will change too. The density change of the traffic flow will spread like a wave. This kind of echo of traffic flow is just like sound wave and water wave. Li et al. [1] proposed an approximation considering the velocity effect of the preceding car (VE) model; Knospe et al. [2] proposed a CA model, which takes into account the driver's hopes for a smooth and comfortable driving; Kerner et al. [3] proposed a fully discrete KKW model; and Lee et al. [4], Kang [5], and Dion et al. [6] accurately and realistically depicted vehicle queues by comparing conventional methods of the queuing analysis and deeply analyzed traffic wave's transmission characteristics on vehicles queuing process.

Most of the existing traffic flow theories use the continuum model to describe the traffic state. The time and space variables are included in the equation, and the flow characteristics of traffic flow are taken into account. In the reality, the traffic flow is not a real continuous fluid medium. There are lots factors that can have a great impact on the motion characteristics of traffic flow. The factors include the interaction of individual vehicles, the reaction of drivers, the randomness of information judgment of drivers and others. Driving in the subsea tunnel, there are something different from the ordinary highway such as special scene of long-term claustrophobia, lack of light and easy fatigue of drivers. There is no good theoretical model been established on how vehicles react and conduct after the traffic events occur in the subsea tunnel. Based on the existing continuous traffic flow model, considering the dispersion and randomization of traffic flow, and combining with the special scene in subsea tunnel, a new traffic flow model is built for better describing the traffic flow in subsea tunnel.

3 Main Characters of the Subsea Tunnel

The whole traffic flow in subsea tunnel is the result of all the individual vehicle driving state. Drivers adjust their driving behavior according to their own driving state, surrounding traffic environment and surrounding vehicle running state.

The density of traffic flow in subsea tunnel increases with the increase of traffic volume. When drivers enter the tunnel and encounter an event, the overall speed of traffic flow will inevitably decrease. A reverse direction echo of traffic flow will produce and spread to the upstream at a certain speed. When the density of traffic flow reaches a certain value, this event will decelerate the vehicles speed entering the upstream of the tunnel.

When the abnormal events are eliminated, the fleet gradually returns to normal state. Because the impact of echo caused by emergencies is considered, the process of congestion dissipation is not considered for the time being.

4 Echo Model of Saturated Traffic Flow in Subsea Tunnel

The traffic flow change of the subsea tunnel due to event is described as follows: assuming that when the traffic flow of the subsea tunnel at a certain time is running at state A with traffic volume Q_1, speed V_1 and density K_1. When the vehicle encounters the front driving behavior abnormality, it begins to slow down, which will lead to the state change. The state changes to B with traffic volume Q_2, speed V_2 and density K_2 respectively. With the time passing, the subsequent vehicles will slow down or even stop in state C. At this time, $Q_3 = 0$, $V_3 = 0$, $\mathrm{K}_3 = \mathrm{K}_\mathrm{J}$ and K_J are the blocking parking density.

When the traffic events are eliminated, the fleet gradually returns to normal state. Because the impact of echo caused by emergencies is considered, the process of congestion dissipation is not considered for the time being.

For vehicles causing emergency, the specific micro description is that: assuming that the vehicle body length of vehicles passing in the tunnel is L_0, the average distance of safe braking for free driving under submarine immersed tunnel is L, the free flow speed is V_f, and the distance between the two adjacent vehicles which collides at t time is assumed to be l_0, and the vehicle speed after driving is v_1. The traffic density at the rear car is $\mathrm{k} = 1/l$.

According to the characteristics subsea tunnel, the consistency of the common characters of the drivers in the tunnel is the fundamental reason for the echo velocity transmission. The driving behavior of drivers in subsea tunnel is summarized as follows:

(1) The drivers can judge the traffic flow condition in front of the driver by their ideal distance H.
(2) The average time distance of mental vehicle head is upper limit H_0. When the actual headway beyond the upper limit, the flow speed is affected.
(3) Acceleration or deceleration behavior in driving behavior is the driver's judgment according to the traffic information transmitted by the front vehicle. If the headway of the front vehicle is lower than the lower limit of the driver's mental headway, the driver takes deceleration measures. The overall speed of the traffic flow decreases and the echo transmits back.
(4) The drivers mental vehicle head H is relative to drivers' reaction. But in the subsea, there is a comprehensive influence factor g.

The process of the traffic fleet change when there is an event. Assuming that the vehicle length is L_0. The average distance of safe braking under the subsea tunnel is

L and the velocity of free flow is v_f. The distance between the two adjacent vehicles with collision at the time t is assumed to be l_1. The vehicle speed after driving is v_1. The traffic density at the rear car is $k_1 = \frac{1}{l_1}$.

$$v_1 = \frac{l_1 - L_0}{\tau} \tag{1}$$

Rear car eager flow Q_1 is defined as the product of velocity and density, and can also be expressed as the sum of echo flow and other information flow.

$$Q_1 = v_1 * k_1 = Q_c + Q_0 = \frac{v_1}{l_1} \tag{2}$$

where, Q_c is the flow generated by echo velocity, Q_0 is the undisturbed flow. Assuming c is the echo velocity, then:

$$Q_c = \frac{(l_1 - l)v_1}{l_1^2} = \pm\frac{c}{l} \tag{3}$$

where ' + ' means the forward acceleration echo; '−' Corresponding to reverse deceleration echo.

Define Q_0 is the traffic generated for other information,

$$Q_0 = \frac{lv_1}{l_1^2} \tag{4}$$

If $\mathrm{x} = l/L$, then l_1 can be expressed as a function of x, and it is recorded as $l_1 = f(x) * L$. The headway space of the free flow and the saturated flow is l_f. When the traffic flow is close to saturation, according to the characteristics of driving behavior, the function $f(x)$ can meet the following conditions:

(1) $\mathrm{x} \geq \mathrm{L}_0/L$, and $\mathrm{x} \leq \mathrm{L}_f/L$;
(2) $\mathrm{f(x)} \geq \mathrm{x}$, and f(x) monotonous;
(3) f(x) has upper bound, that is, when the headway L is large to a certain extent, it is free flow state. The driver can drive at his own speed without being affected by the front traffic flow, at this time $l_1 = l_f$.

Assuming that the echo velocity of backward propagation is c, the above formula is brought into the expression of echo velocity, and then it can be obtained:

$$\mathrm{c} = \pm Q_c * l_1 = \pm\frac{l(l_1 - l)v_1}{l_1^2} = \pm\frac{x(f(x) - x)(f(x)L - L_0)}{\tau(f^2(x))} \tag{5}$$

According to the above formula, the echo speed c is the driver's reaction speed τ According to the existing research τ. The change process is different when the vehicle accelerates and decelerates, τ is a process of growing from small to large and keeping unchanged when the headway reaches a certain value. If an appropriate function f(x) is chosen, c can be made to get the maximum value in a certain distance space. Due to the group effect of vehicle driving in the whole traffic flow, the maximum echo velocity c should be the farthest value that can be observed to the upstream of the traffic flow. In addition, when the headway is the same, the echo velocity curve is not the same, that is, the echo velocity curve is not the same when accelerating and decelerating. The critical headway l corresponding to the maximum value of echo velocity curve during deceleration l_c is the dividing point between stable and unstable traffic flow.

Through analysis and comparison, f(x) is defined as:

$$\mathrm{f(x)} = \begin{cases} \frac{1}{cthx-\mu} & \frac{L_0}{L} \le x \le \frac{l_f}{L} \\ l_f & x > \frac{l_f}{L} \end{cases} \tag{6}$$

μ is the free velocity v_f. $cthx = (e^x + e^{-x})/(e^x - e^{-x})$.

When $l = l_f$, that is $x_f = l_f/L$. The expected speed is equal to the free speed, i.e. $v_1 = v_f$. So μ can be calculated as:

$$\mu = cth\left|\frac{\tau v_f + L_0}{L}\right| - \frac{L}{\tau v_f + L_0} \tag{7}$$

5 Conclusions

The subsea tunnel has important location and heavy traffic volume, the measures after traffic event should be considered ahead. Under the saturated traffic flow, little incident will lead to disastrous consequences such as huge casualties, social losses, and environmental hazards. The theoretical echo model is established in the paper. The characters of the subsea is considered such as the different reaction time and the broadcast process. The speed of the echo model in the subsea is given. The drivers' reaction after the event, which will give great technique support to the better operation of subsea tunnel.

References

1. Li XB, Wu QS, Jiang R (2001) Cellular automaton model considering the velocity effect of a car on the successive car. Phys Rev E 64(6):066128
2. Knospe W, Schadschneider A, Schreckenberg M, Santen L (2000) Towards a realistic microscopic description of highway traffic. J Phys A 33(48):477–485
3. Kerner BS, Klenov SL, Wolf DE (2002) Cellular automata approach to three-phase traffic theory. J Phys A: Math Gen 35(47):9971–10013
4. Lee HK, Barlovic R, Schreckenberg M, Kim D (2004) Mechanical restriction versus human overreaction triggering congested traffic states. Phys Rev Lett 92(23):238702

5. Kang YS (2000) Delay, stop and queue estimation for uniform and random traffic arrivals at fixed-time signalized intersections. The Faculty of the Virginia Polytechnic Institute and State University, Blacksburg
6. Dion F, Rakha H, Kang YS (2004) Comparison of delay estimates at under-saturated and over-saturated pre-timed signalized intersections. Transp Res Part B: Methodol 38(2):99–122

Prediction for Taxi-Hailing Demand—An Adaptive Multi-view Deep Learning Model

Xin Tang, Yongfeng Ma, Zhuopeng Xie, and Shuyan Chen

Abstract The emergence of taxi requesting service have changed the situation of traditional taxis. Nowadays, it has gained great popularity all over the world. As more people use this service, some problems spring up gradually and the imbalance between supply and demand is one of the most serious ones. It is urgent for researchers to solve this problem because it affects greatly the service quality of taxi system. In summary, traditional papers involving the demand prediction of taxi-hailing are mainly divided into two directions. Some researchers tend to use regional attributes (e.g., land use variables) for spatial modeling and analysis (e.g., using geographically weighted regression, GWR) and then interpret the model. The others mainly depend on spatiotemporal correlation to predict the demand (e.g., autoregressive integrated moving average, ARIMA) and then focus on the improvement of prediction. Based on previous researches, this paper proposes an adaptive multi-view deep learning model which comprehensively integrates the focuses of previous studies. The framework of the model includes four views, namely feature views (including various features of zones such as the conditions of weather, transportation, land use and so on), semantic views (division of zones that functionally similar), spatial views (searching for adaptive spatial neighbors), temporal views (searching for adaptive time windows). By embedding these views into our model, the accuracy and extension of prediction are greatly improved. To evaluate the prediction performance of our model, it is respectively compared with models which use different

X. Tang · Y. Ma · Z. Xie · S. Chen (✉)
Jiangsu Key Laboratory of Urban ITS, School of Transportation, Jiangsu Province Collaborative Innovation Center of Modern Urban Traffic Technologies, Southeast University, Southeast University Road, Jiangning District No.2, Nanjing 211189, China
e-mail: chenshuyan@seu.edu.cn

X. Tang
e-mail: 213151836@seu.edu.cn

Y. Ma
e-mail: mayf@seu.edu.cn

Z. Xie
e-mail: 220203282@seu.edu.cn

W. Wang et al. (eds.), *Green Transportation and Low Carbon Mobility Safety*,
Lecture Notes in Electrical Engineering 944,
https://doi.org/10.1007/978-981-19-5615-7_51

algorithms or have different model frameworks. Validation based on large amounts of data shows the superiority of our proposed model.

Keywords Taxi-hailing · Demand estimation · Multi-view deep learning · Dynamic time warping

1 Introduction

As an important part of urban transportation system, the mode of the taxi keeps changing with the penetration of Internet technology. Nowadays, the online taxi-hailing service plays a vital role in urban transportation system. Compared with the blindly searching for passengers of traditional taxis in the early days, taxi drivers nowadays are guided by an efficient system. They are quickly allocated target orders which are mostly in nearby locations. As a result, the empty rate of taxis is greatly reduced and the passengers can just wait for a minute instead of looking hard for a taxi.

However, taxi-hailing service also has some problems. With the promotion of this new service, more and more drivers and passengers join it. It gradually causes the imbalance between supply and demand. If in a zone, supply is more than demand, it will make it difficult for some drivers to get an order nearby, thus making it difficult for drivers to earn more money. Similarly, if demand is much more than supply, many passengers need to line up for a taxi.

Taxi demand forecasting is fast becoming a key instrument to solve the problem of supply–demand imbalance. By predicting the future demand of a zone, the taxis will be guided by the real-time information and the gap between supply and demand will narrow greatly. There is no doubt that this study will bring effective changes to the taxi-hailing service.

Recently, many studies of transport planning experts have given some ideas and methods on the prediction of taxi-hailing demand.

Some researchers have proposed innovative taxi demand models and efficient algorithms that show good prediction performance. Chen et al. [1] tried to predict the gap between the supply and demand of ride-hailing services in a certain region in the future. In this study, a framework based on deep neural network structure is adopted and multiple data sources (such as car-hailing orders, weather condition and traffic data) are used. Several models such as linear regression, gradient boosting decision tree (GBDT), random forest (RF) are used to achieve high-precision prediction. Kamga et al. [2] studied the impact of time and weather conditions in New York City. The study used 147 million taxi travel records covering 10 months. The results show that, under different time periods and weather conditions, the number of passengers, the supply of taxis, the trip distance and pickup frequency are substantially various, which in turn lead to the differences of drivers' income. Rong et al. [3] studied the problem of taxi call prediction and proposed a prediction framework using several algorithms, combined with basic attributes such as order, meteorological information and personalized historical profile. The effectiveness of this method is verified by the evaluation on the data of an online taxi-calling service platform. In addition, some deep learning prediction models used in the field of traffic flow prediction also provide references for our prediction model [4–8].

Some other papers provide ideas of prediction methods. Cheng et al. [9] proposed an adaptive spatiotemporal KNN model which considers adaptive spatial neighbors, time windows, spatiotemporal weights and other parameters. We take this model for reference and apply it to taxi demand forecasting effectively. Cardozo et al. [10] made a reasonable use of different candidate variables and completed the spatial analysis logically. Yao et al. [11] proposed a deep learning model considering multiple views. The model framework is established from different views, which greatly improves the prediction performance. Some studies have put forward some valuable suggestions on the division of traffic zones which is worth studying [12–14].

To sum up, previous researchers have achieved many valuable results and their studies are deep and extensive. However, most of these papers focus on their own research direction and are not comprehensive. For example, some researchers make the prediction just in spatiotemporal views. Some use socioeconomic attributes of different regions to predict and visualize. Others are focused on the improvement and comparison of different models. Thus, on the basis of previous studies, this paper proposes an innovative model—an adaptive multi-view deep learning model, which are based on the feature views, temporal view, spatial view and semantic view. In the feature view, we extract various related variables such as land use variables, weather conditions, transportation system variables from multi-source data. In the temporal and spatial view, cross-correlation and autocorrelation functions are utilized to find the adaptive spatial neighbors and time windows. In the semantic view, since the functional similarity of some zones is represented by the fluctuation of demand sequence, we use dynamic time warping (DTW) [15–18] to cluster these zones and predict respectively. By comparison with big data, our model shows a better prediction result.

The remainder of this paper is organized as follows: The following section describes the construction of our model. We introduce the feature view, semantic view, spatial and temporal view respectively and explain how to embed them in the model. In Sect. 3, the prediction performance of different models is evaluated and their results are compared. Section 4 concludes this paper and suggest the future work (Fig. 1).

2 Methods

In this section, we introduce the construction of our demand prediction model. Data is provided by Didi Chuxing GAIA Initiative (https://gaia.didichuxing.com). Firstly, from the feature view, candidate variables in a zone are extracted from multi-source data. Next, from the semantic view, we use DTW to cluster the functionally close zones which mostly have the similar fluctuation in demand sequence. Then, from the

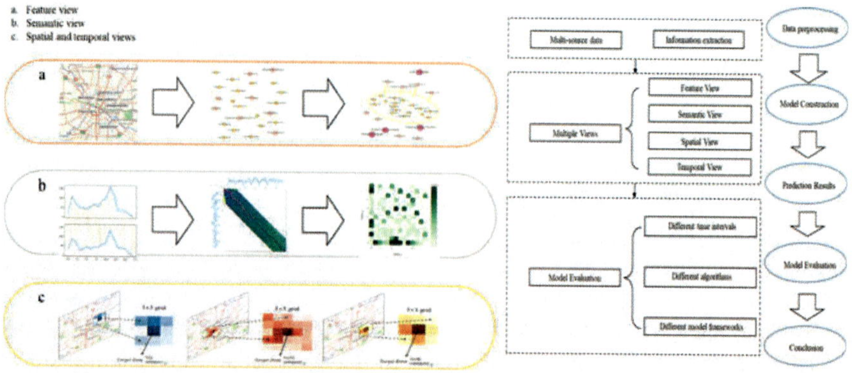

Fig. 1 Flow diagram of the model framework

perspective of the spatial and temporal view, we achieve adaptive spatial neighbors and time windows for each category respectively based on the results of DTW. Finally, we build a deep learning framework with views embedded and show the prediction performance.

2.1 *Feature View*

Different zones have different characteristics of taxi demand partly because of the spatial heterogeneity [19, 20]. In order to extract the potential variables that affect taxi-hailing demand, we combine different data sources in our paper. It includes various types of point-of-interest (POI) data obtained through Baidu API, road information and weather conditions of Chengdu. According to the data, we select different methods to extract the variables that are related to taxi-hailing demand. The variables extracted are shown in Fig. 2.

(1) Land use variables

POI data can be used to evaluate the situation of land use in a zone [21, 22]. For example, if there are many residential buildings and dormitories in a zone, we can think that the living services in this area are relatively significant. However, there are correlations between different types of POI. For example, residential buildings and dormitories regularly have similarity in spatial distribution. Since there are too many categories of POI, we apply Exploratory Factor Analysis (EFA) [23, 24] to simplify data and extract key variables.

In this paper, Principal Component Analysis is selected as the Extraction Method and Varimax with Kaiser Normalization is used as Rotation Method. After many tests, we choose the number of factors as four.

The KMO test is used to test partial correlations between variables. Generally, if the value is greater than 0.5, it indicates that the data is suitable for factor analysis.

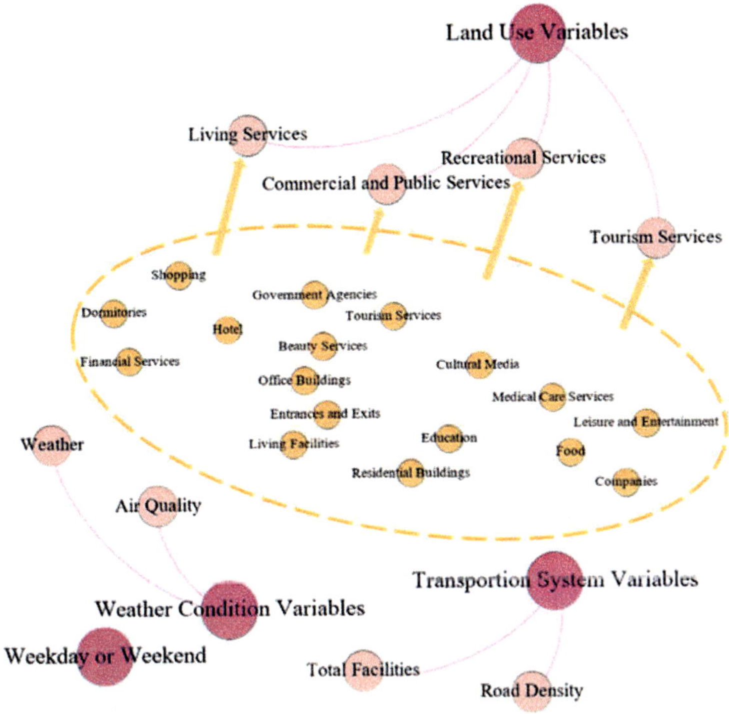

Fig. 2 The visualization of extracted variables

As shown in Table 1, the value of KMO test reaches 0.939, representing that EFA is suitable for POI data. The significance level of Bartlett's Test of Sphericity is less than 0.05, suggesting that there is significant correlation between variables at this confidence level. Therefore, it is reasonable to extract the influencing factors through EFA. When the number of candidate factors is four, 77.484% of total variance of original variables is explained. Considering the data is complicated, this result is acceptable.

Table 1 EFA testing results

Index		Value
KMO measure of sampling adequacy		0.939
Bartlett's test of sphericity	Approx. Chi-square	3618.737
	Df	171
	Sig	0.000
Explanation of total variance		77.484

The result in Table 2 shows that factor 1 has a high factor loading in Residential buildings, Medical Care Services, Dormitories and Living Facilities, so factor 1 can be interpreted as Living services, representing the basic life of residents. Factor 2 has a high factor loading in Office Buildings, Companies, Entrances and Exits, Government Agencies, Cultural Media, Financial Services and other categories. So it can be interpreted as Commercial and public services, involving residents' work and public activities. Factor 3 has higher factor loading at Beauty Services, Food, Sports, Hotel, Leisure and Entertainment, which can be interpreted as Recreational services. Factor 4 has a very high factor loading in tourist attractions. So it represents Tourism services.

(2) Transportation system variables

The road accessibility and the coverage of public transport facilities directly affect the travel in this region [25]. Public transport stations tend to be the peak taxi-hailing areas. And the accessibility of roads also affect the generation of taxi-hailing demand. In this paper, we take the road density and total transport facilities as the Transportion system variables.

Table 2 Rotated component matrix

Point of interest	Component			
	1	2	3	4
Beauty services	0.710	0.263	0.504	0.000
Leisure and entertainment	0.749	0.354	0.335	0.083
Residential buildings	0.885	0.157	0.069	−0.090
Companies	0.360	0.689	0.201	−0.095
Office buildings	0.219	0.760	0.410	0.019
Entrances and exits	0.607	0.701	0.240	0.098
Medical care Services	0.833	0.221	0.046	0.060
Dormitories	0.811	0.315	−0.102	−0.009
Government agencies	0.439	0.755	−0.081	0.103
Education	0.228	0.575	0.432	−0.135
Cultural media	0.093	0.832	0.232	0.079
Tourism services	0.021	0.078	0.033	0.969
Living facilities	0.804	0.288	0.303	0.012
Food	0.766	0.338	0.408	0.111
Shopping	0.551	0.155	0.552	−0.067
Sports	0.067	0.387	0.724	0.057
Hotel	0.170	0.571	0.528	0.149
Financial services	0.337	0.768	0.266	0.051

Table 3 Description of transportation system variables

Variable	Description	Max	Min	Mean	S.D
Total facilities	The number of transportation	109	0	28.78	23.22
Road density	Road length/Area of a zone (km/km^2)	28.34	4.53	9.95	7.74

Table 4 Description of weather condition variables

Variable	Description	Value	Days
Weather	Rain or no rain	0 or 1	31
Air quality	Value of AQI	56–105	31

We use Baidu API to obtain all public transport stations, parking lots, charging stations and other important transport facilities in Chengdu. Then we divide them into each zone and count the number of transport facilities within them.

In addition, we also obtain length, grades and coordinates of roads within the study area from OpenStreetMap. By these, we can calculate the density of the road network in each zone. It represents the accessibility to the target zone and consequently affects the taxi-hailing demand. The description of Transportation system variables is shown in Table 3.

(3) Weather condition variables

When it rains in an area, taxi-hailing demand tend to be higher than usual. Because there are fewer people walking or cycling and many people choose to commute by taxi. Similarly, poor air quality often changes the choices of residents' travel [2].

We obtain weather data in Chengdu in October 2016 from the official meteorological website. Weather data includes the weather conditions and air quality of a zone. The weather conditions includes sunny, cloudy, light rain, heavy rain and so on. Because rain is the main factor influencing choices of people's travel, we divide the weather variables into rain or no rain (1 means rain and 0 means no rain). Air Quality Index (AQI) is directly used to measure air quality.

The Weather condition variables are shown in Table 4.

2.2 Semantic View

Generally speaking, functionally similar zones often have similar fluctuation on demand sequences [11]. For example, residential zones tend to have higher demand during rush hours, while recreational communities tend to have higher demand at night or on weekends. Based on such a hypothesis, we cluster zones which share similar functions and make predictions according to different characteristics of them.

On the one hand, the selected time interval leads to a lack of data so that it is impossible to make the prediction based on each zone. On the other hand, if the prediction is based on the historical demand data of all zones within the research scope, the differences will reduce the accuracy of the prediction model. Therefore, it is necessary to extract zones with similar functions for unified prediction. In the study of similarity matching of sequences, the drift of time series leads to the failure of the traditional Euclidean distance measurement method even though the fluctuation of two sequences are very similar. DTW can find the non-linear alignment for the two time series (Fig. 3a). Euclidean distances between locations are less susceptible to negative similarity measurements caused by distortions in the direction of time axis.

DTW is used to calculate the similarity between sequences $Q\ (q_1,q_2,\ldots,q_i,\ldots,q_n)$ and $C\ (c_1,c_2,\ldots,c_i,\ldots,c_n)$, we construct an n-by-m matrix where the (ith, jth) element of the matrix contains the distance $d(q_i, c_j)$ between the two points q_i and c_j (Typically the Euclidean distance is used, so $d(q_i, c_j) = (q_i - c_j)^2$). Each matrix element (i, j) corresponds to the alignment between the points q_i and c_j. This is illustrated in Fig. 3a. A warping path W, is a contiguous (in the sense stated below) set of matrix elements that defines a mapping between Q and C. The kth element of W is defined as $w_k = (i, j)_k$. Therefore, $W = w_1, w_2,..,w_k,..,w_n$ ($\max(m, n) \le K \le m + n - 1$). We are interested only in the path which minimizes the warping cost:

$$DTW(Q, C) = \sqrt{\sum_{k=1}^{K} (q_i - c_j)^2} \tag{1}$$

The K in the denominator is used to compensate for the fact that warping paths may have different lengths. This path can be found very efficiently using dynamic programming to evaluate the following recurrence which defines the cumulative distance $\gamma(i, j)$ as the distance d(i, j) found in the current cell and the minimum of the cumulative distances of the adjacent elements:

$$\gamma(i, j) = d(qi, cj) + \min\{\gamma(i-1, j-1), \gamma(i-1, j), \gamma(i, j-1)\} \tag{2}$$

As shown in Fig. 3b, DTW compresses any two sequences at some points to obtain the shortest Euclidean distance between two sequences in another way, and takes this distance as the distance basis of hierarchical clustering. The steps are shown as follows:

(1) Considering each zone as a class and calculate the cumulative distance of demand sequence based on DTW algorithm.
(2) Merging the two categories with the smallest cumulative distance into one.
(3) Recalculating the cumulative distance between the new classes.
(4) Repeating step 2 and 3 until all categories are finally combined into the specified number of categories. Figure 3c illustrates the result of hierarchical clustering based on DTW.

Through hierarchical clustering based on DTW, traffic zones with similar demand sequences within the research scope are considered as a group (Fig. 3d). The demand sequences in the same group have similar characteristics, which is convenient for

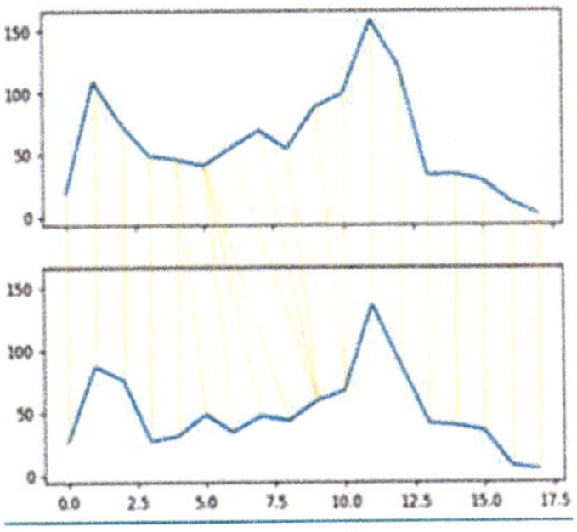

(a) Non-linear alignment

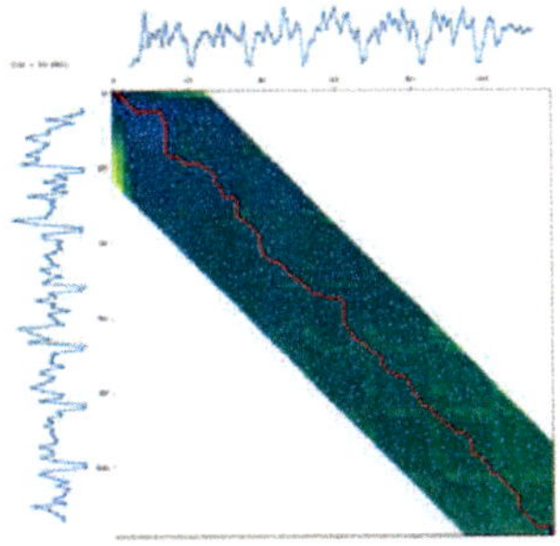

(b) Cumulative distance

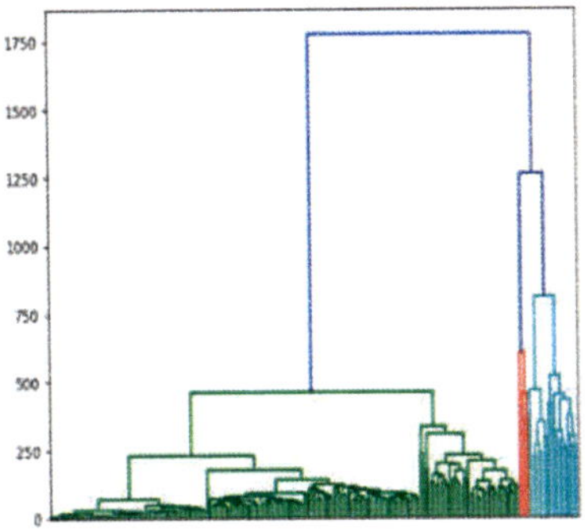

(c) Hierarchical clustering

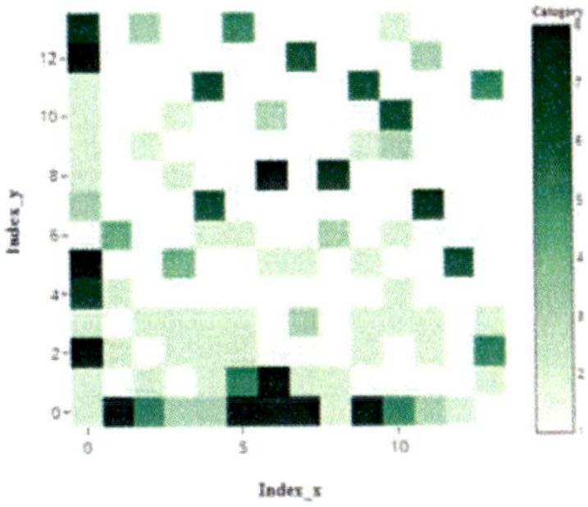

(d) Categories of all traffic zones

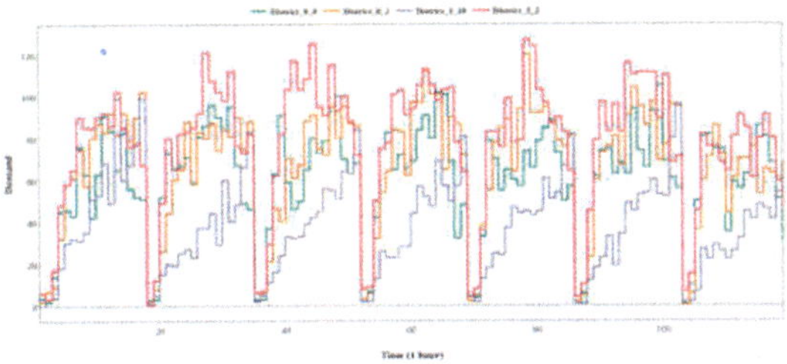

(e) Demand series of several sones in clustering I

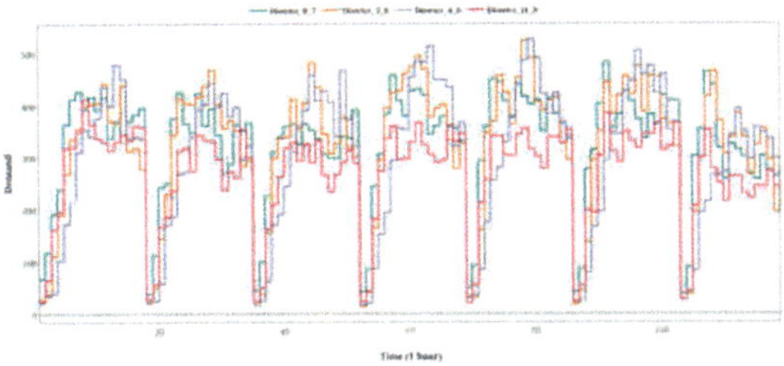

(f) Demand series of several zones in clustering II

Fig. 3 DTW-hierarchical clustering

prediction model to capture the variation characteristics of fluctuation (Fig. 3e and f).

2.3 Spatial View and Temporal View

As Tobler's First Law of Geography says that everything is related to everything else, but near things are more related to each other. The influence of zones nearby on the target zone cannot be ignored. We propose the method of cross-correlation to obtain the relevant spatial neighbors. Based on DTW clustering, adaptive spatial neighbors are used for different types of zones (Fig. 4).

Adaptive spatial neighbors are used to determine how the traffic demand of the predicted zone are affected by the surrounding zones and are determined by calculating the correlations among traffic zones. Most traffic zones are affected by different internal and external factors (for example, driving distance between zones) and this cause a lag in the impact of the surrounding zones on target zone. These delayed spatiotemporal relationships cannot be exactly expressed by correlation coefficients.

Assume that the demand series can be indicated as $\{D_t^{i,j}, i \in [1,N], j \in [1,N], \in [t_0, t_c]\}$, where t_0 and t_c denote the start time point and the current point of the time series, i represent the ith zone in the longtitude direction and j represent the jth zone in the latitude direction. $D_t^{i,j}$ is the demand series of zone$_{ij}$ at time point t, and N is the total number of traffic zones. The time series of central and surrounding traffic zones at time point t can be expressed as $C = \{D_t^{i,j}\}$, $S = \{D_t^{m,n}\}$, $i \in [1, N]$, $j \in [1, N]$, $i \in [1, N]$, $j \in [1, N]$, and their correlation at lag φ is defined as Eq. 3:

$$\begin{cases} ccf_{c,s}(\varphi) = \dfrac{\gamma_{c,s}(\varphi)}{\sigma_c \sigma_s}, \varphi = 0, \pm 1, \pm 2, \ldots, \\ \gamma_{c,s}(\varphi) = \sum \left[(u_t - \mu_c)(z_{t+\varphi} - u_s) \right] \\ \sigma_c = \sqrt{\sum (u_t - \mu_c)^2} \\ \sigma_s = \sqrt{\sum (z_{t+\varphi} - \mu_s)^2} \end{cases} \quad (3)$$

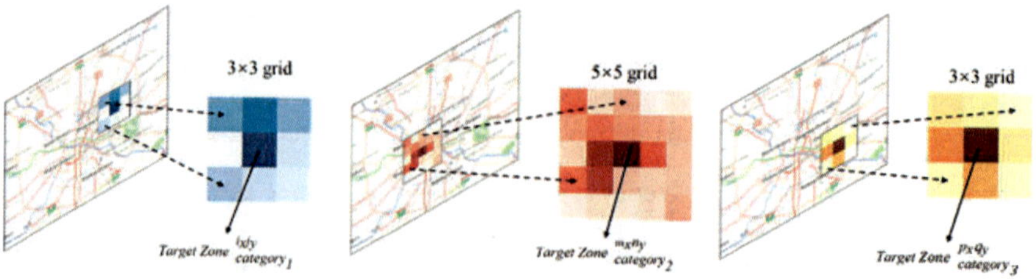

Fig. 4 Spatio-temporal correlation

In this definition, the lag can be regarded as a dependent variable by the cross-correlation function and the lag value that maximizes the cross-correlation function is the average delay time of the surrounding zones to the target zone [9]. The formal definition is expressed as follows:

$$\varphi_{m,n} = |argmin(ccf_{cd}(\varphi))|m, n \in [1, N], ccf_{cd}(\varphi) > threshold \tag{4}$$

where $\varphi_{m,n}$ is the lag value that get the correlation of the surrounding traffic zones to the target zone lower than a reasonable threshold, and $\varphi_{m,n}$ represents the maximum impact time of the surrounding traffic zones to the target zone.

As shown in Fig. 5, we set $Zone_{2,1}$ as the predicted road segment, select the traffic demand data sampled at 1 h, and set the 1-h prediction interval (same as one lag). At first, Fig. 5a illustrates the auto- correlation of $Zone_{2,1}$ under various lag. Next, the cross-correlation between $Zone_{2,1}$ and its surrounding traffic $Zone_{2,3}$ are calculated (Fig. 5b).

Under the predetermined time window, the Pearson correlation coefficients between the demand sequence of target zone and surrounding zones are calculated iteratively. The neighboring zones that meet the threshold condition are added to the spatial neighbors set of the target zone. When the Pearson correlation coefficient between the demand sequence of all zones in the m th round area and the demand sequence of the target round area is lower than a certain threshold, the iterative calculation will be stopped, and all zones in the *m* th round area are regarded as spatial neighbors.

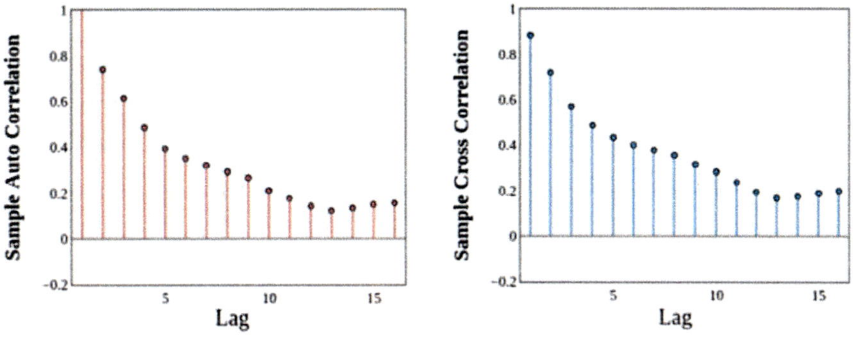

(a) Zone $_{2,1}$ (b) Zone $_{3,3}$

Fig. 5 Visualization of correlation

2.4 Prediction Part

The prediction models are based on results of DTW-Hierarchical Clustering algorithm and four different views are embedded into this model. Considering the spatio-temporal correlation of traffic demand data, the results of the adaptive spatial neighbors and time windows are input to the prediction model.

In order to evaluate the prediction results of the model, we mainly compare the models from two aspects.

(1) Time intervals and algorithms
Based on the estimated variables, we compare our model with several normal prediction models, including autoregressive integrated moving average model (ARIMA), deep neural network (DNN), support vector machine (SVM), decision tree (DT), gradient boosting decision tree (GBDT) and long short-term memory (LSTM). We also compared the results of different models using different time intervals which may significantly affect the prediction accuracy.
(2) Model frameworks
In this part, we consider the influence of four different views including the feature view, semantic view, spatial and temporal views on prediction. Our improved model integrates all views and other models just consider part of the views. To evaluate the effectiveness of this framework, as shown in Table 5, we compare it with different model frameworks.

- Feature view + Semantic view + Spatial view + Temporal view
The prediction model LSTM in this part considers the spatio-temporal features of traffic demand, the heterogeneity and regionally social and environmental factors. All traffic zones are clustered by DTW-Hierarchical Clustering. The input of this framework contains demand series of target zones and its surrounding zones having cross relationship, and regionally social and environmental variables.
- Semantic view + Spatial view + Temporal view
The prediction model LSTM in this part considers the spatio-temporal features of traffic demand, the heterogeneity. The input of this framework is the same as the first framework except regionally social and environmental variables.
- Feature view + Semantic view + Temporal view

Table 5 Models with different frameworks

Model	Multiple views			
	Feature view	Semantic view	Spatial view	Temporal view
Improved	1	1	1	1
Non-feature	0	1	1	1
Non-spatial	1	1	0	1
Non-semantic	1	0	1	1

The prediction model LSTM in this part considers the temporal features of traffic demand, the heterogeneity and regionally social and environmental factors. This framework regards the prediction as a simple time series problem and disregard the influence of spatial factors.

- Feature view + Spatial view + Temporal view
  The prediction model LSTM in this part considers the spatio-temporal features of traffic demand, and regionally social and environmental factors. The model ignores the spatial heterogeneity of the taxi-hailing demand. Therefore, demand series of all traffic zones are trained together.

3 Results

In this section, to show the superiority of our proposed model, we compare and evaluate the models from the following two perspectives. The first is from the time intervals and algorithms. The second is in the perspective of model frameworks. The results fully reflect the excellent performance of our model.

(1) Evaluation of different time intervals and algorithms.

In order to compare the prediction performance of seven algorithms at different time periods, we set time intervals as 0.5 h, 1 h, 1.5 h, 2 h, 3 h. Table 6 shows the prediction performance of the different models. The ARIMA model, regard the problem of the traffic demand prediction as a simple time series problem and disregard of the influence of regional attribute variables on the predicted traffic district. Therefore, its prediction performance is lower, based on the accuracy (1-MAPE) values, than the prediction performance of other models at all time periods.

As can be seen from the Table 6, although the accuracy of six algorithms almost fall in the range of 0.7–0.85, LSTM has the highest stability, followed by ARIMA. In addition, LSTM mostly has the highest accuracy than other six models for all time intervals.

Figure 6a and b compares the prediction results of the approaches at different time intervals, in which the best ones are presented in bold. It can easily found with the increase of time interval, MAPE and RMSE of LSTM, DNN, SVM, DTR and RF all increase, with those of SVM and DTR increasing most obviously. To predict the

Table 6 Prediction performance of different models

Interval (hour)	Index	ARIMA	SVM	DTR	RF	GBR	DNN	LSTM
0.5 h	MAPE	63.66%	80.65%	82.87%	88.65%	82.77%	86.45%	87.42%
	RMSE	21.75	19.84	15.13	12.98	15.54	11.39	10.51

(continued)

Table 6 (continued)

Interval (hour)	Index	ARIMA	SVM	DTR	RF	GBR	DNN	LSTM
1 h	MAPE	65.86%	77.77%	82.52%	86.81%	84.56%	85.30%	88.50%
	RMSE	23.36	21.41	16.99	11.85	13.64	12.67	9.82
1.5 h	MAPE	64.67%	78.45%	81.76%	84.17%	81.76%	81.87%	86.78%
	RMSE	26.95	23.18	19.52	16.69	16.78	14.88	12.95
2 h	MAPE	62.65%	74.75%	78.57%	85.75%	82.76%	82.13%	85.68%
	RMSE	28.72	28.48	22.87	21.35	19.03	18.70	15.11
3 h	MAPE	67.77%	68.87%	70.71%	81.05%	74.88%	80.63%	80.67%
	RMSE	33.35	30.98	23.19	22.45	24.09	23.80	19.32

traffic demand in 0.5-h and 2-h, RF outperforms the other six approaches, followed by LSTM. In other prediction intervals, LSTM has the best prediction accuracy. It shows that LSTM is superior to other models, suggesting that it is important to consider the heterogeneity of traffic districts for taxi-hailing demand prediction.

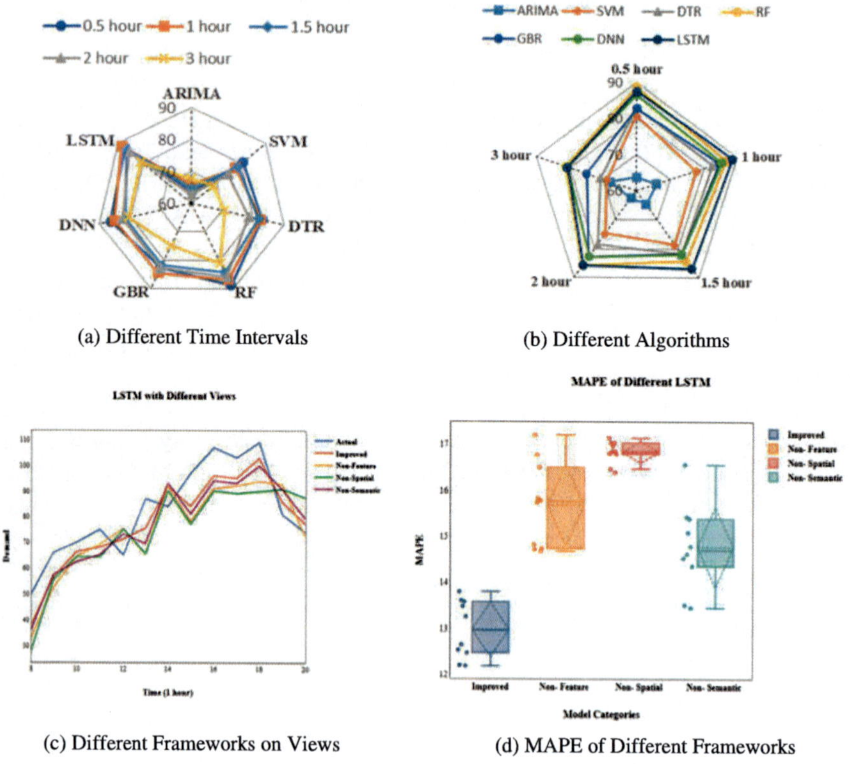

(a) Different Time Intervals

(b) Different Algorithms

(c) Different Frameworks on Views

(d) MAPE of Different Frameworks

Fig. 6 Comparison accuracy of different models and intervals

(2) Evaluation of different model frameworks

To further investigate the effectiveness of LSTM, we consider four different prediction factors, including Spatial view, Temporal view, Semantic view, and Feature view. As shown in Fig. 6c and d, temporal LSTM regard the prediction as a simple time series problem and disregard the influence of spatial factors. Therefore, its prediction performance is lower than spatial–temporal LSTM.

The model non-Semantic LSTM introduces the spatiotemporal state matrix, which improves the prediction performance. However, the model ignores the spatial heterogeneity of the taxi-hailing demand and cannot describe the difference in traffic patterns of different traffic districts. The spatial–temporal LSTM introduces the adaptive spatial neighbors, adaptive time windows and regional attributes under DTW- Hierarchical clustering during the modeling process; these parameters adapt to changes in traffic conditions, which much outperforms than spatial–temporal LSTM without Feature view. The experimental results indicate that the proposed spatial–temporal LSTM is greater than other models, suggesting that it is important to consider Feature view, Semantic view (the heterogeneity of traffic zones for taxi-hailing demand prediction) and Spatial–temporal views.

4 Conclusions and Discussion

The importance of taxi demand forecasting in transportation is indisputable. It can optimize the resource allocation of the taxi system so as to alleviate the gap between supply and demand. This paper deepens the previous researches and establishes an adaptive multi-view deep learning model based on multi-source data. It integrates Feature view, Semantic view, Temporal view and Spatial view. In feature view, we extract Land use variables, Transportation system variables and Weather condition variables which may affect taxi-hailing demand. In Semantic view, we propose to use DTW algorithm to classify the zones with similar functions (showing similar fluctuations of demand sequence) into different categories. In Spatial view, we use cross correlation to obtain adaptive spatial neighbors. In temporal view, we use autocorrelation to obtain adaptive time windows. Then, we compare this model with other models of different frameworks or different prediction algorithms. The result shows that our model have a better prediction performance.

The future researches on taxi-hailing demand prediction will continue. Our model is very flexible and extendible and we can optimize our model under this paper's prediction framework. For example, we can introduce other excellent algorithms (such as CNN-LSTM neural networks) and embed them into the model. In addition, we can extend the length and breadth of the data so that we can study in a longer time range and a larger area to improve the adaptability of the model. Last but not least, we need a better interpretation of our model so that it can be applied to practical transportation planning. The authors suggest that future researches can focus on these aspects.

Acknowledgements This work is supported by grants from the National Key R&D Program of China (2018YFB1601600) and the Fundamental Research Funds for the Central Universities (2242019K40215). The taxi-hailing data is mainly provided by Didi Chuxing GAIA Initiative.

References

1. Wang D, Cao W, Li J, Ye J (2017) DeepSD: supply-demand prediction for online car-hailing services using deep neural networks. In: 2017 IEEE 33rd international conference on data engineering (ICDE), San Diego, CA, USA, 17/04/19 - 17/04/22. IEEE, 17/04/19–17/04/22, pp 243–254
2. Kamga C, Yazici MA, Singhal A (2015) Analysis of taxi demand and supply in New York City: implications of recent taxi regulations, August, 2015
3. Rong L, Cheng H, Wang J (2017) Taxi call prediction for online taxicab platforms. In: Song S, Renz M, Moon Y-S (eds) Web and big data. Springer International Publishing, Cham, pp 214–224
4. Xiao J, Wei C, Liu Y (2018) Speed estimation of traffic flow using multiple kernel support vector regression. Phys A 509:989–997
5. Wu Y, Tan H, Qin L, Ran B, Jiang Z (2018) A hybrid deep learning based traffic flow prediction method and its understanding. Transp Res Part C: Emerg Technol 90:166–180
6. Nie L, Jiang D, Guo L, Y6u S (2016) Traffic matrix prediction and estimation based on deep learning in large-scale IP backbone networks. J Netw Comput Appl 76:16–22
7. Cheng S, Lu F, Peng P, Wu S (2019) Multi-task and multi-view learning based on particle swarm optimization for short-term traffic forecasting. Knowl-Based Syst 180:116–132
8. Chen W, An J, Li R, Fu L, Xie G, Bhuiyan MZA, Li K (2018) A novel fuzzy deep-learning approach to traffic flow prediction with uncertain spatial–temporal data features. Futur Gener Comput Syst 89:78–88
9. Cheng S, Lu F, Peng P, Wu S (2018) Short-term traffic forecasting: an adaptive ST-KNN model that considers spatial heterogeneity. Comput Environ Urban Syst 71:186–198
10. Cardozo OD, García-Palomares JC, Gutiérrez J (2012) Application of geographically weighted regression to the direct forecasting of transit ridership at station-level. Appl Geogr 34:548–558
11. Yao H, Wu F, Ke J, Tang X, Jia Y, Lu S, Gong P, Ye J, Li Z (2018) Deep multi-view spatial-temporal network for taxi demand prediction. In: Thirty-second AAAI conference on artificial intelligence
12. Dong H, Wu M, Ding X, Chu L, Jia L, Qin Y, Zhou X (2015) Traffic zone division based on big data from mobile phone base stations. Transp Res Part C: Emerg Technol 58:278–291
13. Sun L, Ling X, He K, Tan Q (2016) Community structure in traffic zones based on travel demand. Phys A 457:356–363
14. Ke J, Yang H, Zheng H, Chen X, Jia Y, Gong P, Ye J (2018) Hexagon-based convolutional neural network for supply-demand forecasting of ride-sourcing services. IEEE Trans Intell Transp Syst 1–14
15. Łuczak M (2016) Hierarchical clustering of time series data with parametric derivative dynamic time warping. Expert Syst Appl 62:116–130
16. Suominen T (2018) Applying MERIS time series and dynamic time warping for delineating areas with similar temporal behaviour in the northern Baltic Sea. Ecol Ind 95:794–804
17. Izakian H, Pedrycz W, Jamal I (2015) Fuzzy clustering of time series data using dynamic time warping distance. Eng Appl Artif Intell 39:235–244
18. Jain B (2019) Revisiting inaccuracies of time series averaging under dynamic time warping. Pattern Recogn Lett 125:418–424
19. Li Z, Wang W, Liu P, Bigham JM, Ragland DR (2013) Using geographically weighted Poisson regression for county-level crash modeling in California. Saf Sci 58:89–97

20. Wang C-H, Chen N (2017) A geographically weighted regression approach to investigating the spatially varied built-environment effects on community opportunity. J Transp Geogr 62:136–147
21. Li B, Cai Z, Jiang L, Su S, Huang X (2019) Exploring urban taxi ridership and local associated factors using GPS data and geographically weighted regression. Cities 87:68–86
22. Li M, Kwan M-P, Wang F, Wang J (2018) Using points-of-interest data to estimate commuting patterns in central Shanghai, China. J Transp Geogr 72:201–210
23. Pugachev I, Kulikov Y, Markelov G, Sheshera N (2017) Factor analysis of traffic organization and safety systems. Transp Res Procedia 20:529–535
24. Jackson ES, Rodgers NH, Rodgers DB (2019) An exploratory factor analysis of action responses to stuttering anticipation. J Fluency Disord 60:1–10
25. Calvo F, Eboli L, Forciniti C, Mazzulla G (2019) Factors influencing trip generation on metro system in Madrid (Spain). Transp Res Part D: Transp Environ 67:156–172

Recognition and Comparison of Driving Styles of Heavy-Duty Truck Drivers Under Different Scenarios

Linghua Yu, Yongfeng Ma, Shuyan Chen, Hong Yao, and Muxiong Zhou

Abstract With the acceleration of urbanization, the demand for heavy-duty trucks has increased and transportation safety and management issues are facing large challenges. The heavy-duty truck driver's behavior is characterized by his or her driving style and plays an important role in driving safety. Consequently, this paper proposes a novel framework to classify driving styles of heavy-duty trucks and make comparision under different scenarios. On rural road and urban road, 11 heavy-duty truck drivers were chosen to conduct experiments under no load or full load. VBOX device was applied to collect data including speed, acceleration and location information. K-means clustering was used to divide driving style into three categories including aggressive, normal and calm. The results show that load and road environment have a great influence on the driving style of heavy-duty truck drivers. It is worth noting that heavy-duty truck drivers are more aggressive with full load than no load when on urban roads. The empirical results demonstrate that the proposed method has efficiency in recognizing the driving style and reveal the variations of the driving style of heavy-duty truck drivers under different scenarios. Moreover, it is meaningful and practical to analyze the driving style in improving road construction, traffic safety and reducing energy consumption.

Keywords Driving style · K-Means clustering · Traffic safety · Heavy-duty truck

L. Yu · Y. Ma (✉) · S. Chen · H. Yao
Jiangsu Key Laboratory of Urban ITS, Jiangsu Province Collaborative Innovation Center of Modern Urban Traffic Technologies, School of Transportation, Southeast University, Southeast University Road, Jiangning District, No.2, Nanjing 211189, China
e-mail: mayf@seu.edu.cn

S. Chen
e-mail: chenshuyan@seu.edu.cn

M. Zhou
Traffic Management Research Institute of the Ministry of Public Security, No.88, Qianrong Road, Jiangning District, Wuxi 214063, China

W. Wang et al. (eds.), *Green Transportation and Low Carbon Mobility Safety*,
Lecture Notes in Electrical Engineering 944,
https://doi.org/10.1007/978-981-19-5615-7_52

1 Introduction

In urban construction projects, heavy-duty trucks are extremely important material tools traveling between urban and rural areas. Both transportation efficiency and transportation security can greatly increase the transportation profit of the transportation company, which is focus of consideration for the driver. It is necessary to take a series of influence factors such as load and road conditions into consideration for the heavy-duty truck drivers when they are operating. Therefore, in face of different scenarios the truck driver presents different driving behaviors and different driving styles. With the advancement of technology and requirements of realistic technology, Advanced Driver Assistance Systems (ADAS) are gradually applied to ensure vehicle driving safety. Therefore, driving style as one of the important parameters has gradually aroused more concern for managers. More and more scholars [1, 2] have begun to explore judgment methods and evaluation criterias of the driving style. Among the current researches on the recognition of driving style, the most empirical data comes from car experiments or driving simulators, but least from truck experiments. However, in order to improve construction efficiency and ensure traffic safety, it is also necessary to recognize driving styles of heavy-duty trucks. Considering the differences in the performance of heavy-duty trucks and cars, the data and analysis concluded from the cars are not suitable for trucks instead. Under aggressive driving style, traffic accidents caused by heavy-duty trucks are correspondingly increasing. Therefore, in this study we conducted a new recognition experiment to recognize the driving style of trucks.

In fact, many studies have been carried out to shed light on the recognition of driving style recently. Many scholars have used novel methods to judge driving style based on their own data, which give us a lot of new ideas. Bejani and Ghatee [3] relied on smart phone sensors to study the impact of traffic levels and vehicle types on driving behavior, and then established an evaluation system. Brombacher et al. [4] used the fuzzy logic to recognize the driving style online on Matlab/Simulink, and then verified the results through vehicle dynamics simulation. Vaitkus et al. [5] divided the driving style into radical and normal according to the accelerometer data, exacted from the same signal on the same route rather than segments. However, it is unreasonable to think that the driver maintains a constant driving style on the whole trip. So some scholars divided each trip into multiple events, on which driving style of the trip could be classified depending. The event is the driver's trip unit, and the relevant parameters of the event can accurately reflect the driver's operation process. The related treatments for events in many researches also vary from each other. For example, Brombacher et al. [6] had used artificial neural networks to recognize the driving style of events on the basis of driving maneuvers and classifications of driving style. Then threshold endpoint was detected to extract maneuvers so that the score of the entire trip was achieved. In contrast, Feng et al. [7] calculated statistical and spectral features of each event on each trip. They adopted Principal Component Analysis (PCA) to treat the calculated parameters and select important parameters among them, and then support vector machine clustering was performed.

Chen et al. [8] developed driving habit graph (DHG) to model the driving styles. The root mean square of vehicle acceleration, the Time-to-Start and the Time-To-Follow were all used to conduct style recognition via Multi-dimension Gaussian Hidden Markov Process (MGHMP). 21 drivers were chosen to take part in experiments and the results were efficient at the same time [9]. In order to recognize the safe driving mode, the smartphone data was used to perform two-stage clustering on events such as harsh events occurrence and acceleration profile. The first stage was to distinguish between aggressive and non-aggressive, and the second stage was to distinguish normal trips [10]. Eren et al. [11] used smartphone sensors to gather relevant information for understanding driver behavior, thus avoiding external sensors. In this paper, two kinds of topic models such as mLDA and mHLDA were used to exact hidden factors that affecting driving style from the actual data mining, and on the process of data pre-processing, KFCM (Kernel Fuzzy C-Means) clustering method was used [12]. Meseguer et al. [13] used data mining and neural networks to classify driving styles and explained the impact of driving style on fuel consumption. Moreover, driving style can also be evaluated and corrected after recognition. Rogavichene and Garmonnikov [14] proposed some application algorithms to reveal this possibility. Javanmardi et al. [15] also modeled the driver and determined the driving behavior through simulation experiments, which providing a good solution to experiment without getting enough driving test samples. Karginova et al. [16] used the data mining to divide the driving style into normal and hard style. For example, high acceleration and braking were considered hard driving style. And Lee and Jang [17] have developed a framework to recognize aggressive driving behaviors with large-scale driving records. Meanwhile abrupt changes and feature were extracted from events using an auto-encoder and finally two-level clustering helped reduce dimension for classification of driving style.

The above discussion emphasizes the classification of events on the trip, but there are also many studies on the classification of events, which is generally divided into two categories based on continuous time window and threshold. Threshold determination is based on a large amount of experimental data, so it is easier to use a threshold to determine the event for cars. The duration of each trip is divided into different lengths. Johnson and Trivedi [18] used the dynamic time warping algorithm to classify the events.

Moreover, how to choose the appropriate parameters is also the focus of study nowadays. Velocity and acceleration as basic parameters are used to characterize driving behavior. Based on the strict correlation between velocity and acceleration, Eboli et al. [19] used two parameters to distinguish between safe and unsafe driving behavior. Later, some researchers suggested that the change of acceleration and deceleration reflects the driver's personality characteristics and is more representative. Therefore, this variable is defined as jerk to judge driving style. Murphey et al. [20] considered the driving style to be the reaction of the driver's behavior on the road, so the jerk statistic, road type and traffic congestion level can be used to predict the driving style. Driving acceleration and jerk were used as measurements in this study to analyze driving performance, and were represented by S and $\bar{a}$ respectively [21]. In addition, the driving style is not only related to the driving parameters, but also related to the physical and emotional conditions, and the structural equation model (SEM) was established to explain the relationship between them [22]. In order to mine other parameters with analytical value, Li et al. [23] used the maneuver transition probabilities to identify the transition patterns between them, and then used the random forest algorithm to classify the selected features. The results showed that there are five types of features that can be used to classify driving styles. Miyajima et al. [24] used his longitudinal acceleration, lateral acceleration and velocity to record driver's risk information in their research. Kalsoom and Halim [25] compared k-means clustering results with hierarchical clustering results by using vehicle's driving characteristics such as ratio of indicators to turns, number of brakes and found that k-means clustering is better.

What can be concluded from the above-mentioned literature, the methods for determining driving style are various but can mainly be classified into two catogories: clustering and neural network. The latter is applied to tagged data while the former is more applicable to unlabeled data. The thresholds used for the segmentation of events are also known through many experiments. There are few studies on urban and rural freight transport vehicles such as heavy-duty trucks. Therefore, under the premise of unclassified data, the appropriate time window can contain more events, which is also a good method for classification. In the parameter selection, the velocity is the basis, but later research found that the acceleration can reflect the driver's driving style better. In general, driving style recognition is a new research direction, especially in traffic safety and road construction as a focus of many transportation companies nowadays.

The purpose of this study is to explore the difference in driving styles of heavy-duty truck drivers in rural and urban area with full or no load. According to the driving style recognition method related above, in this study truck experiments are conducted to collect data and acceleration and jerk are selected as driving parameters to determine the driving style of heavy-duty truck drivers. In the second section, we give some details of the methods used in this paper. The third part introduces the data collection and experimental procedures. Finally, the fourth part analyzes the results of classification and compares the difference of four scenarios. Furthermore future work is discussed.

2 Methods

2.1 *Measurements*

2.1.1 Driving Style

As can be seen from above, the current driving style could be an additional input for Advanced Driver Assistance Systems (ADAS) to alert the driver to adjust the vehicle dynamics in a dangerous environment or unsafe driving style.

In this paper, the driving style is classified into three types: aggressive driving, normal driving and calm driving [20]. The driving behavior of the aggressive driver is severe and fuel consumption often increases while the normal driver is moderately mild and deals with emergencies more carefully. The calm driver will comprehensively consider more information and keep cautious during the driving. And in this case, fuel consumption is the least.

2.1.2 Time Window

The driving style is not static but dynamic throughout the entire trip, which is inseparable from some external factors such as the driver's personality, mood, family, etc. In one trip, specific events such as acceleration, deceleration and turn can be used to describe the driver's driving style. This article divides the trip into multiple events according to unequal time size [7]. Since the threshold is unknown due to the lack of a large number of existing researches, the way of divide trip into several fixed events is not suitable for heavy-duty trucks. In order to capture the exact event information, a fixed time window is applied. And this also plays a role in smoothing abnormal parameters within one trip by calculating parameters in each time window. The length of the time window has a certain influence on the extraction of the driver's characteristic value. The time window with an appropriate size will contain more complete events. If the time window is too short, it cannot capture the complete acceleration or deceleration events, thus making the events more radical. On the contrary, if the time widow is too long, it may capture multiple events, thus making the events more gradual. This narrows the difference between events resulting in inaccurate classification. Referring to the previous literature [20], a window size of six is reasonable for driving style recognition, which has a three second overlap between the two adjacent windows.

2.1.3 Describing Features

In order to effectively judge the driving style with specific parameters, this paper selects acceleration and jerk (the first derivation of acceleration) as the analytical parameters. Statistical features of the parameter in each time window are calculated to represent the characteristics of each event.

Acceleration reflects the change of velocity while jerk reflects the change of acceleration, both characterizing the tendency of driving style. In general, the driver's driving style should be described by the driver's rapid deceleration, and jerk as the changing rate of acceleration plays an important role verified in many studies. However, the driving style cannot be determined by one of the factors alone. For example, when the vehicle is at high or low acceleration, the jerk may be the same. In this case, it is difficult to judge the driving style. So considering the four parameters on different scenarios, the learned driving style is more credible.

The vehicle is subject to the vector acceleration represented by $\overline{a}$ during driving, which includes a_{long} (longitudinal acceleration in the same direction of the motion) and a_{lat} (lateral acceleration in the direction transverse to the direction of the motion) components [19]. The modulus $|\overline{a}|$ is expressed as:

$$|\overline{a}| = \sqrt{a_{long}^2 + a_{lat}^2} \tag{1}$$

A trip is represented by the speed function of time t, i.e., DC(t), $t = 0,\ldots, t_e$. The jerk function following is expressed as:

$$J_{long}(t) = \frac{d(a_{long})}{dt} \tag{2}$$

$$J_{lat}(t) = \frac{d(a_{lat})}{dt} \tag{3}$$

$$J(t) = \sqrt{J_{long}^2 + J_{lat}^2} \tag{4}$$

X represents the given N sequential data (X = {x_1, x_2,..., x_N}). The mean $\overline{m}$ and standard s can be calculated as follows: mean acceleration ($|\overline{a}|_{mean}$), standard deviation of acceleration ($|\overline{a}|_{std}$), mean acceleration (J_{mean}), standard deviation of acceleration (J_{std}).

$$\overline{m} = \frac{1}{N}\sum_{i=1}^{N} f(x_i) \tag{5}$$

$$s = \sqrt{\frac{1}{N-1}\sum_{i=1}^{N}(f(x_i) - \overline{m})^2} \tag{6}$$

where X_i is ith the acceleration and spot jerk in four scenarios respectively.

2.2 *Clustering and Classification*

As one of the unsupervised learning algorithms, k-means clustering is inclined to implement. It mainly refers to dividing some unlabeled samples into several clusters, where the samples in each cluster are the closest, and the data of each cluster are as far away from that of other cluters as possible. The main idea of this algorithm is to first define k centroids from each other as far as possible, then associate each point of the dataset with the nearest centroid, and finally recalculate the new centroids as the clustering center. Based on this, a new round of association is made until the position of the centroid is no longer changed.

The goal of the entire process is to minimize a squared error function. The objective function is expressed as follows:

$$E = \sum_{i=1}^{k} \sum_{x \in C_i} \|x - \mu_i\|^2 \tag{7}$$

where μ_i is the mean vector of the cluster center C_i and expressed as:

$$\mu_i = \frac{1}{|C_i|} \sum_{x \in C_i} x \tag{8}$$

Clustering is applied to many different situations, such as data mining, data compression, image recognition and image classification.

3 Data Collection

3.1 Apparatus

In order to obtain parameters such as the latitude and longitude position and acceleration of the experimental truck, VBOX-IISX10 in Fig. 1a is used to collect data. The test truck chosen in this experiment is a HOWO concrete mixer truck with a total mass of 25,000 kg, the dimensions of which are 9100 × 2500 × 3900 (mm). Figure 1b shows the experimental data of VBOX recorded, which was processed in VBOX Tools software.

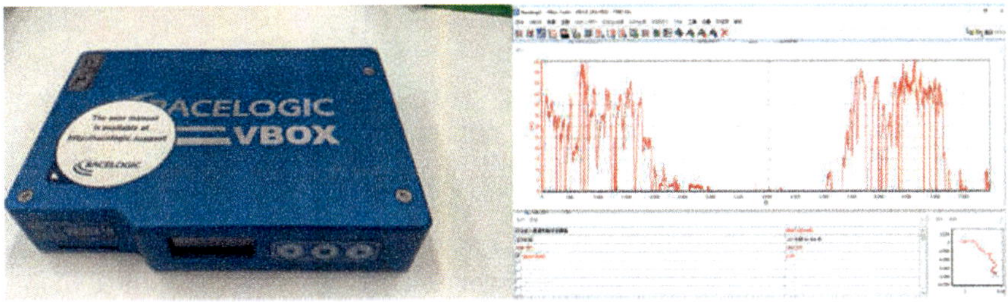

(a) VBOX-IISX10 (b) VBOX Tools

Fig. 1 Experimental apparatus

3.2 Participants

Chinese Truck Driver Survey Report shows that truck drivers are predominantly male (95.8% for men). Therefore, this study only focuses on male heavy-duty truck drivers. Professional heavy-duty truck drivers of Lanye Construction Group in Pukou distinct, Nanjing, China were recruited for the study. And all participants meet the occupational standards and hold a valid driving license. A total of 11 truck drivers participated in the study, ranging in age from 31 to 53 years (mean ± SD: 40.8 ± 6.6 years). The average driving age of the tested drivers is 12.8 years, with a standard deviation of 5.1 years. No serious traffic accidents occurred in the previous three years for all participants. Each subject signed an informed consent agreement prior to participation and received compensation when they finished the study.

Subjects were not allowed to drink alcohol and eat ingesting foods that affect the nervous system, such as coffee and tea, for twenty-four hours prior to the start of the study. They are also supervised to ensure adequate sleep to avoid affecting the nervous system. In order to reduce the impact of work stress, their working time and intensity are guaranteed to be consistent from start to finish. The study adheres to China's laws on scientific research on healthy human volunteers and each participant was paid for their participation who signed an agreement prior to participation.

3.3 Experimental Procedure

The heavy-duty trucks traveled between rural and urban roads every day, and each driver switched for a total of three days with and without load on their work to ensure adequate samples. The 11 heavy-duty truck drivers participating in the test were required to arrive at the begining point before 8:00 am. After finishing preparation of the test truck and equipment, drivers departed the begining point at approximately 8:30 am to complete the transportation handover every day. The route comprised one four-lane, first-class highway with low-density traffic flow, and the road geometry was well controlled in that 90% of the test route consisted of long, straight roads. At noon, the driver will take a 10-min lunch break to get enough physical strength. Each driver finished their trips on the route. In this study, the ambient temperature and noise were also controlled to avoid influence of these parameters on the physiological state of the driver. Since this experiment is an uncontrolled one, the working hours of the participants are not strictly required. But the total driving time did not exceed 8 h, but was no less than 5 h, in compliance with the Chinese Road Traffic Safety Law Implementation Regulations that driving time cannot exceed 8 h in a 24-h period. The purpose of this paper is to analyze the differences of driving styles in four scenarios, so it is only necessary to ensure that each driver's driving route is consistent. The part route and field procedure are illustrated in the figure (Fig. 2). The VBOX system was installed in the test truck to collect the data of velocity, longitudinal acceleration and lateral acceleration. In order to ensure the accuracy and stability of the extracted

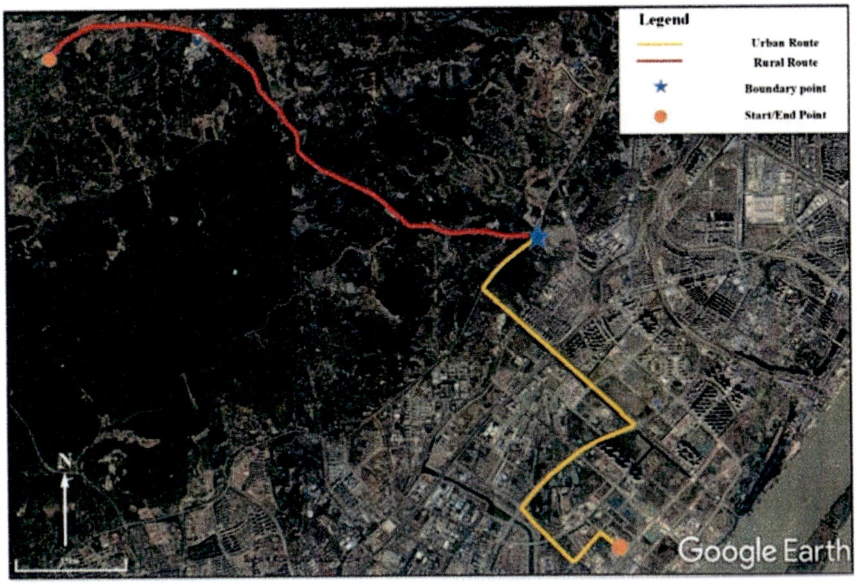

Fig. 2 The driving route in Nanjing, China

data, all experiments were carried out under good weather conditions. There was no crash or near crash occurred during transportation. The experiments were conducted from April 2019 to June 2019.

4 Results

The research results are divided into two parts. The first is the preprocessing of data and extraction of related features, and the second is the custering of data with time windows of specific size and driving style of truck drivers under four scenarios are classified based on this.

4.1 Data Pre-Processing and Analysis

The objective of this paper is to compare and analyze the driving styles of trucks with or without load in rural and urban roads. The data were processed with python and the classification of the data was shown on VBOX tools software. The main task is to calculate acceleration and jerk and their mean and standard deviation over a time window of 6 s which has a 3 s overlap between the two adjacent windows. The description statistics for pre-processing data are in Table 1 (Fig. 3).

Table 1 Description statistics for pre-processing data

variable	count	mean	std	min	25%	50%	75%	max
v_avg	71,415	35.157	14.532	1.130	24.339	36.142	46.500	66.783
v_std	71,415	1.299	1.174	0.015	0.448	0.915	1.757	7.745
acc_avg	71,415	0.556	0.529	0.000	0.166	0.389	0.777	2.756
acc_std	71,415	0.650	0.345	0.034	0.393	0.574	0.831	1.993
jerk_avg	71,415	0.739	0.635	0.000	0.267	0.573	1.027	3.539
jerk_std	71,415	6.513	3.223	0.166	4.179	5.763	8.103	18.385
rural_urban	71,415	0.359	0.480	0	0	0	1	1
empty_full	71,415	0.570	0.495	0	0	1	1	1

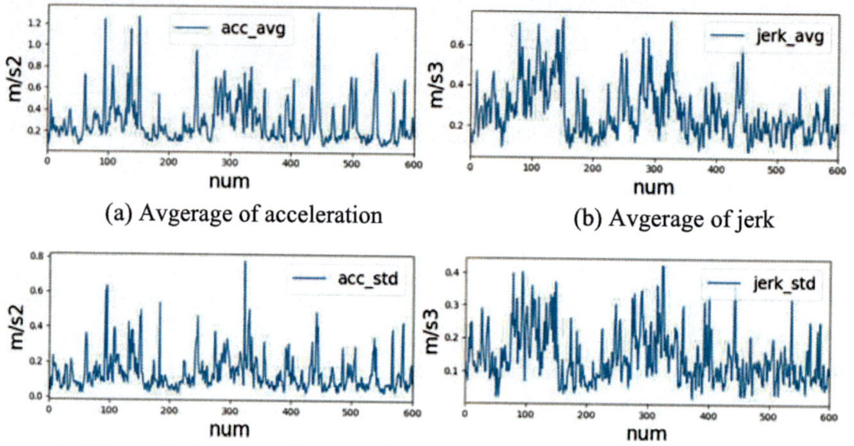

(a) Avgerage of acceleration (b) Avgerage of jerk

(c) Standard deviation of acceleration (d) Standard deviation of jerk

Fig. 3 Sample dataset collected

4.2 *Classification Results and Comparison*

4.2.1 Driving Style Classification

In this part k-means algorithm was used to divide the records into three categories with the four parameters obtained above as variables. According to the coordinates of the centroid, each type was classified into aggressive driving, normal driving and calm driving. In this paper, eleven drivers were selected from the experimental participants, and the driving styles in four cases were obtained. As is shown in the Fig. 4a, trucks with full load in the rural are divided into three classes and Fig. 4b shows indicators of features in each category. From it we can know that the indicators of lable 1 are larger than others generally. The value of lable 0 is more concentrated and this means the data in this class has less degree of change in the Fig. 5. While indicators

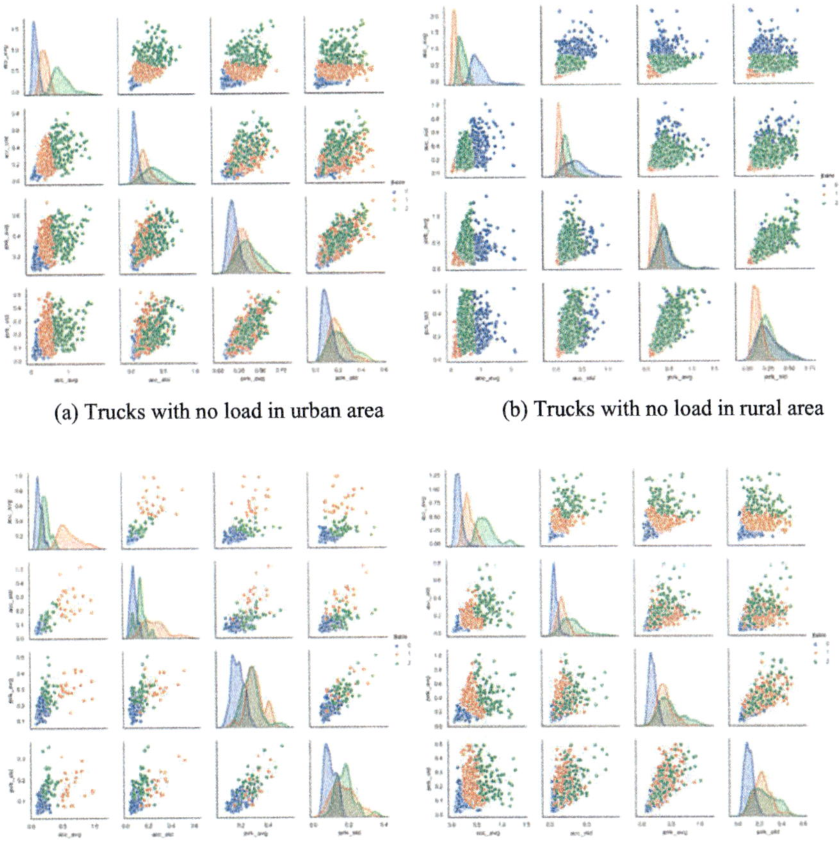

(a) Trucks with no load in urban area

(b) Trucks with no load in rural area

(c) Trucks with full load in urban area

(d) Trucks with full load in rural area

Fig. 4 Cluster results

of lable 2 lie in the middle without too big or too small change. Accordingly, we can deduce that blue points represent calm type, green points norm type and yellow points aggressive type.

In order to analyze the differences in driving styles in four scenarios, experimental data of 11 drivers in 33 days was processed and the processing results are shown in the Table 1. It can be observed (Table 1) that different drivers have corresponding changes in the face of environmental changes. All eleven drivers can take the calm driving style, and the possibility of adopting the aggressive driving style is no more than 12%. However, there is still a difference in the driver's tendency to drive calmly. For example, when a large car is fully loaded on a country road, the possibilities of the eleven drives with calm driving style are 55%, 69%, and 58% respectively. The proportions of the eleven people adopting the aggressive driving style are similar.

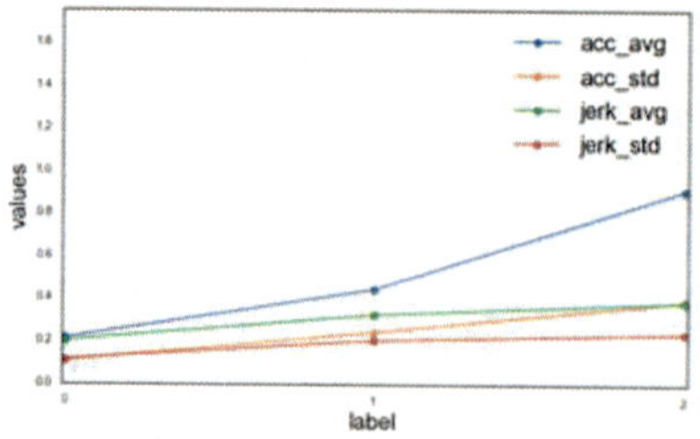

(a) Trucks with no load in urban area

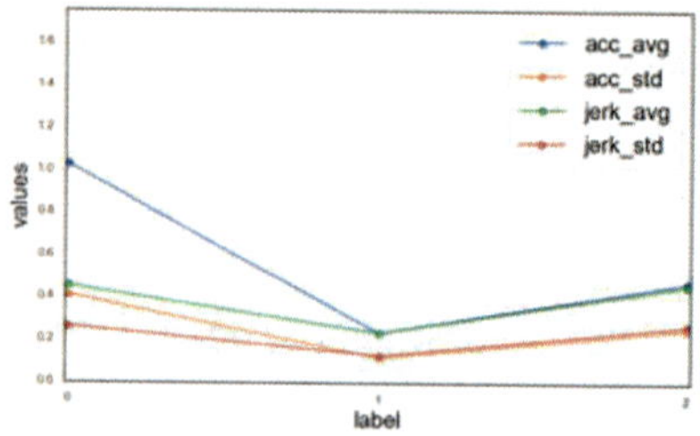

(b) Trucks with no load in rural area

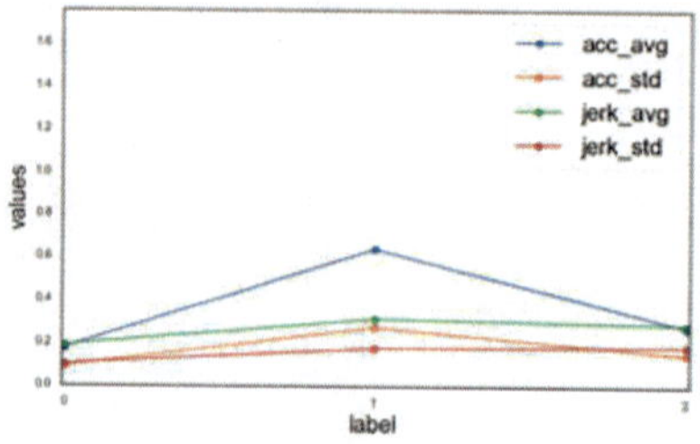

(c) Trucks with full load in urban area

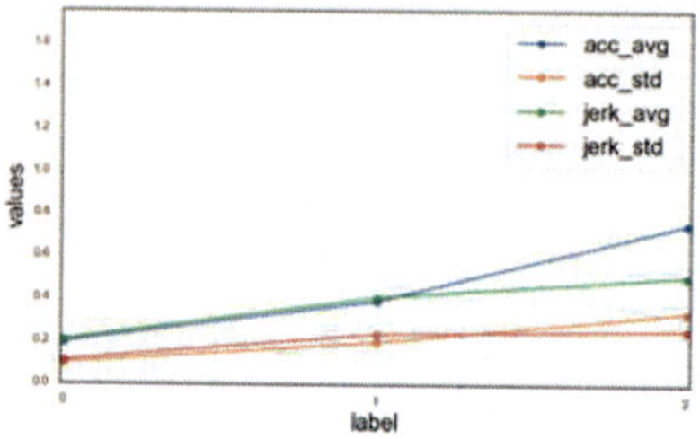

(d) Trucks with full load in rural area

Fig. 5 Parameters of cluster centroids

4.2.2 Comparison of Trucks in Rural and Urban Area

As we all know, there are big differences between urban and rural area. The rural area is sparsely populated, where vehicles are rarely interfered by other vehicles. The urban area is full of crowded buildings with many intersections and intermittent flows. In theory, the urban road will hinder drivers because of its heavy traffic flow. The driver needs to pass safely and cautiously. While in rural area drivers will be calmer and they can drive more freely with less attention due to low traffic flow.

Drivers who drive the truck with full load in the urban are more aggressive than doing so in rural area, and the proportions of taking aggressive and calm styles in the urban are higher from the figure (Fig. 6). These tendencies indicate that drivers are more susceptible to environmental stimuli. That is to say, when transportation time is sharply increased because of the road and traffic condition, the drivers are more eager to transport the goods to the destination as soon as possible, thus accelerating frequently. On the other hand, they have to keep serious to ensure the integrity of the materials.

Drivers who drive the truck with no load in the rural are more aggressive than doing so in urban area without responsibility to guarantee the integrity of the materials. However, unlike the first situation, the proportion of taking the calm style in the rural is still lower than that in the urban because drivers are more susceptible to the environment and feel safe on spacious rural area. So in rural area the driving style is more aggressive, and the proportion of taking the calm style is decreasing.

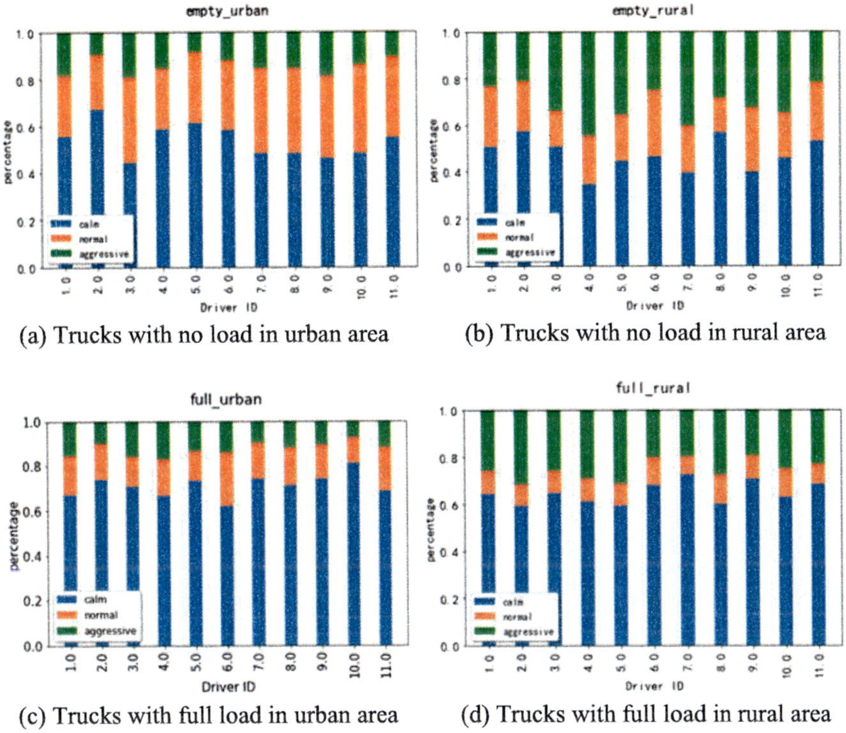

(a) Trucks with no load in urban area (b) Trucks with no load in rural area

(c) Trucks with full load in urban area (d) Trucks with full load in rural area

Fig. 6 Proportion of different driving styles

4.2.3 Comparison of Trucks with Full Load and no Load

In the daily operation, trucks should switch between full loaded and empty-loaded modes due to the need of the operation. It is helpful for job evaluation to analyze the driving styles in both cases. In theory, the driver is more aggressive with no load because of the reduced pressure. While when the truck is fullly loaded, the driver needs to pay attention to the safety of materials. Correspondingly, the driver is inclined to be calm in that scenario. A further analysis will be carried out in this section.

In rural area, drivers who drive the truck with no load are more aggressive than driving the truck with full load and the result is consistent with the theory above. Because the driver needs to consider the safety of materials and the weight of the truck, it is not appropriate to take violent actions. Otherwise, the vehicle will be slow to respond to sudden situation due to inertia. Drivers need to be cautious and calm because they want to transport goods to their destination completely.

In urban area, it is worth noting that drivers who drive the truck with full load are more aggressive than that with no load. This just reflects a psychological characteristic of the driver. When in urban area, heavy-duty trucks stop frequently because of more intersections, thus delaying the delivery of materials. In order to transport materials faster, the driver is more anxious and presents a more aggressive driving style. Therefore, in the analysis of driving style, in addition to considering theoretical common sense, it is necessary to explore the psychology of the driver, which is easily affected by the environment.

5 Conclusions

The purpose of this study is to recognize and compare the driving style of heavy-duty truck drivers in rural and urban area with full or no load. The driving parameters of the heavy-duty truck drivers were collected through truck experiments, in which acceleration and jerk were selected as the evaluation indicators. The average and standard deviation of the two parameters respectively were calculated within the fixed time window as the processing unit, and then the k-means algorithm was used to classify the driving styles in the four discussed circumstances. The main conclusions of the article are as follows.

The acceleration and jerk can be used as the basis for judging the driving style, and the clustering effect is obviously reasonable. In the case of the same road condition, the difference in driving style between drivers is not significant, indicating that the environment has a greater impact on driving behavior. In the four scenarios, the aggressive truck with no load in rural area accounts for a large proportion and the calm one accounts for the least. It shows that under the no-transport pressure, driver in the rural is less cautious and behaves more aggressively. In the rural, the increase in load is inversely proportional to the ratio of aggressive driving and positively proportional to that in the urban. With no load, the greater the road disturbance, the calmer the driver; but the more aggressive the driver is when fully loaded. The experimental results are different from the conventional cognition, indicating that there are still other factors such as driver psychological characteristics, weather and road emergency during the actual driving process of the truck.

The data used in this paper is mainly from the VBOX device collector, and the research carried out under the assumption that other factors are always consistent. In the future work, we will use datasets that includes high-precision GPS equipment, eye tracker data, and their combination to analyze influencing factors of driving style. At the same time, the weight of each factors in the dataset should also be considered. Morever since the driver style was not marked during the experiment, the data obtained was unlabeled, which cannot verify the actual clustering effect. This will be considered in future research that we can determine which classification is better. These studies can be used as important input for Advanced Driver Assistance Systems, greatly promoting the development of Autonomous Driving System.

Acknowledgements This work is supported by grants from the National Key R&D Program of China (2018YFB1601600) and Open Project of Key Laboratory of Ministry of Public Security for Road Traffic Safety(2021ZDSYSKFKT02).

References

1. Marina Martinez C, Heucke M, Wan FY, Gao B, Cao D (2018) Driving style recognition for intelligent vehicle control and advanced driver assistance: a survey. IEEE Trans Intell Transp Syst 19(3):666–676
2. Meiring GAM, Myburgh HC (2015) A review of intelligent driving style analysis systems and related artificial intelligence algorithms. Sensors (Basel, Switzerland) 15(12):30653–30682
3. Bejani MM, Ghatee M (2018) A context aware system for driving style evaluation by an ensemble learning on smartphone sensors data. Transp Res Part C: Emerg Technol 89:303–320
4. Dorr D, Grabengiesser D, Gauterin F (2014) Online driving style recognition using fuzzy logic. In: 17th International Ieee Conference On Intelligent Transportation Systems (Itsc). Qingdao, China, 2014/10/8–2014/10/11. IEEE, pp 1021–1026
5. Vaitkus V, Lengvenis P, Zylius G (2014) Driving style classification using long-term accelerometer information. In: Driving Style Classification Using Long-Term Accelerometer Information, Miedzyzdroje, 2014/9/2–2014/9/5. IEEE, pp 641–644
6. Brombacher P, Masino J, Frey M, Gauterin F (2017) Driving event detection and driving style classification using artificial neural networks. In: 2017 Ieee International Conference On Industrial Technology (Icit), Toronto, On, 2017/3/22–2017/3/25. IEEE, pp 997–1002
7. Feng Y, Pickering S, Chappell E, Iravani P, Brace C (2018) Driving style analysis by classifying real-world data with support vector clustering. In: 2018 3rd Ieee International Conference On Intelligent Transportation Engineering (Icite), Singapore, 2018/9/3 2018/9/5. IEEE, pp 264–268
8. Chen SW, Fang CY, Tien CT (2013) Driving behaviour modelling system based on graph construction. Transp Res Part C: Emerg Technol 26:314–330
9. Sun B, Deng W, Wu J, Li Y, Zhu B, Wu L (2017) Research on the classification and identification of driver's driving style. In: 2017 10th International Symposium On Computational Intelligence And Design (Iscid), Hangzhou, 2017/12/9–2017/12/10. IEEE, pp 28–32
10. Mantouka EG, Barmpounakis EN, Vlahogianni EI (2019) Identification of driving safety profiles from smartphone data using machine learning techniques. Saf Sci. https://doi.org/10.4271/2012-01-0744
11. Eren H, Makinist S, Akin E, Yilmaz A (2012) Estimating driving behavior by a smartphone. In: 2012 Ieee Intelligent Vehicles Symposium, Alcal De Henares, Madrid, Spain, 2012/6/3–2012/6/7. IEEE, pp 234–239
12. Qi G, Wu J, Zhou Y, Du Y, Jia Y, Hounsell N, Stanton NA (2019) Recognizing driving styles based on topic models. Transp Res Part D: Transp Environ 66:13–22
13. Meseguer JE, Toh CK, Calafate CT, Cano JC, Manzoni P (2017) Driving styles: a mobile platform for driving styles and fuel consumption characterization. J Commun Netw 19(2):162–168
14. Rogavichene L, Garmonnikov I (2017) Innovative technologies for assessment and correction of the driving style. Transp Res Procedia 20:564–570
15. Javanmardi S, Bideaux E, Trégouët JF, Trigui R, Tattegrain H, Bourles EN (2017) Driving style modelling for eco-driving applications. Ifac-Papersonline 50(1):13866–13871
16. Karginova N, Byttner S, Svensson M (2012) Data-driven methods for classification of driving styles in buses. In: Sae Technical Paper Series, Apr. 24, 2012. Sae International400 Commonwealth Drive, Warrendale, Pa, United States. https://doi.org/10.4271/2012-01-0744
17. Lee J, Jang K (2017) A framework for evaluating aggressive driving behaviors based on in-vehicle driving records. Transp Res Part F: Traffic Psychol Behav. https://doi.org/10.1016/J.Trf.2017.11.021
18. Johnson DA, Trivedi MM (2011) Driving style recognition using a smartphone as a sensor platform. In: 2011 14th International IEEE Conference On Intelligent Transportation Systems (Itsc), Washington, Dc, Usa, 2011/10/5–2011/10/7. IEEE, pp 1609–1615
19. Eboli L, Mazzulla G, Pungillo G (2016) Combining speed and acceleration to define car users' safe or unsafe driving behaviour. Transp Res Part C: Emerg Technol 68, pp 113–125

20. Murphey YL, Milton R, Kiliaris L (2009) Driver's style classification using jerk analysis. In: 2009 Ieee Workshop On Computational Intelligence In Vehicles And Vehicular Systems, Nashville, Tn, Usa, 2009/3/30–2009/4/2. IEEE, pp 23–28
21. Perez MA, Sudweeks JD, Sears E, Antin J, Lee S, Hankey JM, Dingus TA (2017) Performance Of basic kinematic thresholds in the identification of crash and near-crash events within naturalistic driving data. Accid; Anal Prev 103:pp 10–19
22. Eboli L, Mazzulla G, Pungillo G (2017) How drivers' characteristics can affect driving style. Transp Res Procedia 27:945–952
23. Li G, Li SE, Cheng B, Green P (2017) Estimation of driving style in naturalistic highway traffic using maneuver transition probabilities. Transp Res Part C: Emerg Technol 74:113–125
24. Miyajima C, Ukai H, Naito A, Amata H, Kitaoka N, Takeda K (2011) Driver risk evaluation based on acceleration, deceleration, and steering behavior. In: 2011 IEEE International Conference On Acoustics, Speech And Signal Processing (Icassp), Prague, Czech Republic, 2011/5/22–2011/5/27. IEEE, pp 1829–1832
25. Kalsoom R, Halim Z. Clustering the driving features based on data streams. In: Inmic, Lahore, Pakistan, 2013/12/19–2013/12/20. IEEE, pp 89–94

Author Index

W. Wang et al. (eds.), *Green Transportation and Low Carbon Mobility Safety*, Lecture Notes in Electrical Engineering 944, https://doi.org/10.1007/978-981-19-5615-7

Y

Z